Principles of
Electron Optics

Principles of Electron Optics

Volume 1
Basic Geometrical Optics

by

P. W. HAWKES

*CNRS Laboratory of Electron Optics,
Toulouse, France*

and

E. KASPER

*Institut für Angewandte Physik
Universität Tübingen,
Federal Republic of Germany*

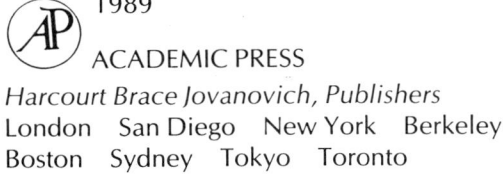

1989
ACADEMIC PRESS
Harcourt Brace Jovanovich, Publishers
London   San Diego   New York   Berkeley
Boston   Sydney   Tokyo   Toronto

ACADEMIC PRESS LIMITED
24/28 Oval Road,
London NW1 7DX

United States Edition published by
ACADEMIC PRESS, INC.
San Diego, CA 92101

Copyright © 1989
by ACADEMIC PRESS LIMITED

All rights reserved. No part of this publication may
be reproduced, stored in a retrieval system or
transmitted, in any form or by any means, electronic,
mechanical, photocopying or otherwise, without the
prior permission of the publishers.

British Library Cataloguing in Publication Data

Hawkes, P. W.
 Principles of electron optics.
 Vol 1: Basic Geometrical Optics
 1. Electron Optics
 I. Title II. Kasper, E.
 537.5'6

ISBN 0-12-333351-2

Printed in Great Britain by St Edmundsbury Press Limited,
Bury St Edmunds, Suffolk.

Contents of Volume 1
Basic Geometrical Optics

Preface
Chapter	1		Introduction	1
	1.1		Organization of the subject	3
	1.2		History	8

PART I – CLASSICAL MECHANICS

Chapter	2		Relativistic Kinematics	17
	2.1		The Lorentz equation and general considerations	17
	2.2		Conservation of energy	18
	2.3		The acceleration potential	19
	2.4		Definition of coordinate systems	22
	2.5		Conservation of axial angular momentum	24
Chapter	3		Different Forms of Trajectory Equations	27
	3.1		Parametric representation in terms of the arc-length	27
	3.2		Relativistic proper-time representation	29
	3.3		The cartesian representation	30
	3.4		Scaling rules	33
Chapter	4		Variational Principles	35
	4.1		The Lagrange formalism	35
	4.2		General rotationally symmetric systems	38
	4.3		The canonical formalism	41
	4.4		The time-independent form of the variational principle	43
	4.5		Static rotationally symmetric systems	44
Chapter	5		Hamiltonian Optics	46
	5.1		Introduction of the characteristic function	46
	5.2		The Hamilton–Jacobi equation	48
	5.3		The analogy with light optics	49
	5.4		The influence of vector potentials	51
	5.5		Gauge transformations	53

5.6	Poincaré's integral invariant	54
5.7	The problem of uniqueness	57
5.8	Résumé	58

PART II – CALCULATION OF STATIC FIELDS

Chapter	6	Basic Concepts and Equations	61
	6.1	General considerations	61
	6.2	Field equations	62
	6.3	Variational principles	65
	6.4	Rotationally symmetric fields	67
	6.5	Planar fields	69
Chapter	7	Series Expansions	73
	7.1	Azimuthal Fourier series expansions	73
	7.2	Radial series expansions	78
	7.3	Rotationally symmetric fields	85
	7.4	Multipole fields	88
	7.5	Planar fields	90
	7.6	Fourier–Bessel series expansions	91
Chapter	8	Boundary-Value Problems	94
	8.1	Boundary-value problems in electrostatics	94
	8.2	Boundary conditions in magnetostatics	96
	8.3	Examples of boundary-value problems in magnetostatics	101
Chapter	9	Integral Equations	107
	9.1	Integral equations for scalar potentials	107
	9.2	Problems with interface conditions	111
	9.3	Reduction of the dimensions	113
	9.4	Important special cases	117
	9.5	Résumé	124
Chapter	10	The Boundary-Element Method	125
	10.1	Evaluation of the Fourier integral kernels	125
	10.2	Numerical solution of one-dimensional integral equations	131
	10.3	Superposition of aperture fields	143
	10.4	Three-dimensional Dirichlet problems	149
	10.5	Examples of applications of the boundary-element method	158
Chapter	11	The Finite-Difference Method (FDM)	159
	11.1	The choice of grid	159

	11.2	The Taylor series method	160
	11.3	The integration method	162
	11.4	Nine-point formulae	165
	11.5	Iterative solution techniques	170
Chapter 12		The Finite-Element Method (FEM)	175
	12.1	Formulation for round magnetic lenses	175
	12.2	Formulation for self-adjoint elliptic equations	179
	12.3	Solution of the finite-element equations	182
	12.4	Improvement of the finite-element method	183
	12.5	Comparison and combination of different methods	184
Chapter 13		Field-Interpolation Techniques	188
	13.1	One-dimensional differentiation and interpolation	188
	13.2	Two-dimensional interpolation	194

PART III – THE PARAXIAL APPROXIMATION

Chapter 14		Introduction	201
Chapter 15		Systems with an Axis of Rotational Symmetry	202
	15.1	Derivation of the paraxial ray equations from the general ray equations	207
	15.2	Variational derivation of the paraxial equations	212
	15.3	Forms of the paraxial equations and general properties of their solutions	213
	15.4	The Abbe sine condition and Herschel's condition	219
	15.5	Some other transformations	222
Chapter 16		Gaussian Optics of Rotationally Symmetric Systems: Asymptotic Image Formation	225
	16.1	Real and asymptotic image formation	225
	16.2	Asymptotic cardinal elements and transfer matrices	226
	16.3	Gaussian optics as a projective transformation	235
	16.4	Use of the angle characteristic to establish the optical quantities	237

	16.5	The existence of asymptotes	239
Chapter 17		Gaussian Optics of Rotationally Symmetric Systems: Real Cardinal Elements	242
	17.1	Real cardinal elements for high magnification and high demagnification	242
	17.2	Osculating cardinal elements	246
	17.3	Inversion of the principal planes	253
	17.4	Approximate formulae for the cardinal elements: the thin-lens approximation and the weak-lens approximation	257
Chapter 18		Electron Mirrors	261
	18.1	Introduction	261
	18.2	A time-like parameter as independent variable	264
	18.3	The cartesian representation	271
	18.4	A quadratic transformation	274
Chapter 19		Quadrupole Lenses	276
	19.1	Paraxial equations for quadrupoles	277
	19.2	Transaxial lenses	286
Chapter 20		Cylindrical Lenses	290

PART IV – ABERRATIONS

Chapter 21		Introduction	297
Chapter 22		Perturbation Theory: General Formalism	303
Chapter 23		The Relation Between Permitted Types of Aberration and System Symmetry	315
	23.1	Introduction	315
	23.2	$N = 1$	322
	23.3	$N = 2$	325
	23.4	$N = 3$	329
	23.5	$N = 4$	330
	23.6	$N = 5$ and 6	333
	23.7	Systems with an axis of rotational symmetry	334
	23.8	Note on the classification of aberrations	336
Chapter 24		The Geometrical Aberrations of Round Lenses	339
	24.1	Introduction	339

	24.2	Derivation of the real aberration coefficients	339
	24.3	Spherical aberration	350
	24.4	Coma	365
	24.5	Astigmatism and field curvature	369
	24.6	Distortion	378
	24.7	The variation of the aberration coefficients with aperture position	382
	24.8	Reduced coordinates	384
	24.9	Seman's transformation of the characteristic function	387
Chapter	25	Asymptotic Aberration Coefficients	393
Chapter	26	Chromatic Aberrations	409
	26.1	Real chromatic aberrations	409
	26.2	Asymptotic chromatic aberrations	415
Chapter	27	Aberration Matrices and the Aberrations of Lens Combinations	418
Chapter	28	The Aberrations of Mirrors and Cathode Lenses	425
	28.1	The parametric form of the theory	425
	28.2	Systems with curved cathodes	429
	28.3	Structure of the aberrations	430
	28.4	The cartesian form of the aberration theory	432
Chapter	29	The Aberrations of Quadrupole Lenses and Octopoles	434
	29.1	Introduction	434
	29.2	Geometrical aberration coefficients	434
	29.3	Aperture aberrations	453
	29.4	Chromatic aberrations	460
	29.5	Quadrupole multiplets	461
Chapter	30	The Aberrations of Cylindrical Lenses	466
Chapter	31	Parasitic Aberrations	470
	31.1	Small deviations from rotational symmetry; axial astigmatism	470
	31.2	Classification of the parasitic aberrations	472
	31.3	Numerical determination of parasitic aberrations	475
	31.4	The isoplanatic approximation	477

PART V – DEFLECTION SYSTEMS

Chapter 32		Deflection Systems and their Aberrations	483
	32.1	Introduction	483
	32.2	The paraxial optics of deflection systems	487
	32.3	The aberrations of deflection systems	497
	32.4	Stigmators	516

PART VI – COMPUTER-AIDED ELECTRON OPTICS

Chapter 33		Numerical Calculation of Trajectories, Paraxial Properties and Aberrations	525
	33.1	Introduction	525
	33.2	Numerical solution of ordinary differential equations	526
	33.3	Standard applications in electron optics	533
	33.4	Differential equations for the aberrations	537
	33.5	Least-squares-fit methods in electron optics	543
	33.6	Determination and evaluation of aberration discs	547
	33.7	Optimization procedures	558
Chapter 34		The Use of Computer Algebra Languages	565
	34.1	Introduction	565
	34.2	Computer algebra, its role in electron optics	566
	34.3	Two practical examples	569

Notes and References
    Preface and Chapter 1   575
    Part I, Chapters 2–5   586
    Part II, Chapters 6–13   586
    Part III, Chapters 14–20   591
    Part IV, Chapters 21–31   601
    Part V, Chapter 32   619
    Part VI, Chapters 33 and 34   622
    Conference Proceedings   1192
Index

Contents of Volume 2
Applied Geometrical Optics

Preface

PART VII – INSTRUMENTAL OPTICS   627

Chapter 35	Electrostatic Lenses	629
Chapter 36	Magnetic Lenses	687
Chapter 37	Electron Mirrors	796
Chapter 38	Cathode Lenses and Field-Emission Microscopy	799
Chapter 39	Quadrupole Lenses	801
Chapter 40	Deflection Systems	823

PART VIII – ABERRATION CORRECTION AND BEAM INTENSITY DISTRIBUTION (CAUSTICS)   855

Chapter 41	Aberration Correction	857
Chapter 42	Caustics and their Applications	879

PART IX – ELECTRON GUNS   905

Chapter 43	General Features of Electron Guns	907
Chapter 44	Theory of Electron Emission	918
Chapter 45	Pointed Cathodes without Space Charge	934
Chapter 46	Space Charge Effects	953
Chapter 47	Brightness	971
Chapter 48	Emittance	989
Chapter 49	The Boersch Effect	1004
Chapter 50	Complete Electron Guns	1017

PART X – SYSTEMS WITH A CURVED OPTIC AXIS   1037

Chapter 51	General Curvilinear Systems	1039
Chapter 52	Magnetic Sector Fields	1058

| Chapter 53 | Unified Theories of Ion Optical Systems | 1080 |

Notes and References 1101
Index

Principles of Electron Optics

Volume 3

Topics to be covered in this volume are:
- Derivation of the laws of electron propagation from Schrödinger's equation.
- Image formation and the notion of resolution in electron microscopes.
- Electron–specimen interactions (scattering theory and dynamic theory).
- Image processing:
    - discretization and coding;
    - enhancement;
    - restoration;
    - analysis, description and pattern recognition.
- Electron holography and interference.
- Coherence, brightness and the spectral functions.

Preface

The last attempt to cover systematically the whole of electron optics was made by the late Walter Glaser, whose *Grundlagen der Elektronenoptik* appeared in 1952; although a revised abridgement was published in the *Handbuch der Physik* four years later, we cannot but recognise that those volumes are closer to the birth of the subject, if we place this around 1930, than to the present day.

Furthermore, electron optics has been altered dramatically during these intervening decades by the proliferation of large fast computers. Analytic expressions for the aberration coefficients of superimposed deflection and round magnetic lens fields, for example, have been derived only recently, partly because the latest generation of microlithography devices required them but also because they could only be evaluated by numerical methods: the earlier practice of seeking models permitting hand calculation could never have served here. Again, computer calculations have shed considerable light on electron gun behaviour, as the length of Part IX testifies convincingly; in 1952, Glaser was able to condense his account of gun theory into four pages!

The growth of electron optics is not, however, solely due to the computer. Many systems that had not been thoroughly explored have now been analysed in detail and, in many cases, we have had to renounce the attempt to reproduce in detail new results, however interesting, to keep the number of pages within reasonable limits. This work should therefore be regarded as both a textbook and a source-book: the fundamentals of the subject are set out in detail, and there the student should find everything needed to master the basic ideas or to begin the analysis of some class of systems not yet explored; the principal electron optical components are likewise dealt with in great detail. Where optical elements that are not quite so common are concerned, however, we have felt at liberty to direct the reader to original articles and reviews, or specialist texts, to leave space for topics of wider interest.

The following chapters are, moreover, limited to geometric optics: wave optics is to be covered in a companion volume. With the Schrödinger equation as starting point, we shall there examine the propagation of electron waves in electrostatic and magnetic fields and study image formation and resolution in the principal electron optical instruments. This demands some discussion of electron–specimen interactions. A chapter will be devoted to the four broad themes of image processing: discretization and

PREFACE

coding; enhancement; restoration; and analysis, description and pattern recognition. In another, we shall give an account of the steadily growing field of electron holography. Finally, we shall return to the optics of electron sources in order to understand the concept of coherence and we shall show how the notions of brightness, partial coherence and various associated spectral functions are inter-connected.

Students of electron optics have been fortunate in that many excellent textbooks on the subject have appeared over the years, the first when the subject was still young (Brüche and Scherzer, 1934; Myers, 1939; Klemperer, 1939; Picht, 1939); these were followed in the 1940s by the encyclopaedic Zworykin et al. (1945), Cosslett (1946) and Gabor (1945). Many books on the subject appeared in the 1950s, of which the texts by Glaser already mentioned, Sturrock (1955) Grivet et al. (1955, 1958) and Kel'man and Yavor (1959) are the most important for our present purposes. Subsequently, however, the flow has shrunk to a trickle, new editions and short introductory texts dominating, with the exception of the multi-author volumes edited by Septier (1967, 1980, 1983); conversely, monographs on limited topics have become more common. Although certainly 'standing on the shoulders of giants', the present volumes do differ considerably from their many predecessors in that the developments of the past twenty years are accorded ample space.

For whom is this work intended? A knowledge of physics and mathematics to first degree level is assumed, though many reminders and brief recapitulations are included. It would be a suitable background text for a post-graduate or final year course in electron optics, and much of the material has indeed been taught for some years in the University of Tübingen; a course in the University of Cambridge likewise covered many of the principles. Its real purpose is, however, to provide a self-contained, detailed and above all modern account of electron optics for anyone involved with particle beams of modest current density in the energy range up to a few mega-electronvolts. Such a reader will find all the basic equations with their derivations, recent ideas concerning aberration studies, extensive discussion of the numerical methods needed to calculate the properties of specific systems and guidance to the literature of all the topics covered.

Composition of volumes such as these puts us in debt to a host of colleagues: many have permitted us to reproduce their results; the librarians of our institutes and the Librarian and Staff of the Cambridge Scientific Periodicals Library have been unflagging in their pursuit of recondite and elusive early papers; Mrs. Ströer has uncomplainingly word-processed hundreds of pages of mathematical and technical prose; Mrs. Maczkiewicz and Mr. Inial have taken great pains with the artwork as have Mrs. Bret and her colleagues with the references; Academic Press and Professor Dr K.-H.

PREFACE

Herrmann, Director of the Institut für Angewandte Physik der Universität Tübingen, have generously supported this work; the Zentrum für Datenverarbeitung has provided the text-editing facilities needed for TEX. To all of these we are extremely grateful. We also thank the many authors and publishers who have been good enough to allow us to reproduce published drawings. The details are as follows.

SPRINGER VERLAG: Figs 17.4–5, 36.6–7, 36.11 and 42.1–7 from *Grundlagen der Elektronenoptik* by W. Glaser; Fig. 43.2 from the article by D. Kamke in *Handbuch der Physik* **33**; and Figs 36.3–4, 36.13, 36.17 and 36.36–37 from the chapters by F. Lenz, W.D. Riecke and T. Mulvey in *Magnetic Electron Lenses*.

WISSENSCHAFTLICHE VERLAGSGESELLSCHAFT: Figs 35.5, 36.8–9, 36.15, 36.18–26, 36.27–33, 36.35, 41.3–4, 41.7, 41.11, 42.15, 49.3 and 50.6–8 from *Optik*.

INSTITUTE OF PHYSICS: Figs 35.10, 36.38–40 and 36.43 from *J. Phys. E: Sci. Instrum.*; Fig. 36.16 from *Repts Prog. Phys.*; and Fig. 35.3 from the *Proceedings of the Fifth European Congress on Electron Microscopy* (Manchester).

JAPANESE SOCIETY OF ELECTRON MICROSCOPY: Figs 36.12, 36.34 and 41.5–6 from *J. Electron Microsc.*

NORTH–HOLLAND PUBLISHING CO.: Figs 36.41–42 from *Ultramicroscopy*; and Fig. 42.17 from *Nucl. Instrum. Meth.*

JAPANESE JOURNAL OF APPLIED PHYSICS: Figs 35.13–14 and 36.44–45.

VEB DEUTSCHER VERLAG DER WISSENSCHAFTEN: Fig. 36.14 from *Exp. Tech. Phys.*

IEEE: Figs 40.24–25 from *IEEE Trans Electron Devices*.

ACADEMIC PRESS: Figs. 45.8–9 and 48.7–8 from *Adv. Opt. Electron Microsc.*; Figs 40.14–19 from *Microcircuit Engineering 83*.

"*Now, soldiers, march away; And how thou pleasest, God, dispose the day!*"
   Henry V, *on the eve of the battle of Agincourt*...

Introduction

1
Introduction

1.1 Organization of the subject

The properties of beams of free electrons, released from a material source and propagating through a vacuum region in some device, are of interest in many diverse fields of instrumentation and technology. The study of such electron beams forms the subject of electron optics, which divides naturally into geometrical optics, when the wavelength is negligible, and wave optics, in which effects due to the finite wavelength are considered. This first volume is concerned with geometrical optics, a knowledge of which is needed to analyse an extremely wide range of instruments: cathode-ray tubes; the family of electron microscopes, which now includes the fixed-beam and scanning transmission instruments, the scanning electron microscope and the emission microscope; electron spectrometers and mass spectrographs if we include charged particles other than electrons; image converters; electron interferometers and diffraction devices; electron welding machines; and electron-beam lithography devices. We could indeed include electron accelerators, such as betatrons and electron synchrotrons, but a rather different approach is often more useful in those machines. This list is by no means complete but it already demonstrates the great diversity of the possible applications of electron optics.

Over the years, a vast amount of knowledge about the many branches of electron optics has been accumulated and we have therefore had to be selective. The main emphasis is on the principles of electron optics, and technical details are only included to bridge the gap to the practical application of these principles. This seems justified, for the principles remain unaffected by the passage of time whereas instrumental development is so rapid that surveys and review articles are the best means of charting its progress.

The physical properties of electrons in a free beam may be classified as follows:
(a) corpuscular properties;
(b) wave properties;

(c) macroscopic interactions;
(d) microscopic or atomic interactions;
(e) radiative properties.

A similar classification is given by Sturrock (1955).

The *corpuscular properties* are described by classical (relativistic) mechanics, the electron being regarded as a charged particle acted on by electromagnetic forces. For almost all electron optical devices, extensive studies must be made on the assumption that these corpuscular properties alone are important and the present volume is almost wholly confined to the corresponding *geometrical* optics. Geometrical particle optics is very similar for all charged particles, ions in particular, and Parts I and X are written in such a way that many of the relations derived are valid for ions and electrons, or can be converted straightforwardly.

The rest-mass of the electron is extremely small, a characteristic that has important consequences for the technology associated with electron beams. Only quite modest voltages are needed to accelerate electrons to a very high velocity, and the time of flight between the departure of an electron from the cathode and its arrival at its destination in a typical device is so small that it can almost invariably be ignored completely. It is therefore quite sufficient to study the purely *geometrical* shape of the electron motion within the beam, although a time-like curve parameter may prove to be advantageous in numerical calculations.

A further consequence of the extremely small inertia of the beam electrons is that deflection by suitably placed magnetic or electrostatic fields occurs virtually instantaneously, in synchronism with the applied voltages or currents, unless the frequency involved is very high indeed. The performance of many devices relies upon this property. We shall consider electron motion only in static, that is, *time-independent* fields. This is justified even when studying the deflector in a scanning device, the time of flight being so short that the applied field is quasistatic; the time dependence is then merely a common amplitude factor.

A knowledge of the *wave properties* of the electron is essential to understand the concept of resolution in electron microscopes, to analyse the interactions between electron beams and targets of all kinds, and to analyse the behaviour of electron interferometers and diffraction devices and of course to comprehend electron holography. These topics will occupy much of Volume 3.

The *macroscopic interactions* in an electron beam are a consequence of the fact that the latter may be regarded as a cloud of negative charges, which creates an electric field; this is superimposed on the external applied field and can thus alter the focusing properties of the device. In principle, of course, this occurs for every electron beam, but in reality such space charge effects are of importance only when the local beam intensity is very high. The space charge density can be treated as a macroscopic observable and the associated field calculation remains within the framework of classical electrostatics; we therefore call these interactions macroscopic. Such effects occur mainly in electron guns where the beam intensity can be high, and are therefore dealt with in Part IX, devoted to guns. Such effects are also extremely important in accelerators, but these are not within the scope of this book.

The *microscopic* or atomic interactions are the various scattering processes that occur within the beam on the atomic scale. Such processes arise when an electron beam encounters a specimen or target and electron–electron collisions may also occur within the beam. The latter give rise to the Boersch effect, an anomalous broadening of the electron spectrum at beam waists, which are themselves enlarged. Classical collision theory is capable of providing an approximate explanation of this effect, which is examined briefly in Chapter 49. Collisions between beam electrons and the atoms in a target can only be properly understood in terms of quantum mechanics; some space is devoted to this topic in Volume 3.

Finally, we come to the *radiative properties* of the electron, essentially the emission of bremsstrahlung when the acceleration is very high. This occurs mainly in high-voltage electron microscopes where particles with an energy of 1 MeV or more collide with the specimen placed in the path of the beam. Although the staff around the microscope must be protected from this radiation, the damage to the specimen is negligible in comparison with that inflicted by the mechanical bombardment. The bremsstrahlung caused by the acceleration of the electrons *in vacuo* only becomes important at the energies encountered in high-energy physics, which are beyond the scope of these volumes; we therefore ignore bremsstrahlung throughout.

Our theme is thus the study of the motion of electrons, regarded as classical charged particles of negligible extent, through static electric or magnetic fields. We begin with the derivation of the conservation laws for the electron motion and cast these into a form particularly well suited to electron optics. Various forms of the trajectory equations are established but these are not at all satisfactory for our purpose, which is the study not of single trajectories but of whole families of electron paths: not ballistics but optics. It is *Hamiltonian theory* that enables us to make the transition. As early as 1827, Hamilton drew his famous analogy between geometrical optics and classical mechanics; this tells us that, just as in optics, there must exist a mechanical characteristic function, or *eikonal*, with the property that the trajectories are always locally orthogonal to the surfaces of constant value of this function. This is true only in the absence of magnetic vector potentials; in the general case, when magnetic fields are present, the correct form of this orthogonality relation emerges from the Hamilton–Jacobi theory presented in Chapter 5. This theory is very important, for it forms the cornerstone of geometrical electron optics.

Most instrumental research is concerned with the design of new or improved electron optical systems, for which an accurate knowledge of the properties of families of rays traced through such systems is indispensable. This proceeds in two stages: first, the field distribution must be established, after which rays can be traced and quantities characteristic of the system calculated. A knowledge of the field distribution is usually needed only in the immediate vicinity of a curve in space, frequently a symmetry axis, known as the optic axis; the beam is required to remain close to this axis to prevent the aberrations from degrading the performance of the device. Unfortunately, the required information about the field can rarely be obtained without solving a *boundary-value problem*, since the field will be generated by electrodes and magnetic materials, such as polepieces, at some distance from this axis.

In practice, field calculation is the most complicated part of numerical design and the principal methods are presented in considerable detail in Part II. *Series expansions* for electrostatic potentials and hence fields and for magnetic scalar and vector potentials are also listed since these are repeatedly needed in later chapters where the trajectory equations and aberration coefficients of various types of system are derived.

This thorough presentation of the physical and mathematical fundamentals leads naturally to the systematic investigation of electron optical components: how are these to be characterized, how can we code complex behaviour in terms of a few easily calculated parameters? Parts III and IV provide the traditional answers in terms of the paraxial approximation and the aberrations that measure departures from it. In the *paraxial approximation*, it is assumed that the electron trajectories remain so close to the optic axis that equations of motion linear in the off-axis coordinates describe them satisfactorily. Although this is an excellent first-order approximation, it is clearly an idealization, a consequence of which is that some electron optical systems appear to be free of any image defects and hence capable of producing a stigmatic, unblurred image or a sharp focus.

In reality, no system is free of aberrations. One of the major tasks of electron optics is to establish what types of aberrations can occur in any given system and then to reduce the most deleterious as far as possible. A long Part is therefore devoted to the *theory of aberrations*. Since all wave optical considerations are excluded from this volume, only *geometrical* and *chromatic aberrations* are investigated. The former are those that measure the discrepancy between the true point of arrival of an electron at its destination and the point predicted by the paraxial approximation, due to the inadequacies of the latter and to small imperfections in the system; the chromatic aberrations are those caused by the presence of electrons with different energies in the beam, arising from the small spread of the initial energies at the cathode surface or from the loss of various amounts of energy when traversing a thin specimen.

A separate Part is devoted to a similar analysis of deflection systems, of great practical importance for microlithography in the current quest for miniaturization. Such systems may be very complex, magnetic and electric deflection fields occupying the same region as a magnetic round lens field, and the number of degrees of freedom becomes very large. Both the theory and the experimental adjustment of such combinations reflect this complexity but it has proved necessary to resort to such intricate arrangements in modern electron beam lithography machines, which are used to produce the semiconducting integrated circuits required in computers.

In the next four Parts, VI–IX, the principles set out in the first half of the book are applied to the many different types of electron optical components—round lenses, quadrupoles, mirrors, cathode lenses—for each of which the general theory takes a special form. We have preferred to organize this material by component rather than by instrument, since the latter must inevitably be understood as an assembly of individual modules. In Part VI, the numerical techniques needed for tracing trajectories and computing aberrations in any system are presented, together with an introductory account of computer algebra, a tool that is very useful for establishing aberration integrals and evaluating these for the few models that permit a result to be obtained in closed form.

Part VII indicates what practical information is available in the literature concerning the optical properties of the various elements analysed in theory in earlier chapters, while in Part VIII we examine two special topics of sufficient importance to warrant separate treatment: aberration correction, essentially for systems of round lenses, and the theory of caustics.

Another topic of great complexity is the study of electron sources, usually known as electron guns. An entire Part (IX) is devoted to these, for although the degrees of freedom are not unduly large in number, the theoretical description requires concepts that are of little importance elsewhere and depend essentially on the purpose to which the gun is to be put. Thus the gun of an electron interferometer is very different from that of an electron welding machine. We have tried to impose a pattern on this complex and many-faceted topic by following the flight of the electron through the gun. First we examine the principal emission processes and the focusing effects in the neighbourhood of the cathode. This is followed by the theory of space charge, which may of course be important elsewhere than in the cathode region. We next introduce a number of quantities that are employed to characterize the beam farther from the cathode, and, in particular, brightness, emittance and the energy spectrum, which are very important when considering the suitability of the gun for specific tasks. The Part ends with a few remarks about the design of complete guns.

The final Part is devoted to systems in which the optic axis is curved, though in practice almost always a plane curve, a situation that arises in the electric or magnetic prisms of electron spectrometers. With this, most aspects of geometrical electron optics have been covered. The reader will notice that the emphasis throughout is on physical principles and on their theoretical formulation. Technical details are included only when they seem necessary to render the practical applications of these principles comprehensible. Inclusion of technological details would have made the book impossibly large and rapidly obsolete, for there are few branches of the subject that are not in continual development. The lifetime of the underlying principles is, however, substantially longer.

The subject has acquired a very voluminous literature over the decades, so that a full bibliography would alone fill many pages. We have adopted a compromise towards these many publications: papers of especial relevance are mentioned in the body of the text but, in order not to interrupt the flow, the majority are grouped in annotated bibliographic appendices at the end of each book. In this way, the reader is directed to the literature of each topic but is not continually distracted by notes and references. Even so, we have made no attempt to trace the history of the subject in these appendices and we therefore complete this introduction with a succinct account of the principal stages through which the subject has passed.

1.2 History

Electron optics was born in the 1920s. In 1925, Louis de Broglie argued convincingly that a wavelength should be associated with moving particles, electrons in particular; and in 1927, Hans Busch demonstrated that the action of an axially symmetric coil on electrons can be described in the language of geometrical optics, in terms of a focal length: "Eine kurze Spule hat also die Eigenschaft, die Kathodenstrahlen nach der Achse zu um einen Winkel γ abzulenken, der proportional der Achsenentfernung ... des Strahles ist. Genau die gleiche Eigenschaft besitzt aber für Lichtstrahlen eine Sammellinse"; this was an explicit statement of his conclusions adumbrated a year earlier (Busch, 1926). De Broglie's paper soon led to the experiments on electron diffraction of Davisson and Germer (1927) and of Thomson and Reid (1927). Busch's idea of associating a lens-like character with a short magnetic field was tested by Max Knoll and his young student,

Ernst Ruska (Ruska and Knoll, 1931), who went on to combine such lenses into the first electron microscope, built in the Electrotechnical Institute of the Berlin Technological University (Knoll and Ruska, 1932a,b).

An electron microscope has much in common with its light optical ancestor. It consists of a source of illumination, condenser lenses to direct the illuminating beam onto a suitably sized region of the specimen with an appropriate angular spread, an objective lens to provide a first magnification and projector lenses to magnify the intermediate images still further. In appearance and nature, however, each of these optical elements is very different from those of the familiar compound microscope. The source of illumination is now an electron gun, of which the commonest type is the simple thermionic triode structure. A filament is heated, thus releasing electrons which are accelerated to the desired energy by a suitably polarized anode. A third electrode, the wehnelt, placed between filament and anode, improves the performance of such sources considerably.

The lenses are short stretches of rotationally symmetric magnetic field, created by a current-carrying coil enclosed in an iron yoke. The interior of the microscope must be evacuated to a pressure typically of the order of 10^{-6} Torr ($\sim 10^{-4}$ Pa) since electrons are scattered or halted by a very small amount of matter in their path. For the same reason, the specimen must be exceedingly thin (at most tens or hundreds of nanometres thick for a 100 kV instrument). In these conditions, the electrons are deflected or 'scattered' within the specimen but almost no electrons fail to emerge from the far side. The specimen is thus a 'phase object' and contrast is created at the image by various mechanisms analogous to those encountered in the phase-contrast microscope. This image is rendered visible by allowing the electrons to fall on a fluorescent screen or a recording medium, a photographic emulsion. One aspect of electron lenses deserves special mention: their optical quality is astonishingly poor! They suffer from two lens defects that have been virtually eliminated from glass lenses: spherical aberration, a defect that severely limits the numerical aperture at which they can be operated and hence the resolution attainable; and chromatic aberration, by which we mean that their focusing power varies rapidly with the velocity of the incoming electrons. The high spherical aberration has the practical consequence of deteriorating the resolution of an electron microscope by some two orders of magnitude: with perfect lenses, the resolution limit might be expected to be of the order of picometres, whereas it is in reality of the order of hundreds of picometres (that is, of the order of ångströms). The harmful effects of chromatic aberration are avoided by using nearly monoenergetic electrons and stabilizing the lens currents to a very high degree, typically to one part in a million.

The first tentative studies of Ruska and Knoll, with which Bodo von

Borries was soon associated, were sufficiently encouraging to initiate a decade of theoretical and empirical electron optics, during which the foundations of the theory were laid, largely by Walter Glaser and Otto Scherzer, and the magnetic electron microscope was perfected to such a point that a commercial model was put on the market by the German company of Siemens in 1938. The British Metropolitan-Vickers company can, however, claim to have been the first commercial firm to supply a microscope, the custom-built EM1 instrument ordered by L.C. Martin for Imperial College, London, where it was installed in 1936; the resolution of the EM1 was not, however, superior to that of a light microscope (Mulvey, 1985).

Meanwhile, comparable work on an electrostatic instrument was being actively pursued in the research department of the Allgemeine Elektrizitätswerke-Gesellschaft (AEG). For full details of these activities, see Ruska's historical volume (1979, 1980) and the 'Selfportrait' of the AEG Research Institute prepared by Ramsauer (1941) with further editions in 1942 and 1943. The early development of the theory is fully chronicled in Glaser's *Grundlagen der Elektronenoptik* (1952).

Outside Germany, many electron microscope projects were launched in the 1930s, though it was not until the end of the Second World War that commercial production began on any scale. The prototypes built in England and Canada are described in various historical articles, especially Gabor (1957), Ruska (1957) and Mulvey (1962, 1967, 1973), and many references and reminiscences are to be found in Hawkes (1985). We must, however, make particular mention of the work of Ladislaus Marton, who constructed a series of simple instruments in Brussels, with which he obtained the earliest osmium-stained biological micrographs, the specimens being the long-leafed sundew and the root of the bird's-nest orchid (Marton, 1934, 1935). Soon after, first Driest and Müller (1935) and then Krause (1936) obtained biological electron micrographs with one of Ruska's microscopes that foreshadowed, albeit faintly, modern biological electron microscopy; Driest and Müller's images of the wing and leg of the common housefly were the first micrographs of unprepared biological specimens.

It was during the 1930s too that the field-emission microscope was developed by E.W. Müller, in one of the Siemens research laboratories in Berlin. In this instrument, a high electric field is maintained at a tip and highly magnified details of the surface are visible in the image as a result of the differences in emission from point to point. Müller's first papers appeared in 1936 and 1937 and a historical account is to be found in Good and Müller (1956).

By the 1950s, electron microscopes were being produced in West Germany, England, France, Holland, Switzerland, Czechoslovakia and the Soviet Union, with more modest activity in other European countries, particularly Scandinavia. In the United States, RCA began manufacturing electron microscopes during the war years, and in Japan commercial production commenced in the late 1940s, though many prototypes were built during the first half of the decade; the Hitachi HU-4 was put on the market in 1947, for example, and the JEOL JEM-1 in 1949 (Sugata, 1968; Fujita, 1986).

Although our subject is not electron microscopy but electron optics, we must digress here to mention an important development in the years 1948–1952, which had a major effect on electron microscope design (Gettner and Ornstein, 1956). The accelerating voltages of the early microscopes were then of the order of 50–80 kV, which was too low to form a sharp image of a biological specimen if the latter was one or more micrometres in thickness. An increase in voltage therefore seemed imperative until in 1948, Pease and Baker succeeded in cutting sections only 0.3–0.5 μm thick with a modified Spencer 820 microtome, and by 1950 the figure had fallen to 0.2 μm. In 1949, Newman et al. introduced methacrylate as an embedding medium, the mechanical properties of which greatly facilitated section cutting. In 1952, Sjöstrand designed a new ultramicrotome (Sjöstrand, 1953) with which sections 20 nm in thickness could be cut reproducibly and "the problem of high resolution electron microscopy of sectioned material had been solved " (Sjöstrand, 1967); the immediate need for high-voltage microscopes in biology vanished and the first megavolt instruments were not built for about another decade. The first of these was a 1.5 MV instrument constructed in Toulouse (Dupouy et al., 1960; Dupouy, 1968, 1985) and this was rapidly followed by a 750 kV machine in Cambridge (Smith et al., 1966; Cosslett, 1981) and the commercial high-voltage microscopes of AEI, GESPA, Hitachi and JEOL. These were all giant versions of the familiar 100 kV instruments, however, and essentially represented only technological progress; their optics was distinctly conservative. Their great bulk and the need for special buildings to house them did, however, furnish one of the reasons for the interest in superconducting lenses that sprang up in the mid-1960s, another being the perfect stability of the magnetic field generated by a coil in the persistent-current mode (Laberrigue and Levinson, 1964; Fernández-Morán, 1965; Boersch et al., 1966; Siegel et al., 1966; Ozasa et al., 1966). Of the various designs, the shielding lens introduced by Dietrich et al. (1969) is now used when it is important that the specimen and its immediate environment be at very low temperature (see Weyl et al., 1972; Hardy, 1973; Dietrich, 1976; Hawkes and Valdrè, 1977; Riecke, 1982; Lefranc et al., 1982). Students of superconducting lenses were not, however, alone in enquiring whether the focusing power of the monster lenses in traditional high-voltage electron microscopes could not be obtained in some other way. We draw attention to the numerous 'unconventional' designs introduced by Mulvey and colleagues, compared and contrasted in Mulvey (1982, 1984), and to the laminated lenses of Murillo (Balladore and Murillo, 1977), in which the yoke is made of highly inhomogeneous material in order to maintain the flux density constant over its cross-section.

The idea of forming an image not by irradiating a comparatively large specimen area and focusing this onto the image plane after suitable magnification but by scanning the specimen with a small probe and collecting a signal from the resulting interaction and using this signal to modulate the intensity of the spot of a cathode-ray tube scanned in synchronism goes back to the late 1930s. In 1938, von Ardenne described a primitive ancestor of the modern scanning electron microscope, in which a probe size of some 10 nm was achieved but at the cost of a very small current indeed ($\sim$ 1 pA); the beam traversed the specimen and struck a photographic film attached to a drum which rotated and advanced appropriately (see von Ardenne, 1978, 1985). In 1942, an instrument in which secondary electrons from a thick target provided the image signal was developed by Zworykin et al. but it was not until 1953 that McMullan described the first of the series of scanning electron microscopes to be built under Charles Oatley's direction in Cambridge, which culminated in the commercial 'Stereoscan', marketed by the Cambridge Instrument Company in 1965 (Oatley et al., 1965, 1985; Oatley, 1982, 1985). More recent versions of these instruments combine the properties of the X-ray microanalyser introduced by Castaing (1951) and perfected by Cosslett and Duncumb (1956) and Duncumb (1958), who added beam scanning, with the host of signals available in a scanning electron microscope with the result that analytical electron microscopy (AEM) has become a discipline in its own right.

The next major instrumental development occurred in the early 1960s. A celebrated set of curves (Oatley et al., 1965) relating probe size, number of lines in the image and the time needed to record an image of acceptable quality had shown vividly that the resolution of the scanning microscope could never rival that of the transmission microscope owing to the inadequate performance of the thermionic gun. It was known that a field-emission gun would change this situation dramatically, making it possible to compress a useful current into a probe only a few ångströms in diameter, and it was in 1965 that Crewe first described a scanning transmission electron microscope (STEM) with a field-emission gun (Crewe et al., 1968; Crewe, 1970, 1973). Instrumental development began in three companies, AEI, Vacuum Generators (VG) and Siemens, of which only VG continues to market STEMs; their first commercial instrument was installed in 1974. Those of Siemens and AEI were not pursued and the present tendency is to offer a field-emission gun and a STEM attachment as options with conventional transmission microscopes.

The foregoing account evokes the main steps in electron optical instrumentation for image formation. Many other innovations might have been listed: the development of new types of gun (lanthanum hexaboride and field emission) and the introduction of various types of energy filter and analyser, in particular the focusing device of Castaing and Henry (1962). We now turn to the theory of the subject and single out the principal contributions. We have already mentioned the role of Busch in the founding of geometrical electron optics. The ideas of Hamiltonian mechanics were applied to electron motion by Walter Glaser, who derived expressions for the aberration coefficients with the aid of a characteristic function or eikonal, while Otto Scherzer obtained similar formulae by the 'trajectory method', in which equations of motion including aberration terms are derived and solved by the method of variation of parameters. In 1936, Scherzer published formulae for the coefficients of spherical and chromatic aberration that showed that these can never be made to change sign by skilful lens design; this result and the reactions to it are discussed in detail in Part IV. Eleven years later, it was again Scherzer (1947) who described several types of aberration correctors, capable in principle of cancelling these troublesome coefficients. The history of the 1950s is the history of attempts to use such correctors (see Chapter 41). Such attempts continue today, though this activity diminished considerably with the realization that digital processing of aberrated images might, indeed should, provide an easier solution. Scherzer's were not the only suggestions for aberration correction, however; in 1948, Gabor described a method of correcting spherical aberration by an optical reconstruction technique, which he called holography. For technical reasons, this was unsuccessful at that time (the laser was yet to be invented) but with the development of bright electron sources and coherent light sources, both in-line and off-axis holography have subsequently been extensively developed, as we shall see in Volume 3.

We must return to the 1940s and 1950s to draw attention to some other landmarks in electron optics. In 1943, Grinberg published a very general form of the equations of motion of electrons in electric and magnetic fields and this was later extended to include aberrations by Vandakurov (1955, 1956). Similar general equations were derived by Sturrock (1952), who developed and perfected Glaser's Hamiltonian approach to electron optics in many ways (Sturrock, 1955), and by Rose (1968), who analysed a more limited class of systems. The labour of transforming aberration integrals was reduced by Seman (1951, 1954, 1955, 1958), who introduced a very ingenious method that replaces partial integration by differentiation. At about the same period, Lenz (1956, 1957) clarified the distinction between real and asymptotic aberration coefficients, already examined briefly by

Sturrock (1955). It was not until a decade later that the possibility of writing these asymptotic coefficients as polynomials of at worst fourth order in reciprocal magnification was noticed (Hawkes, 1968), an observation that renders computer-aided design of complex systems less arduous. The 1970s and 1980s have principally seen the application of sophisticated computing techniques in electron optics, and in particular to branches of electron optics that had perforce been almost completely neglected by earlier theoreticians: electron guns are the most striking example. Parts II, VI and IX bear witness to the progress that has been made, much of it in the theoretical electron optics group of the University of Tübingen (Lenz, 1973; Kasper and Lenz, 1980; Kasper, 1982, 1984).

This short account of the history of electron optics and electron microscopes cannot but be invidious: we could have included all the other electron optical instruments and we could have charted the progress of the theory, equation by equation. This would, however, have left all too little space for the Principles of Electron Optics, to which we now turn. References to many other accounts of the history of the subject are to be found in *The Beginnings of Electron Microscopy* (Hawkes, 1985).

Part I

Classical Mechanics

2
Relativistic Kinematics

In the following sections we shall examine the motion of a charged particle of rest mass m_0 and charge Q in an electromagnetic field characterized by the electric and magnetic field vectors $\boldsymbol{E}(\boldsymbol{r},t)$ and $\boldsymbol{B}(\boldsymbol{r},t)$, respectively. Whenever the specialization $Q = -e < 0$ for electrons is not made explicitly, the analysis is valid for the motion of any charged particle.

The derivation of useful trajectory equations for the motion of charged particles—and of all conservation laws satisfied by them—can be performed in a very general and elegant manner by means of variational calculus, as we shall show in Chapter 4. First, however, we give a short introduction to relativistic kinematics, because this offers a better understanding of many of the optical aspects of the variational calculations.

2.1 The Lorentz equation and general considerations

The trajectory equation for the motion of charged particles is most simply represented in its Newtonian form

$$\frac{d\boldsymbol{g}}{dt} = \frac{d}{dt}(m\boldsymbol{v}) = Q\left\{\boldsymbol{E}(\boldsymbol{r},t) + \boldsymbol{v} \times \boldsymbol{B}(\boldsymbol{r},t)\right\} \tag{2.1}$$

$\boldsymbol{v} = d\boldsymbol{r}/dt$ being the velocity and $\boldsymbol{g} = m\boldsymbol{v}$ the *kinetic* momentum, which must be clearly distinguished from the *canonical* momentum, defined in Chapter 4. The Lorentz force, given by the right-hand side of (2.1), is expressed in SI units, which will be used consistently throughout this volume. It is convenient to introduce the familiar abbreviations

$$\beta := \frac{v}{c} = \frac{|\boldsymbol{v}|}{c} \quad , \quad \gamma = \frac{1}{\sqrt{1-\beta^2}} \tag{2.2}$$

whereupon (2.1) has the correct relativistic form if we represent the relativistic mass m by the well-known expression

$$m = \frac{m_0}{\sqrt{1-\beta^2}} \equiv \gamma m_0 \tag{2.3}$$

The trajectory equation does not depend on the values of Q and m_0 separately but only on the ratio Q/m_0 as can be seen by rewriting (2.1) thus:

$$\frac{d}{dt}\left(\frac{v}{\sqrt{1-\beta^2}}\right) = \frac{Q}{m_0}(E + v \times B)$$

In the special case of electron motion, we have

$$Q = -e = -1.602 \times 10^{-19} \text{C} \quad , \quad \frac{e}{m_0} = 1.759 \times 10^{11} \text{C kg}^{-1} \qquad (2.4)$$

The absolute value of the electron charge will always be denoted by e.

The practical evaluation of (2.1) requires the calculation of the field vectors $E(r, t)$ and $B(r, t)$ for arbitrary values of the position vector r and of the time t. The corresponding computer programs have to be written and executed before embarking on trajectory calculations. In practice, field calculation is the most complicated part of theoretical electron and ion optics. We shall deal with this subject in detail in Part II. In the present chapter we shall assume the fields to be known.

In the most general case of arbitrary electron optical devices, (2.1) can only be solved numerically, from given initial conditions. In order to derive more detailed laws, it is necessary to introduce simplifying symmetry conditions, which are assumed to be exactly valid. The inevitable departures from exact symmetry in practical devices will be discussed in later chapters.

2.2 Conservation of energy

In most electron and ion optical devices the applied fields are static, independent of time: $E = E(r)$, $B = B(r)$. It is then possible to derive a law for the conservation of particle energy. This is most easily done by scalar multiplication of both sides of (2.1) with v. We shall denote derivatives with respect to time by dots. Using (2.2) and (2.3) we obtain first

$$v \cdot \frac{d}{dt}\left(\frac{m_0 v}{\sqrt{1-\beta^2}}\right) = Qv \cdot E(r) \quad , \quad (v = \dot{r})$$

Using the identity $v \cdot \dot{v} = v\dot{v}$, the left-hand side can be transformed to a total derivative:

$$v \cdot \frac{d}{dt}(\gamma m_0 v) = \gamma^3 m_0 v \dot{v} = \frac{d}{dt}(\gamma m_0 c^2)$$

In order to transform the right-hand side, we introduce the electrostatic potential $\Phi(\mathbf{r})$ and write

$$\mathbf{E}(\mathbf{r}) = -\text{grad } \Phi(\mathbf{r}) \tag{2.5}$$

The right-hand side then becomes a total derivative too:

$$Q\mathbf{v} \cdot \mathbf{E}(\mathbf{r}) = -Q\dot{\mathbf{r}} \cdot \text{grad } \Phi(\mathbf{r}) = -\frac{d}{dt}(Q\Phi(\mathbf{r}))$$

Integration with respect to time and substitution for the factor γ results in:

$$E_0 := m_0 c^2 \left(\frac{1}{\sqrt{1-\beta^2}} - 1\right) + Q\Phi(\mathbf{r}) = \text{const} \tag{2.6}$$

The first term is the kinetic energy

$$T(v) := m_0 c^2 \left(\frac{1}{\sqrt{1-\beta^2}} - 1\right) \tag{2.7}$$

A power series expansion in v gives

$$T(v) = \frac{m_0}{2} v^2 \left(1 + \frac{3}{4}\beta^2 + \frac{5}{8}\beta^4 + \cdots\right)$$

the first term being the familiar nonrelativistic approximation. The second term in (2.6) is the usual potential energy of classical mechanics,

$$V(\mathbf{r}) := Q\Phi(\mathbf{r}) \tag{2.8}$$

The functions $\Phi(\mathbf{r})$ and hence $V(\mathbf{r})$ are unique apart from the choice of an arbitrary additive constant. The total energy E_0 depends on the choice of this constant and on the initial conditions of the trajectory in question, a trivial point in theory but important in many practical situations.

2.3 The acceleration potential

For practical calculations, it is of great importance that virtually all *scalar* kinetic quantities can be represented as unique functions in space, the constant E_0 being a free parameter. The electrostatic potential $\Phi(\mathbf{r})$ is uniquely defined by its boundary values at the surfaces of the electrodes. In electron optics, the cathode surface in the electron gun is usually chosen to be the equipotential surface $\Phi(\mathbf{r}) = 0$. In this volume we shall adopt this most convenient gauge. The constant E_0 then has a very concrete physical meaning: it is the initial kinetic energy of the corresponding electron trajectory at the cathode surface. This is a small *positive* quantity of the order of an electronvolt. The simplification $E_0 = 0$, common in the literature on electron optics, is too strong a restriction, since this excludes the treatment of energy distributions in electron beams. Here, therefore, the convention $E_0 = 0$ will be adopted in all practical calculations in which the electron energy distribution is not in question. In the remainder of the present chapter, however, the analysis will be completely general.

Once the function $\Phi(r)$ and the constant E_0 have been specified, all the other scalar functions are uniquely defined. The kinetic energy is given by

$$T(\mathbf{r}, E_0) = E_0 - Q\Phi(\mathbf{r}) \tag{2.9}$$

Since T can be rewritten as $T = m_0 c^2 (\gamma - 1)$, the dilatation factor γ is a function of position:

$$\gamma(\mathbf{r}, E_0) = \frac{1}{\sqrt{1-\beta^2}} = 1 + \frac{E_0 - Q\Phi(\mathbf{r})}{m_0 c^2} \tag{2.10}$$

Equation (2.9) also determines the absolute value of the kinetic momentum, $|\mathbf{g}| = g(\mathbf{r})$. In order to find the corresponding function, we first solve $g = m_0 v (1-\beta^2)^{-1/2}$ for $v = v(g)$ and then substitute the resulting expression in (2.7), which yields the well-known formula

$$m_0 c^2 + T = c\sqrt{(m_0 c)^2 + g^2} \equiv mc^2 \tag{2.11}$$

This expression will prove to be of great importance in the Hamiltonian theory. Here we solve (2.11) for g and obtain

$$g \equiv |\mathbf{g}| = \sqrt{2m_0 T \left(1 + \frac{T}{2m_0 c^2}\right)} \tag{2.12}$$

Substituting for T from (2.9), we find

$$g(\mathbf{r}, E_0) = \sqrt{2m_0 \left(E_0 - Q\Phi(\mathbf{r})\right)\left(1 + \frac{E_0 - Q\Phi(\mathbf{r})}{2m_0 c^2}\right)} \tag{2.13}$$

By means of (2.10) and (2.13), the absolute value of the velocity can also be determined as a function, $v(\mathbf{r}, E_0) = g/m_0\gamma$. In the nonrelativistic approximation (2.10), (2.13) and the expression for v simplify to the well-known formulae

$$\gamma \approx 1 \quad , \quad g \approx \sqrt{2m_0 (E_0 - Q\Phi)} \quad , \quad v \approx \sqrt{\frac{2}{m_0}(E_0 - Q\Phi)} \tag{2.14}$$

2.3 THE ACCELERATION POTENTIAL

Since the kinematic functions will be required very often in electron optical calculations, we introduce certain quantities to simplify the notation:

$$\Phi_0 := E_0/e \tag{2.15}$$

$$\epsilon := \frac{e}{2m_0 c^2} = 0.9785 \text{ MV}^{-1} \tag{2.16}$$

$$\eta := \sqrt{\frac{e}{2m_0}} = 2.965 \times 10^5 \text{ C}^{\frac{1}{2}} \text{ kg}^{-\frac{1}{2}} \tag{2.17}$$

Equation (2.13) now becomes

$$g(\mathbf{r}, E_0) = \sqrt{2m_0 e(\Phi_0 + \Phi)\{1 + \epsilon(\Phi_0 + \Phi)\}}$$

Since the radicand will be required very frequently, it is convenient to introduce a new function

$$\hat{\Phi}(\mathbf{r}, \Phi_0) := (\Phi_0 + \Phi(\mathbf{r}))\{1 + \epsilon(\Phi_0 + \Phi(\mathbf{r}))\} \geq 0 \tag{2.18}$$

called the acceleration potential. The circumflex ˆ will be added to all functions and constants defined in terms of $\hat{\Phi}$ rather than Φ. It is now possible to express all the other kinematic functions in terms of $\hat{\Phi}$. Obviously

$$g = \sqrt{2m_0 e \hat{\Phi}} \tag{2.19}$$

On substituting this in (2.11), we find

$$T = m_0 c^2 \left(\sqrt{1 + 4\epsilon \hat{\Phi}} - 1 \right) \equiv \frac{2e\hat{\Phi}}{1 + \sqrt{1 + 4\epsilon \hat{\Phi}}} \tag{2.20}$$

Since $T = m_0 c^2 (\gamma - 1)$, the dilatation factor γ can be expressed as a function of $\hat{\Phi}$ or Φ:

$$\gamma = \sqrt{1 + 4\epsilon \hat{\Phi}} = 1 + 2\epsilon\{\Phi(\mathbf{r}) + \Phi_0\} \tag{2.21}$$

Using (2.19), (2.21) and (2.17), the velocity becomes

$$v \equiv \frac{g}{m_0 \gamma} = 2\eta \sqrt{\frac{\hat{\Phi}}{1 + 4\epsilon \hat{\Phi}}} \tag{2.22}$$

Even the electric field strength $\mathbf{E}$ can be expressed in terms of $\hat{\Phi}$ and its gradient: differentiation of (2.18) results first in

$$\nabla \hat{\Phi} = \left\{ 1 + 2\epsilon(\Phi_0 + \Phi) \right\} \nabla \Phi$$

From (2.21), we see that the factor multiplying $\nabla \Phi$ is equal to γ, so that finally

$$\boldsymbol{E} = -\frac{1}{\gamma}\nabla\hat{\Phi} = -\left(1+4\epsilon\hat{\Phi}\right)^{-\frac{1}{2}}\nabla\hat{\Phi} = \frac{v}{Q}\nabla g \qquad (2.23)$$

the last form being *always* valid in static fields. Thus no independent kinematic function other than $\hat{\Phi}(\boldsymbol{r}, \Phi_0)$ is needed. The gain obtained by this simplification will be obvious later. In order to lighten the notation, we shall omit the argument Φ_0 whenever we are not concerned with energy distributions and chromatic effects.

2.4 Definition of coordinate systems

Many of the subsequent calculations are most favourably carried out by explicit representation in some suitably chosen coordinate system. In order to avoid repetition, we introduce general definitions and standard notations here, which we shall use consistently provided there is no risk of confusion. Any necessary changes of the notation will be mentioned explicitly.

Cartesian coordinates are denoted by (x, y, z) or by subscripts (x_1, x_2, x_3); the cartesian components of any vector and the unit vectors $\boldsymbol{i}$ in the three cartesian directions will always be indicated by the corresponding subscripts. Even when using orthogonal curvilinear coordinates, the unit vector indicating the direction of the gradient of the coordinate in question will be denoted by $\boldsymbol{i}$ and the corresponding subscript. Examples are given below. In sums of cartesian products, the familiar summation convention will be used wherever this does not cause confusion.

Any vector $\boldsymbol{a}$ is thus represented by the equivalent notations

$$\boldsymbol{a} = a_x\boldsymbol{i}_x + a_y\boldsymbol{i}_y + a_z\boldsymbol{i}_z = \sum_{j=1}^{3} a_j\boldsymbol{i}_j = a_j\boldsymbol{i}_j \qquad (2.24)$$

and scalar products are written

$$\boldsymbol{a}\cdot\boldsymbol{b} = \sum_{j=1}^{3} a_j b_j = a_j b_j \qquad (2.25)$$

As well as cartesian coordinates, we shall need cylindrical coordinates (z, r, φ) and spherical coordinates $(R, \underline{G}, \varphi)$. All these coordinate systems are related by the familiar transformations

$$\begin{aligned} x &= R\sin\vartheta\cos\varphi = r\cos\varphi \\ y &= R\sin\vartheta\sin\varphi = r\sin\varphi \\ z &= R\cos\vartheta \quad , \quad r = R\sin\vartheta \\ 0 &\leq r \leq R \quad , \quad 0 \leq \vartheta < \pi \quad , \quad 0 \leq \varphi < 2\pi \end{aligned} \qquad (2.26)$$

2.4 DEFINITION OF COORDINATE SYSTEMS

The choice of notation for the spherical coordinate R is unusual but spherical coordinates are very rarely used in electron optics. The only important case is the treatment of cathode tips (Chapter 45). Cylindrical coordinates are very frequently used and the corresponding notation is as simple as possible.

The element of length, ds, is given by

$$ds^2 = dx^2 + dy^2 + dz^2$$
$$= dz^2 + dr^2 + r^2 d\varphi^2$$
$$= dR^2 + R^2 d\vartheta^2 + R^2 \sin^2\vartheta \, d\varphi^2$$

The position vector in particular is given by

$$\boldsymbol{r} = x\boldsymbol{i}_x + y\boldsymbol{i}_y + z\boldsymbol{i}_z = z\boldsymbol{i}_z + r\boldsymbol{i}_r = R\boldsymbol{i}_R \qquad (2.27a)$$

and the velocity by

$$\boldsymbol{v} = \dot{\boldsymbol{r}} = \dot{x}\boldsymbol{i}_x + \dot{y}\boldsymbol{i}_y + \dot{z}\boldsymbol{i}_z = \dot{z}\boldsymbol{i}_z + \dot{r}\boldsymbol{i}_r + r\dot{\varphi}\boldsymbol{i}_\varphi$$
$$= \dot{R}\boldsymbol{i}_R + R\dot{\vartheta}\boldsymbol{i}_\vartheta + R\sin\vartheta\,\dot{\varphi}\boldsymbol{i}_\varphi \qquad (2.27b)$$

The components of arbitrary vectors, characterized by the corresponding subscripts, have the value of the corresponding projections; for example, if $\boldsymbol{a} = a_R\boldsymbol{i}_R + a_\vartheta\boldsymbol{i}_\vartheta + a_\varphi\boldsymbol{i}_\varphi$, then $a_R = \boldsymbol{a}\cdot\boldsymbol{i}_R$. This is clearly different from the familiar, but less convenient, covariant formalism, which we do not need here. A simple consequence of our definitions is that

$$|\boldsymbol{a}|^2 = a_j a_j = a_z^2 + a_r^2 + a_\varphi^2 = a_R^2 + a_\vartheta^2 + a_\varphi^2 \qquad (2.28)$$

No metric tensor is needed in this context. To facilitate the evaluation of vector products, it is useful to remember that the basic vectors, $\boldsymbol{i}_z$, $\boldsymbol{i}_r$, $\boldsymbol{i}_\varphi$ and $\boldsymbol{i}_R$, $\boldsymbol{i}_\vartheta$, $\boldsymbol{i}_\varphi$ respectively, form positively orientated orthonormalized bases.

In order to avoid giving explicit cartesian representations of vector differentiations in different vector spaces, we introduce the familiar abbreviation

$$\frac{\partial}{\partial \boldsymbol{u}} := \boldsymbol{i}_x\frac{\partial}{\partial u_x} + \boldsymbol{i}_y\frac{\partial}{\partial u_y} + \boldsymbol{i}_z\frac{\partial}{\partial u_z} = \boldsymbol{i}_j\frac{\partial}{\partial u_j} \qquad (2.29)$$

for differentiation in the space in question. Ordinary vector differentiation is denoted by $\nabla := \partial/\partial \boldsymbol{r}$. Some common representations in cylindrical coordinates are as follows:

$$\nabla = \boldsymbol{i}_z\frac{\partial}{\partial z} + \boldsymbol{i}_r\frac{\partial}{\partial r} + \boldsymbol{i}_\varphi\frac{1}{r}\frac{\partial}{\partial \varphi} \qquad (2.29a)$$

$$\nabla^2 = \frac{\partial^2}{\partial z^2} + \frac{\partial^2}{\partial r^2} + \frac{1}{r}\frac{\partial}{\partial r} + \frac{1}{r^2}\frac{\partial^2}{\partial \varphi^2} \qquad (2.29b)$$

These are extensively used in Part II. Partial derivatives are denoted in various different ways, as shown in the following examples:

$$\frac{\partial^2 V}{\partial x^2} \equiv \partial^2_{xx} V \equiv V_{|xx}$$

$$\frac{\partial^2 V}{\partial x \partial y} \equiv \partial^2_{xy} V \equiv V_{|xy}$$

Expressions of this kind will appear intermittently in later chapters.

2.5 Conservation of axial angular momentum

This conservation law may be obtained very elegantly by use of the Lagrange formalism but the subsequent elementary presentation brings out clearly its physical meaning.

We now assume that the electron optical device is rotationally symmetric about an optic axis. This assumption is valid in all perfect round lenses and mirrors. The fields in these devices are usually also static, but this additional assumption is not necessary. The following considerations remain valid in time-dependent systems and in ion optics. As everywhere in physics, the conservation of axial angular momentum is a direct consequence of the assumption of rotational symmetry.

On forming the vector product of r with (2.1), we see that the expression on the left-hand side is the derivative of the familiar mechanical angular momentum vector:

$$\frac{d}{dt}(m\mathbf{r} \times \mathbf{v}) = Q\mathbf{r} \times (\mathbf{E} + \mathbf{v} \times \mathbf{B}) \qquad (2.30)$$

On the right-hand side, only the component parallel to the axis can be represented as a total derivative, and we therefore confine our attention to the evaluation of this component.

In view of the assumed symmetry, we may anticipate that cylindrical coordinates (z, r, φ) will be most appropriate, the z-axis coinciding with the optic axis. The axial component of (2.30) then becomes

$$\frac{d}{dt}(mr^2\dot{\varphi}) = Qr\{E_\varphi + (\mathbf{v} \times \mathbf{B})_\varphi\} = Qr(E_\varphi + \dot{z}B_r - \dot{r}B_z) \qquad (2.31)$$

In order to transform the expression on the right-hand side into a total derivative with respect to time, it is necessary to introduce the *magnetic flux function* $\Psi(z, r, t)$. This is defined as the magnetic flux through a

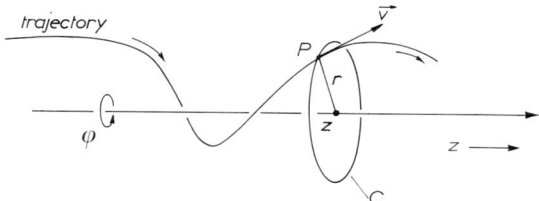

Fig. 2.1: Trajectory of a charged particle; a coaxial circular disc C is associated with an arbitrary point P with coordinates $(z, r, \varphi; t)$. This circle travels along the axis and its radius varies as the point P moves along the trajectory.

coaxial circular disc C, located in a plane z=const and of radius r. The coordinates (z, r, φ) specify the instantaneous position of the charged particle on its trajectory. This is illustrated in Fig. 2.1.

This function $\Psi(z, r, t)$ is easily evaluated by expressing the surface element $d\boldsymbol{a} = da\, \boldsymbol{i}_z = 2\pi r' dr' \boldsymbol{i}_z$ in cylindrical coordinates, giving

$$\Psi(z,r,t) := \int_C \boldsymbol{B} \cdot d\boldsymbol{a} = 2\pi \int_0^r r' B_z(z, r', t)\, dr' \tag{2.32}$$

Differentiation with respect to r gives immediately

$$B_z(z, r, t) = \frac{1}{2\pi r} \frac{\partial \Psi}{\partial r} \tag{2.33}$$

Using the condition div $\boldsymbol{B} = 0$, the second representation

$$B_r(z, r, t) = -\frac{1}{2\pi r} \frac{\partial \Psi}{\partial z} \tag{2.34}$$

can be derived (see Section 6.4). Finally the integral induction law $\oint \boldsymbol{E} \cdot d\boldsymbol{r} = -\partial \Psi / \partial t$, applied to the circumference of the disc C, results in

$$E_\varphi(z, r, t) = -\frac{1}{2\pi r} \frac{\partial \Psi}{\partial t} \tag{2.35}$$

Introducing (2.33), (2.34) and (2.35) into (2.31) we obtain:

$$\frac{d}{dt}(mr^2 \dot{\varphi}) = -\frac{Q}{2\pi} \left(\frac{\partial \Psi}{\partial t} + \dot{z} \frac{\partial \Psi}{\partial z} + \dot{r} \frac{\partial \Psi}{\partial r} \right) = -\frac{Q}{2\pi} \frac{d\Psi}{dt} \tag{2.36}$$

Integration with respect to time leads to the conservation law for the axial angular momentum:

$$N := \frac{m_0 r^2 \dot{\varphi}}{\sqrt{1-\beta^2}} + \frac{Q}{2\pi} \Psi(z, r, t) = \text{const} \tag{2.37}$$

This expression differs from that familiar in classical mechanics by the presence of the important second term, which contains the electromagnetic interaction in integral form. Its physical origin and meaning appear very clearly on comparing (2.31) with (2.36). The *electric* torque, QrE_φ, is an induction effect in a *fixed* loop (specified by z and r). The *magnetic* part of the torque, $Qr(\boldsymbol{v} \times \boldsymbol{B})_\varphi$, is the corresponding induction effect in a loop *moving* with the particle and is hence a consequence of the implicit variation of Ψ with time: $-Q\boldsymbol{v} \cdot \nabla\Psi/2\pi$. This latter part remains nonzero even in static magnetic round lenses where it forces the particle beam to rotate about the optic axis, a fact of great importance in the physics of magnetic lenses.

In the vicinity of the optic axis, this rotation is of most interest and can be easily calculated. In a sufficiently small paraxial domain, the magnetic field can be considered as radially homogeneous; the magnetic flux is then $\Psi = \pi r^2 B(z,t)$, the function $B(z,t)$ being the axial flux density. Furthermore, we have $\dot{r}^2 + r^2\dot{\varphi}^2 \ll \dot{z}^2$. From (2.37), with $m(z,t) = m_0(1 - \dot{z}^2/c^2)^{-1/2}$, we then obtain

$$N = r^2\left\{m(z,t)\dot{\varphi} + \frac{1}{2}QB(z,t)\right\}$$

We now confine our considerations to trajectories that intersect the optic axis at some point. For such trajectories, N vanishes and we obtain

$$\dot{\varphi} = -\frac{QB(z,t)}{2m(z,t)} \tag{2.38}$$

This is the local and instantaneous value of the familiar *Larmor* frequency.

In the important case of electron trajectories in static round lenses, it is convenient to represent the azimuth as a function of z. On the optic axis the relation

$$m\dot{z} = \sqrt{2m_0 e\hat{\phi}_0(z)}$$

is satisfied, $\hat{\phi}_0(z)$ being the axial acceleration potential. Using $d\varphi/dz = \dot{\varphi}/\dot{z}$ and integrating with respect to z we find

$$\varphi(z) = \eta \int_{z_0}^{z} \hat{\phi}_0^{-\frac{1}{2}}(z')B(z')dz' + \varphi(z_0) \tag{2.39}$$

In Part III we shall show that the trajectory equations simplify considerably in a coordinate system that is twisted round the optic axis by this local angle of rotation.

3
Different Forms of Trajectory Equations

The Lorentz equation (2.1) is generally valid but not always convenient. In very many practical applications, expressions for the trajectories of the form $r = r(t)$ are of no interest; it is their geometrical shape that is required. In devices in which the fields E and B depend explicitly on the time, it is rarely possible to eliminate the latter from the trajectory equation. We therefore confine the following discussion to systems with static fields. In order to avoid repetition, we shall give here only those forms of the trajectory equations that cannot (or at least, only with greater difficulty) be derived from variational principles but are yet of practical importance.

3.1 Parametric representation in terms of the arc-length

The differential arc-length is given by $ds := |dr| = v\,dt$. The transformation of the differential operator is hence given by $d/dt = v\,d/ds$. Introducing this into (2.1) and noting that $v/v = dr/ds$, we obtain

$$\frac{dg}{ds} = \frac{QE}{v} + Q\frac{dr}{ds} \times B$$

It is now of great importance that v and $g = mv$ are unique functions of r, as explained in Chapter 2. Thus the time is already eliminated. By means of (2.23), the electric field vector can also be eliminated, giving

$$\frac{d}{ds}\left\{g(r)\frac{dr}{ds}\right\} = \operatorname{grad} g(r) + Q\frac{dr}{ds} \times B(r) \qquad (3.1)$$

In the absence of the magnetic term, this trajectory equation is even valid for the propagation of light in matter provided that the geometric approximation is adequate (see e.g. Born and Wolf, 1959 eq.(3.2.2); Kasper, 1972). In this case the kinetic momentum is given by $g = \hbar k_0 n(r, k_0)$, k_0 being the mean wave number (2π/wavelength) in the vacuum and $n(r, k_0)$ the corresponding index of refraction in the material; $\hbar = h/2\pi$ is Planck's constant. Hence (3.1) is a very general trajectory equation.

3. DIFFERENT FORMS OF TRAJECTORY EQUATIONS

Some useful relations can be derived from (3.1) by calculating its components with respect to the orthonormal basis vectors

$$t := \frac{d\boldsymbol{r}}{ds} \quad , \quad \boldsymbol{n} := R\frac{dt}{ds} \quad , \quad \boldsymbol{b} = \boldsymbol{t} \times \boldsymbol{n} \tag{3.2}$$

the local tangent, principal normal and binormal respectively, R being the (positive) radius of curvature. Scalar multiplication of (3.1) with each of these unit vectors in turn, using $dg/ds = \boldsymbol{t} \cdot \mathrm{grad}\, g$ and some elementary vector operations, yields:

$$\frac{g}{R} = \boldsymbol{n} \cdot \mathrm{grad}\, g - Q\boldsymbol{b} \cdot \boldsymbol{B} \equiv Q(\boldsymbol{n} \cdot \boldsymbol{E}/v - \boldsymbol{b} \cdot \boldsymbol{B}) \tag{3.3}$$

$$0 = \boldsymbol{b} \cdot \mathrm{grad}\, g + Q\boldsymbol{n} \cdot \boldsymbol{B} \equiv Q(\boldsymbol{b} \cdot \boldsymbol{E}/v + \boldsymbol{n} \cdot \boldsymbol{B}) \tag{3.4}$$

In a purely *magnetic* field ($\boldsymbol{E} \equiv 0$), (3.4) shows that the vector $\boldsymbol{n}$ is always orthogonal to $\boldsymbol{B}$. From (3.3) the (absolute) curvature is found to be

$$\frac{1}{R} = -\frac{Q}{g}\boldsymbol{b} \cdot \boldsymbol{B} \equiv \frac{1}{g}|Q\boldsymbol{t} \times \boldsymbol{B}| \tag{3.5}$$

For electrons the curvature can be rewritten with the aid of the formulae of Section 2.3 as

$$\frac{1}{R} = \frac{\eta}{\sqrt{\hat{U}}}|\boldsymbol{t} \times \boldsymbol{B}| \tag{3.6}$$

where $\hat{U} := \hat{\Phi}$ is now a *constant* acceleration potential. Equation (3.1) then simplifies to:

$$\frac{d^2\boldsymbol{r}}{ds^2} = \frac{\eta}{\sqrt{\hat{U}}}\boldsymbol{B}(\boldsymbol{r}) \times \frac{d\boldsymbol{r}}{ds} \tag{3.7}$$

For the circular motion of charged particles in a homogeneous magnetic field the familiar relation

$$BR = g|Q|^{-1} \tag{3.7a}$$

is a simple consequence of (3.5). This 'BR-product' is of great importance in the design of spectrometers and analysers (see Part X).

In a purely *electrostatic* field the binormal $\boldsymbol{b}$ is always orthogonal to $\boldsymbol{E}$, and (3.3) can then be rewritten in the familiar self-evident form

$$\frac{mv^2}{R} = |QE_n| = Q|\boldsymbol{t} \times \boldsymbol{E}| \tag{3.8}$$

The dependence of the left-hand side on the acceleration potential can be obtained by use of the kinematic functions, given in Section 2.3.

3.2 Relativistic proper-time representation

The arc-length is a highly unsuitable trajectory parameter for numerical trajectory computations in electron guns and mirrors, since the radius of curvature varies over an extremely wide range of values. This implies that the integration step length Δs must also vary considerably. The time would be a much better parameter but has the disadvantage that the range of values is now too small. Other parametrizations such as that of Kel'man et al. (1972, 1973) are of specific interest for mirrors. These will be dealt with in Chapter 18.

A favourable parameter, which exhibits none of the disadvantages mentioned above and is very advantageous, particularly in the numerical investigation of electron guns, is defined by (Kasper, 1985):

$$d\sigma = u\sqrt{1-\beta^2}\,dt = u\,dt' \qquad (u = \text{const}) \tag{3.9}$$

dt' being the relativistic proper-time element, observable in the frame of reference of the electron. The observable nature of the proper-time element is, of course, purely abstract, but the trajectory equation can be slightly simplified by introducing (3.9). A particularly suitable choice of the constant factor u is

$$u = \sqrt{\frac{2e}{m_0}\hat{U}} = 2\eta\sqrt{\hat{U}} \tag{3.10}$$

the second constant $\hat{U}$ being the acceleration potential at some suitably chosen fixed point of reference. From (3.9) and (3.10), it is obvious that u has the dimensions of a speed and thus $d\sigma$ that of a length, though $d\sigma$ is proportional in magnitude to dt'.

Equation (3.1) can now be straightforwardly transformed by means of the operator relation

$$\frac{1}{\sqrt{1-\beta^2}}\frac{d}{dt} = u\frac{d}{d\sigma}$$

The kinematic momentum is given by

$$\boldsymbol{g} = m_0 u\frac{d\boldsymbol{r}}{d\sigma} = \sqrt{2m_0 e\hat{U}}\,\frac{d\boldsymbol{r}}{d\sigma} \tag{3.11}$$

which is already a simplification, since the factor before the derivative is now a constant, rather than a complicated function. Multiplying by the factor $\gamma = (1-\beta^2)^{-1/2}$ and using the relations given above, the Lorentz equation (2.1) for electrons ($Q = -e$) now transforms into

$$m_0 u^2\frac{d^2\boldsymbol{r}}{d\sigma^2} = -e\gamma\boldsymbol{E} + e\boldsymbol{B}\times\frac{d\boldsymbol{r}}{d\sigma}$$

The electric term is eliminated by means of (2.23). Using (3.10) we obtain finally

$$\frac{d^2 \boldsymbol{r}}{d\sigma^2} = \frac{1}{2}\mathrm{grad}\left(\frac{\hat{\Phi}(\boldsymbol{r})}{\hat{U}}\right) + \frac{\eta}{\sqrt{\hat{U}}}\boldsymbol{B}(\boldsymbol{r}) \times \frac{d\boldsymbol{r}}{d\sigma} \qquad (3.12)$$

The normalization constant $\hat{U}$ can be set arbitrarily to any positive value. In devices having an asymptotically field-free domain, we may choose any reference point $\boldsymbol{R}_0$ located in this domain: $\hat{U} = \hat{\Phi}(\boldsymbol{R}_0)$. This implies that $d\sigma = ds$ in that domain. In an electron microscope a good choice for the reference point $\boldsymbol{R}_0$ is the centre of the recording screen. Then $\hat{\Phi}(\boldsymbol{R}_0)$ is usually the maximum of $\hat{\Phi}$. Without loss of generality, the starting point of each electron trajectory at the cathode surface may be the point with $\sigma = 0$. The final value of σ is then slightly longer than the length L of the device, roughly $\sigma_{max} \sim 1.5L$. There is thus no need for any special precautions to avoid inconvenient scales.

The conservation laws for energy and axial angular momentum can also be represented in a very convenient form. By scalar multiplication of (3.12) with $d\boldsymbol{r}/d\sigma$ and integration with respect to σ we find

$$\left|\frac{d\boldsymbol{r}}{d\sigma}\right|^2 = \frac{\hat{\Phi}(\boldsymbol{r})}{\hat{U}} \qquad (3.13)$$

With the conventions adopted in Section 2.3, the constant of integration must be zero.

For motion in static fields, (2.37) can be simplified with the aid of kinematic functions, the result being

$$r^2\frac{d\varphi}{d\sigma} \equiv x\frac{dy}{d\sigma} - y\frac{dx}{d\sigma} = \frac{\eta}{\sqrt{\hat{U}}}\left(\frac{N}{e} + \frac{\Psi(z,r)}{2\pi}\right) \qquad (3.14)$$

These conservation laws are useful as additional checks of the accuracy in numerical computations. In practice, the evaluation of (3.12) has proved to be the most successful method of calculating Lorentz trajectories.

3.3 The cartesian representation

The various representations of the trajectory equation derived hitherto are suitable for calculating individual trajectories from given initial conditions but they are hardly suitable for developing a systematic theory of focusing and aberrations. For this, a cartesian representation $x = x(z)$, $y = y(z)$ is preferable. Such a formalism is possible if the electron optical device in

3.3 THE CARTESIAN REPRESENTATION

question has a straight optic axis, if $\hat{\Phi} \gtrsim 20$ eV and if the slopes $x'(z)$, $y'(z)$ of *all* trajectories remain finite. The last two conditions are not satisfied in electron guns and mirrors and it is for this reason that there is at present no entirely satisfactory theory of the aberrations in these devices.

In what follows, we shall use the explicit cartesian representation (2.27) with $x = x(z)$, $y = y(z)$. Differentiation with respect to z will be denoted by a prime, thus

$$\varrho t(z) \equiv r'(z) = x'(z)i_x + y'(z)i_y + i_z \tag{3.15}$$

Since $|t| = 1$, the absolute value of r' is:

$$\varrho := |r'| = \sqrt{1 + x'^2 + y'^2} \tag{3.16}$$

This function has a very simple geometric meaning: $\varrho^{-1} = t \cdot i_z = \cos\alpha$, α being the angle between the local tangent and the optic axis. This holds even for skew trajectories.

The required cartesian representation of the trajectory equation is most easily obtained by substituting

$$\frac{d}{ds} = \frac{dz}{ds}\frac{d}{dz} = \frac{1}{\varrho}\frac{d}{dz}$$

in (3.1), giving

$$\frac{1}{\varrho}\frac{d}{dz}\left(\frac{g}{z}\frac{dr}{dz}\right) = \operatorname{grad} g + \frac{Q}{\varrho} r' \times B$$

Expanding the derivative on the left-hand side yields

$$\frac{g}{\varrho^2}r'' + \frac{r'}{\varrho}\frac{d}{dz}\left(\frac{g}{\varrho}\right) = \operatorname{grad} g + \frac{Q}{\varrho} r' \times B \tag{3.17}$$

These are three scalar differential equations for the two functions $x(z)$, $y(z)$; the third equation

$$\frac{1}{\varrho}\frac{d}{dz}\left(\frac{g}{\varrho}\right) = \frac{\partial g}{\partial z} + \frac{Q}{\varrho} i_z \cdot (r' \times B) \tag{3.18}$$

is therefore dependent on the first two and may be omitted. In fact, it is possible to derive the x- and y- components of (3.17) directly from a two-dimensional variational principle but not (3.18). Here we shall use (3.18) to simplify (3.17) by eliminating the second term. Multiplying (3.18) by r' and subtracting the result from (3.17), we find

$$\frac{g}{\varrho^2}r'' = \operatorname{grad} g - r'\frac{\partial g}{\partial z} + \frac{Q}{\varrho}\left\{r' \times B - i_z \cdot (r' \times B) r'\right\}$$

3. DIFFERENT FORMS OF TRAJECTORY EQUATIONS

The third component of this is a trivial identity. The two components of interest are given explicitly by

$$x'' = \frac{\varrho^2}{g}\left(\frac{\partial g}{\partial x} - x'\frac{\partial g}{\partial z}\right) + \frac{Q\varrho}{g}\left\{y'(B_z + x'B_x) - B_y(1 + x'^2)\right\}$$

$$y'' = \frac{\varrho^2}{g}\left(\frac{\partial g}{\partial y} - y'\frac{\partial g}{\partial z}\right) + \frac{Q\varrho}{g}\left\{-x'(B_z + y'B_y) + B_x(1 + y'^2)\right\}$$

The magnetic terms can be rewritten in a more compact form by introducing the tangential component of $\boldsymbol{B}$:

$$B_t := \boldsymbol{t} \cdot \boldsymbol{B} = \frac{1}{\varrho}(B_z + x'B_x + y'B_y) \tag{3.19}$$

and we finally obtain

$$\begin{aligned} x'' &= \frac{\varrho^2}{g}\left(\frac{\partial g}{\partial x} - x'\frac{\partial g}{\partial z}\right) + \frac{Q\varrho^2}{g}(y'B_t - \varrho B_y) \\ y'' &= \frac{\varrho^2}{g}\left(\frac{\partial g}{\partial y} - y'\frac{\partial g}{\partial z}\right) + \frac{Q\varrho^2}{g}(-x'B_t + \varrho B_y) \end{aligned} \tag{3.20}$$

These trajectory equations are valid for all charged particles, provided that the conditions mentioned above are satisfied. With $Q = 0$ and $g = \hbar k_0 n(\boldsymbol{r})$, they are even valid for light rays, $n(\boldsymbol{r})$ being the optical index of refraction. The vacuum momentum $\hbar k_0$ cancels out and we obtain the ray equations

$$x'' = \frac{\varrho^2}{n}\left(\frac{\partial n}{\partial x} - x'\frac{\partial n}{\partial z}\right) \quad , \quad y'' = \frac{\varrho^2}{n}\left(\frac{\partial n}{\partial y} - y'\frac{\partial n}{\partial z}\right) \tag{3.21}$$

For electron trajectories, (2.19) and (2.17) may be used, whereupon (3.20) become

$$\begin{aligned} x'' &= \frac{\varrho^2}{2\hat{\Phi}}\left(\frac{\partial \hat{\Phi}}{\partial x} - x'\frac{\partial \hat{\Phi}}{\partial z}\right) + \frac{\eta\varrho^2}{\sqrt{\hat{\Phi}}}(\varrho B_y - y'B_t) \\ y'' &= \frac{\varrho^2}{2\hat{\Phi}}\left(\frac{\partial \hat{\Phi}}{\partial y} - y'\frac{\partial \hat{\Phi}}{\partial z}\right) + \frac{\eta\varrho^2}{\sqrt{\hat{\Phi}}}(-\varrho B_x + x'B_t) \end{aligned} \tag{3.22}$$

These trajectory equations are equally well suited for the numerical computation of individual trajectories and for the development of a systematic theory of focusing and aberrations. It must be emphasized that only the existence of a straight optic axis is required, not rotational symmetry about this axis. Thus not only can round lenses be considered here, but also stigmators, systems of multipole lenses and deflection units such as saddle coils.

3.4 Scaling rules

A number of simple scaling rules can be derived for trajectories in static fields. Since the time-dependent form of the trajectories is now of no interest, we confine the discussion to the time-independent form. Scale changes are most easily performed on (3.1). We shall consider only two very important special cases.

For purely *electrostatic* fields, (3.1) becomes linear in $g(\mathbf{r})$, which means that (3.1) is unaffected by a scale-transform $g(\mathbf{r}) = g_0 g^*(\mathbf{r})$ of the kinematic momentum. Any experimental change of scales, however, is made by alteration of the electrode potentials and this affects the kinematic momentum only indirectly. From (2.13), it is obvious that a simple rule can only be expected in the nonrelativistic case $|E_0 - Q\Phi(\mathbf{r})| \ll m_0 c^2$. Since the initial energy E_0 is not constant in a particle beam, a unique change of scale common to all the particles is only possible if $E_0 \ll |Q\Phi|$. For electrons and negatively charged ions, (2.13) then simplifies to:

$$g(\mathbf{r}) = \sqrt{2m_0|Q|\Phi(\mathbf{r})} \quad , \quad \Phi \geq 0 \tag{3.23}$$

which is only valid sufficiently far from the cathode. In these circumstances, a linear scale transform $\Phi(\mathbf{r}) = U_0 \Phi^*(\mathbf{r})$ of the potential is equivalent to such a transform of the kinematic momentum, the relation between the scaling factors being given by $g_0 = U_0^{1/2}$. Introducing (3.23) into (3.1), the factor $(2m_0|Q|)^{1/2}$ is seen to cancel; the trajectory equation then simplifies to

$$\frac{d}{ds}\left(\sqrt{\Phi(\mathbf{r})}\,\frac{d\mathbf{r}}{ds}\right) = \mathrm{grad}\,\sqrt{\Phi(\mathbf{r})} \tag{3.24}$$

which can be rewritten as

$$2\Phi(\mathbf{r})\frac{d^2\mathbf{r}}{ds^2} = \left\{\frac{d\mathbf{r}}{ds} \times \nabla\Phi(\mathbf{r})\right\} \times \frac{d\mathbf{r}}{ds} \tag{3.25}$$

Here the relation $d\Phi/ds = (d\mathbf{r}/ds) \cdot \nabla\Phi$ has been used. This trajectory equation is *linear* in $\Phi(\mathbf{r})$. A linear scale change $\Phi = U_0 \Phi^*$ can be performed most easily by alteration of all the electrode potentials by the same factor, the cathode being $\Phi^* = 0$. From (3.25) it is obvious that the geometric shape of the trajectories depends neither on Q and m_0 nor on the scale factor U_0. These constants affect only the time of propagation, which is of little interest.

These conclusions are together known as the *electrostatic principle*, which also holds for positively charged particles, provided that $\Phi(\mathbf{r})$ is replaced by $|\Phi(\mathbf{r})|$ and $\Phi(\mathbf{r}) = 0$ is now the emitting anode surface. In practice, a consequence of this principle is that an electrostatic microscope can be operated with different kinds of particles and fixed *ratios* of the voltage differences between the acceleration electrodes and the source.

The consequences of a geometric scale change can also be derived from (3.25). A transform

$$r = ar^* \quad , \quad s = as^* \quad , \quad \Phi^*(r^*) = \Phi(r)U_0 \tag{3.26}$$

does not affect the geometric shape of the trajectories, since (3.25) is invariant with respect to (3.26). The magnitude of the trajectories is proportional to the distances between the electrodes if the *shape* of the field is unaltered.

In purely *magnetostatic* devices the solutions of the trajectory equations depend on Q/m_0. We shall confine our considerations to electron motion; we must then investigate the effect of scale changes on (3.7). The constant $\hat{U}$ will now be the relativistic acceleration potential. Introducing the scale transforms

$$r = ar^* \quad , \quad s = as^* \quad , \quad \hat{U} = \hat{U}_0\hat{U}^* \quad , \quad B(r) = B_0 B^*(r^*) \tag{3.27}$$

into (3.7) we find that this trajectory equation remains invariant if the condition

$$aB_0 = \sqrt{\hat{U}_0} \tag{3.28}$$

is satisfied. This can be put into a more practical form by introducing scales for the field-producing currents. The magnetic field strength B is related to the electric current I by an expression of the form $B = \mu I/l$, l being some typical length and μ a permeability. An appropriate scale transform is now given by

$$B = B_0 B^* \quad , \quad I = I_0 I^* \quad , \quad l = al^* \quad , \quad aB_0 = I_0 \tag{3.29}$$

which must be compatible with (3.27). Equation (3.28) now simplifies to

$$I_0^2 = \hat{U}_0 \tag{3.30}$$

This simple scaling rule is of great importance in the practical design of magnetic lenses.

4
Variational Principles

All our analysis has so far been based on the Lorentz equation (2.1). This equation is, however, identical with the Euler–Lagrange equations of Hamilton's variational principle which may hence be regarded as more fundamental. This principle may be stated thus:

$$W := \int_{t_0}^{t_1} L(\boldsymbol{r}, \boldsymbol{v}, t)\, dt = \text{extr.} \tag{4.1}$$

where 'extr.' denotes an extremum or at least a stationary value. The necessary constraints are that t_0, t_1, $\boldsymbol{r}(t_0)$ and $\boldsymbol{r}(t_1)$ remain fixed and that the variable of integration must not be varied: $\delta t = 0$.

4.1 The Lagrange formalism

In charged particle dynamics the integrand L, the Lagrangian, takes the form

$$L = m_0 c^2 (1 - \sqrt{1 - \beta^2}) + Q(\boldsymbol{v} \cdot \boldsymbol{A} - \Phi) \tag{4.2}$$

$\Phi(\boldsymbol{r}, t)$ and $\boldsymbol{A}(\boldsymbol{r}, t)$ being the electromagnetic potentials (e.g. Goldstein, 1959). These are related to the field vectors $\boldsymbol{E}$ and $\boldsymbol{B}$ by

$$\boldsymbol{B}(\boldsymbol{r}, t) = \operatorname{curl} \boldsymbol{A}(\boldsymbol{r}, t) \tag{4.3}$$

$$\boldsymbol{E}(\boldsymbol{r}, t) = -\operatorname{grad} \Phi(\boldsymbol{r}, t) - \frac{\partial}{\partial t} \boldsymbol{A}(\boldsymbol{r}, t) \tag{4.4}$$

These relations do *not* provide a unique definition of the potentials; in other words, the same field vectors $\boldsymbol{E}$ and $\boldsymbol{B}$ may be obtained from *different* sets of potentials. The consequences of adopting different gauges for Φ and $\boldsymbol{A}$ are discussed in Section 5.5. Since only $\boldsymbol{E}$ and $\boldsymbol{B}$ have physical significance, but not Φ and $\boldsymbol{A}$, the results of all calculations should be presented in a gauge-invariant form.

It is convenient to rewrite the Lagrangian, L (4.2), as $L = T^* - V^*$ with

$$T^*(v) := m_0 c^2 (1 - \sqrt{1 - \beta^2}) \tag{4.5}$$

$$V^*(\boldsymbol{r}, \boldsymbol{v}, t) := Q\{\Phi(\boldsymbol{r}, t) - \boldsymbol{v} \cdot \boldsymbol{A}(\boldsymbol{r}, t)\} \tag{4.6}$$

since this closely resembles the familiar classical form $L = T - V$. The function $T^*(v)$, known as the kinetic potential, is closely related to the kinetic energy T, by $T^* = T(1 - \beta^2)^{1/2}$ (see 2.7). The function V^* is a generalization of the familiar potential energy V (2.8), since it is now velocity-dependent.

Since there are no geometric constraints in charged particle dynamics, we adopt cartesian coordinates. The Euler–Lagrange equations

$$\frac{d}{dt}\left(\frac{\partial L}{\partial \dot{x}_j}\right) - \frac{\partial L}{\partial x_j} = 0 \qquad (j = 1, 2, 3) \tag{4.7a}$$

are obviously equivalent to

$$\frac{d}{dt}\left(\frac{\partial T^*}{\partial v_j}\right) = \frac{d}{dt}\left(\frac{\partial V^*}{\partial v_j}\right) - \frac{\partial V^*}{\partial x_j} \tag{4.7b}$$

Performing the differentiations on the left-hand side, we first obtain

$$\frac{dT^*}{dv} = \frac{m_0 v}{\sqrt{1 - \beta^2}} \equiv g(v) \tag{4.8}$$

But $\partial v/\partial v_j = v_j/v$ and so

$$\frac{\partial T^*}{\partial v_j} = \frac{dT^*}{dv}\frac{\partial v}{\partial v_j} = \frac{m_0 v_j}{\sqrt{1 - \beta^2}} = g_j \quad , \quad j = 1, 2, 3$$

This result is obviously the cartesian representation of the well-known expression for the kinematic momentum. By using (2.29) with $\boldsymbol{u} = \boldsymbol{v}$, the kinematic momentum can be rewritten as

$$\boldsymbol{g}(\boldsymbol{v}) = \frac{\partial T^*}{\partial \boldsymbol{v}} = \frac{m_0 \boldsymbol{v}}{\sqrt{1 - \beta^2}} \tag{4.9}$$

and (4.7b) as

$$\dot{\boldsymbol{g}} = \frac{d}{dt}\left(\frac{\partial V^*}{\partial \boldsymbol{v}}\right) - \frac{\partial V^*}{\partial \boldsymbol{r}} =: \boldsymbol{F} \tag{4.10}$$

We now evaluate the expression $\boldsymbol{F}$ on the right-hand side:

$$\frac{\partial V^*}{\partial \boldsymbol{v}} = -Q\boldsymbol{A} \quad , \quad \frac{\partial V^*}{\partial \boldsymbol{r}} = Q\nabla\Phi - Q\frac{\partial}{\partial \boldsymbol{r}}(\boldsymbol{v} \cdot \boldsymbol{A})$$

and thus

$$\boldsymbol{F} = Q\left(-\frac{d\boldsymbol{A}}{dt} - \nabla\Phi + \frac{\partial}{\partial \boldsymbol{r}}(\boldsymbol{v} \cdot \boldsymbol{A})\right)$$

4.1 THE LAGRANGE FORMALISM

The total derivative $d\boldsymbol{A}/dt$ is given by

$$\frac{d}{dt}\boldsymbol{A}(\boldsymbol{r}(t),t) = \frac{\partial \boldsymbol{A}}{\partial t} + \dot{x}_j \frac{\partial}{\partial x_j}\boldsymbol{A} \equiv \frac{\partial \boldsymbol{A}}{\partial t} + (\boldsymbol{v}\cdot\nabla)\boldsymbol{A}$$

We now make use of the vector identity

$$\boldsymbol{v}\times\operatorname{curl}\boldsymbol{A}(\boldsymbol{r}) = \operatorname{grad}\boldsymbol{v}\cdot\boldsymbol{A}(\boldsymbol{r}) - (\boldsymbol{v}\cdot\nabla)\boldsymbol{A}(\boldsymbol{r})$$

valid for any constant $\boldsymbol{v}$ and any vector function $\boldsymbol{A}(\boldsymbol{r})$. (This may easily be verified by writing the terms out in cartesian coordinates or, more simply still, by introducing tensor notation.) Recalling that $\boldsymbol{v}$ is to be treated as a constant in differentiations with respect to $\boldsymbol{r}$, we find

$$\boldsymbol{F} = Q\left(-\nabla\Phi - \frac{\partial \boldsymbol{A}}{\partial t} + \boldsymbol{v}\times\operatorname{curl}\boldsymbol{A}\right)$$

Substituting for $\boldsymbol{E}$ and $\boldsymbol{B}$ from (4.4) and (4.3), we obtain finally

$$\boldsymbol{F} := \frac{d}{dt}\left(\frac{\partial V^*}{\partial \boldsymbol{v}}\right) - \frac{\partial V^*}{\partial \boldsymbol{r}} = Q(\boldsymbol{E} + \boldsymbol{v}\times\boldsymbol{B}) \tag{4.11}$$

which means that the Lorentz force is the *functional* derivative of the generalized potential V^* given by (4.6). Equation (4.10) in combination with (4.11) is identical with (2.1).

This calculation is only slightly longer than the usual derivation of the Lorentz equation by direct evaluation of (4.7a) without making the separation $L = T^* - V^*$, but it brings out clearly the physical meaning of the various expressions since it uses as few gauge-dependent quantities as possible. The direct evaluation of (4.7a) leads, as a first step, to the definition of the *canonical* momentum

$$p_j := \frac{\partial L}{\partial \dot{x}_j} \quad , \quad j = 1,2,3$$

or, in vector notation and using (4.2)

$$\boldsymbol{p} = \frac{\partial L}{\partial \boldsymbol{v}} = \boldsymbol{g} + Q\boldsymbol{A}(\boldsymbol{r},t) \tag{4.12}$$

Though familiar and frequently used in theoretical physics, this quantity has *no* physical meaning as an observable, since it is gauge-dependent.

A great advantage of the variational calculus is the fact that the value of the integral appearing in (4.1) is invariant with respect to transformations of the coordinates and of the variable of integration. Since it is usually easier to perform these transformations on the Lagrangian than on the corresponding Lorentz equation, new forms of trajectory equations can be derived straightforwardly. Not every useful form can be obtained in this way, however. For instance, it is impossible to derive (3.1) as Euler–Lagrange equations, since the arc-length does not satisfy the necessary constraints and is thus not a permissible variable of integration in the action integral W. An excellent example of the beneficial use of coordinate transformations is given below.

4.2 General rotationally symmetric systems

We assume that the system in question is rotationally symmetric about an optic axis and we introduce the corresponding cylindrical coordinates. The components of $\boldsymbol{E}$ and $\boldsymbol{B}$ must not depend on the azimuth φ, and it may be assumed that the same is true of A_z, A_r, A_φ and Φ. In cylindrical coordinates, the Lagrangian (4.2) is given explicitly by

$$L = m_0 c^2 \left[1 - \left\{ 1 - (\dot{z}^2 + \dot{r}^2 + r^2 \dot{\varphi}^2) c^{-2} \right\}^{\frac{1}{2}} \right]$$
$$+ Q \left(\dot{z} A_z + \dot{r} A_r + r\dot{\varphi} A_\varphi - \Phi \right) \tag{4.13}$$

A_z, A_r, A_φ and Φ being functions of z, r and t only. Hence L does not depend explicitly on φ, which means that φ is a cyclic variable. The corresponding canonical momentum is therefore a constant of motion

$$p_\varphi = \partial L / \partial \dot{\varphi} = \text{const}$$

or

$$p_\varphi = \frac{m_0 r^2 \dot{\varphi}}{\sqrt{1 - \beta^2}} + Q r A_\varphi(z, r, t) = \text{const}$$

Integrating (4.3) over the circular disk C introduced in Section 2.5 and using Stokes's theorem we find

$$\oint \boldsymbol{A} \cdot d\boldsymbol{r} = 2\pi r A_\varphi = \int_C \boldsymbol{B} \cdot d\boldsymbol{a} = \Psi(z, r, t)$$

Thus the constant p_φ is identical with the axial angular momentum N of (2.37). More suggestively, we can use (4.12) to represent the axial angular momentum as

$$N = (\boldsymbol{r} \times \boldsymbol{p})_z \tag{4.14}$$

4.2 GENERAL ROTATIONALLY SYMMETRIC SYSTEMS

but this does *not* mean that p has become an observable quantity, even though we can measure N. We note that although the rotationally symmetric gauge adopted for the potentials is most convenient, it is not absolutely necessary. If some other (unsymmetric) gauge were used, (4.14) would not hold, whereas (2.37) always remains valid. Moreover, the derivation of the conservation law given in Section 2.5 gives more physical insight than does the formalism presented here. The evaluation of the remaining two Lagrange equations

$$\frac{d}{dt}\left(\frac{\partial L}{\partial \dot{r}}\right) - \frac{\partial L}{\partial r} = 0 \quad , \quad \frac{d}{dt}\left(\frac{\partial L}{\partial \dot{z}}\right) = \frac{\partial L}{\partial z} \qquad (4.15)$$

is straightforward. Expressing (4.3) and (4.4) in cylindrical coordinates, we find

$$\frac{d}{dt}(m\dot{r}) - mr\dot{\varphi}^2 = Q(E_r + r\dot{\varphi}B_z - \dot{z}B_\varphi)$$

$$\frac{d}{dt}(m\dot{z}) = Q(E_z - r\dot{\varphi}B_r + \dot{r}B_\varphi) \qquad (4.16)$$

$$m = \frac{m_0}{\sqrt{1-\beta^2}} = m_0\left\{1 - (\dot{z}^2 + \dot{r}^2 + r^2\dot{\varphi}^2)c^{-2}\right\}^{-\frac{1}{2}} \qquad (4.17)$$

The remaining calculations are left to the reader. The same results can, of course, be obtained by transforming the Lorentz equation in the appropriate fashion.

The conservation law $p_\varphi = N =$ const may be used to eliminate $\dot{\varphi}$ from (4.16). Better still, $\dot{\varphi}$ may be eliminated from (4.13) before evaluating (4.15). Owing to the dependence of the mass on $\dot{\varphi}$ in (4.17), the resulting formulae are highly complicated. It is thus advantageous to eliminate $\dot{\varphi}$ from (4.13) only in the nonrelativistic approximation $m = m_0$. Moreover, we make the simplifying assumption that all the terms in B_φ can be neglected, since this field component is only produced by the particle beam itself and is always very weak in comparison with the external magnetic field. $B_\varphi \equiv 0$ is most easily satisfied by the gauge $A_z \equiv A_r \equiv 0$, which will henceforward be adopted. From (4.4), we see that E_z and E_r are now represented by a quasistationary approximation:

$$E_z = -\frac{\partial \Phi}{\partial z} \quad , \quad E_r = -\frac{\partial \Phi}{\partial r} \quad \text{with} \quad \Phi = \Phi(z, r, t)$$

a simplification which is justified even in technical applications involving high-frequency devices. The essential induction effect is incorporated in the components E_φ and A_φ and thus in the dependence of Φ on time.

4. VARIATIONAL PRINCIPLES

With all these simplifications the Lagrangian now becomes

$$L = \frac{1}{2}m_0(\dot{z}^2 + \dot{r}^2 + r^2\dot{\varphi}^2) + Q(r\dot{\varphi}A_\varphi - \Phi) \tag{4.18}$$

Solving (2.37) for $\dot{\varphi}$ in the nonrelativistic limit, we obtain

$$\dot{\varphi} = \frac{N - Q\Psi(z,r,t)/2\pi}{m_0 r^2} \tag{4.19}$$

On inserting this in (4.18) we obtain a Lagrangian that is a function of z, r, t, $\dot{z}$ and $\dot{r}$ only but is a very cumbersome expression. We can get a more compact form by making the Legendre transform

$$L^* := L - \dot{\varphi}p_\varphi \equiv L - \dot{\varphi}N \tag{4.20}$$

before substituting for $\dot{\varphi}$. This transform does not change the final equations of motion since the corresponding action integrals differ from each other only by a fixed constant and have thus the same extremal trajectories:

$$W^* = \int_{t_0}^{t_1} L^* \, dt = \int_{t_0}^{t_1} L \, dt - N(\varphi_1 - \varphi_0)$$

Substitution of (4.19) in L^* now results in

$$L^* = \frac{m_0}{2}(\dot{z}^2 + \dot{r}^2) - X(z,r,t) \tag{4.21}$$

with the *effective* potential energy

$$X(z,r,t) = Q\Phi(z,r,t) + \frac{\{N - Q\Psi(z,r,t)/2\pi\}^2}{2m_0 r^2} \tag{4.22}$$

The latter contains two contributions, the familiar electric term $Q\Phi$ and a *centrifugal* potential. The latter contains the terms involving $\dot{\varphi}$ and has the value $m_0 r^2 \dot{\varphi}^2/2$. It differs from the familiar classical form in possessing a contribution from the magnetic flux Ψ. The final form of the trajectory equations is now obtained by writing down the Euler–Lagrange equation of (4.21) (Störmer, 1904, 1906, 1933):

$$m_0 \ddot{r} = -\frac{\partial X}{\partial r} \quad , \quad m_0 \ddot{z} = -\frac{\partial X}{\partial z} \tag{4.23}$$

4.3 THE CANONICAL FORMALISM

The third equation is (4.19), which can be rewritten in compact form if the quantity N, though constant with respect to time, is regarded as a free parameter. It is readily seen that

$$\dot{\varphi} = -\frac{\partial L^*}{\partial N} = \frac{\partial X}{\partial N} \qquad (4.24)$$

In this sense L^* is a Routhian function.

In static fields, for which X is a function of z and r only, the law of conservation of energy can be simplified to

$$E_0 = \frac{m_0}{2}(\dot{z}^2 + \dot{r}^2) + X(z,r) = \text{const} \qquad (4.25)$$

as can be easily verified.

This example demonstrates that the use of the Lagrange formalism in electron optics can be quite advantageous, since the elementary derivation of (4.23) turns out to be more complicated if the electromagnetic fields are time-dependent. These equations are very useful in studies of particle motion in high-frequency devices, for which alternative simple forms of the trajectory equations are not available.

4.3 The canonical formalism

As well as the Lagrangian, the Hamiltonian function is of great importance. The latter is needed in Hamilton–Jacobi theory, which is outlined in Chapter 5; the Hamiltonian itself and the associated canonical equations of motion are needed in the theory of electron emission from cathodes (see Chapter 44).

In vector notation, the Legendre transform between the Lagrange function $L(\boldsymbol{r}, \boldsymbol{v}, t)$ and the Hamilton function $H(\boldsymbol{r}, \boldsymbol{p}, t)$ has the form

$$H(\boldsymbol{r},\boldsymbol{p},t) := \boldsymbol{p}\cdot\boldsymbol{v} - L(\boldsymbol{r},\boldsymbol{p},t) \qquad (4.26)$$

in which the velocity $\boldsymbol{v}$ has to be expressed in terms of the canonical momentum $\boldsymbol{p}$. This expression is obtained by solving (4.12) for $\boldsymbol{v}$, the first step being

$$\boldsymbol{v} = \frac{1}{m}\boldsymbol{g} = \frac{1}{m}(\boldsymbol{p} - Q\boldsymbol{A})$$

The relativistic mass can be expressed as a function of $\boldsymbol{p}$ with the aid of (2.11) and we finally obtain

$$\boldsymbol{v} = \frac{c(\boldsymbol{p} - Q\boldsymbol{A})}{\sqrt{(m_0 c)^2 + (\boldsymbol{p} - Q\boldsymbol{A})^2}} \qquad (4.27)$$

Introducing this into (4.26), we obtain the Hamilton function

$$H = c\sqrt{(m_0 c)^2 + (\boldsymbol{p} - Q\boldsymbol{A})^2} - m_0 c^2 + Q\Phi \qquad (4.28)$$

This expression is always valid, even in systems with time-dependent electromagnetic potentials. The canonical equations of motion are given by

$$\boldsymbol{v} \equiv \dot{\boldsymbol{r}} = \frac{\partial H}{\partial \boldsymbol{p}} \qquad (4.29)$$

$$\dot{\boldsymbol{p}} = -\frac{\partial H}{\partial \boldsymbol{r}} \qquad (4.30)$$

Equations (4.29) and (4.30) can easily be verified by performing the necessary differentiations. Equation (4.29) proves to be identical with (4.27), while (4.30) is equivalent to the Lagrange equation $\dot{\boldsymbol{p}} = \partial L/\partial \boldsymbol{r}$. More generally, the canonical equations can be derived from a variational principle of least action in *phase space* (defined as the union of the vector spaces of $\boldsymbol{r}$ and $\boldsymbol{p}$). This variational principle will not be investigated here.

An important law can be established concerning the variation of the Hamiltonian with time. Taking the total derivative with respect to time of both sides of (4.26) we have first

$$\frac{dH}{dt} = \frac{\partial H}{\partial t} + \dot{\boldsymbol{r}} \cdot \frac{\partial H}{\partial \boldsymbol{r}} + \dot{\boldsymbol{p}} \cdot \frac{\partial H}{\partial \boldsymbol{p}} = \dot{\boldsymbol{p}} \cdot \boldsymbol{v} + \boldsymbol{v} \cdot \dot{\boldsymbol{p}} - \frac{\partial L}{\partial t} - \dot{\boldsymbol{r}} \cdot \frac{\partial L}{\partial \boldsymbol{r}} - \dot{\boldsymbol{v}} \cdot \frac{\partial L}{\partial \boldsymbol{v}}$$

Using the canonical equations (4.29) and (4.30), the first part of this equation reduces to $dH/dt = \partial H/\partial t$. Definition (4.12) together with the Lagrange equation shows that the expression on the far right-hand side reduces to $-\partial L/\partial t$, thus

$$\frac{dH}{dt} = \frac{\partial H}{\partial t} = -\frac{\partial L}{\partial t} \qquad (4.31)$$

The most important consequence of this relation is that the value of H is *conserved* in all systems with *static* electromagnetic fields, since the only explicit dependence of H and L on time occurs in the potentials $\Phi(\boldsymbol{r}, t)$ and $\boldsymbol{A}(\boldsymbol{r}, t)$. By comparison of (4.28) with (2.6), (2.7), (2.8) and (2.11), it can be seen that the first two terms on the right-hand side represent the kinetic energy T as a function of $\boldsymbol{p}$ and $\boldsymbol{r}$, while the last term $Q\Phi$ is the potential energy $V(\boldsymbol{r})$, so that H is the same as the total energy E_0 of the motion. According to the conventions adopted in Section 2.3, this quantity $H = E_0$ is the kinetic starting energy of the electron at the cathode surface; it thus has a very concrete and important meaning, since this quantity will be used in the statistical analysis of the emission process.

4.4 The time-independent form of the variational principle

Since the value of H is conserved in all static systems, we can cast the variational principle (4.1) into a very attractive form by eliminating L. Solving (4.26) for L and introducing the resulting expression into (4.1), we find

$$W = \int_{t_0}^{t_1} (\boldsymbol{p} \cdot \boldsymbol{v} - H)\, dt = \text{extr.}$$

In all *static* systems, the contribution

$$\int_{t_0}^{t_1} H\, dt = E_0(t_1 - t_0)$$

is a fixed constant, which does not affect the equation of motion and may hence be omitted. We thus obtain the (reduced) principle of least action:

$$\overline{S} := \int_{t_0}^{t_1} \boldsymbol{p} \cdot \boldsymbol{v}\, dt = W + E_0(t_1 - t_0) = \text{extr.} \tag{4.32}$$

Elimination of the time and the introduction of any other legitimate variable of integration u is now straightforward:

$$\overline{S} = \int_{u_0}^{u_1} \boldsymbol{p} \cdot \frac{d\boldsymbol{r}}{du}\, du = \int_{\boldsymbol{r}_0}^{\boldsymbol{r}_1} \boldsymbol{p} \cdot d\boldsymbol{r} = \text{extr.} \tag{4.33}$$

The second representation shows that $\overline{S}$ is *invariant* with respect to the choice of the parameter.

In systems with a straight optic axis, the axial coordinate z is always the most convenient choice of variable of integration. Using (4.12) and (3.15) we obtain

$$\overline{S} = \int_{z_0}^{z_1} (\boldsymbol{g} + Q\boldsymbol{A}) \cdot \boldsymbol{r}'(z)\, dz \tag{4.33a}$$

Since the direction of the kinematic momentum vector $\boldsymbol{g}$ must be the same as that of the local tangent $\boldsymbol{t}$, we may use

$$\boldsymbol{g} \cdot \boldsymbol{r}' = g(\boldsymbol{r})\, \boldsymbol{t} \cdot \boldsymbol{r}' = \varrho g(\boldsymbol{r})$$

where ϱ is given by (3.16). Bringing all this together and writing out the expression for $\overline{S}$ explicitly, we obtain finally

$$\overline{S} = \int_{z_0}^{z_1} \overline{M}(x, y, z, x', y') \, dz \qquad (4.34)$$

with

$$\overline{M}(x, y, z, x', y') = \sqrt{1 + x'^2 + y'^2} \, g(\mathbf{r}) + Q(x' A_x + y' A_y + A_z) \qquad (4.35)$$

For electron motion, (2.19) may be used. The corresponding Euler equations

$$\frac{d}{dz}\left(\frac{\partial \overline{M}}{\partial x'}\right) = \frac{\partial \overline{M}}{\partial x} \quad , \quad \frac{d}{dz}\left(\frac{\partial \overline{M}}{\partial y'}\right) = \frac{\partial \overline{M}}{\partial y} \qquad (4.36)$$

are the x- and y-components of (3.17) and will not be discussed here. An important way of developing a theory of aberrations consists in expanding (4.35) as a Taylor series before evaluating (4.36) and then developing an appropriate perturbation calculus.

4.5 Static rotationally symmetric systems

A good example of the simplification achieved by eliminating the time as a trajectory parameter is the static rotationally symmetric system. Moreover, the Störmer equation of motion thus obtained is of some importance. It is advantageous to transform (4.35) into cylindrical coordinates. In static systems, the only surviving component of $\mathbf{A}$ is $A_\varphi = \Psi(z,r)/2\pi r$, Ψ being the magnetic flux. Thus we immediately obtain

$$\overline{M}(z, r, r', \varphi') = g(z,r)\sqrt{1 + r'^2 + r^2 \varphi'^2} + Q\varphi' \Psi(z,r)/2\pi$$

Since φ is cyclic, the conservation of axial angular momentum is now expressed by

$$N = p_{\varphi'} = \frac{\partial \overline{M}}{\partial \varphi'} = \frac{g(z,r) r^2 \varphi'}{\sqrt{1 + r'^2 + r^2 \varphi'^2}} + \frac{Q\Psi(z,r)}{2\pi} = \text{const} \qquad (4.37)$$

On comparison with (2.37), it is clear that the value of N is *invariant* with respect to the parametric transform. Every canonical momentum can be shown to be invariant with respect to every parametric transform in variational theory.

4.5 STATIC ROTATIONALLY SYMMETRIC SYSTEMS

Even for relativistic motion, it is easy to eliminate φ' by solving (4.37) for φ', the result being

$$\varphi' = \frac{\sqrt{1+r'^2}}{r^2\mu(z,r)}\{N - Q\Psi(z,r)/2\pi\} \tag{4.38}$$

with

$$\mu(z,r) := \{g^2(z,r) - r^{-2}(N - Q\Psi/2\pi)^2\}^{\frac{1}{2}} \tag{4.39}$$

Elementary calculations show that

$$\varrho = \sqrt{1+r'^2+r^2\varphi'^2} = g(z,r)\mu^{-1}\sqrt{1+r'^2}$$

The integrand of (4.34) can be simplified by means of the Legendre transform $M^* = \overline{M} - \varphi'N$ (cf. 4.20), the result being

$$M^*(z,r,r') = \sqrt{1+r'^2}\,\mu(z,r)$$

Evaluation of the Euler equation

$$\frac{d}{dz}\left(\frac{\partial M^*}{\partial r'}\right) = \frac{\partial M^*}{\partial r}$$

yields

$$\frac{d}{dz}\left(\frac{r'\mu}{\sqrt{1+r'^2}}\right) = \sqrt{1+r'^2}\,\frac{\partial\mu}{\partial r}$$

which can be further simplified to the Störmer equation

$$r''(z) = \frac{1+r'^2}{\mu(z,r)}\left(\frac{\partial\mu}{\partial r} - r'\frac{\partial\mu}{\partial z}\right) \tag{4.40}$$

This trajectory equation is similar in structure to (3.21) and to the electric terms of (3.20). The function $\mu(z,r)$ is the length of the meridional projection of $\boldsymbol{g}$, since (4.39) can be rewritten as $\mu = |\boldsymbol{i}_z g_z + \boldsymbol{i}_r g_r|$. When considering the motion of electrons, it is helpful to introduce the acceleration potential by means of (2.19). We find

$$r'' = \frac{1+r'^2}{2\tilde{\Phi}(z,r)}\left(\frac{\partial\tilde{\Phi}}{\partial r} - r'\frac{\partial\tilde{\Phi}}{\partial z}\right) \tag{4.41}$$

with the effective potential

$$\tilde{\Phi}(z,r) = \hat{\Phi}(z,r) - \frac{\{N + e\Psi(z,r)/2\pi\}^2}{2m_0 e r^2} \tag{4.42}$$

the second term being again a centrifugal contribution.

On comparing these equations with the corresponding formulae in Section 4.2, the reader will notice that no approximations have been necessary here and the exact calculation has not become at all complicated. The essential simplification is a consequence of the fact that, in the time-independent representation, the relativistic mass is a simple function $m(z,r)$; the need to use (4.17) in the time-dependent situation highly complicates the calculations.

5
Hamiltonian Optics

In the previous chapter we investigated the various types of variational principles and showed how they can be made to yield suitable forms of trajectory equations. These trajectory equations are to be solved by methods that are explained in later chapters and the solutions will provide us with a certain measure of physical understanding. Though this is a possible way of investigating electron optical devices, it is not entirely satisfactory. In geometric light optics, Hamilton's theory of characteristic functions, in which the rays of light are treated as trajectories orthogonal to eikonal or characteristic functions, was a major advance, making it possible to investigate whole bundles of trajectories instead of individual ones. This is the main difference between optical and purely mechanical or ballistic treatments.

In charged particle optics, the analogue of the eikonal theory is well-known under the name of Hamilton–Jacobi theory, which we shall now consider in detail. In the standard textbooks on classical mechanics, this theory is derived by means of canonical transformations, which is a very general but undeniably elaborate method (see e.g. Goldstein, 1959, Chapter VIII). We shall not follow their example. The presentation that follows is considerably simpler.

5.1 Introduction of the characteristic function

We again set out from (4.1). For given boundary values, t_0 and t_1, of t and hence of the vectors $\boldsymbol{r}_0 := \boldsymbol{r}(t_0)$ and $\boldsymbol{r}_1 := \boldsymbol{r}(t_1)$, a function $W(\boldsymbol{r}_0, t_0; \boldsymbol{r}_1, t_1)$ can be defined to be the stationary value of (4.1), obtained by integration over a *physical* trajectory, a solution of the corresponding Euler equations. This definition may be complicated by the presence of singularities and ambiguities, which will be investigated later. For the moment, we shall assume that the value of the integral expression is a unique and differentiable function of its arguments. We shall see that it is appropriate to regard this function W as a characteristic of the system and we shall indeed refer to it as a *characteristic function*.

We now consider differential variations of the endpoint, the starting point being unaltered. In order to distinguish these variations from the notation used for integration, we denote them by Δt_1 and $\Delta \boldsymbol{r}_1$ rather than

5.1 INTRODUCTION OF THE CHARACTERISTIC FUNCTION 47

dt and $d\boldsymbol{r}$. Since the curves connecting the fixed starting point and shifted endpoint must correspond to physical trajectories, the condition

$$\Delta \boldsymbol{r}_1 = \boldsymbol{v}(t_1)\Delta t_1 = \boldsymbol{v}_1 \Delta t_1$$

must be satisfied. The corresponding variation of W is then given by $\Delta W = L_1 \Delta t_1 = L(\boldsymbol{r}_1, \boldsymbol{v}_1, t_1)\Delta t_1$. The Lagrangian L_1 can be expressed in terms of the Hamiltonian H_1 at the terminal point 1, the result being

$$\Delta W = (\boldsymbol{p}_1 \cdot \boldsymbol{v}_1 - H_1)\Delta t_1 = \boldsymbol{p}_1 \cdot \Delta \boldsymbol{r}_1 - H_1 \Delta t_1 \qquad (5.1)$$

On the other hand we can expand the difference

$$\Delta W := W(\boldsymbol{r}_0, t_0; \boldsymbol{r}_1 + \Delta \boldsymbol{r}_1, t_1 + \Delta t_1)$$

as a Taylor series. Retaining only the first order terms, we obtain

$$\Delta W = \Delta \boldsymbol{r}_1 \cdot \frac{\partial W}{\partial \boldsymbol{r}_1} + \Delta t_1 \frac{\partial W}{\partial t_1} \qquad (5.2)$$

The two expressions for ΔW must be identical for all increments Δt_1 and $\Delta \boldsymbol{r}_1$ that represent a physical motion. Certainly we can choose arbitrary values of Δt_1 provided that a continuous range of values leads from t_1 to $t_1 + \Delta t_1$. It is now of importance that (5.1) must be identical with (5.2) in every respect, that is, for *all* acceptable configurations of $\boldsymbol{r}_0$, t_0, $\boldsymbol{r}_1$ and t_1. This implies that $\Delta \boldsymbol{r}_1$ must be regarded as *independent* of Δt_1, even though $\boldsymbol{r}_1 = \boldsymbol{v}_1 \Delta t_1$; the velocity $\boldsymbol{v}_1$ may be an arbitrary vector only subject to $|\boldsymbol{v}_1| < c$. From (5.1) and (5.2), we then obtain the necessary and sufficient conditions

$$\boldsymbol{p}_1 = \frac{\partial W}{\partial \boldsymbol{r}_1} \quad , \quad H_1 = -\frac{\partial W}{\partial t_1} \qquad (5.3)$$

By considering variations of the starting point, we likewise find

$$\boldsymbol{p}_0 = -\frac{\partial W}{\partial \boldsymbol{r}_0} \quad , \quad H_0 = \frac{\partial W}{\partial t_0} \qquad (5.4)$$

Thus the variation ΔW due to alterations of both sets of arguments is given by Hamilton's central equation:

$$\Delta W = \Big[\boldsymbol{p} \cdot \Delta \boldsymbol{r} - H \Delta t\Big]_0^1 \equiv \boldsymbol{p}_1 \cdot \Delta \boldsymbol{r}_1 - \boldsymbol{p}_0 \cdot \Delta \boldsymbol{r}_0 - H_1 \Delta t_1 + H_0 \Delta t_0 \quad (5.5)$$

Since W is a continuously differentiable function of its arguments, (5.5) must hold even for completely arbitrary differential increments.

5.2 The Hamilton–Jacobi equation

In the following discussion we shall assume that the starting coordinates r_0, t_0 are uniquely specified and remain unaltered, while the terminal coordinates r_1, t_1 may vary within physically allowed domains. In order to simplify the notation, we shall omit the arguments r_0 and t_0 whenever they are not explicitly needed. Furthermore, we shall omit the subscript 1 of the terminal coordinates, and the ordinary vector notation for the gradient will refer to differentiation with respect to r_1.

Introducing (5.3) into the Hamiltonian $H(r, p, t)$, we immediately obtain the Hamilton–Jacobi equation

$$H(r, \text{grad } W, t) = -\frac{\partial W}{\partial t} \tag{5.6}$$

W now being a function of r and t. Recalling (4.28), (5.6) is given explicitly by

$$c\left\{(m_0 c)^2 + (\text{grad } W - Q\mathbf{A})^2\right\}^{\frac{1}{2}} - m_0 c^2 + Q\Phi + \frac{\partial W}{\partial t} = 0 \tag{5.7}$$

Φ and $\mathbf{A}$ being functions of r and t, like W. This is a partial differential equation of first order for the function $W(r, t)$. Since only the derivatives of W appear in (5.7) and not the function itself, the solution of (5.7) may contain an arbitrary additive constant. A reasonable and simple normalization is $W(r_0, t_0; r_0, t_0) = 0$, since the solution then fits the original definition of W, namely as the integral representation of the action. From among the many solutions of (5.7), we shall select all those that satisfy this condition. They correspond to the paths of all trajectories that start at the point r_0 at the time t_0.

Even after imposing this restriction on the set of solutions, the practical solution of (5.7) will be extremely complicated in the general case; further simplification is therefore necessary. In all systems with static potentials $\Phi(r)$, $\mathbf{A}(r)$, the separation (in full notation)

$$W(r_0, t_0, r, t) = \overline{S}(r_0, r) - E_0(t - t_0) \tag{5.8}$$

is possible and advantageous, (5.7) then simplifying to

$$c\left\{(m_0 c)^2 + (\text{grad } \overline{S} - Q\mathbf{A}(r))^2\right\}^{\frac{1}{2}} - m_0 c^2 + Q\Phi(r) = E_0 \tag{5.9}$$

The *point characteristic function* $\overline{S}(\mathbf{r}_0, \mathbf{r})$ introduced by this separation is exactly the same as the function $\overline{S}$ appearing in the integral representations (4.32) and (4.33) if the normalization condition for W is satisfied. Thus, instead of solving (5.9) directly with $\overline{S}(\mathbf{r}_0, \mathbf{r}_0) = 0$, $\overline{S}$ may in practice be obtained by evaluating (4.33).

Familiar though it is in theoretical physics, (5.9) is an inconvenient form of the reduced Hamilton–Jacobi equation. By means of (2.13) we can simplify it to

$$\left\{ \operatorname{grad} \overline{S}(\mathbf{r}) - Q\mathbf{A}(\mathbf{r}) \right\}^2 = g^2(\mathbf{r}) \tag{5.10}$$

which is still exact. In the case of electron propagation, we can rewrite it as

$$\left\{ \operatorname{grad} \overline{S}(\mathbf{r}) + e\mathbf{A}(\mathbf{r}) \right\}^2 = 2m_0 e \hat{\Phi}(\mathbf{r}) \tag{5.11}$$

In all practical calculations we shall use this form. Despite this simplification, the practical solution may still be very arduous.

5.3 The analogy with light optics

Having formally introduced the characteristic function, we now discuss its physical meaning. For simplicity, we first assume that the vector potential $\mathbf{A}(\mathbf{r})$ vanishes. In the static case, the first equation in (5.3) then reduces to

$$\mathbf{p}(\mathbf{r}) \equiv \mathbf{g}(\mathbf{r}) = \operatorname{grad} \overline{S}(\mathbf{r}) \tag{5.12}$$

and (5.11) to

$$\left\{ \operatorname{grad} \overline{S}(\mathbf{r}) \right\}^2 = \left\{ \mathbf{g}(\mathbf{r}) \right\}^2 = g^2(\mathbf{r}) = 2m_0 e \hat{\Phi}(\mathbf{r}) \tag{5.13}$$

the last term of this equation being valid for electrons. The physical meaning of (5.12) and (5.13) is quite clear: (5.12) expresses the vector $\mathbf{g} = m\mathbf{v}$ as a function of $\mathbf{r}$, while (5.13) is the condition that the length of this vector is in agreement with relativistic kinematics. The truly new result, going beyond relativistic kinematics, is that the local direction of the vector $\mathbf{g}$ is always orthogonal to the corresponding surface $\overline{S}(\mathbf{r}) = \text{const}$ (5.12). Since the vector $\mathbf{g} = g(\mathbf{r})\mathbf{t}(\mathbf{r})$ always points in the same direction as the local tangent $\mathbf{t}(\mathbf{r})$ of the trajectory in question, we can draw the following conclusion: the particle trajectories are orthogonal to the set of surfaces $\overline{S}(\mathbf{r}) = \text{const}$.

This statement is illustrated in Fig. 5.1. The analogy with geometric light optics is now complete. We may use the relation $\mathbf{g} = \hbar k_0 n(\mathbf{r})\mathbf{t}(\mathbf{r})$,

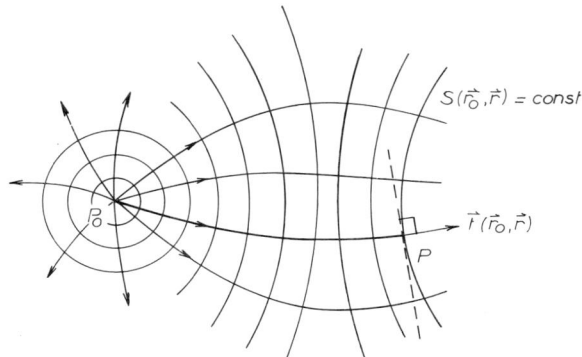

Fig. 5.1: Simplified representation of the point characteristic function $S(r_0, r)$ in an isotropic medium. The point P_0, with position vector r_0, may be regarded as a point source, from which the various rays emanate. A number of surfaces $S = $ const are shown in a two-dimensional section. The value of S at any arbitrary point P is equal to the variational integral along the trajectory from P_0 to P. The local tangent t at P is orthogonal to the corresponding surface $S = $ const.

$t(r)$ being the local tangent vector of a ray passing through the point r. The vacuum momentum $\hbar k_0$ cancels out if we introduce an *eikonal function* $L(r)$ by writing $\overline{S}(r) =: \hbar k_0 L(r)$, and we obtain

$$n(r) t(r) = \text{grad } L(r) \tag{5.14}$$

$$\{\text{grad } L(r)\}^2 = n^2(r) \tag{5.15}$$

Equation (5.14) expresses Hamilton's statement that light rays are the normals to a family of surfaces $L(r) = $ const known as wavefronts or eikonal surfaces (see e.g. Born and Wolf (1959) eq. (3.1.15b), where (5.15) is referred to as the eikonal equation, recalling the work of Bruns (1895) on 'Das Eikonal'). The eikonal itself is an optical length. Explicitly, the point eikonal

$$L(r_0, r_1) = \int_{P_0}^{P_1} n(r)\, ds \tag{5.16}$$

is the optical length between the points P_0 and P_1 with position vectors r_0 and r_1, respectively. The integration is to be performed along a physical trajectory. According to Fermat's principle this function takes the same value along all trajectories connecting P_0 and P_1 that are continuously deformable into one another, and is stationary when evaluated along a trajectory. For rotationally symmetric dioptric systems, this stationary value is a true minimum (Sturrock, 1955 p.60). This result is exactly analogous to (4.32).

It is possible and of potential interest to introduce an electron optical index of refraction in such a way that (5.12) and (5.13) are in formal agreement with (5.14) and (5.15), respectively. Since the index of refraction may be defined in light optics as the ratio of the momentum in the medium in question and in vacuo, we may similarly define the electron optical index of refraction to be

$$n_E(\mathbf{r}) = \frac{g(\mathbf{r})}{G} = \sqrt{\frac{\hat{\Phi}(\mathbf{r})}{\hat{U}}} \qquad (5.17)$$

$G = (2m_0 e\hat{U})^{1/2}$ being some suitable constant momentum. We now have complete formal agreement if we write $\overline{S} = GL$ instead of $\overline{S} = \hbar k_0 L$. There is, however, an important difference between light optics and electron optics. With respect to the propagation of light, the vacuum is the privileged medium of reference as it is the only medium free of dispersion and absorption. In electron optics there is no privileged medium of reference, the choice of the positive constant G being arbitrary; different choices are to be found in the literature. A definition of an electron optical index of refraction such as (5.17) thus offers no particular advantage and it seems more sensible to use (5.12) and (5.13) in their original form, since the kinematic momentum has a direct experimental significance.

A more serious difference between light optics and electron optics is that, in the nonrelativistic approximation, the condition $\nabla^2(n_E^2) = 0$ is satisfied in all source-free domains, whereas no such condition obtains in light optics. Although the index $n(\mathbf{r})$ for the refraction of light is a piecewise constant function in all the principal optical devices, the free choice of the lens surfaces makes aberration correction possible. The fact that Laplace's equation must be satisfied is, however, such a strong restriction that some of the aberrations in electron optical devices are rigorously incapable of correction (see Part IV). The relativistic terms in (5.17) do not alter this situation.

5.4 The influence of vector potentials

The analogy between light optics and electron optics breaks down completely when vector potentials have to be considered. We then have

$$\mathbf{p}(\mathbf{r}) = \mathbf{g}(\mathbf{r}) + Q\mathbf{A}(\mathbf{r}) = \operatorname{grad} \overline{S}(\mathbf{r}) \qquad (5.18)$$

Since the vectors $\mathbf{g}$ and $\mathbf{A}$ are in general *not* parallel to each other and $\mathbf{g}$ has the direction of the local tangent, $\mathbf{p}$ does not always point in this

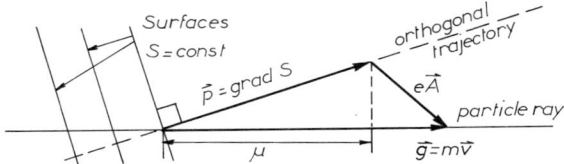

Fig. 5.2: Influence of the vector potential A on the characteristic function S.

tangential direction. The rays of particles are hence *no longer* orthogonal trajectories of surfaces $\overline{S}(r) = $ const. This is shown in Fig. 5.2.

The principle of least action (4.33) can be rewritten as $\delta \overline{S} = 0$ or

$$\overline{S}(r_0, r_1) = \int_{r_0}^{r_1} p \cdot dr = \int_{P_0}^{P_1} \mu(r, t)\, ds = \text{extr}. \tag{5.19}$$

with

$$\mu(r, t) := t \cdot p = g(r) + Q\, t \cdot A(r) \tag{5.20}$$

and for electrons:

$$\mu(r, t) = \sqrt{2m_0 e \hat{\Phi}(r)} - e\, t \cdot A(r) \tag{5.21}$$

The meaning of this quantity is shown in Fig. 5.2. The generalization of (5.17) is now

$$n_E(r, t) := \sqrt{\frac{\hat{\Phi}(r)}{\hat{U}}} - \frac{\eta}{\hat{U}^{\frac{1}{2}}} t \cdot A(r) \tag{5.22}$$

Expressions of this form or proportional to it are commonly encountered in the literature. The square-root term is regarded as an isotropic contribution to the index of refraction, while the vector potential term—due to its dependence on t—is an anisotropic contribution. It must be emphasized that all these considerations are of a purely formal character and have no experimental significance, as will soon be obvious. Nevertheless, the point characteristic function $\overline{S}(r_0, r_1)$, defined by (5.19), does retain—disregarding any constant normalization factor—the character of the optical distance between the points r_0 and r_1. Various choices of the normalization factor (essentially $\hat{U}$ in (5.22)) are to be found in the literature (see Picht, 1939, 1957, eq. (3.10–11) or Picht, 1963, Section 3.8; Glaser, 1952 Sections 9–10; Grivet, 1965, Chapter 6; Kel'man and Yavor, 1959, 1968, Section 4 of Chapter 1). A system of units—not adopted here—has been devised by Sturrock (1955, Section 1.2) to eliminate e, m_0 and c from the equations. We emphazise that the choice of any particular normalization is only a question of convenience and has no physical meaning.

It is possible and obviously sensible to choose a gauge for $A(r)$ such that $A(r)$ vanishes in all domains in which B is zero. Thus at least in the field-free domains in front of and behind magnetic devices, we can make use of the orthogonality between trajectories and eikonal surfaces.

5.5 Gauge transformations

In the previous chapters we have frequently used electromagnetic potentials but, apart from the purely electrostatic potential, never specified their gauge. We now discuss the influence of different gauges on various physical quantities.

Equations (4.3) and (4.4) can be satisfied by different pairs of potentials Φ, A and Φ', A', say, provided that these pairs are related by

$$\Phi = \Phi' - \frac{\partial}{\partial t} F(r,t) \quad , \quad A = A' + \operatorname{grad} F(r,t) \tag{5.23}$$

This is called a gauge transformation. The function $F(r,t)$ is arbitrary so long as it is sufficiently differentiable. The other Maxwell equations impose further restrictions but do not completely eliminate the freedom of choice of F. Since only the field vectors have experimental significance and not the potentials, all quantities that depend in any way on $F(r,t)$ have *no* experimental significance.

The kinematic functions, introduced in Chapter 2, are gauge-invariant since they are essentially related to the kinetic energy and not to the potential. In the variational formalism, the gauge dependent quantities are the generalized potential V^*, the canonical momentum p, the Lagrangian L and the Hamiltonian H, the corresponding transforms being given by

$$V^* = V^{*\prime} - Q\left(\frac{\partial F}{\partial t} + v \cdot \operatorname{grad} F\right) \equiv V^{*\prime} - Q\frac{dF}{dt} \tag{5.24}$$

$$L = L' + Q\frac{dF}{dt} \tag{5.25}$$

$$p = p' + Q\operatorname{grad} F \tag{5.26}$$

$$H = H' - Q\frac{\partial F}{\partial t} \tag{5.27}$$

The transform of the variational integral (4.1) involves integration of a total derivative, the result being

$$W = W' + \Big[QF(r,t)\Big]_0^1 \tag{5.28}$$

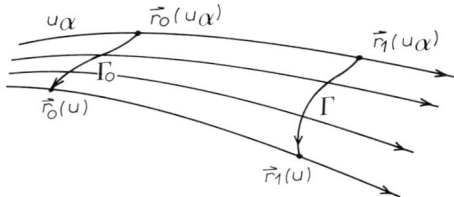

Fig. 5.3: A one-parameter family of rays, each ray being uniquely characterized by a specific value of the parameter u. Two paths of integration, Γ_0 and Γ, are also shown.

If the endpoints are kept fixed, W and W' differ only by a constant and thus have the same extremals as solutions of the corresponding Euler–Lagrange equations. If, however, the endpoints are regarded as variables, (5.28) shows that the action W is essentially gauge dependent. The characteristic functions are therefore gauge-dependent and are not observable quantities. Only for the propagation of particles in purely electrostatic fields can (5.12) and (5.13) be regarded as having a physical meaning, as they have a gauge-invariant form. In (5.20), however, the gauge-dependence is now obvious and so definition (5.22) has no particular advantage.

5.6 Poincaré's integral invariant

We now return to (5.5). In time-independent systems, variations of the time are of no interest and we thus choose $\Delta t_1 = \Delta t_0 = 0$. Since spatial variations involve only the static point characteristic function $\overline{S}(\boldsymbol{r}_0, \boldsymbol{r}_1)$, (5.5) then simplifies to

$$\Delta \overline{S} = \boldsymbol{p}_1 \cdot \Delta \boldsymbol{r}_1 - \boldsymbol{p}_0 \cdot \Delta \boldsymbol{r}_0 \qquad (5.29)$$

We now consider a one-parameter family of non-intersecting rays, each ray being uniquely characterized by a well-defined value of some parameter u, as shown in Fig. 5.3. This means that the points with vectors $\boldsymbol{r}_0(u)$ and $\boldsymbol{r}_1(u)$ are located on the same ray. It is useful to introduce derivatives with respect to u, for instance $\Delta \boldsymbol{r}_0 = \Delta u \, d\boldsymbol{r}_0/du$, with similar expressions for $\Delta \boldsymbol{r}_1$ and $\Delta \overline{S}$. The quantity Δu then cancels out from (5.29), so that

$$\frac{d\overline{S}}{du} = \boldsymbol{p}_1(u) \cdot \frac{d\boldsymbol{r}_1}{du} - \boldsymbol{p}_0(u) \cdot \frac{d\boldsymbol{r}_0}{du} \qquad (5.30)$$

is exactly valid.

5.6 POINCARÉ'S INTEGRAL INVARIANT

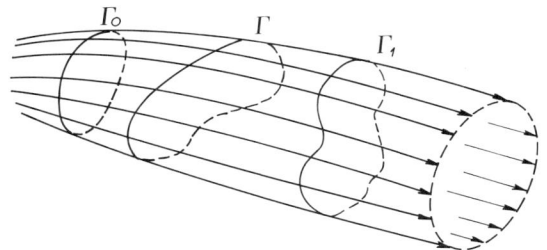

Fig. 5.4: Tube of non-intersecting rays and three closed paths of integration, Γ_0, Γ_1 and Γ round its surface.

In the next step we consider a tube of non-intersecting rays, its mantle surface now being a one-parameter family, as shown in Fig. 5.4. On this surface we choose two closed loops Γ_0 and Γ_1 with parametric representations $\boldsymbol{r}_0(u)$ and $\boldsymbol{r}_1(u)$ for $u_\alpha \leq u \leq u_\beta$, respectively. Integration of (5.30) over the whole interval of u results in

$$\int_{u_\alpha}^{u_\beta} \frac{dS}{du}\, du = \int_{u_\alpha}^{u_\beta} \boldsymbol{p}_1(u) \cdot \frac{d\boldsymbol{r}_1}{du}\, du - \int_{u_\alpha}^{u_\beta} \boldsymbol{p}_0(u) \cdot \frac{d\boldsymbol{r}_0}{du}\, du$$

Since $\boldsymbol{r}_j(u_\beta) = \boldsymbol{r}_j(u_\alpha)$, $(j = 0,1)$, and $\overline{S}$ is a unique function of its arguments, the expression on the left-hand side vanishes. Thus

$$I := \int_{u_\alpha}^{u_\beta} \boldsymbol{p}_1(u) \cdot \frac{d\boldsymbol{r}_1}{du}\, du = \int_{u_\alpha}^{u_\beta} \boldsymbol{p}_0(u) \cdot \frac{d\boldsymbol{r}_0}{du}\, du$$

is invariant. The parametric representation facilitates the evaluation of these integrals but is not absolutely necessary; the value of I is invariant with respect to parametric transforms. Since Γ_0 and Γ_1 are arbitrary loops, the expression

$$I = \oint_\Gamma \boldsymbol{p} \cdot d\boldsymbol{r} \tag{5.31}$$

has the same value for any closed loop Γ on the surface of the tube. This is *Poincaré's integral invariance theorem*. This quantity is even invariant with respect to gauge transformations, since it can be rewritten as

$$I = \oint_\Gamma \boldsymbol{g} \cdot d\boldsymbol{r} + Q\Psi_\Gamma \tag{5.32}$$

where Ψ_Γ is the magnetic flux through Γ:

$$\Psi_\Gamma = \oint_\Gamma \boldsymbol{A} \cdot d\boldsymbol{r} = \int_{(\Gamma)} \boldsymbol{B} \cdot d\boldsymbol{a}$$

The flux term in (5.32) gives rise to a phase shift of magnitude $Q\Psi_\Gamma/\hbar$ in the wave-optical interference patterns produced by electron optical 'biprism devices'. This phase shift, known as the Bohm–Aharonov effect (Ehrenberg and Siday, 1949; Aharonov and Bohm, 1959), will be discussed in detail in Volume 3.

Another interesting consequence of (5.31), is the existence of the point characteristic function. The invariance of the integral expression given by (5.31) can be derived by following an alternative route, using canonical transforms (Goldstein, 1959), and may thus be regarded as fundamental. Let us now assume that the whole bundle of rays shown in Fig. 5.4 and also all rays propagating laminarly in the interior of the tube intersect at some point $\boldsymbol{r}_0$. Then any closed loop Γ may be contracted to this point $\boldsymbol{r}_0$, and hence the integral I vanishes. From $\oint \boldsymbol{p} \cdot d\boldsymbol{r} \equiv 0$ (even in the interior), we can deduce that curl $\boldsymbol{p} \equiv 0$. Hence there must be a function $U(\boldsymbol{r}) \not\equiv 0$ such that $\boldsymbol{p} = \text{grad } U$. With the reasonable assumption $U(\boldsymbol{r}_0) = 0$ we recover the point characteristic function

$$U(\boldsymbol{r}_1) =: \overline{S}(\boldsymbol{r}_0, \boldsymbol{r}_1) = \int_{\boldsymbol{r}_0}^{\boldsymbol{r}_1} \boldsymbol{p}(\boldsymbol{r}) \cdot d\boldsymbol{r}$$

Since the value of this integral is independent of the path of integration between $\boldsymbol{r}_0$ and $\boldsymbol{r}_1$, it must be identical with that of (5.19).

It is often preferable to use the invariance theorem in its differential form. This can be easily obtained in the following way. We now consider *congruences* of rays. These are two-parameter manifolds or families of rays, represented by functions $\boldsymbol{r}(u,v;s)$, u and v being the parameters in question and s the arc-length. For instance, all monoenergetic rays emerging from a 'point source' at $\boldsymbol{r}_0$ form a congruence, the parameters u and v then being angles characterizing the starting direction. The definition is, however, more general. It is easily seen that the generalization of (5.30) for a congruence is given by

$$\begin{aligned}\frac{\partial S}{\partial u} &= \boldsymbol{p}_1 \cdot \frac{\partial \boldsymbol{r}_1}{\partial u} - \boldsymbol{p}_0 \cdot \frac{\partial \boldsymbol{r}_0}{\partial u} \\ \frac{\partial S}{\partial v} &= \boldsymbol{p}_1 \cdot \frac{\partial \boldsymbol{r}_1}{\partial v} - \boldsymbol{p}_0 \cdot \frac{\partial \boldsymbol{r}_0}{\partial v}\end{aligned} \quad (5.33)$$

since all vector quantities are now functions of u and v (neglecting the irrelevant dependence on s). From the condition that $\partial^2 \overline{S}/\partial u \partial v$, calculated in different ways, must be the same continuous function, we obtain

$$\frac{\partial \boldsymbol{p}_1}{\partial v} \cdot \frac{\partial \boldsymbol{r}_1}{\partial u} - \frac{\partial \boldsymbol{p}_0}{\partial v} \cdot \frac{\partial \boldsymbol{r}_0}{\partial u} = \frac{\partial \boldsymbol{p}_1}{\partial u} \cdot \frac{\partial \boldsymbol{r}_1}{\partial v} - \frac{\partial \boldsymbol{p}_0}{\partial u} \cdot \frac{\partial \boldsymbol{r}_0}{\partial v}$$

or rearranging

$$\frac{\partial \boldsymbol{r}_1}{\partial u} \cdot \frac{\partial \boldsymbol{p}_1}{\partial v} - \frac{\partial \boldsymbol{p}_1}{\partial u} \cdot \frac{\partial \boldsymbol{r}_1}{\partial v} = \frac{\partial \boldsymbol{r}_0}{\partial u} \cdot \frac{\partial \boldsymbol{p}_0}{\partial v} - \frac{\partial \boldsymbol{p}_0}{\partial u} \cdot \frac{\partial \boldsymbol{r}_0}{\partial v}$$

Since the points $\boldsymbol{r}_0$ and $\boldsymbol{r}_1$ may be chosen arbitrarily so long as both are located on the same trajectory specified by the values of u and v, the expressions on each side do not depend on position but only on u and v, and are thus constant along each ray. This constant is the familiar Lagrange bracket

$$\{u,v\} := \frac{\partial \boldsymbol{r}}{\partial u} \cdot \frac{\partial \boldsymbol{p}}{\partial v} - \frac{\partial \boldsymbol{p}}{\partial u} \cdot \frac{\partial \boldsymbol{r}}{\partial v} = \text{const} \tag{5.34}$$

A congruence is said to be *normal* if $\{u,v\} \equiv 0$. This is equivalent to $I = 0$, since the Poincaré invariant can be obtained from (5.34) by integration (Sturrock, 1955 Section 2.3). In every normal congruence, there is thus a family of surfaces $U(\boldsymbol{r}) = \text{const}$ such that $\boldsymbol{p} = \text{grad } U$, U usually being the point characteristic function $\overline{S}$. As is illustrated in Fig. 5.2, this does not automatically imply orthogonality with respect to the trajectories themselves.

5.7 The problem of uniqueness

In the preceding considerations we have tacitly assumed that the solutions of the Hamilton–Jacobi equation are unique and regular. This is very often not the case. A simple example is shown in Fig. 5.5. In a beam, there may exist an envelope surface, formed by a one-parameter family of rays. This surface, which is known as a caustic, usually has sharp edges, and its extension depends on the positions and the shapes of any apertures confining the beam. The caustic represents a singularity of the point characteristic function, since it separates the domain of no solution from that with two solutions, where the rays intersect. The example illustrated is highly simplified; in realistic electron optical devices, caustics may have a very complicated structure. We shall therefore not investigate them here in a general manner; instead we shall treat some concrete examples later,

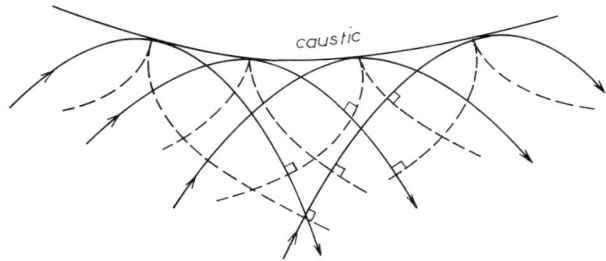

Fig. 5.5: Particle trajectories (full lines) forming a caustic and wavefronts $S(r) =$ const (broken lines) in a two-dimensional section through a beam in an isotropic medium. The domain beyond the caustic is inaccessible to the particles; within the caustic, the trajectories may intersect. At the caustic, the lines $S =$ const form cusps.

see Chapter 42. The foregoing theory remains valid in domains accessible to the beam, the vicinity of caustics being excluded.

In domains in which different functional branches of the point characteristic function overlap, each branch is to be treated separately.

5.8 Résumé

In this part we have dealt with the general theoretical fundamentals of electron optics. We have derived various forms of trajectory equations, kinematic functions and conservation laws. We have investigated various kinds of variational principles and these have permitted us to derive the general theory of Hamiltonian optics. All the relations obtained are essential tools in the investigation of particular aspects of electron optical devices. Since a knowledge of the applied electromagnetic fields is needed before the trajectory equations can be solved, we must now interrupt the purely electron optical discussions and deal with field calculations. In Part III, we shall return to the theory of electron propagation.

Part II

Calculation of Static Fields

6
Basic Concepts and Equations

6.1 General considerations

In this Part we shall deal with the calculation of electrostatic or magnetostatic fields, that is to say, time-independent fields. This major simplification is justified, since in the vast majority of practical electron optical devices the applied fields are static. Even in electric and magnetic deflection units, the frequencies of the time-dependent fields are so low that a quasistatic approximation is entirely justified. By this, we mean that all field functions can be separated in the form $F(r,t) = f_1(r)f_2(t)$, where $f_1(r)$ is practically independent of the frequency and may be calculated as a static field. Special high-frequency devices for which these assumptions do not hold are not treated in this book.

Charged particles usually propagate *in vacuo* in a very narrow domain far distant from any material walls. The only exceptions are the immediate vicinity of emitting surfaces in electron guns, the surfaces of mirrors, specimens, apertures and recording devices. Specimens, apertures and recording media are of little interest in the present context, because they are usually located in field-free domains or are assumed to have no effect on the field distribution. The vicinities of cathodes and reflecting surfaces will be excluded from the following discussion. With these exceptions, the space through which particles travel will be referred to as the *extended paraxial domain*. Usually this is a narrow but long tube around the optic axis of the system in question, see Fig. 6.1. One of the aims of this Part is to derive suitable series expansions for the field in this paraxial domain, since these are fundamental for the investigation of trajectories, focusing properties and aberration effects in practical devices.

Although a concrete knowledge of the field in the appropriate extended paraxial domain would be quite sufficient for all further electron optical considerations (always excluding cathodes and mirrors), this knowledge cannot be obtained without making a complete calculation of the field within the whole device. The reason for this is that a static field within a given domain can only be calculated as the solution of a boundary-value problem, as is further outlined in Chapter 8. In electron optics, the boundaries are the surfaces of electrodes or polepieces, which are thus of great importance

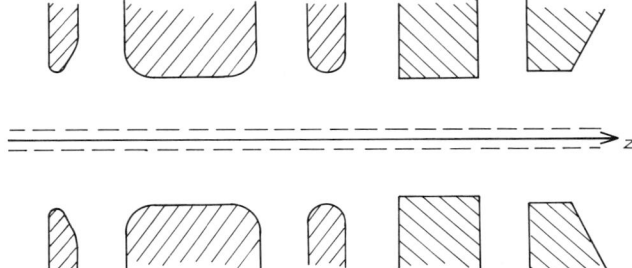

Fig. 6.1: Axial section through part of some electron optical device; the paraxial region is indicated by dashed lines.

though they are usually far from the paraxial domain. Thus the second important aim of this Part is to present techniques for solving boundary-value problems.

Throughout this Part we shall assume that electrodes and magnetic polepieces have isotropic material properties. The corresponding material coefficients may be functions of position. This is mainly the case when saturation effects arise in ferromagnetic polepieces. We shall adopt the following standard electrodynamic notation:

E : electric field strength; $\qquad D$: displacement vector;
H : magnetic field strength; $\qquad B$: magnetic flux density;
ϵ, ϵ_0 : permittivity; $\qquad \mu, \mu_0$: permeability;
$\nu = 1/\mu$: magnetic reluctance; $\qquad A$: vector potential;
ϱ : space charge density; $\qquad j$: electric current density;
σ : surface charge density; $\qquad \omega$: surface current density.

Scalar potential functions are denoted in different ways, as they will appear frequently in different contexts; very often they have only a formal mathematical meaning.

6.2 Field equations

In the case of stationary fields, Maxwell's equations reduce to

$$\begin{aligned} \operatorname{curl} \boldsymbol{E} = 0 \quad &, \quad \operatorname{curl} \boldsymbol{H} = \boldsymbol{j} \\ \operatorname{div} \boldsymbol{D} = \varrho \quad &, \quad \operatorname{div} \boldsymbol{B} = 0 \quad , \quad (\operatorname{div} \boldsymbol{j} = 0) \end{aligned} \qquad (6.1)$$

6.2 FIELD EQUATIONS

These are to be complemented by the material equations

$$\boldsymbol{D} = \epsilon \boldsymbol{E} \quad , \quad \boldsymbol{B} = \mu \boldsymbol{H} \quad , \quad (\text{or} \quad \boldsymbol{H} = \nu \boldsymbol{B}) \tag{6.2}$$

In ferromagnetic materials, the reluctance ν is a function of $B = |\boldsymbol{B}|$, hence

$$\boldsymbol{H} = \nu(B)\boldsymbol{B} \tag{6.3}$$

In (6.1) the space charge density $\varrho(\boldsymbol{r})$ and the current density $\boldsymbol{j}(\boldsymbol{r})$ are regarded as given functions of position. The determination of space charge distributions will be treated in Chapter 46. Electric fields in the interior of conducting materials are not considered here. The source-free Maxwell equations permit us to introduce electromagnetic potentials,

$$\boldsymbol{E} = -\text{grad } \Phi(\boldsymbol{r}) \quad , \quad \boldsymbol{B} = \text{curl } \boldsymbol{A}(\boldsymbol{r}) \tag{6.4}$$

these equations being special cases of (4.3) and (4.4). Combining (6.4) with (6.1) and (6.2), we obtain partial differential equations of Poisson's type. For the electrostatic potential we have

$$\text{div } \left\{ \epsilon(\boldsymbol{r}) \text{grad } \Phi(\boldsymbol{r}) \right\} = -\varrho(\boldsymbol{r}) \tag{6.5}$$

In homogeneous dielectric media, where ϵ is constant, (6.5) reduces to

$$\nabla^2 \Phi(\boldsymbol{r}) = -\varrho(\boldsymbol{r})/\epsilon \tag{6.6}$$

and in domains free of space charge, to Laplace's equation

$$\nabla^2 \Phi(\boldsymbol{r}) = 0 \tag{6.7}$$

The derivation of Poisson's equation for the vector potential proceeds as follows. In the general case, combination of $\nabla \times \boldsymbol{H} = \boldsymbol{j}$, $\boldsymbol{H} = \nu \boldsymbol{B}$ with $\nu = \nu(B)$ and $\boldsymbol{B} = \nabla \times \boldsymbol{A}$ results in

$$\text{curl } \left\{ \nu(|\text{curl } \boldsymbol{A}|) \cdot \text{curl } \boldsymbol{A}(\boldsymbol{r}) \right\} = \boldsymbol{j}(\boldsymbol{r}) \tag{6.8}$$

This complicated non-linear vector Poisson equation has to be solved in ferromagnetic domains with saturation effects. When the permeability $\mu = \nu^{-1}$ is constant, this differential equation can be simplified considerably. Using the vector differential identity

$$\text{curl curl } \boldsymbol{A} = \text{grad div } \boldsymbol{A} - \nabla^2 \boldsymbol{A}$$

valid only in cartesian representations, we obtain first

$$-\text{grad div } \boldsymbol{A} + \nabla^2 \boldsymbol{A} = -\mu \boldsymbol{j}(\boldsymbol{r})$$

A further important simplification can be achieved by choosing a gauge for $\boldsymbol{A}$ that satisfies

$$\text{div } \boldsymbol{A}(\boldsymbol{r}) \equiv 0 \tag{6.9}$$

We then finally obtain the vector Poisson equation

$$\nabla^2 \boldsymbol{A}(\boldsymbol{r}) = -\mu \boldsymbol{j}(\boldsymbol{r}) \tag{6.10}$$

Yet another simplification is possible in current-free domains. We then have $\nabla \times \boldsymbol{H} = 0$ and it is always permissible to write

$$\boldsymbol{H}(\boldsymbol{r}) = -\text{grad } \chi(\boldsymbol{r}) \tag{6.11}$$

$\chi(\boldsymbol{r})$ being the scalar magnetic potential. Since μ is constant, we have div $\boldsymbol{H} = 0$ and χ satisfies the Laplace equation

$$\nabla^2 \chi(\boldsymbol{r}) = 0 \tag{6.12}$$

The simplification thus achieved lies in the fact that only one scalar differential equation is to be solved instead of three coupled by (6.9). The representation (6.11), (6.12) is of course less general than that in terms of the vector potential, since the condition $\boldsymbol{j}(\boldsymbol{r}) = 0$ cannot be true throughout the whole space. The domain of solution of (6.12) is thus to be confined in such a way that $\boldsymbol{j} = 0$ and that the solution for χ remains unique. Important examples are given in Chapters 8 and 9.

Although (6.11) is the correct form of the gradient representation of $\boldsymbol{H}$, it is rather inconvenient, since it is the flux density $\boldsymbol{B}$, and not $\boldsymbol{H}$, that figures in the trajectory equations. After solving the boundary-value problem for the magnetic field, we are only interested in the source-free vacuum field. We may therefore introduce a new potential $W(\boldsymbol{r})$, writing

$$\boldsymbol{B} = -\text{grad } W \quad , \quad \nabla^2 W = 0 \quad , \quad W = \mu_0 \chi \tag{6.12a}$$

We shall use this representation whenever this is possible and causes no confusion.

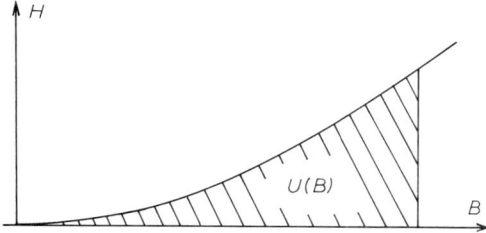

Fig. 6.2: A branch of the magnetization curve, in which $U(B)$ denotes the area under the curve. As $B \to \infty$, the gradient tends asymptotically to a constant value.

6.3 Variational principles

In connection with the finite-element method (see Chapter 12), it is of importance that the partial differential equations given above can also be derived from a variational principle, the integrand being the stored field energy. This principle takes different forms for scalar and vector potentials.

The general form of variational principle for m coupled functions of position $y_1(\mathbf{r}) \ldots y_m(\mathbf{r})$ in a three-dimensional domain S is given by

$$\delta W = \delta \int_S \Lambda(\mathbf{r}, y_1 \ldots y_m, \nabla y_1 \ldots \nabla y_m) \, d^3r = 0 \qquad (6.13)$$

The boundary ∂S and the boundary values $y_1 \ldots y_m$ on it must not vary. The corresponding Euler equations are

$$\frac{\partial \Lambda}{\partial y_i} = \sum_{k=1}^{3} \frac{\partial}{\partial x_k} \left[\frac{\partial \Lambda}{\partial (\partial y_i / \partial x_k)} \right] \equiv \nabla \cdot \frac{\partial \Lambda}{\partial (\nabla y_i)} \qquad (6.14)$$

The cartesian components of the gradients are now treated as independent variables.

In applications involving electrostatic fields, we have $m = 1$, $y_1(\mathbf{r}) = \Phi(\mathbf{r})$ being the electrostatic potential. The corresponding Lagrange density Λ is given by

$$\Lambda = \frac{\epsilon}{2} E^2 - \varrho \Phi \quad , \quad (\mathbf{E} = -\nabla \Phi) \qquad (6.15)$$

Applying (6.14), we obtain (6.5).

When source-free magnetic field domains are to be studied, we may choose scalar potential representations; as before $m = 1$ and now $y_1(\mathbf{r}) = \chi(\mathbf{r})$. The Lagrange density Λ is now the familiar energy density:

$$\Lambda = \frac{\mu}{2} H^2 \quad , \quad (\mathbf{H} = -\nabla \chi) \qquad (6.16)$$

The corresponding Euler equation is (6.12).

6. BASIC CONCEPTS AND EQUATIONS

In the case of general magnetostatic fields, we have to choose $m = 3$, and $y_i(\boldsymbol{r}) = A_i(\boldsymbol{r})$, $i = 1, 2, 3$, are then the three cartesian components of the vector potential $\boldsymbol{A}$. We shall need the absolute flux density

$$B = |\boldsymbol{B}| = |\operatorname{curl} \boldsymbol{A}| \tag{6.17}$$

The source-free term of the Lagrange density Λ is then a function of the form

$$U(B) = \int_0^B H(B')\,dB' \tag{6.18}$$

provided that any magnetic media are isotropic, that is, their permeability may vary with B but not with the direction of $\boldsymbol{B}$ —it is a scalar, not a tensor. For a non-linear medium with saturation effects, but without hysteresis effects, a graph of such a function is shown in Fig. 6.2. Differentiation of (6.18) with respect to $\boldsymbol{B}$ gives

$$\boldsymbol{H}(\boldsymbol{B}) = \frac{\partial U}{\partial \boldsymbol{B}} = \frac{\boldsymbol{B}}{B}\frac{dU(B)}{dB} =: \nu(B)\boldsymbol{B} \tag{6.19a}$$

in accordance with (6.3) and hence

$$\nu(B) = \frac{1}{B}\frac{dU}{dB} \tag{6.19b}$$

The complete Lagrange density is now given by

$$\Lambda = U(B) - \boldsymbol{j} \cdot \boldsymbol{A} \tag{6.20}$$

Evaluation of (6.14) for the three cartesian components of $\boldsymbol{A}$ and expression of the result in vector notation leads to (6.8). These tedious but elementary calculations are not reproduced here.

In the case of linear (unsaturated) media, (6.18) simplifies to the familiar energy density

$$U(B) = \frac{1}{2\mu}B^2 = \frac{1}{2\mu}(\nabla \times \boldsymbol{A})^2 \tag{6.21}$$

$\mu = 1/\nu$ now being a constant. With the additional constraint (6.9), the Euler equations reduce to (6.10).

In all the cases considered here, the Lagrange density Λ has the physical meaning of an energy density, which generally contains an additional interaction term. The functional $F = \int \Lambda d^3r$ is thus an energy. Since this has no upper bound, the concrete evaluation of the variational principle always results in a *minimum* of F. This is of importance for the finite-element method. If the boundary ∂S is extended to infinity, the functional F must remain finite. This implies that any electrodynamic quantities must satisfy the 'natural' boundary conditions, which means that as $|\mathbf{r}| \to \infty$ they must converge to zero in such a way that all the integrals involved remain finite.

6.4 Rotationally symmetric fields

Rotationally symmetric fields are of particular interest in electron optics, as the most common electron lenses are round, by which we mean they are built up from rotationally symmetric fields. Rotationally symmetric electrostatic fields are a simple special case of the Fourier series expansion treated in Chapter 7 and will therefore not be considered here. Magnetic fields, however, require special attention, as we now explain.

For rotational symmetry it is advantageous to introduce cylindrical coordinates (z, r, φ), as defined in Section 2.4. It is necessary to assume that the current density $\mathbf{j}(\mathbf{r})$ is circular and it is further convenient, though not necessary, to assume that the vector potential $\mathbf{A}(\mathbf{r})$ is likewise circular. This means that both vector functions have only azimuthal components:

$$\mathbf{j}(\mathbf{r}) = j(z, r) \mathbf{i}_\varphi \qquad (6.22)$$

$$\mathbf{A}(\mathbf{r}) = A(z, r) \mathbf{i}_\varphi \qquad (6.23)$$

This already implies that div $\mathbf{j} = 0$ and div $\mathbf{A} = 0$. The cylindrical components of $\mathbf{B} = \text{curl } \mathbf{A}$ are given by

$$B_z(z, r) = \frac{1}{r} \frac{\partial}{\partial r}(rA) \quad , \quad B_r = -\frac{\partial A}{\partial z} \quad , \quad B_\varphi = 0 \qquad (6.24)$$

which represent a magnetic field, the direction of which always lies in a *meridional* plane. The opposite case of a circular magnetic field produced by a meridional current distribution is of little interest in electron optics but is of importance in plasma physics. This case will not be treated here.

It is advantageous to introduce the magnetic flux function $\Psi(z, r)$, as in Section 2.5:

$$\Psi(z, r) = 2\pi \int_0^r r' B_z(z, r') \, dr' \qquad (6.25)$$

This is the static special case of (2.32). An immediate consequence is that $B_z(r, z) = (2\pi r)^{-1} \partial \Psi / \partial r$ is the static special case of (2.33). The condition div $B = 0$, expressed in cylindrical coordinates, now becomes

$$\text{div } B = \frac{\partial B_z}{\partial z} + \frac{1}{r}\frac{\partial}{\partial r}(rB_r) = 0 \qquad (6.26)$$

in which we have used $B_\varphi \equiv 0$ (6.24).

Introducing the expression obtained above for B_z into (6.26), we find

$$\frac{\partial}{\partial r}(rB_r) = -\frac{1}{2\pi}\frac{\partial^2 \Psi}{\partial z \partial r} \equiv \frac{\partial}{\partial r}\left(-\frac{1}{2\pi}\frac{\partial \Psi}{\partial z}\right)$$

Integration with respect to r yields first

$$rB_r = -\frac{1}{2\pi}\frac{\partial \Psi}{\partial z} + C(z)$$

$C(z)$ being an arbitrary differentiable function of z. At the optic axis ($r = 0$), B_r must remain finite. Since Ψ and $\partial \Psi/\partial z$ are proportional to r^2 in the vicinity of the optic axis, we can conclude that $C(z) \equiv 0$. Since the restriction to static fields is irrelevant in this context, we thus obtain (2.34).

By applying Stokes's integral theorem to the circle $z = \text{const}$, $r = \text{const}$, $r' \leq r$ shown in Fig. 2.1 we find immediately

$$\Psi(z,r) = 2\pi r A(z,r) \qquad (6.27)$$

A corresponding relation has already been used in Section 4.2. In connection with the paraxial properties of magnetic round lenses, it is helpful to introduce an auxiliary potential function

$$\Pi(z,r) = \frac{2}{r}A(z,r) \qquad (6.28)$$

We can now write the different representations of B_z and B_r as follows:

$$B_z(z,r) = \frac{\partial A}{\partial r} + \frac{A}{r} = \Pi + \frac{r}{2}\cdot\frac{\partial \Pi}{\partial r} = \frac{1}{2\pi r}\cdot\frac{\partial \Psi}{\partial r} \qquad (6.29)$$

$$B_r(z,r) = -\frac{\partial A}{\partial z} = -\frac{r}{2}\cdot\frac{\partial \Pi}{\partial z} = -\frac{1}{2\pi r}\cdot\frac{\partial \Psi}{\partial z} \qquad (6.30)$$

Comparing the different representations and noting that $\Pi(z,r)$ must remain finite as $r \to 0$, we see that

$$B_z(z,0) = \Pi(z,0) \qquad (6.31)$$

Furthermore, we see that, in the immediate vicinity of the optic axis, B_r and A are proportional to r. These facts are of great importance in the physics of magnetic lenses.

When we come to discuss the boundary conditions, it will be useful to write (6.29) and (6.30) in vector form:

$$\boldsymbol{B}(\boldsymbol{r}) = \frac{1}{2\pi r}\boldsymbol{i}_\varphi \times \mathrm{grad}\ \Psi(z,r) \tag{6.32}$$

This is easily verified by writing out (6.32) in cylindrical components.

The partial differential equation to be satisfied by $A(z,r)$ is most easily obtained by substituting (6.24) into (6.8). In cylindrical coordinates, only the azimuthal component fails to vanish:

$$\frac{\partial}{\partial z}\left(\nu\frac{\partial A}{\partial z}\right) + \frac{\partial}{\partial r}\left(\nu\frac{\partial A}{\partial r} + \nu\frac{A}{r}\right) = -j(z,r) \tag{6.33}$$

In unsaturated materials $\mu = \nu^{-1}$ is a constant and (6.33) then simplifies to

$$\frac{\partial^2 A}{\partial z^2} + \frac{\partial^2 A}{\partial r^2} + \frac{1}{r}\frac{\partial A}{\partial r} - \frac{A}{r^2} = -\mu j \tag{6.34}$$

Introducing (6.27) and (6.28) into this equation, more convenient partial differential equations, having no term in r^{-2}, are obtained:

$$\frac{\partial^2 \Pi}{\partial z^2} + \frac{\partial^2 \Pi}{\partial r^2} + \frac{3}{r}\frac{\partial \Pi}{\partial r} = -\frac{2\mu j}{r} \tag{6.35}$$

$$\frac{\partial^2 \Psi}{\partial z^2} + \frac{\partial^2 \Psi}{\partial r^2} - \frac{1}{r}\frac{\partial \Psi}{\partial r} = -2\pi\mu r j \tag{6.36}$$

In Chapters 7 and 8, these equations will be encountered again as formal special cases of a more general differential equation.

6.5 Planar fields

Planar fields are such that the field components are independent of one of the three cartesian coordinates. In practice, they are idealizations of three-dimensional fields, obtained by neglecting the fringe-field domains in one direction. Fields of this type are approximately realized in such devices as electrostatic deflection units, slit lenses, secondary emission multipliers and deflection magnets with plane surfaces.

Without loss of generality, we may choose the cartesian coordinate system in such a manner that the field does not depend on the coordinate y. In the $z - x$ plane, we introduce the complex variable

$$w = z + \mathrm{i}x = r\mathrm{e}^{\mathrm{i}\varphi} \tag{6.37}$$

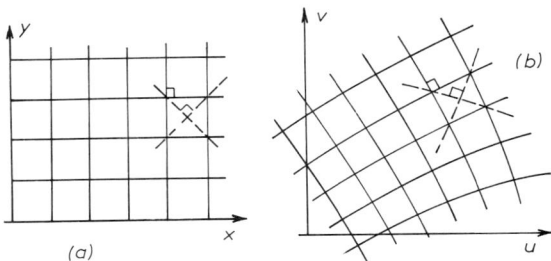

Fig. 6.3: (a) Square grid in the (x,y)-plane. b) Conformal map in the (u,v)-plane; the mapping is such that angles are conserved.

Any analytic function

$$f(w) = u(z,x) + iv(z,x) \qquad (6.38)$$

is a solution of Laplace's equation. This is a consequence of differentiability in the complex plane. As is well-known from the theory of analytic functions, u and v satisfy the Cauchy–Riemann equations

$$\frac{\partial}{\partial z}u(z,x) = \frac{\partial}{\partial x}v(z,x) \quad , \quad \frac{\partial}{\partial z}v(z,x) = -\frac{\partial}{\partial x}u(z,x) \qquad (6.39)$$

From these, the orthogonality relation

$$\frac{\partial u}{\partial z}\frac{\partial v}{\partial z} + \frac{\partial u}{\partial x}\frac{\partial v}{\partial x} \equiv \operatorname{grad} u \cdot \operatorname{grad} v = 0, \qquad (6.40)$$

the conformity relation

$$|\operatorname{grad} u| = |\operatorname{grad} v| = |f'(w)| \qquad (6.41)$$

(with $f'(w) := df/dw$) and the Laplace equations

$$\nabla^2 u = 0 \quad , \quad \nabla^2 v = 0 \quad , \quad \left(\nabla^2 = \frac{\partial^2}{\partial z^2} + \frac{\partial^2}{\partial x^2}\right) \qquad (6.42)$$

can be derived. Equations (6.40) and (6.41) are characteristic of a conformal mapping, as shown in Fig. 6.3. The function $f(w)$ defined by (6.38) can be interpreted as a transform from the square-shaped map in the z–x-plane (Fig. 6.3a) to a curvilinear map in the u–v-plane (Fig. 6.3b). As the size of the curved cells is decreased, these cells collapse to squares.

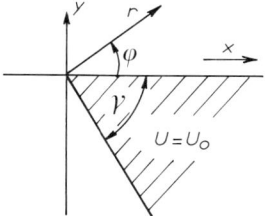

Fig. 6.4: Coordinate system adapted to a sharp edge.

A standard method of calculating planar fields consists in finding the inverse function $[f(w)]^{-1}$ that transforms the given boundary in the z-x-plane into a pair of lines $u =$ const. All lines $u(z,x) =$ const are then equipotentials of the field and all lines $v(z,x) =$ const are orthogonal flux lines. Such a transform is again a conformal mapping. This method of field calculation, commonplace in the mathematical literature, will not be outlined here since it is more complicated than the standard numerical techniques described later. It continues to be employed in the analysis of mass spectrometers (Boerboom and Chen, 1984; Wallington, 1970, 1971). An extremely detailed account is given by Durand (1966).

Some interesting analytical functions and the corresponding potential fields are the following:

(i) Arbitrary convergent power series expansions:

$$f(w) = \sum_m a_m w^m = \sum_m a_m r^m e^{im\varphi} \tag{6.43}$$

Non-negative integral exponents m lead to regular Taylor series expansions, while negative integers generate singular Laurent series expansions. The potential functions obtained from Taylor series expansions are approximately realized in the central zones of multipole devices, see Chapter 7. Non-integral values of m give rise to potential fields in the vicinity of sharp edges. The potential of the sharp edge shown in Fig. 6.4, for example, is given by

$$u(r,\varphi) = u_0 + \sum_{n=1}^{\infty} A_n r^{n\mu} \sin n\mu\varphi \quad , \quad \mu = \frac{\pi}{2\pi - \gamma} \tag{6.44}$$

Here n is a positive integer. It is easy to verify that $u = u_0 =$ const on the lines $\varphi = 0$ and $\varphi = 2\pi - \gamma$ for all sets of coefficients $A_1, A_2, \ldots$

(ii) Complex Fourier integral expansions:

$$f(w) = \int_0^{\infty} A(k) \exp(ikw) \, dk \quad (k \text{ real}, \geq 0) \tag{6.45}$$

With $A(k) = A'(k) - iA''(k)$ and $A(0) = 0$ we obtain a potential distribution that satisfies $u(z,x) \to 0$ as $x \to \infty$:

$$u(z,x) = \int_0^\infty e^{-kx}\{A'(k)\cos kz + B'(k)\sin kz\}\,dk \qquad (6.46)$$

This is of importance in investigations on electron mirrors.

(iii) Logarithmic singularity

$$f(w) = \log Cw \quad , \quad u = \log Cr \quad , \quad v = \varphi \qquad (6.47)$$

(C is a positive constant.) This is of great importance in the practical application of the integral equation method, see Chapter 10. The inverse function

$$w = z + ix = r\exp(i\varphi) = \frac{1}{C}\exp(u+iv) \qquad (6.48a)$$

$$r = \frac{1}{C}\exp(u) \quad , \quad \varphi \equiv v \qquad (6.48b)$$

describes the transform from quadratic to polar grids, see also Section 11.4.2.

7
Series Expansions

As has been mentioned in Section 6.1, the electromagnetic field in the extended paraxial domain is of paramount interest in electron optics. In this domain, the field will be represented by series expansions, the general structures of which can already be derived without explicit solution of the corresponding boundary-value problem. As will be obvious later, the field in the vicinity of the optic axis can be obtained by analytic continuation of the axial distribution.

In this chapter we shall assume that the optic axis is straight. Series expansions adapted to curvilinear axes are required in the treatment of devices with sector fields and will be derived in that context (see Part X). We shall further assume that the paraxial domain is usually source-free. This is certainly true for all current distributions, since the windings of coils are always far from the optic axis. Electric space charge may accumulate in the extended paraxial domain. Its distribution is, however, so inhomogeneous that the corresponding series expansions are of little practical value. Here, therefore, we shall mostly exclude them; they will be considered in detail when we come to treat electron guns (see Part IX).

7.1 Azimuthal Fourier series expansions

The aim of the following considerations is to decompose a three-dimensional field into a sequence of uncoupled two-dimensional fields. This is advantageous, since two-dimensional fields are far more easily calculated than three-dimensional ones. The required decoupling is obtained by expanding the field as a series of complete orthogonal functions, most favourably as a Fourier series (Glaser, 1952, Section 35). Since the present discussion is quite general, source distributions are not excluded.

7.1.1 Scalar potentials
Let us now consider a general Poisson equation

$$\nabla^2 V(\boldsymbol{r}) = -S(\boldsymbol{r}) \qquad (7.1)$$

regardless of the special meanings of the functions $V(\boldsymbol{r})$ and $S(\boldsymbol{r})$. In cylindrical coordinates (z, r, φ) this Poisson equation takes the form

$$\frac{\partial^2 V}{\partial z^2} + \frac{\partial^2 V}{\partial r^2} + \frac{1}{r} \cdot \frac{\partial V}{\partial r} + \frac{1}{r^2} \cdot \frac{\partial^2 V}{\partial \varphi^2} = -S(z, r, \varphi) \qquad (7.2)$$

Both functions V and S are periodic with respect to the azimuth φ, the period being 2π, which suggests that we should introduce their Fourier series expansions. These can be represented in different equivalent forms. One essential requirement is that it should be easily possible to transform these series expansions and the corresponding ones for the gradient into *regular* power series expansions with respect to x and y. With these transforms in mind, we introduce the variables

$$w := x + iy = r \exp(i\varphi) \qquad (7.3)$$
$$s := x^2 + y^2 = r^2 = ww^* \qquad (7.4)$$

the asterisk denoting complex conjugation. The series expansions in question then have the form

$$V(z, r, \varphi) = \sum_{m=0}^{\infty} r^m \Re\left\{U_m(z, s) e^{im\varphi}\right\} \qquad (7.5)$$

$$S(z, r, \varphi) = \sum_{m=0}^{\infty} r^m \Re\left\{g_m(z, s) e^{im\varphi}\right\} \qquad (7.6)$$

and in the cartesian representation

$$V(\mathbf{r}) = \sum_{m=0}^{\infty} \Re\left\{U_m(z, s) w^m\right\} \qquad (7.7)$$

with a corresponding expression for $S(\mathbf{r})$. Differentiation of (7.7) with respect to z immediately results in:

$$\frac{\partial V}{\partial z} = \sum_{m=0}^{\infty} \Re\left\{w^m \frac{\partial U_m(z, s)}{\partial z}\right\} \qquad (7.8)$$

The remaining derivatives are most favourably expressed in complex form:

$$\frac{\partial V}{\partial x} + i\frac{\partial V}{\partial y} = \sum_{m=0}^{\infty} \left\{w^{m+1} \frac{\partial U_m}{\partial s} + (w^*)^{m-1}\left(s\frac{\partial U_m^*}{\partial s} + mU_m^*\right)\right\} \qquad (7.9)$$

For $m = 0$ this series expansion contains *no* singularity, since $(w^*)^{-1}s = w$. Recalling that U_0 must be a real function, we obtain the expression $2w\partial U_0/\partial s$ for the zero-order term of this series expansion.

7.1 AZIMUTHAL FOURIER SERIES EXPANSIONS

Introducing (7.5) and (7.6) into (7.2), we first find a sequence of uncoupled differential equations

$$\frac{\partial^2 U_m}{\partial z^2} + \frac{\partial^2 U_m}{\partial r^2} + \frac{2m+1}{r}\frac{\partial U_m}{\partial r} = -g_m(z,r) \quad m = 0,1,2,\ldots \quad (7.10)$$

It is possible to introduce the variable $s = r^2$ (Glaser, 1952, Section 35); this is, however, unfavourable with respect to the numerical solution, see Chapter 11.

In practice, it is impossible to evaluate infinite series expansions numerically; instead, we have to truncate them in a suitable manner. From (7.5) we can get an idea of how this is to be done. Since U_m and g_m may be regarded as slowly varying functions within the extended paraxial domain, the factor r^m is of greatest importance. Let R be some characteristic bore radius of the device in question. Then the exponent $m = M$ at which the series is truncated should be chosen in such a way that all terms $(r/R)^m$ with $m > M$ can be neglected. In practice $M = 5$ is usually sufficient, since $r/R \lesssim 10^{-1}$. In order to avoid any misunderstanding, we emphasize that in concrete calculations, (7.10) (with $0 \leq m \leq M$) are to be solved as a sequence of boundary-value problems, which implies that r may extend far beyond the paraxial domain. Only the values of the solutions obtained inside the extended paraxial domain are to be retained.

In (7.10), the coefficients of the partial derivatives are simple real factors. A further simplification is therefore possible by considering only real source terms $g_m(z,s)$ and real solutions $U_m(z,s)$. Complex solutions can easily be obtained by forming linear combinations with appropriate complex factors. In Section 7.2, we shall thus assume that the functions $U_m(z,s)$ are *real*.

7.1.2 Vector potentials

In Chapters 4 and 5, we have seen that the vector potential contributes to the canonical momentum and to the characteristic function. Though the vector potential itself is not an observable quantity, it still plays an important role in theoretical considerations. We now derive appropriate series expansions for it. Equations (6.9) and (6.10) must be satisfied and furthermore we impose the natural boundary condition $\boldsymbol{A}(\boldsymbol{r}) \to 0$ for $|\boldsymbol{r}| \to \infty$. Then $\boldsymbol{A}(\boldsymbol{r})$ is uniquely defined. For technical reasons we have $j_x = j_y = 0$ on the optic axis, which implies that $A_x = A_y = 0$ for $x = y = 0$.

The desired series expansions will be similar in form to (7.7). Here it is more convenient to consider the imaginary parts instead of the real ones, which is only a minor difference. Furthermore it is convenient to introduce complex transverse components, thus

$$A_T := A_x + iA_y = \frac{i}{2} \sum_{m=1}^{\infty} C_m(z,s) w^m \tag{7.11}$$

$$A_z := -\Im \left(\sum_{m=0}^{\infty} D_m(z,s) w^m \right) \tag{7.12}$$

In a similar way the current density $\boldsymbol{j}(\boldsymbol{r})$ is represented by

$$j_T := j_x + ij_y = \frac{i}{2} \sum_{m=1}^{\infty} J_m(z,s) w^m \tag{7.13}$$

$$j_z := -\Im \left(\sum_{m=0}^{\infty} L_m(z,s) w^m \right) \tag{7.14}$$

In the functions introduced on the right-hand side, the variable s is again defined by (7.4).

A straightforward differentiation in cartesian coordinates results in

$$\operatorname{div} \boldsymbol{A} = -\Im \left\{ \sum_{m=1}^{\infty} \left(D_{m-1|z} + sC_{m|s} + mC_m \right) w^{m-1} \right\} = 0$$

Here the subscripts behind vertical bars denote partial differentiations with respect to the corresponding variable (see Section 2.4). Since different powers w^m of w must be linearly independent, this condition $\operatorname{div} \boldsymbol{A} = 0$ can only be satisfied if

$$D_{m-1|z} + sC_{m|s} + mC_m = 0 \quad (m \geq 1) \tag{7.15}$$

is valid for all integers m. Similarly the condition $\operatorname{div} \boldsymbol{j} = 0$ leads to

$$L_{m-1|z} + sJ_{m|s} + mJ_m = 0 \quad (m \geq 1) \tag{7.16}$$

The determination of $\nabla^2 \boldsymbol{A}$ in cartesian coordinates is also straightforward; we obtain

$$\nabla^2 A_T = \frac{i}{2} \sum_{m=1}^{\infty} \left(C_{m|zz} + 4sC_{m|ss} + 4(m+1)C_{m|s} \right) w^m$$

$$\nabla^2 A_z = -\Im \left\{ \sum_{m=0}^{\infty} \left(D_{m|zz} + 4sD_{m|ss} + 4(m+1)D_{m|s} \right) w^m \right\}$$

7.1 AZIMUTHAL FOURIER SERIES EXPANSIONS

These series expansions are to be matched to (7.13) and (7.14). Using the linear independence of different powers of w and the vector Poisson equation $\nabla^2 \boldsymbol{A} = -\mu \boldsymbol{j}$ we obtain

$$C_{m|zz} + 4sC_{m|ss} + 4(m+1)C_{m|s} = -\mu J_m \tag{7.17}$$
$$D_{m|zz} + 4sD_{m|ss} + 4(m+1)D_{m|s} = -\mu L_m \tag{7.18}$$

With the transformations

$$\frac{\partial}{\partial s} = \frac{1}{2r}\frac{\partial}{\partial r} \quad , \quad \frac{\partial^2}{\partial s^2} = \frac{1}{4s}\frac{\partial^2}{\partial r^2} - \frac{1}{4sr}\frac{\partial}{\partial r}$$

we obtain the more suitable differential equations

$$C_{m|zz} + C_{m|rr} + \frac{2m+1}{r}C_{m|r} = -\mu J_m(z,s) \tag{7.19}$$
$$D_{m|zz} + D_{m|rr} + \frac{2m+1}{r}D_{m|r} = -\mu L_m(z,s) \tag{7.20}$$

which have the same mathematical structure as (7.10). Here, however, their solutions are additionally coupled by (7.15).

The magnetic flux density is obtained by differentiation, $\boldsymbol{B} = \nabla \times \boldsymbol{A}$. After some elementary calculations we find:

$$B_T := B_x + iB_y$$
$$= -\sum_{m=0}^{\infty}\left(sD_{m|s}^* + mD_m^*\right)(w^*)^{m-1} - \sum_{m=0}^{\infty}\left(\frac{1}{2}C_{m+1|z} - D_{m|s}\right)w^{m+1}$$
$$\tag{7.21}$$

In *source-free domains*, where $\boldsymbol{j} = 0$, this expression can be further simplified. In order to show this, we differentiate (7.15) with respect to s and make use of (7.17) with $J_m = 0$, obtaining

$$-D_{m-1|sz} = sC_{m|ss} + (m+1)C_{m|s} = -\frac{1}{4}C_{m|zz}$$

This can be integrated with respect to z. Since $\boldsymbol{A}$ has to vanish asymptotically, no additive constant can appear, and so

$$4D_{m|s} = C_{m+1|z} \tag{7.22}$$

Introducing this into (7.21) and eliminating $C_{m+1|z}$, we find now

$$B_T = -\sum_{m=0}^{\infty}\left\{\left(sD_{m|s}^* + mD_m^*\right)(w^*)^{m-1} + D_{m|s}w^{m+1}\right\}$$

This equation has the same structure as (7.9). Therefore with $\mu = \mu_0$ *in vacuo* we can introduce a scalar potential $W(\mathbf{r})$ by writing

$$W(\mathbf{r}) = \Re \sum_{m=0}^{\infty} \left(D_m(z,s) w^m \right) \qquad (7.23)$$

Calculation of the axial component B_z confirms this result. A straightforward differentiation yields

$$B_z = A_{y|x} - A_{x|y} = \Re \sum_{m=1}^{\infty} \left(sC_{m|s} + mC_m \right) w^{m-1}$$

On the right-hand side the expression $D_{m|z}$ can be introduced by means of (7.15) and we then find

$$B_z = -\frac{\partial}{\partial z} \Re \left(\sum_{m=0}^{\infty} D_m(z,s) w^m \right) = -W_{|z}$$

since the permeability μ is constant. The results obtained are in agreement with (6.11) and (6.12), as they should be.

Comparison of (7.12) and (7.23) shows that W and A_z are interrelated as the real and imaginary parts of a complex function:

$$W - \mathrm{i} A_z = \sum_{m=0}^{\infty} D_m(z,s) w^m \qquad (7.24)$$

This relation holds, of course, only in source-free domains.

The representation of the vector potential given above is a generalization of Sturrock's (1951) formula. Other gauges, which do not satisfy div $\mathbf{A} = 0$, have been introduced by Glaser (1952, Section 36) and by Schwertfeger and Kasper (1974). The present procedure is convenient, since we always arrive at the same class of partial differential equations.

7.2 Radial series expansions

7.2.1 Scalar potentials

In accordance with the assumption that the paraxial domain is source-free, we shall now investigate solutions of (7.10) with source terms vanishing for sufficiently small values of r. From (7.5) and (7.7), it can be seen that a power series expansion with respect to $s = r^2$ will be the most suitable. We therefore introduce

$$U_m(z,s) = \sum_{n=0}^{\infty} \frac{(-1)^n}{(2n)!} c_n(z,m) s^n \qquad (7.25)$$

7.2 RADIAL SERIES EXPANSIONS

into (7.10), the factor $(-1)^n/(2n)!$ being included for reasons of convenience. The coefficients c_n, unknown at the moment, are related by recurrence formulae:

$$\left(1 + \frac{2m+1}{2n+1}\right) c_{n+1}(z,m) = c_n''(z,m) \quad , \quad n = 0, 1, 2, \ldots$$

where primes denote differentiation with respect to z. The coefficient $c_0(z, m)$ can be chosen arbitrarily provided that all its derivatives remain finite for all values of z ($|z| \to \infty$ included). This function is the *axial* value of U_m:

$$c_0(z,m) = U_m(z,0) =: u_m(z) \tag{7.26}$$

the subscript 0 being omitted to lighten the notation. In terms of r^2 the required power series expansion is now given by

$$U_m(z, r^2) = \sum_{n=0}^{\infty} \frac{m!}{n!(m+n)!} \left(-\frac{r^2}{4}\right)^n u_m^{(2n)}(z) \tag{7.27}$$

$$= u_m(z) - \frac{r^2 u_m''(z)}{4(m+1)} + \left(\frac{r^2}{4}\right)^2 \frac{u_m^{(4)}(z)}{2!(m+1)(m+2)} + O(r^6)$$

Apart from the notation, this radial series expansion is identical with that given by Glaser (1952); it is equivalent to that given by Kasper (1982) if α is identified with $2m + 1$. Equation (7.27) shows that each function $U_m(z, s)$ is already uniquely determined by its axial distribution $u_m(z)$ (called the axial harmonic), provided that the series expansion converges. This property of the solution is very similar to that of analytic functions in the complex plane and by analogy we shall refer to (7.27) as an analytic continuation of the axial values.

In the most general case, the convergence of (7.27) cannot be proven but must be assumed, at least in the extended paraxial domain. In this context there may arise problems since in fact (7.27) does *not* converge for all values of r. Unfortunately even reliable estimates for the radius of convergence are not known (apart from some special examples).

In principle, it should be possible to compute the field in an entire device by analytic continuation of the appropriate axial harmonics $u_m(z)$, since the solution as a boundary-value problem shows clearly that the singularities must be located outside or at the boundaries. But in practice this is impossible for various reasons. A first problem is that the analytical continuation is numerically unstable. Even if one starts with the correct functions $u_m(z)$, one will *not* obtain the correct boundary values of $U_m(z, r^2)$, since rounding and truncation errors, initially very small, may increase dramatically. Another serious difficulty is that it will be impossible to obtain reasonable shapes of electrodes or polepieces if the functions $u_m(z)$ are only slightly different from those corresponding to a realistic field. This problem is further discussed in Chapter 34.

In a realistic field computation, the first step is the solution of a boundary-value problem. In this way the axial harmonics $u_m(z)$ can be determined uniquely. These have then to be differentiated numerically. Finally (7.27) can be evaluated for arbitrary values of z and sufficiently small values of r ($r/R \lesssim 10^{-1}$, see Section 7.1).

A very important practical application of the radial series expansion arises in general theoretical calculations, where no concrete numerical evaluations are required. The purpose of such calculations is the derivation of general rules for focusing properties and aberration coefficients in classes of devices. In this context, it is often helpful to use simple analytic models of the axial field distributions. These models must contain some free parameters with which a fit to a correct numerical solution for the $u_m(z)$ is possible.

7.2.2 Vector potentials

Since the functions $C_m(z,s)$ and $D_m(z,s)$, introduced in Section 7.1.2, satisfy differential equations of the same basic type as (7.10), their radial series expansions must be similar to (7.27), always assuming that the domain of solution is source-free. Here we have to consider *two* series expansions, one for each of the two functions $C_m(z,s)$ and $D_m(z,s)$, but due to (7.15) and (7.22) these are linearly dependent. The consequence is that, though the three components of $\mathbf{A}(\mathbf{r})$ are different functions of position, only *one* axial harmonic $\Pi_m(z)$ can be introduced independently for each Fourier component. Since the scalar potential W itself, given by (7.23), has little significance, we define the axial harmonics by means of the relations

$$\Pi_m(z) := -D_{m|z}(z,0) \quad , \quad (m \geq 0) \tag{7.28}$$

and hence

$$D_m(z,0) = -\int_{-\infty}^{z} \Pi_m(z')\,dz' \tag{7.29}$$

Applying (7.20) (with $L_m(z,s) = 0$), the radial series expansion of $D_m(z,s)$ is then given by

$$D_m(z,r^2) = -\sum_{n=0}^{\infty} \frac{m!}{n!(m+n)!}\left(-\frac{r^2}{4}\right)^n \Pi_m^{(2n-1)}(z)$$

$$= -\int_{-\infty}^{z} \Pi_m(z')\,dz' + \frac{r^2 \Pi'_m(z)}{4(m+1)} - \left(\frac{r^2}{4}\right)^2 \frac{\Pi'''_m(z)}{2!(m+1)(m+2)} + O(r^6)$$

(7.30)

The axial values of the function $C_m(z, r^2)$ can be determined from (7.15). On the optic axis, the term $sC_m|_s$ vanishes and using (7.28), we find

$$C_m(z, 0) = \Pi_{m-1}(z)/m \quad (m \geq 1) \tag{7.31}$$

The series expansion for $C_m(z, r^2)$ is given by

$$C_m(z, r^2) = \sum_{n=0}^{\infty} \frac{m!}{n!(m+n)!} \left(-\frac{r^2}{4}\right)^n C_m^{(2n)}(z, 0)$$

which must be similar to (7.27), since in the homogeneous case the corresponding differential equations (7.10) and (7.17) have the same formal structure. With the aid of (7.31), we find

$$C_m(z, r^2) = \sum_{n=0}^{\infty} \frac{(m-1)!}{n!(m+n)!} \left(-\frac{r^2}{4}\right)^n \Pi_{m-1}^{(2n)}$$

$$= \frac{\Pi_{m-1}(z)}{m} - \frac{r^2 \Pi''_{m-1}(z)}{4m(m+1)} + \left(\frac{r^2}{4}\right)^2 \frac{\Pi_{m-1}^{(4)}(z)}{2!m(m+1)(m+2)} - O(r^6)$$

(7.32)

It is straightforward to prove, by carrying out the necessary differentiations, that (7.15) and (7.22) are satisfied.

In order to obtain the complete series expansions of the vector potential $\mathbf{A}$, the expansions, derived above, have to be substituted in (7.11) and (7.12). Practical expressions will be given in the next sections. Apart from the different notation, our results are identical with those given by Sturrock (1951). They differ from the formulae of Schwertfeger and Kasper (1974), which do not satisfy div $\mathbf{A} = 0$, and also from those of Glaser (1952), in which $A_z \equiv 0$ is assumed and which do not necessarily satisfy $\mathbf{A} \to 0$ for $z \to \infty$. The particular gauge that does satisfy the natural boundary conditions is most convenient in practical applications.

7.2.3 Explicit representations

In almost all practical applications, it is quite sufficient to truncate the power series expansions of the potentials after the terms of fourth order in x and y, and consequently those of the field strength after the third order. These series expansions play an important role in the theory of electron optical aberrations.

In order to introduce a comparatively simple and easily remembered *real* representation of the series expansions, we define real axial harmonics $p_m(z)$, $q_m(z)$, $P_m(z)$ and $Q_m(z)$ as follows:

$$U_m(z,0) =: \frac{(-1)^m}{m!}\Big(p_m(z) - iq_m(z)\Big) \quad m = 0,1,2,\ldots \quad (7.33)$$

$$D_m(z,0) =: \frac{(-1)^m}{m!}\Big(P_m(z) - iQ_m(z)\Big) \quad m = 0,1,2,\ldots \quad (7.34)$$

Equation (7.33) will be used exclusively for the expansion of the electrostatic potential $\Phi(\boldsymbol{r})$, while (7.34) will refer exclusively to the vector potential $\boldsymbol{A}(\boldsymbol{r})$ and the flux density $\boldsymbol{B}(\boldsymbol{r})$. The Fourier coefficients with $m = 0$ and $m = 1$ have a special meaning which will be encountered frequently. We therefore introduce a special notation for these coefficients apart from q_0, which is identically zero:

$$p_0(z) = \phi(z) \quad , \quad p_1(z) = F_1(z) \quad , \quad q_1(z) = F_2(z) \quad (7.35)$$

Here $\phi(z) = \Phi(z,0,0)$ is the familiar axial potential, $F_1(z) = -\Phi_{|x}(z,0,0)$ and $F_2(z) = -\Phi_{|y}(z,0,0)$ are the transverse components of the field strength $\boldsymbol{E}$ on the optic axis. The electric potential is then given by

$$\begin{aligned}\Phi(\boldsymbol{r}) = {}& \phi(z) - \frac{1}{4}(x^2+y^2)\phi''(z) + \frac{1}{64}(x^2+y^2)^2\phi^{(4)}(z) \\ & - xF_1(z) - yF_2(z) + \frac{1}{8}(x^2+y^2)(xF_1'' + yF_2'') \\ & + \frac{1}{2}(x^2-y^2)p_2(z) + xyq_2(z) - \frac{1}{24}(x^4-y^4)p_2'' - \frac{1}{12}(x^3y+xy^3)q_2'' \\ & - \frac{1}{6}p_3(z)(x^3-3xy^2) + \frac{1}{6}q_3(z)(y^3-3x^2y) \\ & + \frac{1}{24}p_4(z)(x^4-6x^2y^2+y^4) + \frac{1}{6}q_4(z)(x^3y-xy^3) \end{aligned} \quad (7.36)$$

The corresponding series expansion in cylindrical coordinates reveals the structure better:

$$\begin{aligned}\Phi = {}& \phi - \frac{r^2}{4}\phi'' + \frac{r^4}{64}\phi^{(4)} \\ & - r(F_1\cos\varphi + F_2\sin\varphi) + \frac{r^3}{8}(F_1''\cos\varphi + F_2''\sin\varphi) \\ & + \frac{r^2}{2}(p_2\cos 2\varphi + q_2\sin 2\varphi) - \frac{r^4}{24}(p_2''\cos 2\varphi + q_2''\sin 2\varphi) \\ & - \frac{r^3}{6}(p_3\cos 3\varphi + q_3\sin 3\varphi) + \frac{r^4}{24}(p_4\cos 4\varphi + q_4\sin 4\varphi) \end{aligned} \quad (7.37)$$

7.2 RADIAL SERIES EXPANSIONS

For magnetic fields, we introduce the following special notation for the coefficients:

$$\Pi_0(z) =: B(z) \quad , \quad P_0(z) = -\int_{-\infty}^{z} B(z)\,dz \quad , \quad Q_0 \equiv 0 \quad (7.38)$$

$$P_1(z) =: B_1(z) \quad , \quad Q_1(z) =: B_2(z) \quad (7.39)$$

Again P_0 is essentially a scalar axial potential, but here this has little physical meaning. The functions $B(z)$, $B_1(z)$ and $B_2(z)$ are, however, very important, since they represent the axial value of $\mathbf{B}(\mathbf{r})$:

$$\mathbf{B}(0,0,z) = \mathbf{i}_x B_1(z) + \mathbf{i}_y B_2(z) + \mathbf{i}_z B(z) \quad (7.40)$$

The *scalar magnetic potential* is then given by:

$$W(\mathbf{r}) = -\int B\,dz + \frac{1}{4}(x^2+y^2)B'(z) - \frac{1}{64}(x^2+y^2)^2 B'''(z)$$
$$- xB_1(z) - yB_2(z) + \frac{1}{8}(x^2+y^2)(xB_1'' + yB_2'')$$
$$+ \frac{1}{2}(x^2-y^2)P_2(z) + xyQ_2(z) - \frac{1}{24}(x^4-y^4)P_2'' - \frac{1}{12}(x^3 y + xy^3)Q_2''$$
$$- \frac{1}{6}(x^3 - 3xy^2)P_3(z) + \frac{1}{6}(y^3 - 3x^2 y)Q_3(z)$$
$$+ \frac{1}{24}(x^4 - 6x^2 y^2 + y^4)P_4(z) + \frac{1}{6}(x^3 y - xy^3)Q_4(z) \quad (7.41)$$

$$W = -\int B\,dz + \frac{r^2}{4}B'(z) - \frac{r^4}{64}B'''(z)$$
$$- r(B_1 \cos\varphi + B_2 \sin\varphi) + \frac{r^3}{8}(B_1'' \cos\varphi + B_2'' \sin\varphi)$$
$$+ \frac{r^2}{2}(P_2 \cos 2\varphi + Q_2 \sin 2\varphi) - \frac{r^4}{24}(P_2'' \cos 2\varphi + Q_2'' \sin 2\varphi)$$
$$- \frac{r^3}{6}(P_3 \cos 3\varphi + Q_3 \sin 3\varphi) + \frac{r^4}{24}(P_4 \cos 4\varphi + Q_4 \sin 4\varphi) \quad (7.42)$$

The *vector potential* is given by

$$A_x = -\frac{y}{2}\left(B - \frac{1}{8}(x^2+y^2)B''\right)$$
$$+ \frac{1}{4}(x^2 - y^2)B_2' - \frac{1}{48}(x^4 - y^4)B_2''' - \frac{xy}{2}B_1' + \frac{1}{24}(x^3 y + xy^3)B_1'''$$
$$- \frac{1}{12}(x^3 - 3xy^2)Q_2' - \frac{1}{12}(y^3 - 3x^2 y)P_2'$$
$$+ \frac{1}{48}(x^4 - 6x^2 y^2 + y^4)Q_3' - \frac{1}{12}(x^3 y - xy^3)P_3' \quad (7.43)$$

$$A_y = \frac{x}{2}\left(B - \frac{1}{8}(x^2 + y^2)B''\right)$$
$$+ \frac{1}{4}(x^2 - y^2)B_1' - \frac{1}{48}(x^4 - y^4)B_1''' + \frac{xy}{2}B_2' - \frac{1}{24}(x^3y + xy^3)B_2'''$$
$$- \frac{1}{12}(x^3 - 3xy^2)P_2' + \frac{1}{12}(y^3 - 3x^2y)Q_2'$$
$$+ \frac{1}{48}(x^4 - 6x^2y^2 + y^4)P_3' + \frac{1}{12}(x^3y - xy^3)Q_3' \qquad (7.44)$$

$$A_z = -xB_2(z) + yB_1(z) + \frac{1}{8}(x^2 + y^2)(xB_2'' - yB_1'')$$
$$+ \frac{1}{2}(x^2 - y^2)Q_2 - xyP_2 - \frac{1}{24}(x^4 - y^4)Q_2'' + \frac{1}{12}(x^3y + xy^3)P_2''$$
$$- \frac{1}{6}(x^3 - 3xy^2)Q_3 - \frac{1}{6}(y^3 - 3x^2y)P_3$$
$$+ \frac{1}{24}(x^4 - 6x^2y^2 + y^4)Q_4 - \frac{1}{6}(x^3y - xy^3)P_4 \qquad (7.45)$$

Differentiation using $\boldsymbol{B} = -\nabla W = \nabla \times \boldsymbol{A}$ and truncation of the resulting expressions beyond the third-order terms results in

$$B_x = -\frac{x}{2}B'(z) + \frac{x}{16}(x^2 + y^2)B'''$$
$$+ B_1(z) - \frac{1}{8}(3x^2 + y^2)B_1'' - \frac{1}{4}xyB_2''$$
$$- xP_2 - yQ_2 + \frac{1}{6}x^3P_2''' + \frac{1}{12}(3x^2y + y^3)Q_2''$$
$$+ \frac{1}{2}(x^2 - y^2)P_3 + xyQ_3$$
$$- \frac{1}{6}(x^3 - 3xy^2)P_4 + \frac{1}{6}(y^3 - 3x^2y)Q_4 \qquad (7.46)$$

$$B_y = -\frac{y}{2}B'(z) + \frac{y}{16}(x^2 + y^2)B'''$$
$$+ B_2(z) - \frac{1}{8}(3y^2 + x^2)B_2'' - \frac{1}{4}xyB_1''$$
$$- xQ_2 + yP_2 - \frac{1}{6}y^3P_2''' + \frac{1}{12}(x^3 + 3xy^2)Q_2''$$
$$+ \frac{1}{2}(x^2 - y^2)Q_3 - xyP_3$$
$$- \frac{1}{6}(x^3 - 3xy^2)Q_4 - \frac{1}{6}(y^3 - 3x^2y)P_4 \qquad (7.47)$$

7.3 ROTATIONALLY SYMMETRIC FIELDS

$$B_z = B - \frac{1}{4}(x^2 + y^2)B''$$
$$+ xB_1' + yB_2' - \frac{1}{8}(x^2 + y^2)(xB_1''' + yB_2''')$$
$$- \frac{1}{2}(x^2 - y^2)P_2' - xyQ_2'$$
$$+ \frac{1}{6}(x^3 - 3xy^2)P_3' - \frac{1}{6}(y^3 - 3x^2y)Q_3' \qquad (7.48)$$

In these formulae, general cartesian components B_x, B_y, B_z are clearly distinguished from their axial values B_1, B_2, B by the notation of the subscripts. The expressions for the electric field strength are obtained by interchanging the symbols as follows:

$$\begin{aligned} B_{1,2} \to F_{1,2} &\quad,\quad B \to -\phi' \quad,\quad B_{x,y,z} \to E_{x,y,z} \\ P_j \to p_j &\quad,\quad Q_j \to q_j \end{aligned} \qquad (7.49)$$

The resulting formulae are not given explicitly here.

7.3 Rotationally symmetric fields

In Sections 7.3–7.5, we give the paraxial series expansions for the most important applications to be studied in later chapters in the notation that will be used there.

7.3.1 Electrostatic fields

Since it describes round lenses, the rotationally symmetric scalar potential field is the most important special case. For electrostatic fields, the following identifications are necessary in Sections 7.1.1 and 7.2.1:

$$m = 0 \quad,\quad U_0(z,s) = \Phi(r) =: \Phi(z,r) \quad,\quad u_0(z) = \Phi(z,0) =: \phi(z)$$

With this notation, we obtain

$$\Phi(z,r) = \sum_{n=0}^{\infty} \frac{1}{(n!)^2} \left(-\frac{r^2}{4}\right)^n \phi^{(2n)}(z)$$
$$= \phi(z) - \frac{r^2}{4}\phi''(z) + \frac{r^4}{64}\phi^{(4)}(z) - \frac{r^6}{2304}\phi^{(6)}(z) + O(r^8) \qquad (7.50)$$

The cartesian components of $\mathbf{E} = -\mathrm{grad}\,\Phi$ are most rapidly obtained by direct differentiation of (7.50):

$$E_z(z,r) = -\phi'(z) + \frac{r^2}{4}\phi^{(3)}(z) - \frac{r^4}{64}\phi^{(5)}(z) + \frac{r^6}{2304}\phi^{(7)}(z) - O(r^8) \qquad (7.51)$$

$$E_r = -\frac{r}{2}R_E \quad,\quad E_x = -\frac{x}{2}R_E \quad,\quad E_y = -\frac{y}{2}R_E \qquad (7.52a)$$

with

$$R_E(z,r^2) := \phi''(z) - \frac{r^2}{8}\phi^{(4)}(z) + \frac{r^4}{192}\phi^{(6)}(z) - \frac{r^2}{9216}\phi^{(8)}(z) + O(r^8) \quad (7.52b)$$

It is possible to express (7.50) in closed form as a complex integral (Glaser, 1952)

$$\Phi(z,r) = \frac{1}{2\pi}\int_0^{2\pi} \phi(z + ir\cos\alpha)\,d\alpha$$

This has, however, little practical value and does not circumvent the difficulties described above associated with the analytic continuation.

In some electron optical devices, the space charge of the beam is important. We therefore present here the corresponding series expansions. It is convenient to expand the space charge density $\varrho(z,r)$ as in (7.50):

$$\begin{aligned}\epsilon_0^{-1}\varrho(z,r) &= \sum_{n=0}^{\infty}\frac{1}{(n!)^2}\left(-\frac{r^2}{4}\right)^n a_n(z) \\ &= a_0(z) - \frac{r^2}{4}a_1(z) + \frac{r^4}{64}a_2(z) - \dots\end{aligned} \quad (7.53)$$

The coefficients a_n are here *independent* functions of z. Substituting (7.53) into the axisymmetric form of Poisson's equation,

$$\frac{\partial^2\Phi}{\partial z^2} + \frac{\partial^2\Phi}{\partial r^2} + \frac{1}{r}\frac{\partial\Phi}{\partial r} = -\frac{\varrho}{\epsilon_0}$$

and introducing for Φ a series expansion similar to (7.50) but with unknown coefficients still to be determined, we find after some elementary calculations

$$\begin{aligned}\Phi(z,r) = \phi(z) &- \frac{r^2}{4}\left(\phi'' + a_0(z)\right) + \frac{r^4}{64}\left(\phi^{(4)} + a_0'' + a_1(z)\right) \\ &- \frac{r^6}{2304}\left(\phi^{(6)} + a_0^{(4)} + a_1'' + a_2(z)\right) + \dots\end{aligned} \quad (7.54)$$

The coefficient $a_0(z)$, representing the most important space charge term, is related to the axial space charge density by $\varrho(z,0) = \epsilon_0 a_0(z)$.

7.3.2 Magnetic fields

The series expansions for round magnetic fields are obtained by recognizing that he formulae of Section 6.4 are a special case of those given in Sections 7.1.2 and 7.2.2. We need to consider here only fields in source-free vacuum domains, where a scalar potential W can be applied.

Equations (6.23), (6.28) and (7.11), (7.12) describe the *same* field in different ways if

$$\Pi(z,r) \equiv C_1(z,s) \quad , \quad \Im(D_0(z,s)) = 0$$

so that recalling (6.31) and setting $m = 1$ in (7.31), we have

$$\Pi(z,0) \equiv C_1(z,0) = \Pi_0(z) = B(z)$$

$B(z)$ being the axial flux density. From (7.32) we now have the series expansion

$$\Pi(z,r) = \sum_{n=0}^{\infty} \frac{1}{n!(n+1)!} \left(-\frac{r^2}{4}\right)^n B^{(2n)}(z)$$

$$= B(z) - \frac{r^2}{8} B''(z) + \frac{r^4}{192} B^{(4)}(z) - \frac{r^6}{9216} B^{(6)}(z) + O(r^8) \quad (7.55a)$$

Using this, we obtain series expansions for the various components of $\boldsymbol{A}$ and related quantities:

$$A_\varphi = A = \frac{r}{2}\Pi(z,r) \qquad \Psi = \pi r^2 \Pi(z,r) \qquad (7.55b)$$

$$A_x = -\frac{y}{2}\Pi(z,r) \quad , \quad A_y = \frac{x}{2}\Pi(z,r) \quad , \quad A_z = 0 \qquad (7.55c)$$

The components of $\boldsymbol{B}$ can be obtained in two different but equivalent ways, from $\boldsymbol{B} = \nabla \times \boldsymbol{A}$ and from $\boldsymbol{B} = -\nabla W$. In the latter case we have to start from (7.30) with $W = D_0(z,r^2)$. In both cases, we arrive at

$$B_z(z,r) = B(z) - \frac{r^2}{4} B''(z) + \frac{r^4}{64} B^{(4)}(z) - \frac{r^6}{2304} B^{(6)}(z) + O(r^8) \quad (7.56a)$$

$$B_r = -\frac{r}{2} R_M \quad , \quad B_x = -\frac{x}{2} R_M \quad , \quad B_y = -\frac{y}{2} R_M \qquad (7.56b)$$

with $R_M = \partial \Pi/\partial z$ given by

$$R_M(z,r^2) = B'(z) - \frac{r^2}{8} B'''(z) + \frac{r^4}{192} B^{(5)}(z) - \frac{r^6}{9216} B^{(7)}(z) + O(r^8) \quad (7.56c)$$

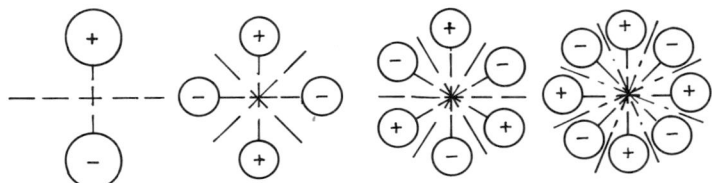

Fig. 7.1: From left to right, multipoles of order $m = 1$ (dipole), $m = 2$ (quadrupole), $m = 3$ (sextupole) and $m = 4$ (octopole). The optic axis is always perpendicular to the plane of the diagram.

These expressions are of particular interest in connection with the physics of round lenses.

For completeness, we state the series expansions of the scalar potential:

$$W(z,r) = -\int B\, dz + \frac{r^2}{4}B' - \frac{r^4}{64}B''' + \frac{r^6}{2304}B^{(5)} - O(r^8) \qquad (7.57)$$

On comparing all these series expansions, there are seen to be only two sets of denominators, which appear frequently in different contexts.

7.4 Multipole fields

In electron optics the meaning of this term is slightly different from that familiar in electrodynamics. In the present context, we do not consider series expansions of fields in terms of spherical harmonics but only those in terms of the azimuth in cylindrical coordinates, as given by (7.5) and (7.9). Multipole fields are then those that have well-defined symmetry properties with respect to the azimuth φ, as is illustrated schematically in Fig. 7.1.

In practice, such fields are often created by a suitable configuration of electrodes and polepieces, their major axes being parallel to the optic axis, see Fig. 7.2. Since the extent of these elements must be finite, fringe fields are inevitable. It is thus impossible to create 'pure' multipole fields in the sense that their dependence on the azimuth φ corresponds to a single harmonic (or finite number of them). In practice this is of no consequence; the only essential requirement is that a well-defined symmetry exists. Consequently each physical multipole field consists of a superposition of different harmonics having the same symmetry properties. The field is then classified by its lowest order harmonic component. Here we state explicitly the electrostatic multipole potentials of the lowest orders $m = 1$ and $m = 2$.

The *dipole field* ($m = 1$) is characterized by having only one plane of even symmetry and a perpendicular one of odd symmetry. Commonly, two

7.4 MULTIPOLE FIELDS

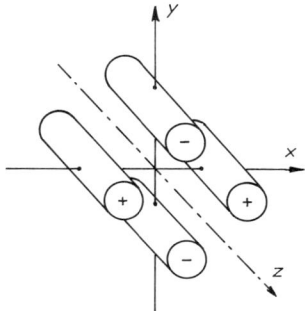

Fig. 7.2: Simplified diagram of a real quadrupole (see also Fig. 19.1).

such fields, rotated at 90° with respect to each other, are superimposed; the potential is then given up to terms in r^5 by

$$-\Phi_D = r\cos\varphi\left\{F_1(z) - \frac{r^2}{8}F_1''(z) + \frac{r^4}{192}F_1^{(4)}(z)\right\}$$
$$+ \frac{1}{6}r^3\cos 3\varphi\left\{p_3(z) - \frac{r^2}{16}p_3''(z)\right\}$$
$$+ \frac{1}{120}r^5\cos 5\varphi\, p_5(z) + O(r^7)$$
$$+ r\sin\varphi\left\{F_2(z) - \frac{r^2}{8}F_2''(z) + \frac{r^4}{192}F_2^{(4)}(z)\right\}$$
$$+ \frac{1}{6}r^3\sin 3\varphi\left\{q_3(z) - \frac{r^2}{16}q_3''(z)\right\}$$
$$+ \frac{1}{120}r^5\sin 5\varphi\, q_5(z) + O(r^7) \tag{7.58}$$

Such fields are employed in *deflection units*, see Chapters 32 and 40.

The *quadrupole field* ($m=2$) is characterized by two perpendicular planes of even symmetry and two planes of odd symmetry, inclined at 45° relative to the former. More generally, two such fields, inclined at 45° with respect to each other, may be superimposed. The electrostatic potential is then given up to terms in r^6 by

$$\Phi_Q = \frac{1}{2}r^2\cos 2\varphi\left\{p_2(z) - \frac{r^2}{12}p_2''(z) + \frac{r^4}{384}p_2^{(4)}(z)\right\}$$
$$+ \frac{1}{2}r^2\sin 2\varphi\left\{q_2(z) - \frac{r^2}{12}q_2''(z) + \frac{r^4}{384}q_2^{(4)}(z)\right\}$$
$$+ \frac{1}{720}r^6\left\{p_6(z)\cos 6\varphi + q_6(z)\sin 6\varphi\right\} + O(r^8) \tag{7.59}$$

Fields of this type occur in multiplets of quadrupole lenses (Chapter 29) and in stigmators (Chapter 32).

The radial series expansions of the multipole fields of higher multiplicity ($m > 2$) can be terminated after the nonvanishing term of lowest order. The reason for this is that, within the paraxial domain, these fields represent only weak perturbations or corrections. The potential is then given by

$$\Phi_m(\mathbf{r}) = \frac{(-r)^m}{m!} \left\{ p_m(z) \cos m\varphi + q_m(z) \sin m\varphi \right\} \tag{7.60}$$

So far we have given explicit expressions only for the electrostatic potential Φ. The corresponding magnetostatic potential W is obtained by replacing the symbols as follows:

$$\Phi \to W \quad , \quad F_1 \to B_1 \quad , \quad F_2 \to B_2 \quad , \quad p_j \to P_j \quad , \quad q_j \to Q_j \tag{7.61}$$

for each subscript j.

7.5 Planar fields

In Section 6.5 we introduced planar solutions of Laplace's equation as analytic functions of a complex variable $z + ix$. This slightly unorthodox choice was adopted for the purposes of electron optics, where the z-axis is almost always made to coincide with the optic axis. We now reconsider planar fields in the (z, x)-plane, confining the discussion to fields with well-defined symmetry properties.

The power series expansion of potentials with *odd* mirror symmetry with respect to the plane $x = 0$ is given by

$$\begin{aligned}-\Phi(z, x) &= \sum_{n=0}^{\infty} \frac{(-1)^n}{(2n+1)!} x^{2n+1} F^{(2n)}(z) \\ &= x \left\{ F(z) - \frac{x^2}{6} F''(z) + \frac{x^4}{120} F^{(4)}(z) + O(x^6) \right\}\end{aligned} \tag{7.62}$$

It is easily verified that this is a solution of Laplace's equation. The function $F(z) = -\partial V/\partial x$, $(x = 0)$, has the meaning of an axial field strength.

Fields of this type are approximately realized in the electric deflection units of oscillographs (Chapter 32) and (as the analogous magnetic potential) in the fringe-field domains of sector magnets with plane fronts (see

Chapter 52). Equation (7.62) can be transformed into a special case of (7.58), the nonzero axial harmonics then being

$$p_1(z) \equiv F(z) \quad , \quad p_3(z) = -\frac{1}{4} F''(z) \quad , \quad p_5(z) = \frac{1}{16} F^{(4)}(z) \qquad (7.63)$$

This shows that the planar deflection field is a special type of dipole field.

The paraxial series expansion of potentials with *even* mirror symmetry with respect to the plane $x = 0$ is given by

$$\Phi(z,x) = \phi(z) - \frac{x^2}{2} \phi''(z) + \frac{x^4}{24} \phi^{(4)}(z) - \frac{x^6}{720} \phi^{(6)}(z) + O(x^8) \qquad (7.64)$$

$\phi(z)$ being the potential in the symmetry plane. These planar fields are a special case of multipole fields with $p_2 = -\phi''/2$, $p_4 = \phi^{(4)}/8$. They occur in electrostatic slit lenses.

7.6 Fourier–Bessel series expansions

In Section 7.1 we introduced azimuthal Fourier series expansions with coefficient functions depending on z and r. The evaluation of the paraxial series expansions, derived in the subsequent sections, is not the only way of calculating the coefficient functions. An alternative procedure is to separate the general scalar potential $V(\mathbf{r})$ into two functions, one in z only, the other in r only:

$$V_m(z, r) = r^m U_m(z, r^2) =: Z_m(z) R_m(r) \qquad (7.65)$$

When this is introduced—together with a factor $\exp(im\varphi)$—into Laplace's equation, ordinary differential equations are obtained:

$$Z_m''(z) + k^2 Z_m(z) = 0 \qquad (7.66a)$$

$$R_m''(z) + \frac{1}{r} R_m'(z) - \left(k^2 + \frac{m^2}{r^2} \right) R_m(r) = 0 \qquad (7.66b)$$

The separation constant here has an arbitrary positive value k^2. The general solution of (7.66a) is

$$Z_m(z) = C_m(k) e^{ikz} \quad , \quad (-\infty < k < \infty) \qquad (7.67a)$$

$C_m(k)$ being any regular function of k. Equation (7.66b) is the differential equation for modified Bessel functions, its regular solution being given by

$$R_m(r) = I_m(kr) \qquad (7.67b)$$

Putting all this together, we obtain a solution of Laplace's equation in the form of a Fourier–Bessel series expansion:

$$V(z,r,\varphi) = \sum_{m=0}^{\infty} \Re \left\{ e^{im\varphi} \int_{-\infty}^{\infty} C_m(k) e^{ikz} I_m(kr) \, dk \right\} \quad (7.68)$$

The paraxial series expansion can now be obtained by introducing the well-known Taylor series expansion

$$I_m(x) = \left(\frac{x}{2}\right)^m \sum_{n=0}^{\infty} \frac{(x^2/4)^n}{n!(m+n)!} \quad , \quad (m = 0, 1, 2, \ldots) \quad (7.69)$$

with $x = kr$ into (7.68). The resulting expression will not be given here. The main difference between it and (7.27) is that repeated differentiations are replaced by Fourier integrals.

The most important special case is that of rotational symmetry, $m = 0$, (7.68) then simplifying to

$$V(z,r) = \int_{-\infty}^{\infty} C_0(k) e^{ikz} I_0(kr) \, dk \quad (7.70)$$

The reality of this expression is guaranteed by requiring that

$$C_0^*(-k) = C_0(k) \quad (7.71)$$

This Fourier coefficient $C_0(k)$ is the Fourier transform of the axial potential $u_0(z)$. The relation between differentiations and Fourier transforms is very simple here:

$$u_0^{(\nu)}(z) = \int_{-\infty}^{\infty} (ik)^\nu C_0(k) e^{ikz} \, dk \quad (7.72)$$

Furthermore, it is easy to determine $C_0(k)$ from the boundary values $V(z,a)$ of the potential on the surface $r = a$ of an infinitely long cylinder. Applying the inverse Fourier transform to (7.70) with $r = a$, we obtain rapidly

$$C_0(k) = \left\{ 2\pi I_0(ka) \right\}^{-1} \int_{-\infty}^{\infty} V(z,a) e^{-ikz} \, dz \quad (7.73)$$

This expression satisfies (7.71).

7.6 FOURIER–BESSEL SERIES EXPANSIONS

Historically, (7.70) and (7.73) have played an important role in the development of simple analytic field models for round electron lenses. A few details are given in Chapters 35 and 36. Nowadays, interest in these models has dwindled, for it is as easy to calculate field distributions exactly as to match parameters to a model. One or two models remain useful for teaching purposes and to gain a rapid qualitative understanding of the dependence of the properties of some device on various parameters. Nevertheless, the Fourier–Bessel series expansion is still of some importance. For instance, van der Merwe (1978a,b, 1979, 1980) used it to calculate rotationally symmetric lenses and Franzen (1984) applied it to quadrupole lenses in cathode-ray tubes.

8
Boundary-Value Problems

Hitherto we have concentrated on fields in the extended paraxial domain. We are in a position to carry out the field calculation once a sequence of axial harmonics $u_m(z)$, $m = 0, 1, 2, \ldots$ is known, but these functions are so far unspecified. The potential inside a domain of solution is specified by its boundary values at the surfaces of this domain and by its source distribution. We elaborate on this in the following sections.

8.1 Boundary-value problems in electrostatics

In electron optics, the electric fields inside insulators and in current-carrying metal conductors are of very little interest and will not be considered here. The domain of solution is the vacuum part of the device in question. This may be multiply connected but it always contains the optic axis. Its boundary is formed by the surfaces of all surrounding metallic electrodes or at least by relevant parts of these. It may prove to be convenient to assume that parts of the boundary are located in the vacuum and even to extend these to infinity, though this is clearly an idealization.

In almost all cases of practical interest, the electric field exhibits simplifying symmetry properties, since a completely unsymmetric field serves no practical purpose. Imperfections in the machining of the electrodes will not be considered here; this topic is treated in Section 9.4.6. Any symmetry properties of the field can be exploited to reduce the relevant domain Γ of solution; the field obtained is subsequently completed by means of symmetry operations.

These remarks are illustrated in the example shown in Fig. 8.1. The appropriate choice of the domain Γ and its boundary $\partial\Gamma$ does, of course, depend strongly on the particular properties of the device in question.

Whenever it is sufficient to consider the field in a planar axial section through the device, we shall adopt the notation presented in Fig. 8.2. The vectors n and t are unit vectors. The surface normal n is directed outwards from medium 1, even in the general three-dimensional case. The contour of the boundary in the axial section will always be oriented positively in the sense shown in Figs 8.1 and 8.2. This choice will be adopted throughout this Part. In the case of electrostatic fields, the medium 1 will be identical with the domain Γ of the desired solution.

8.1 BOUNDARY-VALUE PROBLEMS IN ELECTROSTATICS

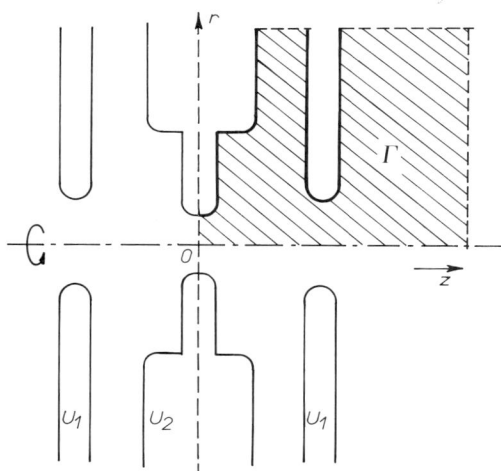

Fig. 8.1: Round symmetric electrostatic lens. The domain Γ within which the solution is sought can be confined to the vacuum region for which $z \geq 0$, $r \geq 0$. Whether the domain Γ must be closed or can be extended to infinity depends on the method of calculation.

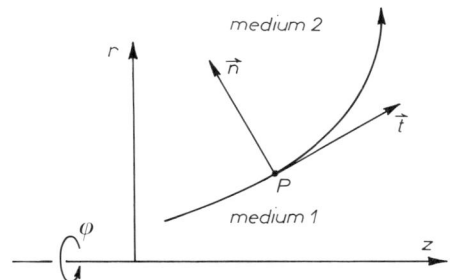

Fig. 8.2: Unit vectors normal to the surface (n) and tangent to it in the meridional section (t) at an arbitrary point P of the boundary. The unit vector i_φ is perpendicular to the meridional section shown and satisfies $i_\varphi = t \times n$.

The boundary-value problem for the electrostatic field is now defined in the following manner:
(i) At all electrode surfaces the electrostatic potentials $\Phi(r)$ must have a constant value equal to the known potential of the corresponding electrode.
(ii) If the boundary consists of several separate parts located in the vacuum, the corresponding surface potential is generally not constant and

is to be defined reasonably, for instance by means of linear, quadratic or logarithmic interpolations. This situation often arises in narrow gaps between electrodes.

(iii) If a plane of negative mirror symmetry of the potential forms a part of the boundary, the potential over this plane is constant, usually zero.

(iv) At all infinite parts of the boundary, the potential is constant. These constant values must be chosen consistently.

(v) At all planes or at an axis of positive mirror symmetry, the normal component of the field strength (normal derivative of the potential) vanishes. The optic axis in every rotationally symmetric device is certainly such an axis.

The boundary-value problem, specified in this manner, has a unique solution, and later we shall describe computational methods for obtaining this solution. In this context, the following relations are very useful. Since all electrode surfaces must be equipotentials, the field strength on their vacuum side is given by

$$\boldsymbol{E}(\boldsymbol{r}) = -\frac{1}{\epsilon_0}\sigma(\boldsymbol{r})\boldsymbol{n}(\boldsymbol{r}) \tag{8.1}$$

Its magnitude is then

$$-\boldsymbol{E}\cdot\boldsymbol{n} = \frac{\partial \Phi}{\partial n} = \sigma(\boldsymbol{r})/\epsilon_0 \tag{8.2}$$

Throughout this Part the symbol $\partial/\partial n = \boldsymbol{n}\cdot\nabla$ denotes the familiar normal derivative. The function $\sigma(\boldsymbol{r})$, defined for all metallic surfaces, is the surface charge density. Initially, this function is unknown, but once the boundary-value problem has been solved it may be used with advantage in the ensuing field computations, see Chapter 9.

8.2 Boundary conditions in magnetostatics

Whereas the material properties, of the metallic electrodes are unimportant in electrostatics, since the electrostatic field vanishes inside any conductor, the situation in magnetostatics is far more complicated. Apart from the case of perfect superconductors, the magnetic field inside polepieces does not vanish. In consequence, it is not always possible to confine the domain of solution to the vacuum part of the field. It is of course this region that is of greatest interest for calculating the optical properties, but a knowledge of the field distribution in the yoke is often needed when the shape of the latter is being designed. In the most general case, the field

8.2 BOUNDARY CONDITIONS IN MAGNETOSTATICS

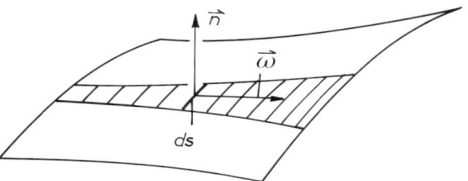

Fig. 8.3: Representation of a surface current distribution; the direction of the vector $\boldsymbol{\omega}$ is the same as that of the tangent. The current increment $dI = \omega \cdot ds$ is constant along the stripe indicated.

computation will become extremely complicated; instead of treating this, therefore, we shall consider some classes of important field configurations with simplifying properties. First, however, we shall formulate the general boundary conditions, which must always be satisfied.

We shall use the notation explained in Fig. 8.2; here medium 1 is the vacuum while medium 2 is any ferromagnetic or superconducting material. In the general case, at all surfaces of materials, the *interface conditions*

$$\boldsymbol{n} \cdot (\boldsymbol{B}_2 - \boldsymbol{B}_1) = 0 \tag{8.3}$$
$$\boldsymbol{n} \times (\boldsymbol{H}_2 - \boldsymbol{H}_1) = \boldsymbol{\omega}(\boldsymbol{r}) \tag{8.4}$$

must be satisfied, the subscript referring to the material in which the field vectors are defined. The function $\boldsymbol{\omega}(\boldsymbol{r})$ is the *surface current density*. This is a vector function defined only on surfaces. It must always have the same direction as the local tangent: $\boldsymbol{\omega}(\boldsymbol{r}) = \omega(\boldsymbol{r})\boldsymbol{t}_c(\boldsymbol{r})$, $\boldsymbol{t}_c(\boldsymbol{r})$ being a normalized tangential vector which may differ from the vector $\boldsymbol{t}$ introduced earlier. The physical meaning is as follows: $dI = \omega(\boldsymbol{r})ds$ is the electric current flowing through a surface line-element ds oriented perpendicularly to $\boldsymbol{\omega}(\boldsymbol{r})$, see Fig. 8.3. The whole distribution of surface currents must, of course, satisfy the requirements for the conservation of electric current.

Such surface current distributions arise in superconducting devices. They are caused by induction effects, when the field in the vacuum domain is switched on. Furthermore, surface current distributions offer a convenient way of describing flat layers of current-conducting windings located in the vicinity of the surfaces of magnetic shielding tubes. Further examples are given below.

In very many cases, the function $\boldsymbol{\omega}(\boldsymbol{r})$ vanishes identically, (8.4) then simplifying to

$$\boldsymbol{n} \times (\boldsymbol{H}_2 - \boldsymbol{H}_1) = 0 \tag{8.5}$$

Equation (8.3) expresses the continuity of the normal component of $\boldsymbol{B}$, while (8.5) implies that the tangential component of $\boldsymbol{H}$ is continuous. Even

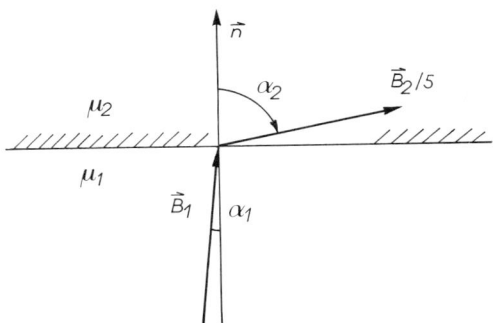

Fig. 8.4: Refraction of lines of magnetic flux for $\mu_2/\mu_1 = 50$.

when saturation effects occur, we may introduce material coefficients (6.3 or 6.19). Writing $B_j = \mu_j H_j$, $\mu_j = 1/\nu_j$ for $j = 1, 2$, we can now derive the familiar law for the refraction of flux lines (Fig. 8.4):

$$\frac{\tan \alpha_2}{\tan \alpha_1} = \frac{\mu_2}{\mu_1} \tag{8.6}$$

The continuity laws break down at sharp edges, where no local surface normal n can be defined. We shall therefore assume that such edges are slightly rounded off, as is the case in all practical devices.

These interface conditions are very simple but refer to vector fields. The computation of vector fields is possible in principle but is usually complicated, requiring much computing time and memory capacity. It is therefore advantageous to use *scalar* potentials from which the field can be determined by differentiation. Unfortunately, the scalar potentials $\chi(\mathbf{r})$ and $\Pi(\mathbf{r})$, introduced in (6.11–6.12) and (6.28) respectively, are of only very limited applicability. In order to circumvent this difficulty, it is usual to separate the magnetic field strength $\mathbf{H}(\mathbf{r})$ into the contribution $\mathbf{H}_0(\mathbf{r})$ of the isolated coils in vacuum and the contribution $\mathbf{H}_M(\mathbf{r})$ of the ferromagnetic parts,

$$\mathbf{H}(\mathbf{r}) = \mathbf{H}_0(\mathbf{r}) + \mathbf{H}_M(\mathbf{r}) \tag{8.7}$$

By definition the following conditions are to be satisfied in the whole space:

$$\operatorname{div} \mathbf{H}_0(\mathbf{r}) = 0 \quad , \quad \operatorname{curl} \mathbf{H}_0(\mathbf{r}) = \mathbf{j}(\mathbf{r}) \tag{8.8}$$

Together with the natural boundary conditions at infinity, this is already sufficient to calculate $\mathbf{H}_0(\mathbf{r})$ uniquely by means of Biot–Savart's law:

$$\mathbf{H}_0(\mathbf{r}) = \int \frac{(\mathbf{r}' - \mathbf{r}) \times \mathbf{j}(\mathbf{r}')}{4\pi |\mathbf{r}' - \mathbf{r}|^3} d^3 r' \tag{8.9}$$

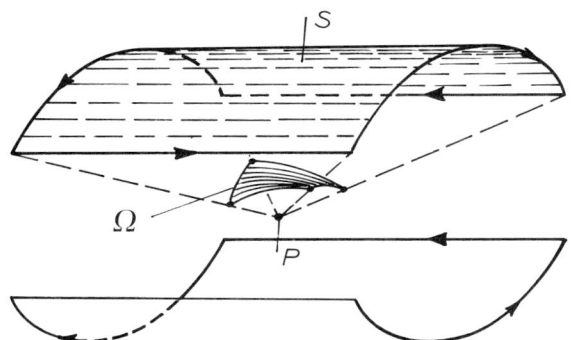

Fig. 8.5: Pair of saddle coils. The arrows indicate the local direction of the electric current. The solid angle subtended by the upper surface at some arbitrary point P is denoted by Ω. The solid angle subtended at P by the lower surface can likewise be found and is to be subtracted from Ω.

This function can be obtained by differentiation of potentials. In the general case, the vector potential is appropriate:

$$\boldsymbol{H}_0 = \frac{1}{\mu_0}\operatorname{curl}\boldsymbol{A}_0 \quad , \quad \boldsymbol{A}_0(\boldsymbol{r}) = \frac{\mu_0}{4\pi}\int \frac{\boldsymbol{j}(\boldsymbol{r}')d^3r'}{|\boldsymbol{r}' - \boldsymbol{r}|} \qquad (8.10)$$

In simply connected domains excluding any sources, we may also use (6.11–6.12). For a single closed winding carrying an electric current I, this representation takes the familiar form

$$\boldsymbol{H}_0 = -\operatorname{grad}\chi_0 \quad , \quad \chi_0(\boldsymbol{r}) = \frac{I}{4\pi}\int_S \frac{(\boldsymbol{r} - \boldsymbol{r}')\cdot d\boldsymbol{a}'}{|\boldsymbol{r} - \boldsymbol{r}'|^3} \qquad (8.11)$$

The two-dimensional surface integral is to be evaluated over any surface S enclosed by the windings. An example of this is shown in Fig. 8.5. Since the integral in (8.11) is equal to the solid angle Ω under which the winding would be seen from the point $\boldsymbol{r}$ (see Fig. 8.5), $\chi_0(\boldsymbol{r})$ is known as the 'solid angle potential'. As the point $\boldsymbol{r}$ passes through the surface S, $\chi_0(\boldsymbol{r})$ varies discontinuously, the jump being $\pm I$. In the case of several closed windings the contributions of all the windings are to be summed up appropriately. Important practical applications of this integral are given in Chapter 40.

The representation (8.11) is most convenient in the extended paraxial domain, since it is always possible to define the surfaces S of integration in such a way that (8.11) is unique in this domain. Apart from some special cases, (8.10) is less convenient. The evaluation of (8.9) is always possible. In the case of surface current distributions the three-dimensional integration collapses to a surface integration, which means that $j(r')d^3r'$ is to be replaced by $\omega(r')da'$. In every case, the resulting function $H_0(r)$ is unique in the whole space. In the following considerations we shall assume that these integrations can be carried out for all points of reference r that are to be considered.

Let us now focus our attention on the second term $H_M(r)$ in (8.7). Since the whole current density j is already associated with H_0, the field H_M must satisfy curl $H_M = 0$, and hence

$$H_M(r) = -\text{grad } \chi_M(r) \qquad (8.12)$$

$\chi_M(r)$ being a *unique* scalar potential in the whole space and defined as the *reduced* magnetic scalar potential. In every unsaturated medium ($\mu = $ const) it can be concluded from div $B = 0$, div $H_0 = 0$ and div $H = 0$ that div $H_M = 0$ so that

$$\nabla^2 \chi_M(r) = 0 \quad \text{for} \quad \mu = \text{const} \qquad (8.13)$$

This is invalid at the material surfaces, where formal scalar surface charge distributions must be introduced; these are the analogue of electrostatic surface charges. Magnetic surface charges, however, have no physical meaning, but are only a convenience in calculating, as will become obvious in Section 9.2.

The interface conditions (8.3), (8.4) and (8.5) are considerably simplified by the separation in (8.7) combined with (8.12). Since $\chi_M(r)$ must be a unique function in the whole space, this potential itself and the tangential components of its gradient must be continuous at all material surfaces, while the normal component will be discontinous. The field contribution H_0 has the opposite behaviour: its normal component is continous at interfaces, while the tangential components are discontinuous, the corresponding jump being obtained from (8.4):

$$n \times \left\{ (H_0)_2 - (H_0)_1 \right\} = \omega \qquad (8.14)$$

This jump is already considered in the Biot–Savart integration over the surface currents and hence (8.4) and (8.5) contain no further information.

Introducing (8.7) and (8.12) into (8.13) and using the fact that

$$B_j = \mu_j H_j = \mu_j \left(H_0 - \nabla \chi_M \right)_j \quad (j = 1, 2)$$

8.3 EXAMPLES OF BOUNDARY-VALUE PROBLEMS

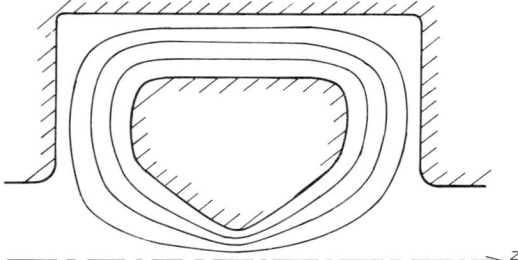

Fig. 8.6: Magnetic flux lines in a superconducting device; the upper half of an axial section is shown.

on both sides of the corresponding interface, we obtain the fundamental interface condition

$$\mu_2 \left(\frac{\partial \chi_M}{\partial n} \right)_2 - \mu_1 \left(\frac{\partial \chi_M}{\partial n} \right)_1 = (\mu_2 - \mu_1) \, \boldsymbol{n} \cdot \boldsymbol{H}_0 \qquad (8.15)$$

The implementation of this condition will be examined in Section 9.2.

8.3 Examples of boundary-value problems in magnetostatics

In this section we shall confine our considerations to important classes of boundary-value problems in the proper sense, by which we mean cases in which it is possible to confine the necessary field calculation entirely to *one* medium (1). The subscript 1 will then be omitted when this does not cause confusion, and we set $\mu_1 = \mu_0$ when medium (1) is the vacuum.

8.3.1 Devices with superconducting yokes
Owing to the Meissner–Ochsenfeld effect, the magnetic field is completely expelled from the interior of any superconductor. On the vacuum side of its surface the magnetic field must have a locally tangential direction, see Fig. 8.6. This condition can be satisfied only by the presence of appropriate surface current distributions $\boldsymbol{\omega}(\boldsymbol{r})$, which are *unknown* prior to the solution of the corresponding boundary-value problem. Since $\boldsymbol{\omega}(\boldsymbol{r})$ must be known in order to calculate $\boldsymbol{H}_0(\boldsymbol{r})$, the separation (8.7) is unhelpful in this context.

In rotationally symmetric devices, the necessary boundary conditions are most simply satisfied by use of the flux potential Ψ. From (6.32) it is obvious that $\boldsymbol{B}$ is tangential if the boundary contour C is a line $\Psi = \text{const}$. The Dirichlet problem for $\Psi(z, r)$ is then very simple:

$$\begin{cases} \Psi(z, r) = \Psi_B = \text{const} , & (z, r) \in C \\ \Psi(z, 0) = 0 , & -\infty < z < \infty \end{cases} \qquad (8.16)$$

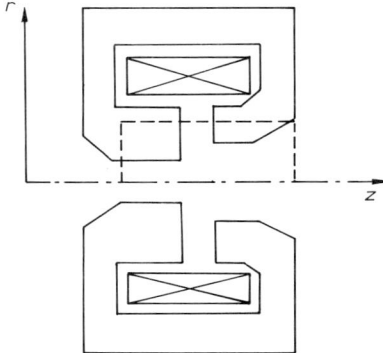

Fig. 8.7: Highly simplified axial section through a conventional magnetic lens.

For the field in the paraxial domain, the potential $\Pi(z,r)$ is more suitable. From (6.27), (6.28) and from the regularity requirements at the optic axis, the boundary conditions

$$\begin{cases} \Pi(z,r) = \Psi_B/\pi r^2 &, \quad (z,r) \in C \\ \partial\Pi/\partial r = 0 & \text{for } r = 0 \end{cases} \quad (8.16a)$$

can be derived. The solution of the corresponding boundary-value problem for Π is not unduly complicated. The constant Ψ_B has the physical meaning of the total flux through the bore of the superconducting polepiece.

An analogous Dirichlet problem can be formulated for planar fields but this will not be treated here since the simplification to planar field structures is generally not satisfactory in superconducting devices.

8.3.2 Conventional round magnetic lenses

Figure 8.7 shows an axial section through a typical magnetic lens, and Fig. 8.8 the relevant vacuum domain of the magnetic field. The contours of the casing are schematically simplified. It is only approximately possible to confine the field calculation to the domain Γ; for this the following assumptions must be made:

(i) The permeability of the casing material must be extremely high, $\mu_2 \gtrsim 10^4 \mu_0$, and saturation effects must nowhere occur.
(ii) The cross-section of the casing must be large enough to ensure that practically all the magnetic flux flows through the gap.
(iii) The gap has to be long (in the radial direction) and narrow, so that the field between the pole faces may be regarded as practically homogeneous.

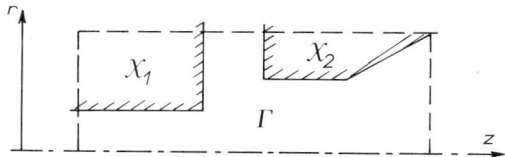

Fig. 8.8: Enlarged view of the region of Fig. 8.7 enclosed in dashed lines; the scalar potential $\chi(r)$ is defined within the domain Γ.

These assumptions cannot, of course, be satisfied precisely in a mathematical sense. The following considerations represent a technical simplification.

Since the currents in the coils are located completely outside the domain Γ, we may use the total scalar potential $\chi(r)$. From (8.6) we can conclude that the angle α_1 on the vacuum side of the casing surfaces must be extremely small. Thus the flux lines intersect these surfaces practically *orthogonally*. This implies that the surfaces may be regarded as equipotentials $\chi(r) = $ const. One surface potential χ_1 may be chosen arbitrarily, for instance $\chi_1 = 0$. The other is determined by Ampère's law

$$\oint \boldsymbol{H} \cdot d\boldsymbol{r} = NI = \chi_1 - \chi_2 \qquad (8.17)$$

In this relation NI is the total number of ampère-turns of the coil; the integration loop must enclose all the windings and must pass through the gap.

The boundary-value problem to be solved now takes the following form:
 (i) Inside Γ the potential $\chi(z, r)$ is to be calculated by solving $\nabla^2 \chi = 0$.
 (ii) At the surfaces of the polepieces and in asymptotic regions of the bores the potential is constant, χ_1 or χ_2 respectively.
(iii) At the upper part of the boundary, inside the gap, the potential is to be interpolated linearly.
(iv) On the optic axis, $\partial \chi / \partial r = 0$ must be satisfied.

The simplifying assumptions reduce this boundary-value problem to the analogue of an electrostatic problem and it can be solved by means of corresponding techniques. As in (8.1) and (8.2), it will be convenient to introduce formal magnetic surface charge densities, satisfying

$$\boldsymbol{H}(\boldsymbol{r}) = -\sigma_M(\boldsymbol{r})\boldsymbol{n}(\boldsymbol{r}) \qquad (8.18)$$
$$-\boldsymbol{H} \cdot \boldsymbol{n} = \partial \chi / \partial n = \sigma_M(\boldsymbol{r}) \qquad (8.19)$$

Another problem that can be solved by employing the potential $\chi(r)$ is the fringe field of sector magnets with screening plates (Part X).

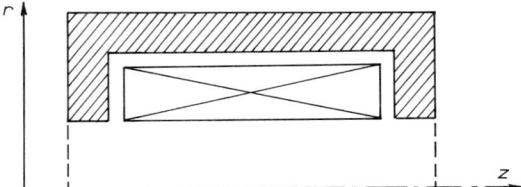

Fig. 8.9: Axial section through a very simple unconventional magnetic lens.

8.3.3 Unconventional round magnetic lenses

The approximations described above are of limited application and may break down even when comparatively simple configurations are considered. A typical example is shown in Fig. 8.9. The 'gap' in the shield of this lens is so wide that a simple linear interpolation is unreasonable and the distribution of the electric current in the coil is now of importance. This is clearly a case for the flux potential $\Psi(z,r)$, since the interface conditions are then strongly simplified.

The continuity of the magnetic flux requires that $\Psi(z,r)$ be continuous at any material surface not conducting surface currents. Consequently the tangential component of $\nabla\Psi$ is also continuous. Hence (8.3) is already satisfied: introducing (6.32) into (8.3) and recalling that $\bm{n} \times \bm{i}_\varphi = \bm{t}$ for the tangential vector $\bm{r}$ (Fig. 8.2), we do indeed obtain

$$(\bm{t}\cdot\nabla\Psi)_2 = (\bm{t}\cdot\nabla\Psi)_1$$

The second interface condition is obtained by introducing (6.32) into (8.5). After some elementary calculations, we find

$$\nu_2\left(\frac{\partial\Psi}{\partial n}\right)_2 = \nu_1\left(\frac{\partial\Psi}{\partial n}\right)_1 \tag{8.20}$$

In the case of an unsaturated casing with extremely high permeability, $\mu_2 \gg \mu_1$, $\nu_2 \ll \nu_1$, it is reasonable to make the approximation $\nu_2 \to 0$, (8.20) then simplifying to

$$\left(\frac{\partial\Psi}{\partial n}\right)_1 = 0 \tag{8.21}$$

We now have to solve the following boundary-value problem: the domain Γ of solution is the whole of space excluding all ferromagnetic parts. Inside this domain (6.36) is to be solved. At the optic axis and at infinity, Ψ must vanish, while at the iron surfaces the Neumann condition (8.21) must be satisfied.

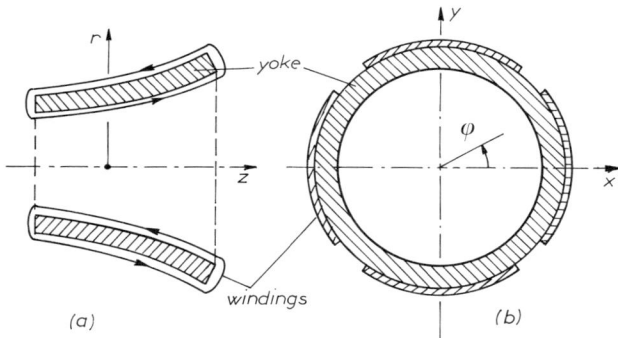

Fig. 8.10: Simplified representation of a toroidal deflection system. (a) Axial section, (b) cross-section.

The simplifications introduced above are invalid if parts of the iron become seriously saturated. In cases where this is liable to occur, another approximate field calculation is to be used, which is described in Chapter 12.

8.3.4 Toroidal magnetic deflection systems

Such systems are frequently employed as scanning units, for instance in television tubes and scanning electron microscopes. A simplified diagram is given in Fig. 8.10. Two pairs of coils, rotated at 90° with respect to each other, are wound round a rotationally symmetric ferrite shield in such a way that each winding remains in a meridional plane.

It is convenient to regard the coils as surface current distributions. Following Schwertfeger and Kasper (1974), the H-field inside the yoke may be neglected, since the permeability is very high. The deflection currents are never strong enough to cause saturation effects and we can therefore confine the following discussion to the vacuum domain Γ of the device. We now omit the subscript 1. In view of these assumptions, (8.4) simplifies to

$$H(r) \times n(r) = \omega(r) \quad (r \in \partial\Gamma) \tag{8.22}$$

$\partial\Gamma$ denoting the surfaces of the coils on their vacuum side. Since any integration contour which remains completely in the vacuum never encloses any current lines, $\oint H \cdot dr = 0$ is always valid and hence the scalar potential χ is unique in the whole vacuum domain. Forming the vector product of (8.22) with n and introducing $H = -\nabla\chi$ we obtain

$$\operatorname{grad} \chi - (n \cdot \operatorname{grad} \chi)n = \omega \times n \quad \text{on} \quad \partial\Gamma \tag{8.23}$$

The expression on the left-hand side is the *tangential* component of $\nabla\chi$; this now has a given value for all surface points.

So far these considerations are quite general and are also applicable to deflection systems with saddle coils. The characteristic feature of toroidal systems is that the direction of the vector function $\boldsymbol{\omega}$ is meridional. This implies that $\boldsymbol{\omega}(\boldsymbol{r})$ may be written

$$\boldsymbol{\omega}(\boldsymbol{r}) = J(\varphi)\boldsymbol{t}(\boldsymbol{r})/r \tag{8.24}$$

φ being the azimuth with respect to the optic axis and $\boldsymbol{t}(\boldsymbol{r})$ the local tangential vector in the meridional direction; $J(\varphi)$ is the azimuthal current distribution function, which means $J(\varphi)d\varphi$ is the total electric current flowing through the windings located between φ and $\varphi + d\varphi$.

It is now possible to integrate (8.23), since (8.23) is consistent with the assumption that the boundary values of χ are only dependent on φ. Introducing $\nabla\chi = r^{-1}\chi'(\varphi)\boldsymbol{i}_\varphi$ and (8.24) into (8.23), and recalling that $\boldsymbol{t} \times \boldsymbol{n} = \boldsymbol{i}_\varphi$, we obtain $\chi'(\varphi) = J(\varphi)$ and hence

$$\chi(\varphi) = \chi_0 + \int_0^\varphi J(\alpha)\,d\alpha \quad \text{on} \quad \partial\Gamma \tag{8.25}$$

This is essentially the same as the formula of Schwertfeger and Kasper (1974); here the derivation is more general, since (8.23) may also be applied to more general types of deflection systems, see Section 9.4.4.

The boundary-value problem to be solved is now comparatively simple: at the surface $\partial\Gamma$ of a rotationally symmetric shield the potential $\chi(\boldsymbol{r})$ has uniquely determined boundary values which are *not* rotationally symmetric. At infinity the potential must vanish, and in the vacuum domain Γ, Laplace's equation is satisfied.

This presentation of the important classes of boundary-value problems is by no means complete. We cannot devote more space to them here but we hope that the reader has some impression of the complexity of the problems to be considered.

9
Integral Equations

As is well-known in classical electrodynamics, it is possible to reduce the problem of solving a boundary-value problem to that of solving an integral equation. This is very advantageous since methods of field calculation based upon integral equations have gained great importance. In this chapter we shall present the general theory; details of numerical procedures are given in Chapter 10.

9.1 Integral equations for scalar potentials

In the following account, we consider a domain Γ in three-dimensional space and its boundary $\partial\Gamma$. Inside the domain Γ we attempt to solve a uniquely specified boundary-value problem for Poisson's equation (7.1)

$$\nabla^2 V(\mathbf{r}) = -S(\mathbf{r}) \qquad (9.1)$$

9.1.1 General theory

In order to obtain an integral equation, we start from Green's theorem for a modified domain Γ' with boundary $\partial\Gamma'$ and for a variable of integration $\mathbf{r}'$

$$\int_{\Gamma'} \left(G \nabla'^2 V - V \nabla'^2 G \right) d^3 r' = \int_{\partial\Gamma'} \left(G \frac{\partial V}{\partial n'} - V \frac{\partial G}{\partial n'} \right) da' \qquad (9.2)$$

valid for any differentiable functions $G(\mathbf{r}')$ and $V(\mathbf{r}')$, regardless of their special meanings. The operator $\partial/\partial n' = \mathbf{n}' \cdot \nabla'$ is the so-called normal derivative, the derivative in the direction of the *outward* oriented surface normal $\mathbf{n}'$ on $\partial\Gamma'$. The boundary $\partial\Gamma'$ itself may consist of several distinct closed surfaces and the integral on the right-hand side of (9.2) is then the sum of the contributions arising from the different surfaces; this summation is implicit in the notation.

The function G in (9.2) can be chosen arbitrarily; the most suitable choice is the free-space Green's function, defined by

$$G(\mathbf{r}, \mathbf{r}') = (4\pi |\mathbf{r} - \mathbf{r}'|)^{-1} \qquad (9.3)$$

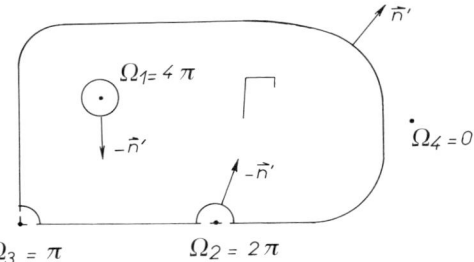

Fig. 9.1: Two-dimensional section through a three-dimensional domain Γ, showing various positions of reference points and the associated solid angles.

and satisfying the differential equation

$$\nabla^2 G = \nabla'^2 G' = -\delta(\boldsymbol{r} - \boldsymbol{r}')$$

where $\delta(\boldsymbol{r}-\boldsymbol{r}')$ denotes Dirac's distribution. The use of the latter is familiar but not always favourable. Difficulties arise if the reference point $\boldsymbol{r}$ is located on the boundary. In order to circumvent these difficulties, we do *not* employ the δ-function formalism but instead, we modify the given domain Γ as sketched in Fig. 9.1. An internal point $\boldsymbol{r}$ is completely enclosed by a small sphere of solid angle $\Omega_1 = 4\pi$. Around a regular boundary point, a small hemisphere with solid angle $\Omega_2 = 2\pi$ is excluded; at a sharp line-edge of intersection angle α, a spherical segment with solid angle $\Omega_3 = 2\alpha$ is removed; finally, an external point $\boldsymbol{r}$ needs no exclusion surface and hence $\Omega_4 = 0$. In the domain Γ', obtained after excluding the immediate neighbourhood of the reference point, (9.2) is valid.

In order to simplify the notation, we introduce the abbreviation

$$\epsilon(\boldsymbol{r}) := \frac{1}{4\pi}\Omega(\boldsymbol{r}) = \begin{cases} 1 & \text{for } \boldsymbol{r} \in \Gamma, \\ \alpha/2\pi & \text{for } \boldsymbol{r} \in \partial\Gamma, \\ 0 & \text{for } \boldsymbol{r} \notin \Gamma, \end{cases} \qquad (9.4)$$

which implies that $\alpha = \pi$, $\epsilon = 1/2$ for *regular* boundary points. We also write

$$\sigma(\boldsymbol{r}') := \frac{\partial V}{\partial n'} \equiv \boldsymbol{n}' \cdot \nabla' V(\boldsymbol{r}') \qquad (9.5)$$

$$P(\boldsymbol{r}, \boldsymbol{r}') := \frac{\partial G}{\partial n'} \equiv \frac{\boldsymbol{n}' \cdot (\boldsymbol{r} - \boldsymbol{r}')}{4\pi|\boldsymbol{r} - \boldsymbol{r}'|^3} \qquad (9.6)$$

At the boundary these normal derivatives will be discontinuous; they are then defined as the limits obtained on approaching from the interior of Γ'.

9.1 INTEGRAL EQUATIONS FOR SCALAR POTENTIALS

Introducing all this into (9.2), we soon notice that only the terms involving P are critical so far as the limit of vanishing radius ϱ of the sphere or sector of exclusion is concerned. We may then approximate the slowly varying function $V(r')$ by $V(r)$. Recalling that $|r - r'| = \varrho = n' \cdot (r - r')$ on the spherical surfaces S, we obtain

$$\int_{\partial \Gamma'} V(r') \frac{\partial G}{\partial n'} da' \to \fint_{\partial \Gamma} V(r') P(r, r') da' + \lim_{\varrho \to 0} V(r) \int_S \frac{da'}{4\pi \varrho^2}$$

The final, surface integral is just the expression $\epsilon(r)$ defined by (9.4). The symbol $\fint$ denotes the principal value of the corresponding integral, defined as the value obtained by proceeding to the limit $\varrho \to 0$ in the integration over the not-excluded parts of $\partial \Gamma$. For reference points r outside $\partial \Gamma$ this is straightforward; for $r \in \partial \Gamma$ this limit exists, since in this case we have $n' \cdot (r - r') \to 0$ due to the orthogonality between tangents and surface normals. In the subsequent presentation we shall not indicate the principal value explicitly since all improper surface integrals are to be evaluated in this way.

Putting all this together, we arrive finally at

$$\epsilon(r)V(r) = \int_\Gamma G(r, r') S(r') d^3r'$$
$$+ \int_{\partial \Gamma} \Big(G(r, r') \sigma(r') - P(r, r') V(r') \Big) da' \quad (9.7)$$

The expressions on the right-hand side can be interpreted in the following way: the first is a space-source term, the second a surface-source term and the last is a surface-polarization term. Since the boundary functions $V(r')$ and $\sigma(r')$ are still independent, (9.7) alone does not suffice to determine the potential distribution uniquely. We may prescribe an additional boundary condition

$$a(r')V(r') + b(r')\sigma(r') = c(r') \quad , \quad r' \in \partial \Gamma \quad (9.8)$$

the surface functions $a(r')$, $b(r')$ and $c(r')$ being known with $a^2 + b^2 > 0$. It is then possible to solve (9.7). We now discuss the two most familiar special cases.

9.1.2 Dirichlet problems

Here the boundary values $V(r')$ are specified uniquely, while $\sigma(r')$ is unknown. In (9.8) we may choose $a \equiv 1$, $b \equiv 0$; $c(r') \equiv \overline{V}(r')$ are then the given boundary values. Equation (9.7) is now a two-dimensional integral equation of Fredholm's first kind for the unknown σ. After solving it, the same equation (9.7) can be used to evaluate the potential at any point r in Γ.

Although this procedure is perfectly correct, it is rather inconvenient for numerical solutions since the factor $\epsilon(r)$ is discontinuous and the polarization term requires careful handling in order to obtain the appropriate principal value. It is therefore highly desirable to find alternative forms of the integral equation that do not contain these terms.

If the boundary $\partial\Gamma$ consists of a number of *closed* surfaces with *constant* boundary values $\overline{V}$ on them, the difficult terms can be eliminated completely. For simplicity, we consider only one such surface $\partial\Gamma$, the domain Γ being its *exterior*. The second boundary is an infinite sphere; since the field is required to satisfy the natural boundary condition, this sphere need not be considered here. We now have to evaluate the integral term

$$I := \int_{\partial\Gamma} P(r,r')V(r')\,da' = V \int_{\partial\Gamma} P(r,r')\,da'$$

Since (9.7) is quite generally valid, we are at liberty to set V equal to a constant; for the moment, therefore, we introduce

$$V \equiv 1 \quad , \quad \sigma \equiv 0 \quad , \quad S \equiv 0$$

and recalling that Γ is now the exterior, we obtain the mathematical identity

$$\int_{\partial\Gamma} P(r,r')\,da' = 1 - \epsilon(r) \tag{9.9}$$

The integral expression I now simplifies to

$$I = \{1 - \epsilon(r)\}V$$

Introducing this into (9.7) we find

$$V(r) = \int_{\Gamma} G(r,r')S(r')\,d^3r' + \int_{\partial\Gamma} G(r,r')\sigma(r')\,da' \tag{9.10}$$

This Fredholm equation for σ is now quite generally applicable without exception. When the solution is still unknown, the reference point r must be located at the boundary and $V(r)$ on the left-hand side is then the corresponding boundary value. After obtaining the solution for σ, the same equation (9.10) may be employed to compute the potential at any point r

in the space, even on $\partial\Gamma$ or outside Γ. The potential itself is *continuous* if the reference point crosses the boundary.

We can generalize (9.10) to include configurations with a boundary consisting of several closed surfaces and even boundary values of V that are *not* constant on these surfaces. The basic form of (9.10) remains unaltered, but now the surface charge density σ is the difference between the normal derivatives of V on each side of the corresponding surface. Physically, (9.10) can be interpreted as a Coulomb integral over space charges and over surface charges as a degenerate case of the former. When it comes to numerical evaluation, (9.10) is far more convenient than (9.7).

9.1.3 Neumann problems

It is now the boundary values of $\partial V/\partial n$ that are uniquely specified while those of $V(\mathbf{r})$ are unknown. This boundary value problem has a solution only if $\oint \sigma(\mathbf{r}')\,da'$ vanishes on $\partial\Gamma$. In (9.8) we may specify $a \equiv 0$, $b \equiv 1$, so that $c(\mathbf{r}') = \sigma(\mathbf{r}')$ are given boundary values. Equation (9.7) is now an integral equation of Fredholm's *second* kind for the boundary values of $V(\mathbf{r})$. Apart from an unimportant additive constant, this integral equation has a unique solution. After this has been found, (9.7) can be used to compute the potential $V(\mathbf{r})$ at arbitrary points inside Γ.

The polarization term cannot be eliminated from (9.7). Since this term contains a strong singularity and is discontinuous at the boundary $\partial\Gamma$, great care must be taken in numerical computations. Thus the concrete evaluation of such expressions should be avoided whenever possible by appropriate transformation of the boundary-value problem.

9.2 Problems with interface conditions

The theory outlined in Section 9.1.1 is quite standard in classical electrodynamics and can be applied to boundary-value problems in electrostatics and magnetostatics. The integral equation, given below, is less familiar. In connection with electron optical applications it has been mentioned by Kasper (1982) and explicitly derived by Scherle (1983), who also demonstrated that it can be applied in practical numerical computations. Alternative formulations will be given at the end of this section.

Since interface conditions are most important in *magnetostatic* problems, we shall confine the discussion to these, although it would be no problem to establish an integral equation for electric fields. It is necessary to assume unsaturated (linear) media. For simplicity, we consider here only two different domains, the vacuum domain Γ_1 and a ferromagnetic shield Γ_2, $\Gamma_1 \cup \Gamma_2$ being the whole space. The convention concerning the choice of the surface normal, represented in Fig. 8.2, then holds. Generalizations to more than two different domains have been worked out (Scherle, 1983); apart from the introduction of an iterative solution technique, they contain nothing essentially new; some results will be given in Chapter 10.

We start from (9.7). The appropriate potential is here $\chi_M(\boldsymbol{r})$, introduced in (8.7) and (8.12). From (8.13), we see that there is no space-source term. On the boundary $\partial\Gamma = \Gamma_1 \cap \Gamma_2$ we have $\epsilon = 1/2$ and hence (9.7) specializes to

$$\frac{\chi_M(\boldsymbol{r})}{2} = (-1)^j \oint_{\partial\Gamma} \left\{ \chi_M(\boldsymbol{r}')P(\boldsymbol{r},\boldsymbol{r}') - G(\boldsymbol{r},\boldsymbol{r}')\sigma_j(\boldsymbol{r}') \right\} da' \qquad (9.11)$$

$$(j = 1, 2 \quad , \quad \boldsymbol{r} \in \partial\Gamma)$$

The index j indicates the domain from which the surface $\partial\Gamma$ is approached; we recall that in the polarization term the principal value of the integral is to be taken. Since neither the potential itself nor its normal derivatives σ_j are known, a second relation is needed, namely (8.15). This enables us to eliminate the normal derivative from (9.11) by forming appropriate linear combinations. In this context we need to use (9.5) with $\chi_M \equiv V$. The result of these elementary calculations is as follows:

$$-\lambda\chi_M(\boldsymbol{r}) + \oint_{\partial\Gamma} P(\boldsymbol{r},\boldsymbol{r}')\chi_M(\boldsymbol{r}')\, da' = \oint_{\partial\Gamma} G(\boldsymbol{r},\boldsymbol{r}')\, \boldsymbol{n}' \cdot \boldsymbol{H}_0(\boldsymbol{r}')\, da' \qquad (9.12)$$

with $\boldsymbol{r} \in \partial\Gamma$ and

$$\lambda := \frac{1}{2}\frac{\mu_2 + \mu_1}{\mu_2 - \mu_1} \qquad (9.13)$$

This is an integral equation of Fredholm's second kind for the surface values of the potential. Once it has been solved, the problem reduces to an ordinary Dirichlet problem. We can hence introduce the calculated surface values of χ_M in the left-hand side of (9.10) and solve this equation (with vanishing space-source S) for σ. It can be shown that $\sigma = \sigma_1 - \sigma_2$, but this is of little use, since the solution of (9.10) gives σ directly.

The derivation presented here differs from Scherle's method but is equivalent to it. In the literature on magnetic field computation, many other forms of integral equation are derived, which are essentially equivalent to (9.12), but not always so suitable for numerical evaluation. Many technical points are elucidated in the proceedings of COMPUMAG (1976) and the subject has been reviewed by Iselin (1981). Besides Scherle's thesis, the publications of Lucas (1976) and Kuroda (1983) are particularly concerned with field calculation for electron optical designs by means of integral equations. Scherle's method has the advantage that only one scalar integral equation for a potential is needed instead of three coupled ones for a vector field and that the singularity of the integral kernel $P(\mathbf{r},\mathbf{r}')$ is the weakest possible. The field can be evaluated everywhere in space.

9.3 Reduction of the dimensions

The derivation of a two-dimensional integral equation means that three-dimensional unknown functions have already been reduced to two dimensions. Very often the integral equations obtained are soluble only numerically. In electron optics, however, there is an important class of configurations that can be treated by means of one-dimensional integral equations. This is the class of all devices the electrodes or polepieces of which have rotationally symmetric surfaces. It is not necessary to assume that the boundary values are also rotationally symmetric. In Chapter 7 we have shown that the three-dimensional Poisson equation can be reduced to a sequence of *uncoupled* two-dimensional equations by means of Fourier series expansions. Here we shall show that integral equations can also be simplified in an analogous manner. For Dirichlet problems, an approximate theory has been developed by Kasper and Scherle (1982) and by Kasper (1984a,b); the theory of a method of evaluating (9.12) has been developed by Scherle, who has also demonstrated that it can be applied in practice.

9.3.1 Dirichlet problems

The space-source term in (9.10) can now be omitted without loss of generality. This term alone produces a particular solution $V_s(\mathbf{r})$. If $V_s \neq 0$, then the subsequent reasoning is valid for $V - V_s$ instead of V. Since this adds nothing new, we assume that $S(\mathbf{r}) \equiv 0$, (9.10) then simplifying to a pure surface integral

$$V(\mathbf{r}) = \frac{1}{4\pi} \int_{\partial \Gamma} \frac{\sigma(\mathbf{r}')\,da'}{|\mathbf{r}-\mathbf{r}'|} \tag{9.14}$$

There is no advantage to be gained by separating the factor r^m in (7.5) and we thus introduce the notation

$$V(\mathbf{r}) = \sum_{m=0}^{\infty} \Re\left\{V_m(z,r)e^{im\varphi}\right\} \tag{9.15}$$

The boundary $\partial \Gamma$ is suitably represented in parametric form in terms of the azimuth φ and the arc-length s along the meridional line C passing through the reference point $\mathbf{r}$ (see Fig. 9.2). Since $\partial \Gamma$ is assumed to be rotationally symmetric, this line C can be represented by $z = z(s)$, $r = r(s)$. Whenever

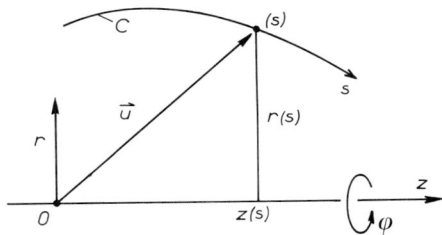

Fig. 9.2: Meridional section through a rotationally symmetric boundary with contour C, along which the arc-length s is adopted as parameter.

this causes no confusion, we shall omit the argument s and introduce the simplifying notation $z' = z(s')$, $r' = r(s')$. When z, z' and r, r' appear as arguments of functions, the vectors $\boldsymbol{u} = (z,r)$, $\boldsymbol{u}' = (z',r')$ will be used.

The values of the functions V_m, occurring in (9.15), on the boundary are given by the inverse Fourier transform of $V(\boldsymbol{u}(s),\varphi)$ and hence depend on s:

$$v_m(s) := V_m(\boldsymbol{u}(s)) = \frac{\kappa_m}{2\pi} \int_0^{2\pi} V(\boldsymbol{u}(s),\varphi) e^{-im\varphi} \, d\varphi \qquad (9.16)$$

with

$$\kappa_m = 2 - \delta_{m,0} = \begin{cases} 1 & \text{for } m = 0 \\ 2 & \text{for } m \neq 0 \end{cases} \qquad (9.16a)$$

It is helpful to expand the surface charge density σ as a Fourier series with respect to φ:

$$\sigma(\boldsymbol{r}) = \sum_{m=0}^{\infty} \Re\left\{\sigma_m(s) e^{im\varphi}\right\} \qquad (9.17)$$

Introducing (9.17) and (9.15) with the coefficients $v_m(s)$ into (9.14), we obtain

$$\sum_{m=0}^{\infty} \Re\left\{v_m(s) e^{im\varphi}\right\} = \sum_{m=0}^{\infty} \Re \int_{\partial \Gamma} \frac{\sigma_m(s') e^{im\varphi'}}{4\pi |\boldsymbol{r}-\boldsymbol{r}'|} \, da'$$

The distance $D := |\boldsymbol{r}-\boldsymbol{r}'|$ is explicitly given by

$$D(\boldsymbol{u},\boldsymbol{u}',\alpha) = \left\{(z-z')^2 + r^2 + r'^2 - 2rr' \cos\alpha\right\}^{\frac{1}{2}} \qquad (9.18)$$

9.3 REDUCTION OF THE DIMENSIONS

$\alpha = \varphi' - \varphi$ being the difference of azimuth. The element of surface area can be written as $da' = r' ds' d\alpha$. Eliminating φ', we find

$$\sum_{m=0}^{\infty} \Re\left\{v_m(s)e^{im\varphi}\right\} = \sum_{m=0}^{\infty} \Re\left\{e^{im\varphi} \int_C G_m(\boldsymbol{u}, \boldsymbol{u}') r' \sigma_m(s')\, ds'\right\}$$

where the abbreviation

$$G_m(\boldsymbol{u}, \boldsymbol{u}') = \frac{1}{4\pi} \int_0^{2\pi} \frac{e^{im\alpha}\, d\alpha}{D(\boldsymbol{u}, \boldsymbol{u}', \alpha)}$$

has been introduced. This expression is a *real* function of its arguments, since it can be rewritten as

$$G_m(\boldsymbol{u}, \boldsymbol{u}') = \frac{1}{2\pi} \int_0^{\pi} \frac{\cos(m\alpha)\, d\alpha}{D(\boldsymbol{u}, \boldsymbol{u}', \alpha)} \tag{9.19}$$

From (9.18) the *symmetry* relation

$$G_m(\boldsymbol{u}, \boldsymbol{u}') = G_m(\boldsymbol{u}', \boldsymbol{u}) \tag{9.20}$$

is obvious. These functions are essentially the Fourier coefficients of the free-space Green's function $G(\boldsymbol{r}, \boldsymbol{r}')$.

From the uniqueness of Fourier series expansions, we now obtain a sequence of *uncoupled* one-dimensional Fredholm equations, given in explicit notation by

$$v_m(s) = \int_C G_m\!\left(\boldsymbol{u}(s), \boldsymbol{u}(s')\right) r(s') \sigma_m(s')\, ds' \tag{9.21}$$

$$\text{for} \quad m = 0, 1, 2, \ldots$$

These have the formal structure

$$v_m(s) = \int_C K_m(s, s') \sigma_m(s')\, ds' \tag{9.22}$$

In Chapter 10 we shall show that the kernel functions G_m of (9.19) can be evaluated analytically and that the resulting expressions contain complete elliptic integrals. We shall further show that there are convenient techniques for numerical solution of (9.22). Thus the solution of Dirichlet problems with rotationally symmetric boundaries can be regarded as a standard technique.

9.3.2 Problems with interface conditions

In the case of rotationally symmetric boundaries, Scherle's integral equation (9.12) can again be decomposed into a sequence of uncoupled integral equations for the Fourier potentials $V_m(z,r)$. As before, we start from (9.15), with $V \equiv \chi_M$. It is now necessary to introduce a Fourier series expansion for the component $\boldsymbol{n} \cdot \boldsymbol{H}_0$ appearing on the right-hand side of (9.12):

$$\boldsymbol{n}(\boldsymbol{r}) \cdot \boldsymbol{H}_0(\boldsymbol{r}) = \Re \sum_{m=0}^{\infty} N_m(z,r) e^{im\varphi} \qquad (9.23)$$

Then, by arguments similar to those described above, we obtain first

$$\oint_{\partial \Gamma} G(\boldsymbol{r},\boldsymbol{r}')\boldsymbol{n}' \cdot \boldsymbol{H}_0(\boldsymbol{r}') \, da' = \sum_{m=0}^{\infty} \Re\left\{ q_m(s) e^{im\varphi} \right\} \qquad (9.24)$$

with

$$q_m(s) := \oint_C G_m(\boldsymbol{u},\boldsymbol{u}') r' N_m(\boldsymbol{u}') \, ds' \qquad (9.25)$$

This is a known function of s which can be evaluated by numerical integration over s'.

In order to evaluate the integral expression on the left-hand side of (9.12), we need the Fourier coefficients of the function $P(\boldsymbol{r},\boldsymbol{r}')$ defined by (9.6). Since the normal derivative involves only the coordinates z' and r', the operator $\partial/\partial n'$ can be applied *after* the integration over φ' has been carried out. We thus obtain

$$\begin{aligned} P_m(\boldsymbol{u},\boldsymbol{u}') &:= \int_0^{2\pi} \frac{\partial}{\partial n'} G(\boldsymbol{r},\boldsymbol{r}') e^{im(\varphi-\varphi')} \, d\varphi' \\ &= \frac{\partial}{\partial n'} G_m(\boldsymbol{u},\boldsymbol{u}') \end{aligned} \qquad (9.26)$$

Like the integral expressions (9.19), these functions take only *real* values but the symmetry properties do not hold.

Introducing the Fourier series expansions of $V(\boldsymbol{r})$ into the left-hand side of (9.12) and considering (9.26), we finally obtain a sequence of uncoupled Fredholm equations of the second kind:

$$-\lambda v_m(s) + \oint_C P_m\Big(\boldsymbol{u}(s),\boldsymbol{u}(s')\Big) r(s') v_m(s') \, ds' = q_m(s) \qquad (9.27)$$

$$m = 0, 1, 2, \ldots$$

Since G_m, P_m and λ are real, it is sufficient to investigate solution techniques for *real* integral equations. Complex solutions can easily be obtained by forming linear combinations of real solutions with constant complex factors. We have thus achieved a major simplification of the original boundary value problem.

9.3.3 Planar fields
In this case the integral equations are already familiar. We shall only examine briefly the Dirichlet problem in the (z, x) plane. Let C be a boundary line represented parametrically in terms of the arc-length s: $z = z(s)$, $x = x(s)$. This line may consist of several distinct curves. Let $v(s) = V(z(s), x(s))$ be given boundary values. Then the integral equation takes the form

$$v(s) = \int_C q(s') \ln\left[a\left\{(z-z')^2 + (x-x')^2\right\}^{-\frac{1}{2}}\right] ds' \qquad (9.28)$$

where $z' = z(s')$, $x = x(s')$ has been introduced ; a is an arbitrary positive normalization constant and $q(s')$ the source distribution function on C. Since the logarithmic kernel function is singular at infinity, either $\int q(s')\, ds' \equiv 0$ or the entire field must be enclosed within a closed loop on which V is zero.

9.4 Important special cases

In this section, we shall examine some field calculation problems that can be solved by numerical evaluation of the integral equations derived above. Technical details are discussed in Chapter 10.

9.4.1 Rotationally symmetric scalar potentials
Here it is not necessary to carry out the Fourier series expansions; the required results are obtained directly for $m = 0$. The function $G_0(\boldsymbol{u}, \boldsymbol{u}')$ is the scalar potential of a uniformly charged ring of radius r' or r and is treated in detail in Chapter 10. Equation (9.21) with $m = 0$ was the starting point for the development of the boundary-element method (BEM) of field calculation, first introduced by Cruise (1963) and extensively studied by Harrington (Harrington, 1967, 1968; Harrington et al., 1969; Mautz and Harrington, 1970). This version of the BEM can be directly applied to electrostatic round lenses of various shapes (Lewis, 1966; Singer and

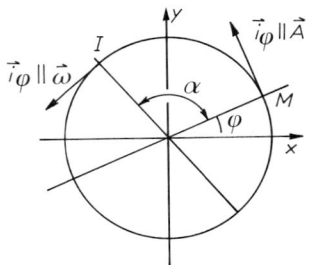

Fig. 9.3: The relation between vectors and angles in rotationally symmetric vector-potential fields. M is a fixed point and I is the integration point.

Braun, 1970; Read et al., 1971; Adams and Read, 1972; Harting and Read, 1976), electron guns with arbitrary rotationally symmetric cathodes (Rauh, 1971; Kuroda and Suzuki, 1972), round electron mirrors and conventional magnetic lenses.

9.4.2 Rotationally symmetric vector potentials

Though the general theory was developed for scalar potentials, it can easily be modified to make it applicable to vector potentials. Since each cartesian component of $\boldsymbol{A}$ separately satisfies a Poisson equation like (9.1) with $S \to \mu_0 j_k$ ($k = 1, 2, 3$), the general conclusions leading to (9.10) must hold for each component of $\boldsymbol{A}$ separately. Collecting the three integral equations into a single vector expression, we now find

$$\boldsymbol{A}(\boldsymbol{r}) = \mu_0 \int_\Gamma G(\boldsymbol{r},\boldsymbol{r}')\boldsymbol{j}(\boldsymbol{r}')\,d^3r' + \mu_0 \int_{\partial\Gamma} G(\boldsymbol{r},\boldsymbol{r}')\boldsymbol{\omega}(\boldsymbol{r}')\,da' \qquad (9.29)$$

The condition div $\boldsymbol{A} = 0$ is satisfied if the surface currents are conserved, as must be the case for physical reasons.

In rotationally symmetric devices, the vectors $\boldsymbol{A}$, $\boldsymbol{j}$ and $\boldsymbol{\omega}$ have only azimuthal components (see also Section 6.4). In the important case of superconducting round lenses, the space currents $\boldsymbol{j}$ are absent, (9.29) then simplifying to

$$A(\boldsymbol{u}) = \mu_0 \oint_C \int_0^{2\pi} \frac{\omega(\boldsymbol{u}')\boldsymbol{i}_\varphi \cdot \boldsymbol{i}_{\varphi'}\, r'\, ds'\, d\alpha}{4\pi D(\boldsymbol{u},\boldsymbol{u}',\alpha)}$$

where the notation introduced in Section 9.3 ($\boldsymbol{u} = (z,r)$ etc.) has been used. The factor $\cos\alpha = \boldsymbol{i}_\varphi \cdot \boldsymbol{i}_{\varphi'}$ arises from the projection of $\boldsymbol{\omega}$ on the direction of $\boldsymbol{A}$, see Fig. 9.3. Using (9.19) with $m = 1$, (6.27) and the boundary conditions (8.16), we finally obtain the integral equation:

9.4 IMPORTANT SPECIAL CASES

$$2\pi r \mu_0 \oint G_1(\boldsymbol{u}, \boldsymbol{u}') r' \omega(\boldsymbol{u}') \, ds' = \Psi_B = \text{const} \quad (9.30)$$

which may formally be considered as a special case of (9.21) with $v(s) = \Psi_B / 2\pi\mu_0 r(s)$.

It is also of some interest that the azimuthal vector potential of a rotationally symmetric coil in the absence of magnetic materials can be represented by

$$A_0(\boldsymbol{u}) = \mu_0 \iint_F G_1(\boldsymbol{u}, \boldsymbol{u}') r' j(\boldsymbol{u}') \, dr' dz' \quad (9.31)$$

F being the domain $r' > 0$ of the axial section through the coil. This is a special case of (8.10).

9.4.3 Unconventional magnetic lenses

In Section 8.3.3 we have indicated that such lenses can be treated by obtaining the solution of a Neumann problem for the rotationally symmetric flux potential Ψ. Alternatively, they can be calculated as a special case of (9.25) and (9.27) with $m = 0$. In this case, no Fourier series expansions are necessary, the quantities of interest being the zero-order Fourier coefficients themselves. In this way really complicated devices with very open structures like those investigated by Mulvey (1982) can be calculated, provided that saturation effects need not be considered.

9.4.4 Magnetic deflection coils

The following considerations (Kasper, 1984a) are valid for both toroidal coils and saddle coils. If the windings are so close to the shield surface that the surface current distribution approximation is reasonable, the total scalar potential $\chi(\boldsymbol{r})$ may be used. The boundary conditions on $\chi(\boldsymbol{r})$ are given by (8.23). In the case of toroidal systems, we could integrate this equation, but in more complicated cases the corresponding integration becomes very complicated. The application of Fourier series expansions, however, makes this integration unnecessary, as we now show.

We again represent each surface function as a function of the azimuth φ and the arc-length s along the axial contour C. The meridional tangent vector is $\boldsymbol{t} = \partial \boldsymbol{r}/\partial s$. The surface potential and its gradient are then given by

$$\chi = \chi(\varphi, s) \quad , \quad \nabla \chi = \frac{1}{r} \frac{\partial \chi}{\partial \varphi} \boldsymbol{i}_\varphi + \frac{\partial \chi}{\partial s} \boldsymbol{t} + \frac{\partial \chi}{\partial n} \boldsymbol{n} \quad (9.32)$$

The surface current distribution may be written

$$\boldsymbol{\omega}(\varphi, s) = \omega_\varphi(\varphi, s) \boldsymbol{i}_\varphi + \omega_s(\varphi, s) \boldsymbol{t} \quad (9.33)$$

Introducing (9.33) and (9.32) into (8.23) and recalling that $t \times n = i_\varphi$ and $i_\varphi \times n = -t$, we obtain

$$\frac{\partial \chi}{\partial \varphi} = r\omega_s \quad , \quad \frac{\partial \chi}{\partial s} = -\omega_\varphi \qquad (9.34)$$

The condition obtained by evaluating $\partial^2 \chi / \partial \varphi \partial s$ from each of these,

$$\frac{\partial^2 \chi}{\partial \varphi \partial s} = \frac{\partial}{\partial s}(r\omega_s) = -\frac{\partial}{\partial \varphi}\omega_\varphi \qquad (9.35)$$

is identical with the continuity equation for surface currents and thus imposes a restriction on the choice of surface current distributions, which is automatically satisfied by any real distribution of wires.

For a single pair of deflection coils, we can choose the origin of φ in such a way that the appropriate Fourier series expansions can be written as

$$\omega_s(\varphi, s) = \sum_m M_m(s) \cos(m\varphi) \qquad (9.36)$$

$$\omega_\varphi(\varphi, s) = -\sum_m A_m(s) \sin(m\varphi) \qquad (9.37)$$

$$\chi(\varphi, s) = \sum_m v_m(s) \sin(m\varphi) \qquad (9.38)$$

m always being an *odd* integer. Introduction of these series expansions into (9.34) and (9.35) results in

$$v_m(s) = \frac{1}{m} r(s) M_m(s) \quad , \quad m = 1, 3, 5, \ldots \qquad (9.39)$$

and

$$v'_m(s) = \frac{d}{ds}\left(\frac{r}{m} M_m\right) = A_m(s) \quad , \quad m = 1, 3, 5, \ldots \qquad (9.40)$$

This is an important simplification. The boundary values $v_m(s)$ of the Fourier potentials are already uniquely determined by the Fourier coefficients of the meridional component, while the azimuthal components A_m can be obtained by mere differentiation and are not needed explicitly for the solution of the boundary-value problem: the only major computations needed are the Fourier transformation of $\omega_s(\varphi, s)$ and the subsequent numerical solution of (9.21). Thereafter the integral expression in (9.21) can be evaluated to give $V_m(z, r)$ at any point (z, r) outside the surface.

9.4 IMPORTANT SPECIAL CASES

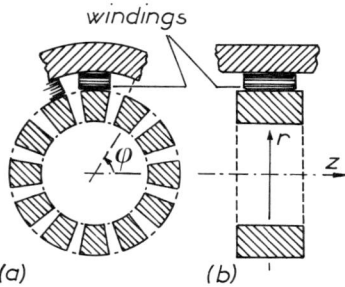

Fig. 9.4: Simplified representation of a magnetic multipole system. (a) Cross-section, (b) meridional section. Only parts of the coils and the outer screening ring are shown.

In the case of a purely toroidal system of coils the surface current density is given by

$$\omega_\varphi \equiv 0 \quad , \quad \omega_s = I\,W(\varphi)/r(s) \tag{9.41}$$

I being the electric current through the windings and $W(\varphi)$ the winding density, by which we mean that $W(\varphi)d\varphi$ is the number of windings between φ and $\varphi + d\varphi$. In this case (9.39) simples to

$$v_m = \frac{4I}{\pi m}\int_0^{\pi/2} W(\varphi)\cos(m\varphi)\,d\varphi \quad , \quad m = 1,3,5,\ldots \tag{9.42}$$

(Schwertfeger and Kasper, 1974). We cannot expect any further simplification, as we now have to solve Dirichlet problems with constant boundary values.

If the windings are not close to the shield surface, as is frequently the case in systems of saddle coils, the one-dimensional integral equations arising from Scherle's equation, namely (9.25) and (9.27), must be evaluated. Then, of course, the necessary computation is considerably greater but the results will be very accurate. The details are discussed in Chapter 40.

9.4.5 Multipole systems

Electric or magnetic multipole systems are commonly used as stigmators or as strong focusing lenses in electron optical devices. Nowadays such systems are also used as deflectors in scanning devices. Since these systems are in no sense rotationally symmetric, the theory of one-dimensional integral equations cannot be properly applied to them. If, however, parts of the surfaces form a rotationally symmetric face and the gaps between

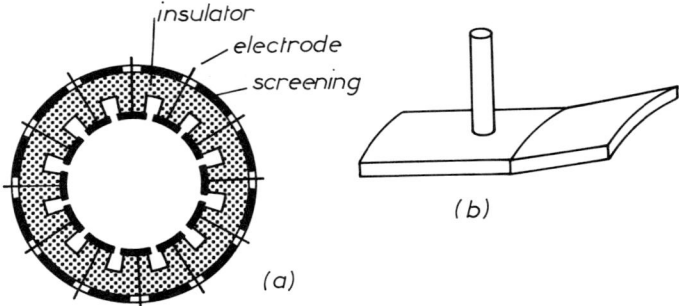

Fig. 9.5: Simplified view of an electric multipole device. (a) Cross-section, (b) perspective view of a single electrode.

adjacent poles are comparatively narrow, as shown in Figs 9.4 and 9.5, the theory is approximately applicable. Magnetic systems like the one shown in Fig. 9.4 are in use in the devices developed by the Darmstadt group (see Rose, 1971; Rose and Plies, 1973) to compensate the third-order spherical aberration and the axial chromatic aberration in an electron microscope of very high resolution. Electric systems of the kind shown in Fig. 9.5 have been investigated by Munro and Chu (1982) and Chu and Munro (1982) and are used as deflection units in electron lithography devices.

The approximation underlying the treatment of these systems is that in the gaps, it must be possible to make a reasonable interpolation for the potential with respect to the azimuth φ, usually by a linear expression. Then, on a rotationally symmetrical surface—consisting of the cylindrical bore, the gaps, the ring-shaped parts of the end-planes and sometimes parts of the outer surfaces of the poles and a screening surface enclosing the whole systems as well—reasonable values of the potential can be defined. These are then to be introduced into (9.16) after which the uncoupled one-dimensional integral equations (9.21) can be solved. The errors caused by the incorrect boundary values do not seriously influence the field in the paraxial domain. This approximation is therefore satisfactory.

9.4.6 Small perturbations of the rotational symmetry

Since it is impossible to build rotationally symmetric devices perfectly, the effects of small perturbations such as shifts, tilts or ellipticity of the pole-pieces or electrodes on the field in the paraxial domain are of interest. Such effects determine the tolerance limits for the machining of electron optical devices, and they have attracted considerable attention (see Chapter 31). The first successful attempts to calculate them numerically were made by

9.4 IMPORTANT SPECIAL CASES

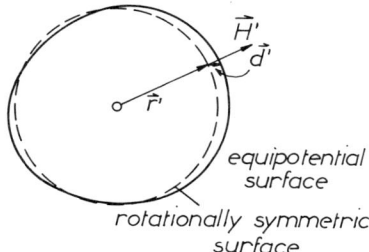

Fig. 9.6: Simplified cross-section through an imperfect lens, showing the real and the ideal contour of a polepiece. The vectors r', d' and H' are the projections of the position vector r, the shift $d(r)$ and the field strength $H(r)$, respectively.

Janse (1971). Here, we describe how his method can be used to calculate the field in an imperfect round magnetic lens.

Janse's method is a perturbation calculus. First, as a zero-order approximation, the ideal round system is calculated; along the contour C in the axial section through the system the boundary values $H(s)$ of the field strength are then determined (s being the arc-length along C). This is illustrated in Fig. 9.6. Owing to the rotational symmetry of the perfect system, the field strength $H(r) = H(\varphi, s)$ at any surface point r with coordinates φ and s is then known.

In the next step a deviation function $d(r)$ is defined as the local shift from the ideal surface to the real one. In the parametrization adopted here, this is a function $d = d(\varphi, s)$. The shift may be in any direction but must be very small. This is illustrated in Fig. 9.6.

Since the real distorted pole surface coincides with the equipotentials $\chi(r + d) = $ const, the potential at the ideal round surface is perturbed by a quantity

$$\delta\chi(r) = \chi(r) - \chi(r + d) \approx -d \cdot \mathrm{grad}\,\chi = d(r) \cdot H(r) \quad (9.43)$$

or

$$\delta\chi(\varphi, s) = d(\varphi, s) \cdot H(\varphi, s) \quad (9.43a)$$

This function represents the boundary values of a perturbation potential and is to be introduced into (9.16). The integral equations derived above can then be applied to this problem.

Janse did not in fact use integral equation techniques but solved the Dirichlet problems corresponding to (7.10) with vanishing source terms by means of the finite-difference method. By solving (9.12) with $V \equiv \chi$, however, the required normal derivative $H = -n\sigma$ at the equipotential surfaces is obtained directly, thus saving subsequent differentiation. These simplifications are, of course, not possible if $\sigma \neq \partial \chi / \partial n$.

After solution of the corresponding boundary-value problems, the rotationally symmetric field and the perturbation field are to be superimposed. It is sufficient to limit this to the paraxial domain, whereupon the influence of various kinds of perturbations on the electron trajectories can be studied.

9.5 Résumé

Without entering into the details of concrete numerical calculations, we have developed a general theory of field calculation in systems having a straight optic axis. Two basic ideas, the introduction of azimuthal Fourier series expansions and the formulation of integral equations, have been worked out in some detail, since these are particularly well adapted to the needs of electron optical field calculations. The use of Fourier series expansions results in a sequence of uncoupled mathematical structures of lower dimensions. Since we shall finally be interested in the field in the paraxial domain, we may terminate the calculation of the Fourier coefficients (axial harmonics) after the first few orders, which are of most importance. This is thus a very economic technique. The use of integral equations rather then partial differential equations further reduces the number of dimensions, since parts of the necessary integrations have already been carried out. In all cases in which the material properties of the polepieces or electrodes are constant, integral equation methods have proved to be very powerful and efficient. There are, of course, problems that cannot be solved in this way. These will be treated in the context in which they arise.

10

The Boundary-Element Method

In Chapter 9 we derived various types of integral equations. We now turn to their concrete numerical solution. The corresponding procedure, called the integral-equation method (IEM) or the boundary-element method (BEM), has proved to be very powerful. Since the early investigations (Cruise, 1963; Lewis, 1966; Harrington, 1967, 1968; Singer and Braun, 1970), many presentations have been published, which often differ only in minor details. The list of references given in the bibliography makes no pretence at completeness. We recall that the physical idea behind this method is to calculate the charge density distribution (in the electrostatic case) corresponding to the voltages applied to the electrodes and then calculate the potential distribution in space created by this charge distribution. The complexity of the method in some practical situations is a consequence of the complicated nature of the boundary conditions. We first consider one-dimensional integral equations.

10.1 Evaluation of the Fourier integral kernels

The numerical solution of (9.21) and (9.27) requires the evaluation of the Fourier integral kernels G_m, defined by (9.18) and (9.19), and of their partial derivatives. The special case $m = 0$ is already familiar in classical electrodynamics, since $G_0(z, r; z', r')$ is the potential at (z, r) due to a uniformly charged ring located at (z', r'). Cases for which $m \neq 0$ have been investigated by Kasper and Scherle (1982).

10.1.1 Introduction of moduli

The meaning of some of the geometric variables appearing in the subsequent theory is presented in Fig. 10.1. We assume that a ring of radius r' is located in the plane $z' = $ const. The coordinates (z, r) define an arbitrary point in the axial section through the field. It is convenient to introduce the distances

$$d_{1,2} := \sqrt{(z - z')^2 + (r \mp r')^2} \qquad (10.1)$$

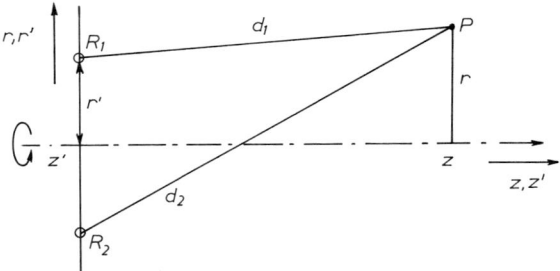

Fig. 10.1: In an axial section through an axisymmetric configuration, a single ring is seen as the circles R_1 and R_2. P is an arbitrary reference point.

and the quantities

$$p := d_1 d_2 = \left[\{(z-z')^2 + r^2 + r'^2\}^2 - 4r^2 r'^2\right]^{\frac{1}{2}} \tag{10.2}$$

and

$$S := d_1 + d_2 = \sqrt{2}\left\{p + (z-z')^2 + r^2 + r'^2\right\}^{\frac{1}{2}} \tag{10.3}$$

These quantities have a very simple geometric meaning: d_1 and d_2 are the minimal and the maximal distance from the ring, respectively, as shown in Fig. 10.1, and S is the sum of these distances. The surfaces $S = $ const are oblate spheroids with the ring as their common focal line. It is helpful to introduce dimensionless moduli:

$$k := \frac{2\sqrt{rr'}}{d_2} \quad , \quad k' := \sqrt{1-k^2} = \frac{d_1}{d_2} \tag{10.4}$$

$$\varrho := \frac{4rr'}{S^2} = \frac{d_2 - d_1}{d_2 + d_1} = \frac{1-k'}{1+k'} \tag{10.5}$$

These moduli are defined only in the interval [0,1]. It is of great importance that the relation between k and ϱ is a Landen transformation of the arguments of elliptic integrals (Whittaker and Watson, 1927, Section 22.42), see Section 10.1.3.

10.1.2 Radial series expansions
Again using the abbreviations $\boldsymbol{u} = (z,r)$, $\boldsymbol{u}' = (z',r')$, we obtain from (9.18) and (9.21) the explicit integral representation

$$G_m(\boldsymbol{u};\boldsymbol{u}') = \frac{1}{2\pi}\int_0^{2\pi} \frac{\cos m\alpha \, d\alpha}{\{(z-z')^2 + r^2 + r'^2 - 2rr'\cos\alpha\}^{\frac{1}{2}}} \tag{10.6}$$

10.1 EVALUATION OF THE FOURIER INTEGRAL KERNELS

Introducing the variables S and ϱ, defined by (10.3) and (10.5), we soon find the simpler representation

$$G_m(\boldsymbol{u}; \boldsymbol{u}') = \frac{1}{\pi S} \int_0^{2\pi} \frac{\cos m\alpha \, d\alpha}{\left(1 - 2\varrho \cos \alpha + \varrho^2\right)^{\frac{1}{2}}} \qquad (10.7)$$

The required radial series expansion is now most easily obtained by expanding the denominator in terms of Legendre polynomials:

$$\left(1 - 2\varrho \cos \alpha + \varrho^2\right)^{-\frac{1}{2}} = \sum_{l=0}^{\infty} \varrho^l P_l(\cos \alpha)$$

Integration of each term of the resulting sum yields

$$G_m(\boldsymbol{u}; \boldsymbol{u}') = \frac{1}{S} \sum_{j=0}^{\infty} a_{mj} \varrho^{m+2j} \qquad (10.8)$$

the coefficients being given by

$$a_{mj} = \frac{(2j)! \, (2m + 2j)!}{2^{2m+4j} \, (j!)^2 \, \{(j+m)!\}^2} \qquad (10.9)$$

Since ϱ is proportional to r as long as $r \ll r'$, (10.8) essentially represents a series expansion with respect to r. This agrees with our former results that the mth Fourier coefficient of the potential must be proportional to r^m if r is very small (see Section 7.1).

The series expansion (10.8) converges for all values of ϱ in the interval $0 \leq \varrho < 1$. In practice it can be evaluated from a reasonable number of terms in the interval $0 \leq \varrho \leq 0.5$. The number of terms necessary depends on the acceptable error limit and on the actual value of ϱ. With $j_{max} = 20$, excellent accuracy is obtained. Since $\varrho \leq 0.5$ usually covers the domain occupied by the electron trajectories, (10.8) is very useful in practical field computations.

10.1.3 Recurrence relations

The solution of the integral equation derived in Chapter 9 requires the Fourier kernels G_m to be evaluated in the interval $0 \leq \varrho < 1$. In the interval $0.5 < \varrho < 1$ this can be performed by means of appropriate recurrence relations. The latter can take different forms, which have been investigated by Kasper and Scherle (1982). For the sake of brevity, we shall treat here only one of these forms.

10. THE BOUNDARY-ELEMENT METHOD

Introducing the more familiar modulus k (10.4) into (10.6) and substituting $\beta = \alpha/2$, we find the representation

$$G_m(\boldsymbol{u};\boldsymbol{u}') = \frac{1}{\pi d_2} \int_0^{\pi/2} \frac{\cos 2m\beta \, d\beta}{\left(1 - k^2 \cos^2 \beta\right)^{1/2}}$$

which is clearly a generalization of the familiar complete elliptic integral of the first kind. In fact, the special case $m = 0$ is the well-known formula for the potential of a uniformly charged ring.

For the subsequent discussion, we introduce the complete elliptic integrals

$$\left.\begin{aligned}
K(k) &:= \int_0^{\pi/2} \left(1 - k^2 \sin^2 \beta\right)^{-\frac{1}{2}} d\beta \\
E(k) &:= \int_0^{\pi/2} \left(1 - k^2 \sin^2 \beta\right)^{\frac{1}{2}} d\beta \\
D(k) &:= \int_0^{\pi/2} \sin^2 \beta \left(1 - k^2 \sin^2 \beta\right)^{-\frac{1}{2}} d\beta \equiv (K - E)k^{-2}
\end{aligned}\right\} \quad (10.10)$$

For all values $0 < k < 1$, these can be computed very accurately by means of the algorithm of repeated algebraic and geometric means (see Abramowitz and Stegun, 1965, p. 598). In terms of (10.10), the lowest order Fourier kernels may be written

$$G_0 = \frac{1}{\pi d_2} K(k) \quad, \quad G_1 = \frac{2D(k) - K(k)}{\pi d_2}$$

As it is necessary to introduce the modulus ϱ for the evaluation of the radial series expansion, it is convenient to introduce ϱ into the integral expression as well; this implies that we have to establish the functions $I_m(\varrho)$:

$$G_m(\boldsymbol{u};\boldsymbol{u}') =: S^{-1} I_m(\varrho) \quad, \quad 0 \leq \varrho < 1 \quad (10.11)$$

As far as G_0 and G_1 are concerned, this is easily done by means of the Landen transformation

$$K(k) = (1 + \varrho)K(\varrho) \quad, \quad E(k) = \frac{2}{1 + \varrho} E(\varrho) - (1 - \varrho)K(\varrho)$$

10.1 EVALUATION OF THE FOURIER INTEGRAL KERNELS

After some elementary calculations, it is easy to confirm that (10.11) is satisfied for $m = 0$ and $m = 1$ by

$$I_0(\varrho) = \frac{2}{\pi}K(\varrho) \quad , \quad I_1(\varrho) = \frac{2}{\pi}\varrho D(\varrho) \tag{10.12}$$

The higher order integral expressions are given by the linear recurrence relation ($m \geq 1$)

$$(2m+1)I_{m+1}(\varrho) = 2m\left(\varrho + \frac{1}{\varrho}\right)I_m(\varrho) - (2m-1)I_{m-1}(\varrho) \tag{10.13}$$

This differs from the expressions given by Kasper and Scherle: here the more convenient normalization $I_0(0) = 1$ has been chosen. Equations (10.13) can be easily evaluated in ascending sequence. This recursive procedure is slightly unstable. For practical purposes it is stable enough if $\varrho \geq 0.5$ and $m \leq 12$. There are ways of extending the region of allowable orders m, but these will not be treated here since $m = 12$ is quite sufficient. Combined with the paraxial series expansion (10.8) for $\varrho \leq 0.5$, the recursive procedure defined by (10.12) and (10.13) provides a convenient way of computing generalized elliptic integrals.

An interesting complete integral representation of the function $I_m(\varrho)$ is the expression

$$I_m(\varrho) = \frac{2\varrho^m}{\pi} \int_0^{\pi/2} \sin^{2m}\beta \left(1 - \varrho^2 \sin^2\beta\right)^{-\frac{1}{2}} d\beta$$

Obviously (10.12) are satisfied. By expanding the integrand as a power series and performing the integrations before the summation, it is possible to verify (10.8) and (10.9). Finally, (10.13) can be proved by induction.

10.1.4 Analytic differentiation

It is of great importance in connection with field computation that the derivatives with respect to z and r can also be calculated quite easily. Differentiation of the expressions for S and ϱ gives

$$\frac{\partial S}{\partial z} = -\frac{\partial S}{\partial z'} = S(z-z')/p \tag{10.14}$$

$$\frac{\partial \varrho}{\partial z} = -\frac{\partial \varrho}{\partial z'} = -2\varrho(z-z')/p \tag{10.15}$$

$$\frac{\partial S}{\partial r} = \frac{r}{pS}(S^2 - 4r'^2) \tag{10.16}$$

$$\frac{\partial \varrho}{\partial r} = \frac{4r'}{S^2}\left(1 - \frac{2r^2}{p} + \frac{8r^2r'^2}{pS^2}\right) \tag{10.17}$$

The derivatives with respect to r' are obtained by interchanging r and r'. On the optic axis ($r = 0$ or $r' = 0$) all these derivatives remain regular. The differentiation of (10.8) is now a straightforward procedure. For $\varrho < 0.5$ the derivatives of (10.12) and (10.13) are to be used. From the standard formulae for the derivatives of complete elliptic integrals we have

$$I_0'(\varrho) = \frac{\varrho I_0(\varrho) - I_1(\varrho)}{1 - \varrho^2} \quad , \quad I_1'(\varrho) = I_0'(\varrho)/\varrho \qquad (10.18)$$

The differentiation of (10.13) results immediately in

$$(2m+1)I_{m+1}'(\varrho) = 2m\left(\varrho + \frac{1}{\varrho}\right)I_m'(\varrho) - (2m-1)I_{m-1}'(\varrho) - 2m\left(\frac{1}{\varrho^2} - 1\right)I_m(\varrho) \qquad (10.19)$$

For $m \geq 1$ this is an inhomogeneous linear recurrence relation, which can easily be solved after evaluating (10.13).

The derivatives of $G_m(\boldsymbol{u}, \boldsymbol{u}')$ can now be put into a very convenient form. From (10.14) and (10.15) we find easily

$$\frac{\partial G_m}{\partial z} = -\frac{\partial G_m}{\partial z'} = \frac{z'-z}{pS} Q_m(\varrho) \quad , \quad Q_m(\varrho) := I_m + 2\varrho I_m'(\varrho) \qquad (10.20a)$$

After some elementary calculations we obtain from (10.16) and (10.17) the result

$$\frac{\partial G_m}{\partial r} = \frac{1}{pS}\left\{\frac{\varrho}{r}I_m'(\varrho) + (r'\varrho - r)Q_m(\varrho)\right\} \qquad (10.20b)$$

As $r \to 0$ this formula gives a finite result since ϱ/r remains finite.

Altogether we obtain a comparatively simple numerical procedure for the computation of the kernel functions $G_m(\boldsymbol{u}, \boldsymbol{u}')$ and their derivatives. This procedure can be carried out for arbitrary values of the arguments with the exception of singular combinations, the latter being defined by $d_1 = 0$ or $k = \varrho = 1$.

Computer programs embodying this procedure have been tested and its practical applicability is now well established. In the literature it is usually the special case of rotationally symmetric fields ($m = 0$) that is investigated. Apart from the fact that there is no need to evaluate the recurrence formulae, this specialization brings no major simplification.

10.2 NUM. SOLUTION OF ONE-DIMEN. INTEGRAL EQUATIONS

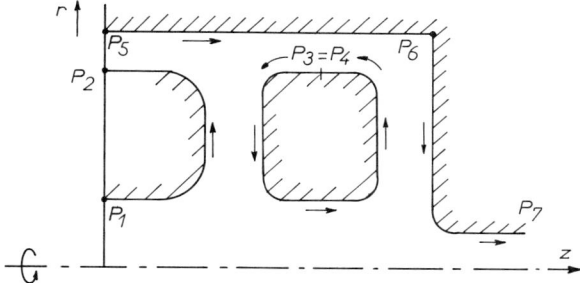

Fig. 10.2: Upper-right quadrant of an axial section through a device with axial and mirror symmetry. The contour C now consists of three separate curves, one closed, the other open. The orientation has been chosen in such a way that the field domain is always on the right as one follows the contour; this makes it easy to distinguish between interior and exterior domains.

10.2 Numerical solution of one-dimensional integral equations

We now discuss the solution of integral equations such as (9.21) and (9.27). These are special cases of a Fredholm equation of the second kind:

$$\int_C K(s,s')Y(s')\,ds' = U(s) + \lambda Y(s) \qquad (10.21)$$

Depending on the particular application, the variables appearing in (10.21) have different physical meanings. In applications to Dirichlet problems (9.21), Y is a surface charge density, U is a surface potential, while λ vanishes. In applications to problems with interface conditions (9.27), Y is a surface potential, U arises from the boundary values of an external field (9.23 and 9.25), while λ is given by (9.14). The variables s and s' are usually arc-lengths in an axial section through the device, but may also be other parameters, such as angles, if these are more favourable. In most applications the contour line C of the boundary consists of two or more separated loops. As a result of using any symmetry properties of the device, these loops may be open or closed, as is shown in Fig. 10.2. In the transformation to a linear scale for s and s', the contours are mapped onto a sequence of disjoint intervals, as is shown in Fig. 10.3. On closed loops the functions appearing in (10.21) must satisfy cyclic conditions, since they must be unique.

10.2.1 Conventional solution techniques
The structure of the kernel function $K(s,s')$, is so complicated that (10.21)

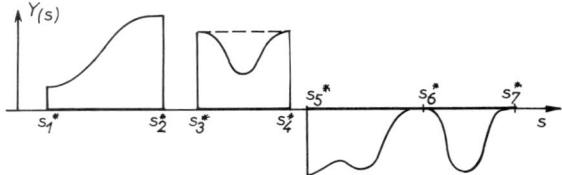

Fig. 10.3: The intervals of s and the surface charge distribution corresponding to Fig. 10.2. The three branches of C now correspond to three separated intervals. The surface charge density $Y(s)$ must satisfy $Y(s_3^*) = Y(s_4^*)$ (cyclicity) and $Y'(s_1^*) = Y'(s_2^*) = Y'(s_5^*) = 0$ (even or mirror symmetry).

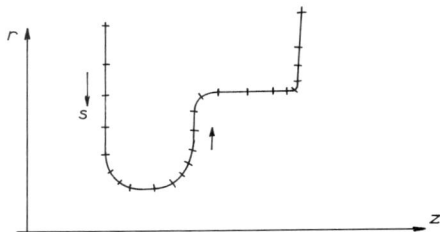

Fig. 10.4: Part of a boundary and the appropriate discretization. The line elements must be short in the vicinity of edges and in regions where the boundary curvature is large. Abrupt large changes in the length of the elements should be avoided.

can only be solved numerically. In every case, this requires a suitable discretization of the boundary, as is shown in Figs. 10.4 and 10.5. Usually, the unknown function $Y(s')$ is assumed to vary very slowly, so that in each integration subinterval it can be replaced by a constant, the value at the centre of the subinterval. With these step functions, (10.21) is then approximated by

$$\sum_{k=1}^{N} H_{ik} \tilde{Y} = U_i \quad , \quad i = 1 \ldots N \qquad (10.22)$$

where the notation

$$\tilde{Y}_k := Y(\overline{s}_k) \quad , \quad U_i := U(\overline{s}_i) \quad , \quad i, k = 1 \ldots N \qquad (10.23)$$

and the matrix elements

$$H_{ik} := \int_{-h_k}^{h_k} K(\overline{s}_i, \overline{s}_k + \tau) \, d\tau - \lambda \delta_{ik} \qquad (10.24)$$

10.2 NUM. SOLUTION OF ONE-DIMEN. INTEGRAL EQUATIONS

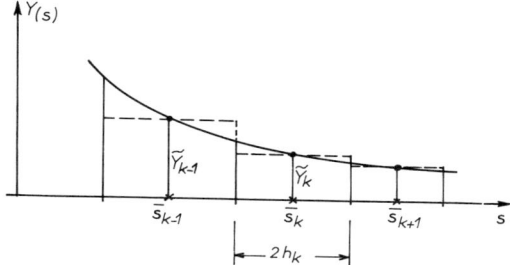

Fig. 10.5: Some sub-intervals of integration and the corresponding approximation of a function $Y(s)$ by a step-function. The quantity $\bar{s}_k$ denotes the midpoint of the k-th subinterval. The area of the column of the histogram is not exactly identical with that under the curve $Y(s)$.

have been introduced. The quantities $\bar{s}_i$ and $\bar{s}_k$ refer to the midpoints of the subintervals in question and h_k is the half-width of the subinterval with label k.

The integration in (10.24) can be performed numerically without problem for $i \neq k$. For $i = k$, however, the integral becomes improper, since the kernel then contains a logarithmic singularity. In the case of rotational symmetry, the approximation

$$r'G_0(\boldsymbol{u}, \boldsymbol{u}') = \frac{1}{2\pi}\ln(8r'/d_1) + R_0(d_1) \quad , \quad d_1 \ll r' \tag{10.25}$$

can be derived from (10.6), $R_0(d_1)$ being a bounded function. Then for $i = k$ the integrand of (10.24) becomes

$$K(\bar{s}_i, \bar{s}_i + \tau) = \frac{1}{2\pi}\ln(8\bar{r}_i/|t|) + R_0(|t|) \tag{10.25a}$$

since $d_1 = |t|$ is the distance from the singularity. From the recurrence formula (10.13), the Fourier integral kernels of higher orders clearly have the same singularity; they differ from (10.25) only in the nonsingular term $R_0(d_1)$.

The numerous applications of the BEM published in the literature (see list of references) differ essentially in the numerical procedure adopted for the integration over the singularity. Once this problem has been solved, the implementation of the BEM raises no further problem. The linear system of equations (10.22) can be solved directly by means of standard techniques, such as the Gaussian elimination algorithm, the number N usually being between 50 and 500 in order of magnitude. Thereafter, the surface charge density is known and can be introduced into the surface Coulomb integral in (9.10) or (9.14), giving the value at any point $\boldsymbol{r}$ in space.

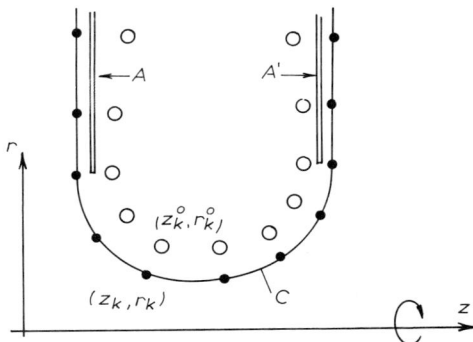

Fig. 10.6: Part of an electrode, showing the boundary curve C and a possible configuration of control points (z_k, r_k) and ring positions (z_k^0, r_k^0). A and A' are the upper halves of meridional sections through thin charged plane apertures.

In order to circumvent the complicated evaluation of the improper integrations in the diagonal elements H_{ii}, Hoch et al. (1978) have replaced the surface charge distribution by a sequence of charged rings located *inside* the electrodes, as shown in Fig. 10.6. In addition, they introduced charged thin planar apertures. We postpone consideration of these to Section 10.3. The rings are characterized by coordinates $z' =: z_k^0$, $r' =: r_k^0$ and charges Q_k ($k = 1...N$). At the boundary C an equal number of control points with coordinates $z =: z_i$, $r =: r_i$ and boundary potentials U_i are introduced. Apart from an unimportant constant factor, the conditions that the rotationally symmetric potential $V(z,r)$ has the prescribed boundary values are of the form

$$\sum_{k=1}^{N} Q_k G_0(\boldsymbol{u}_i, \boldsymbol{u}_k^0) = U_i \quad , \quad i = 1...N \qquad (10.26)$$

(we again use $\boldsymbol{u} = (z,r)$, $\boldsymbol{u}_i = (z_i, r_i)$, $\boldsymbol{u}_k^0 = (z_k^0, r_k^0)$). Once the solution of this linear system of equations is known, the potential and its gradient are simply given by

$$V(\boldsymbol{u}) = \sum_{k=1}^{N} Q_k G_0(\boldsymbol{u}; \boldsymbol{u}_k^0) \qquad (10.27a)$$

$$\nabla V(\boldsymbol{u}) = \sum_{k=1}^{N} Q_k \nabla G_0(\boldsymbol{u}; \boldsymbol{u}_k^0) \qquad (10.27b)$$

This is certainly the simplest version of the BEM. Kasper and Scherle (1982) have generalized it for Fourier series expansions like (9.15). In many

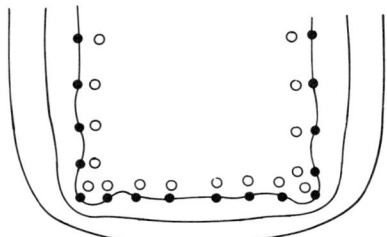

Fig. 10.7: Wavy structure of equipotential lines produced by disjoint rings. This effect, exaggerated here for clarity, decreases rapidly with increasing distance from the surface.

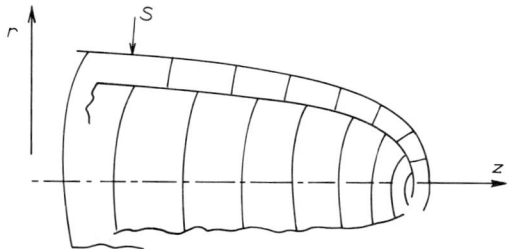

Fig. 10.8: Simplified perspective drawing of the discretization adopted for the cathode in an electron gun. Each control point at the physical surface S is associated with an interior conical mantle over which the corresponding part of the surface charge is to be distributed uniformly or, better, continously.

practical applications to conventional round lenses and to deflection units this method has worked quite satisfactorily, the relative error in the field in its paraxial domain being roughly 10^{-3} to 5×10^{-3}. Difficulties may arise in the vicinity of sharp edges and especially at the surfaces of cathode tips in electron guns. Depending on the shape of the boundaries and on the prescribed boundary values, the equipotential over which the expression (10.27a) assumes a constant electrode potential U has a wavy structure, as shown in Fig. 10.7. This causes a deviation of the local field strength from its correct direction parallel to the local surface normal of the electrode. In order to avoid this weakness, the charge can be distributed over an inner surface, as shown in Fig. 10.8.

It is sufficient to assume constant charge density on each surface element. Replacing the summation in (10.26) by the appropriate integration, one obtains quite an accurate solution. Weysser (1983) has applied this method to electron guns and found that it works reasonably well.

In spite of its wide field of applications, the refined method of Hoch *et al.* is not quite satisfactory in every respect because it cannot be employed to solve problems with interface conditions, since the singularities must then be located at the interfaces themselves. In this case, the corresponding integration over the singularities cannot be circumvented. In the next section we shall present a better solution technique.

10.2.2 Combination with interpolation techniques

If the singularities are assumed to be located at the surfaces of electrodes or yokes, it is necessary to approximate $Y(s')$ in (10.21) by a *smooth* function, since in applications to Dirichlet problems this is essentially the normal component of the field strength. Then, of course, the integration becomes far more complicated. For reasons of conciseness we consider only one integration domain, $s_0 \leq s \leq s_N$; the generalization to more domains raises no difficulties.

A fairly accurate solution is already obtained if the values $\overline{Y}_1 \ldots \overline{Y}_N$, found by solving (10.22), are introduced into a smoothing program based upon a suitable interpolation technique. This has been noticed by Uchikawa *et al.* (1981). Another possibility consists in using more than one point of reference in each integration subinterval and applying Lagrange interpolation techniques. The system of linear equations generated by the discretization of (10.21) does, of course, become more complicated. The best approximation that can be reached in this way is a function $Y(s)$ that is continuous at the internal endpoints of the chosen subintervals. The derivative dY/ds will, however, be discontinuous at these points.

In order to avoid these difficulties, Scherle (1983) has approximated $Y(s)$ by a cubic spline function (see Section 13.1). The points of reference with abscissae $s_0 \ldots s_N$ must then be the *endpoints* of the integration subintervals and not the midpoints. The integration over the singularities was carried out in a correct numerical manner, after these singularities had been separated from regular terms by appropriate series expansions. The whole set of conditions to be satisfied forms a complicated system of linear equations. The accuracy of the solution is, however, very high, as has been shown by applying it to suitable test examples. In these investigations the appropriate integration over singular functions turned out to be the most complicated part of the practical application of the BEM. It is advantageous to choose the *midpoints* of the subintervals for reference, as was done in (10.24). The problem of performing the integration is deferred to Section 10.2.5.

In order to combine these requirements with those of continuous differentiability of the solution $Y(s)$ at the endpoints $s_1 \ldots s_{N-1}$, Kasper (1983) developed a solution technique based on quadratic spline functions. This

10.2 NUM. SOLUTION OF ONE-DIMEN. INTEGRAL EQUATIONS

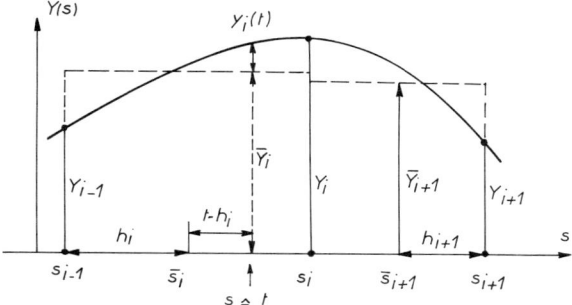

Fig. 10.9: Discrete and mean values of some function $Y(s)$, represented by different interpolation parabolae. At the points (the coordinates of which are written without bars), these parabolae are to be joined together smoothly. The midpoints ($\bar{s}_i$, $\bar{s}_{i+1}$) are introduced to facilitate the integration.

method can be still further generalized and improved, as we now show (Kasper, 1984a).

In each subinterval $s_{k-1} \leq s \leq s_k$, $k = 1\ldots N$, the required function $Y(s)$ is represented in the form

$$Y(s) = \overline{Y}_k + y_k(t) \qquad (10.28a)$$
$$t := (s - \bar{s}_k)/h_k \quad , \quad |t| \leq 1 \qquad (10.28b)$$

which implies

$$\bar{s}_k = (s_k + s_{k-1})/2 \quad , \quad h_k = (s_k - s_{k-1})/2 \quad , \quad (k = 1\ldots N) \qquad (10.29)$$

In the corresponding subinterval the function $y_k(t)$ is the departure of $Y(s)$ from its mean value $\overline{Y}_k$, as is shown in Fig. 10.9. This quantity y_k can be considered as a small perturbation; in the simple conventional approximation (10.22–10.24), it is completely ignored, but here we shall determine it approximately.

Introducing (10.28) into (10.21), we rapidly find that it is convenient to define a perturbation vector with the components

$$u_i := \sum_{k=1}^{N} h_k \int_{-1}^{1} K(\bar{s}_i, \bar{s}_k + h_k t) y_k(t)\, dt - \lambda y_k(0)\delta_{ik} \quad (i = 1\ldots N) \qquad (10.30)$$

eqs. (10.22) then taking the improved form

$$\overline{Y}_k = \sum_{j=1}^{N} \left(H^{-1}\right)_{kj}(U_j - u_j) \quad , \quad k = 1\ldots N \qquad (10.31)$$

This is now to be combined with an appropriate interpolation technique. In the present context this means that for given mean values $\overline{Y}_1 \ldots \overline{Y}_N$, a program is to be set up that supplies the deviations $y_k(t)$ in such a way that the resulting solution $Y(s)$ is continuously differentiable and assumes exactly the values $\overline{Y}_1 \ldots \overline{Y}_N$ as integral means in the corresponding subintervals. This problem can be solved in different ways.

10.2.3 Polynomial series expansions

A fairly simple algorithm, based on series expansions in terms of Legendre polynomials $P_l(t)$, has been proposed by Kasper (1984a). We start from

$$Y(s) = \sum_{l=0}^{L} c_k^{(l)} P_l(t) \quad , \quad s_{k-1} \le s \le s_k \tag{10.32}$$

t being given by (10.28b) and L a fixed integer. Comparing with (10.28a) we notice that

$$\overline{Y}_k = c_k^{(0)} = \frac{1}{2}\int_{-1}^{1} Y\,dt = \frac{1}{2h_k}\int_{-h_k}^{h_k} Y(s)\,ds \tag{10.33}$$

and hence the terms for which $l > 0$ represent the perturbation $y_k(t)$. Introducing (10.32) into (10.30), we are led to define new matrix elements by

$$M_{ik}^{(l)} = h_k \int_{-1}^{1} K(\overline{s}_i, \overline{s}_k + th_k) P_l(t)\,dt - \lambda P_k(0)\delta_{ik}$$

$$(i = 1 \ldots N, \quad k = 1 \ldots N, \quad l = 0 \ldots M) \tag{10.34}$$

(10.24) being thus a special case $H_{ik} = M_{ik}^{(0)}$. The improved form of (10.22) is now given by

$$\sum_{l=0}^{L}\sum_{k=1}^{N} M_{ik}^{(l)} c_k^{(l)} = U_i \quad (i = 1 \ldots N) \tag{10.35}$$

In matrix-vector notation (matrices being denoted here by sans-serif type) this may be rewritten as

$$\mathbf{c}^{(0)} = \mathsf{I}^{(0)}\left(\boldsymbol{U} - \sum_{l=1}^{L} \mathsf{M}^{(l)}\mathbf{c}^{(l)}\right) \equiv \mathsf{I}^{(0)}(\boldsymbol{U} - \boldsymbol{u}) \tag{10.36}$$

10.2 NUM. SOLUTION OF ONE-DIMEN. INTEGRAL EQUATIONS

$I^{(0)} = (M^{(0)})^{-1} = H^{-1}$ denoting the inverse of the matrix of (10.24). Equation (10.36) is a more concrete form of (10.31), as the summation term makes it clear how the perturbation u is to be computed.

In order to solve the whole system (10.36) uniquely, we need additional conditions. These arise from the requirements of continuity at the endpoints $s_0 \ldots s_N$ of the subintervals. In this context it is helpful to introduce the integral function

$$F(s) := \int_{s_0}^{s} Y(s') \, ds' \qquad (10.37)$$

From (10.33) it is obvious that its values $F_k := F(s_k)$ satisfy the relations

$$F_0 = 0 \quad , \quad F_k = F_{k-1} + \Delta F_k \qquad (10.38a)$$
$$\Delta F_k = 2h_k \overline{Y}_k \equiv 2h_k c_k^{(0)} \qquad (10.38b)$$

Hence, by means of suitable numerical differentiation procedures, we can now calculate the sets $Y_0 \ldots Y_N$ and $Y_0' \ldots Y_N'$ and even derivatives of higher order, referring to $s_0 \ldots s_N$. In order to limit the rounding errors, it is preferable to apply a differentiation technique that uses the increments ΔF_k directly instead of the accumulated values F_k. The differentiation routine must be set up with some care as the resulting error may otherwise increase with decreasing step-width. A suitable procedure will be presented in Section 13.1.

With increasing order L, the accuracy should become better, but numerical instabilities also tend to increase and the memory requirements soon become prohibitive. A reasonable compromise is $L = 4$. The values $\overline{Y}_k$, Y_k, Y_k', Y_{k-1}, Y_{k-1}' uniquely specify a polynomial of fourth order that is consistent with these data. The required relations for the Legendre coefficients with $l > 0$ are then given by

$$\sum_{l=1}^{4} c_k^{(l)} P_k(\mp 1) = [Y_{k-1} - \overline{Y}_k; \; Y_k - \overline{Y}_k]$$

$$\sum_{l=1}^{4} c_k^{(l)} \dot{P}_k(\mp 1) = [h_k Y_{k-1}' - \overline{Y}_k; \; h_k Y_k]$$

where the first term on the right-hand side refers to the coordinate $s = s_{k-1}$, $t = -1$ while the second term refers to $s = s_k$, $t = 1$; the dot denotes differentiation with respect to t. By forming suitable linear combinations, this linear system of equations can be easily split up into two uncoupled subsystems of rank 2 and then solved, the results being

$$c_k^{(1)} = \left\{ 6(Y_k - Y_{k-1}) \qquad - h_k(Y'_k + Y'_{k-1}) \right\}/10$$

$$c_k^{(2)} = \left\{ 10(Y_k + Y_{k-1} - 2\overline{Y}_k) \quad - h_k(Y'_k - Y'_{k-1}) \right\}/14$$

$$c_k^{(3)} = \left\{ -(Y_k - Y_{k-1}) \qquad + h_k(Y'_k + Y'_{k-1}) \right\}/10$$

$$c_k^{(4)} = \left\{ -3(Y_k + Y_{k-1} - 2\overline{Y}_k) + h_k(Y'_k - Y'_{k-1}) \right\}/14$$

(10.39)

10.2.4 Iterative solution

We now have enough conditions to determine all the quantities uniquely. The resulting linear system of equations could be solved directly but this would be most inefficient because the rank of the coefficient matrix is very high and the matrix is sparse, large regions containing only zero elements. By using a suitable iteration procedure, the memory requirements can be reduced drastically.

In the *initialization* step, we evaluate and store the matrix elements $M_{ik}^{(l)}$, given by (10.34). The matrix $\mathsf{M}^{(0)}$ is then inverted and afterwards its memory location is occupied by the inverse $\mathsf{I}^{(0)}$. In practice, it has been found that in many cases, essentially in applications to systems with thick electrodes or polepieces, the matrix elements $M_{ik}^{(l)}$ with $l \geq 2$ and $|i-k| > 2$ can be ignored without significant loss of accuracy. Thus in favourable cases the memory requirement is little more than $2N^2$ for these matrices. After storing the matrices, the inhomogeneity vector U is determined and vanishing start-vectors $c^{(l)}$, $(l = 1\ldots 4)$ are assumed.

In the *iteration loop*, (10.36) is first solved for $c^{(0)}$; the increments ΔF_k are then known from (10.38b). Next the sets $Y_0\ldots Y_N$ and $Y'_0\ldots Y'_N$ are determined by numerical differentiation. These are now introduced on the right-hand side of (10.39). In this way we obtain improved vectors $c^{(1)}\ldots c^{(4)}$. The norms of the differences relative to the vectors of the preceding iteration provide a convergence control. The procedure is stopped when all the difference norms have fallen below given tolerance limits; otherwise a fresh loop starts with (10.36).

This procedure usually converges after about 20 (or less) iterations. Finally we can obtain a continuously twice-differentiable solution $Y(s)$ by numerical calculation of the second order derivatives $Y''_0\ldots Y''_N$ at $s_0\ldots s_N$ and application of the quintic spline technique outlined in Section 13.1. This gives a very smooth solution. It is highly accurate whenever the

10.2 NUM. SOLUTION OF ONE-DIMEN. INTEGRAL EQUATIONS

inhomogeneity $U(s)$ varies slowly in every subinterval. This is, for instance, true in applications to electrostatic Dirichlet problems, where $U(s)$ is the locally constant electrode potential. A more general numerical technique, which is also accurate for rapidly varying inhomogeneities $U(s)$, has been devised by Ströer (1987), to whose work we refer the interested reader.

10.2.5 Evaluation of improper integrals

A careful inspection of the kernel functions to be evaluated shows that these contain logarithmic singularities; the derivatives of the kernels may have singularities of first order. For the sake of brevity, we assume the abscissa of the singularity in question to be $x = 0$.

A singular function of the type

$$f_1(x) = \ln|\varphi_1(x)| \quad \text{with} \quad \varphi_1(0) = 0 \quad , \quad \varphi_1'(0) \neq 0$$

can be reduced to the form

$$f_1(x) = \ln|x| + \ln|\varphi_1(x)/x|$$

where the second term on the right is a *regular* function, which can be expanded as a Taylor series about $x = 0$. Likewise, a function of the type

$$f_2(x) = 1/\varphi_2(x) \quad \text{with} \quad \varphi_2(0) = 0 \quad , \quad \varphi_2'(0) \neq 0$$

can be reduced to

$$f_2(x) = \frac{1}{x\varphi_2'(0)} + \frac{x\varphi_2'(0) - \varphi_2(x)}{x\varphi_2'(0)\,\varphi_2(x)}$$

where again the second term is regular, since both numerator and denominator have a common zero of second order, which cancels out. It is therefore sufficient to consider functions of the form

$$F(x) = f_1(x)\ln|x| + f_2(x)/x + f_3(x) \tag{10.40}$$

$f_1(x)$, $f_2(x)$, and $f_3(x)$ being arbitrary *regular* functions.

Kasper (1983) has developed a simple method for integration over such functions. The corresponding formula is similar to a Gauss quadrature and takes the basic form

$$\int_{-h}^{h} F(x)\,dx = h \sum_{\mu=1}^{N} w_\mu \left\{ F(p_\mu h) + F(-p_\mu h) \right\} + O(h^{2N+1}) \tag{10.41}$$

with positive abscissae p_μ and weight factors w_μ. Thanks to the symmetric arrangement of these, the contributions from the singularity in $f_2(x)/x$ and from all antisymmetric terms cancel out, and (10.41) automatically furnishes the principal value of the integral. The parameters p_μ and w_μ are obtained by numerical solution of the non-linear equations

$$\left. \begin{array}{l} \displaystyle\int_0^1 u^{2\nu}\, du = \sum_{\mu=1}^N w_\mu p_\mu^{2\nu} = \frac{1}{2\nu+1} \\[2ex] \displaystyle\int_0^1 u^{2\nu} \ln\frac{1}{u}\, du = \sum_{\mu=1}^N w_\mu p_\mu^{2\nu} \ln\frac{1}{p_\mu} = \frac{1}{(2\nu+1)^2} \end{array} \right\} \quad \nu = 0, 1 \ldots N-1$$

The results for the case $N = 4$, which are adequate in most cases, are as follows:

$$\begin{array}{ll} p_1 = 0.0399\ 4596\ 2203 & w_1 = 0.1270\ 7679\ 2574 \\ p_2 = 0.2801\ 7249\ 6204 & w_2 = 0.3267\ 4417\ 6078 \\ p_3 = 0.6361\ 2394\ 4954 & w_3 = 0.3523\ 4912\ 8452 \\ p_4 = 0.9223\ 6045\ 1138 & w_4 = 0.1938\ 3290\ 3896 \end{array}$$

It is now easy to integrate all kernel functions $K(s, s')$ and their normal derivatives over arbitrary smooth boundary contours C. It is necessary neither to approximate the contour by a polygon, nor to choose the arc-length as the variable of integration. As (10.41) makes clear, there is no need to express the integrand explicitly in the form (10.40), which would be extremely tedious. The only knowledge required is the *implicit* singular character of the integrand. The user of this method need only set up a program that supplies correctly the value of the integrand for arbitrary arguments.

In conclusion, it emerges that the BEM, at least in its one-dimensional version, is a highly attractive procedure. Practical examples demonstrating this are given in Section 10.5.

10.3 Superposition of aperture fields

Though the general method described above can be applied to any axisymmetric Dirichlet problem, this is not always the best way of obtaining the solution. Problems may arise in systems with very narrow apertures in electrodes with plane fronts, as shown in the example presented in Fig.

10.3 SUPERPOSITION OF APERTURE FIELDS

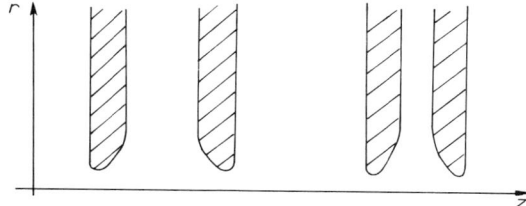

Fig. 10.10: Upper half of the axial section through an accelerator.

10.10. Here many integration subintervals are necessary in order to obtain the electric field correctly in its asymptotic domains. The computation of surface charge distributions can, however, be confined to the principal inhomogeneous domains of the field if the asymptotic field is represented correctly by appropriate superpositions of the fields of thin plane circular apertures. A combination of the BEM with such fields has been proposed by Hoch et al. (1978). Pure superposition of aperture fields had been investigated earlier by Regenstreif (1951), Lenz (1956) and Dommaschk (1965). In the following analysis, we shall first study the field of one single aperture and then superpositions of such fields.

10.3.1 Electric field of a single aperture

We consider now a thin plane circular aperture with radius $r' = R$. Without loss of generality, we can choose the coordinate system in such a manner that the aperture plane is given by $z' = 0$. It is appropriate to introduce oblate spheroidal coordinates $(u, v, \varphi,)$, φ being the usual azimuth and u, v defined by the transformation

$$z = Ruv \quad , \quad r = R\Big\{(1+u^2)(1-v^2)\Big\}^{\frac{1}{2}}$$
$$(-\infty < u < \infty \quad , \quad 0 \le v \le 1) \tag{10.42}$$

These are shown in Fig. 10.11. The surfaces $u = $ const are confocal oblate spheroids, the surfaces $v = $ const confocal orthogonal hyperboloids. Among the latter, the optic axis $(v = 1)$ and the surface of the aperture itself $(v = 0)$ are degenerate special cases.

In these new coordinates, Laplace's equation takes the form

$$\frac{\partial}{\partial u}\left\{(1+u^2)\frac{\partial \Phi}{\partial u}\right\} + \frac{\partial}{\partial v}\left\{(1-v^2)\frac{\partial \Phi}{\partial v}\right\} = 0$$

This can be solved by separation of variables,

$$\Phi(u,v) = F(u)P(v)$$

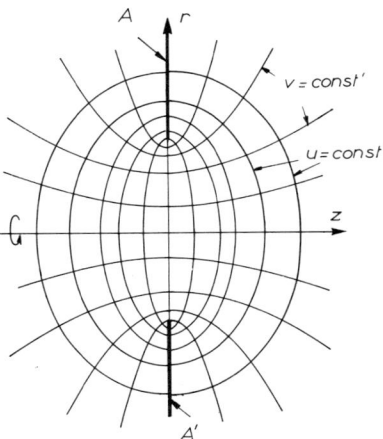

Fig. 10.11: Coordinate lines $u = \text{const}$ and $v = \text{const}'$ in a system of oblate spheroidal coordinates (u, v, φ). The azimuth φ is constant in this axial section. A and A' denote the closed parts of the aperture plane; these are singularities of the coordinate system.

Writing the separation constant in the familiar form $l(l+1)$, we obtain

$$\frac{d}{dv}\{(1-v^2)P'(v)\} = -l(l+1)P(v) \qquad (10.43)$$

$$\frac{d}{du}\{(1+u^2)F'(u)\} = l(l+1)F(u) \qquad (10.44)$$

Equation (10.43) is Legendre's differential equation, its *regular* solutions being the well-known Legendre polynomials $P_l(v)$ for *integral values* of the subscript l. Equation (10.44) reduces to (10.43) if we write $u' = iu$ and is hence solved by general Legendre functions with imaginary argument. For physical reasons the resulting solution for Φ must correspond asymptotically to a homogeneous field, which means that $\Phi \sim z = Ruv$ for $|u| \gg 1$. With this constraint, the general solution is

$$\Phi(u,v) = Auv + Bv(1 + u \arctan u) + C$$

The coefficients A, B and C are uniquely specified by the conditions $\Phi(u,0) = \Phi_0$, $\partial\Phi/\partial z = -E_l$ for $z \to -\infty$ and $\partial\Phi/\partial z = -E_r$ for $z \to +\infty$, the result being

$$\Phi(u,v) = \Phi_0 + Rv\left\{-\frac{E_l + E_r}{2}u + \frac{E_l - E_r}{\pi}(1 + u \arctan u)\right\} \qquad (10.45)$$

10.3 SUPERPOSITION OF APERTURE FIELDS

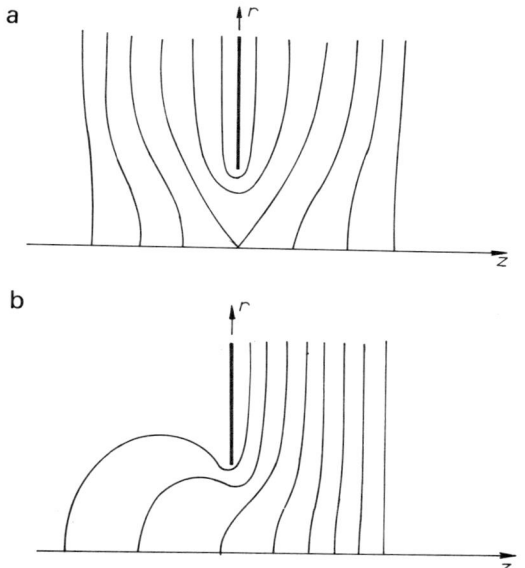

Fig. 10.12: Equipotentials in the upper half of an axial section through the field of a thin charged aperture.
(a) Symmetric field, $E_1 = -E_2$; (b) $E_1 = 0$, the field vanishes asymptotically as $z \to -\infty$.

The constants Φ_0, E_l and E_r have the physical meaning of the aperture potential and the asymptotic field strengths, respectively. Examples of such solutions, differing in the choice of the constants, are shown in Figs 10.12a,b.

In order to represent the potential and the asymptotic field strength in cylindrical coordinates, we need the inverse transformation corresponding to (10.42). This can be expressed in terms of the distances d_1 and d_2, defined by (10.1) and shown in Fig. 10.1; in this context, the singular ring is the edge of the aperture, $z' = 0$, $r' = R$. We then have

$$u^2 = \frac{(d_1 + d_2)^2}{4R^2} - 1 \quad , \quad v^2 = 1 - \frac{(d_2 - d_1)^2}{4R^2} \tag{10.46}$$

For the later computations, it is more convenient to express the transform in *relative* coordinates $\bar{z} = z/R$, $\bar{r} = r/R$; we find

$$\xi := \bar{z}^2 + \bar{r}^2 - 1 \equiv u^2 - v^2 \tag{10.47a}$$

$$\eta := \sqrt{\xi^2 + 4\bar{z}^2} \equiv u^2 + v^2 \tag{10.47a}$$

$$v = \begin{cases} \sqrt{(|\xi|+\eta)/2} & \text{for } \xi < 0 \\ |\bar{z}|/\sqrt{(\xi+\eta)/2} & \text{for } \xi \geq 0 \end{cases} \quad (10.47c)$$

$$u = \bar{z}/v \quad (10.47d)$$

The matrix of the partial derivatives can then be computed very easily:

$$w := (R\eta)^{-1} \equiv \left\{R(u^2+v^2)\right\}^{-1} \quad (10.48a)$$

$$\frac{\partial u}{\partial z} = wv(1+u^2) \quad , \quad \frac{\partial u}{\partial r} = wu\bar{r} \quad (10.48b)$$

$$\frac{\partial v}{\partial z} = wu(1-v^2) \quad , \quad \frac{\partial v}{\partial r} = -wv\bar{r} \quad (10.48c)$$

By means of these relations it is easy to compute $\Phi(z,r)$ and its derivatives $\partial \Phi/\partial z$, $\partial \Phi/\partial r$ or even derivatives of higher orders. The corresponding expressions are not given here.

10.3.2 Superposition procedure

We now consider configurations of coaxial thin apertures and their fields. Any such configuration of N apertures is uniquely specified by the aperture positions z_i, the radii R_i and the physical surface potentials U_i, $i = 1, 2, \ldots N$, and also by the asymptotic field strengths E_l for $z \to -\infty$ and E_r for $z \to \infty$. The linear superposition of the corresponding single-aperture fields can be represented in many different but equivalent ways. Here we choose a superposition of N *symmetric* single-aperture potentials and the potential Φ_H of one homogeneous field:

$$\Phi_A(z,r) = \sum_{i=1}^{N} C_i v_i (1 + u_i \arctan u_i) + A + Bz \quad (10.49)$$

the quantities u_i and v_i being spheroidal coordinates referring to the aperture with subscript i. This representation is the most simple. The coefficients $C_1 \ldots C_N$ are to be determined from the asymptotic field for large values of r:

$$C_i v_i (1 + u_i \arctan u_i) \to \frac{\pi C_i}{2R_i} |z - z_i| \quad \text{for} \quad r \gg R_i \quad (10.50)$$

Since the asymptotic field strength must be

$$F_i := \frac{U_i - U_{i+1}}{z_{i+1} - z_i} \quad \text{for} \quad z_i < z < z_{i+1} \quad , \quad i = 1 \ldots N-1 \quad (10.51a)$$

$$F_0 := E_l \quad \text{for} \quad z \to -\infty \quad , \quad F_N := E_r \quad \text{for} \quad z \to \infty \quad (10.51b)$$

at large off-axis distances r, we find

$$C_k = (F_{k-1} - F_k) R_k/\pi \quad , \quad k = 1 \ldots N \tag{10.52}$$

The total contribution of the homogeneous field can be represented as

$$A + Bz = \frac{1}{2} \Big\{ U_1 - E_l(z - z_1) + U_N - E_r(z - z_N) \Big\} \tag{10.53}$$

The algorithm corresponding to these formulae is very easy to program.

10.3.3 Combination with the BEM

In earlier investigations, prior to the publication of Hoch *et al.* (1978), mere superposition of aperture fields was used to approximate the electric field in devices with plane electrodes containing circular bores. This is entirely reasonable for very thin electrodes separated by large distances, $|z_{i+1} - z_i| \gg R_i + R_{i+1}$. Generally speaking, this method fails when applied to devices with thick electrodes although Regenstreif modelled the latter by two thin electrodes at the same potential. A typical example is shown in Fig. 10.13a, which represents equipotentials in an axial section through an electron gun with a thick wehnelt electrode.

In such a situation we suggest the following procedure. As a first step the appropriate aperture parameters are determined, so that the potentials Φ_A, given by (10.49), can be computed for any point in space. The chosen apertures may be located in the front planes of the electrodes, or inside the electrodes as suggested by Hoch *et al.* (1978). In the latter case the appropriate aperture potentials are to be obtained by linear extrapolation.

The contribution Φ_A alone, of course, will not give the full result. We therefore consider additional surface charge distributions. These can be confined to the vicinity of the bores, where the surface values of Φ_A differ strongly from the prescribed boundary values. Thus, as a second step, we introduce a reasonable discretization of those parts B of the boundaries where such surface charges are to be applied. We can then solve numerically the integral equation

$$\int_B \frac{1}{\epsilon_0} G_0(z, r; z', r') \sigma(s') \, ds' = U(s) - \Phi_A(z, r) \tag{10.54}$$

where $z = z(s)$, $z' = z(s'), \ldots$, are parametric representations of surface points and $U(s)$ is the given boundary value function.

After solving (10.54), the potential $\Phi(z, r)$ and the field strength $\boldsymbol{E} = -\nabla \Phi$ at any point of reference can be computed by numerically superposing Φ_A and $-\nabla \Phi_A$ and the corresponding terms arising from the surface

10. THE BOUNDARY-ELEMENT METHOD

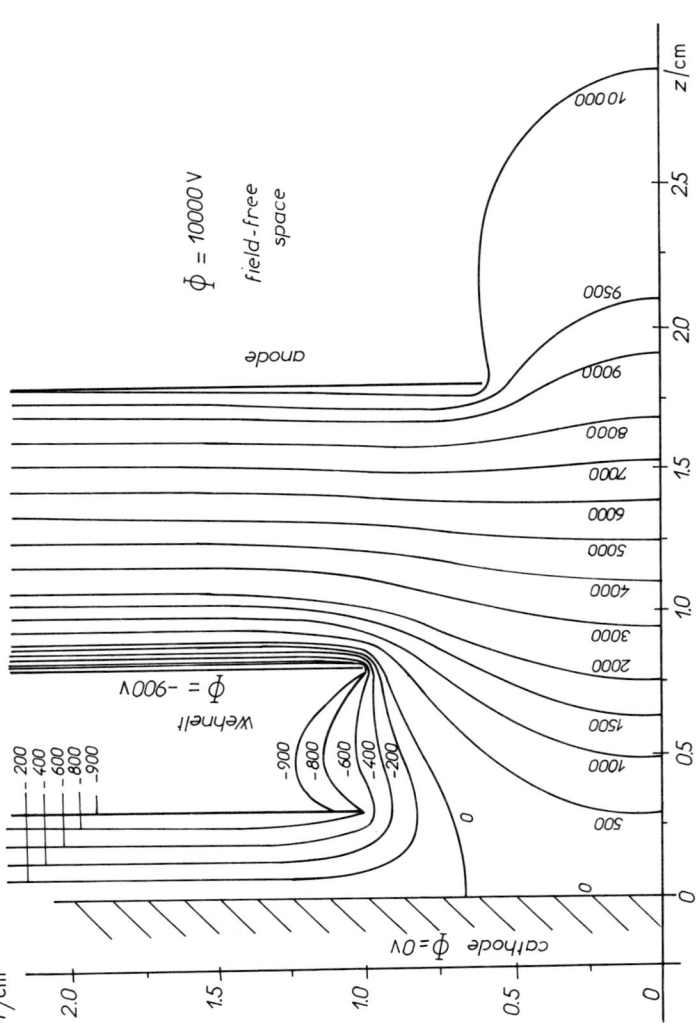

Fig. 10.13: Equipotentials $\Phi(z,r) = $ const in the upper half of an axial section through an electron gun with a plane cathode, a thick wehnelt and a thin anode.
(a) Attempt to solve the Dirichlet problem by mere superposition of aperture potentials. The equipotential $\Phi(z,r) = -900$V does not fit the cylindrical bore of radius 1 cm of the wehnelt at all well. The examples makes it very clear that the potentials of additional surface charge distributions are indispensable.

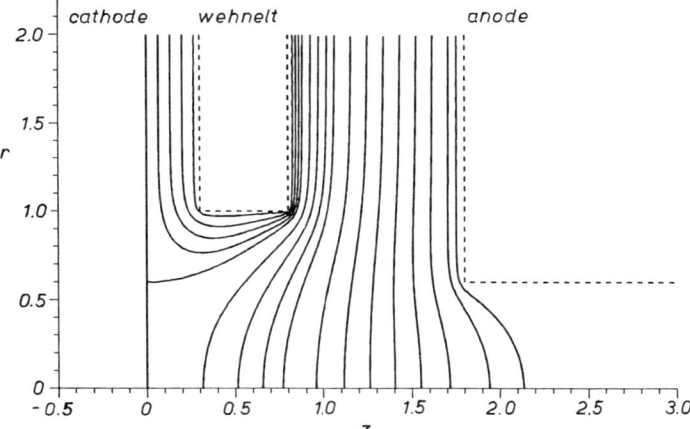

Fig. 10.13b: Equipotentials $\Phi(z,r) = $ const in the upper half of an axial section through an electron gun with a plane cathode, a thick wehnelt and a thin anode. The improvement achieved by the introduction of a suitable surface charge distribution.

Coulomb integrals. The construction of this field is such that it satisfies the boundary conditions with a high degree of accuracy. The results for the chosen example are presented in Fig. 10.13b.

The consideration of aperture fields is only one of many possible ways of extending the BEM. Another possibility is the superposition of the fields produced by axial charge distributions. This can be helpful for field calculation in systems with pointed cathodes and will therefore be dealt with in Chapter 45.

10.4 Three-dimensional Dirichlet problems

There are three-dimensional Dirichlet problems that cannot be reduced to a sequence of two-dimensional ones. Such problems arise in most situations with non-rotationally symmetric boundaries. A typical example is the field in the vicinity of the hairpin of an electron gun, see Figs 10.19 and 10.20a,b. This case has been investigated by Eupper (1985) in order to estimate the influence of the electric field perturbation on the astigmatism in the electron beam.

The extreme complexity of general three-dimensional boundary-value problems renders their concrete numerical solution much more complicated than that of two-dimensional problems. Here we can deal only with one family of problems, the three-dimensional Dirichlet problem for Laplace's

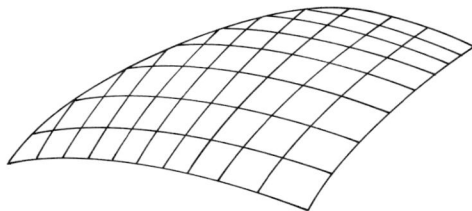

Fig. 10.14a: Dissection of a curved surface into surface elements by means of two families of surface curves. In at least one of these families, neighbouring curves are locally parallel (provided we consider small enough regions).

equation $\nabla^2 V(\mathbf{r}) = 0$.

The general method of solving such problems may appear to be straightforward. The boundary $\partial\Gamma$ is dissected into a set of N sufficiently small finite surface elements Δ_k, the centroids of which are located at $\mathbf{r}_k$ ($k = 1 \ldots N$). In each such element, the surface charge density σ_k is assumed to be constant. Then (9.14) is approximated by a linear system of equations:

$$V(\mathbf{r}_j) = \frac{1}{4\pi\epsilon_0} \sum_{k=1}^{N} \sigma_k \int_{\Delta_k} \frac{da'}{|\mathbf{r}_j - \mathbf{r}'|} \quad , \quad j = 1 \ldots N \qquad (10.55)$$

This system can be solved for $\sigma_1 \ldots \sigma_N$, after which the surface Coulomb integral can be evaluated for any position $\mathbf{r}$ of reference.

This is, in fact, the usual way of solving such problems. For instance, Munro and Chu (1982) have applied this method to an electrostatic deflection unit. Such devices are still very simple, since all the surface elements Δ_k can be chosen to be rectangles. In more general cases, when triangular surface elements cannot be avoided, the method may become very tedious. It is certainly possible to calculate the potential of any triangular surface element with a uniform or even a linear charge distribution in a completely analytical manner (Durand, 1966; Eupper, 1985), but this is very laborious. Although the corresponding expression for $V(\mathbf{r})$ can be built up entirely from elementary functions, its evaluation is very slow. Since the number of surface elements must be large in order to achieve good accuracy, the whole procedure is extremely inefficient. The problem of saving unnecessary operations is far more important than in the case of two-dimensional field calculations.

With a view to improving the efficiency, Eupper (1982, 1985) has made an unconventional proposal. In order to avoid the evaluation of improper integrals, the charges are assumed to be located on surfaces $\partial\Gamma'$ chosen

10.4 THREE-DIMENSIONAL DIRICHLET PROBLEMS

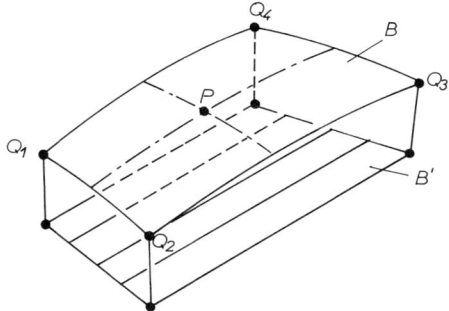

Fig. 10.14b: Perspective view of a single surface element and the associated trapezoid. The lines in the latter indicate the charged bars from which the surface charge element is built up. P and $Q_1, \ldots Q_4$ are control points for the potential.

close to the corresponding electrode surfaces $\partial \Gamma$, but in the *interior* of the corresponding electrode as is shown in Figs 10.14a and 10.14b. These surfaces are to be chosen reasonably, in the sense that the distance between parallel parts of $\partial \Gamma'$ must be larger than the distance to the corresponding material surface $\partial \Gamma$. Difficulties thus arise with this method for very thin electrodes and in the vicinity of sharp edges.

The interior surfaces $\partial \Gamma'$ are now dissected into general trapezoidal elements (with parallelograms and triangles as special cases). Each such element Δ_k is associated with one control point r_k located on the true electrode surface $\partial \Gamma$. The conditions that the potentials $V(r_1), V(r_2) \ldots V(r_N)$ assume their prescribed values are now set up; this is straightforward. The new idea here is that one part of the necessary integrations, that in the longitudinal direction of each trapezium, can be carried out analytically, resulting in the potential of a charged bar; the corresponding expression will be given below. The remaining integration in the transverse direction is then carried out numerically. This procedure is comparatively simple and yet much faster than entirely analytic integration. The discretization by trapezia is so flexible that even complicated problems like those shown in Figs 10.19 and 10.20 can be solved satisfactorily.

Let us now consider a charged bar of length $2a$, the direction of which is indicated by a unit vector t. Let the charge per length unit be $q(s)$ for $-a \leq s \leq a$. The origin of the coordinate system may be chosen to coincide with the centroid of the bar. The potential is then given by

$$V(r) = \frac{1}{4\pi\epsilon_0} \int_{-a}^{a} \frac{q(s)ds}{|r - ts|} \qquad (10.56)$$

152   10. THE BOUNDARY-ELEMENT METHOD

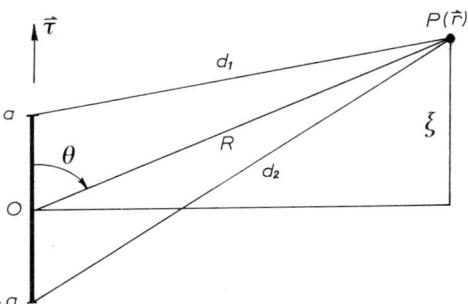

Fig. 10.15: Notation employed to characterize the position of an arbitrary point P relative to a bar of length $2a$.

There are two ways of carrying out the integration, direct analytic integration and integration after an appropriate series expansion. In the most important case of a linear charge distribution $q(s) = q_0 + q_1 s$, direct integration results in

$$4\pi\epsilon_0 V(\boldsymbol{r}) = (q_0 + q_1 \xi)\ln\left(\frac{d_1 + d_2 + 2a}{d_1 + d_2 - 2a}\right) + q_1(d_1 - d_2) \qquad (10.57)$$

with

$$d_{1,2} = |\boldsymbol{r} \mp a\boldsymbol{t}| \quad , \quad \xi = \boldsymbol{t}\cdot\boldsymbol{r} \equiv (d_2^2 - d_1^2)/4a \qquad (10.57a)$$

The quantities d_1 and d_2 are the distances of the point $\boldsymbol{r}$ from the endpoints of the bar, as shown in Fig. 10.15.

Alternatively, we may use asymptotic multipole series expansion. After introducing spherical coordinates R, ϑ, defined by

$$R := |\boldsymbol{r}| \quad , \quad \mu := \cos\vartheta := \xi/R \qquad (10.58)$$

we first write down the series expansion

$$|\boldsymbol{r} - \boldsymbol{t}s|^{-1} = \left(R^2 - 2R\mu s + s^2\right)^{-\frac{1}{2}} = R^{-1}\sum_{l=0}^{\infty}\left(\frac{s}{R}\right)^l P_l(\mu) \quad , \quad s < R$$

Introducing this into (10.56) and defining moments of the charge distribution by

$$M_l := a^{-(l+1)}\int_{-a}^{a} q(s)s^l\,ds \qquad (10.59)$$

10.4 THREE-DIMENSIONAL DIRICHLET PROBLEMS

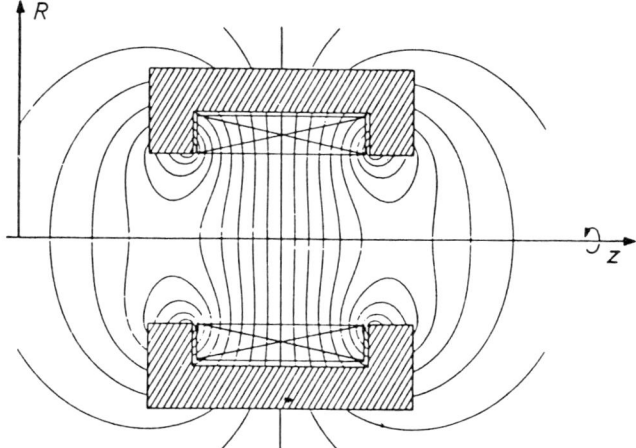

Fig. 10.16: Equipotentials of the reduced magnetic scalar potential $\chi_M(\mathbf{r})$ of a magnetic lens with a very wide gap. Courtesy of W. Scherle (1983).

we obtain

$$4\pi\epsilon_0 V(\mathbf{r}) = \sum_{l=0}^{\infty} \left(\frac{a}{R}\right)^{l+1} M_l P_l(\mu) \tag{10.60}$$

where the $P_l(\mu)$ are Legendre polynomials. Using the relation $\mu P_l'(\mu) + (l+1)P_l(\mu) = P_{l+1}'(\mu)$, the gradient of this potential can be written

$$4\pi\epsilon_0 \nabla V = \frac{1}{a}\sum_{l=0}^{\infty} \left(\frac{a}{R}\right)^{l+2} M_l \left\{ t P_l'(\mu) - \frac{\mathbf{r}}{R} P_{l+1}'(\mu) \right\} \tag{10.61}$$

In the case of a linear charge distribution the moments M_l are given by

$$M_l = \begin{cases} q_0/(l+1/2), & l \text{ even} \\ q_1 a/(1+l/2), & l \text{ odd} \end{cases}$$

The series expansions (10.60) and (10.61) have a very simple structure and are hence attractive if they can be terminated after a very few terms. This is the case when $R/a \gtrsim 5$. In conclusion, the calculation of such a potential field can be made fast enough to be useful as a basic routine in the numerical solution of three-dimensional Dirichlet problems.

Practical tests have shown that it is quite sufficient to assume constant surface charge density in each trapezoidal element if the field is needed only in domains far from all boundaries, as is the case in deflection units. When

154 10. THE BOUNDARY-ELEMENT METHOD

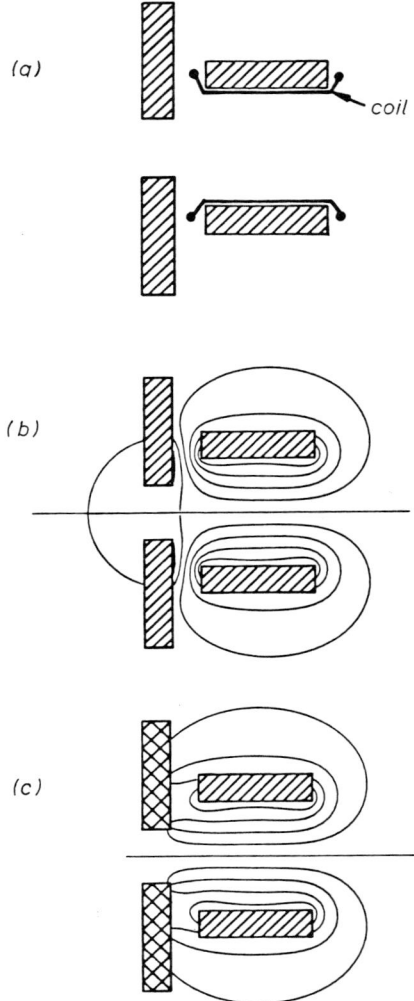

Fig. 10.17: Equipotentials of the reduced magnetic scalar potential $\chi_m(r)$ in an axial section through a system of saddle coils, a ferromagnetic yoke and a pierced shielding plate. (a) The position of the saddle coils (b) Ferromagnetic shielding plate (c) Superconducting shielding plate. (Note: the coils and their field $H_0(r)$ are omitted from (b) and (c) to prevent confusion.) Courtesy of W. Scherle (1983).

10.4 THREE-DIMENSIONAL DIRICHLET PROBLEMS

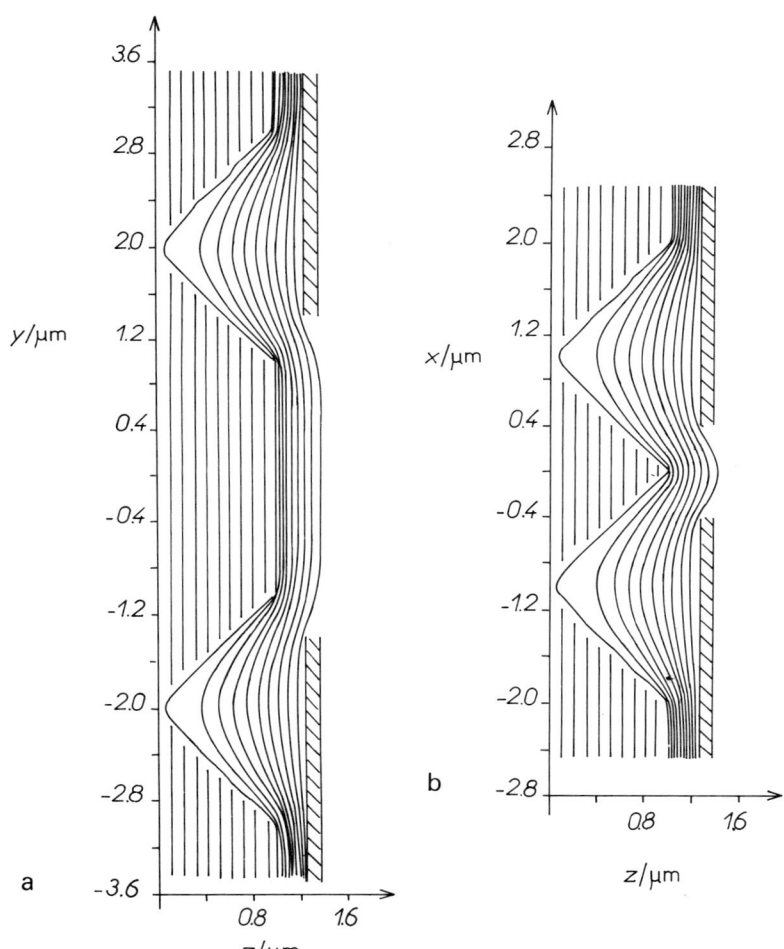

Fig. 10.18a,b: (a)–(c) Equipotentials in different sections through a field electron emission source consisting of a cathode with a hipped roof and an anode with a rectangular bore. Courtesy of M. Eupper (1982).

10. THE BOUNDARY-ELEMENT METHOD

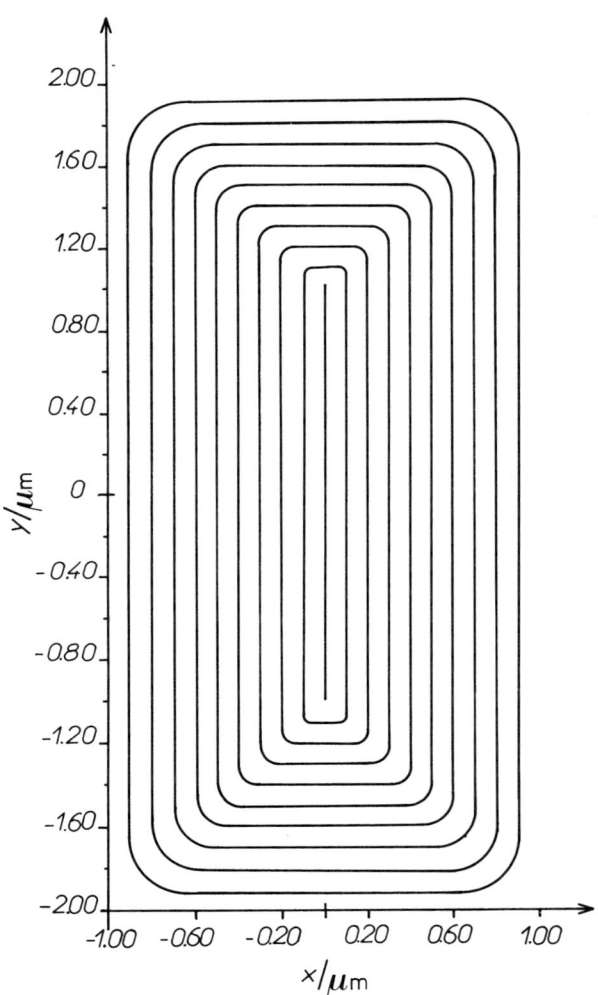

Fig. 10.18c

10.4 THREE-DIMENSIONAL DIRICHLET PROBLEMS

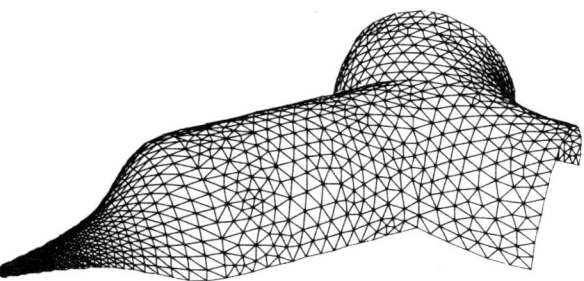

Fig. 10.19: Perspective view of a pointed cathode welded on a hairpin support; Only parts of the surface discretization are shown for reasons of clarity. Courtesy of M. Eupper (1983).

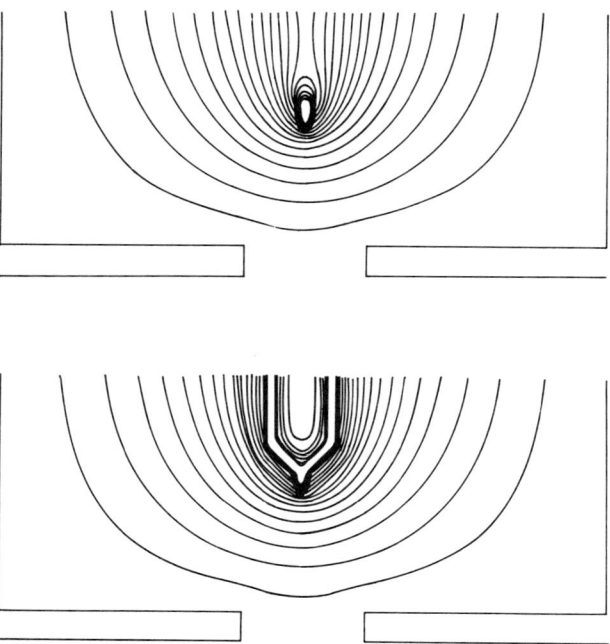

Fig. 10.20: (a), (b) Equipotentials of the electrostatic potential in two perpendicular sections through the system shown in Fig. 10.19. The position of the anode is indicated. The total width is 10 mm and the anode–cathode distance is 2 mm.

the method is applied to asymmetric electron guns, higher accuracy—especially of the field strength—can be achieved by choosing the charge distribution in each surface element as a bilinear function of the corresponding coordinates, so that the whole distribution becomes continuous (in the vicinity of the cathode). The structure of the linear system of equations for the surface charges does, of course, become more complicated.

Alternative techniques for the solution of three-dimensional Dirichlet problems are the three-dimensional versions of the finite-difference and finite-element methods. These are generally so dependent on the problem in question that they will not be treated in any detail here. For an example, see the detailed description of the use of the three-dimensional FDM to calculate electrostatic focusing fields in cathode ray tubes given by Franzen (1984).

10.5 Examples of applications of the boundary-element method

The rotationally symmetric solution of (9.11), corresponding to the order $m = 0$ in (9.27), is useful for calculating the magnetic field of a round lens with a very wide gap (Scherle, 1983) as is shown in Fig. 10.16. For clarity, only the equipotentials of the reduced scalar potential $\chi_M(\boldsymbol{r})$ are presented. A typical example of field calculation in deflection systems (Scherle, 1983) is shown in Figs 10.17a–c. Again, only the equipotentials of $\chi_M(\boldsymbol{r})$ in one section through the field are shown. The case in which there are three different domains of solution can be analysed by iterative solution of two coupled integral equations of the form (9.11) (Scherle, 1983).

The three-dimensional BEM becomes a very powerful method when problems with very large differences in the geometrical dimensions are to be solved. Some typical examples are field-electron emission sources with a hipped-roof cathode (Figs 10.18a–c; Eupper, 1982) and with a rotationally symmetric tip welded on a hairpin-shaped support (Figs 10.19 and 10.20a,b). More details of field calculations in electron sources are given in Chapter 45.

11
The Finite-Difference Method (FDM)

The finite-difference method (FDM), usually combined with an iterative technique to solve the corresponding linear system of equations, is a standard procedure for field computation. It was introduced by H. Liebmann as early as 1918 and is thus often called 'Liebmann's method'. The associated mathematical theory is exhaustively studied in the literature, for instance by Varga (1962), Forsythe and Wasow (1960), Ames (1969) and Jacobs (1977). Survey articles on the application of the FDM to electron optical problems have been published by Weber (1967), Bonjour (1980) and Kasper (1982). In recent years, however, more powerful techniques have been developed for solving boundary-value problems, and the original form of the FDM has lost much of its earlier importance; we shall therefore discuss it only briefly.

11.1 The choice of grid

The basic idea of the FDM is to cover the entire domain of solution of a boundary-value problem by a finite rectangular grid. In order to obtain the greatest possible simplification, it is usual to specialize to square-shaped grids. We have to distinguish between regular internal points (A), irregular internal points (B), regular axial points (C, D), irregular axial points (E) and boundary points (F), as shown in Fig. 11.1. In the practical organization of a FDM program, each point has to be assigned to one of the classes and handled accordingly, which complicates the actual application of the FDM. Since the boundary of the domain is in general curved, these complications cannot be avoided.

In the subsequent presentation we shall consider a two-dimensional Dirichlet problem associated with a general elliptic differential equation:

$$A(u,v)V_{|uu} + B(u,v)V_{|vv} + a(u,v)V_{|u} + b(u,v)V_{|v}$$
$$= C(u,v)V(u,v) + G(u,v) \quad (11.1)$$

(For an explanation of the notation, see Section 2.4.) The coordinate system (u, v) may be curvilinear, but is almost invariably chosen to be *orthogonal*. A term with $V_{|uv}$ cannot then appear and is hence omitted from (11.1).

160   11. THE FINITE-DIFFERENCE METHOD (FDM)

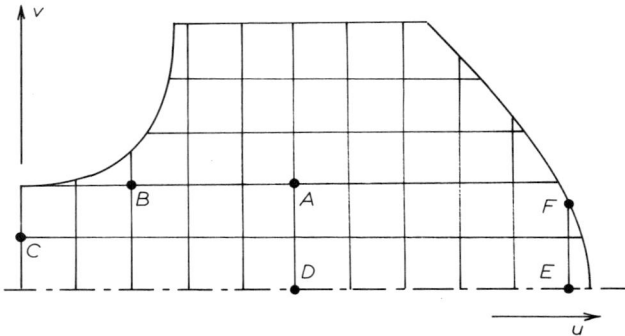

Fig. 11.1: Grid with square meshes and distinction between different types of node.

There are essentially two different ways of deriving discrete formulae, the Taylor series method and the integral method. Both are in very widespread use and equivalent in the sense that they differ only in higher order terms of the discretization errors.

11.2 The Taylor series method

Since irregular configurations are to be treated in the vicinity of the boundary at least, we now consider a general five-point configuration, as shown in Fig. 11.2a. In order to establish a discrete form of (11.1), we expand $V(u,v)$ as a Taylor series with respect to the coordinate differences $u - u_0$ and $v - v_0$. In applications to five-point configurations, we have to truncate this after the second order terms. This implies that, along the lines $u = u_0 = $ const and $v = v_0 = $ const, we can approximate the potential by Lagrange interpolation parabolae. For instance, the parabola that fits the potential at the points P_3, P_0 and P_1 (see Fig. 11.2a) is given by

$$V(u, v_0) = V_0 + (u - u_0)V_{|u} + \frac{1}{2}(u - u_0)^2 V_{|uu}$$

The derivatives $V_{|u}$ and $V_{|uu}$ refer to the central point P_0 and are given by

$$V_{|u} = \frac{1}{h_1 h_3 (h_1 + h_3)} \left\{ h_3^2 V_1 - h_1^2 V_3 + (h_1^2 - h_3^2)V_0 \right\}$$

$$V_{|uu} = \frac{2}{h_1 h_3 (h_1 + h_3)} \left\{ h_3 V_1 + h_1 V_3 - (h_1 + h_3)V_0 \right\}$$

Similar expressions are obtained for the derivatives $V_{|v}$ and $V_{|vv}$. Introducing all these into (11.1), we obtain a finite-difference approximation for

11.2 THE TAYLOR SERIES METHOD

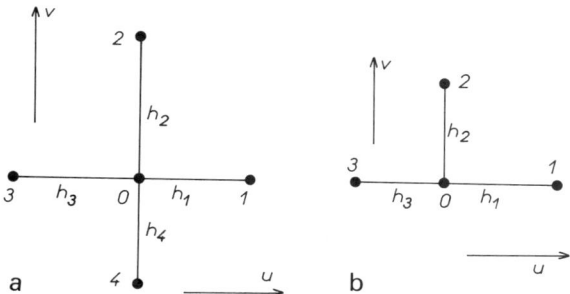

Fig. 11.2: (a) Five-point configuration for internal nodes 0. (b) Four-point configuration to be used when the node 0 lies on the axis.

the latter. This is to be solved for the value V_0 at the grid point P_0 in question. The resulting linear relation has the general form

$$V_0 = \beta_0 + \beta_1 V_1 + \beta_2 V_2 + \beta_3 V_3 + \beta_4 V_4 \qquad (11.2a)$$

the coefficients $\beta_0 \ldots \beta_4$ depending on the position of the grid point in question and being given by

$$\beta_0 = G_0/N$$
$$\beta_1 = \frac{2A_0 + a_0 h_3}{h_1(h_1 + h_3)N} \quad , \quad \beta_2 = \frac{2B_0 + b_0 h_4}{h_2(h_2 + h_4)N}$$
$$\beta_3 = \frac{2A_0 - a_0 h_1}{h_3(h_1 + h_3)N} \quad , \quad \beta_4 = \frac{2B_0 - b_0 h_2}{h_4(h_2 + h_4)N} \qquad (11.2b)$$
$$N = C_0 + \frac{2A_0 + a_0(h_3 - h_1)}{h_1 h_3} + \frac{2B_0 + b_0(h_4 - h_2)}{h_2 h_4}$$

A relation of this kind holds for all regular and irregular internal grid points. On the axis of symmetry, however, only four-point configurations can be evaluated, as is shown in Fig. 11.2b. In such a case the symmetry condition $V(-u, v) = V(u, v)$ leads to

$$V_{|v} = 0 \quad , \quad \lim_{v \to 0}\left[\frac{V_{|v}(u, v)}{v}\right] = V_{|vv}(u, 0)$$

The second-order derivative is then given by

$$V_{|vv}(u, 0) = 2(V_2 - V_0)h_2^{-2}$$

Introducing these approximations into (11.1) we obtain the four-point formula. It is necessary to assume that $b(u, v) = \tilde{b}(u, v)/v$, where $\tilde{b}(u, v)$ is an even function with respect to v and may also vanish. The coefficients are then given by

$$\beta_4 = 0 \quad , \quad \beta_2 = 2(B_0 + \tilde{b}_0)/Nh_2^2$$
$$N = C_0 + \frac{2A_0 + a_0(h_3 - h_1)}{h_1 h_3} + \frac{2(B_0 + \tilde{b}_0)}{h_2^2} \quad (11.2c)$$

while β_0, β_1 and β_3 remain the same as in (11.2b), apart from the different normalization factor N.

As a comparatively simple example, we shall now consider the differential equation

$$V_{|zz} + V_{|rr} + \frac{\alpha}{r} V_{|r} = -g(z, r) \quad (11.3)$$

which includes (7.10) as a special case with $\alpha = 2m+1$. We limit the discussion to regular grid points, for which the finite-difference approximations can be given easily in explicit notation. With

$$V_{i,k} := V(ih, kh) \quad (i, k \text{ integers}) \quad (11.4)$$

and a similar notation for $g(z, r)$ we obtain

$$V_{i,k} = \frac{1}{4} \left(V_{i+1,k} + V_{i-1,k} + V_{i,k+1} + V_{i,k-1} + h^2 g_{i,k} \right)$$
$$+ \frac{\alpha}{8k} \left(V_{i+1,k} - V_{i-1,k} \right) \quad , \quad k > 1 \quad (11.5a)$$

for internal mesh points and

$$V_{i,0} = \frac{1}{2(\alpha + 2)} \left\{ V_{i+1,0} + V_{i-1,0} + 2(\alpha + 1)V_{i,1} + h^2 g_{i,0} \right\} \quad (11.5b)$$

for axial mesh points (if $\alpha \neq -2$). The resulting discretization error is of fourth order in the mesh-length h. The set of relations (11.5a,b) forms a linear system of equations specified by two subscripts i and k.

11.3 The integration method

In the subsequent presentation, we assume that the differential equation to be solved is self-adjoint:

$$\frac{\partial}{\partial u}\left(PV_{|u}\right) + \frac{\partial}{\partial v}\left(PV_{|v}\right) = QV + S \quad (11.6)$$

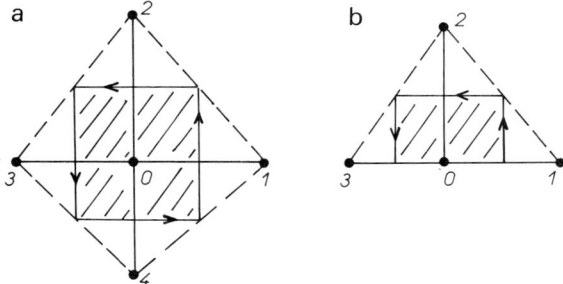

Fig. 11.3: Rectangular loops and areas of integration (a) for internal nodes and (b) for axial nodes. In the latter case, both the loop and the area can be doubled by exploiting the mirror symmetry.

the coefficients P, Q and S being regular functions of u and v. Applying Gauss's integral theorem to any domain R and its closed boundary C we obtain

$$\oint_C P \frac{\partial V}{\partial n} ds = \iint_R (QV + S) \, du \, dv \qquad (11.7)$$

$\partial V/\partial n$ denoting the normal derivative of V in the outward direction. This integral relation is exact. Its practical evaluation, however, requires several simplifying assumptions. For instance, when applying (11.7) to the configuration shown in Fig. 11.3a, we can make the approximation $\partial V/\partial n = (V_i - V_0)/h_i$ ($i = 1 \ldots 4$) on the corresponding side s_i of the rectangular contour. Furthermore we make the simplification $V \approx V_0$ under the double integral; we then obtain the finite-difference equation

$$\sum_{i=1}^{4} \frac{V_i - V_0}{h_i} \int_{s_i} P \, ds = \iint_R (QV_0 + S) \, du \, dv \qquad (11.8a)$$

Provided that the integrands are sufficiently simple analytic functions, it is possible to perform the remaining integrations analytically, but this brings no essential gain in accuracy, since the approximation made for $\partial V/\partial n$ is then too inaccurate. Thus, in order to obtain a practical form of the discretization, we assume $Q \to Q_0 = Q(u_0, v_0)$, $S \to S_0 = S(u_0, v_0)$ on the right-hand side of (11.8a). On the left-hand side, we assume that the integrand P is piecewise constant, for instance $P(u,v) \to P(u_0 + h_1/2, v_0) =: \overline{P}_1$ on the right-hand side of the rectangular integration path shown in Fig. 11.3a, the other parts of this path being treated analogously. We then find immediately

$$(V_1 - V_0)\frac{h_2 + h_4}{2h_1}\overline{P}_1 + (V_2 - V_0)\frac{h_1 + h_3}{2h_2}\overline{P}_2$$
$$+ (V_3 - V_0)\frac{h_2 + h_4}{2h_3}\overline{P}_3 + (V_4 - V_0)\frac{h_1 + h_3}{2h_4}\overline{P}_4 \quad (11.8b)$$
$$= \frac{1}{4}(V_0 + Q_0 S_0)(h_1 + h_3)(h_2 + h_4)$$

Solving this for V_0, we obtain a linear relation which differs from (11.2) only in the values of the coefficients $\beta_0 \ldots \beta_4$. The two discretizations are equivalent in the sense that they differ only in discretization errors of third or fourth order in the mesh-length. These error terms are to be neglected in any case. For axial nodes O, some special considerations are necessary, which are not given here. For the most important special case mentioned below, the reader will find them in Janse (1971) and Kasper (1976).

The above considerations can be applied to (11.3), since this differential equation can be rewritten as

$$\frac{\partial}{\partial z}\left(r^\alpha V_{|z}\right) + \frac{\partial}{\partial r}\left(r^\alpha V_{|r}\right) = -r^\alpha g(z, r) \quad (11.9)$$

so that in (11.6) we have $P = r^\alpha$, $Q \equiv 0$, $S = -r^\alpha g$. The corresponding discretization formulae have been published by Janse (1971) and Kasper (1976, 1982) and will not be repeated here. The discretization differs from (11.5a) essentially in the fact that here the integrations in (11.8a) are carried out analytically and all the coefficients remain strictly positive, whereas in (11.5a), the coefficient of $V_{i,k-1}$ becomes negative for $\alpha > 2k$. Positive coefficients mean increased stability of the entire system of equations when these are solved by iterative techniques (see Section 11.5), but the final accuracy of the solution obtained is not better than that given by (11.5a). It is interesting to note that the discretization formula obtained by integration over the configuration shown in Fig. 11.3b is identical with (11.5b).

The integral method may be further generalized to discretization in general triangular grids. The corresponding algorithms have been derived by Colonias (1974) and Winslow (1967). These will not be treated here, since they seem to be less favourable than the finite-element method presented in Chapter 12. Some other refinements of the FDM are possible; see, for example, Lenz (1973) and Kasper (1982) for further details.

11.4 Nine-point formulae

The accuracy of the FDM can be improved considerably by the use of nine-point formulae (Durand, 1966; Kasper, 1976, 1984a,b). These are advantageous in the case of a regular grid; the solution of the problem that arises for irregular grid points is given below.

11.4.1 General formalism

We reconsider the differential equation of the general form (11.6). More particularly we assume that the coefficient functions have a common factor v^α with $v \geq 0$, $\alpha \geq -1$, so that

$$\frac{\partial}{\partial u}\left(p^2 v^\alpha V_{|u}\right) + \frac{\partial}{\partial v}\left(p^2 v^\alpha V_{|v}\right) + p^2 v^\alpha(\hat{q}V + s) = 0 \qquad (11.10)$$

$p(u,v)$, $\hat{q}(u,v)$ and $s(u,v)$ being *finite* analytical functions of their variables and $p > 0$. There is a wide class of differential equations that fit (11.10). One practical example is (11.9) with $p \equiv 1$, $\hat{q} \equiv 0$, $v = r$, $s = g$; we shall meet others below. The same type of discretization can be applied to all these equations, as will be obvious from the following considerations.

We first note that by writing

$$V(u,v) =: \frac{U(u,v)}{p(u,v)} \qquad (11.11)$$

(11.10) collapses to the simpler form

$$\Delta_\alpha U = -g(u,v) := -\left(q(u,v)U + ps\right) \qquad (11.12)$$

with

$$\Delta_\alpha := \frac{\partial^2}{\partial u^2} + \frac{\partial^2}{\partial v^2} + \frac{\alpha}{v}\frac{\partial}{\partial v} \qquad (11.13)$$

and

$$q(u,v) = \hat{q}(u,v) - \frac{\Delta_\alpha p}{p} \qquad (11.14)$$

If we consider cylindrical coordinates, $u = z$ and $v = r$, all the differential equations derived in Chapter 7 are seen to be special cases of (11.12). There are, however, important examples of representations in other coordinate systems, as we shall see later.

Just as for the five-point discretization, we have to distinguish between on-axis formulae ($v = 0$) and off-axis formulae ($v > 0$), and again these

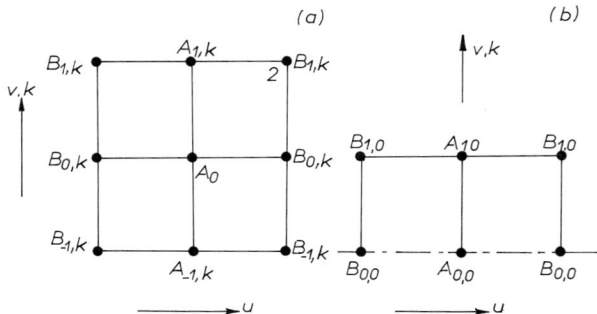

Fig. 11.4: Notation employed for the points and coefficients in nine-point configurations. (a) General case and (b) on-axis node.

can take different forms. Here we shall present only the results derived by Kasper (1984a), since these seem to be most favourable for practical applications. For reasons of space we cannot reproduce here their lengthy derivation, to be found in the corresponding publication.

We again use a notation with two subscripts $i = u/h$, $v = k/h$; the notation for the coefficients is explained in Figs 11.4a,b. The discretization formulae are most conveniently given by an *implicit* representation in terms of a new array W, defined by

$$W_{i,k} := U_{i,k} + \frac{h^2}{12} g_{i,k} \equiv p_{i,k} \left\{ V_{i,k} + \frac{h^2}{12} \left(q_{i,k} V_{i,k} + s_{i,k} \right) \right\} \qquad (11.15)$$

The *off-axis* discretization ($k \neq 0$) is found to be

$$W_{i,k} = A_{-1,k} W_{i,k-1} + A_{1,k} W_{i,k+1} + A_{0,k} h^2 g_{i,k}$$
$$+ \sum_{j=-1}^{1} B_{j,k} \left(W_{i-1,k+j} + W_{i+1,k+j} \right) + O(h^6) \qquad (11.16)$$

The coefficients are given by

$$\gamma_k := \frac{3\alpha(\alpha - 2)}{10(24k^2 + \alpha^2 - 2\alpha - 6)}$$
$$A_{\pm 1,k} = \frac{1}{5} \pm \frac{\alpha}{10k} + \gamma_k \left(1 \mp \frac{2\alpha + 5}{6k} \right)$$
$$A_{0,k} = \frac{3}{10} - \gamma_k \quad , \quad B_{0,k} = \frac{1}{5} - \gamma_k \qquad (11.17)$$
$$B_{\pm 1,k} = \frac{1}{20} \pm \frac{\alpha}{40k} \mp \gamma_k \frac{\alpha + 1}{12k}$$

11.4 NINE-POINT FORMULAE

This set of coefficients is obviously independent of the mesh-length h. It is sensible to compute it at the beginning of the program and leave it in store.

On the optic axis ($v = 0, k = 0$) a slightly different discretization is necessary. Equation (11.15) remains valid but in the discretization not only are the coefficients different but some other terms appear:

$$W_{i,0} = A_{1,0}W_{i,1} + \sum_{j=0}^{1} B_{j,0}\left(W_{i-1,j} + W_{i+1,j}\right)$$
$$+ h^2\left\{A_{0,0}g_{i,0} + C(g_{i,0} - g_{i,1})\right\} + O(h^6) \tag{11.18}$$

the set of coefficients being given by

$$\beta := \frac{(1+\alpha)(6+\alpha)}{6(3+\alpha)} \quad , \quad \gamma_0 := \frac{1}{2(2+\alpha-\beta)}$$
$$A_{1,0} = 2\gamma_0(1+\alpha-\beta)$$
$$B_{0,0} = \gamma_0(1-\beta) \quad , \quad B_{1,0} = \gamma_0\beta \tag{11.19}$$
$$A_{0,0} = \gamma_0 \quad , \quad C = \gamma_0\frac{\alpha(1+\alpha)}{6(3+\alpha)}$$

For $\alpha < 0$ ($\alpha = -1$ for flux fields), the axial discretization fails, but then we have simply $W_{i,0} = U_{i,0} = g_{i,0} = s_{i,0} = 0$, so that (11.18) is no longer needed.

In order to reduce the necessary amount of calculations and storage locations, we rewrite (11.15) in the form

$$h^2 g_{i,k} = \frac{(q_{i,k}W_{i,k} + p_{i,k}s_{i,k})}{\left(h^{-2} + \frac{1}{12}q_{i,k}\right)} =: C_{i,k}W_{i,k} + S_{i,k} \tag{11.20a}$$

from which the field V has been eliminated. The sets of coefficients $C_{i,k}$ and $S_{i,k}$ are calculated once in the beginning and stored. Next, the boundary values of W are determined from (11.15) and stored. The boundary-value problem for the array W can now be solved using (11.16) and (11.18) and with (11.20a) as source terms. In this major calculation only three arrays, C, S and W, are needed simultaneously. Finally the required function V is obtained by solving (11.15) for $V_{i,k}$, or equivalently from

$$V_{i,k} = p_{i,k}^{-1}\left\{W_{i,k} - \frac{1}{12}(C_{i,k}W_{i,k} + S_{i,k})\right\} \tag{11.20b}$$

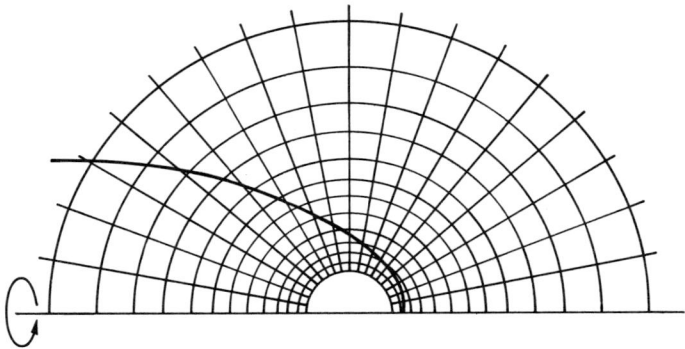

Fig. 11.5: An expanding spherical-mesh grid and part of a curved boundary that does not fit this grid ($h = \pi/20$).

This can be performed without any increase in memory requirement, since there is no iteration and the coefficients C, S and W are no longer needed once the expressions on the right-hand side have been evaluated.

11.4.2 Expanding spherical-mesh grid
The formalism developed above fits very many different forms of the equations of Laplace, Poisson and Helmholtz. In electron optics, they are most frequently expressed in cylindrical coordinates but this is not always advantageous. In order to calculate the field in field-emission electron guns (see Part IX), Kang et al. (1981, 1983) introduced an expanding spherical grid which they called SCWIM (spherical coordinates with increasing mesh-width). This can be regarded as a conformal mapping of an originally cylindrical grid with square-shaped meshes:

$$z + ir = R_0 \exp(u + iv) = R(\cos\vartheta + i\sin\vartheta)$$

with an arbitrary positive constant R_0, and hence

$$R = R_0 \exp u \quad , \quad \vartheta \equiv v \qquad (11.21)$$

If the mesh-length h in the coordinates (u, v) is constant, the corresponding grid in the real space expands exponentially, as shown in Fig. 11.5. (This interpretation is not mentioned by Kang et al.).

Introducing (11.21) into the rotationally symmetric Poisson equation in spherical coordinates (R, ϑ):

$$\frac{\partial^2 V}{\partial R^2} + \frac{2}{R}\frac{\partial V}{\partial R} + \frac{1}{R^2}\frac{\partial^2 V}{\partial \vartheta^2} + \frac{\cot\vartheta}{R^2}\frac{\partial V}{\partial \vartheta} = -\frac{\varrho(R,\vartheta)}{\epsilon_0} \qquad (11.22)$$

we find after some elementary calculations

$$\frac{\partial}{\partial u}\left(e^u \sin v \frac{\partial V}{\partial u}\right) + \frac{\partial}{\partial v}\left(e^u \sin v \frac{\partial V}{\partial v}\right) = -\frac{R_0^2 e^{3u} \sin v}{\epsilon_0} \varrho(R_0 e^u, v) \quad (11.23)$$

This can be brought into the form (11.10) with

$$\alpha = 1 \quad , \quad p^2 = e^u \frac{\sin v}{v} \quad , \quad \hat{q} \equiv 0 \quad , \quad (0 \leq v < \pi) \quad (11.24a)$$

$$s = \frac{R_0^2 e^{2u}}{\epsilon_0} \varrho(R_0 e^u, v) = \frac{R^2}{\epsilon_0} \varrho(R, \vartheta) \quad (11.24b)$$

The evaluation of (11.14) results in

$$q = \frac{1}{4}\left(\frac{1}{\sin^2 v} - \frac{1}{v^2}\right) = \frac{1}{12}\left(1 + \frac{v^2}{5}\right) + O(v^4) \quad (11.24c)$$

The derivation of the corresponding nine-point discretization is now a straightforward matter. It has been proposed by Kasper (1984a) and worked out by Killes (1985). In comparison with a five-point discretization with equal meshes, the gain in accuracy is considerable, so that it is certainly worthwhile to use the nine-point discretization whenever this is possible. All that is required is the determination of the coefficients in (11.20) from (11.24), the result being

$$p_{j,k} = \exp(\frac{1}{2}jh)\left(\frac{\sin hk}{hk}\right)^{\frac{1}{2}}$$

$$q_{j,k} =: q_k = \frac{1}{4}\left(\frac{1}{\sin^2 hk} - \frac{1}{h^2 k^2}\right)$$

$$C_{j,k} =: C_k = \frac{12 q_k}{q_k + 12 h^{-2}}$$

$$S_{j,k} = \frac{12 p_{j,k} s_{j,k}}{q_k + 12 h^{-2}}$$

where j and k are integers. Equations (11.16) and (11.18) with (11.20a) and finally (11.20b) can then be solved.

As Killes (1985) pointed out, this method is useful for $0 \leq \vartheta = hk \leq \pi/2$. For larger values of ϑ, it is better to discretize the variable $\vartheta' := \pi - \vartheta$ and to join the fields in the two domains together smoothly at $\vartheta = \vartheta' = \pi/2$. The concept of conformal mapping can be generalized further but we shall not pursue this here.

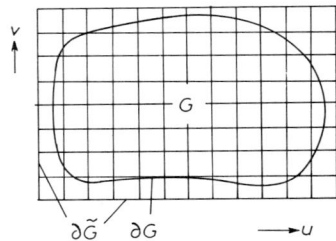

Fig. 11.6: Extension of a regular grid beyond the curved boundary ∂G of a given domain G.

11.4.3 Combination with the BEM

It frequently happens that a boundary-value problem is to be solved in which the boundary contours do not fit the grid. For every irregular internal node the general five-point formulae then have to be applied. This complicates the FDM very considerably and diminishes its accuracy. This disadvantage can be circumvented in different ways. In the case treated above and for the solution of other Dirichlet problems for Poisson's equation $\nabla^2 V = -\varrho(\boldsymbol{r})/\epsilon_0$, combination with the boundary-element method (BEM) is effective (Kasper, 1984a,b; Killes, 1985). First, the regular grid is extended beyond the boundaries so that only *regular* points are obtained, as is demonstrated in Figs 11.5 and 11.6. In this grid, Poisson's equation is solved with arbitrary reasonable boundary values. The values at the true curved boundary are then determined by interpolation. This can be done very accurately since interpolation in regular grids raises no problems (Killes, 1985). The values obtained are now subtracted from the prescribed boundary values, and with these differences the Dirichlet problem $\nabla^2 V' = 0$ is solved using the BEM. The required total solution is then obtained by superposition. The advantage lies in the fact that the solution of Poisson's equation by means of the FDM is much faster than the evaluation of Coulomb integrals, while the BEM can be applied easily to configurations with curved boundaries.

11.5 Iterative solution techniques

These will be briefly treated here; the same techniques are also used to solve the equations resulting from finite-element approximations. In order to obtain a highly accurate final solution, the mesh-length must be sufficiently small. Then, of course, the rank N of the system of equations for the potential at the internal nodes becomes very large, typically $N \sim 10^4$. It is then practically impossible to obtain the solution directly. However, a

11.5 ITERATIVE SOLUTION TECHNIQUES

solution can be found iteratively, since the coefficient matrix of the system is sparse.

In the past decades, numerous iterative techniques for solving large linear systems of equations have been developed. The corresponding mathematical literature is very extensive; some of the major works are cited in the bibliographic listing.

The first step in the application of any of these methods is the choice of an appropriate numbering of the grid points. All the internal grid points must be counted exactly once in a one-dimensional sequence and no boundary point must be counted as the whole domain of solution is scanned. For instance, an appropriate numbering of the internal points with indices $I_1 \leq i \leq I_2$, $K_1 \leq k \leq K_2$ is given by

$$\mu := i - I_1 + 1 + (k - K_1)(I_2 - I_1 + 1)$$
$$1 \leq \mu \leq N := (K_2 - K_1 + 1)(I_2 - I_1 + 1)$$

but permutations of this sequence are also allowed. With this linear sequence of numbering, the whole system of finite-difference equations can be brought into the general form

$$V_\mu = \sum_{\nu=1}^{N}{}' C_{\mu\nu} V_\nu + Q_\nu \quad , \quad \mu = 1 \ldots N \quad (11.25)$$

the prime indicating that the case $\mu = \nu$ is to be excluded. The matrix on the right-hand side is large but sparse, its nonzero elements can be easily calculated and it is therefore not necessary to store them. The inhomogeneous terms Q_ν arise from the boundary values of the potential and from the source terms.

The standard iterative technique for solving linear systems of equations like (11.25) is the successive overrelaxation method (SOR). The corresponding procedure is defined by

$$S_\mu^{(n+1)} := \sum_{\nu=1}^{\mu-1} C_{\mu\nu} V_\nu^{(n+1)} + \sum_{\nu=\mu+1}^{N} C_{\mu\nu} V_\nu^{(n)} + Q_\nu \quad (11.26a)$$

$$V_\mu^{(n+1)} = V_\mu^{(n)} + \omega \left(S_\mu^{(n+1)} - V_\mu^{(n)} \right) \quad , \quad \mu = 1 \ldots N \quad (11.26b)$$

Here the superscript in parentheses denotes the iteration number. The starting values $V_1^{(0)} \ldots V_N^{(0)}$ can be chosen arbitrarily; a sensible guess for these is quite sufficient. The constant ω, the relaxation parameter, must satisfy $1 < \omega < 2$. It is very important for the convergence of the iteration

11. THE FINITE-DIFFERENCE METHOD (FDM)

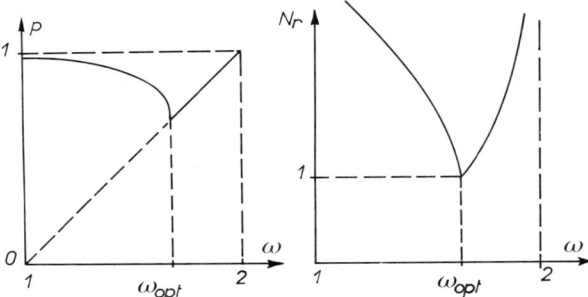

Fig. 11.7: Asymptotic behaviour of the damping factor p and relative iteration number N_r as functions of the relaxation parameter ω.

procedure to choose it suitably, as will be discussed below. The practical meaning of (11.26a,b) is as follows. The quantity $S_\mu^{(n+1)}$ is an approximation for the expression on the right-hand side of (11.25). This Gauss-Seidel value is calculated from the newest values V_ν ($\nu = 1\ldots N$); it is, however, not accepted as the next approximation, but instead the difference between it and the preceding value $V_\mu^{(n)}$ is magnified by a factor ω, as is obvious from (11.26b).

In the practical setup of a SOR program, each successive approximation $V_\mu^{(l)}$, $l = 0, 1, 2, \ldots$ for the same variable V_μ is stored in a single location assigned to V_μ, and for S only one storage location is necessary. Since the array $\{Q_\nu, \nu = 1\ldots N\}$ is usually sparse, the necessary number of storage locations is in practice not much greater than N.

The theory of the convergence of the SOR is investigated exhaustively in the mathematical literature; see Ames (1969), Varga (1962) and Weber (1967), for example. A sufficient and often necessary criterion for the convergence of SOR is that

$$\sum_{\nu=1}^{N}{}' |C_{\mu\nu}| \leq 1 \quad , \quad \mu = 1\ldots N \qquad (11.27)$$

At least one of these sums must be less then 1. For $\alpha > 1$, criterion (11.27) is violated by (11.5a), so that SOR cannot be applied to (11.5a). When the process does converge, its rate depends essentially on the choice of ω, as is shown in Fig. 11.7. The number of iterations N_{it} needed to reduce the iteration error Δ below a given error limit ϵ does of course depend on the definition of Δ, the choice of ϵ and the initialization $V_\mu^{(0)}$, $\mu = 1\ldots N$. But the value of ω_{opt}, where N_{it} has its sharp minimum, and the relative iteration number, $N_r = N_{it}/N_{min}$, do *not* depend on these quantities.

11.5 ITERATIVE SOLUTION TECHNIQUES

According to the general theory of SOR, the optimum value of ω is given by

$$\omega_{opt} = \frac{2}{1 + \sqrt{1 - \lambda^2}} \quad , \quad |\lambda| < 1 \tag{11.28}$$

λ being the largest (real) eigenvalue of the matrix C in (11.25). For any value of ω this quantity λ is related to the asymptotic damping factor p, $(p < 1)$ by

$$\lambda = \frac{(p + \omega - 1)^2}{\omega^2 p} \tag{11.29}$$

Since λ and p are very difficult to calculate exactly, Carré (1961) has proposed a method of estimating these quantities, and hence ω_{opt}, from the actual rates of convergence during the computations. Winslow (1967) has modified this method. Both versions work adequately in many applications but not in every case. For further information the reader is referred to the original papers.

A further refinement of the SOR is the familiar successive line overrelaxation (SLOR) method. Here SOR is combined with Gaussian elimination for tridiagonal subsystems. For instance, in applications to (11.5a,b), these equations are first solved directly along each radial row of the grid, the values in the neighbouring rows being regarded as known for the moment; thereafter the values obtained are modified by overrelaxation and the algorithm proceeds to the next row. The whole grid is scanned repeatedly in this way until sufficient convergence has been achieved.

The main advantage of the SLOR method is that it removes the instabilities of the simple SOR when (11.27) is not satisfied. This has been reported by Kasper and Lenz (1980), who applied the SLOR to (11.5a,b).

Still more refined techniques for solving large but sparse systems of linear equations are the alternating direction implicit methods (ADI: Peaceman and Rachford , 1955; Varga, 1962; Jacobs, 1977), the strongly implicit methods (Stone, 1968) and the cyclic reduction methods (Buneman, 1971, 1973). With the rapid increase of memory capacity, direct solution techniques are also becoming more and more attractive. Special procedures that order the corresponding matrix in such a way that only its nonzero part is stored then become necessary. Munro (1971, 1973) has applied such methods to systems of equations arising in electron optical field computations.

In summary, we wish to emphasize that by the introduction of refined techniques, much computation time can be saved, but, of course, the development of the corresponding computer program then becomes a major task. In any given case, one has to make a reasonable compromise between the time spent on writing an adequate computer program (or the expense of a commercial one) and the computation time (CPU) saved. Furthermore, we again stress that these techniques are not restricted to FDM but can also be used in combination with the FEM, treated in the next chapter.

12
The Finite-Element Method (FEM)

Though the basic ideas and equations of the FDM are very simple, the practical application of this method to boundary-value problems can become extremely tedious if the boundaries are of an irregular shape. In the finite-element method (FEM), this difficulty is removed by the use of general triangular grids, as is shown in Fig. 12.1. Such grids can be fitted to any shape of boundary, once the latter has been represented approximately by a polygon. Since the numerical differentiation now becomes very complicated, partial differential equations are not considered here. Instead, the equations governing the values of the potential at the nodes of the grid are derived directly from an appropriate variational principle (see Section 6.3).

The FEM was proposed by Courant (1943). It came into practical use with the development of modern computers and has found widespread application in mechanical and electrical engineering. Typical examples are problems in fluid dynamics and aerodynamics, elasticity, heat conduction and magnetic field computations for electric machines (Chari and Silvester, 1980). For further details we refer to the books of Zienkiewicz (1967, 1971). Mathematical problems associated with the FEM are treated in detail by Norrie and de Vries (1973). In electron optics, the FEM was first used by Munro (1971), who applied it to the computation of magnetic fields in round lenses. Since this is of especial interest in electron optics, we now deal with this application. The presentation of the FEM given below differs from Munro's version in the introduction of form functions and in the unification of the methods for saturated and unsaturated lenses.

12.1 Formulation for round magnetic lenses

The appropriate variational principle for magnetic field calculation is (6.13) in combination with (6.17), (6.18) and (6.29). Though magnetization curves are usually presented in the form $B = B(H)$ and thus (6.18) is quite familiar, it is more convenient for the FEM to write

$$\beta := B^2(z,r) \quad , \quad V(\beta) := U(B) \tag{12.1}$$

Then (6.19) is in agreement with

$$\frac{1}{\mu} = \nu(\beta) = 2V'(\beta) \tag{12.2}$$

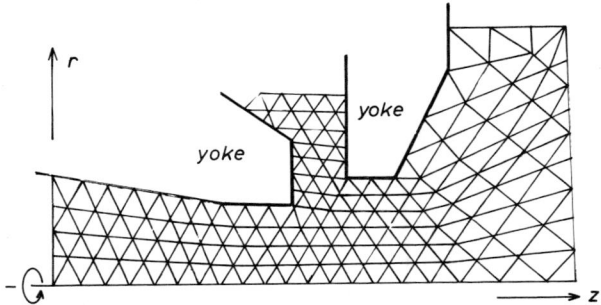

Fig. 12.1: Finite-element discretization; this example concerns the vacuum part of a meridional section through a round magnetic lens.

We now introduce cylindrical coordinates (z, r, φ) and perform the integration over φ. The variational principle (6.13) now takes the form

$$W = 2\pi \iint_S \{V(\beta) - j(z,r)A(z,r)\}\, r\,dr\,dz = \text{minimum} \qquad (12.3)$$

S being the domain of integration in the axial section and

$$\beta = (\nabla \times \boldsymbol{A})^2 = \left(\frac{\partial A}{\partial z}\right)^2 + \left(\frac{\partial A}{\partial r} + \frac{A}{r}\right)^2 \qquad (12.4)$$

In order to solve this variational equation, some simplifying assumptions are necessary.

The expression (12.3) must be minimized with respect to any permitted variations of the vector potential $A(z,r)$. In the FEM these are variations of the values $A_j := A(z_j, r_j)$, $j = 1...N$, assumed at the *internal* nodes of a triangular grid, N being the total number of such nodes. In this context it is convenient to introduce dimensionless form functions $f_j(z,r)$ associated with the node with the corresponding subscript; a detailed definition will be given below.

We start now from a series expansion

$$A(z,r) = \sum_{k=1}^{N} A_k f_k(z,r) \qquad (12.5)$$

and with (12.4) in mind, we introduce the abbreviation

$$F_{l,k}(z,r) = F_{k,l}(z,r) = f_{l|z}f_{k|z} + (f_{l|r} + r^{-1}f_l)(f_{k|r} + r^{-1}f_k) \qquad (12.6)$$

12.1 FORMULATION FOR ROUND MAGNETIC LENSES

The function β may now be written as the quadratic form

$$\beta(z,r) = \sum_{l=1}^{N}\sum_{k=1}^{N} F_{l,k}(z,r) A_l A_k \qquad (12.7)$$

Introducing (12.5) and (12.7) into (12.3) we obtain a discretization of this functional. The minimization condition now takes the form

$$\frac{\partial W}{\partial A_i} = 2\pi \iint_S \left\{ \nu'(\beta) \frac{\partial \beta}{\partial A_i} - f_i(z,r) j(z,r) \right\} r\, dr\, dz = 0 \quad , \quad (i=1\ldots N) \qquad (12.8)$$

Evaluating this expression and recalling (12.2), we soon notice that it is favourable to introduce the matrix elements

$$L_{i,k} = 2\pi \iint_S \nu(\beta) F_{i,k}(z,r)\, r\, dr\, dz = L_{k,i} \qquad (12.9)$$

$$M_i = 2\pi \iint_S j(z,r) f_i(z,r)\, r\, dr\, dz \qquad (12.10)$$

Equation (12.8) then takes the concise form

$$\sum_{k=1}^{N} L_{i,k}(\beta) A_k = M_i \quad , \quad i=1\ldots N \qquad (12.11)$$

which represents a *non-linear* system of equations in the general case, since in saturated media ν is a function of β and hence depends implicitly on $A_1 \ldots A_N$.

So far, the discussion has been quite general. The choice of the form functions is quite arbitrary, except that these must remain linearly independent so that the matrix L in (12.9) is invertible for fixed values of ν. In order to perform the necessary numerical integrations in (12.9) and (12.10), however, a reasonably simple choice for the form functions is necessary. In the simplest choice, equivalent to Munro's version of the FEM, they are piecewise linear functions which are joined together continuously at the nodes. Even the first-order derivatives are then discontinuous on the grid lines, however. With an alteration of the numbering, the form function corresponding to an arbitrary internal node 0 is sketched in Fig. 12.2a,b. It is nonzero only in the configuration shown, consisting of $n=6$ triangles with the common node 0, and there it is a pyramid of unit height. It is defined in one particular subdomain ($i=1$), see Fig. 12.3, by

12. THE FINITE-ELEMENT METHOD (FEM)

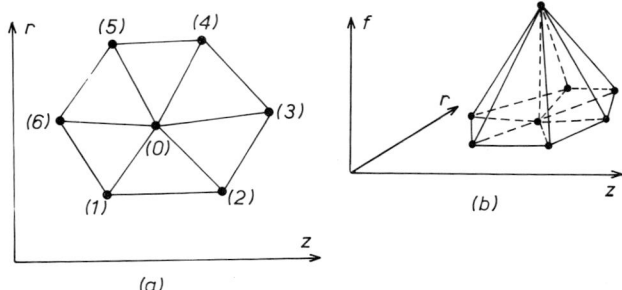

Fig. 12.2: (a) Hexagonal configuration ($n = 6$) of nearest neighbours of an arbitrary internal node 0. (b) Perspective view of the corresponding linear form function; outside the hexagonal domain, this function vanishes.

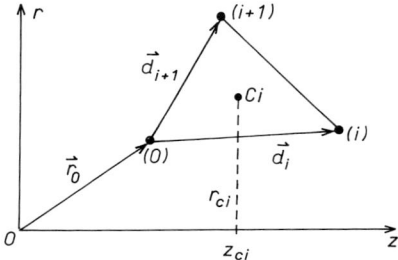

Fig. 12.3: Notation for one of the triangular elements of which the hexagonal configuration of Fig. 12.2 is composed; $d_j = r_j - r_0$ ($j = i, i+1$) denote the side vectors and C_i the centroid. In the text, the area is denoted by a_i.

$$f_{012}(z,r) = \frac{z_1 r_2 - z_2 r_1 + z(r_1 - r_2) + r(z_2 - z_1)}{z_1 r_2 - z_2 r_1 + z_0(r_1 - r_2) + r_0(z_2 - z_1)} \qquad (12.12a)$$

which assumes the nodal values

$$f_{012}(z_0, r_0) = 1 \quad , \quad f_{012}(z_1, r_1) = f_{012}(z_2, r_2) = 0 \qquad (12.12b)$$

Linear functions of this type are said to be barycentric.

The remaining calculation is elementary but very lengthy. We have to differentiate the linear form functions, substitute these into (12.9) and (12.10) and then integrate. In this context some simplifying assumptions are necessary, essentially concerning the factors $\nu(\beta)$ and $j(z,r)$. In order to facilitate the integration, these factors are assumed to be piecewise linear or

12.2 FORMULATION FOR SELF-ADJOINT ELLIPTIC EQUATIONS

even piecewise constant functions. The results of these considerations will be given at the end of Section 12.2, since we need the notation introduced there.

12.2 Formulation for self-adjoint elliptic equations

A variational principle is certainly a very common starting point but is not the only possible one. The following method is equivalent to it but can be made more general so that it remains applicable in cases where no variational principle is known.

We now regard (z, r) as quasi-cartesian coordinates in a meridional plane and consider a self-adjoint elliptic differential equation of the form

$$\frac{\partial}{\partial z}\left(P\frac{\partial \Psi}{\partial z}\right) + \frac{\partial}{\partial r}\left(P\frac{\partial \Psi}{\partial r}\right) + Q(z,r) = 0 \qquad (12.13)$$

$P = P(z, r)$ being a positive coefficient function. As in (12.5), we expand the potential Ψ in terms of form functions:

$$\Psi(z,r) = \sum_{k=1}^{N} \Psi_k f_k(z,r) \qquad (12.14)$$

Here M is the total number of nodes; the numbering can always be chosen so that $i = 1\ldots N$ refer to internal nodes, while $j = N+1\ldots M$ refer to boundary nodes. In Dirichlet problems, the boundary values $\Psi_{N+1}\ldots\Psi_M$ are kept fixed. Substituting this in (12.13), multiplying throughout by $f_i(z,r)$, $i = 1\ldots N$, and integrating over the whole domain S of solution, we find

$$\iint_S \left[\sum_k \Psi_k f_i\left\{\frac{\partial}{\partial z}\left(Pf_{k|z}\right) + \frac{\partial}{\partial r}\left(Pf_{k|r}\right)\right\} + Qf_i\right] dz\,dr = 0$$

With the necessary continuity conditions, integration by parts now leads to

$$\iint_S \left\{-\sum_k P\Psi_k\left(f_{i|z}f_{k|z} + f_{i|r}f_{k|r}\right) + Qf_i\right\} dz\,dr$$
$$+ \oint_{\partial S} P\Psi_k f_i\left(n_z f_{k|z} + n_r f_{k|r}\right) ds = 0 \quad, \quad i = 1\ldots N$$

n_z, n_r denoting the cylindrical components of the outward-directed boundary normal. The contour integral vanishes, since the form functions referring to inner nodes must vanish at the boundary in the case of a Dirichlet problem. Introducing the arrays

$$G_{i,k}(z,r) = f_{i|z} f_{k|z} + f_{i|r} f_{k|r} = \nabla f_i \cdot \nabla f_k \qquad (12.15a)$$

$$\overline{P}_{i,k} = \iint_S P(z,r) G_{i,k}(z,r) \, dz \, dr \qquad (12.15b)$$

$$\overline{Q}_i = \iint_S Q(z,r) f_i(z,r) \, dz \, dr \qquad (12.16)$$

we obtain the finite-element equations

$$\sum_{k=1}^N \overline{P}_{i,k} \Psi_k = \overline{Q}_i \quad , \quad i = 1 \ldots N \qquad (12.17)$$

which are identical with those obtained by evaluating the corresponding variational principle. From the latter, it might be concluded that (12.17) with (12.15b) remains valid even when the normal derivatives of the form functions are discontinuous at the grid lines, but this is not always true. In (12.17) the contributions of the boundary values to the inhomogeneity are incorporated on the left-hand side, as the summation covers *all* the nodes.

The matrix elements (12.15a) are considerably simpler than (12.6), as they are scalar products. Consequently, the results of the discretization using linear form functions can be cast into a fairly simple explicit form. Assuming that $P(z,r)$ and $Q(z,r)$ are constant and refer to the centroid c_i in each triangular element, we find, after some lengthy elementary calculations, for a configuration of n triangular elements with common node 0 like that shown in Fig. 12.2:

$$\Psi_0 \sum_{i=1}^n P_{ci}(\boldsymbol{d}_{i+1} - \boldsymbol{d}_i)^2 / a_i$$

$$= \sum_{i=1}^n \frac{1}{a_i} \left[P_{ci} \{ d_i^2 \Psi_{i+1} + d_{i+1}^2 \Psi_i - \boldsymbol{d}_i \cdot \boldsymbol{d}_{i+1} (\Psi_i + \Psi_{i+1}) \} \right. \qquad (12.18)$$

$$\left. + \frac{4}{3} \sum_{i=1}^n a_i Q_{ci} \right]$$

Here the notation of Figs 12.2 and 12.3 has been adopted; it is cyclic in the sense that $\boldsymbol{d}_{i+n} = \boldsymbol{d}_i$, $\Psi_{i+n} = \Psi_i$; a_i is the area of the element with the side vectors $\boldsymbol{d}_i$ and $\boldsymbol{d}_{i+1}$.

12.2 FORMULATION FOR SELF-ADJOINT ELLIPTIC EQUATIONS

This theory can be applied to the field in round magnetic lenses. The potential Ψ is then to be identified with the flux function $\Psi = 2\pi r A$, introduced in Section 6.4, and (12.13) must be identified with the flux equation arising from (6.33):

$$\frac{\partial}{\partial z}\left(\frac{\nu}{r}\frac{\partial \Psi}{\partial z}\right) + \frac{\partial}{\partial r}\left(\frac{\nu}{r}\frac{\partial \Psi}{\partial r}\right) = -2\pi j(z,r) \qquad (12.19)$$

The coefficients are hence

$$P(z,r) = \nu(z,r)/r \quad , \quad Q(z,r) = 2\pi j(z,r) \qquad (12.20)$$

The differential equation and the corresponding finite-element discretization remain applicable even in the non-linear case; we then have $\nu = \nu(\beta)$ with

$$\beta = |B|^2 = \frac{1}{4\pi^2 r^2}\left[\Psi_{|z}^2 + \Psi_{|r}^2\right] = \frac{1}{4\pi^2 r^2}\sum_i\sum_k \Psi_i \Psi_k G_{i,k} \qquad (12.21)$$

In the vicinity of the optic axis at least, quadratic form functions are necessary, since we know that $\Psi \propto r^2$ in the paraxial domain. Recalling that Ψ must vanish at the outer boundary and at the axis, we perceive that all the summations run only over the internal nodes ($i = 1 \ldots N$), as in Section 12.1.

We now state briefly the corresponding formulae for the vector potential, which result from the considerations in Section 12.1. Although not identical with Munro's formulae, they are equivalent to them.

Again adopting the notation introduced in Fig. 12.3, we find for the value of β at the centroid C_i

$$\beta_{ci} = \frac{1}{4a_i^2}\{d_{i+1}(A_0 - A_i) + d_i(A_{i+1} - A_0) + s_i(A_0 + A_i + A_{i+1})\}^2$$

with an additional shift

$$s_i = \frac{2a_i}{3r_{ci}}i_z$$

Then with $\nu_{ci} := \nu(\beta_{ci})$ and $j_{ci} = j(z_{ci}, r_{ci})$, both referring to the centroid c_i of the element with label i, we obtain

$$A_0 \sum_{i=1}^n \frac{r_{ci}\nu_{ci}}{a_i}(d_{i+1} - d_i + s_i)^2 - \frac{4}{3}\sum_{i=1}^n r_{ci}a_i j_{ci}$$
$$= \sum_{i=1}^n \frac{r_{ci}\nu_{ci}}{a_i}\{A_i(d_{i+1} - s_i) - A_{i+1}(d_i + s_i)\}\cdot(d_{i+1} - d_i + s_i) \qquad (12.22)$$

Apart from different material coefficients, the essential difference between (12.22) and (12.18) lies in the appearance of the shift s_i which results from the term A/r in (12.4).

Another interesting application of the FEM to magnetic lenses is the calculation of magnetic circuits made of anisotropic material. In this case the reluctance $\nu(\beta)$ is to be replaced by a symmetric tensor. Its components depend directly on the position r as a consequence of the variable crystallographic orientation in the material and indirectly due to saturation effects. Such calculations are extremely complicated; nevertheless, magnetic circuits with anisotropic material can be advantageous. Balladore et al. (1981, 1984) have shown that the size and weight of the yoke can be appreciably reduced in this way.

12.3 Solution of the finite-element equations

In the case of linear (unsaturated) media, the matrix elements $L_{i,k}$ in (12.11) and $\overline{P}_{i,k}$ in (12.17) are constants; the corresponding systems of equations are therefore linear and can be solved by means of standard techniques. It is usual to employ direct solution techniques, in which case it is essential to make use of an ordering that minimizes the bandwidth of the corresponding sparse matrix. These techniques cannot be outlined here; the reader is referred to the corresponding literature (Cuthill and McKee, 1969; Gibbs et al., 1976; Duff, 1977). A very fast iterative procedure using the preconditioned conjugate-gradient method has been developed by Lencová and Lenc (1986).

Equation (12.18) has already been cast into a form which is suitable for iterative techniques such as SOR and SLOR. These are efficient if the coefficients in (12.18), referring to each internal node (0), are computed once and for all at the beginning and then stored. In the absence of source terms Q_{ci} and with $n = 6$, the total memory requirement is $7N$.

When saturation or other non-linear effects are present, the situation becomes more complicated, as (12.11) and (12.17) are now non-linear systems of equations: iterative procedures are unavoidable. Direct techniques must be combined with Newton's iterative procedure (Munro, 1973). When the SOR is employed, a quasi-linearization is necessary; (12.11) and (12.17) already have the appropriate form if $\nu(\beta)$ is first treated as a linear coefficient during each iteration over the field and then recalculated before the next cycle according to (12.2) together with (12.7) or (12.21).

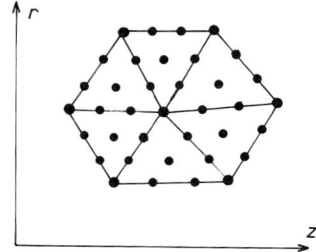

Fig. 12.4: Configuration of finite elements and reference points to be used in a third order approximation.

12.4 Improvement of the finite-element method

The form of the FEM presented above is only the very simplest version. We have chosen it in order to display the basic ideas clearly. The FEM can, of course, be improved in many ways. The corresponding theory has been developed in so much detail that it is impossible to treat it adequately here. It is even impossible to present a complete list of references. We must confine our considerations to some essential points.

The linear form functions given by (12.12) are those of the lowest permissible order. For the azimuthal component of the vector potential $A(z,r)$ such a choice is reasonable, since in the most important, paraxial domain the function $A(z,r)$ is proportional to r. There are also presentations of the FEM in which a linear approximation is made for scalar potentials (Munro, 1973) and flux functions (Bonjour, 1980). This is clearly a locally very inaccurate approximation, which has been criticized by Kasper and Lenz (1980), who showed that the FEM is then less accurate than the FDM, at least in the paraxial domain. This weakness can be avoided by using higher order approximations for the potential. The corresponding theory has been presented by Silvester and Konrad (1973). It is then necessary to introduce additional points of reference along the mesh lines of the grid and also in the interior of the elements (see Fig. 12.4). The values of the potentials at these additional points are unknown. The rank of the system of FE equations and, of course, the complexity of its structure increase accordingly but a gain in accuracy may be achieved.

A second important aspect is the appropriate choice of the grid. In practice it is too tedious to let the user of a program choose every node individually. Suitable algorithms for the automatic generation of the grid have been proposed by Winslow (1967) and Munro (1973). Whether these algorithms work efficiently or not depends essentially on the shape of the boundary in question. More recently, Hermeline (1982) has proposed a new method which works well but is highly sophisticated. The user of a commercial FE program is usually supplied with a method for generating the grid, even a curved grid, but setting up one's own program will be generally very laborious.

A third aspect of the FEM—and indeed of the FDM—is the need to confine the spatial extension of the grid. Very often the fringe fields of a particular configuration spread out over a much larger region than can be covered by the grid. In order to keep the error introduced by cutting off the field at the boundary of the grid sufficiently small, the size of the grid must often be very large. This drawback of the FEM can be removed by the introduction of additional infinite elements. The trial functions for the potential, to be evaluated in such elements, must be consistent with the asymptotic form of the real potential. Such infinite elements have been proposed by Bettess (1977) and their use in electron optics has been investigated by Lencová and Lenc (1982, 1984).

The most serious drawback of the FEM becomes apparent when we come to compute the field strength. This problem will be discussed in Chapter 13.

12.5 Comparison and combination of different methods

In the previous sections, we have dealt with three major methods of calculating potentials, the BEM, the FDM and the FEM. The question that now arises is which one should be preferred in a given case. The answer depends on the details of the particular problem to be solved.

In all cases in which a one-dimensional linear integral equation can be derived, the BEM is the most advantageous means of obtaining a solution. The necessary discretization can be easily fitted to arbitrary boundaries, regardless of whether these are curved or piecewise straight with sharp edges. Even extreme differences in the dimensions of boundaries, as in field-emission electron guns, for example, are no obstacle to this method. There is no need to cut off fringe-fields, as theoretically the domain of solution is the whole space. With a comparatively modest memory capacity, high accuracy can be achieved. The linear system of equations to be solved is well-conditioned and can hence be solved directly by means of a simple Gaussian elimination without any pivoting. After determining the appropriate surface-source distributions, the analytic expressions for the field strength can be evaluated at any point of reference. In principle, there is no need for additional interpolation and numerical differentiation techniques, though these may be helpful in some cases.

12.5 COMPARISON AND COMBINATION OF DIFFERENT METHODS   185

The same conclusions hold when the two-dimensional BEM is applied to three-dimensional boundary-value problems (Section 10.4). Of course, this method is then more complicated than the one-dimensional BEM, but this is an inevitable consequence of the greater complexity of the problem to be solved; alternative methods such as the FDM and the FEM will also become more complicated.

The FDM is suitable only when a regular grid fits the boundary, since the inclusion of irregularities, though quite elementary, is very tedious. Since highly regular domains of solution are very rare, the FDM is not advantageous, in electron optics at least, unless the improvement outlined in Sections 11.4.2 and 11.4.3 is incorporated.

The FEM is theoretically applicable to any kind of boundary-value problem, even in three dimensions. This method may be very effective if a completely worked-out and tested program package is available but if this is not the case, we should prefer the BEM, since the latter can be easily programmed by a single scientist, at least in the one-dimensional version. Only when saturation effects become important in ferromagnetic materials does the FEM seem to be really necessary. Moreover, after the potentials have been determined at the nodes, the serious problem of calculating the field strength will arise.

Taking into account all the points discussed above, we come to the conclusion that in most cases the BEM is very advantageous and it is for this reason that we have dealt extensively with this method.

There have been several investigations on combinations of the different methods of field calculation, and one has already been outlined in Section 11.4.3. In electron optics this possibility has been found valuable for field computation in field-emission electron guns (see Part IX). More generally, the combination of different methods for the solution of Dirichlet problems in electrostatics has been investigated by Schaefer (1982, 1983), who has proved quite generally that iterative solutions of Dirichlet problems in two and more overlapping domains converge. He has developed a suitable technique for the solution of such problems, which he calls Schwarz's alternating method. Though this method can be very powerful, we cannot describe it here for reasons of space. Unfortunately, this method does not work for problems with interface conditions or for non-linear problems, where a suitable coupling of different methods is particularly interesting. In these cases a combination of the FEM with the BEM is possible, as has been proposed by McDonald and Wexler (1972) and McDonald et al. (1973) and by Lencová and Lenc (1982, 1984). A similar method has been proposed by Kasper (1984a,b); we now outline this briefly.

A typical example of the application of a hybrid method is presented in Fig. 12.5, which shows a half-axial section through an open magnetic

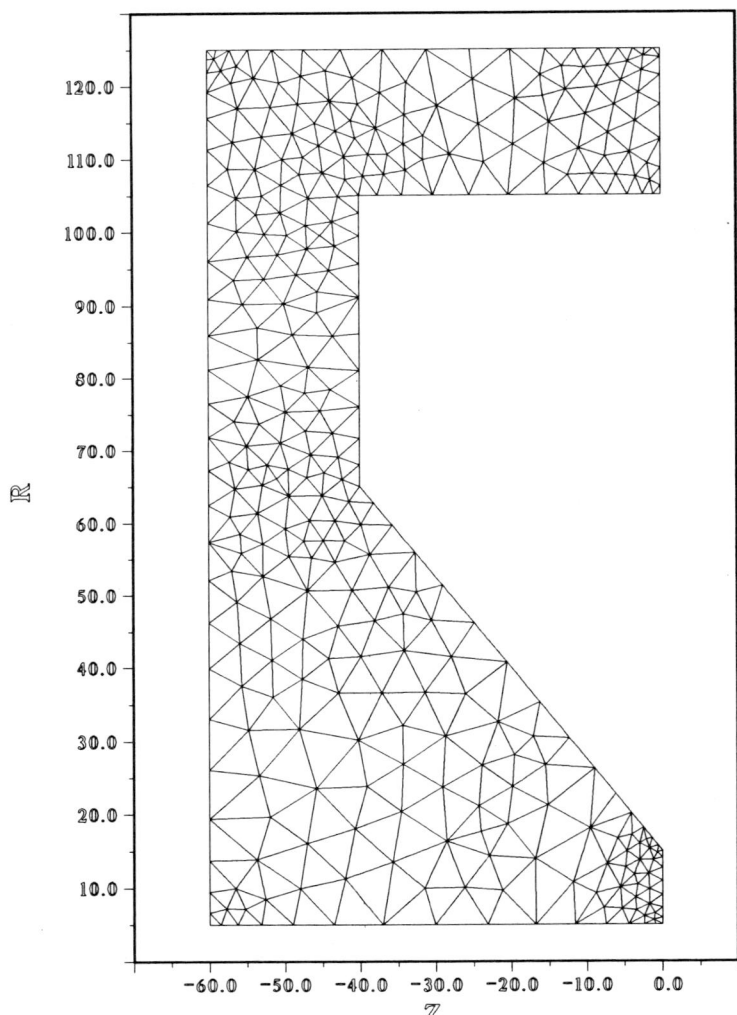

Fig. 12.5a: Upper half of a meridional section through an open, round, magnetic lens with a ferromagnetic core and a rectangular distribution of windings. Only the interior of the yoke is discretized by a triangular-mesh grid.

12.5 COMPARISON AND COMBINATION OF DIFFERENT METHODS   187

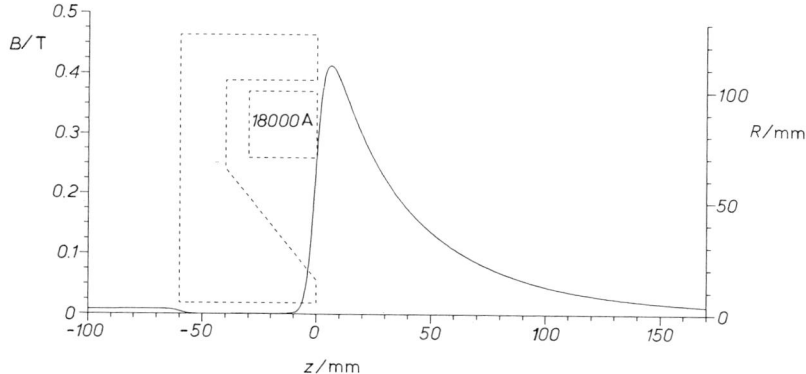

Fig. 12.5b: Axial field strength for an excitation of 18000 A-turns.

lens. Owing to this open structure the magnetic fringe-field extends so far that it becomes very impractical to apply the FEM in the vacuum domain V. On the other hand, the saturation effects in the polepieces make the application of Scherle's method impossible. We hence apply the FEM only to the polepieces and use the BEM in the outer domain.

The finite-element equations can be solved if the boundary values of $A(z,r)$ at the surface contour C are known. Then by means of suitable interpolation techniques the normal derivative $(\partial A/\partial n)_i$ on the inner side can be determined. When this has been done, we also know the normal derivative $(\partial A/\partial n)_v$ on the vacuum side from

$$\frac{1}{\mu_i}\left(\frac{\partial \hat{A}}{\partial n}\right)_i = \frac{1}{\mu_0}\left(\frac{\partial \hat{A}}{\partial n}\right)_v \quad , \quad \hat{A} := rA \qquad (12.23)$$

We can now solve the integral equation

$$\frac{1}{2}A(\boldsymbol{u}) = \oint_C \left\{ A(\boldsymbol{u}')\frac{\partial G_1}{\partial n'} - G_1(\boldsymbol{u},\boldsymbol{u}')\left(\frac{\partial A}{\partial n'}\right)_v \right\} r' ds'$$
$$+ \mu_0 \int_V G_1(\boldsymbol{u},\boldsymbol{u}') j(\boldsymbol{u}') r' dr' dz' \qquad (12.24)$$

in which we have written $\boldsymbol{u} = (z,r)$, $\boldsymbol{u}' = (z',r')$ and G_1 is defined by (9.21) with $m = 1$.

When the whole system of coupled equations has been solved iteratively, we have the appropriate solution in the partly saturated iron together with an unbounded and smooth vacuum field. Thus the drawbacks of using each of the individual methods separately have been circumvented.

13
Field-Interpolation Techniques

The finite-difference and the finite-element methods yield the values of a potential at the nodes of a discrete grid. This is only the first step in a full field calculation, since the computation of electron trajectories requires a knowledge of the field strength at arbitrary points in the field. This implies that suitable techniques for interpolation and numerical differentiation will be required.

The application of predictor-corrector methods to the computation of trajectories requires that the field strength be a smooth function, especially on the grid lines separating two adjacent meshes. With respect to the FDM, this problem has been solved satisfactorily. The irregular grid employed in the FEM, however, makes accurate computation of the field strength practically impossible. As far as we know, a very accurate tracing of Lorentz trajectories through an irregular grid has never been reported. In practice, the use of the FEM limits us to interpolations and differentiations along the optic axis.

In this respect, the boundary-element method is clearly superior, since the field strength can be computed as a continous superposition of analytic functions once the surface-source distributions have been determined. Nevertheless, in order to save computation time, interpolation techniques may still be valuable since the BEM is slow.

13.1 One-dimensional differentiation and interpolation

Numerical differentiation and interpolation in one dimension are the basis for all the corresponding procedures in two and three dimensions. Moreover they are of immediate importance in electron optics, since a knowledge of the axial potential and of its derivatives is sufficient for the determination of most electron optical properties.

We consider here an arbitrary analytical function $Y(z)$, which may be an axial potential, an axial deflection field strength or any other function of interest. Let us assume now that only the discrete values

$$Y_i := Y(z_i) \quad , \quad i = 0 \ldots N \quad , \quad (z_i > z_{i-1}) \qquad (13.1)$$

13.1 ONE-DIMENSIONAL DIFFERENTIATION AND INTERPOLATION

are known. They may, for instance, result from a field calculation program using the FDM or the FEM. We now wish to calculate $Y(z)$ and some of its derivatives for arbitrary values of z with $z_0 \leq z \leq z_N$.

This is a standard problem in numerical analysis and a wide variety of methods has been devised to solve it but not all of these are suitable. If problems are not to arise in ray-tracing programs (see Chapter 33), all the required derivatives of $Y(z)$ must be continuous at $z_0 \ldots z_N$. Most of the familiar interpolation techniques do not satisfy this requirement, and the best technique proves to be Hermite interpolation, which will now be outlined.

13.1.1 Hermite interpolation

We assume for the moment that the derivatives Y' and Y'' at $z_0 \ldots z_N$ are known; their determination will be discussed further below. We can then apply cubic or quintic Hermite interpolation. In each interval $z_{i-1} \leq z \leq z_i (i = 1 \ldots N)$, *cubic* interpolation consists in the procedure

$$h_i := z_i - z_{i-1} \quad , \quad t := (2z - z_i - z_{i-1})/h_i \quad , \quad |t| \leq 1 \quad (13.2)$$

$$f_{1,2}(t) = \frac{1}{4}(2 \mp 3t \pm t^3) \quad , \quad g_{1,2}(t) = \frac{1}{8}(t^3 \mp t^2 - t \pm 1) \quad (13.3)$$

$$Y(z) = Y_{i-1}f_1(t) + Y_i f_2(t) + h_i\{Y'_{i-1}g_1(t) + Y'_i g_2(t)\} \quad (13.4)$$

while *quintic* interpolation is given by (13.2) in combination with

$$F_{1,2}(t) = \frac{1}{2}(1 \mp t) \mp \frac{t}{16}(7 - 10t^2 + 3t^4)$$

$$G_{1,2}(t) = -\frac{t}{32}(7 - 10t^2 + 3t^4) \pm \frac{1}{32}(5 - 6t^2 + t^4) \quad (13.5)$$

$$H_{1,2}(t) = \frac{1}{64}(1 - t^2)^2(1 \mp t)$$

$$Y(z) = Y_{i-1}F_1(t) + Y_i F_2(t) + h_i\{Y'_{i-1}G_1(t) + Y'_i G_2(t)\} \quad (13.6)$$
$$+ h_i^2\{Y''_{i-1}H_1(t) + Y''_i H_2(t)\}$$

The *form functions* $f_{1,2}(t)$, $g_{1,2}(t)$ or $F_{1,2}(t)$, $G_{1,2}(t)$, $H_{1,2}(t)$, respectively, are defined in such a way that $Y(z)$ and certain of its derivatives assume the prescribed values at z_i and z_{i-1}. Since each internal endpoint z_i is common to the intervals $[z_{i-1}, z_i]$ and $[z_i, z_{i+1}]$, $Y(z)$ and $Y'(z)$ are continuous in the cubic Hermite procedure, while the quintic procedure also ensures the continuity of $Y''(z)$.

13.1.2 Cubic splines

The formulae given above require that the derivatives at $z_0 \ldots z_N$ have been calculated and stored prior to the actual interpolation. We now discuss the determination of these derivatives. In connection with (13.3) and (13.4), the *cubic spline* technique is very convenient and is in widespread use. Cubic spline functions are Hermite interpolation functions (13.4) that remain continuous after *two* differentiations. This requirement imposes conditions on $Y_0' \ldots Y_N'$, which can be cast into the form of a tridiagonal system of equations. With the abbreviation $k_i = h_i^{-1}$, this is given by

$$k_i Y_{i-1}' + 2(k_i + k_{i+1})Y_i' + k_{i+1}Y_{k+1}'$$
$$= 3k_i^2(Y_i - Y_{i-1}) + 3k_{i+1}^2(Y_{i+1} - Y_i) \qquad (13.7)$$
$$i = 1 \ldots N - 1$$

The terminal values Y_0' and Y_N' can be chosen independently, provided that they are not determined uniquely by such constraints as symmetries or periodicity. If there is apparently no reasonable way of determining Y_0' and Y_N', the linear equations

$$k_1 Y_0' + (k_2 + k_1)Y_1' = 2D_1 + \frac{k_1}{k_1 + k_2}(D_1 + D_2)$$
$$(3k_{N-1} - k_N)Y_{N-1}' + k_N Y_N' = 2D_N + \frac{k_N}{k_N + k_{N-1}}(D_N + D_{N+1}) \qquad (13.8)$$
$$D_\nu := (Y_\nu - Y_{\nu-1})k_\nu^2 \quad , \quad \nu = 1,2; N-1, N$$

can be combined with (13.7); these equations are obtained if $Y'''(z)$ is assumed to be continuous at $z = z_1$ and $z = z_{N-1}$, respectively. The complete tridiagonal system of equations can be solved directly by means of the Gauss algorithm without pivoting. A generalization to quintic spline functions is possible but very delicate and we do not recommend this.

In cubic splines, the second derivative is only a piecewise linear function and hence not very accurate; one should thus not use cubic splines if $Y''(z)$ is needed explicitly. In order to obtain high accuracy in such cases, an improved differentiation technique is necessary and quintic Hermite interpolation should then be used.

13.1.3 Differentiation using difference schemes

Among the many ways of performing numerical differentiations, the technique outlined below has proved very effective. The explicit use of the unequal spacing of the abscissae $z_0 \ldots z_N$ makes the formulae cumbersome and should be avoided. This can be achieved in the following way.

We choose a parametric representation of the function in question, the parameter being denoted by x:

$$z = z(x) \quad , \quad Y = Y(x) \qquad (13.9a)$$
$$x = ih \quad , \quad z_i = z(ih) \quad , \quad Y_i = Y(ih) \quad , \quad i = 0, 1, \ldots N \qquad (13.9b)$$

13.1 ONE-DIMENSIONAL DIFFERENTIATION AND INTERPOLATION

Without loss of generality we can choose $h = 1$, as we do below. Since the two functions $z(x)$ and $Y(x)$ are to be treated in the same manner, it is sufficient to deal only with the differentiation of $Y(x)$; the corresponding derivatives will be denoted by dots.

We now introduce finite differences

$$\Delta Y_i = Y_{i+1} - Y_i \qquad (13.10a)$$
$$\delta^2 Y = Y_{i+1} - 2Y_i + Y_{i-1} = \Delta Y_i - \Delta Y_{i-1} \qquad (13.10b)$$
$$\delta^{2n+2} Y_i = \delta^{2n} Y_{i+1} - 2\delta^{2n} Y_i + \delta^{2n} Y_{i-1} \quad , \quad n \geq 1 \qquad (13.10c)$$

Derivatives with respect to x are then given by

$$S_j := \frac{1}{2} Y_j - \frac{1}{12} \delta^2 Y_j + \frac{1}{60} \delta^4 Y_j - \frac{1}{280} \delta^6 Y_j + \frac{1}{1260} \delta^8 Y_j - \cdots$$
$$\dot{Y}_i = S_{i+1} - S_{i-1} \qquad (13.11)$$
$$\ddot{Y}_i = \delta^2 Y_i - \frac{1}{12} \delta^4 Y_i + \frac{1}{90} \delta^6 Y_i - \frac{1}{560} \delta^8 Y_i + \frac{1}{3150} \delta^{10} Y_i - \cdots$$

If the highest order is chosen reasonably, these formulae give accurate results, since they are highly symmetric. In the vicinity of the margins they are not directly applicable. In order to avoid special asymmetric formulae, it is preferable to extrapolate the function $Y(x)$ a certain distance beyond the interval in which the derivatives are actually needed. This can be done with the aid of symmetries, of periodicities or of well-known asymptotic properties. If none of these is applicable, a polynomial extrapolation can be made. For a polynomial of degree n, this extrapolation takes the simple form

$$Y_{j+1} = \sum_{k=0}^{n} \binom{n+1}{k+1} (-1)^k Y_{j-k} \qquad (13.12)$$

Analogous formulae with correspondingly lower degree hold for the differences ΔY_{j+1} and $\delta^{2m} Y_{j+1}$.

Sometimes the first-order increments $\Delta Y_1 \ldots \Delta Y_k$ are given directly, for instance in the procedure outlined in Section 10.2.3. It is then possible to set up the differentiation procedure directly in terms of these increments. This provides additional numerical stability, as the subtraction of large Y-values is avoided. The corresponding elementary manipulations are not given here.

Finally the required derivatives with respect to the coordinate z are given by

$$Y_i' = \dot{Y}_i / \dot{z}_i \quad , \quad Y_i'' = (\ddot{Y}_i \dot{z}_i - \dot{Y}_i \ddot{z}_i)/\dot{z}_i^3 \quad , \quad (i = 0 \ldots N) \qquad (13.13)$$

Derivatives of higher orders can be computed easily by applying this procedure to these sets of derivatives instead of to the function itself. If the abscissae $z_0 \ldots z_N$ are equidistant, the differentiations of $z(x)$ can be omitted since we have simply

$$\dot{z}_i = z_{i+1} - z_i = h = \text{const} \quad , \quad \ddot{z}_i = 0$$

The procedure is thus very economic.

For the interpolation of derivatives $Y^{(n)}$, the following procedure is efficient: (13.2), (13.5) and (13.6) are completed by

$$\begin{aligned}
Y^{(n)}(z) &= Y_{i-1}^{(n)} F_1(t) + Y_i^{(n)} F_2(t) \\
&+ h_i \left\{ Y_{i-1}^{(n+1)} G_1(t) + Y_i^{(n+1)} G_2(t) \right\} \\
&+ h_i^2 \left\{ Y_{i-1}^{(n+2)} H_1(t) + Y_i^{(n+2)} H_2(t) \right\} \quad , \quad n \geq 1
\end{aligned} \tag{13.14}$$

This has the advantage that only the form functions themselves need to be computed and not their derivatives, so that computation time is saved. Furthermore, even the derivatives of higher orders remain twice continuously differentiable, and hence are very smooth. A very high accuracy can be achieved.

13.1.4 Evaluation of radial series expansions

In Chapter 7 we have derived a variety of radial series expansions, which are of particular interest in electron optics. These are all determined uniquely by certain axial functions, the axial harmonics. With the technique outlined above, their higher order derivatives can be computed numerically for a sequence of abscissae $z_0 \ldots z_N$ and then stored. Using the interpolation formula (13.14), it is now easy to evaluate the radial series expansions for the potential, the field strengths and even for derivatives of second order at any point (z, r) of reference within the domain of convergence. This is straightforward and is undoubtedly the fastest method of field computation.

The analytical character of the solution obtained with the BEM allows *analytical* differentiation of the axial potential, which is clearly preferable if the corresponding procedure remains reasonably simple. For the functions involved in the calculation of rotationally symmetric fields, this is certainly the case.

For a single charged ring, specified by its position (z', r') and the normalized charge 2π, the axial potential, here denoted by γ, can be easily calculated from (10.6):

$$\gamma(z - z', r') = G_0(z, 0; z'r') = (2R)^{-1} \tag{13.15a}$$

13.2 TWO-DIMENSIONAL INTERPOLATION

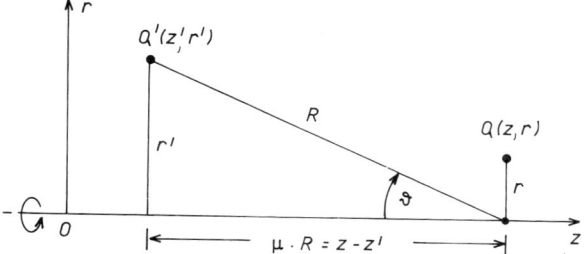

Fig. 13.1: Notation used in the extended paraxial series expansion. Q is an arbitrary reference point and Q' is the trace of a ring in this meridional section.

with
$$R = \sqrt{(z-z')^2 + r'^2} \qquad (13.15b)$$

These geometric quantities are sketched in Fig. 13.1. The derivatives of γ with respect to z can be brought into a very convenient form:

$$\gamma^{(n)}(z-z',r') = \frac{(-1)^n n! P_n(\mu)}{2R^{n+1}} \quad (n \geq 0) \qquad (13.16a)$$

$P_n(\mu)$ being Legendre polynomials with argument

$$\mu = (z-z')/R = \cos\vartheta \qquad (13.16b)$$

where the angle ϑ is shown in Fig. 13.1. These formulae can be evaluated efficiently.

The order in which the differentiation with respect to z and the integration over the boundary C are performed in the integral equation can be exchanged. Once the source distribution $\sigma(s)$ is known, the axial potential $\phi(z)$ and its derivatives can hence be calculated from

$$\phi^{(n)}(z_i) = \oint_C \gamma^{(n)}\Big(z_i - z(s), r(s)\Big)\sigma(s)r(s)\,ds \qquad (13.17)$$
$$(i = 0, 1, 2 \ldots N \quad, \quad n \geq 0)$$

After these values have been computed and stored, (13.14) can be employed for the calculation of $\phi^{(n)}(z)$ for arbitrary z, after which the evaluation of the radial series expansions is straightforward.

This concept can be generalized to include the superposition of aperture fields (Section 10.3) and of various multipole fields, but this will not be dealt with here.

194   13. FIELD-INTERPOLATION TECHNIQUES

13.2 Two-dimensional interpolation

Here we consider two-dimensional functions $P(u,v)$, known at the nodes of a rectangular grid. The coordinates u and v will usually be the cylindrical coordinates z and r in a meridional section through an axisymmetric system, though this special meaning is not absolutely necessary. We now describe algorithms for calculating $P(u,v)$ and its partial derivatives at an arbitrary point Q with coordinates (u,v).

This problem is of importance for the computation of equipotentials and Lorentz trajectories at large off-axis distances. The accurate tracing of a Lorentz trajectory through an electron optical system may require as many as 2000 calls of the field program, this number rapidly increasing with worsening smoothness of the field strength at the grid lines. When the analytic fields supplied by the boundary-element method are used, this problem does not arise, but each single call of the field program may then take so much time that it is preferable to store the values of the potential and the components of the field strength at the nodes of a suitably chosen square-shaped grid. The frequent evaluations at arbitrary points can subsequently be performed very fast by means of interpolation. This is particularly important when several Lorentz trajectories are to be computed, for instance in electron guns.

This interpolation problem has been solved in many different ways. In electron optics, different proposals have been made by Weber (1967), Lenz (1973), Kern (1978) and Kasper (1982). Two- and three-dimensional Hermite interpolation has been used, for instance by Eupper (1985). In the subsequent presentation we shall first examine simple two-dimensional Hermite interpolation, after which we consider possible improvements.

13.2.1 Hermite interpolation
Our object is to calculate a function $P(u,v)$ at some point Q, located arbitrarily in the grid, as shown in Fig. 13.2. It is convenient to denote the partial derivatives by

$$U := P_{|u} \equiv \frac{\partial P}{\partial u} \quad , \quad V := P_{|v} \equiv \frac{\partial P}{\partial v} \qquad (13.18)$$

We assume that the nodal values of P, U and V have been computed prior to the interpolation stage and stored in two-dimensional arrays having two subscripts. The array elements $P_{i,k}$, $U_{i,k}$ and $V_{i,k}$ refer to the node with coordinates (u_i, v_k).

With this information, bivariate cubic Hermite interpolation, which is based on the form factors (13.3), can be applied. The interpolation

13.2 TWO-DIMENSIONAL INTERPOLATION

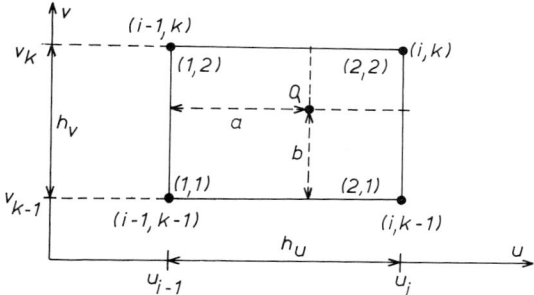

Fig. 13.2: Values of the subscripts used in two-dimensional interpolation. Outside the rectangle, the subscripts J and L are those of the potential; inside the rectangle, the subscripts j and l are those of the coefficients. The lengths a and b are given by $a = h_u(1+s)/2$ and $b = h_v(1+t)/2$.

formulae can be written explicitly in terms of these functions:

$$P(u,v) = \sum_{j=1}^{2}\sum_{l=1}^{2}\Big\{ f_j(s)f_l(t)P_{J,L}$$

$$+ h_u g_j(s)f_l(t)U_{J,L} + h_v f_j(s)g_l(t)V_{J,L} \Big\} \tag{13.19a}$$

with the auxiliary quantities

$$\begin{aligned} h_u &= u_i - u_{i-1} & s &= (2u - u_i - u_{i-1})/h_u \\ h_v &= v_k - v_{k-1} & t &= (2v - v_k - v_{k-1})/h_v \\ J &= i + j - 2 & L &= k + l - 2 \end{aligned} \tag{13.19b}$$

The values of U and V at the point Q are obtained by the appropriate differentiations; the corresponding elementary expressions will not be given here. The derivatives U and V are still continuously differentiable on the grid lines if the arrays $[U_{i,k}], [V_{i,k}]$ are calculated by applying the cubic spline technique to the potentials in the corresponding rows and columns of the grid. This method can easily be generalized to three-dimensional problems.

13.2.2 The use of derivatives of higher order

As in the one-dimensional case, the accuracy and smoothness can be improved by the use of derivatives of higher orders at the nodes of the grid. Such a proposal has been made by Kasper (1982) but this requires a particular partial differential equation to be satisfied, which is not always the case. Here we treat the most general case.

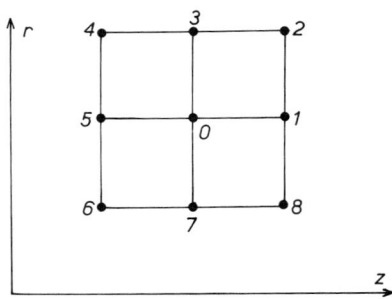

Fig. 13.3: Simplified provisional numbering of the nodes in nine-point formulae for higher derivatives at the central node.

The necessary procedure is reasonably simple only if the grid is square-shaped: $u_i = ih$, $v_k = kh$ with i and k integers. We assume again that the arrays $[P_{i,k}]$, $[U_{i,k}]$ and $[V_{i,k}]$ are known beforehand. Owing to the higher accuracy required, the cubic spline technique is inadequate and a more accurate differentiation technique must be employed.

It is now of great importance that the derivatives of higher orders can be calculated in the form of *local* finite differences, so that there is no need to store them all. For conciseness, we temporarily introduce the notation presented in Fig. 13.3. The appropriate Taylor series expansions about the central node 0 yield the formulae

$$\begin{aligned} P_{|uu} &= 2(P_1 - 2P_0 + P_5)/h^2 - 0.5(U_1 - U_5)/h \\ P_{|vv} &= 2(P_3 - 2P_0 + P_7)/h^2 - 0.5(V_3 - V_7)/h \\ P_{|uv} &= -0.25(P_2 - P_4 + P_6 - P_8)/h^2 \\ &\quad + 0.5(V_1 - V_5 + U_3 - U_7)/h \end{aligned} \qquad (13.20)$$

These derivatives refer to the central node, and their remainder is of fourth order in the mesh-length.

This method can be extended to the determination of derivatives of still higher orders. Even some of the derivatives of fifth order can be calculated in this way, but the finite differences involved then become quite numerous. For reasons of space we must confine this account to the simplest non-trivial case, which is sufficient in very many practical applications. The mixed derivatives of third order are given by fairly simple expressions:

$$\begin{aligned} P_{|uuv} &= 0.25(U_2 - U_4 + U_6 - U_8)/h^2 + O(h^2) \\ P_{|uvv} &= 0.25(V_2 - V_4 + V_6 - V_8)/h^2 + O(h^2) \end{aligned} \qquad (13.21)$$

On an *axis of symmetry*, typically the optic axis ($v = 0$), some of the neighbours of the node 0 are missing. We can either extend the arrays

13.2 TWO-DIMENSIONAL INTERPOLATION

beyond this axis and fill them up according to the symmetry or make explicit use of the symmetry. In the case of positive symmetry, $P(u, -v) = P(u, v)$, the finite differences for an axial point ($v_0 = 0$) take the simple form

$$P_{|vv} = 4(P_3 - P_0)/h^2 - V_3/h$$
$$P_{|uv} = P_{|uuv} = 0 \qquad (13.22)$$
$$P_{|uvv} = 0.5(V_2 - V_4)/h^2$$

while $P_{|uu}$ remains unaltered.

All these finite differences are simple enough to be recalculated in every new call of the field-calculation program. We have to identify the central node 0 of Fig. 13.3 with each of the four corners in Fig. 13.2 in turn and then calculate the corresponding derivatives. This results in a 16-point configuration and the evaluation of 20 simple finite differences.

The necessary interpolations are written most concisely in the form

$$X(u, v) = \sum_{j=1}^{2} \sum_{l=1}^{2} \Big\{ f_j(s) g_l(t) X_{J,L} + h g_j(s) f_l(t) X_{J,L|u}$$
$$+ h f_j(s) g_l(t) X_{J,L|v} + h^2 g_j(s) g_l(t) X_{J,L|uv} \Big\} \qquad (13.23)$$
$$(J = i + j - 2 \quad , \quad L = k + l - 2)$$

where the symbol X denotes P, U or V, respectively, and the subscripts J and L refer to the four corners of the mesh cell in question. Only the arrays for the potential and the first-order derivatives are stored; the rest are recalculated, but the time spent on the latter is compensated for by the saving in the computation of differentiated form functions. Equation (13.23) implies that the same procedure is to be carried out three times, but with different coefficients. The design of this interpolation scheme is such that the field strengths—that is, the derivatives U and V—are continuously differentiable. Even the second-order derivatives, needed in a procedure to be outlined in Chapter 34, are fairly smooth.

In practice, many operations can be saved if the calculations are performed with a mesh-length $h = 1$, to which all stored derivatives and calculated finite differences must refer. Each computed result is finally multiplied only once by the appropriate power of the actual mesh-length. We have not presented this version here for pedagogic reasons but we recommend it for any real program. Moreover, some computation time can be saved by calculating the finite differences referring to the four corners of the *same* mesh cell only once even though these quantities are needed several times. This situation can arise if several subsequent points Q of reference in a very accurate ray-tracing program are located in the same cell. Such points can easily be identified by comparing the subscripts (i, k) with those of the previous call and by skipping the corresponding parts of the procedure when they are the same.

In conclusion, the field interpolation can be made sufficiently accurate and fast for the purposes of ray tracing. A still more accurate but also more sophisticated method of interpolation has been worked out by Killes (1985), to which we refer for the details.

Part III

The Paraxial Approximation

14
Introduction

The general form of the trajectory equations in electromagnetic fields has been derived in Part I (3.22) but in many practical situations these equations are unnecessarily complicated. In a very large class of electron optical instruments, the electrons remain in the vicinity of a curve, frequently a straight line, which we call the optic axis. The behaviour of the various optical elements can then be characterized by simpler equations, obtained by expanding the fields and potentials about this axis and retaining only the terms of lowest order. We shall see that these equations are usually second-order, linear, homogeneous differential equations and their solutions describe the linear imaging properties of lenses of various kinds. With a little care, mirrors can also be included and some aspects of electron guns and cathode lenses can even be characterized in this way.

We shall give two derivations of the paraxial equations for systems with an axis of rotational symmetry, since these are of such importance. First, we simply insert the series expansions for the components of the magnetic flux B and the electrostatic potential Φ into the general equations (3.22) and neglect all but the terms of lowest order. In the alternative derivation, we expand the characteristic function $\overline{M}$ (4.25) as a power series in the off-axis coordinates and their derivatives; the Euler equations (4.26) of the variational relation (4.33) then yield the paraxial trajectory equation if we retain only quadratic terms in the expansion. For systems of lower symmetry, we employ only one of these methods, usually the latter. The function $\overline{M}$ will almost invariably be scaled with respect to $(2m_0 e)^{\frac{1}{2}}$, as in (15.23). The momentum then scales to $\hat{\phi}^{\frac{1}{2}}$ and we shall often refer to this quantity as the momentum, with components $\hat{\phi}^{\frac{1}{2}} x'$ and $\hat{\phi}^{\frac{1}{2}} y'$, though it of course does not have the proper dimensions.

The presence of a magnetic field leads us to introduce a new coordinate system, twisted about the z-axis with respect to the cartesian system in terms of which the field expansions are given in Part II. In this Part therefore, we denote the 'fixed' cartesian system by (X, Y, z), reserving the lower-case (x, y, z) for the twisted or 'rotating' coordinate system, in terms of which all later calculations will be performed. Only static fields will be considered here. Dynamic fields are more conveniently treated separately.

15
Systems with an Axis of Rotational Symmetry

Round lenses are by far the most common in electron optical instruments and we now examine their paraxial properties in detail. A typical electrostatic lens consists of two, three or more electrodes, in the form of plates in which round holes have been cut, their centres lying on a common axis (Fig. 15.1a), or of cylinders, again with the same axis (Fig. 15.1b). Although in theory the field extends indefinitely, in practice it rapidly becomes negligibly small and we speak of the field region and the field-free space outside it. Electrostatic lenses may have an overall accelerating or retarding effect, in which case the constant potential in front of the lens is not the same as that behind it (Fig. 15.1c); they are then often known as *immersion lenses* even though a real object is rarely immersed in the electrode field. An exception is the cathode lens, which is terminated by an unperforated electrode, the properties of which are to be studied. If the lens has no overall accelerating effect, in practice it very often has three electrodes (Fig. 15.1d) and is then known as an *einzel lens* or *unipotential lens*. We reserve the term einzel lens for a three-electrode design such as that illustrated in Fig. 15.1d. A special case of the electrostatic lens is the electron gun (Fig. 15.1e), in which electrons are generated by a filament or cathode, in the form of a point or hairpin, and rapidly accelerated to the operating voltage of the instrument in which they are employed. Guns need special treatment, however, and are discussed in detail in Part IX.

Round magnetic lenses are devices that generate a rotationally symmetric magnetic field, effectively confined to a narrow region. The traditional design, which has changed little since its introduction by Ruska in the early 1930s (Knoll and Ruska, 1932a,b; Ruska, 1934a,b), consists of a large number of windings enclosed in an iron casing; a slot in the latter, finished with circular polepieces, concentrates the field as shown schematically in Fig. 15.2a. In some designs, the windings are in the superconducting state and carry persistent currents. In others, the entire lens is in the superconducting state and the field is confined by a diamagnetic shield (Fig. 15.2b). More radical departures from this geometry are employed for special purposes; two extreme shapes are illustrated in Figs 15.2c and d.

All these types of magnetic lens rely on current-carrying conductors to provide the magnetomotive force. Permanent magnets may be used instead,

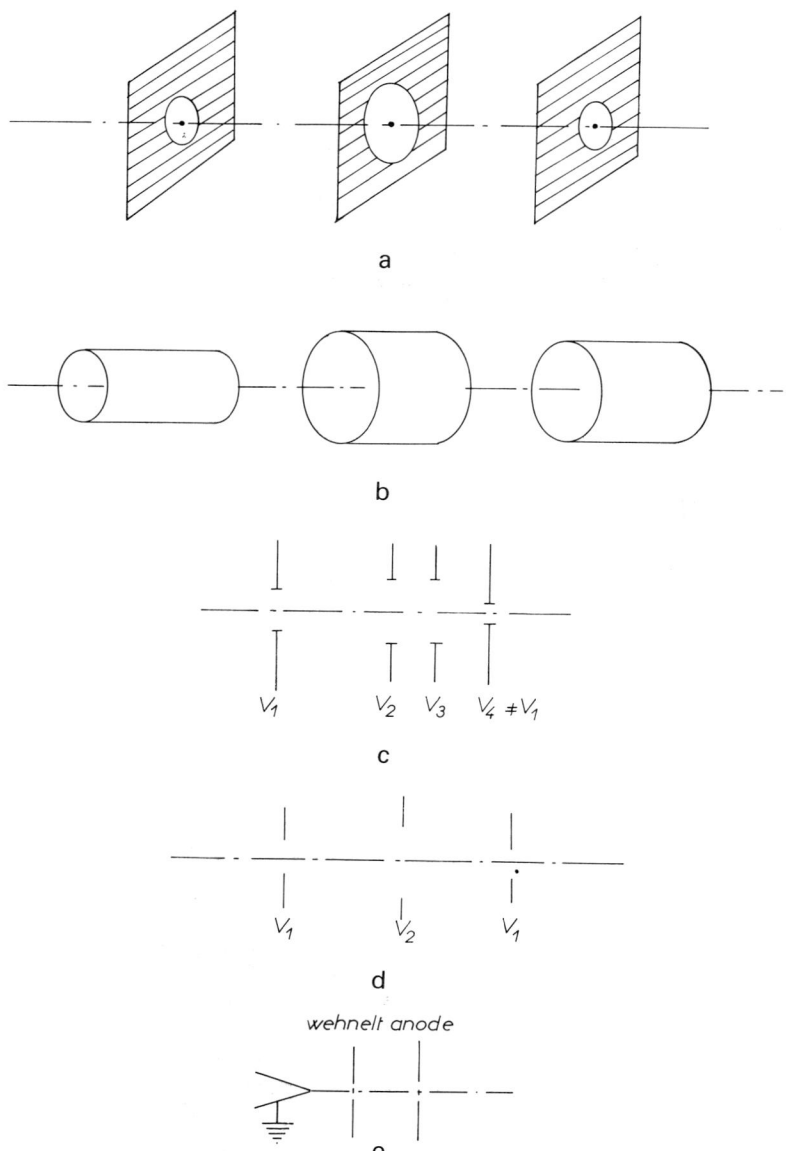

Fig. 15.1: Forms of electrostatic lenses. (a) Plates with circular openings. (b) Cylinders along a common axis. (c) Lens with an overall accelerating $(V_4 < V_1)$ or retarding ($V_4 > V_1$) effect. (d) Einzel or unipotential lens. (e) Gun nomenclature.

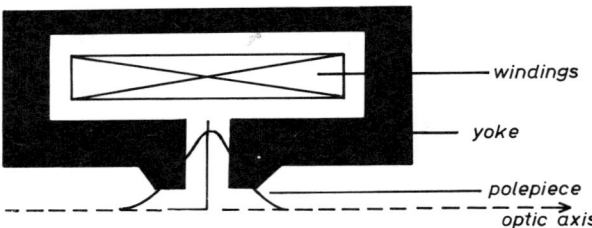

a

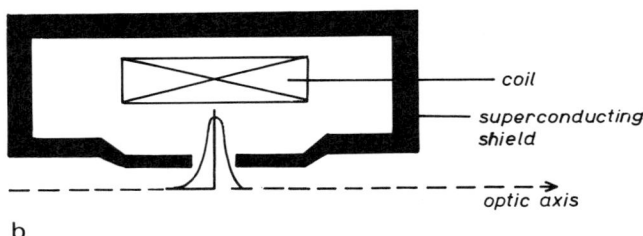

b

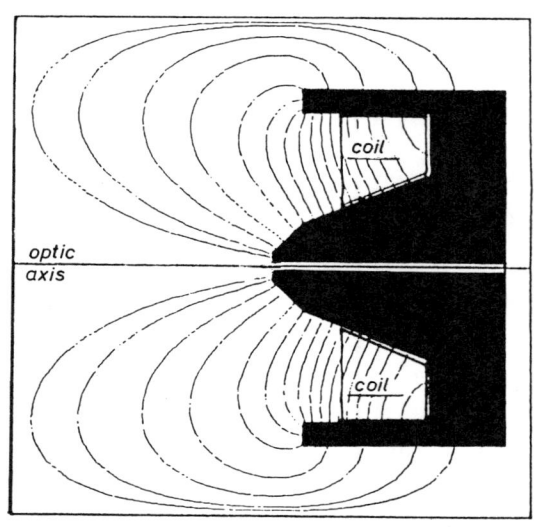

c

205

d

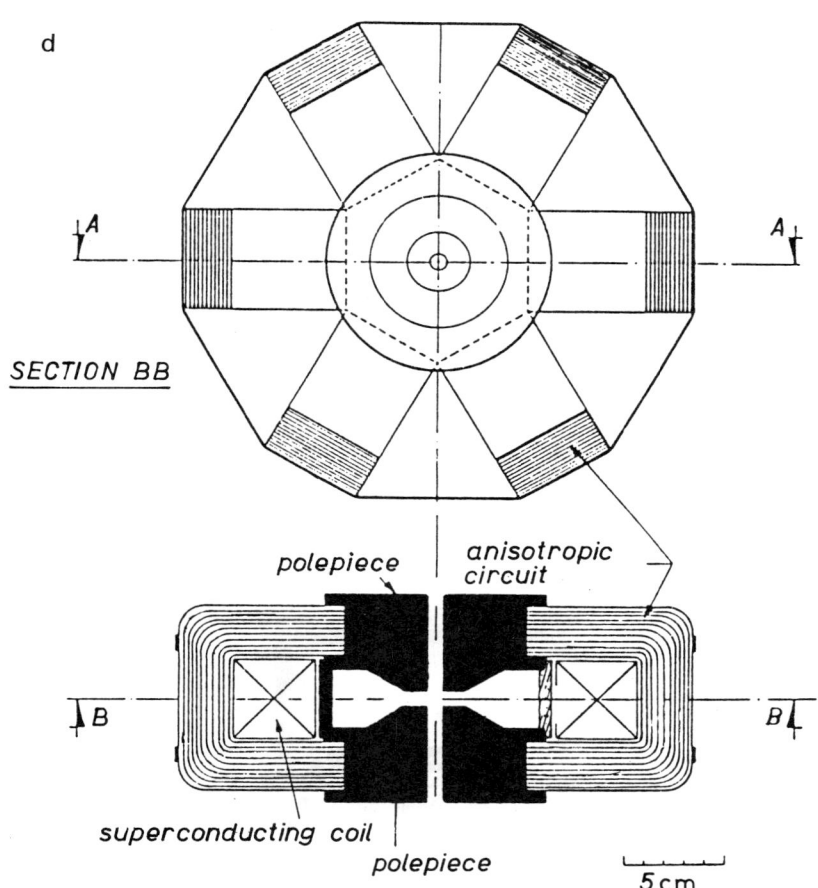

SECTION BB

polepiece
anisotropic circuit

superconducting coil
polepiece

5 cm

SECTION AA

e

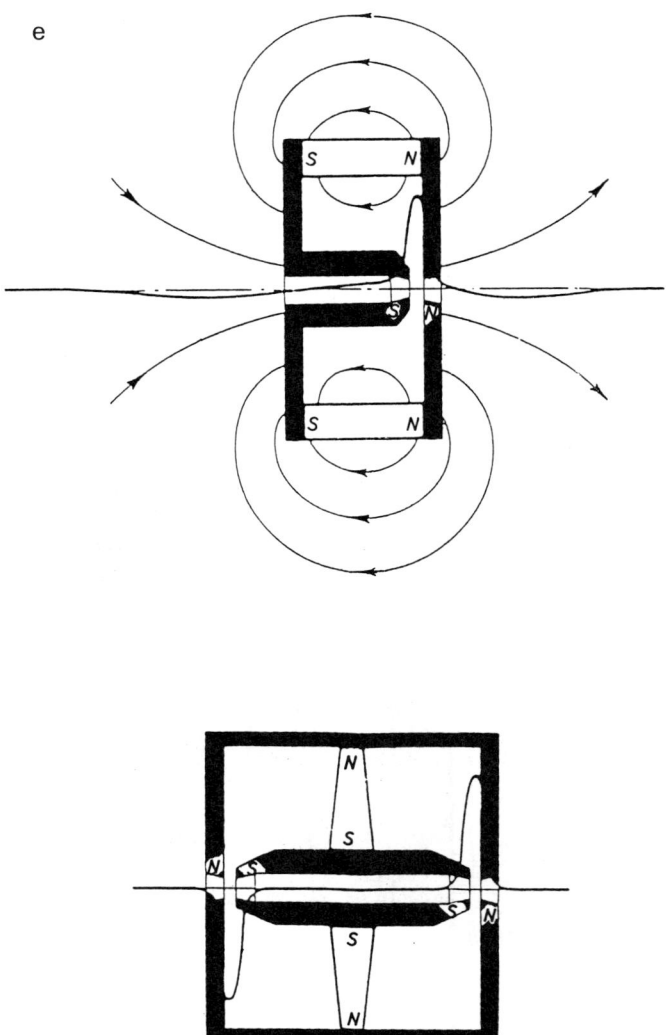

Fig. 15.2: Forms of magnetic lenses. (a) Conventional lens. (b) Superconducting shielding lens. (c) Single-pole lens (with bore). (d) Laminated lens. (e)–(f) Permanent-magnet lenses.

and have been incorporated into commercial electron microscopes, but their inflexibility is a severe handicap. Figures 15.2e and f show permanent-magnet lenses and the axial fields within them.

Real objects are regularly immersed deeply within the field of magnetic lenses, unlike electrostatic lenses. When discussing their properties, we need distinguish only two types of lens, those in which a real object or image is situated within the field and those in which the lens transfers an intermediate image from one plane to another. Nevertheless it is usual to describe lenses in terms of the role they play. Thus we speak of condenser lenses if the 'intermediate image' being transferred is the image of the source of a microscope but of intermediates or projectors if a genuine image is in question. Geometrically, these might be very similar. Likewise, the same lens may be an objective, if the specimen is immersed in it, or a probe-forming lens, if it forms a fine probe within (or indeed outside) the field. These distinctions will become more clear in Part VII.

15.1 Derivation of the paraxial ray equations from the general ray equations

We now derive the paraxial trajectory equations for electrostatic and magnetic lenses. Into the general trajectory equations (3.22), we substitute the expansions for $\Phi(X, Y, z)$ and $B(X, Y, z)$ given in Part II. We now set Φ_0 equal to zero, thereby disregarding chromatic effects. From (7.14), we find

$$\frac{\partial \hat{\Phi}}{\partial X} \approx -\frac{1}{2}(1 + 2\epsilon\phi)X\phi'' = -\frac{1}{2}\gamma X\phi''$$
$$\frac{\partial \hat{\Phi}}{\partial z} \approx (1 + 2\epsilon\phi)\phi' = \gamma\phi' \quad (15.1)$$

where as usual $\gamma = m/m_0 = (1 - v^2/c^2)^{-1/2} = 1 + 2\epsilon\Phi$ (2.2, 2.21). Neglecting quadratic and higher order terms in X, Y and their derivatives, the electrostatic terms on the right-hand side of (3.22a) become

$$\frac{1 + X'^2 + Y'^2}{2\hat{\Phi}}\left(\frac{\partial \hat{\Phi}}{\partial X} - X'\frac{\partial \hat{\Phi}}{\partial z}\right) \approx -\frac{\gamma}{4\hat{\phi}}(2X'\phi' + X\phi'') \quad (15.2)$$

with a similar expression for (3.22b).

For the magnetic term, we substitute the appropriate expansions (7.18–19); noting that B_t can be replaced by B_z since we are neglecting quadratic terms, we find

$$\frac{\eta\varrho^2}{\sqrt{\hat{\Phi}}}(\varrho B_Y - Y'B_t) \approx -\frac{\eta}{2\sqrt{\hat{\phi}}}(YB' + 2Y'B) \quad (15.3a)$$

and
$$\frac{\eta\varrho^2}{\sqrt{\hat{\Phi}}}(-\varrho B_X + X'B_t) \approx \frac{\eta}{2\sqrt{\hat{\phi}}}(XB' + 2X'B) \tag{15.3b}$$

The pair of trajectory equations (3.22) thus collapse to the following in the paraxial approximation:

$$X'' + \frac{\gamma\phi'}{2\hat{\phi}}X' + \frac{\gamma\phi''}{4\hat{\phi}}X + \frac{\eta B}{\sqrt{\hat{\phi}}}Y' + \frac{\eta B'}{2\sqrt{\hat{\phi}}}Y = 0$$
$$Y'' + \frac{\gamma\phi'}{2\hat{\phi}}Y' + \frac{\gamma\phi''}{4\hat{\phi}}Y - \frac{\eta B}{\sqrt{\hat{\phi}}}X' - \frac{\eta B'}{2\sqrt{\hat{\phi}}}X = 0 \tag{15.4}$$

This pair of coupled linear differential equations can be cast into a simpler form by replacing the coordinate system (X, Y, z) by a new system, rotated with respect to the former by a *variable* angle $\theta(z)$. In order to see this, we introduce the complex coordinate (7.3)

$$w = X + iY \tag{15.5}$$

so that (15.4) become

$$w'' + \frac{\gamma\phi'}{2\hat{\phi}}w' + \frac{\gamma\phi''}{4\hat{\phi}}w - i\frac{\eta B}{\sqrt{\hat{\phi}}}w' - i\frac{\eta B'}{2\sqrt{\hat{\phi}}}w = 0 \tag{15.6}$$

The final two terms containing i explicitly can be removed by introducing a new complex coordinate, u, such that

$$w =: u \exp i\theta(z) \tag{15.7}$$

Equation (15.6) becomes

$$u'' + u'\left(2i\theta' + \frac{\gamma\phi'}{2\hat{\phi}} - \frac{i\eta B}{\sqrt{\hat{\phi}}}\right)$$
$$+ u\left\{i\theta'' - \theta'^2 + i\theta'\left(\frac{\gamma\phi'}{2\hat{\phi}} - \frac{i\eta B}{2\sqrt{\hat{\phi}}}\right) + \frac{\gamma\phi''}{4\hat{\phi}} - \frac{i\eta B'}{2\sqrt{\hat{\phi}}}\right\} = 0 \tag{15.8}$$

and the terms explicitly involving i vanish if we choose

$$\theta' = \frac{\eta B}{2\hat{\phi}^{\frac{1}{2}}} \tag{15.9}$$

15.1 DERIVATION OF THE PARAXIAL RAY EQUATION

(so that $\theta'' = \eta B'/2\hat{\phi}^{1/2} - \eta B\gamma\phi'/4\hat{\phi}^{3/2}$) and we obtain

$$u'' + \frac{\gamma\phi'}{2\hat{\phi}} u' + \frac{\gamma\phi'' + \eta^2 B^2}{4\hat{\phi}} u = 0 \tag{15.10}$$

We note that (15.9) is essentially the same as (2.39). Explicitly, writing

$$u = x + iy \tag{15.11}$$

we have

$$x'' + \frac{\gamma\phi'}{2\hat{\phi}} x' + \frac{\gamma\phi'' + \eta^2 B^2}{4\hat{\phi}} x = 0$$

$$y'' + \frac{\gamma\phi'}{2\hat{\phi}} y' + \frac{\gamma\phi'' + \eta^2 B^2}{4\hat{\phi}} y = 0 \tag{15.12}$$

or again

$$\frac{d}{dz}\left(\hat{\phi}^{\frac{1}{2}} x'\right) + \frac{\gamma\phi'' + \eta^2 B^2}{4\hat{\phi}^{\frac{1}{2}}} x = 0$$

$$\frac{d}{dz}\left(\hat{\phi}^{\frac{1}{2}} y'\right) + \frac{\gamma\phi'' + \eta^2 B^2}{4\hat{\phi}^{\frac{1}{2}}} y = 0 \tag{15.13}$$

Physical significance of the coordinate rotation

This transformation to the rotating coordinate system (x, y, z) is of great importance. We therefore consider it in more detail before proceeding. The complex transformation (15.7) may be written

$$\begin{aligned} X &= x\cos\theta - y\sin\theta \\ Y &= x\sin\theta + y\cos\theta \end{aligned} \tag{15.14}$$

so that in any plane $z = $ const, the axes X-Y are inclined at an angle $\theta(z)$ to x-y (Fig. 15.3). This angle increases monotonically provided that the sign of $B(z)$ does not change and the x-y-axes therefore twist round the z-axis like the blades of a propeller or the ridge of a screw of variable pitch. Figure 15.3a gives a perspective view of this and 15.3b shows a view along the z-axis. We shall see in Part VII that $\int_{-\infty}^{\infty} B(z)\,dz = 0$ in any permanent-magnet lens and the total rotation in such a lens is hence zero.

Unlike a conventional cartesian coordinate system the coordinate *surfaces* are not planes: the surfaces $x = 0$ and $y = 0$ are curved, though everywhere normal to each other, intersecting along the z-axis. The element of length ds in (x, y, z) is not equal to $(dx^2 + dy^2 + dz^2)^{1/2}$ but is given by

$$\begin{aligned} ds^2 &= dX^2 + dY^2 + dz^2 \\ &= dx^2 + dy^2 + dz^2\{1 + (x^2 + y^2)\theta'^2\} + 2(x\,dy - y\,dx)\theta'\,dz \end{aligned} \tag{15.15}$$

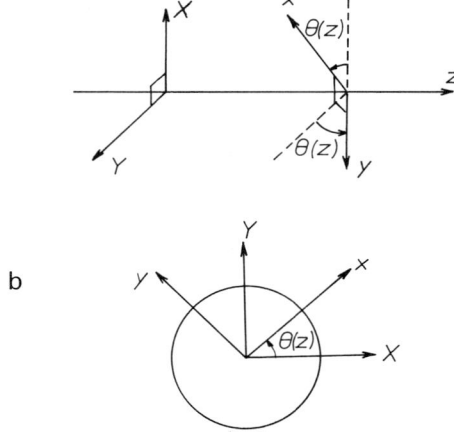

Fig. 15.3: Fixed coordinates (X, Y, Z) and rotating coordinates (x, y, z). (a) Perspective view. (b) View along the optic axis.

This rotation about the axis is closely related to the phenomenon of Larmor precession; if we express the rate of change of θ as a function of time rather than axial distance, using $d\theta/dt = \theta' dz/dt$ and $dz/dt = 2\eta \hat{\phi}/\gamma$, we find, in agreement with (2.38)

$$\frac{d\theta}{dt} = \frac{e}{2m_0} \frac{B}{\gamma} = \frac{\eta^2 B}{\gamma} \tag{15.16}$$

or using (3.9)

$$\frac{d\theta}{d\tau} = \eta^2 B \tag{15.16a}$$

which is indeed the Larmor precession frequency.

The fact that the paraxial equations separate in the rotating coordinates implies that an electron initially travelling on one of the coordinate surfaces, or on any surface $\alpha x + \beta y = 0$, remains on this surface; this leads us to ask what becomes of the angular momentum, and in particular of its axial component N (4.14). We have

$$N = (\bm{r} \times \bm{p})_z \tag{15.17}$$

in which we recall that $\bm{p}$ is the canonical momentum (4.12), $\bm{p} = \bm{g} - e\bm{A}$ and $\bm{g}$ is the kinetic momentum (2.12, 2.19). In the paraxial approximation,

$$\bm{g} := (2m_0 e \hat{\Phi})^{\frac{1}{2}} \bm{t} \approx (2m_0 e \hat{\phi})^{\frac{1}{2}} (\bm{i}_z + r' \bm{i}_r + r\varphi' \bm{i}_\varphi)$$

15.1 DERIVATION OF THE PARAXIAL RAY EQUATION

for $\Phi_0 = 0$ and

$$-e\boldsymbol{A} = -eAi_\varphi \approx -\frac{1}{2}erBi_\varphi$$

so that

$$\boldsymbol{p} = \boldsymbol{g} - e\boldsymbol{A} = (2m_0 e\hat{\phi})^{\frac{1}{2}}(i_z + r'i_r) + \left\{(2m_0 e\hat{\phi})^{\frac{1}{2}}\varphi' - \frac{1}{2}eB\right\}ri_\varphi \quad (15.18)$$

Hence

$$N = \left\{(2m_0 e\hat{\phi})^{\frac{1}{2}}\varphi' - \frac{1}{2}eB\right\}r^2 \quad (15.19)$$

or writing $XY' - X'Y = r^2\varphi'$

$$N = (2m_0 e\hat{\phi})^{\frac{1}{2}}(XY' - X'Y) - \frac{1}{2}eBr^2 \quad (15.20)$$

From (15.14) we have

$$XY' - X'Y = xy' - x'y + (x^2 + y^2)\theta' = xy' - x'y + \frac{\eta Br^2}{2\hat{\phi}^{\frac{1}{2}}}$$

and hence

$$N = \sqrt{2m_0 e\hat{\phi}}\,(xy' - x'y) \quad (15.21)$$

so that in the rotating coordinate system, the magnetic field does not appear explicitly in N. If N vanishes, the azimuthal angle φ remains at a constant angular distance from θ, $\varphi = \theta + \text{const}$ and the corresponding trajectories are said to be *meridional*. They lie in the curved surfaces defined by $\alpha x + \beta y = 0$, which intersect along the z-axis and are inclined at some fixed angle to the coordinate surfaces $x = 0$ and $y = 0$. Clearly any trajectory that intersects the axis at some point is a meridional trajectory, and vice versa. Rays that are not meridional are said to be *skew*. *Note:* we have used cartesian coordinates (X, Y, z) and rotating pseudo-cartesian coordinates (x, y, z) above in preference to polar coordinates, since the calculation becomes complicated when skew rays are considered in the latter system. Extensive discussion of the correct way of handling skew rays when polars are used is to be found in most of the earlier texts on electron optics (e.g. de Broglie, 1950; Rusterholz, 1950; Picht, 1963); the problem vanishes when cartesians are employed (Glaser, 1952, Section 42).

15.2 Variational derivation of the paraxial equations

We now take as our starting point (4.34–4.36), expanding the function $\overline{M}$ that plays the role of refractive index as a power series in x and y and retaining only quadratic terms. From (4.35) and (2.13), we have

$$\overline{M}(X,Y,X',Y',z) = \left\{2m_0 e\Phi(1+\epsilon\Phi)(1+X'^2+Y'^2)\right\}^{\frac{1}{2}} \quad (15.22)$$
$$- e(X'A_X + Y'A_Y + A_z)$$

Substituting for A_X, A_Y and A_z from (7.43–7.45) and for Φ from (7.36) into

$$M := \frac{\overline{M}}{(2m_0 e)^{\frac{1}{2}}} \quad (15.23)$$
$$= \left\{\hat{\Phi}(1+X'^2+Y'^2)\right\}^{\frac{1}{2}} - \eta(X'A_X + Y'A_Y + A_z)$$

we obtain a power series in X,Y and their derivatives, the quadratic terms of which, $M^{(2)}$, are given by

$$M^{(2)} = -\frac{\gamma\phi''}{8\hat{\phi}^{\frac{1}{2}}}(X^2+Y^2) + \frac{1}{2}\hat{\phi}^{\frac{1}{2}}(X'^2+Y'^2) - \frac{1}{2}\eta B(XY' - X'Y) \quad (15.24)$$

It is already clear that the Euler equations of $\delta \int M^{(2)} dz = 0$ will be coupled and we therefore attempt to transform the coordinates in such a way that the mixed term in $XY' - X'Y$, the source of the coupling, is eliminated. From (15.14) we obtain

$$\begin{aligned} X^2 + Y^2 &= x^2 + y^2 \\ X'^2 + Y'^2 &= x'^2 + y'^2 + 2\theta'(xy' - x'y) + \theta'^2(x^2+y^2) \\ XY' - X'Y &= xy' - x'y + \theta'(x^2+y^2) \end{aligned} \quad (15.25)$$

Substituting into $M^{(2)}$, we find

$$M^{(2)} = (x^2+y^2)\left(-\frac{\gamma\phi''}{8\hat{\phi}^{\frac{1}{2}}} + \frac{\hat{\phi}^{\frac{1}{2}}}{2}\theta'^2 - \frac{1}{2}\eta B\theta'\right)$$
$$+ (x'^2+y'^2)\frac{\hat{\phi}^{\frac{1}{2}}}{2} + (xy' - x'y)(\theta'\hat{\phi}^{\frac{1}{2}} - \frac{1}{2}\eta B) \quad (15.26)$$

and the term in $xy' - x'y$ vanishes if we select

$$\theta' = \frac{\eta B}{2\hat{\phi}^{\frac{1}{2}}} \tag{15.27}$$

(see 15.9). The function $M^{(2)}$ becomes

$$M^{(2)} = -\frac{1}{8\hat{\phi}^{\frac{1}{2}}}(\gamma\phi'' + \eta^2 B^2)(x^2 + y^2) + \frac{1}{2}\hat{\phi}^{\frac{1}{2}}(x'^2 + y'^2) \tag{15.28}$$

Hence

$$\frac{\partial M^{(2)}}{\partial x'} = \hat{\phi}^{\frac{1}{2}} x' \quad , \quad \frac{\partial M^{(2)}}{\partial y'} = \hat{\phi}^{\frac{1}{2}} y' \tag{15.29}$$

$$\frac{\partial M^{(2)}}{\partial x} = -\frac{1}{4\hat{\phi}^{\frac{1}{2}}}(\gamma\phi'' + \eta^2 B^2)x$$

$$\frac{\partial M^{(2)}}{\partial y} = -\frac{1}{4\hat{\phi}^{\frac{1}{2}}}(\gamma\phi'' + \eta^2 B^2)y \tag{15.30}$$

and the paraxial equations are thus

$$\frac{d}{dz}\left(\hat{\phi}^{\frac{1}{2}} x'\right) + \frac{\gamma\phi'' + \eta^2 B^2}{4\hat{\phi}^{\frac{1}{2}}} x = 0 \tag{15.31}$$

with an identical equation for $y(z)$, as already found (15.13).

15.3 Forms of the paraxial equations and general properties of their solutions

15.3.1 Reduced coordinates
In the absence of an electrostatic field, the paraxial equations take the form

$$u'' + F(z)u = 0 \tag{15.32}$$

($u = x + iy$) with

$$F(z) := \frac{\eta^2 B^2}{4\hat{\phi}} \tag{15.33}$$

When $\phi(z)$ is not constant, they can again be reduced to this form by a simple transformation of the off-axis coordinates. We write

$$u(z) =: v(z)a(z) \tag{15.34}$$

in which v is a new reduced complex coordinate and $a(z)$ is a real function, chosen so that all terms involving dv/dz disappear. Substituting (15.34) into (15.10), we obtain

$$v'' + \left(2\frac{a'}{a} + \frac{\gamma\phi'}{2\hat{\phi}}\right)v' + \left(\frac{a''}{a} + \frac{a'}{a}\frac{\gamma\phi'}{2\hat{\phi}} + \frac{\gamma\phi'' + \eta^2 B^2}{4\hat{\phi}}\right)v = 0 \quad (15.35)$$

and the coefficient of v' vanishes if

$$\frac{a'}{a} = -\frac{\gamma\phi'}{4\hat{\phi}} = -\frac{\hat{\phi}'}{4\hat{\phi}} \quad (15.36)$$

or

$$a(z) = \hat{\phi}^{-\frac{1}{4}} \quad (15.37)$$

giving

$$v''(z) + G(z)v(z) = 0 \quad (15.38)$$

with

$$G(z) := \frac{3}{16}\left(\frac{\phi'}{\hat{\phi}}\right)^2 \left(1 + \frac{4}{3}\epsilon\hat{\phi}\right) + \frac{\eta^2 B^2}{4\hat{\phi}} \quad (15.39)$$

and

$$u(z) = v(z)/\hat{\phi}^{\frac{1}{4}} \quad (15.40)$$

The substitution (15.40) was introduced into electron optics by Picht (1932), and is widely known as Picht's transformation; see too Glaser (1933a-d) and Cotte (1938). This result is of interest for two reasons. First, it is simpler to perform numerical calculations with (15.38) than (15.10). Secondly, the function $G(z)$ is essentially non-negative, and we shall see that this imposes an interesting restriction on electron lenses: they always exert a converging action.

15.3.2 Stigmatic image formation

The paraxial equations are linear, homogeneous and of second order and their most general solution is therefore of the form

$$u(z) = Au_1(z) + Bu_2(z) \quad (15.41)$$

in which $u_1(z)$ and $u_2(z)$ are any pair of linearly independent solutions of (15.10). We shall find it necessary to introduce several such pairs of solutions and we shall adopt a consistent notation for each in subsequent chapters, but many paraxial properties are quite general and in no way depend on any particular choice. The most important result concerns the

15.3 FORMS OF THE PARAXIAL EQUATION

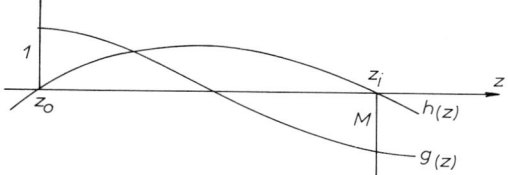

Fig. 15.4: The paraxial solutions $g(z)$ and $h(z)$.

existence of stigmatic image formation. The form of (15.10) or (15.12–15.13) alone is sufficient to predict that pairs of planes can always be found having the properties associated with point-to-point image formation.

Consider a particular solution $h(z)$ of the paraxial equation for the complex coordinate u (15.10) that intersects the axis at $z = z_o$ and $z = z_i$: $h(z_o) = h(z_i) = 0$ (Fig. 15.4). A pencil of rays intersecting the plane $z = z_o$ at some point P_o ($u_o = x_o + iy_o$) may be described by

$$u(z) = u_o g(z) + \lambda h(z) \tag{15.42}$$

in which $g(z)$ is a solution of (15.10) that is linearly independent of $h(z)$; for convenience, we have set $g(z_o) = 1$; λ is a (complex) parameter characterizing the various members of the pencil. In the plane $z = z_i$, we have

$$u(z_i) = u_o g(z_i) \tag{15.43}$$

for all λ and hence for every ray passing through P_o. Since this is true of all points in the plane $z = z_o$, the latter will be stigmatically imaged in $z = z_i$. Moreover the ratio $u(z_i)/u_o$ is constant and so the distribution of points P_i will be identical with the distribution of P_o, apart from a change of scale: the image is a linearly magnified (or reduced) representation of the object.

If we return to the fixed cartesian system ($w = X + iY, z$), we find

$$w(z_i) = u(z_i) \exp\{i\theta(z_i)\}$$
$$= g(z_i) \exp\left[i\{\theta(z_i) - \theta(z_o)\}\right] w(z_o)$$

or

$$w(z_i) = M(z_i, z_o) w(z_o) \tag{15.44}$$

where

$$M(z_i, z_o) = g(z_i) \exp\left[i\{\theta(z_i) - \theta(z_o)\}\right] \tag{15.45}$$

For single-stage image formation, in which a meridional ray from P_o intersects the axis only once between P_o and P_i, $g(z_i)$ is negative and we may write

$$M(z_i, z_o) = -|g(z_i)| \exp\left[i\{\theta(z_i) - \theta(z_o)\}\right]$$
$$= |g(z_i)| \exp\left[i\{\theta(z_i) - \theta(z_o) + \pi\}\right] \quad (15.46)$$

The modulus of $g(z_i)$, and hence $|M(z_i, z_o)|$, is referred to as the transverse magnification and the image rotation is clearly equal to $\arg(M) - \pi$. Nevertheless, the complex magnification $M(z_i, z_o)$ is rarely used and in the remainder of this book we shall reserve the symbol M for the transverse magnification, regarded as an algebraic quantity:

$$M := g(z_i) \quad (15.47)$$

The notion of complex magnification is valuable when we need to consider the reversal of an imaging system. A pencil of rays from P_i to P_o will not retrace the paths of those from P_o to P_i since the direction of rotation will be opposite. This is readily seen from (15.45) which tells us that

$$M(z_o, z_i) = M^{*-1}(z_i, z_o) \quad (15.48)$$

in which the asterisk denotes the complex conjugate.

A number of useful general relations can be deduced from the form of (15.32) or (15.38) alone. Thus the fact that $F(z)$ and $G(z)$ are never negative tells us that all electron lenses have a net converging action although this need not be true of local zones of electrostatic lenses. To see this, we note that the curvature ϱ of any solution of (15.32) or (15.38), given by $\varrho = u''/(1 + u'^2)^{3/2}$ or $v''/(1 + v'^2)^{3/2}$, is always opposite in sign to u or v respectively. Thus a solution of the appropriate paraxial equation that approaches the field parallel to the axis will be bent towards the latter. If the ray crosses the axis in the field, it will again be bent back towards it and if the field is long enough, the ray will oscillate about the axis. Thus the effect of the field is that of a converging lens. Nevertheless, care is needed here since rays can intersect the axis more than once in a strong lens and, as we shall see in the next section, the sign of the focal length will then be that associated with a divergent lens.

We have been basing our argument on the positivity of $F(z)$ or $G(z)$ and it is safe to conclude that a ray incident from field-free space parallel to the axis will intersect the axis at least once before emerging into image space. It is not, however, necessarily true that actual electron trajectories in electrostatic (or mixed electrostatic and magnetic) fields always bend towards the axis: the term in $(\hat{\phi}_o/\phi)^{1/4}$ may be large enough to reverse the curvature locally. Provided the electron is not driven beyond the paraxial region, however, the convergent action will always dominate, as our reasoning based on $G(z)$ shows.

15.3.3 The Wronskian

Another property of the paraxial equations is the existence of an invariant, the Wronskian, from which a number of interesting optical relations can be derived. Let $u_1(z)$ and $u_2(z)$ be a pair of linearly independent solutions of (15.10), so that

$$\frac{d}{dz}\left(\hat{\phi}^{\frac{1}{2}}u_1'\right) + \frac{\gamma\phi'' + \eta^2 B^2}{4\hat{\phi}^{\frac{1}{2}}} u_1 = 0$$
$$\frac{d}{dz}\left(\hat{\phi}^{\frac{1}{2}}u_2'\right) + \frac{\gamma\phi'' + \eta^2 B^2}{4\hat{\phi}^{\frac{1}{2}}} u_2 = 0 \quad (15.49)$$

Multiplying the first equation by u_2 and the second by u_1 and subtracting, it is easy to show that

$$\frac{d}{dz}\left\{\hat{\phi}^{\frac{1}{2}}(u_1 u_2' - u_1' u_2)\right\} = 0 \quad (15.50)$$

or

$$\hat{\phi}^{\frac{1}{2}}(u_1 u_2' - u_1' u_2) = \text{const} \quad (15.51)$$

The same is of course true of any pair of solutions of the separate equations for $x(z)$ and $y(z)$ (15.13).

Suppose we choose $u_1(z) = h(z)$ and $u_2(z) = g(z)$, where as before (15.42) $g(z_o) = 1$, $g(z_i) = M$ and $h(z_o) = h(z_i) = 0$; we find

$$\hat{\phi}_o^{\frac{1}{2}} h_o' = \hat{\phi}_i^{\frac{1}{2}} h_i' M \quad (15.52)$$

But h_i'/h_o' is the angular magnification, M_α, and we have thus shown that

$$M M_\alpha = (\hat{\phi}_o/\hat{\phi}_i)^{\frac{1}{2}} \quad (15.53)$$

or, if $\phi_o = \phi_i$ as in the case of magnetic and electrostatic einzel lenses:

$$M_\alpha = \frac{1}{M} \quad (\phi_i = \phi_o) \quad (15.54)$$

We may rewrite (15.52) as

$$g(z_i) h'(z_i) \hat{\phi}_i^{\frac{1}{2}} = g(z_o) h'(z_o) \hat{\phi}_o^{\frac{1}{2}} \quad (15.55)$$

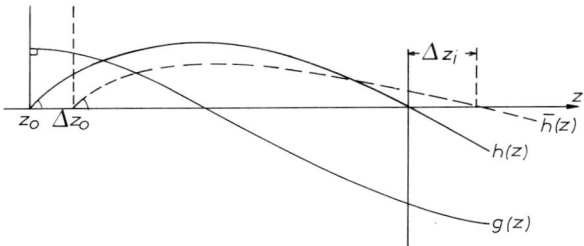

Fig. 15.5: The notion of longitudinal magnification.

which is known in light optics as the Smith–Helmholtz formula (Born and Wolf, eq. 4.4.49); it is also associated with the names of Clausius and Lagrange and was known in a more primitive form to Cotes and Huygens (see Rayleigh, 1886 and Czapski and Eppenstein, 1924 p.116).

A final related quantity is the longitudinal magnification: this tells us how far the image plane moves when the object plane is shifted a small distance. Consider again the pair of rays $g(z)$ and $h(z)$, which we now specify completely as the rays that satisfy the boundary conditions

$$g(z_o) = g'(z_o) = 1$$
$$g'(z_o) = h(z_o) = 0 \tag{15.56}$$

in the original object plane. In the image plane $h(z_i) = 0$ and $g(z_i) = M$. For a neighbouring object plane, distant Δz_o from $z = z_o$ (Fig. 15.5), the corresponding 'h-ray' satisfying

$$\overline{h}(z_o + \Delta z_o) = 0$$
$$\overline{h}'(z_o + \Delta z_o) = 1 \tag{15.57}$$

may be written as a linear combination of $g(z)$ and $h(z)$ since there can be only two linearly independent solutions:

$$\overline{h}(z) = Ah(z) + Bg(z) \tag{15.58}$$

Clearly
$$h(z_o + \Delta z_o) \approx \Delta z_o \quad , \quad g(z_o + \Delta z_o) \approx 1$$

so that
$$A = 1 \quad , \quad B = -\Delta z_o$$

giving
$$\overline{h} = h - g\Delta z_o \tag{15.59}$$

In the shifted image plane, $z = z_i + \Delta z_i$, $\overline{h}(z)$ vanishes and so

$$h(z_i + \Delta z_i) - g(z_i + \Delta z_i)\Delta z_o = 0$$

but $h'(z_i) = M_\alpha$ and $g(z_i) = M$ and hence

$$M_\alpha \Delta z_i - M \Delta z_o = 0$$

or

$$\frac{\Delta z_i}{\Delta z_o} =: M_l = \frac{M}{M_\alpha} = \left(\frac{\hat{\phi}_i}{\hat{\phi}_o}\right)^{\frac{1}{2}} M^2 \qquad (15.60)$$

The quantity M_l is known as the longitudinal magnification and we have

$$M_l M_\alpha = M \qquad (15.61)$$

If $\phi_o = \phi_i$, $M_\alpha = 1/M$ (15.54) and

$$M_l = M^2 = \frac{1}{M_\alpha^2} \qquad (\phi_o = \phi_i) \qquad (15.62)$$

15.4 The Abbe sine condition and Herschel's condition

These two conditions do not strictly belong to paraxial optics, for they are conditions under which particular sets of points are imaged stigmatically irrespective of the ray gradient. They are, however, of interest in electron optics mainly in connection with the foregoing results and we therefore make a short digression to establish them here. They are most easily derived from the invariance of the Lagrange bracket (5.34), as shown by Sturrock (1955,) following the example of Herzberger (1931).

The invariance of the Lagrange bracket $\{u, v\}$ may be translated into concrete terms by considering three neighbouring rays, which we label 0, 1 and 2. The ray zero connects two points A, B as shown in Fig. 15.6; at these points, $\boldsymbol{r} = \boldsymbol{r}_a$, $\boldsymbol{p} = \boldsymbol{p}_a$ and $\boldsymbol{r} = \boldsymbol{r}_b$, $\boldsymbol{p} = \boldsymbol{p}_b$ respectively. The ray 1 is shifted by a small amount from ray 0, so that to its endpoints correspond the values $\boldsymbol{r}_a + \Delta_1 \boldsymbol{r}_a$, $\boldsymbol{p}_a + \Delta_1 \boldsymbol{p}_a$ and $\boldsymbol{r}_b + \Delta_1 \boldsymbol{r}_b$, $\boldsymbol{p}_b + \Delta_1 \boldsymbol{p}_b$; the same is true for the ray 2 except that the increments are now $\Delta_2 \boldsymbol{r}_a$, $\Delta_2 \boldsymbol{p}_a$, $\Delta_2 \boldsymbol{r}_b$ and $\Delta_2 \boldsymbol{p}_b$.

If these rays belong to a congruence, such that ray zero corresponds to the parameters (u, v), ray 1 to $(u + \Delta_1 u, v)$ and ray 2 to $(u, v + \Delta_2 v)$, the invariance of $\{u, v\}$ is equivalent to that of $\Delta_1 \boldsymbol{p} \cdot \Delta_2 \boldsymbol{r} - \Delta_2 \boldsymbol{p} \cdot \Delta_1 \boldsymbol{r}$,

15. SYSTEMS WITH AN AXIS OF ROTATIONAL SYMMETRY

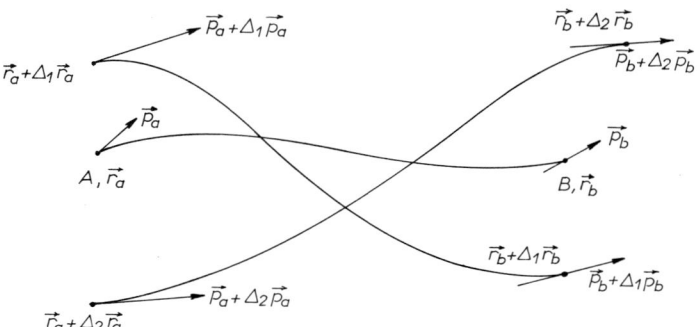

Fig. 15.6: The rays employed in connection with the Lagrange invariant.

which is known as the *Lagrange differential invariant*. We now apply this invariance to two special cases.

Suppose that the points A and B are conjugate and that to every point A' in a plane through A we can find a conjugate point B' in a plane through B. The shift from A to A' and B to B' is chosen to be the displacement Δ_1, while Δ_2 corresponds to the transition to another ray connecting A and B (Fig. 15.7a).

Thus $\Delta_2 \boldsymbol{r}_a = \Delta_2 \boldsymbol{r}_b = 0$ and $\Delta_1 x_a = \Delta_1 x_b = 0$, where the x-axes are taken perpendicular to the planes containing AA' and BB'. The invariance of $\Delta_1 \boldsymbol{p} \cdot \Delta_2 \boldsymbol{r} - \Delta_2 \boldsymbol{p} \cdot \Delta_1 \boldsymbol{r}$ shows that

$$p_a(\Delta_2 t_{ya} \cdot \Delta_1 y_a + \Delta_2 t_{za} \cdot \Delta_1 z_a) = p_b(\Delta_2 t_{yb} \cdot \Delta_1 y_b + \Delta_2 t_{zb} \cdot \Delta_1 z_b) \quad (15.63)$$

where we have written $\boldsymbol{p} = p\boldsymbol{t}$ and $\boldsymbol{t}$ is a unit vector, the components of which are the direction cosines of $\boldsymbol{p}$. The scalar p reduces to $\hat{\phi}^{1/2}$ near the axis.

We now choose the axes Oz_a, Oz_b to coincide with the axes of a rotationally symmetric system and consider points in the planes x_a-z_a, x_b-z_b. Setting $\Delta_1 y_a = \Delta_1 y_b = 0$, $t_{za} = \cos\theta_a$, $t_{zb} = \cos\theta_b$ (Fig 15.7b), we find

$$p_a \cdot \Delta z_a \cdot \sin\theta_a \cdot \Delta_2 \theta_a = p_b \cdot \Delta z_b \cdot \sin\theta_b \cdot \Delta_2 \theta_b \quad (15.64)$$

Writing $\Delta z_b = M_l \Delta z_a$, and integrating with respect to θ, we obtain *Herschel's condition*

$$p_a(\cos\theta_a - 1) = M_l p_b(\cos\theta_b - 1) \quad (15.65)$$

or

$$p_a \sin^2(\theta_a/2) = M_l p_b \sin^2(\theta_b/2) \quad (15.66)$$

15.4 THE ABBE SINE CONDITION AND HERSCHEL'S CONDITION

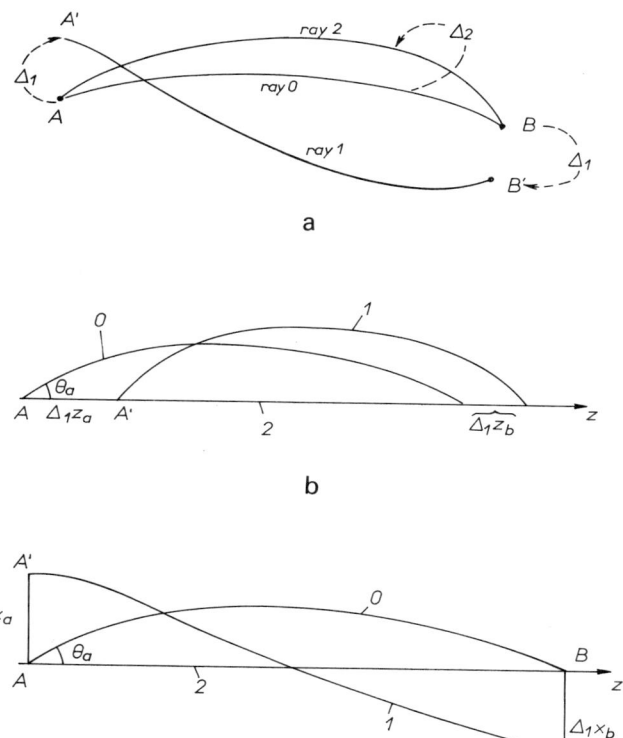

Fig. 15.7: The rays employed in the derivation of (a,b) Herschel's condition and (c) the sine condition.

If this condition is satisfied, an element of the axis close to A will be imaged sharply, even if the ray gradient is not small. If θ_a and θ_b are small, (15.66) becomes

$$p_a \theta_a^2 = M_l p_b \theta_b^2 \qquad (15.67)$$

or

$$p_a/p_b = M_l M_\alpha^2 \qquad (15.68)$$

which is equivalent to (15.61) using $p = \hat{\phi}^{1/2}$ and (15.53).

We may use the Lagrange differential invariant to derive the sine condition; the three rays are now chosen as shown in Fig. 15.7c. From (15.63), we have

$$p_a \cdot \Delta_1 y_a \cdot \cos\theta_a \cdot \Delta_2 \theta_a = p_b \cdot \Delta_1 y_b \cdot \cos\theta_b \cdot \Delta_2 \theta_b$$

(note that $t_{ya} = \sin\theta_a$ and likewise for t_{yb}). Hence

$$p_a \cdot \sin\theta_a \cdot \Delta_1 y_a = p_b \cdot \sin\theta_b \cdot \Delta_1 y_b \tag{15.69}$$

which is known as the *sine condition*, the importance of which was first recognized by Ernst Abbe. When the sine condition is satisfied, a small region around the axis will be imaged sharply irrespective of the ray gradient. Looking ahead to Part IV, this implies that coma must vanish.

For small angles, (15.69) reduces to

$$p_a \theta_a \cdot \Delta_1 y_a = p_b \theta_b \cdot \Delta_1 y_b \tag{15.70}$$

which is equivalent to (15.55).

15.5 Some other transformations

We have seen that the paraxial equations for electrostatic lenses and mixed lenses can be cast into a more convenient form by means of Picht's transformation (15.40). We briefly mention some of the other transformations that have been proposed; we shall meet still others in connection with electron mirrors and with specific field models.

We first enquire whether the term in du/dz in (15.10) can be removed, not by a change of the transverse (dependent) variable as in the Picht transformation but by introducing a different axial (independent) variable. We write

$$\zeta = \zeta(z) \quad , \quad u = u(\zeta) \quad , \quad \phi = \phi(\zeta) \tag{15.71}$$

so that

$$\frac{du}{dz} = \frac{du}{d\zeta}\frac{d\zeta}{dz} \quad \text{and} \quad \frac{d^2u}{dz^2} = \frac{d^2u}{d\zeta^2}\left(\frac{d\zeta}{dz}\right)^2 + \frac{du}{d\zeta}\frac{d^2\zeta}{dz^2} \tag{15.72}$$

The paraxial equation (for electrostatic fields only) thus becomes

$$\ddot{u}\left(\frac{d\zeta}{dz}\right)^2 + \dot{u}\left\{\frac{\dot\phi\gamma}{2\phi}\left(\frac{d\zeta}{dz}\right)^2 + \frac{d^2\zeta}{dz^2}\right\} \\ + \gamma u\,\frac{\ddot\phi\,(d\zeta/dz)^2 + \dot\phi\,(d^2\zeta/dz^2)}{4\hat\phi} = 0 \tag{15.73}$$

in which dots denote differentiation with respect to ζ. The term in $\dot u$ vanishes if

$$\frac{d}{dz}\left(\hat\phi^{\frac{1}{2}}\frac{d\zeta}{dz}\right) = 0$$

15.5 SOME OTHER TRANSFORMATIONS

or, apart from an unimportant multiplicative constant,

$$\zeta(z) = \int^z \hat{\phi}^{-\frac{1}{2}}(z')\,dz' \tag{15.74}$$

giving

$$\ddot{u} + \left\{ \frac{\gamma}{4\hat{\phi}} \ddot{\phi} - \frac{1}{8} \left(\frac{\gamma\dot{\phi}}{\hat{\phi}} \right)^2 \right\} u = 0 \tag{15.75}$$

An incorrect nonrelativistic form of this equation is given by Picht (1963 p.166). Another transformation, also introduced by Picht (1932, 1963 p.167), provides a means of designing lenses for a specific purpose, by generating the potential distribution that will create desired trajectories. We merely indicate the procedure: several examples are worked out in detail by Picht. From (15.10), in which we again set $B = 0$ and consider the nonrelativistic approximation, we see that the paraxial equation can be written

$$3\frac{u''}{u}(u\phi) + (u\phi)'' = 0 \tag{15.76}$$

Setting

$$u\phi =: T(z) \tag{15.77a}$$

and

$$3\frac{u''}{u} =: t(z) \tag{15.77b}$$

(15.76) becomes

$$T''(z) + t(z)T(z) = 0 \tag{15.78}$$

Thus, given $u(z)$ we can calculate $t(z)$, solve (15.78) for $T(z)$ and finally extract $\phi(z)$ from (15.77a). Picht gives another method of solving this problem, which we shall not describe here.

Hitherto, we have discussed the motion of electrons in terms of coordinates of position, deriving the ray gradients by differentiation. Position and canonical momentum are, however, conjugate variables, as explained in Part I, and we should therefore expect to be able to work in terms of either at will. Returning to the equations

$$\frac{d}{dz}\left(\frac{\partial M^{(2)}}{\partial x'}\right) = \frac{\partial M^{(2)}}{\partial x}$$

and writing $p = \partial M^{(2)}/\partial x'$, we see from (15.13) that

$$x = -\frac{4\hat{\phi}^{\frac{1}{2}}}{\gamma\phi'' + \eta^2 B^2} p'$$

so that substituting for x' in $p = x'\hat{\phi}^{1/2}$, we obtain

$$p = -\hat{\phi}^{\frac{1}{2}} \frac{d}{dz}\left(\frac{4p'\hat{\phi}^{\frac{1}{2}}}{\gamma\phi'' + \eta^2 B^2}\right)'$$

or writing

$$G(z) := \frac{\gamma\phi'' + \eta^2 B^2}{4\hat{\phi}^{\frac{1}{2}}}$$

$$p'' - \frac{G'}{G}p' + \frac{G}{\hat{\phi}^{\frac{1}{2}}}p = 0$$

All the rules of Gaussian optics that we shall establish in Chapter 16 could equally well be derived from this equation; this duality is noted in Hawkes (1966).

16
Gaussian Optics of Rotationally Symmetric Systems: asymptotic image formation

16.1 Real and asymptotic image formation

The fact that the paraxial trajectory equations are linear, second order and homogeneous is itself sufficient for us to anticipate that the imaging properties of the corresponding fields can be characterized by a small number of quantities. We discuss this in detail in the following paragraphs but we must first explain the notions of *real* and *asymptotic* image formation; the distinction between these is not the same as that between real and virtual in light optics, despite some similarities.

Since electron lenses consist of regions containing magnetic or electrostatic fields, it is possible, and in practice common, to immerse the specimen of which a magnified image is required within the field itself, particularly in the case of magnetic lenses. The lens field is thus divided into two regions playing different roles (Fig. 16.1). In a light microscope, any lenses preceding the specimen, region I in Fig. 16.1, belong to the condenser system, while the lens immediately after the specimen, region II, is the objective proper. In an electron microscope, different parts of the same lens may thus play different roles. The properties of region I will provide information about the illumination, those of region II about the image formation. In such a situation, it is clearly necessary to study the regions separated by the real object independently and the corresponding characteristics will be referred to as "real".

In a multi-lens system, most of the lenses will simply transfer an intermediate image from one plane to another, with the appropriate magnification, and the entire lens field contributes to this transfer. Here we must study the coordination between incoming and outgoing asymptotes, as shown in Fig. 16.2. If the intermediate image that acts as object for a lens is well outside the lens field, on the object side, the situation is exactly as in light optics. If it falls within the lens field or beyond it, then the *asymptotic object* is analogous to the familiar "virtual object"; similar remarks apply to the image. In this context, we note that when discussing

226   16. GAUSSIAN OPTICS OF ROTATIONALLY SYMMETRIC SYSTEMS

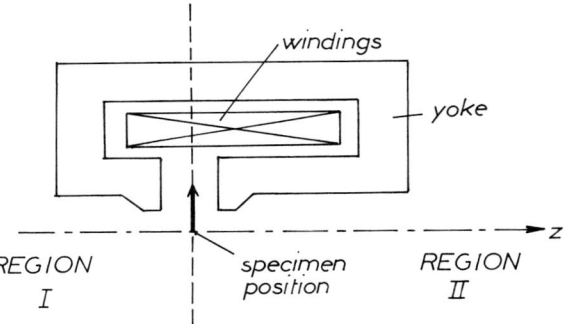

Fig. 16.1: The various parts of a magnetic objective lens.

asymptotic imagery, the notions of object space and image space are used to refer not to regions of physical space—the object may lie anywhere as may the image—but to the space to which the corresponding asymptotes belong. This will become clearer when we discuss asymptotic image formation below.

Finally, we note that both possible hybrid situations may occur: real object–asymptotic image and asymptotic object–real image (formation of a small probe within a field, for example).

16.2 Asymptotic cardinal elements and transfer matrices

We first discuss these matters in terms of specific solutions of the paraxial equations, after which we show that the same results may be obtained from the more abstract notion of bilinear transformations. In this first discussion, we use the form (15.12) of the paraxial equations and for ease of understanding we use the real x-coordinate. The reasoning for y is of course identical and we could naturally have used the complex u.

Consider a field region characterized by $\phi(z)$ and $B(z)$ (Fig. 16.3) and two solutions of the paraxial equation, $G(z)$ and $\overline{G}(z)$, satisfying the boundary conditions

$$\lim_{z \to -\infty} G(z) = 1 \quad , \quad \lim_{z \to \infty} \overline{G}(z) = 1 \qquad (16.1)$$

A general solution thus has the form

$$x(z) = AG(z) + B\overline{G}(z) \qquad (16.2)$$

16.2 ASYMPTOTIC CARDINAL ELEMENTS AND TRANSFER MATRICES

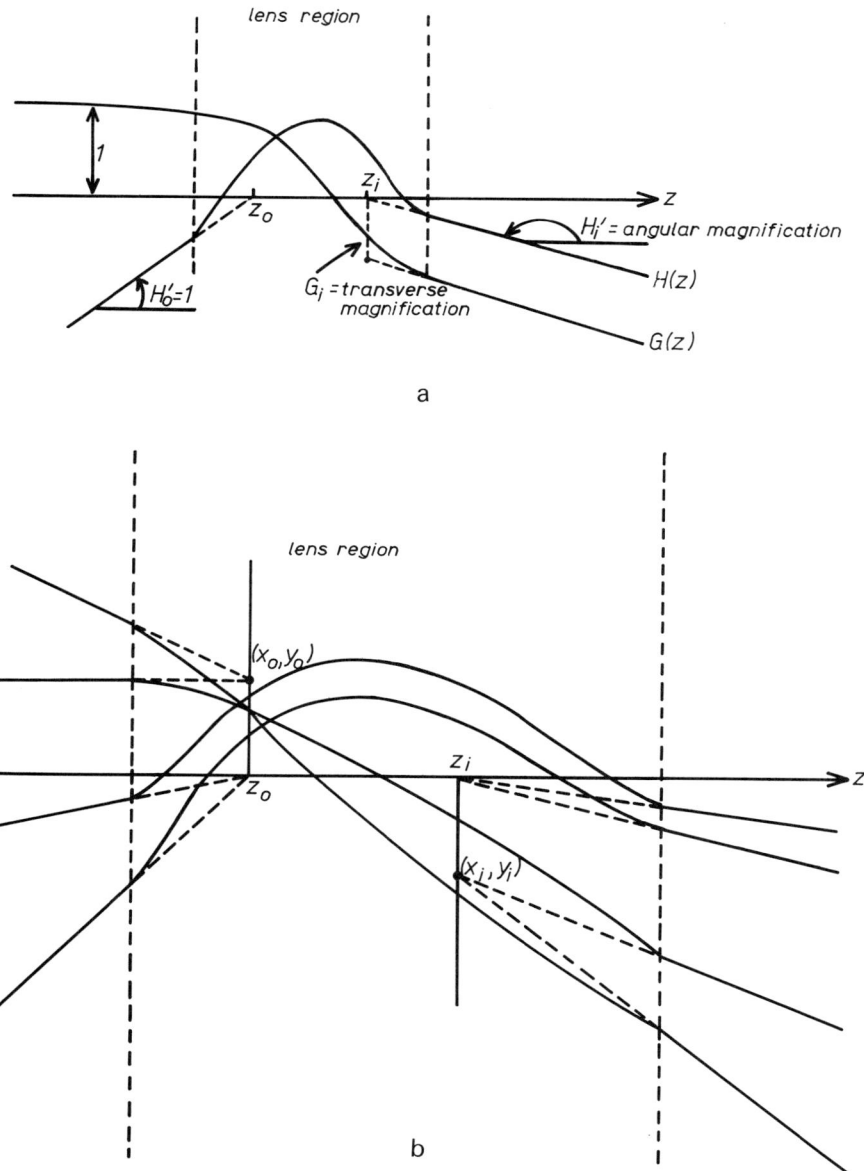

Fig. 16.2: (a) The paraxial solutions $G(z)$ and $H(z)$. (b) Asymptotic image formation.

16. GAUSSIAN OPTICS OF ROTATIONALLY SYMMETRIC SYSTEMS

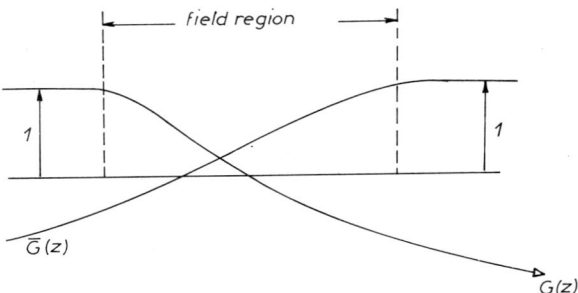

Fig. 16.3: The paraxial solutions $G(z)$ and $\overline{G}(z)$.

The rays $G(z)$ and $\overline{G}(z)$ tend to the following asymptotes:

$$\lim_{z \to \infty} G(z) = (z - \zeta_i)G'_i$$
$$\lim_{z \to -\infty} \overline{G}(z) = (z - \zeta_o)\overline{G}'_o$$
(16.3)

Any ray incident parallel to the axis can be written as $\lambda G(z)$, where λ is a constant, and will have as its emergent asymptote $\lambda(z - \zeta_i)G'_i$. Thus rays incident parallel to the axis generate emergent asymptotes that all intersect the axis at $z = \zeta_i$ and we denote this point by $z = z_{Fi}$ and refer to it as the *asymptotic image focus*. By exactly analogous reasoning, we see that all rays that emerge parallel to the axis correspond to incident asymptotes that intersect the axis at $z = \zeta_o$; we write $\zeta_o = z_{Fo}$ and refer to this as the *asymptotic object focus*.

The incident asymptote to the ray $G(z)$ and its emergent asymptote intersect at a point in the plane $z = z_{Pi}$, such that

$$1 = (z_{Pi} - z_{Fi})G'_i$$

or

$$z_{Pi} = z_{Fi} + \frac{1}{G'_i}$$
(16.4)

Likewise the asymptotes to $\overline{G}(z)$ intersect in a plane $z = z_{Po}$,

$$1 = (z_{Po} - z_{Fo})\overline{G}'_o$$

giving

$$z_{Po} = z_{Fo} + \frac{1}{\overline{G}'_o}$$
(16.5)

16.2 ASYMPTOTIC CARDINAL ELEMENTS AND TRANSFER MATRICES

The planes z_{Pi} and z_{Po} are known as the *asymptotic principal planes*. The distances $1/G'_i$ and $1/\overline{G}'_o$ are known as the *asymptotic focal lengths* (Fig. 16.4). We write

$$f_i := -\frac{1}{G'_i} \quad , \quad f_o := \frac{1}{\overline{G}'_o} \tag{16.6}$$

so that (16.4–16.5) become

$$z_{Pi} = z_{Fi} - f_i \quad , \quad z_{Po} = z_{Fo} + f_o \tag{16.7}$$

This apparent inconsistency in the choice of sign is explained by considering the relation between f_o and f_i. Since (15.51)

$$\hat{\phi}^{\frac{1}{2}}(G\overline{G}' - G'\overline{G}) = \text{const} \tag{16.8}$$

we see that

$$\hat{\phi}_o^{\frac{1}{2}} \overline{G}'_o = -\hat{\phi}_i^{\frac{1}{2}} G'_i \tag{16.9}$$

or using (16.6),

$$f_o \hat{\phi}_o^{-\frac{1}{2}} = f_i \hat{\phi}_i^{-\frac{1}{2}} \tag{16.10}$$

In magnetic lenses, therefore, with the sign convention of (16.6), we have $f_o = f_i$ and shall frequently drop the suffix. In the class of electrostatic lenses that provide no overall acceleration, so that $\hat{\phi}_o = \hat{\phi}_i$, we again have $f_o = f_i$. Furthermore, f_o and f_i will both be positive if the rays $G(z)$ and $\overline{G}(z)$ intersect the axis only once, since electron lenses always have a convergent focusing action. As the lens is made stronger, however, there comes a point at which the emergent asymptotes are parallel to the optic axis as well as the incident asymptotes and the lens then behaves like a telescope (f_o and $f_i \to \infty$). Beyond this point, the rays return towards the axis but now G'_i is positive and $\overline{G}'_o$ negative (Fig. 16.4c). Formally, therefore, lenses operating in these conditions belong to the class of divergent lenses but since they have this character because they are so strongly convergent, this terminology is never used.

Returning to the general solution (16.2), we can express the incident and emergent asymptotes in the following way:

$$\lim_{z \to -\infty} x(z) = A + B \frac{z - z_{Fo}}{f_o} \tag{16.11a}$$

$$\lim_{z \to \infty} x(z) = -A \frac{z - z_{Fi}}{f_i} + B \tag{16.11b}$$

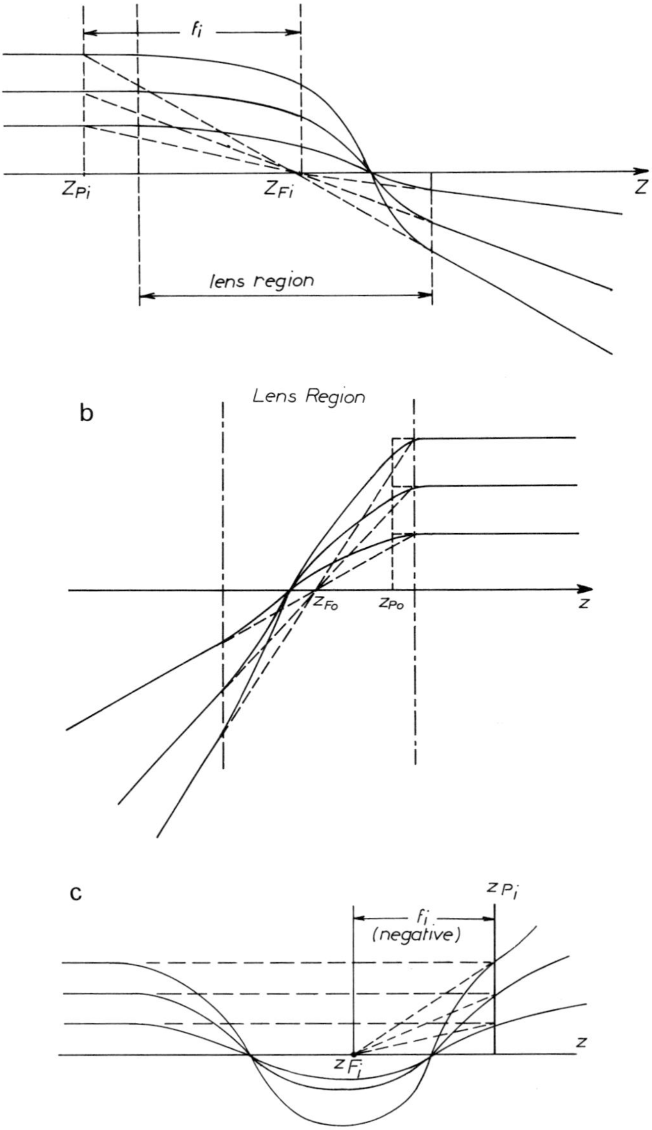

230   16. GAUSSIAN OPTICS OF ROTATIONALLY SYMMETRIC SYSTEMS

Fig. 16.4: The asymptotic cardinal elements. (a) Image focus and principal plane. (b) Object focus and principal plane. (c) Image focus and principal plane for a strong lens; the image focal length has become negative.

16.2 ASYMPTOTIC CARDINAL ELEMENTS AND TRANSFER MATRICES

Eliminating A and B, we find, with $Q_{12} := (z_1 - z_{Fo})(z_2 - z_{Fi})$

$$\begin{pmatrix} x_2 \\ x_2' \end{pmatrix} = \begin{pmatrix} -(z_2 - z_{Fi})/f_i & f_o + Q_{12}/f_i \\ -1/f_i & (z_1 - z_{Fo})/f_i \end{pmatrix} \begin{pmatrix} x_1 \\ x_1' \end{pmatrix} \qquad (16.12)$$

in which x_2 denotes $x(z)$ in some plane $z = z_2$ on the emergent asymptote and x_2' the gradient of the latter ($x_2' = -A/f_i$); x_1 denotes $x(z)$ in some plane z_1 on the incident asymptote and x_1' the gradient ($x_1' = B/f_o$). Writing

$$\boldsymbol{x} = \begin{pmatrix} x \\ x' \end{pmatrix} \qquad (16.13)$$

and

$$T = \begin{pmatrix} -(z_2 - z_{Fi})/f_i & f_o + Q_{12}/f_i \\ -1/f_i & (z_1 - z_{Fo})/f_i \end{pmatrix} \qquad (16.14)$$

(16.12) reduces to

$$\boldsymbol{x}_2 = T\boldsymbol{x}_1 \qquad (16.15)$$

The matrix T is known as the *transfer matrix*, and we shall see that it encapsulates in a convenient way all the paraxial behaviour of the lens. From it, all the familiar imaging relations may be derived straightforwardly. Suppose that the planes $P_o(z_1 = z_o)$ and $P_i(z_2 = z_i)$ are conjugate, that is, that all rays from any point in P_o converge to a point in P_i. For this, the expression for x_i must be independent of x_o' and hence

$$f_o + (z_o - z_{Fo})(z_i - z_{Fi})/f_i = 0 \qquad (16.16)$$

or

$$(z_o - z_{Fo})(z_i - z_{Fi}) = -f_i f_o \qquad (16.17)$$

This is *Newton's lens equation*. Introducing the expression for z_{Po} and z_{Pi} (16.7), (16.17) becomes

$$(z_o - z_{Po})(z_i - z_{Pi}) - f_i(z_o - z_{Po}) + f_o(z_i - z_{Pi}) = 0$$

or

$$\frac{f_o}{z_{Po} - z_o} + \frac{f_i}{z_i - z_{Pi}} = 1 \qquad (16.18a)$$

Writing

$$\tilde{f} := f_o \left(\frac{\hat{\phi}_i}{\hat{\phi}_o}\right)^{\frac{1}{4}} = f_i \left(\frac{\hat{\phi}_o}{\hat{\phi}_i}\right)^{\frac{1}{4}} = (f_o f_i)^{\frac{1}{2}} \qquad (16.19a)$$

so that $\tilde{f}$ is the geometric mean of the focal lengths, and

$$\tilde{\phi} = (\hat{\phi}_o \hat{\phi}_i)^{\frac{1}{2}} \tag{16.19b}$$

this becomes

$$\frac{\hat{\phi}_o^{\frac{1}{2}}}{z_{Po} - z_o} + \frac{\hat{\phi}_i^{\frac{1}{2}}}{z_i - z_{Pi}} = \frac{\tilde{\phi}^{\frac{1}{2}}}{\tilde{f}} \tag{16.18b}$$

This is the thick-lens counterpart of the familiar thin-lens equation.

From (16.19a) we see that, irrespective of z_1 and z_2, the determinant of the transfer matrix T (16.14) has the value

$$\det T = f_o/f_i = (\hat{\phi}_o/\hat{\phi}_i)^{\frac{1}{2}}$$

For conjugate planes, the matrix equation simplifies to

$$\begin{pmatrix} x_i \\ x_i' \end{pmatrix} = \begin{pmatrix} -(z_i - z_{Fi})/f_i & 0 \\ -1/f_i & (z_o - z_{Fo})/f_i \end{pmatrix} \begin{pmatrix} x_o \\ x_o' \end{pmatrix} \tag{16.20}$$

and denoting the transverse magnification by M, we have

$$-(z_i - z_{Fi})/f_i = M \tag{16.21}$$

If $x_o = 0$ the ratio of x_i' to x_o' is the angular magnification M_α (15.52),

$$M_\alpha = (z_o - z_{Fo})/f_i \tag{16.22}$$

so that using (16.17),

$$M_\alpha = -f_o/(z_i - z_{Fi}) = f_o/f_i M \tag{16.23}$$

Hence

$$M M_\alpha = f_o/f_i = (\hat{\phi}_o/\hat{\phi}_i)^{\frac{1}{2}} \tag{16.24}$$

as already shown (15.53).

Thus

$$z_i = z_{Fi} - f_i M \quad , \quad z_o = z_{Fo} + f_o/M \tag{16.25}$$

From (16.20), we see immediately that the principal planes are the pair of conjugate planes with unit magnification:

$$\begin{pmatrix} x_{Pi} \\ x_i' \end{pmatrix} = \begin{pmatrix} 1 & 0 \\ -1/f_i & f_o/f_i \end{pmatrix} \begin{pmatrix} x_{Po} \\ x_i' \end{pmatrix} \tag{16.26}$$

16.2 ASYMPTOTIC CARDINAL ELEMENTS AND TRANSFER MATRICES

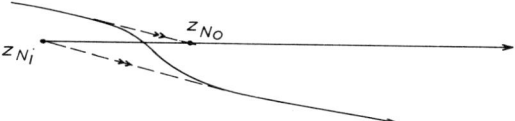

Fig. 16.5: Nodal points.

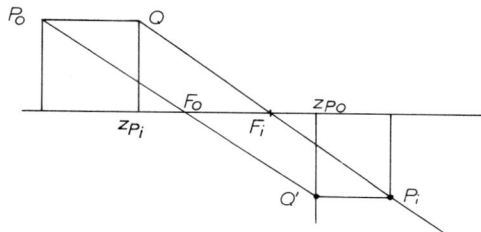

Fig. 16.6: Asymptotic image construction. The image of an object P_o is obtained by first drawing a line through P_o and the object focus F_o, which intersects the object principal plane at Q'. A second line is drawn through the point Q in the image principal plane at the same height as P_o and the image focus F_i. The point of intersection of this second line and a line through Q' parallel to the optic axis is the image P_i of P_o.

A third pair of axial points, the nodal points, is occasionally of interest. These are points having the property that a ray whose object asymptote intersects the axis at the object nodal point has an emergent asymptote intersecting the axis at the image nodal point, these asymptotes being parallel (Fig. 16.5). If these points are denoted by $z = z_{No}$, $z = z_{Ni}$, then

$$(z_{No} - z_{Fo})/f_i = 1 \tag{16.27}$$

so that if $\hat{\phi}_o = \hat{\phi}_i$, the nodal points and principal planes coincide. In general,

$$\begin{aligned} z_{No} &= z_{Fo} + f_i = z_{Po} - f_o + f_i \\ z_{Ni} &= z_{Fi} - f_o = z_{Pi} - f_o + f_i \end{aligned} \tag{16.28}$$

Once the foci and principal points are known, a simple construction enables us to obtain the point P_i conjugate to any object point P_o. First, a line is drawn through P_o parallel to the optic axis, intersecting the plane $z = z_{Pi}$ at Q; a line is then drawn through Q and the image focus, F_i (Fig. 16.6). Next, a line is drawn through P_o and the object focus F_o, intersecting the plane $z = z_{Po}$ at Q'; a line through Q' parallel to the optic axis intersects the line through Q and F_i at the image point P_i.

The various quantities that characterize the imaging properties of a lens—the foci, principal points, nodal points and focal lengths—are known as its *cardinal elements*. In electron optics, only the focal lengths and the positions of the foci are routinely tabulated.

The matrix expression (16.12) is very convenient when we wish to calculate the cardinal elements of a doublet of two lenses, separated by a field-free region. Instead of using the general form of the matrix elements given in (16.12), however, we use the simple form that connects principal planes (16.26). We have

$$\begin{pmatrix} x(z_{Pi}^{(2)}) \\ x_i'^{(2)} \end{pmatrix} = \begin{pmatrix} 1 & 0 \\ -1/f_i^{(2)} & f_o^{(2)}/f_i^{(2)} \end{pmatrix} \begin{pmatrix} x(z_{Po}^{(2)}) \\ x_o'^{(2)} \end{pmatrix} \quad (16.29)$$

$$\begin{pmatrix} x(z_{Po}^{(2)}) \\ x_o'^{(2)} \end{pmatrix} = \begin{pmatrix} 1 & D \\ 0 & 1 \end{pmatrix} \begin{pmatrix} x(z_{Pi}^{(1)}) \\ x_i'^{(1)} \end{pmatrix} \quad (16.30)$$

$$\begin{pmatrix} x(z_{Pi}^{(1)}) \\ x_i'^{(1)} \end{pmatrix} = \begin{pmatrix} 1 & 0 \\ -1/f_i^{(1)} & f_o^{(1)}/f_i^{(1)} \end{pmatrix} \begin{pmatrix} x(z_{Po}^{(1)}) \\ x_o'^{(1)} \end{pmatrix} \quad (16.31)$$

in which the superscripts (1) and (2) characterize the first and second lenses and

$$D = z_{Po}^{(2)} - z_{Pi}^{(1)} \quad (16.32)$$

is the distance between the object principal plane of the second lens and the image principal plane of the first. Multiplying the matrices, we obtain

$$\begin{pmatrix} x(z_{Pi}^{(2)}) \\ x_i'^{(2)} \end{pmatrix} = T_D \begin{pmatrix} x(z_{Po}^{(1)}) \\ x_o'^{(1)} \end{pmatrix} \quad (16.33)$$

where

$$T_D = \begin{pmatrix} -(z_{Pi}^{(2)} - \zeta_{Fi})/\varphi_i & \varphi_o + Q_{oi}/\varphi_i \\ -1/\varphi_i & (z_{Po}^{(1)} - \zeta_{Fo})/\varphi_i \end{pmatrix}$$

$$= \begin{pmatrix} 1 - D/f_i^{(1)} & Df_o^{(1)}/f_i^{(1)} \\ (D - f_i^{(1)} - f_o^{(2)})/f_i^{(1)} f_i^{(2)} & -f_o^{(1)}(D + f_o^{(2)})/f_i^{(1)} f_i^{(2)} \end{pmatrix}$$

(16.34)

in which ζ_{Fo} and ζ_{Fi} denote the object and image foci of the doublet and φ_i its image focal length; we have written $Q_{oi} := (z_{Po}^{(1)} - \zeta_{Fo})(z_{Pi}^{(2)} - \zeta_{Fi})$. We see that

$$\varphi_i = \frac{f_i^{(1)} f_i^{(2)}}{f_i^{(1)} + f_o^{(2)} - D} = \frac{f_i^{(1)} f_i^{(2)}}{D_F} \quad (16.35)$$

16.3 GAUSSIAN OPTICS AS A PROJECTIVE TRANSFORMATION

with

$$D_F := f_i^{(1)} + f_o^{(2)} - D = z_{Fi}^{(1)} - z_{Fo}^{(2)} \tag{16.36}$$

$$\begin{aligned} \zeta_{Fi} - z_{Pi}^{(2)} &= f_i^{(2)}(f_i^{(1)} - D)/D_F \\ \zeta_{Fo} - z_{Po}^{(1)} &= f_o^{(1)}(f_o^{(2)} + D)/D_F \end{aligned} \tag{16.37}$$

We have not exhausted the forms of the transfer matrix that are occasionally useful. Thus we shall encounter the focal transfer matrix between the (non-conjugate) foci, and the Dušek matrix, in which $z_2 = z_1$. We shall discuss these as the need arises.

16.3 Gaussian optics as a projective transformation (collineation)

The reasoning in Section 16.2 has been based on the physical notion of rays and their asymptotes but the characteristic quantities of Gaussian optics, the cardinal elements, also emerge from a more abstract approach. The linearity of the equations of motion is sufficient for us to assert that the object space and image space of a field region are connected by a *projective transformation*, or *collineation*. By this we mean that if (x_o, y_o, z_o) are the cartesian coordinates of a point P_o in object space and (x_i, y_i, z_i) those of a point P_i in image space, then

$$x_i = F_1/F_4 \quad , \quad y_i = F_2/F_4 \quad , \quad z_i = F_3/F_4 \tag{16.38a}$$

where

$$F_j = a_j x_o + b_j y_o + c_j z_o + d_j \quad (j = 1, 2, 3, 4) \tag{16.38b}$$

and solving for x_o, y_o, z_o

$$\begin{aligned} x_o &= F_1'/F_4' \quad , \quad y_o = F_2'/F_4' \quad , \quad z_o = F_3'/F_4' \\ F_j' &= a_j' x_i + b_j' y_i + c_j' z_i + d_j' \end{aligned} \tag{16.39}$$

From (16.38b), it is immediately obvious that the images of all points lying in the plane $F_4 = 0$ are at infinity: $F_4 = 0$ is thus the object focal plane and similarly, the plane $F_4' = 0$ is the image focal plane.

Suppose now that the system has rotational symmetry about the z-axis and suppose too that the coordinates in image space are rotated with respect to those in object space by the appropriate angle θ (15.9 or 15.27). An object point $(0, y_o, z_o)$ will be transformed into $(0, y_i, z_i)$ but z_i will be unaltered if y_o is replaced by $-y_o$ whereas y_i will become $-y_i$. From the relations

$$y_i = \frac{b_2 y_o + c_2 z_o + d_2}{b_4 y_o + c_4 z_o + d_4} \quad , \quad z_i = \frac{b_3 y_o + c_3 z_o + d_3}{b_4 y_o + c_4 z_o + d_4} \tag{16.40}$$

we deduce that $b_4 = b_3 = c_2 = d_2 = 0$, giving

$$y_i = \frac{b_2 y_o}{c_4 z_o + d_4} \quad , \quad z_i = \frac{c_3 z_o + d_3}{c_4 z_o + d_4} \tag{16.41}$$

or

$$y_o = \frac{c_4 d_3 - c_3 d_4}{b_2} \frac{y_i}{c_4 z_i - c_3} \quad , \quad z_o = -\frac{d_4 z_i - d_3}{c_4 z_i - c_3}$$

The focal planes are thus given by the solution of

$$\begin{aligned} c_4 z_o + d_4 = 0 &: z_{Fo} = -d_4/c_4 \\ c_4 z_i - c_3 = 0 &: z_{Fi} = c_3/c_4 \end{aligned} \tag{16.42}$$

On measuring distances from these planes, by writing

$$\begin{aligned} Z_o &:= z_o + d_4/c_4 \\ Z_i &:= z_i - c_3/c_4 \end{aligned} \tag{16.43}$$

and introducing f_o, f_i thus:

$$f_o := b_2/c_4 \quad , \quad f_i := -(c_4 d_3 - c_3 d_4)/b_2 c_4 \tag{16.44}$$

we obtain

$$\frac{y_i}{y_o} = \frac{f_o}{Z_o} = -\frac{Z_i}{f_i} \tag{16.45}$$

This yields Newton's lens equation (16.17)

$$Z_o Z_i = -f_o f_i \tag{16.46}$$

Furthermore, the magnification $M = dy_i/dy_o$ for constant z_o is given by

$$M = y_i/y_o = f_o/Z_o = -Z_i/f_i \tag{16.47}$$

so that for $M = 1$ we have $Z_o = f_o$, $Z_i = -f_i$. The planes thus defined are the principal planes, situated at distances f_o and f_i from the foci as already shown (16.7).

We shall not pursue this further but clearly all the remaining results of Gaussian optics can be derived straightforwardly. For lengthy discussion of the use of projective transformations in this context, see Ollendorff (1955), Born and Wolf (1959), Czapski and Eppenstein (1924) and Carathéodory (1937).

16.4 Use of the angle characteristic to establish the Gaussian optical quantities

Finally, we offer a third method of establishing the relations of Gaussian optics, setting out from the *angle characteristic*. This function is obtained from the point characteristic function by a Legendre transformation, which effectively changes the arguments from point coordinates to momenta. We define a function T as follows:

$$T := S + p_a x_a + q_a y_a - p_b x_b - q_b y_b \qquad (16.48)$$

in which (p, q) are the transverse components of $\mathbf{p}$. From (5.29) we know that

$$\Delta S = p_b \Delta x_b + q_b \Delta y_b - (p_a \Delta x_a + q_a \Delta y_a) \qquad (16.49)$$

when the integral in S is taken along a ray so that

$$\Delta T := -(x_b \Delta p_b + y_b \Delta q_b) + x_a \Delta p_a + y_a \Delta q_a \qquad (16.50)$$

and T must be a function of p_a, q_a, p_b and q_b; the function T is known as the *angle characteristic*. Provided that p_b is not proportional to p_a and q_b to q_a, we see that

$$\begin{aligned} x_b &= -\frac{\partial T}{\partial p_b}, & x_a &= \frac{\partial T}{\partial p_a} \\ y_b &= -\frac{\partial T}{\partial q_b}, & y_a &= \frac{\partial T}{\partial q_a} \end{aligned} \qquad (16.51)$$

Consider now a pair of points A, A' in object space and B, B' in image space. We assume that the angle characteristic between A and B, T_{AB}, is known and we calculate the new value between A', B', assuming asymptotic image formation. We write

$$T_{A'B'} = T_{A'A} + T_{AB} + T_{BB'} \qquad (16.52)$$

From (16.48), we know that

$$\begin{aligned} T_{A'A} &= S_{A'A} + p_{A'} x_{A'} + q_{A'} y_{A'} - p_A x_A - q_A y_A \\ &= (z_A - z_{A'}) \left(\hat{\phi}_A - p_A^2 - q_A^2 \right)^{\frac{1}{2}} \end{aligned} \qquad (16.53)$$

in which we have used $p_A = p_{A'}$, $q_A = q_{A'}$, $S_{AA'} = \hat{\phi}_A^{\frac{1}{2}} \overline{AA'}$ and

$$x_A = x_{A'} + (z_A - z_{A'})x'_A \quad , \quad y_A = y_{A'} + (z_A - z_{A'})y'_A$$

with

$$x'_A = p_A/(\hat{\phi}_A - p_A^2 - q_A^2)^{1/2} \quad , \quad y'_A = q_A/(\hat{\phi}_A - p_A^2 - q_A^2)^{1/2}$$

A similar expression is obtained for $T_{BB'}$. To the paraxial approximation therefore

$$T_{A'B'} = T_{AB} - (z_{A'} - z_A)\left(1 - \frac{p_A^2 + q_A^2}{2\hat{\phi}_A}\right)\hat{\phi}_A^{\frac{1}{2}}$$
$$+ (z_{B'} - z_B)\left(1 - \frac{p_B^2 + q_B^2}{2\hat{\phi}_B}\right)\hat{\phi}_B^{\frac{1}{2}} \qquad (16.54)$$

Since the system has rotational symmetry, the quantities p and q can only appear in the combinations $p_A^2 + q_A^2$, $p_B^2 + q_B^2$, $p_A p_B + q_A q_B$ and we write

$$T_{AB} = \alpha(p_A^2 + q_A^2)/2 + \beta(p_B^2 + q_B^2)/2 - \overline{f}(p_A p_B + q_A q_B) \qquad (16.55)$$

By applying (16.51) to (16.54), with $x_a = x_{A'}$, $p_a = p_{A'} = p_A$ etc., we see that

$$x_{A'} = \left\{\alpha + (z_{A'} - z_A)/\hat{\phi}_A^{\frac{1}{2}}\right\}p_A - \overline{f}p_B$$
$$y_{A'} = \left\{\alpha + (z_{A'} - z_A)/\hat{\phi}_A^{\frac{1}{2}}\right\}q_A - \overline{f}q_B$$
$$x_{B'} = -\left\{\beta - (z_{B'} - z_B)/\hat{\phi}_B^{\frac{1}{2}}\right\}p_B + \overline{f}p_A \qquad (16.56)$$
$$y_{B'} = -\left\{\beta - (z_{B'} - z_B)/\hat{\phi}_B^{\frac{1}{2}}\right\}q_B + \overline{f}q_A$$

If $p_B = 0$, then $x_{A'} = 0$ in the plane $z_{A'} - z_A = -\alpha\hat{\phi}_A^{1/2}$ for all p_A while if $p_A = 0$, then $x_{B'}$ vanishes in $z_{B'} - z_B = \beta\hat{\phi}_B^{1/2}$. These are the foci: we write $z_{Fi} - z_B = \beta\hat{\phi}_B^{1/2}$, $z_{Fo} - z_A = -\alpha\hat{\phi}_A^{1/2}$. Eliminating p_A or p_B between the equations for $x_{A'}$ and $x_{B'}$, we find

$$\overline{f}\, x_B = \left(\beta - \frac{z_{B'} - z_B}{\hat{\phi}_B^{\frac{1}{2}}}\right) x_A$$

or

$$\overline{f}\, x_A = \left(\alpha + \frac{z_{A'} - z_A}{\hat{\phi}_A^{\frac{1}{2}}}\right) x_B$$

provided that

$$(\alpha\hat{\phi}_A^{\frac{1}{2}} + z_{A'} - z_A)(\beta\hat{\phi}_B^{\frac{1}{2}} - z_{B'} + z_B) = \overline{f}^2 (\hat{\phi}_A\hat{\phi}_B)^{\frac{1}{2}} \qquad (16.57)$$

Thus $x_{B'} = x_{A'}$ if $z_{B'} - z_B = \hat{\phi}_B^{1/2}(\beta - \overline{f})$ and $z_{A'} - z_A = \hat{\phi}_A^{1/2}(\overline{f} - \alpha)$. These are therefore the principal planes,

$$\begin{aligned} z_{P_i} - z_B &= \hat{\phi}_B^{\frac{1}{2}}(\beta - \overline{f}) \quad \text{or} \quad z_{P_i} - z_{F_i} = -\hat{\phi}_B^{\frac{1}{2}}\overline{f} \\ z_{P_o} - z_A &= \hat{\phi}_A^{\frac{1}{2}}(\overline{f} - \alpha) \quad \text{or} \quad z_{P_o} - z_{F_o} = \hat{\phi}_A^{\frac{1}{2}}\overline{f} \end{aligned} \qquad (16.58)$$

With the sign convention of (16.7), we recognize that the focal lengths are given by

$$f_o = \hat{\phi}_A^{\frac{1}{2}}\overline{f} \quad , \quad f_i = \hat{\phi}_B^{\frac{1}{2}}\overline{f} \qquad (16.59)$$

Condition (15.57) may then be written in the form of Newton's lens equation.

16.5 The existence of asymptotes

In the foregoing sections, we have assumed that the curved trajectories within the field region tend to asymptotes in object and image space, "outside" the field. In theory, however, the fields continue indefinitely, though they of course become vanishingly small, and we need to be sure that it is legitimate to use the concept of asymptotes. In particular, we need to establish conditions concerning the rate at which the field functions tend to zero for large values of $|z|$. These questions have been explored in detail by Glaser and Bergmann (1950), whom we follow closely.

If a general solution $x(z)$ of the paraxial equation tends to an asymptote, the gradient $x'(z)$ and the intercept of the tangent to $x(z)$ with an arbitrary plane perpendicular to the axis must both tend to constant values:

$$\lim_{z \to \infty} x'(z) = a \quad , \quad \lim_{z \to \infty} \{x(z) - z\, x'(z)\} = b \qquad (16.60)$$

and similarly for $z \to -\infty$. In reduced coordinates (15.40), these conditions become

$$\lim_{z \to \infty} \hat{\phi}^{-\frac{1}{4}} \left\{ \xi'(z) - \frac{1}{4}\frac{\gamma}{\hat{\phi}} \phi'\xi(z) \right\} = a \qquad (16.61a)$$

$$\lim_{z \to \infty} \hat{\phi}^{-\frac{1}{4}} \left\{ \xi(z)\left(1 + \frac{\gamma}{4\hat{\phi}}\phi' z\right) - z\xi'(z) \right\} = b \qquad (16.61b)$$

240   16. GAUSSIAN OPTICS OF ROTATIONALLY SYMMETRIC SYSTEMS

with $\xi = x\hat{\phi}^{1/4}$. Since $\hat{\phi}$ is always finite in real fields, ξ and ξ' must be finite. The condition (16.61a) may therefore be replaced by

$$\lim_{z \to \infty} \xi'(z) = A \quad , \quad \lim_{z \to \infty} \xi\phi' = \overline{A} \tag{16.62}$$

ϕ' must therefore vanish at infinity, and we assume that it falls to zero faster than $1/z$, which is easily justified. The second condition (16.61b) thus becomes

$$\lim_{z \to \infty} (\xi - \xi' z) = B \tag{16.63}$$

which is immediately recognizable as the condition that solutions of the reduced equation (15.38) tend to asymptotes as $z \to \infty$. Multiplying (15.38) by z and integrating, we obtain (with ξ in place of v)

$$\xi - \xi' z = \int G(z) \xi z \, dz$$

so that for any upper bounds α or $\overline{\alpha}$

$$\lim_{z \to \infty} (\xi - \xi' z) = B = \lim_{\alpha \to \infty} \int^{\alpha} G\xi z \, dz \equiv \int^{\overline{\alpha}} G\xi z \, dz + \lim_{\alpha \to \infty} \int_{\overline{\alpha}}^{\alpha} G\xi z \, dz$$

$$= \int^{\overline{\alpha}} G\xi z \, dz + \lim_{\alpha \to \infty} \int_{\overline{\alpha}}^{\alpha} Gz(\xi - \xi' z) \, dz + \lim_{\alpha \to \infty} \int_{\overline{\alpha}}^{\alpha} G\xi' z^2 \, dz \tag{16.64}$$

Choosing $\overline{\alpha}$ very large, we find

$$B = \int^{\overline{\alpha}} G\xi z \, dz + B \lim_{\alpha \to \infty} \int_{\overline{\alpha}}^{\alpha} Gz \, dz - A \lim_{\alpha \to \infty} \int_{\overline{\alpha}}^{\alpha} Gz^2 \, dz \tag{16.65}$$

For the existence of asymptotes in general, therefore, the integrals

$$\int_{-\infty}^{\infty} zG(z) \, dz \tag{16.66}$$

and

$$\int_{-\infty}^{\infty} z^2 G(z) \, dz \tag{16.67}$$

16.5 THE EXISTENCE OF ASYMPTOTES

must converge; the existence of asymptotes parallel to the axis ($A = 0$) is, however, guaranteed by the convergence of $\int^\infty zG(z)\,dz$ alone.

The conditions are *necessary*; we now show that they are also *sufficient*. We can in principle solve (15.38) by an iterative procedure, taking as first approximation a linear expression of the form $\xi_0 = \alpha z + \beta$ so that writing $\xi = \xi_0 + \xi_1 + \ldots + \xi_n + \ldots$,

$$\xi''_{n+1} + G(z)\xi_n = 0 \tag{16.68}$$

or

$$\xi_{n+1} = -\int_z^\infty d\zeta \int_\zeta^\infty G\xi_n \, dX = \int_z^\infty (z + X) G\xi_n \, dX \tag{16.69}$$

Given the convergence of (16.66) and (16.67), we see that for $z > \overline{\alpha}$ and $\overline{\alpha}$ large,

$$\int_{\overline{\alpha}}^z G \, dX < \int_{\overline{\alpha}}^z GX \, dX < \int_{\overline{\alpha}}^\infty GX^2 \, dX =: \epsilon(\overline{\alpha}) < 1 \tag{16.70}$$

and clearly as $\overline{\alpha} \to \infty$, $\epsilon(\overline{\alpha}) \to 0$. From (16.69), we have

$$\xi_1 = \int (\overline{\alpha} + X) G(AX + B) \, dX$$

and since $X > \overline{\alpha}$,

$$\int_\alpha^\infty \overline{\alpha} X G \, dX < \int_{\overline{\alpha}}^\infty X^2 G \, dX = \epsilon(\overline{\alpha}) \tag{16.71}$$

and so

$$|\xi_1| \leq |A| \left\{ \int_{\overline{\alpha}}^\infty \overline{\alpha} X G \, dX + \int_{\overline{\alpha}}^\infty GX^2 \, dX \right\} + |B| \left\{ \int_{\overline{\alpha}}^\infty \overline{\alpha} G \, dX + \int_{\overline{\alpha}}^\infty GX \, dX \right\}$$

From (16.70) and (16.71), we may conclude that

$$|\xi_1| \leq 2(|A| + |B|)\, \epsilon(\overline{\alpha})$$

Iterating, we find

$$|\xi - \xi_0| \leq |\xi_1| + |\xi_2| + \ldots = (|A| + |B|) \frac{2\epsilon(\overline{\alpha})}{1 - 2\epsilon(\overline{\alpha})}$$

As $\overline{\alpha}$ is made larger and hence as $z \to \infty$, $\epsilon(\overline{\alpha}) \to 0$ (16.70) and ξ tends to the linear solution ξ_0. The convergence of (16.66) and (16.67), which we have used in this derivation, is therefore not only a necessary but also a sufficient condition. In practice, the integrals (16.66–16.67) always do converge: since a total system must be electrically or magnetically neutral, $G(z)$ must fall off at least as fast as z^{-6} and any integral of the form $\int G(z) z^n \, dz$ will then converge for $n \leq 4$.

17
Gaussian Optics of Rotationally Symmetric Systems: Real Cardinal Elements

Hitherto we have considered only the coordination between asymptotes. When this is not appropriate, in microscope objective lenses for example, a different set of cardinal elements must be used. After discussing these, we enquire whether fields exist for which fixed real cardinal elements can be defined for a range of object positions. This leads us to the concept of Newtonian imaging fields, to which Glaser attached considerable importance (Glaser and Bergmann, 1950, 1951; also Glaser and Lammel, 1941, 1943); we shall follow his discussion closely. Further contributions were made by Funk (1950) and by Hutter (1945), who wrongly included several non-Newtonian distributions in the family of Newtonian fields. We shall use the rotating coordinate frame (x, y, z) without comment, and it must be remembered that in magnetic lenses, the surfaces $x = 0$ and $y = 0$ are not plane though we shall still speak of 'parallel' rays.

17.1 Real cardinal elements for high magnification and high demagnification

Objective lenses are conventionally operated at high magnification, and to a good approximation we may assume that the image is formed at infinity (Fig. 17.1). The family of rays that emerge parallel to the axis intersect the latter at some point F'_o, which we call the *real object focus*; a family of rays emerging from the lens parallel to one another but not parallel to the axis intersect in the focal plane, the plane through F'_o perpendicular to the axis. This can be seen by introducing the solutions $G(z)$ and $\overline{G}(z)$ of the paraxial equations already used in Section 16.2 (16.1). A family of rays parallel in image space may be written

$$x(z) = \alpha G(z) + c_k \overline{G}(z) \tag{17.1}$$

in which the c_k are constants corresponding to the different rays; since

$$\lim_{z \to \infty} x(z) = \alpha G'_i(z - z_{Fi}) + c_k = -\frac{\alpha}{f_i}(z - z_{Fi}) + c_k \tag{17.2}$$

17.1 REAL CARDINAL ELEMENTS

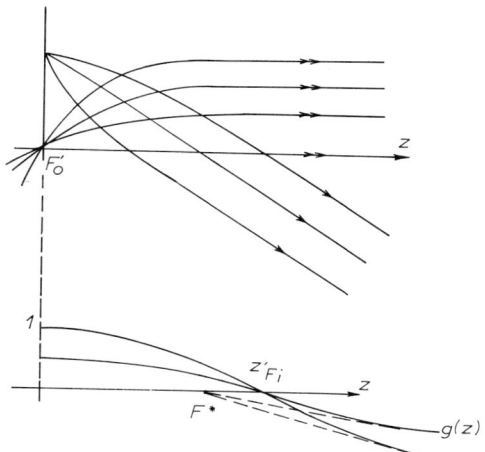

Fig. 17.1: Above: The real object focus F'_o. Rays that intersect in a point in the real object focal plane emerge into image space as a parallel beam, parallel to the axis if the point lies on the axis.
Below: Definition of the ray $g(z)$, which crosses the optic axis at the real image focus ($z = z'_{Fi}$). The corresponding asymptotic focus (F^*) is in practice more useful.

where z_{Fi} and f_i are the asymptotic focus and focal length, we see that (17.1) represents a family of rays all with image gradient $-\alpha/f_i$. In the real object focal plane, $z = z'_{Fo}$, $\overline{G}(z)$ vanishes and hence

$$x(z'_{Fo}) = \alpha G(z'_{Fo}) \quad \text{for all} \quad c_k \tag{17.3}$$

The real image focus is defined to be the point at which rays, parallel to the axis in the plane $z = z'_{Fo}$, intersect the axis. It is now convenient to use a different ray pair: instead of $G(z)$ and $\overline{G}(z)$, we employ $g(z)$ and $\overline{G}(z)$, where

$$g(z'_{Fo}) = 1 \quad , \quad g'(z'_{Fo}) = 0 \tag{17.4}$$

(Fig. 17.1). In the real *image focal plane* $z = z'_{Fi}$, $g(z)$ vanishes:

$$g(z'_{Fi}) = 0 \tag{17.5}$$

It is easy to show that a family of rays, parallel to one another in $z = z'_{Fo}$ but not parallel to the axis, intersect in the real image focal plane $z = z'_{Fi}$. Such a family may be written

$$x(z) = \alpha \overline{G}(z) + c_k g(z)$$

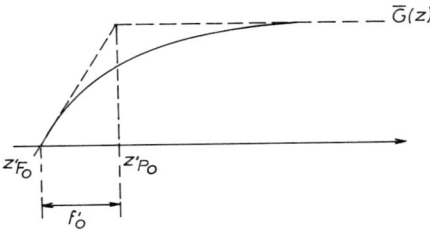

Fig. 17.2: The real object principal plane and the real (objective) focal length.

and in the plane $z = z'_{Fi}$,

$$x(z'_{Fi}) = \alpha \overline{G}(z'_{Fi}) \qquad (17.6)$$

We shall see in Volume 3 that the diffraction pattern of an object placed in the real object focal plane and illuminated with a beam of electrons parallel to the axis in this plane is formed in the real image focal plane; (17.6) expresses a fundamental property of diffraction patterns, that is, that rays that are parallel in the specimen plane intersect in the 'diffraction plane', forming diffraction spots if only isolated directions occur.

In practice, it is more useful to know the location of the *asymptotic* image focus corresponding to the real object focus. This is the point at which the emergent asymptote to $g(z)$ intersects the axis (Fig. 17.1).

Focal lengths are associated with each of these foci. In the case of the real object focus, we define the real object principal plane, $z = z'_{Po}$, to be the plane perpendicular to the axis through the point of intersection of the emergent asymptote to $\overline{G}(z)$ and the tangent to $\overline{G}(z)$ at $z = z'_{Fo}$ (Fig. 17.2). The real focal length f'_o is then given by

$$f'_o = z'_{Po} - z'_{Fo} \qquad (17.7)$$

and since $\lim_{z \to \infty} \overline{G} = 1$, we find

$$f'_o = \frac{1}{\overline{G}'(z'_{Fo})} \qquad (17.8)$$

For the real image focal length f'_i, we have

$$f'_i = -\frac{1}{g'(z'_{Fi})} \qquad (17.9)$$

The Wronskian tells us that for magnetic fields

$$\overline{G}g' - \overline{G}'g = \text{const}$$

17.1 REAL CARDINAL ELEMENTS

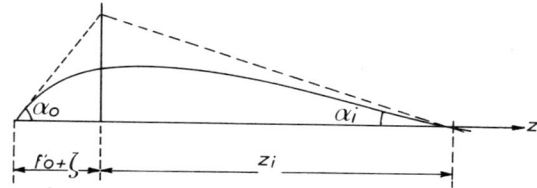

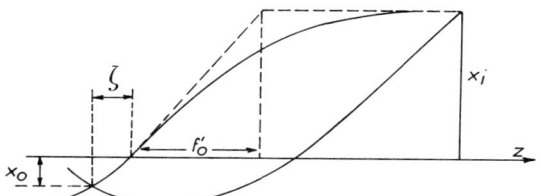

Fig. 17.3: High but not infinite magnification in an objective lens.

and by considering the focal planes, that

$$\frac{1}{f'_o} = \frac{\overline{G}(z'_{Fi})}{f'_i} \tag{17.10}$$

For the hybrid focal length obtained by considering the emergent asymptote to $g(z)$, which we denote f''_i, we have

$$f''_i = -\frac{1}{g'(\infty)} \tag{17.11}$$

and the Wronskian tells us that

$$f''_i = f'_o \tag{17.12}$$

Finally, we consider the practical situation in which a specimen is placed close to but not exactly at $z = z'_{F_o}$, so that the magnification is high but not infinite. From Fig. 17.3a, we see that the magnification is given by

$$|M| = \frac{\alpha_o}{\alpha_i} = \frac{Z_i}{f'_o + \zeta} \approx \frac{Z_i}{f'_o} \tag{17.13}$$

in which $\zeta \ll f'_o$ and Z_i is the distance to the image from z'_{P_o}. From Fig. 17.3b, however, we can express the magnification as

$$|M| = \left|\frac{x_i}{x_o}\right| \approx \frac{f'_o}{\zeta} \tag{17.14}$$

(by similar triangles). Thus $\zeta Z_i = f_o'^2$, in which we once again recognize Newton's lens equation.

17.2 Osculating cardinal elements

The concept of osculating cardinal elements was introduced by Glaser (Glaser and Lammel, 1941, 1943; Glaser and Bergmann, 1950, 1951) in an attempt to establish whether generalized cardinal elements can be defined that collapse to the asymptotic cardinal elements when both object and image lie outside the field and which also allow us to use the lens equation in Newton's form,

$$M = \frac{\overline{f}_o}{z_o - \overline{z}_{Fo}} = -\frac{z_i - \overline{z}_{Fi}}{\overline{f}_i} \tag{17.15}$$

even when the object or the image is located in the field. These new cardinal elements, $\overline{z}_{Fo}$, $\overline{z}_{Fi}$, $\overline{f}_o$ and $\overline{f}_i$, will in general vary with object position z_o and hence with magnification M. From (17.15), we have

$$\frac{1}{\overline{f}_o} = \frac{d}{dz_o}\left(\frac{1}{M}\right) \quad, \quad \frac{1}{\overline{f}_i} = -\frac{dM}{dz_o}\frac{dz_o}{dz_i} \tag{17.16}$$
$$\overline{z}_{Fo} = z_o - \overline{f}_o/M \quad, \quad \overline{z}_{Fi} = z_i + M\overline{f}_i$$

and we regard these as *definitions* of new *osculating* cardinal elements, so called from their geometrical interpretation. From the relation between longitudinal and lateral magnification (15.60), we note that

$$\frac{\overline{f}_i}{\hat{\phi}_i^{\frac{1}{2}}} = \frac{\overline{f}_o}{\hat{\phi}_o^{\frac{1}{2}}} \tag{17.17}$$

The osculating focal lengths and the positions of the osculating foci may be calculated by considering an arbitrary pair of solutions of the paraxial equation, $s(z)$ and $t(z)$, such that the general solution is

$$x(z) = as(z) + bt(z) = \frac{t(z)}{t(z_o)}x_o + a\left\{s(z) - \frac{s(z_o)}{t(z_o)}t(z)\right\} \tag{17.18}$$

with $t(z_o) \neq 0$, $x_o := x(z_o)$ and $b = \{x_o - as(z_o)\}/t(z_o)$. The image plane conjugate to an object plane $z = z_o$ is the plane in which the term multiplied by a vanishes:

$$s(z_i) - \frac{s(z_o)}{t(z_o)}t(z_i) = 0 \tag{17.19}$$

17.2 OSCULATING CARDINAL ELEMENTS

whereupon
$$x(z_i) = \frac{t(z_i)}{t(z_o)} x_o \qquad (17.20)$$

The magnification is thus
$$M = \frac{t(z_i)}{t(z_o)} = \frac{s(z_i)}{s(z_o)} \qquad (17.21)$$

Differentiating, we find
$$\frac{dM}{dz_i} = \frac{1}{t(z_o)} \frac{dt(z_i)}{dz_i} - \frac{t(z_i)}{t^2(z_o)} \frac{dt(z_o)}{dz_o} \frac{dz_o}{dz_i} \qquad (17.22)$$

Equation (17.19) gives
$$\begin{aligned}\frac{dz_o}{dz_i} &= -\frac{t(z_o)s'(z_i) - s(z_o)t'(z_i)}{t'(z_o)s(z_i) - s'(z_o)t(z_i)} \\ &= -\frac{t(z_o)}{t(z_i)} \frac{t(z_o)s'(z_i) - s(z_o)t'(z_i)}{t'(z_o)s(z_o) - s'(z_o)t(z_o)}\end{aligned} \qquad (17.23)$$

where primes denote differentiation with respect to the argument z_o and z_i. From (17.16) and (17.22–17.23), we see that
$$\begin{aligned}\frac{1}{\overline{f}_i} &= \frac{t'(z_o)s'(z_i) - t'(z_i)s'(z_o)}{t(z_o)s'(z_o) - t'(z_o)s(z_o)} \\ \frac{1}{\overline{f}_o} &= \frac{t'(z_o)s'(z_i) - t'(z_i)s'(z_o)}{t(z_i)s'(z_i) - t'(z_i)s(z_i)}\end{aligned} \qquad (17.24)$$

Likewise from (17.16) and (17.24), we obtain
$$\begin{aligned}z_i - \overline{z}_{Fi} &= \frac{t(z_i)s'(z_o) - t'(z_o)s(z_i)}{t'(z_i)s'(z_o) - t'(z_o)s'(z_i)} \\ z_o - \overline{z}_{Fo} &= \frac{t(z_o)s'(z_i) - t'(z_i)s(z_o)}{t'(z_o)s'(z_i) - t'(z_i)s'(z_o)}\end{aligned} \qquad (17.25)$$

The geometrical meanings of these expressions can be extracted by considering the particular ray
$$\xi(z) = t(z)s'(z_i) - s(z)t'(z_i) \qquad (17.26)$$

In the plane $z = z_i$, $\xi'(z_i) = 0$ and so the tangent to $\xi(z)$ is parallel to the axis in this plane. The osculating focal length $\overline{f}_o$ is simply given by
$$\overline{f}_o = \frac{\xi(z_i)}{\xi'(z_o)} \qquad (17.27)$$

and the position of the focus by

$$z_o - \bar{z}_{Fo} = \frac{\xi(z_o)}{\xi'(z_o)} \qquad (17.28)$$

An osculating object principal plane may be defined to be the point of intersection of the tangent to $\xi(z)$ in the object plane and the tangent in the image plane. It is easy to show that

$$\bar{z}_{Po} - \bar{z}_{Fo} = \bar{f}_o \qquad (17.29)$$

Similar relations may be established straightforwardly for the image osculating cardinal elements. These geometrical relations are illustrated in Fig. 17.4a.

These osculating cardinal elements can be used to study image formation in the neighbourhood of a given pair of conjugate points. Consider an object placed in a plane distant Δz_o from z_o; the magnification will change from M to $M + \Delta M$. We see that

$$M + \Delta M = M - \frac{\Delta z_i}{\bar{f}_i} = -\frac{z_i + \Delta z_i - \bar{z}_{Fi}}{\bar{f}_i} \qquad (17.30)$$

Also

$$\frac{1}{M} + \Delta\left(\frac{1}{M}\right) = \frac{1}{M} + \frac{\Delta z_o}{\bar{f}_o} \qquad (17.31)$$

Hence

$$M + \Delta M = -\frac{z_i + \Delta z_i - \bar{z}_{Fi}}{\bar{f}_i} = \frac{\bar{f}_o}{z_o + \Delta z_o - \bar{z}_{Fo}} \qquad (17.32)$$

and so Newton's lens equation is satisfied for the new object position, for small displacements Δz_o.

An example of the variation of the osculating focal length as a function of magnification for the magnetic lens field $B(z) = B_0 \exp(-z^2/a^2)$ is shown in Fig. 17.4b for various values of the lens strength.

From Fig. 17.1, it is immediately clear that the osculating object focus corresponding to an image at infinity is identical with the real object focus discussed in Section 17.1 and the osculating object principal plane likewise coincides with the real object principal plane.

Finally, we consider the question of whether field distributions exist for which the osculating cardinal elements are stationary and thus do not vary with object position (or magnification). This question was very thoroughly studied by Glaser (Glaser and Lammel, 1941), and contributions to the

17.2 OSCULATING CARDINAL ELEMENTS

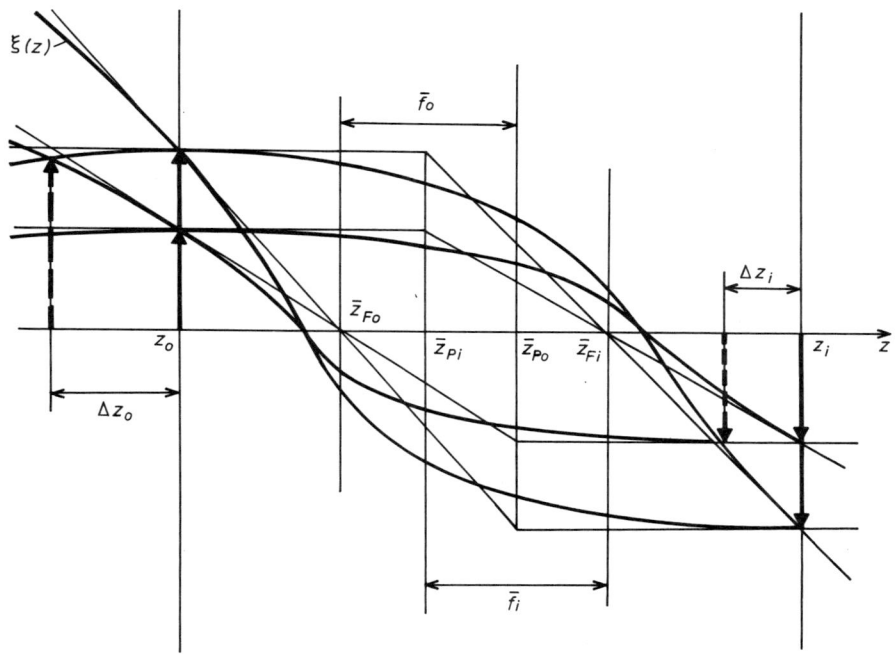

Fig. 17.4a: The osculating cardinal elements.

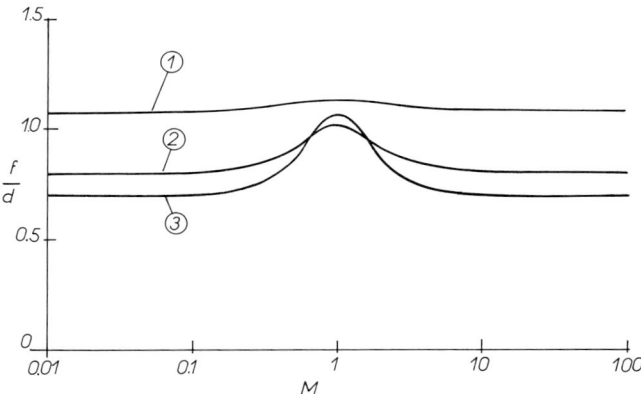

Fig. 17.4b: The osculating focal length as a function of magnification and lens strength for $B(z) = B_0 \exp(-z^2/a^2)$. 1: $J^2/\phi = 142\text{A}^2/\text{V}$; 2: $285\text{A}^2/\text{V}$; 3: $427\text{A}^2/\text{V}$.

topic have also been made by Hutter (1945) and Funk (1950). The list of fields given by Hutter is misleading, however, as it contains several fields that do not possess fixed osculating elements. The fullest account is to be found in Glaser (1952, Section 68), which we follow closely.

We set out from (17.15) and render the axial coordinate dimensionless by writing

$$\zeta := \sigma z/a \tag{17.33}$$

where a is some convenient dimension of the lens and σ is defined below. We no longer add bars to ζ where these were necessary for z. Solving (17.15) for ζ_i we obtain the projective relation

$$\zeta_i = \frac{\zeta_{Fi}\zeta_o - \overline{f}_o\overline{f}_i - \zeta_{Fi}\zeta_{Fo}}{\zeta_o - \zeta_{Fo}} \tag{17.34}$$

and shift the origin to the point midway between F_o and F_i: $\zeta_{Fo} = -\zeta_{Fi}$. Anticipating a result proved in Section 17.3, we set $\zeta_{Fi}^2 - \overline{f}_o\overline{f}_i$ equal to $-\sigma^2$ and (17.34) becomes

$$\zeta_i = \frac{\zeta_{Fi}\zeta_o - \sigma^2}{\zeta_o + \zeta_{Fi}} \tag{17.35}$$

We now map the whole z- or ζ-axis onto a finite domain by writing

$$\zeta =: \sigma \cot \psi \qquad 0 \leq \psi \leq \pi \tag{17.36}$$

and choose ω such that

$$\zeta_{Fi} =: \sigma \cot \frac{\pi}{\omega} \tag{17.37}$$

17.2 OSCULATING CARDINAL ELEMENTS

The projective relation (17.35) then becomes

$$\cot \psi_i = \frac{\cot(\pi/\omega) \cot \psi_o - 1}{\cot \psi_o + \cot(\pi/\omega)} = \cot(\psi_o + \pi/\omega) \tag{17.38}$$

or

$$\psi_i = \psi_o + \pi/\omega \tag{17.39}$$

We now substitute these transformations into the paraxial equations of motion; it will be sufficient to consider the magnetic case (ϕ = const). Replacing z by ψ (17.33 and 17.36) and introducing a new variable ξ,

$$x =: \xi \operatorname{cosec} \psi \tag{17.40}$$

the paraxial equation $x'' + (\eta^2 B^2/4\hat{\phi})x = 0$ becomes

$$\ddot{\xi} + \left\{ 1 + k^2 b^2 (a \cot \psi) \operatorname{cosec}^4 \psi \right\} \xi = 0 \tag{17.41}$$

in which

$$k^2 := \frac{\eta^2 B_0^2 a^2}{4\hat{\phi}} \tag{17.42}$$

and

$$B(z) =: B_0 b(z) \tag{17.43}$$

where B_0 is the greatest value of $B(z)$ so that $b(z) \leq 1$.

Let $s(\psi)$ and $t(\psi)$ be two independent solutions of (17.41), corresponding to the solutions $x_s(\sigma \cot \psi) = s(\psi) \operatorname{cosec} \psi$, $x_t(\sigma \cot \psi) = t(\psi) \operatorname{cosec} \psi$; then (17.15) may be written

$$M = \frac{s(\psi_i)}{s(\psi_o)} \frac{\sin \psi_o}{\sin \psi_i} = \frac{t(\psi_i)}{t(\psi_o)} \frac{\sin \psi_o}{\sin \psi_i} \tag{17.44}$$

Setting

$$r(\psi) := \frac{s}{t} = \frac{x_s(\sigma \cot \psi)}{x_t(\sigma \cot \psi)} \tag{17.45}$$

(17.19) becomes

$$r(\psi_o) = r(\psi_i) \tag{17.46}$$

Since $\psi_i = \psi_o + \pi/\omega$ (17.39), the function $r(\psi)$ must be such that

$$r(\psi_o + \pi/\omega) = r(\psi_o) \tag{17.47}$$

that is, it must be *periodic* with period π/ω. Differentiating (17.45) gives

$$\dot{r} = \frac{\dot{s}t - s\dot{t}}{t^2} \tag{17.48}$$

and using the Wronskian ($\dot{s}t - s\dot{t} = \text{const} = C$), we see that

$$\dot{r} = C/t^2 \tag{17.49}$$

From this we can deduce that $t(\psi)$ must be semi-periodic,

$$t(\psi + \pi/\omega) = -t(\psi) \tag{17.50}$$

and the same is true of $s(\psi)$. The derivatives of these functions are also semi-periodic and returning to (17.41), we see that the field function $b^2(a\cot\psi)\text{cosec}^4\psi$ must be periodic with period π/ω. Writing

$$b^2(a\cot\psi)\text{cosec}^4\psi = F^2(\psi) \tag{17.51}$$

we can conclude that all 'Newtonian' fields, that is, fields for which the osculating cardinal elements are stationary (do not vary with object position), must be of the form

$$b(a\cot\psi) = F(\psi)\sin^2\psi \quad , \quad F(\psi + \pi/\omega) = F(\psi) \tag{17.52}$$

Not all such fields are Newtonian, however: this condition is necessary but not sufficient. Such a condition is provided by (17.47), which we may restate as follows: the solutions $s(\psi)$ and $t(\psi)$ of the Hill differential equation

$$\ddot{\xi} + \left\{1 + k^2 F^2(\psi)\right\}\xi = 0 \quad , \quad F(\psi + \pi/\omega) = F(\psi) \tag{17.53}$$

must be semi-periodic with half-period π/ω. (For further details of Hill equations, see Kamke, 1977, Section C.2 eq. 2.30; Whittaker and Watson, 1927, Chapter XIX).

The simplest function $F(\psi)$ satisfying (17.52) is clearly

$$F(\psi) = 1 \tag{17.54}$$

giving
$$b(a\cot\psi) = \sin^2\psi$$

or
$$B(z) = \frac{B_0}{1 + (z/a)^2} \tag{17.55}$$

This field distribution, introduced by Glaser (1940) and very extensively studied, is known as *Glaser's bell-shaped model*. We shall examine it in detail in Part VII, Chapter 36.

Newtonian fields have been studied by Glaser and Bergmann (1950, 1951); see Glaser (1952, Section 68) for more complicated forms of $F(\psi)$, an example of which is

$$B(a \cot \psi) = B_0 \cos^2 \psi \left\{ 1 + \frac{1+k^2}{k^2} C_1 \frac{1 - C_2 \sin^2 \omega(\pi - \psi)}{\{1 - C_3 \sin^2 \omega(\pi - \psi)\}^2} \right. \\ \left. \cdot \sin^2 \omega(\pi - \psi) \right\}^{\frac{1}{2}} \tag{17.56}$$

where C_1, C_2 and C_3 are constants (some examples are shown in Fig. 17.5). We shall say no more about these fields here, however, because the field shape is a function of the parameter $k^2 \propto B_0^2/\hat{\phi}$. Thus any given member of this family of fields will be Newtonian for only one value $B_0^2/\hat{\phi}$: they are of academic interest only. For some values of the parameters, they appear to represent grossly saturated superconducting lenses quite well (see the field distributions in Bonjour (1974, 1975)) but there is little or no incentive to use models of such restricted utility. The attractive feature of model fields is that they enable us to study the general behaviour of a lens as all the parameters are varied, even if the numerical values are not exact; if a model is not capable of this, there is even less reason to eschew numerical trajectory tracing (Chapter 33).

Another aspect of Newtonian fields, which we shall not consider here, concerns the existence of osculating cardinal elements of higher order, that is, cardinal elements for which (17.15) is valid to better than a linear approximation. The quadratic case is mentioned by Glaser and Bergmann (1951) and the general case has been studied by Putz (1951), whose unpublished work is recapitulated in detail by Glaser (1952, Section 68).

17.3 Inversion of the principal planes

As we have seen, the principal planes may be defined in several ways. For certain definitions, which will emerge from the discussion, the object and image principal planes are crossed in the sense that the object principal plane is on the image-space side of the image principal plane: $z_{Pi} < z_{Po}$. Consider a pair of conjugate planes z_o, z_i; from (16.25),

$$z_i = \frac{z_{Fi} z_o - f_o f_i - z_{Fo} z_{Fi}}{z_o - z_{Fo}} \tag{17.57}$$

254   17. GAUSSIAN OPTICS OF ROTATIONALLY SYMMETRIC SYSTEMS

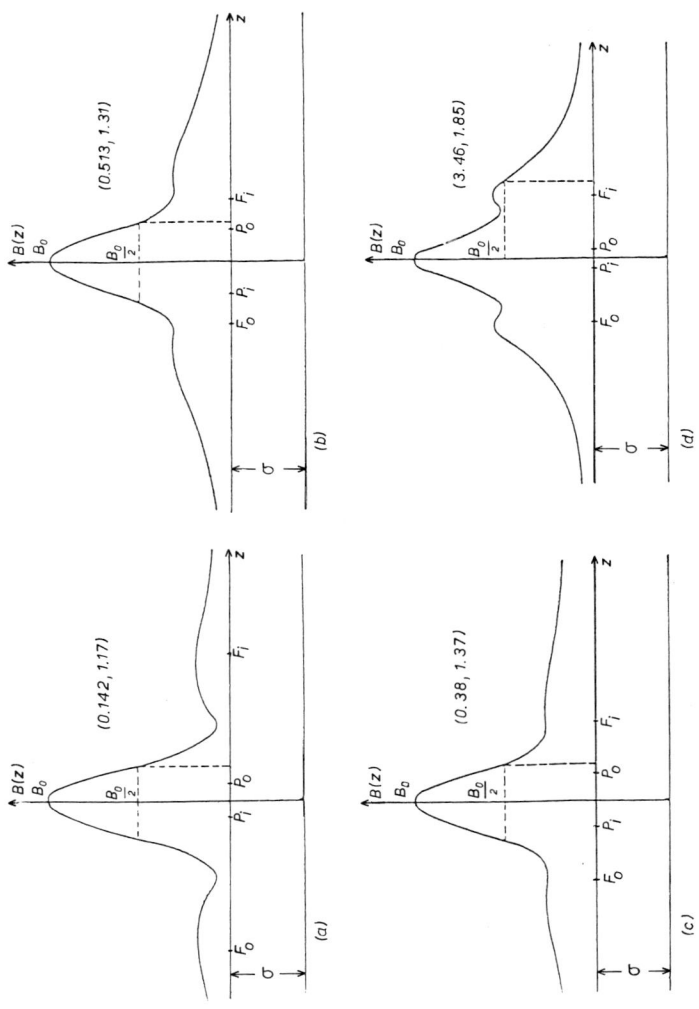

17.3 INVERSION OF THE PRINCIPAL PLANES

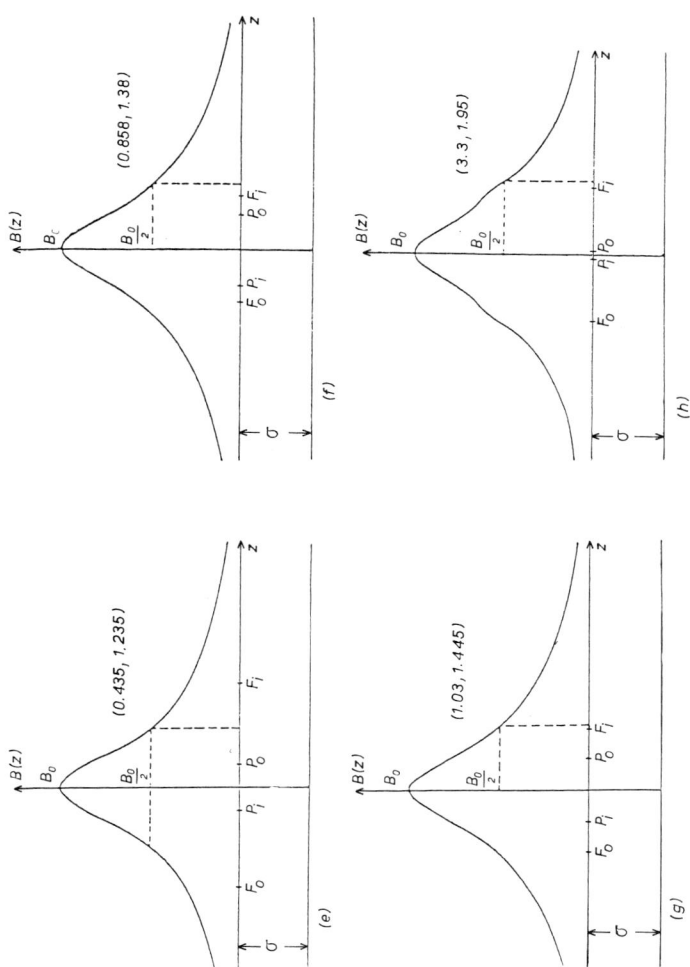

Fig. 17.5: Various Newtonian fields of the form (17.56). (a)–(d), $C_1 = -0.724$, $C_2 = 1.049$, $C_3 = 0.559$; (e)–(h), $C_1 = -0.298$, $C_2 = 1.035$, $C_3 = 0.331$. The parameters (k^2, ω) are as indicated.

as in (17.15). Setting the origin of coordinates midway between z_{Fo} and z_{Fi}, so that

$$-z_{Fo} = z_{Fi} =: z_F > 0 \qquad (17.58)$$

we find

$$z_i = \frac{z_F z_o - (f_o f_i - z_F^2)}{z_o + z_F} \qquad (17.59)$$

This projective transformation must have real or imaginary self-corresponding points, that is, points for which $z_i = z_o =: \zeta$, say. For this

$$\zeta^2 = -f_o f_i + z_F^2 \qquad (17.60)$$

If both object and image are real, there can be no self-corresponding points: a ray setting out from the axial point in $z = z_o$ cannot intersect the axis again in this plane. The self-corresponding points are therefore imaginary and

$$f_o f_i - z_F^2 > 0 \qquad (17.61)$$

If $f_o = f_i =: f$ we may conclude that

$$z_F < f \qquad (17.62)$$

since $z_F > 0$ by hypothesis and the fact that electron lenses are convergent implies that $f > 0$ for single stage imaging. But

$$z_{Pi} = -z_{Po} = z_F - f < 0$$

and so

$$z_{Pi} < z_{Po} \qquad (17.63)$$

If $f_o \neq f_i$, we have

$$z_F^2 < f_o f_i \qquad (17.64)$$

and since

$$\frac{1}{4}(f_o - f_i)^2 > 0$$

we may write

$$z_F^2 < \frac{1}{4}(f_o^2 + 2f_o f_i + f_i^2)$$

or

$$0 < z_F < \frac{1}{2}(f_o + f_i) \qquad (17.65)$$

or

$$z_{Pi} = z_F - f_i < \frac{1}{2}(f_o - f_i)$$

$$z_{Po} = -z_F + f_o > \frac{1}{2}(f_o - f_i)$$

17.4 APPROXIMATE FORMULAE FOR THE CARDINAL ELEMENTS

so that finally $z_{P_o} > z_{P_i}$ as before (17.63).

The above proof is valid only if the object and image are real and if the image is formed in a single stage: there must be no intermediate image, for otherwise we could not take the square roots (17.62–17.65). This inversion of the principal planes was first found experimentally, by Ruska (1934b). A situation which appears to conflict with this has been found by Sturrock (1951, 1955); we return to this in Part VII, where we examine field models.

17.4 Approximate formulae for the cardinal elements: the thin-lens approximation and the weak-lens approximation

Simple expressions for the focal lengths and positions of the foci can be obtained if the lens field is *weak*. We shall find that all weak lenses can be treated as thin lenses, that is, as lenses in which the principal planes coincide when $\phi_i = \phi_o$, but the converse is not necessarily true; when $\phi_i \neq \phi_o$, we are obliged to define thin lenses slightly differently. We set out from the reduced equation (15.38) in the form $v'' = -Gv$ and integrate twice to give

$$v'(z) = a'_0 - \int_{z_0}^{z} Gv \, d\zeta =: a'_0 - G_1(z)$$

$$v(z) = a_0 + a'_0(z - z_0) - \int_{z_0}^{z} G_1(\zeta) \, d\zeta$$

which may be integrated by parts to yield

$$v(z) = a_0 + a'_0(z - z_0) - \int_{z_0}^{z} (z - \zeta) G(\zeta) v(\zeta) \, d\zeta \qquad (17.66)$$

which is a Volterra-type integral equation. A formal solution is given by the Picard–Lindelöff iterative procedure

$$v_{n+1}(z) = a_0 + a'_0(z - z_0) - \int_{z_0}^{z} (z - \zeta) G(\zeta) v_n(\zeta) \, d\zeta$$

$$\text{for} \quad n = 0, 1, \ldots \qquad (17.67)$$

Suppose now that

$$\frac{1}{\tilde{f}} := \int_{-\infty}^{\infty} G(z) \, dz \ll \frac{1}{L} \qquad (17.68)$$

where L is some length characteristic of the lens producing the field $\phi(z)$ or $B(z)$, typically a gap or bore or total length. For convenience, we set the origin at the 'centre of gravity' of $G(z)$, so that its first moment vanishes:

$$\int_{-\infty}^{\infty} zG(z)\, dz = 0 \qquad (17.69)$$

We denote the second moment or "moment of inertia" of $G(z)$ by D:

$$D := \int_{-\infty}^{\infty} z^2 G(z)\, dz \qquad (17.70)$$

Clearly

$$D \ll L \ll \tilde{f} \qquad (17.71)$$

Let us now evaluate $v_1(z)$ from (17.67), taking as zero-order approximation $v_0(z) = a_0 + a_0'(z - z_0)$. In field-free object space, we have

$$v_1(z) = a_0 + a_0'(z - z_0) \qquad (17.72)$$

and in field-free image space,

$$v_1(z) = a_0 + a_0'(z - z_0) + \frac{(a_0' z_0 - a_0)z}{\tilde{f}} + a_0' D \qquad (17.73)$$

Neglecting the final term, we see that incident and emergent rays intersect in the centre-of-gravity plane, $z = 0$. The focal lengths and foci are easily found by expressing the above relation between asymptotes in matrix form and comparing with (16.12). In the plane $z_1 = z_2 = 0$ (Dušek matrix), we have

$$\begin{pmatrix} v_1^{(i)}(0) \\ v_1^{(i)'} \end{pmatrix} = \begin{pmatrix} 1 & 0 \\ -1/\tilde{f} & 1 \end{pmatrix} \begin{pmatrix} v_1^{(0)}(0) \\ v_1^{(0)'} \end{pmatrix}$$

$$\equiv \begin{pmatrix} pz_{Fi}/f_i & p(f_o + z_{Fo} z_{Fi}/f_i) \\ -p/f_i & -pz_{Fo}/f_i \end{pmatrix} \begin{pmatrix} v_1^{(0)}(0) \\ v_1^{(0)'} \end{pmatrix} \qquad (17.74)$$

where $p = (\hat{\phi}_i/\hat{\phi}_o)^{1/4}$, so that

$$\begin{aligned} f_i &= p\tilde{f}\ , & f_o &= -z_{Fo}/p = \tilde{f}/p \\ z_{Fi} &= f_i/p = \tilde{f}\ , & z_{Fo} &= -f_i/p = -\tilde{f} \end{aligned} \qquad (17.75)$$

17.4 APPROXIMATE FORMULAE FOR THE CARDINAL ELEMENTS

If $\phi_i = \phi_o$ ($p = 1$), the lens behaves like a thin lens situated at the centre of gravity of $G(z)$, the principal planes (16.7) coinciding in this plane as expected. If, however, $\phi_i \neq \phi_o$, the object and image focal lengths are related to $\tilde{f}$ explicitly as follows

$$f_o = \left(\frac{\hat{\phi}_o}{\hat{\phi}_i}\right)^{\frac{1}{4}} \tilde{f} \quad , \quad f_i = \left(\frac{\hat{\phi}_i}{\hat{\phi}_o}\right)^{\frac{1}{4}} \tilde{f} \tag{17.76}$$

and so $f_o f_i = \tilde{f}^2$: the intermediate quantity $\tilde{f}$ (16.19a) is the geometric mean of the object and image focal lengths. The foci are equidistant from the centre-of-gravity plane: $z_{Fi} = -z_{Fo} = \tilde{f}$. The principal planes do not, however, coincide in this plane but are separated by a distance

$$z_{Pi} - z_{Po} = z_{Fi} - f_i - z_{Fo} - f_o = 2\tilde{f} - f_i - f_o$$
$$= \tilde{f} \left\{ 2 - \left(\frac{\hat{\phi}_i}{\hat{\phi}_o}\right)^{\frac{1}{4}} - \left(\frac{\hat{\phi}_o}{\hat{\phi}_i}\right)^{\frac{1}{4}} \right\} \tag{17.77}$$

Despite this, the lens is still regarded as thin and we have therefore proved that every weak electron lens can be regarded as a thin lens, where the criterion for weakness is (17.68).

In the cases of magnetic lenses and electrostatic lenses, we obtain the following expressions:

Magnetic lenses
We set $\phi = $ const in $G(z)$ and find

$$\frac{1}{f} = \frac{1}{\tilde{f}} = \frac{\eta^2}{4\hat{\phi}} \int_{-\infty}^{\infty} B^2(z)\, dz \tag{17.78}$$

This expression, first derived by Busch (1927), is known as *Busch's formula* (cf. Ollendorff and Wendt, 1932).

Electrostatic lenses
We now set $B = 0$ in $G(z)$ and obtain

$$\frac{1}{f_o} = \frac{3}{16} \left(\frac{\hat{\phi}_i}{\hat{\phi}_o}\right)^{\frac{1}{4}} \int_{-\infty}^{\infty} \left(\frac{\phi'}{\hat{\phi}}\right)^2 \left(1 + \frac{4}{3}\epsilon\hat{\phi}\right) dz$$
$$\frac{1}{f_i} = \frac{3}{16} \left(\frac{\hat{\phi}_o}{\hat{\phi}_i}\right)^{\frac{1}{4}} \int_{-\infty}^{\infty} \left(\frac{\phi'}{\hat{\phi}}\right)^2 \left(1 + \frac{4}{3}\epsilon\hat{\phi}\right) dz \tag{17.79}$$

17. GAUSSIAN OPTICS OF ROTATIONALLY SYMMETRIC SYSTEMS

or nonrelativistically ($\epsilon\phi \ll 1$, $\hat{\phi} \to \phi$):

$$\frac{1}{f_o} = \frac{3}{16} \left(\frac{\phi_i}{\phi_o}\right)^{\frac{1}{4}} \int_{-\infty}^{\infty} \frac{\phi'^2}{\phi^2} dz$$

$$\frac{1}{f_i} = \frac{3}{16} \left(\frac{\phi_o}{\phi_i}\right)^{\frac{1}{4}} \int_{-\infty}^{\infty} \frac{\phi'^2}{\phi^2} dz$$

(17.80)

We note that it is important to derive (17.79) or (17.80) from the reduced equation. If ordinary coordinates (x, y, z) are employed, it is easy to obtain the wrong result; this point is discussed by Sturrock (1955, p.15ff.), who shows how the confusion arises.

18
Electron Mirrors

18.1 Introduction

In the foregoing discussion of the paraxial properties of rotationally symmetric systems, we have assumed that not only do the electron trajectories remain in the vicinity of the axis but that their gradients also remain small. If, however, the potential barrier in a retarding electrostatic lens is sufficiently high to prevent electrons from passing, the latter will be returned towards object space and the lens will have a mirror action. We shall see that to a first approximation, which it is convenient to call the paraxial approximation, the familiar cardinal elements can again be used for such *contracurrent* or *catoptric* systems, but special precautions must clearly be taken in the vicinity of the turning point, where the gradient is locally very large, passing through infinity.

Several ways of circumventing this difficulty have been proposed, two of which have been studied in detail. In each case, the distance along the optic axis, z, is replaced by some other independent variable, preferably one which, unlike z, increases monotonically as the electron proceeds and hence does not lead to infinite gradients when the electron turns round. The most obvious choice is the time, or a quantity very closely related to it (Recknagel, 1936, 1937; Nicholl, 1938; Regenstreif, 1951; Septier, 1953, 1960; Ehinger and Bernard, 1954; Schiske, 1957). Another possibility is to use a transformation introduced by Hahn (1965) as a means of unifying lens studies, in which the entire z-axis is mapped onto a finite region; its suitability for the study of mirrors is pointed out in Hahn (1971). Another suggestion comes from Bernard (1952), who writes $s^2 = -z$, with the origin of z at the point of reflection of an electron on the axis. The incident part of the trajectory then corresponds to $s < 0$ and the reflected part to $s > 0$; dx/ds and dy/ds do not become infinite.

Another transformation that has been extensively studied employs cartesian coordinates but in such a way that the 'paraxial' equations remain valid. It is still required that the charged particles remain close to the axis but not that their gradients remain small (Kel'man *et al.*, 1971, 1972a,b, 1973a,b; Daumenov *et al.*, 1978; Ximen *et al.*, 1983). We discuss this theory, which has been developed in considerable detail, in Section 18.3.

We assume throughout this Chapter that although the z-component of the electron velocity is reduced to zero by the potential barrier in the mirror, the same is not true of the retarding field: $E_z \neq 0$ in the plane in which the electron is instantaneously stationary. In the exceptional case in which this does occur, the electrons are trapped; this remote contingency is discussed by Recknagel (1937).

Electron mirrors are not the only systems in which the gradients of electron trajectories become too large for the conditions of paraxial imagery to be satisfied. In electron emission structures—guns, cathode lenses and image converters in particular—the electron velocity is locally small and the gradients steep. Guns are dealt with separately in Part IX. The methods of Kel'man et al. and Hahn have been used to study cathode lenses in some detail and any technique that enables mirrors to be analysed is in principle equally suitable for cathode lenses, in which the rays resemble those returning from a mirror except that the equipotentials are determined by the cathode surface. We now say a few words about the instrumental aspect of these various devices.

Electron mirrors are mainly used for reflecting the particle beam in certain types of energy analyser and mass spectrometer. A typical example of the former is the analyser built by Castaing and Henry (1962), which is shown in Fig. 18.1. Here, the direction of the beam is reversed by a repulsive electrostatic field, thus causing the particles to pass twice through the magnetic deflection field. The points of reflection are located somewhere in the field, away from any material surface.

Electron mirrors are also the essential elements of mirror or reflection electron microscopes. Here the beam direction is reversed just in front of the surface of the specimen, and contrast is produced in the image by local variations of some property of the specimen surface: variation in height or magnetic field distribution or electrostatic potential. For a detailed review, see Bok et al. (1971).

In another device (Lichte, 1983), the electron mirror is used as an interferometer (Fig. 18.2). Here, the electrons come so close to the material surface that the reflected beam is modulated by any roughness, as in the mirror microscope. The height distribution over the surface is now determined by analysing the interference pattern obtained by superposing a uniform beam on the reflected beam. Since this effect cannot be understood without going beyond geometrical electron optics, we defer further discussion to Volume 3.

Cathode lenses are employed in electron emission microscopes and in image-converters. *Emission microscopes* are mainly employed in metallurgical investigations. A typical cathode lens is shown schematically in Fig. 18.3. Here, the specimen acts as a (cold) cathode and electrons are

18.1 INTRODUCTION

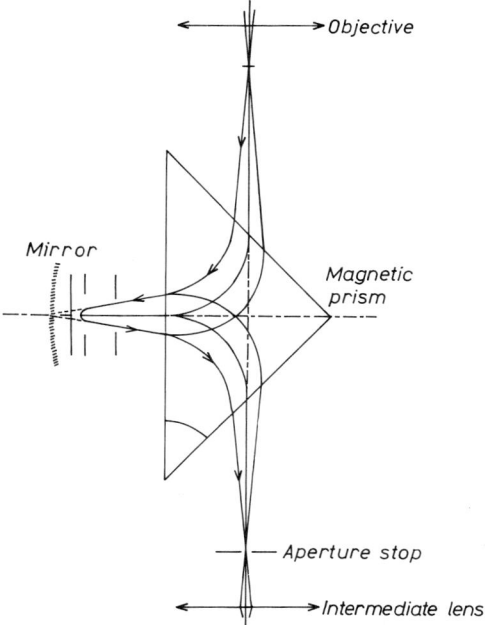

Fig. 18.1: Combination of electrostatic mirror and magnetic prism that permits energy-filtered images or energy-loss spectra to be formed.

ejected from it by lateral bombardment with electrons (Möllenstedt and Düker, 1953), photons (Koch, 1967) or ions. The local intensity of the ensuing electron emission depends to some extent on the intensity of the irradiation, which is assumed to be practically uniform over the area illuminated, and depends principally on material properties such as the work function. The electrons emitted are then used to form an image of the cathode surface on a recording screen, where the distribution of the properties that determined the image is now observable. Here, therefore, the cathode plays two roles: first, it serves as an electron emitter, and in addition, it forms the first electrode of the electrostatic lens in front of it. Such a lens is often known as an 'immersion objective'.

In *image converters* (Fig. 18.4), the function of the cathode is similar. Here, the cathode is a thin spherical layer, transparent to infrared or visible light. The inner surface of the cathode is coated with a layer of conducting material with a low work function so that when light falls on the cathode, photoelectrons are emitted. By imaging the cathode surface

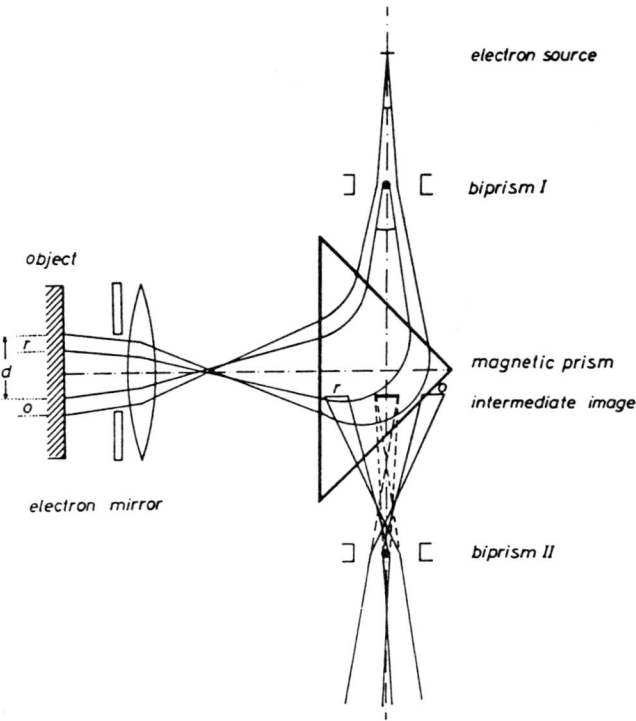

Fig. 18.2: The arrangement of prism, mirror and biprism in an electron mirror interference microscope. The presence of the two biprisms enables phase differences between the partial waves r and o to be measured interferometrically. Courtesy of H. Lichte (1983).

onto a viewing screen with electrons of sufficiently high energy, the feeble light image projected onto the cathode surface can be intensified. These introductory remarks are merely intended to give the reader a general idea of these devices; a few more technical details will be found in Chapters 37 and 38.

18.2 A time-like parameter as independent variable

A common feature of electron mirrors and cathode lenses is the existence of a surface of vanishing acceleration potential. In the vicinity of this surface, the electrons travel very slowly and the angle between their trajectories and the optic axis can take any value. This means that the conventional

18.2 A TIME-LIKE PARAMETER AS INDEPENDENT VARIABLE

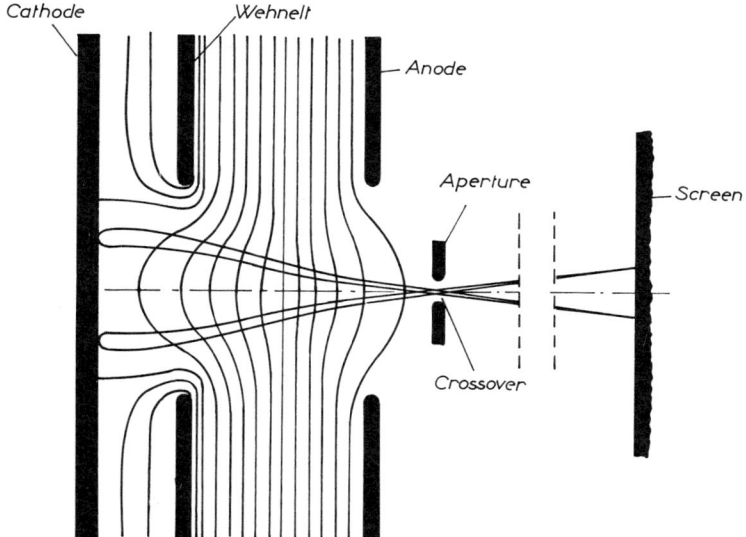

Fig. 18.3: Electrostatic equipotentials and electron trajectories in an electron microscope with a cathode lens.

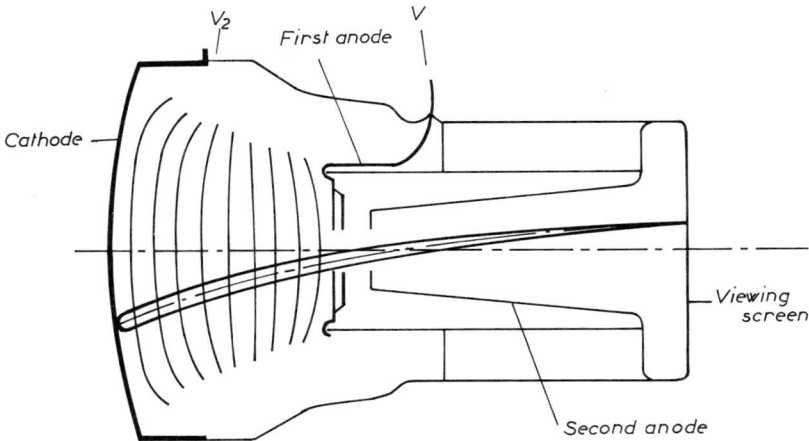

Fig. 18.4: An electrostatic three-electrode image converter for infrared light.

theory of electron lenses, which is based on the assumption that the gradient remains very small, cannot be applied.

One convenient way of overcoming this difficulty is to employ the ray equation in its parametric form; here, this means that we adopt the paraxial form of the Lorentz equation. Similar theories have already been developed by Glaser (1952) and by Zworykin et al. (1945). We shall use as curve parameter not the time t but the variable σ introduced in Section 3.2. We recall that σ is proportional to the proper time t', $d\sigma/dt' = 2\eta \hat{U}^{1/2}$ or $d\sigma/dt = 2\gamma\eta \hat{U}^{1/2}$. Since the kinetic energy of the electrons is low in all systems with mirrors or cathode lenses, a nonrelativistic theory is quite sufficient.

We set out from (3.12). We assume that the beam stays in the vicinity of the optic axis of the rotationally symmetric system, even though its gradient may become large. On substituting the usual paraxial approximations

$$\Phi(X, Y, Z) = \phi(Z) - \frac{1}{4} \phi''(Z)(X^2 + Y^2) \tag{18.1}$$

$$B_Z = B(Z) \quad , \quad B_X = -\frac{1}{2}XB'(Z) \quad , \quad B_Y = -\frac{1}{2}YB'(Z) \tag{18.2}$$

into (3.12) and truncating the resulting equations of motion after the linear terms in X and Y, we find

$$\ddot{X} = -\frac{\phi''}{4U} X - \eta U^{-\frac{1}{2}} \left(\dot{Y}B + \frac{1}{2}YB'\dot{Z} \right) \tag{18.3a}$$

$$\ddot{Y} = -\frac{\phi''}{4U} Y + \eta U^{-\frac{1}{2}} \left(\dot{X}B + \frac{1}{2}XB'\dot{Z} \right) \tag{18.3b}$$

$$\ddot{Z} = \frac{\phi'}{2U} \geq 0 \tag{18.3c}$$

where primes denote differentiation with respect to Z, while dots denote differentiation with respect to the curve parameter σ. Here all coordinates referring to the laboratory frame are denoted by capital letters. The constant U is an arbitrary positive normalization potential, preferably the acceleration potential at the anode of the device. Equation (18.3c) is readily integrated; we first obtain

$$\dot{Z}(\sigma) = \pm\sqrt{(\phi(Z) + \Phi_T)/U} \tag{18.4a}$$

where Φ_T is an arbitrary constant and this can be formally integrated to give

$$\sigma = \sigma_0 \pm \int \left\{ U/(\phi(Z) + \Phi_T) \right\}^{\frac{1}{2}} dZ \tag{18.4b}$$

18.2 A TIME-LIKE PARAMETER AS INDEPENDENT VARIABLE

From the numerical standpoint this expression is unfavourable, since the integral is improper, the radicand being singular at the point of reversal. It is far more convenient to solve (18.3c) directly with the appropriate initial conditions. Equations (18.3a,b) can be further simplified by introducing the familiar rotating frame

$$X + iY = (x + iy)\exp\{i\theta(Z(\sigma))\} \tag{18.5}$$

The necessary calculations are now quite similar to those for ordinary round lenses; the only novel feature is that we have to bear in mind that $Z = Z(\sigma)$ in the necessary differentiations. As in the theory of round lenses, the angle θ of rotation is chosen so that the derivatives $\dot{x}$ and $\dot{y}$ disappear from the equations of motion. This implies

$$\dot{\theta}(Z(\sigma)) = \frac{1}{2}\eta U^{-\frac{1}{2}} B(Z(\sigma)) \tag{18.6}$$

which can be integrated by means of (18.4a):

$$\theta(Z) = \pm \frac{\eta}{2} \int B(Z)\{\phi(Z) + \Phi_T\}^{-\frac{1}{2}} dz \tag{18.7}$$

This result is essentially the same as for round lenses; the normalization constant U has cancelled out, as it must.

After some elementary calculations making use of (18.6), the equations of motion for x and y are found to be

$$\ddot{x}(\sigma) + \frac{1}{4U}\left\{\phi''(Z) + \eta^2 B^2(Z)\right\} x(\sigma) = 0 \tag{18.8a}$$

$$\ddot{y}(\sigma) + \frac{1}{4U}\left\{\phi''(Z) + \eta^2 B^2(Z)\right\} y(\sigma) = 0 \tag{18.8b}$$

These are uncoupled but depend on the solution $Z(\sigma)$ of (18.3c), which has to be obtained first. We now introduce a pair of fundamental solutions, which will enable us to establish the paraxial properties of mirrors and cathode lenses. First, we have to select some nominal value Z_0 for the 'turning point' at which $\dot{Z}$ vanishes, other values then being regarded as aberrations. Without loss of generality we may choose the coordinate system such that $Z_0 = 0$ and the origin of potential such that $\phi(0) = 0$, so that $\Phi_T = 0$ in (18.4a); furthermore we choose the time origin so that $\sigma_0 = 0$ in (18.4b). The corresponding particular solution of (18.3c) will be denoted by $z(\sigma)$. It has the symmetry property

$$z(-\sigma) = z(\sigma) \quad \text{for all} \quad \sigma \tag{18.9}$$

which means that the domain $\sigma < 0$, $\dot{z} < 0$ corresponds to motion towards the reflection surface, while $\sigma > 0$, $\dot{z} > 0$ corresponds to motion after the reflection. In cathode lenses only this latter domain has a physical meaning. In the vicinity of the turning point, the series expansion

$$z(\sigma) = \frac{\phi_0' \sigma^2}{4U} + \frac{\phi_0' \phi_0'' \sigma^4}{96 U^2} + O(\sigma^6) \tag{18.10}$$

is valid.

Substituting $Z = z(\sigma)$ in (18.8a,b), we find that

$$\ddot{x} + F(\sigma)x = 0 \quad , \quad \ddot{y} + F(\sigma)y = 0 \tag{18.11}$$

with a new coefficient function

$$F(\sigma) := \frac{1}{4U} \left\{ \phi''(z(\sigma)) + \eta^2 B^2(z(\sigma)) \right\} \tag{18.12}$$

A convenient pair of linearly independent solutions, which we adopt as the fundamental solutions, are $u(\sigma)$ and $v(\sigma)$, satisfying the initial conditions

$$u(0) = \dot{v}(0) = 1 \quad , \quad \dot{u}(0) = v(0) = 0 \tag{18.13}$$

in the mirror plane. Obviously, besides (18.9) and as a consequence of this choice and of (18.13), the following symmetry properties must hold:

$$F(-\sigma) = F(\sigma) \quad , \quad u(-\sigma) = u(\sigma) \quad , \quad v(-\sigma) = -v(\sigma) \tag{18.14}$$

The general form of these solutions is sketched in Figs 18.5a,b; in the parametric representations (Fig. 18.5b) there is no singularity or ambiguity.

We now investigate image formation by an electron mirror. Consider an object plane, situated at $z = z_o \geq 0$. First, we have to determine the corresponding parameter $\sigma_o \leq 0$ from the inverse function $\sigma = \sigma(z) \leq 0$; there is always exactly one such solution. We next introduce new fundamental solutions $g(\sigma)$, $h(\sigma)$ adapted to the particular object plane. These must satisfy equations of the form of (18.12) with the initial conditions

$$g(\sigma_o) = \dot{h}(\sigma_o) = 1 \quad , \quad \dot{g}(\sigma_o) = h(\sigma_o) = 0 \tag{18.15}$$

instead of (18.13). The linear relations between g, h and u, v are given by

$$g(\sigma) = \dot{v}(\sigma_o) u(\sigma) - \dot{u}(\sigma_o) v(\sigma) \tag{18.16a}$$
$$h(\sigma) = -v(\sigma_o) u(\sigma) + u(\sigma_o) v(\sigma) \tag{18.16b}$$

18.2 A TIME-LIKE PARAMETER AS INDEPENDENT VARIABLE

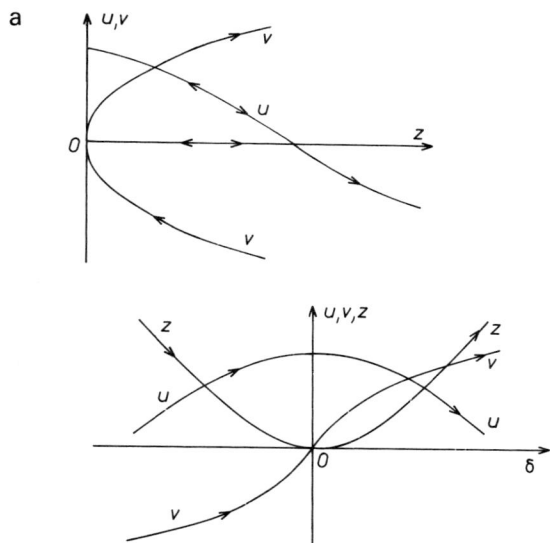

Fig. 18.5: Appearance of the fundamental solutions u and v (a) as functions of the axial coordinate z and (b) in parametric form together with $z(\sigma)$.

The Wronskians of both pairs of solutions are equal to unity, as can easily be verified.

The particular trajectory with the initial velocity components

$$\dot{x}(\sigma_o) := \dot{x}_o \quad , \quad \dot{y}(\sigma_o) := \dot{y}_o \quad , \quad \dot{z}(\sigma_o) := -\sqrt{\phi(z_o)/U} \qquad (18.17a)$$

at the starting point

$$x(\sigma_o) := x_o \quad , \quad y(\sigma_o) := y_o \quad , \quad z(\sigma_o) := z_o \qquad (18.17b)$$

is given by $z = z(\sigma)$ and

$$x(\sigma) = x_o g(\sigma) + \dot{x}_o h(\sigma) \qquad (18.18a)$$
$$y(\sigma) = y_o g(\sigma) + \dot{y}_o h(\sigma) \qquad (18.18b)$$

Just as in the case of ordinary round lenses the position of the conjugate image plane is determined by the next zero of the function $h(\sigma)$: $h(\sigma_i) = 0$ with $\sigma_i > \sigma_o$. Usually the focusing fields are so weak that $\sigma_i > 0$, which means that the image is formed after the reflection. The position of the image plane is then given by $z_i := z(\sigma_i)$ and the lateral image coordinates are (irrespective of $\dot{x}_o$ and $\dot{y}_o$)

$$x_i = x_o g(\sigma_i) \quad , \quad y_i = y_o g(\sigma_i) \qquad (18.19)$$

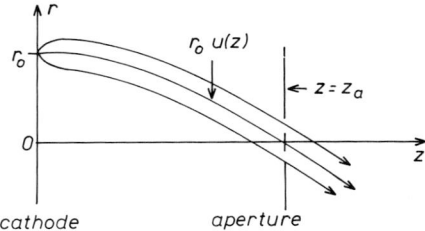

Fig. 18.6: Appropriate location of an aperture in a cathode lens.

Thus $g(\sigma_i)$ is the lateral magnification. In a magnetic field, the image is rotated relative to the object; from (18.6) the corresponding angle is seen to be

$$\theta_i = \frac{\eta}{2} U^{-\frac{1}{2}} \int_{\sigma_o}^{\sigma_i} B(z(\sigma)) \, d\sigma \tag{18.20}$$

From this formula, which is quite unambiguous, it is obvious that the sense of rotation is maintained after the reflection, whereas (18.7) needs more care.

Image formation by a cathode lens is a special case of the foregoing situation. In this case the object is located in the cathode plane $z_o = 0$; hence $\sigma_o = 0$ and consequently $g(\sigma) \equiv u(\sigma)$, $h(\sigma) \equiv v(\sigma)$, as is obvious on comparing (18.13) with (18.15). Other than this simplification, there is nothing new about image formation by a cathode lens. A narrow aperture is often brought into the beam in order to filter out electrons with very large lateral velocity components. In this way the chromatic aberration of the cathode lens can be reduced, but this can only be fully understood after studying the aberrations. Here we merely point out that the appropriate plane for this aperture is given by

$$z_a := z(\sigma_a) \quad \text{with} \quad u(\sigma_a) = 0 \tag{18.21a}$$

as sketched in Fig. 18.6. The lateral coordinates

$$x_a = \dot{x}_o v(\sigma_a) \quad , \quad y_a = \dot{y}_o v(\sigma_a) \tag{18.21b}$$

are then independent of x_o and y_o, so that there is no vignetting.

We could go on to define asymptotic cardinal elements — foci, principal planes and focal lengths — exactly as for round lenses. If we adopt the same sign conventions in the definitions, namely $z_{Fi} - z_{Pi} =: f_i$ and $z_{Po} - z_{Fo} =: f_o$ (16.7), we find that now $f_i = -f_o =: f$, so that with (16.25)

$$z_i - z_{Fi} = -f_i M$$
$$z_o - z_{Fo} = f_o/M$$

we have
$$(z_i - z_{Fi})(z_o - z_{Fo}) = -f_o f_i = f^2$$
If we compare this with the treatment of mirrors in light optics (e.g. Born and Wolf, 1959 Section 4.3, where the sign convention for f_i is the opposite of that adopted here), we see that convergent mirrors correspond to positive values of f_o and hence negative values of f; for divergent mirrors, $f_o < 0$ and hence $f > 0$.

18.3 The cartesian representation

Although the trajectories reverse their direction and have large gradients in the vicinity of the turning point, a paraxial ray equation can nevertheless be derived in the conventional cartesian form. This approach has been investigated by Kel'man et al. (1971, 1972a,b, 1973a,b), who first used it to facilitate the study of cylindrical mirrors (1971, 1972a,b) and subsequently employed it to analyse rotationally symmetric mirrors including magnetic fields (1973a,b). Their theory can be derived from a variational principle, as shown in detail by Daumenov et al. (1978); we shall return to this in Chapter 28 in connection with the aberrations. Here we present a very brief derivation, setting out from (18.8a,b).

Our aim now is to eliminate the time-like parameter σ from (18.8a,b), which can be done with the aid of (18.4a). It is helpful to introduce an axial potential
$$V(Z) = \Phi_T + \phi(Z) \tag{18.22}$$
which depends on the particle energy $e\Phi_T$, as does the turning point Z_0 at which it vanishes. With
$$\frac{d}{d\sigma} = \dot{Z}\frac{d}{dZ} \quad , \quad \frac{d^2}{d\sigma^2} = \ddot{Z}\frac{d}{dZ} + \dot{Z}^2\frac{d^2}{dZ^2}$$
and using (18.3c), (18.4a) and (18.22) we find
$$\frac{d^2}{d\sigma^2} = \frac{V(Z)}{U}\frac{d^2}{dZ^2} + \frac{V'(Z)}{2U}\frac{d}{dZ}$$
and so from (18.8a,b)
$$V(Z)x''(Z) + \frac{1}{2}V'(Z)x'(Z) + \frac{1}{4}\left\{V''(Z) + \eta^2 B^2(Z)\right\}x(Z) = 0$$
$$\tag{18.23a}$$
$$V(Z)y''(Z) + \frac{1}{2}V'(Z)y'(Z) + \frac{1}{4}\left\{V''(Z) + \eta^2 B^2(Z)\right\}y(Z) = 0$$
$$\tag{18.23b}$$

18. ELECTRON MIRRORS

These paraxial ray equations have the same formal structure as those for ordinary round lenses. The novel fact here is that very small values are required only for $x^2 + y^2$, whereas the gradients may be arbitrarily large. The range of validity of the paraxial approximation has clearly been extended.

On the other hand, (18.23) also have disadvantages. A first difficulty is the fact that for a polychromatic beam, it is not the axial potential $\phi(z)$ alone that is involved but the shifted potential $V(Z)$; the turning point Z_0 at which $V(Z_0) = 0$ is energy-dependent and is a singularity. A second drawback is that the familiar numerical solution techniques cannot be employed in the vicinity of a singularity and special series expansions have to be evaluated instead. A third disadvantage is the fact that when tracing rays through mirror fields, great care must be taken over the signs of some of the variables, whereas in the parametric form of the theory the correct signs are obtained automatically. The numerical solution of (18.23) is thus distinctly inconvenient.

We now study the vicinity of the singularity for rays with the *nominal energy*. For these, we have

$$\Phi_T = 0 \quad , \quad Z_0 = 0 \quad , \quad Z = z \quad , \quad V(Z) = \phi(z) \tag{18.24}$$

and since $\phi(0) = 0$, we can introduce a power series expansion

$$\phi(z) = \sum_{n=1}^{\infty} \frac{z^n}{n!} \phi^{(n)}(0) = z\phi'_0 + \frac{z^2}{2}\phi''_0 + \cdots \tag{18.25}$$

For the magnetic field, a power series expansion

$$\beta(z) := \eta^2 B^2(z) = \sum_{n=0}^{\infty} \frac{z^n}{n!} \beta^{(n)}(0) = \beta_0 + z\beta'_0 + \cdots \tag{18.26}$$

is introduced. It is clear that each of the differential equations (18.23a,b) must have two linearly independent solutions and that the general solutions are obtained by appropriate linear superpositions, but it is not possible to represent the latter in terms of initial conditions referring to the singular plane $z = 0$. In view of this situation we can proceed in the following way.

One particular solution $p(z)$ of (18.23) with (18.24), (18.25) and (18.26) is found by introducing an ordinary *regular* power series expansion:

$$p(z) = \sum_{n=0}^{\infty} \frac{z^n}{n!} p^{(n)}(0) = p_0 + zp'_0 + \frac{z^2}{2}p''_0 + \cdots \tag{18.27}$$

18.3 THE CARTESIAN REPRESENTATION

Substituting this together with (18.25) and (18.26) into the differential equation

$$\phi(z)w''(z) + \frac{1}{2}\phi'(z)w'(z) + \frac{1}{4}\left\{\phi''(z) + \beta(z)\right\}w(z) = 0 \qquad (18.28)$$

with $w(z) \equiv p(z)$ and collecting up all terms with equal powers of z, we obtain linear recurrence relations for the coefficients $p^{(n)}(0)$. The beginning of this sequence is

$$\begin{aligned} p_0' &= -\frac{p_0(\phi_0'' + \beta_0)}{2\phi_0'} \\ p_0'' &= -\frac{p_0'(3\phi_0'' + \beta_0) + p_0(\phi_0''' + \beta_0')}{6\phi_0'} \end{aligned} \qquad (18.29)$$

Clearly, only the initial coordinate $p_0 = p(0)$ can be prescribed arbitrarily; the slope p_0' is then already determined.

A second and linearly independent solution $w(z) \equiv q(z)$ of (18.28) is obtained in the form of a *fractional* power series expansion

$$q(z) = z^{\frac{1}{2}} \sum_{n=0}^{\infty} \frac{z^n}{n!} a^{(n)}(0) = z^{\frac{1}{2}}\left(a_0 + za_0' + \frac{z^2}{2}a_0'' + \ldots\right) \qquad (18.30)$$

We again find a sequence of linear recurrence relations, the opening terms of which are

$$\begin{aligned} a_0' &= -\frac{a_0\left(\frac{3}{2}\phi_0'' + \beta_0\right)}{6\phi_0'} \\ a_0'' &= -\frac{a_0'\left(\frac{11}{2}\phi_0'' + \beta_0\right) + a_0\left(\frac{4}{3}\phi_0''' + \beta_0'\right)}{10\phi_0'} \end{aligned} \qquad (18.31)$$

The general solution now has the form

$$w(z) = Ap(z) \pm Bq(z) \qquad (18.32)$$

the two branches of which, distinguished by the sign of the second term, have to be joined together appropriately at $z = 0$.

The physical meaning of this solution is sketched in Fig. 18.7. For very small positive values of z the lowest powers in the series expansions dominate, hence $w(z) = Ap_0 \pm Ba_0 z^{1/2}$. This corresponds to a parabolic trajectory caused by uniform acceleration. In this lowest order approximation the tangents to trajectories in a given plane $z = L > 0$ all intersect in the 'mirror plane' $z = -L$. This is, of course, only true for vanishing axial velocity $\dot{z}$ in the starting plane $z = 0$, which is clearly an idealization. A

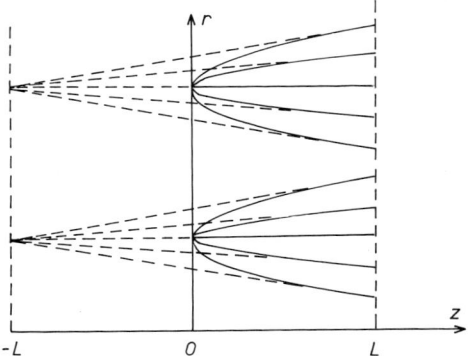

Fig. 18.7: Trajectories approximated by parabolae in the zone $0 \leq z \leq L$. The tangents at $z = L$ all intersect in $z = -L$.

more realistic situation will be studied in Chapter 28, which is concerned with the aberrations of mirrors and cathode lenses.

For numerical evaluation, this form of the trajectory equation is far less convenient than the parametric form dealt with in the previous section. In the case of a mirror, one has first to solve (18.28) for z decreasing from z_0 to a small value L. A system of linear equations for p_0, p_0', p_0'', a_0, a_0', a_0'', arising from the requirement that as many derivatives as possible must be continuous at $z = L$, must then be solved. The trajectory, given in the form of a series expansions, then has to be traced until it again reaches the plane $z = L$. The numerical tracing then has to be resumed. This is a fairly high price to pay for any conceptual advantages gained by avoiding the parametric representation. Moreover we shall see in Chapter 28 that the presence of aberrations renders the original form $x = x(z)$, $y = y(z)$ of the trajectories untenable; the parametric form then becomes clearly advantageous.

18.4 A quadratic transformation

We mention briefly one other transformation designed to avoid the problems arising from the large ray gradients at the turning point. This was introduced by Bernard (1952), who wrote

$$z_v - z =: s^2 \qquad (18.33)$$

for an electron initially travelling in the negative z direction and reflected at $z = z_v$. The negative sign is to be taken before the square root for the

18.4 A QUADRATIC TRANSFORMATION

incident part of the trajectory and the positive sign for the reflected part, which already shows that this transformation is by no means ideal. If we consider only the electrostatic case ($B = 0$), (18.11) becomes, after some calculation,

$$\frac{d^2x}{ds^2} + \left(\frac{1}{2\phi}\frac{d\phi}{ds} - \frac{1}{s}\right)\frac{dx}{ds} + \frac{1}{4\phi}\left(\frac{d^2\phi}{ds^2} - \frac{1}{s}\frac{d\phi}{ds}\right)x = 0 \qquad (18.34)$$

with a similar equation for $y(s)$. The term in dx/ds is eliminated by writing

$$\bar{x} = \phi^{\frac{1}{4}} s^{-\frac{1}{2}} x \qquad (18.35)$$

whereupon (18.34) reduces to

$$\frac{d^2\bar{x}}{ds^2} + G\bar{x} = 0 \qquad (18.36)$$

with

$$G(s) = \frac{3}{16}\left(\frac{d\phi/ds}{\phi}\right)^2 - \frac{3}{4s^2}$$

Equation (18.36) may then be analysed in the usual way and cardinal elements defined.

19
Quadrupole Lenses

Hitherto, we have been studying systems with an axis of rotational symmetry, which are by far the most common in electron optical devices in which the electron accelerating voltage does not exceed one or a few hundred kilovolts or, exceptionally, a few megavolts (in practice, 3 or 4 MV maximum). At higher voltages, focusing elements with lower symmetry are more commonly used, and, in particular, elements with planes of electrical symmetry forming quadrupoles. These possess the property of 'strong focusing', by which we mean that their fields exert a force directly on the electrons, towards or away from the axis, whereas in round magnetic lenses, the focusing force is more indirect, arising from the coupling between B_z and the azimuthal component of the electron velocity. Quadrupoles have also been very thoroughly studied as elements for use at conventional accelerating voltages for a quite different reason. We shall see in Section 24.3 that one of the most undesirable aberrations of round electron lenses cannot be eliminated in any straightforward manner but can in principle be cancelled by introducing quadrupoles and octopoles into the system. This has provided the incentive for exhaustive studies of quadrupole lens properties, comparable in thoroughness with those on round lenses. In this chapter, we derive the paraxial equations of quadrupole systems, and introduce the notion of an orthogonal system.

A brief survey of the history of quadrupole studies is to be found at the beginning of Chapter 39. Here we merely remark that although strong focusing at high energy and aberration correction have been the principal stimuli for research on quadrupoles, their properties were very fully explored long before either of these applications was known: the first study, which was thorough and meticulous, appeared as a Berlin dissertation in 1943 (Melkich, 1947).

In the discussion of Section 7.2.3, the terms in 2φ in the potential expansion (7.37 or 7.59) and their magnetic counterparts (7.43–7.45) were described as quadrupole terms. Here we use the term in a slightly less restrictive sense: a magnetic or electrostatic quadrupole is characterized by the presence of two planes of symmetry in the potential or field (Fig. 19.1), and rotationally symmetric fields as well as octopoles (and higher order $2n$-poles) are not excluded. From (7.36), we see that the term in $p_2(z)$ describes a potential for which the planes $x = 0$ and $y = 0$ are symmetry planes,

19.1 PARAXIAL EQUATIONS FOR QUADRUPOLES

$\Phi(x,y,z) = \Phi(\pm x, \pm y, z)$, and a typical electrostatic quadrupole thus has the form shown in Fig. 7.2 and Fig. 19.1a. The term in $q_2(z)$ simply corresponds to a similar quadrupole inclined at 45° to that of Fig. 19.1a. In the magnetic case, the planes $x = 0$ and $y = 0$ are planes of symmetry for $P_2(z)$ and planes of antisymmetry for $Q_2(z)$. For a field described by Q_2 only, $B_x(x,y,z) = B_x(-x,y,z) = -B_x(x,-y,z) = -B_x(-x,-y,z)$, with analogous relations for B_y (Fig. 19.1b). We shall learn that uncoupled equations can be obtained if the quadrupoles are orientated with respect to the coordinate axes in such a way that $q_2 = 0$ and $P_2 = 0$, that is, as shown in Figs 19.1.

19.1 Paraxial equations for quadrupoles

We set out from the general case in which rotationally symmetric magnetic and electrostatic fields may be present as well as the quadrupole fields themselves. The fields are thus characterized by six axial functions, $\phi(z)$ and $B(z)$ for the round lens components, $p_2(z)$ and $q_2(z)$ for the electrostatic quadrupoles and $Q_2(z)$ and $P_2(z)$ for the magnetic quadrupoles. We substitute the field expansions (7.36) and (7.43–7.45) into the function M, $M = \overline{M}/(2m_0 e)^{1/2}$ (15.23) as in Section 15.2 and expand M as a power series in X, Y and their derivatives. We find

$$
\begin{aligned}
M^{(0)} &= \hat{\phi}^{\frac{1}{2}} \\
M^{(2)} &= -\frac{\gamma \phi''}{8 \hat{\phi}^{\frac{1}{2}}} (X^2 + Y^2) + \left(\frac{\gamma p_2}{4 \hat{\phi}^{\frac{1}{2}}} - \frac{1}{2} \eta Q_2 \right)(X^2 - Y^2) \\
&\quad + \left(\frac{\gamma q_2}{2 \hat{\phi}^{\frac{1}{2}}} + \eta P_2 \right) XY + \frac{1}{2} \hat{\phi}^{\frac{1}{2}} (X'^2 + Y'^2) \\
&\quad - \frac{1}{2} \eta B (XY' - X'Y)
\end{aligned}
\qquad (19.1)
$$

The term in $XY' - X'Y$ can be removed by introducing the rotating coordinate system employed in connection with round magnetic lenses (15.7,

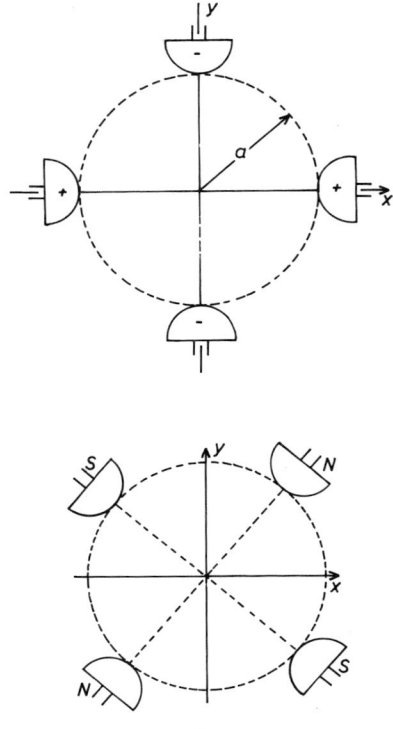

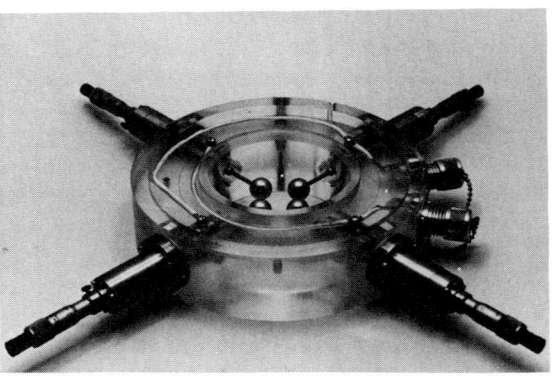

Fig. 19.1: Quadrupoles. (a) Electrostatic quadrupole characterized by $p_2(z)$, with equal and opposite voltages on the electrodes. (b) Magnetic quadrupole characterized by $Q_2(z)$. (c) Photograph of an actual electrostatic quadrupole.

19.1 PARAXIAL EQUATIONS FOR QUADRUPOLES

15.9) whereupon $M^{(2)}$ becomes

$$M^{(2)} = -\left(\frac{\gamma\phi''}{8\hat{\phi}^{\frac{1}{2}}} + \frac{\eta^2 B^2}{8\hat{\phi}^{\frac{1}{2}}}\right)(x^2+y^2)$$

$$+ \left\{\left(\frac{\gamma p_2}{2\hat{\phi}^{\frac{1}{2}}} - \eta Q_2\right)\cos 2\theta + \left(\frac{\gamma q_2}{2\hat{\phi}^{\frac{1}{2}}} + \eta P_2\right)\sin 2\theta\right\}\frac{x^2-y^2}{2}$$

$$+ \left\{-\left(\frac{\gamma p_2}{2\hat{\phi}^{\frac{1}{2}}} - \eta Q_2\right)\sin 2\theta + \left(\frac{\gamma q_2}{2\hat{\phi}^{\frac{1}{2}}} + \eta P_2\right)\cos 2\theta\right\}xy$$

$$+ \frac{1}{2}\hat{\phi}^{\frac{1}{2}}(x'^2 + y'^2) \qquad (19.2)$$

with $\theta' = \eta B/2\hat{\phi}^{1/2}$.
The paraxial equations, $\partial M^{(2)}/\partial x_i = d(\partial M^{(2)}/\partial x'_i)/dz$ ($x_1 = x, x_2 = y$), still do not separate, however, unless

$$\tan\theta(z) = \frac{\gamma q_2/\hat{\phi}^{\frac{1}{2}} + 2\eta P_2}{\gamma p_2/\hat{\phi}^{\frac{1}{2}} - 2\eta Q_2} \qquad (19.3)$$

This is known as the *orthogonality condition* and has been known in various forms for many years (Melkich, 1947; Glaser, 1956, Section 37; Dušek, 1959). This condition can be satisfied in several ways but only one is a practical possibility. Most generally, $\theta(z)$ may vary with z, in which case electrodes and polepieces must be devised and constructed of such shapes that condition (19.3) is everywhere satisfied. More reasonably, $\theta(z)$ may be a constant, not necessarily zero; this requires $B(z) = 0$ and $\gamma q_2 + 2\eta P_2\hat{\phi}^{1/2} \propto \gamma p_2 - 2\eta Q_2\hat{\phi}^{1/2}$. Finally we may set $\theta(z)$ equal to zero, so that $q_2(z) = P_2(z) \equiv 0$, and at least one of $p_2(z)$ and $Q_2(z)$ is not zero everywhere. We retain only this final case.

From now on, then, we consider only orthogonal systems, and we assume that the electrodes and polepieces are disposed so that only $\phi(z)$, $p_2(z)$ and $Q_2(z)$ are allowed to be nonzero. The paraxial equations then take the form

$$\frac{d}{dz}\left(\hat{\phi}^{\frac{1}{2}}x'\right) + \frac{\gamma\phi'' - 2\gamma p_2 + 4\eta Q_2\hat{\phi}^{\frac{1}{2}}}{4\hat{\phi}^{\frac{1}{2}}}x = 0 \qquad (19.4a)$$

$$\frac{d}{dz}\left(\hat{\phi}^{\frac{1}{2}}y'\right) + \frac{\gamma\phi'' + 2\gamma p_2 - 4\eta Q_2\hat{\phi}^{\frac{1}{2}}}{4\hat{\phi}^{\frac{1}{2}}}y = 0 \qquad (19.4b)$$

or nonrelativistically

$$x'' + \frac{\phi'}{2\phi}x' + \frac{\phi'' - 2p_2 + 4\eta Q_2\phi^{\frac{1}{2}}}{4\phi}x = 0 \qquad (19.5a)$$

$$y'' + \frac{\phi'}{2\phi}y' + \frac{\phi'' + 2p_2 - 4\eta Q_2\phi^{\frac{1}{2}}}{4\phi}y = 0 \qquad (19.5b)$$

or in reduced form, $\xi_x := x\hat{\phi}^{1/4}$, $\xi_y := y\hat{\phi}^{1/4}$,

$$\xi_x'' + \left\{ \frac{3}{16}\left(\frac{\phi'}{\hat{\phi}}\right)^2 (1 + \frac{4}{3}\epsilon\hat{\phi}) - \frac{p_2 - 2\eta Q_2 \hat{\phi}^{\frac{1}{2}}}{2\hat{\phi}} \right\} \xi_x = 0 \qquad (19.6a)$$

$$\xi_y'' + \left\{ \frac{3}{16}\left(\frac{\phi'}{\hat{\phi}}\right)^2 (1 + \frac{4}{3}\epsilon\hat{\phi}) + \frac{p_2 - 2\eta Q_2 \hat{\phi}^{\frac{1}{2}}}{2\hat{\phi}} \right\} \xi_y = 0 \qquad (19.6b)$$

Each paraxial equation is a linear, homogeneous, second-order differential equation and by any of the lines of reasoning set out in Chapter 16, we may again establish the existence of cardinal elements. For quadrupoles, it is usually sufficient to list the asymptotic cardinal elements, though real (and osculating) elements can of course be defined if needed. Unlike round lenses, however, two sets of cardinal elements are needed, one to characterize the x–z plane, the other the y–z plane. In the absence of any rotationally symmetric lens field (ϕ = const), the lens action in one of these planes will be convergent while in the other it will be divergent. This is readily seen from (19.4–19.6).

The action of a quadrupole lens on electron rays is most conveniently characterized by a pair of transfer matrices, similar to (16.12) except that the matrix describing the coordination between object and image space is different in the two planes. Before writing down these matrices, we must first introduce the notion of *astigmatic objects and images*. Quadrupoles are commonly employed as multiplets—the quadruplet can have most attractive features—and if the cardinal elements are different in the x–z and y–z planes, the image plane may clearly be different as well: the system will not produce a stigmatic image of a point object. If a further lens follows, this *astigmatic image* will be the object for the subsequent stage, and we must thus expect to have to deal with *astigmatic objects*.

The most general transfer matrix relates position and slope in some plane in object space, $z = z_1$, to the same quantities in a plane in image space, $z = z_2$ (cf. 16.15):

$$\begin{pmatrix} x_2 \\ x_2' \end{pmatrix} = T_x \begin{pmatrix} x_1 \\ x_1' \end{pmatrix} \quad , \quad \begin{pmatrix} y_2 \\ y_2' \end{pmatrix} = T_y \begin{pmatrix} y_1 \\ y_1' \end{pmatrix} \qquad (19.7)$$

and as in (16.12), we write

$$T_x = \begin{pmatrix} -(z_2 - z_{Fi}^{(x)})/f_{xi} & (z_2 - z_{Fi}^{(x)})(z_1 - z_{Fo}^{(x)})/f_{xi} + f_{xo} \\ -1/f_{xi} & (z_1 - z_{Fo}^{(x)})/f_{xi} \end{pmatrix} \qquad (19.8a)$$

$$T_y = \begin{pmatrix} -(z_2 - z_{Fi}^{(y)})/f_{yi} & (z_2 - z_{Fi}^{(y)})(z_1 - z_{Fo}^{(y)})/f_{yi} + f_{yo} \\ -1/f_{yi} & (z_1 - z_{Fo}^{(y)})/f_{yi} \end{pmatrix} \qquad (19.8b)$$

19.1 PARAXIAL EQUATIONS FOR QUADRUPOLES

The cardinal elements are most conveniently defined with the aid of the rays $G_x(z)$, $G_y(z)$, $\overline{G}_x(z)$ and $\overline{G}_y(z)$, which satisfy conditions analogous to (16.1):

$$\lim_{z \to -\infty} G_x(z) = \lim_{z \to -\infty} G_y(z) = 1$$
$$\lim_{z \to \infty} \overline{G}_x(z) = \lim_{z \to \infty} \overline{G}_y(z) = 1 \tag{19.9}$$

The rays $G_x(z)$ and $\overline{G}_x(z)$ satisfy (19.4a) while $G_y(z)$ and $\overline{G}_y(z)$ satisfy (19.4b). The image foci are then the points of intersection of the image asymptotes to G_x and G_y with the axis; the rays $\overline{G}_x$ and $\overline{G}_y$ similarly define the object foci. The focal lengths are given by

$$f_{xi} = -1/G'_{xi} \quad , \quad f_{yi} = -1/G'_{yi}$$
$$f_{xo} = 1/\overline{G}'_{xo} \quad , \quad f_{yo} = 1/\overline{G}'_{yo} \tag{19.10}$$

Let us suppose that the planes $z = z_{xo}$ and $z = z_{xi}$ are conjugate, in the sense that $(T_x)_{12} = 0$, or

$$(z_{xi} - z^{(x)}_{Fi})(z_{xo} - z^{(x)}_{Fo}) = -f_{xi} f_{xo} \tag{19.11}$$

so that

$$T_x = \begin{pmatrix} M_x & 0 \\ -1/f_{xi} & (\hat{\phi}_o/\hat{\phi}_i)^{\frac{1}{2}}/M_x \end{pmatrix} \tag{19.12}$$

in which we have used the Wronskian of (19.4a) to show that $f_{xo}/f_{xi} = (\hat{\phi}_o/\hat{\phi}_i)^{1/2}$; a similar relation is of course true for f_{yo}/f_{yi}. The magnitude M_x is the height of the image asymptote to $G_x(z)$ in the plane $z = z_{xi}$. In general, however, $(T_y)_{12} \neq 0$ when (19.11) is satisfied and so although rays from a point $P(x_o, y_o)$ all have the same x-coordinate $x_i = M_x x_o$ in $z = z_{xi}$, their y-coordinate is a function of both y_o and the gradient y'_o:

$$T_y = \begin{pmatrix} -(z_{xi} - z^{(y)}_{Fi})/f_{yi} & (z_{xi} - z^{(y)}_{Fi})(z_{xo} - z^{(y)}_{Fo})/f_{yi} + f_{yo} \\ -1/f_{yi} & (z_{xo} - z^{(y)}_{Fo})/f_{yi} \end{pmatrix} \tag{19.13}$$

A point $P(x_o, y_o)$ is therefore imaged as a line in the plane $z = z_{xi}$ parallel to the y-axis. Likewise, if we consider a pair of planes $z = z_{yo}$ and $z = z_{yi}$ for which $(T_y)_{12} = 0$, we find that in general $(T_x)_{12}$ does not vanish and again a line image is formed, now parallel to the x-axis. These line images are thus at right-angles to one another and separated by the *astigmatic difference*. One or both may be virtual (Fig. 19.2). If we move an axial point object along the axis, the line foci will move and there will always be

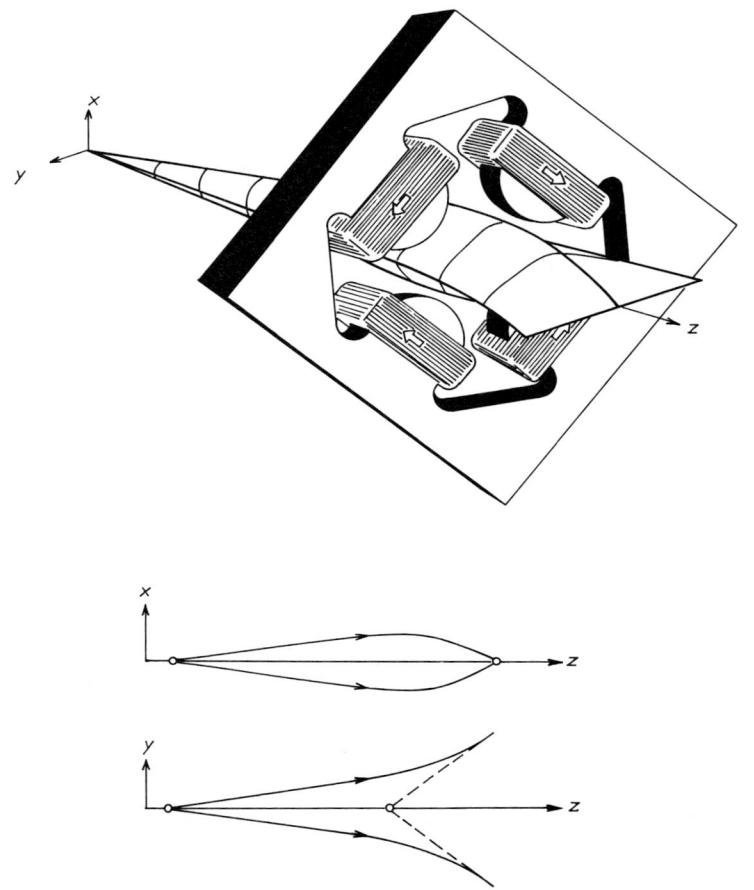

Fig. 19.2: Formation of a line image in a magnetic quadrupole. The arrows show the directions of the currents in the windings.

real or virtual object positions for which the line foci coincide and the image is stigmatic. In general, however, the magnifications in the two planes will not be equal.

The astigmatic difference can be expressed in terms of the cardinal elements and magnification. Using the quadrupole analogue of (16.25),

$$z_{xi} - z_{Fi}^{(x)} = -f_{xi}M_x$$
$$z_{yi} - z_{Fi}^{(y)} = -f_{yi}M_y$$
(19.14)

19.1 PARAXIAL EQUATIONS FOR QUADRUPOLES

and
$$z_{xo} - z_{Fo}^{(x)} = f_{xo}/M_x$$
$$z_{yo} - z_{Fo}^{(y)} = f_{yo}/M_y \tag{19.15}$$

we see that
$$\Lambda_i := z_{xi} - z_{yi} = \Lambda_{Fi} - f_{xi}M_x + f_{yi}M_y$$
$$\Lambda_o := z_{xo} - z_{yo} = \Lambda_{Fo} + f_{xo}/M_x - f_{yo}/M_y \tag{19.16}$$

where
$$\Lambda_{Fi} := z_{Fi}^{(x)} - z_{Fi}^{(y)} = \Lambda_i(M_x = M_y = 0)$$
$$\Lambda_{Fo} := z_{Fo}^{(x)} - z_{Fo}^{(y)} = \Lambda_o(M_x = M_y \to \infty) \tag{19.17}$$

From (19.13), it is readily seen that quadratic equations are obtained for M_x and M_y if we attempt to satisfy the stigmatic imaging condition, $\Lambda_i = \Lambda_o = 0$. The discriminant δ is the same for M_x and M_y and we find

$$M_x = \frac{\Lambda_{Fo}\Lambda_{Fi} - f_{xo}f_{xi} + f_{yo}f_{yi} \pm \delta}{2\Lambda_{Fo}f_{xi}}$$
$$M_y = \frac{-\Lambda_{Fo}\Lambda_{Fi} - f_{xo}f_{xi} + f_{yo}f_{yi} \pm \delta}{2\Lambda_{Fo}f_{yi}} \tag{19.18a}$$

with
$$\delta^2 = (f_{xo}f_{xi} - f_{yo}f_{yi})^2 + \Lambda_{Fo}^2\Lambda_{Fi}^2 + 2(f_{xo}f_{xi} + f_{yo}f_{yi})\Lambda_{Fo}\Lambda_{Fi}$$
$$= (f_{xo}f_{xi} + f_{yo}f_{yi} + \Lambda_{Fo}\Lambda_{Fi})^2 - 4f_{xo}f_{xi}f_{yo}f_{yi} \tag{19.18b}$$

In the usual case in which the signs of f_x and f_y are different, δ^2 is positive and there are two real roots and hence two pairs of stigmatic conjugates.

The cardinal elements of multiplets are established most easily by multiplying the transfer matrices of the individual lenses; these must be separated by transfer matrices corresponding to the *drift spaces*, the spaces between the planes $z = z_2$ for one lens and $z = z_1$ for the next. We recall that these planes may be chosen in various ways: $z_1 = z_2 = 0$, in which case incident position and gradient are related to emergent position and gradient *in the same plane*, conventionally the mid-plane of the lens, is one good choice, thoroughly explored by Dušek (1959). Here we have

$$\begin{pmatrix} x_2 \\ x_2' \end{pmatrix} = \begin{pmatrix} z_{Fi}^{(x)}/f_{xi} & z_{Fi}^{(x)}z_{Fo}^{(x)}/f_{xi} + f_{xo} \\ -1/f_{xi} & -z_{Fo}^{(x)}/f_{xi} \end{pmatrix} \begin{pmatrix} x_1 \\ x_1' \end{pmatrix}$$

$$\begin{pmatrix} y_2 \\ y_2' \end{pmatrix} = \begin{pmatrix} z_{Fi}^{(y)}/f_{yi} & z_{Fi}^{(y)}z_{Fo}^{(y)}/f_{yi} + f_{yo} \\ -1/f_{yi} & -z_{Fo}^{(y)}/f_{yi} \end{pmatrix} \begin{pmatrix} y_1 \\ y_1' \end{pmatrix} \tag{19.19}$$

(We note that Dušek's matrices are trivially different since he used the vectors $(x'\ x)^T$ and $(y'\ y)^T$.)

Another convenient choice involves using different pairs of planes for T_x and T_y, namely the focal planes, since the diagonal matrix elements then vanish:

$$T_x = \begin{pmatrix} 0 & f_{xo} \\ -1/f_{xi} & 0 \end{pmatrix} \quad , \quad T_y = \begin{pmatrix} 0 & f_{yo} \\ -1/f_{yi} & 0 \end{pmatrix} \tag{19.20}$$

This choice has been studied in great detail by Regenstreif (1966, 1967), who has established straightforward rules for writing down the transfer matrices of an arbitrary number of quadrupoles. His procedure can be applied to Dušek matrices (Hawkes, 1970), which we temporarily write as follows:

$$T_i = \begin{pmatrix} a_i & b_i \\ c_i & d_i \end{pmatrix} \tag{19.21}$$

where T_i denotes either T_x or T_y for the i-th quadrupole of a sequence and

$$\begin{aligned} a_i &:= \frac{z_{Fi}}{f_i} \quad , & b_i &:= \frac{z_{Fo} z_{Fi}}{f_i} + f_o \\ c_i &:= -\frac{1}{f_i} \quad , & d_i &:= -\frac{z_{Fo}}{f_i} \end{aligned} \tag{19.22}$$

The separation between the i-th and $(i+1)$-th quadrupoles is denoted by $L_{i,i+1}$ and we write

$$T(L_{i,i+1}) := \begin{pmatrix} 1 & L_{i,i+1} \\ 0 & 1 \end{pmatrix} \tag{19.23}$$

Introducing the distances $X_{i,i+1}$ between the image focus of the i-th quadrupole and the object focus of the $(i+1)$-th quadrupole,

$$X_{i,i+1} := \frac{a_i}{c_i} + L_{i,i+1} + \frac{d_{i+1}}{c_{i+1}} \tag{19.24}$$

we can show (Regenstreif, 1966, 1967; Hawkes, 1970) that the transfer matrix of n quadrupoles separated by $n-1$ drift spaces is given by

$$T^{(n)} = \begin{pmatrix} a^{(n)} & b^{(n)} \\ c^{(n)} & d^{(n)} \end{pmatrix}$$

$$a^{(n)} = a_n \prod_{i=1}^{n-1} c_i \alpha_n \quad , \quad b^{(n)} = d_1 a_n \prod_{i=2}^{n-1} c_i \beta_n$$

$$c^{(n)} = \prod_{i=1}^{n} c_i \gamma_n \quad , \quad d^{(n)} = d_1 \prod_{i=2}^{n} c_i \delta_n \tag{19.25}$$

19.1 PARAXIAL EQUATIONS FOR QUADRUPOLES

in which

$$\alpha_n = \left(X_{n-1,n} - \frac{1}{a_n c_n}\right)\gamma_{n-1} - \frac{\gamma_{n-2}}{c_{n-1}^2}$$

$$\beta_n = \left(X_{n-1,n} - \frac{1}{a_n c_n}\right)\delta_{n-1} - \frac{\delta_{n-2}}{c_{n-1}^2}$$

$$\gamma_n = X_{n-1,n}\gamma_{n-1} - \frac{\gamma_{n-2}}{c_{n-1}^2}$$

$$\delta_n = X_{n-1,n}\delta_{n-1} - \frac{\delta_{n-2}}{c_{n-1}^2} \qquad (19.26)$$

and

$$\gamma_0 = \delta_0 = 0 \quad , \quad \gamma_1 = \delta_1 = 1$$

$$\gamma_2 = X_{1,2} \quad , \quad \delta_2 = X_{1,2} - \frac{1}{c_1 d_1}$$

Another expression for the elements of the transfer matrix between an arbitrary pair of planes, $z = z_n$ and $z = z_0$, may be derived by using the transfer matrix between the principal planes. Writing

$$D^{(i)} := z_{Po}^{(i)} - z_{Pi}^{(i-1)} \qquad 2 \leq i \leq n-1$$
$$D^{(1)} := z_{Po}^{(1)} - z_0 \quad , \quad D^{(n)} := z_n - z_{Pi}^{(n-1)} \qquad (19.27)$$

we form the matrix

$$T(z_0, z_n) = D_n T_{n-1} D_{n-1} \ldots T_2 D_2 T_1 D_1 \qquad (19.28)$$

The elements can be written as Gaussian brackets (Herzberger, 1943, 1958; Hawkes, 1967; Dymnikov, 1968), which are defined as follows:

$$\begin{aligned}
{[x]} &= x \\
{[x_1 x_2]} &= x_1 x_2 + 1 \\
{[x_1 x_2 x_3]} &= x_1 x_2 x_3 + x_1 + x_3 \\
{[x_1 x_2 x_3 \ldots x_n]} &= [x_1 x_2 x_3 \ldots x_{n-2}] + x_n [x_1 x_2 x_3 \ldots x_{n-1}]
\end{aligned} \qquad (19.29)$$

We find

$$T(z_0, z_n) = \begin{pmatrix} [D_n c_{n-1} \ldots D_2 c_1] & [D_n c_{n-1} \ldots D_2 c_1 D_1] \\ [c_{n-1} D_{n-1} \ldots c_1] & [c_{n-1} D_{n-1} \ldots c_1 D_1] \end{pmatrix} \qquad (19.30)$$

For a proof see Hawkes (1967). Gaussian brackets are also employed by Dymnikov (1968) and, in connection with prisms, by Chechulin and Yavor

(1969); renewed interest has been shown in them in light optics too (e.g. Tanaka, 1981, 1982, 1983, 1986).

One common requirement for quadrupole multiplets is that their overall behaviour should be the same as that of a round lens; for this, the cardinal elements in the $x - z$ and $y - z$ planes must coincide and the focal lengths f_x and f_y must be equal. If we consider quadrupoles that may have a round electrostatic lens component, provided that the latter has no overall accelerating or retarding effect ($\phi_i = \phi_o$), we can see on symmetry grounds that one at least of these conditions is satisfied by imposing a certain symmetry on the system. In particular, we perceive that the focal lengths in the $x - z$ and $y - z$ planes are automatically equal in *antisymmetric multiplets*. The latter are defined as follows. If a multiplet consists of $2N$ quadrupoles ($N \geq 1$) such that the central plane is a plane of geometrical symmetry and electrical antisymmetry, we say that the multiplet is *antisymmetric*. The case of $N = 2$ was extensively studied by a group in Leningrad (Yavor, 1962; Dymnikov and Yavor, 1963; Dymnikov et al., 1963a,b, 1964a,b, 1965; Shpak and Yavor, 1964) and has come to be known as *the Russian quadruplet* (Fig. 19.3). Consider two rays, $G_x(z)$ and $\overline{G}_y(z)$. Because of the electrical antisymmetry, the sequence of functions $p_2(z)$ or $Q_2(z)$ encountered by $G_x(z)$ as it proceeds in the positive z-direction will be exactly the same as that traversed by $\overline{G}_y(z)$ if we imagine it travelling in the negative z-direction. The gradients of the emergent asymptotes will hence be equal but opposite in sign, and the focal lengths f_x and f_y will hence be equal (since $\phi_i = \phi_o$, we already have $f_{xi} := f_{xo} =: f_x$ and $f_{yi} := f_{yo} =: f_y$). If we wish to arrange that an antisymmetric multiplet has the same image-forming properties as a round lens, therefore, we have only to ensure that the foci (or principal points) in the $x - z$ and $y - z$ planes coincide. For a given geometry, we need only vary the relative excitation until the condition is satisfied. We therefore obtain a set of pairs of excitations, which is known as the *load characteristic* of the quadruplet. Further details are given in Chapter 39.

19.2 Transaxial lenses

The foregoing discussion has been confined to the optics of quadrupoles in general and we have said little about the shapes of the electrodes and polepieces and hence about the effects of any additional symmetries. Two particular additional symmetry properties are of interest, however; one leads to cylindrical lenses, at least in the electrostatic case, as we see in Chapter 20. A different symmetry characterizes *transaxial lenses*, in which the field or potential is rotationally symmetric but the optic axis is no

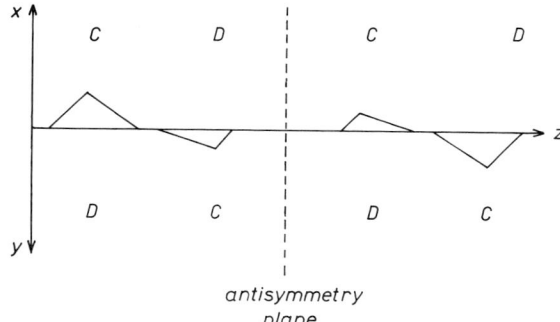

Fig. 19.3: The antisymmetric or "Russian" quadruplet. The centre plane is a plane of geometrical symmetry and electrical antisymmetry. C, D denote convergent and divergent action respectively.

longer the same as the symmetry axis but is perpendicular to it. The electron beam now passes between rotationally symmetric, typically plane electrodes and is focused by the fields in any gaps. Figure 19.4 shows such a system. The electrodes are ideally circular or annular but are in practice reduced to sectors since the electron beam occupies so little of the space available.

It can be seen by comparison with Chapter 20 that such structures bear some resemblance to cylindrical lenses but differ from the latter in that the potential is not independent of one of the transverse cartesian coordinates but is the same for all azimuthal angles φ at a given radial distance from the axis of rotational symmetry. The optical behaviour of such systems was first investigated formally by Strashkevich (1962), to whom we owe the name 'transaxial lenses'; the theory was set out in some detail in his book of 1966. In the early 1970s, it was realized (by V.M. Kel'man and colleagues in Alma-Ata) that certain features of these structures rendered them attractive for use in the collimator of a prism spectrometer, and their properties were investigated in some detail (Glikman et al., 1971; Brodskii and Yavor, 1970, 1971; Karetskaya et al., 1970, 1971a,b,). This work is presented in full in Kel'man et al. (1979), one of the three chapters of which is devoted wholly to these lenses.

The symmetry conditions are now such that

$$\Phi(x, y, z) = \Phi(x, -y, z)$$
$$\Phi(0, y, z) = \Phi(x, y, (z^2 - x^2)^{\frac{1}{2}}) \tag{19.31}$$

19. QUADRUPOLE LENSES

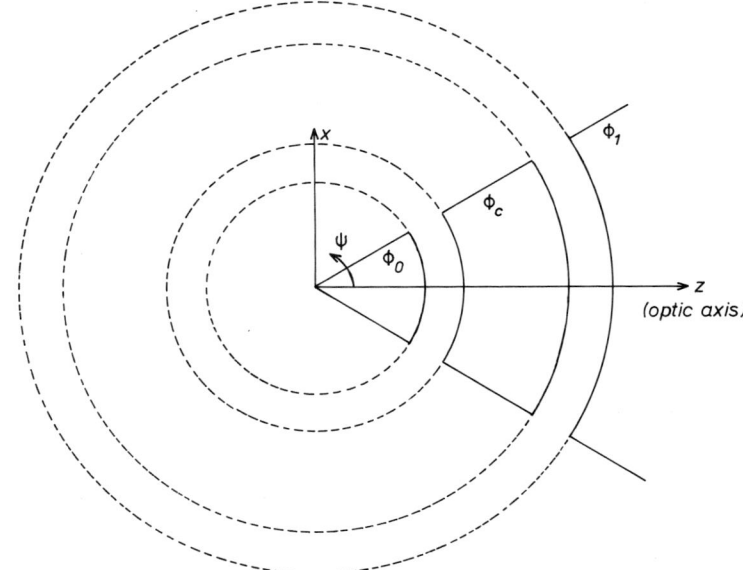

Fig. 19.4: A transaxial lens. The electrodes shown lie in some plane $y = $ const and an identical set lies in the plane $y = -$const.

For small values of x and y, therefore,

$$\Phi(x,y,z) = \phi(z) + \frac{\phi'}{2z} x^2 + \frac{1}{8z^2}\left(\phi'' - \frac{\phi'}{z}\right) x^4 \qquad (19.32)$$
$$+ \phi_2 y^2 + \frac{\phi_2'}{2z} x^2 y^2 + \phi_4 y^4$$

with

$$\phi_2 := -\frac{1}{2}\left(\phi'' + \frac{\phi'}{z}\right) \qquad (19.33)$$
$$\phi_4 := -\frac{1}{12}\left(\phi_2'' + \frac{\phi_2'}{z}\right)$$

Comparing (19.32) with (7.36), we see that

$$p_2(z) = \phi'(z)/z + \phi''/2 \qquad (19.34)$$

The paraxial equations have the form

$$x'' + \frac{\phi'}{2\phi} x' - \frac{\phi'}{2z\phi} x = 0 \qquad (19.35a)$$

$$y'' + \frac{\phi'}{2\phi} x' - \frac{\phi_2}{\phi} y = 0 \qquad (19.35b)$$

All the theory for quadrupoles can hence be employed without further discussion. The form of (19.35a) is, however, such that simple expressions can be obtained for the focal length and foci. Thus on writing

$$\xi := x/z \tag{19.36}$$

(19.35a) becomes

$$\xi'' + \frac{2 + \phi' z / 2\phi}{z} \xi' = 0 \tag{19.37}$$

and after some trivial calculation, we find

$$\begin{aligned}
x &= \frac{x_o z}{z_o} + (x'_o z_o - x_o)\phi_o^{\frac{1}{2}} z \int_{z_o}^{z} \frac{d\zeta}{\phi^{\frac{1}{2}} \zeta^2} \\
&= x'_o z + (x_o - x'_o z_o)\phi_o^{\frac{1}{2}} \left(\frac{1}{\phi^{\frac{1}{2}}} + \frac{z}{2} \int_{z_o}^{z} \frac{\phi' d\zeta}{\phi^{\frac{3}{2}} \zeta} \right)
\end{aligned} \tag{19.38}$$

giving the transfer matrix

$$\begin{pmatrix} x \\ x' \end{pmatrix} = \begin{pmatrix} (\phi_o/\phi)^{\frac{1}{2}} - z/f_i & z - z_o(\phi_o/\phi)^{\frac{1}{2}} + zz_o/f_i \\ -1/f_i & 1 + z_o/f_i \end{pmatrix} \begin{pmatrix} x_o \\ x'_o \end{pmatrix} \tag{19.39}$$

with

$$\frac{1}{f_i} = -\frac{\phi_o^{\frac{1}{2}}}{2} \int_{-\infty}^{\infty} \frac{\phi' d\zeta}{\phi^{\frac{3}{2}} \zeta} \tag{19.40}$$

in which we have extended the limits of integration to infinity in f_i since it is asymptotic imagery that will be of interest. The planes z_o and z_i will be conjugate if

$$z_i - z_o \left(\frac{\phi_o}{\phi_i} \right)^{\frac{1}{2}} + \frac{z_o z_i}{f_i} = 0 \tag{19.41}$$

or

$$\frac{1}{\phi_o^{\frac{1}{2}} z_o} - \frac{1}{\phi_i^{\frac{1}{2}} z_i} = -\frac{1}{f_i \phi_o^{\frac{1}{2}}} \tag{19.42}$$

and the transfer matrix becomes

$$\begin{pmatrix} z_i/z_o & 0 \\ -1/f_i & 1 + z_o/f_i \end{pmatrix} = \begin{pmatrix} z_i/z_o & 0 \\ -1/f_i & z_o \phi_o^{\frac{1}{2}} / z_i \phi_i^{\frac{1}{2}} \end{pmatrix} \tag{19.43}$$

In the converging or y–z plane, there is no such simple solution of the paraxial equation.

20
Cylindrical Lenses

Cylindrical lenses are electrostatic or magnetic devices in which the potential or field is constant in some direction perpendicular to the optic axis. They are the electron optical analogues of glass lenses, the surfaces of which are not spheres, as in rotationally symmetric or 'round' lenses, but cylinders, whence their name. This nomenclature has not always been adhered to in electron optics, and rotationally symmetric lenses are not infrequently referred to as cylindrical, since their central opening is indeed a circular hole and some electrostatic lenses consist of a sequence of metal cylinders.

Cylindrical lenses have a long history, going back to one of the earliest publications on lens properties, the note by Davisson and Calbick (1931) on the lens-like behaviour of round openings and slits. The paraxial properties of such lenses were first discussed by Picht (1939b) and in very much more detail by Gratsiatos (1940). In the same year Strashkevich (1940a,b) gave the ray equations for electrostatic cylindrical lenses, including the reduced form (15.40), and these equations are again to be found in Leitner (1942). Many of the properties were elucidated and rediscovered over the years; the relevant papers are listed in the bibliography to this Chapter. We single out the work of Kel'man *et al.* (1954), Yavor (1955) and Kel'man and Yavor (1955a,b) on magnetic cylindrical lenses, recapitulated in detail in Kel'man and Yavor (1968).

The potential distributions in cylindrical lenses have been calculated or measured in particular, in connection with mass spectrometer design; see in particular Wallington (1970, 1971), Harting and Read (1976), Mulvey and Wallington (1973) and Boerboom (1959, 1960).

We now assume that the potentials Φ (7.36) and W (7.41) are functions of x and z only, so that in the general expansion of (7.36), we have

$$p_2(z) = -\frac{1}{2}\phi''(z) \quad , \quad p_4(z) = \frac{1}{8}\phi^{(4)}(z)$$
$$q_2(z) = q_4(z) = 0$$
(20.1)

and in (7.46-7.48)

$$P_2(z) = \frac{1}{2}B'(z) \quad , \quad P_4(z) = -\frac{1}{8}B'''(z)$$
$$Q_2(z) = Q_4(z) = 0$$
(20.2)

20. CYLINDRICAL LENSES

It is immediately clear from (19.3) that an electrostatic cylindrical lens forms an orthogonal system whereas a magnetic one does not. We have

$$\Phi(x,y,z) = \phi(z) - \frac{1}{2}\phi''x^2 + \frac{1}{24}\phi^{(4)}x^4$$
$$A_x(x,y,z) = -\frac{1}{2}yB + \frac{1}{48}B''y(9x^2 + y^2)$$
$$A_y(x,y,z) = \frac{1}{2}xB - \frac{1}{48}B''y(5x^2 - 3y^2)$$
$$A_z(x,y,z) = -\frac{1}{2}xyB' + \frac{1}{48}B'''xy(3x^2 + y^2)$$
(20.3)

Expanding the function M (15.23), we obtain

$$M^{(0)} = \hat{\phi}^{\frac{1}{2}}$$
$$M^{(2)} = -\frac{\gamma\phi''}{4\hat{\phi}^{\frac{1}{2}}}x^2 + \frac{1}{2}\hat{\phi}^{\frac{1}{2}}(x'^2 + y'^2) + \frac{1}{2}\eta xyB' \quad (20.4)$$
$$- \frac{1}{2}\eta B(xy' - x'y)$$

giving the paraxial equations

$$\frac{d}{dz}(\hat{\phi}^{\frac{1}{2}}x') + \frac{\gamma\phi''}{2\hat{\phi}^{\frac{1}{2}}}x + \eta By' = 0 \quad (20.5a)$$
$$\frac{d}{dz}(\hat{\phi}^{\frac{1}{2}}y' - \eta Bx) = 0 \quad (20.5b)$$

(We note in passing that expression (20.4) for $M^{(2)}$ could have been simplified by the use of Sturrock's partial-integration rule (1955), which tells us that when a term of the form $g(x,y,z)df(x,y,z)/dz$ occurs in a variational function such as $M^{(2)}$, it may be replaced by $-fdg/dz$; here we could have reduced the terms in B to $-\eta Bxy'$.) Integrating (20.5b), we find

$$y' = \frac{C + \eta Bx}{\hat{\phi}^{\frac{1}{2}}} \quad (20.6)$$

in which C is a constant and (20.5a) may then be written

$$x'' + \frac{\gamma\phi'}{2\hat{\phi}}x' + \frac{\gamma\phi'' + 2\eta^2 B^2}{2\hat{\phi}}x = -\frac{\eta CB}{\hat{\phi}} \quad (20.7)$$

This can be recast into reduced form by writing $v = x\hat{\phi}^{1/4}$ (15.40).

Equation (20.7) is a linear, second order, differential equation but is no longer homogeneous when a magnetic field is present. In the absence of such fields ($B = 0$), the paraxial equation for $x(z)$ is almost the same as that for round electrostatic lenses, the only difference being the factor 2 in the denominator of the term in x. The equation for $y(z)$ can be solved immediately, in this case:

$$y(z) = C \int^z \hat{\phi}^{-\frac{1}{2}}(\zeta) \, d\zeta \qquad (20.8)$$

The lens action in the $x - z$ plane is thus described by a transfer matrix of the type (16.12). It is interesting to note that the best thin-lens approximation to the focal length is now obtained by writing

$$v := x\hat{\phi}^{\frac{1}{2}}$$

(and not $v = x\hat{\phi}^{1/4}$ as in 15.40 and 17.79). In the nonrelativistic approximation, (20.7) becomes

$$\frac{d}{dz}\left(\frac{v'}{\phi^{\frac{1}{2}}}\right) = -\frac{v\phi'^2}{2\phi^{\frac{5}{2}}}$$

and the focal length is then

$$\frac{1}{f_i} = \frac{\phi_o^{\frac{1}{2}}}{2} \int \frac{\phi'^2}{\phi^{\frac{5}{2}}} \, dz$$

This expression was obtained by Brodskii and Yavor (1971), who find that it is substantially more accurate than that given by writing $v = x\hat{\phi}^{1/4}$. Glikman et al. (1967a) have demonstrated that electrostatic cylindrical lenses always have a converging action in the $x - z$ plane, using the exact trajectory equation rather than the paraxial equation.

In the $y - z$ plane, we have

$$\begin{pmatrix} y_2 \\ y_2' \end{pmatrix} = \begin{pmatrix} 1 & \int_{z_1}^{z_2}(\hat{\phi}_1/\hat{\phi})^{\frac{1}{2}} \, d\zeta \\ 0 & (\hat{\phi}_1/\hat{\phi}_2)^{\frac{1}{2}} \end{pmatrix} \begin{pmatrix} y_1 \\ y_1' \end{pmatrix} \qquad (20.9)$$

This is the type of transfer matrix that we meet in light optics for a parallel-plane glass plate separating media of different refractive index.

In the general case when $B \neq 0$, (20.7) is solved using the method of variation of parameters. If $g(z)$ and $h(z)$ are two linearly independent

20. CYLINDRICAL LENSES

solutions of the homogeneous equation obtained from (20.7), satisfying the initial conditions

$$g(z_o) = h'(z_o) = 1$$
$$g'(z_o) = h(z_o) = 0 \qquad (20.10)$$

in some object plane, $z = z_o$, the general solution is

$$x(z) = x_o g(z) + x'_o h(z) \qquad (20.11)$$

The solution of the inhomogeneous equation is thus

$$x(z) = x_o g(z) + x'_o h(z)$$
$$- \frac{\eta C h(z)}{\hat{\phi}_o^{\frac{1}{2}}} \int_{z_o}^{z} \frac{g(\zeta) B(\zeta)}{\hat{\phi}^{\frac{1}{2}}(\zeta)} d\zeta + \frac{\eta C g(z)}{\hat{\phi}_o^{\frac{1}{2}}} \int_{z_o}^{z} \frac{h(\zeta) B(\zeta)}{\phi^{\frac{1}{2}}(\zeta)} d\zeta \qquad (20.12)$$
$$=: x_o g(z) + x'_o h(z) + \eta C j(z)$$

From (20.6), we have

$$C = \hat{\phi}_o^{\frac{1}{2}} y'_o + \eta B_o x_o \qquad (20.13)$$

and so

$$x(z) = x_o\{g(z) - \eta^2 B_o j(z)\} + x'_o h(z) + y'_o \eta \hat{\phi}_o^{\frac{1}{2}} j(z) \qquad (20.14)$$

Substituting into (20.6) and integrating, we obtain

$$y(z) = y_o + y'_o \hat{\phi}_o^{\frac{1}{2}} \left(\int_{z_o}^{z} \frac{d\zeta}{\hat{\phi}^{\frac{1}{2}}} + \eta^2 \int_{z_o}^{z} \frac{Bj}{\hat{\phi}^{\frac{1}{2}}} d\zeta \right)$$
$$+ x_o \eta \left(-B_o \int_{z_o}^{z} \frac{d\zeta}{\hat{\phi}^{\frac{1}{2}}} + \int_{z_o}^{z} \frac{Bg}{\hat{\phi}^{\frac{1}{2}}} d\zeta - \eta^2 B_o \int_{z_o}^{z} \frac{Bj}{\hat{\phi}^{\frac{1}{2}}} d\zeta \right) \qquad (20.15)$$
$$+ x'_o \eta \int_{z_o}^{z} \frac{Bh}{\hat{\phi}^{\frac{1}{2}}} d\zeta$$

Systems with optical properties as complicated as (20.14) and (20.15) suggest are unattractive in practice. Their general behaviour has been studied in the context of ophthalmological optics by Gullstrand (1900, 1906, 1908, 1915, 1924) and a more accessible account is given by Carathéodory (1937). The electron optical situation has been explored in detail by Kel'man et al. (1954), Yavor (1955), Kel'man and Yavor (1955) and more recently by Rose (1966/7), and especially (1972).

Part IV

Aberrations

21
Introduction

The paraxial approximation, characterizing the linear coupling between two spaces, object space and image space, describes the dominant behaviour of the various electron optical components but small departures from this can rarely be neglected. In order to assess the magnitude of such non-linear effects, we must proceed to the next higher order approximation, which involves retaining quartic terms in the expression for the refractive index $\overline{M}$ (4.35) for systems with a straight optic axis. We shall find that the results of this calculation can be compactly expressed in terms of coefficients, the geometrical aberration coefficients, which vary in number with the symmetry of the system. The object of most aberration calculations is to obtain expressions for these coefficients in terms of the axial potential or field functions and of particular solutions of the paraxial ray equations.

The calculation may be performed in two very different ways. The most straightforward consists in writing down the ray equations as in Section 15.1 but now retaining higher order terms in the various field and potential expansions. For the most common components, round lenses, quadrupoles and mirrors, this generates linear, second-order, ordinary differential equations, which are now inhomogeneous; the corresponding homogeneous equations are identical with the paraxial equations. These inhomogeneous equations are solved by variation of parameters or by writing down the appropriate Green's function; the difference is purely formal and almost trivial. This procedure is commonly known as the *trajectory method*. The method is best understood by an example, and the reader is referred to Section 24.2, where round lens aberrations are studied in this way. The trajectory method has been used to study systems with arbitrary curved axes in considerable detail. We examine such systems in Part X.

Alternatively, a general perturbation theory may be developed, which enables us to answer the following question: given the solutions of ray equations derived from the paraxial refractive index, $M = M^{(2)}$, what will be the solutions for a slightly different refractive index, $M = M^{(2)} + M^{(p)}$? (Here $M \propto \overline{M}$, see (4.35) and (22.2).) By choosing $M^{(p)} = M^{(4)}$, for example, we obtain the primary geometrical aberrations of systems with a straight axis. Although this second approach may seem more abstract, it has one distinct advantage, which will become apparent below: any interrelations between the various aberration coefficients emerge automatically, whereas they may be by no means obvious when the trajectory method is adopted. This second procedure, which we shall mostly use, is known as the *eikonal method* (from the term introduced by Bruns (1895) in his *Das Eikonal*) or *method of characteristic functions* (following Hamilton, 1828–1837, see Hamilton (1931)).

These *geometrical aberrations*, arising from higher order terms in the field expansions, or more evocatively, from allowing the electrons to stray a little way beyond the truly paraxial region, are not the only type that can arise. The next most important are the *chromatic aberrations*. These are the analogue in electron optics of effects due to variations in refractive index with wavelength in light optics; they can arise in several ways. We have been assuming that the electron energy and any magnetic lens fields and electrostatic lens potentials are static and fixed. In reality, however, the lens excitations will inevitably fluctuate somewhat (except in superconducting lenses operating in the persistent-current mode) and the electrons of the incident beam will never be perfectly monoenergetic: their energies will span a small range, narrow but not negligibly so, for reasons that are examined in Part IX. We shall see in Volume 3 that a wavelength (λ) proportional to $\hat{\Phi}^{-1/2}$ can be associated with electrons accelerated through a voltage drop of Φ, and effects due to variations in electron energy relative to the lens excitation may thus be regarded as due to a change in the focusing properties with λ. They are hence known generically as *chromatic aberrations*. In studying these, we establish *chromatic aberration coefficients*, which characterize the optical consequences of small changes in electron energy, electrostatic lens potential or magnetic lens field strength. In addition, in electron microscopes, some electrons commonly lose energy within the specimen and this too contributes to the energy spread in the beam and hence to the chromatic blur.

Various other perturbations have been characterized by aberration coefficients and calculated by the methods set out in this Part: effects due to the inclusion of relativity, those caused by moderate space charge forces and those provoked by small departures from the assumed symmetry of the system. Most formulae are now available in relativistically correct form and the incentive to explore the difference between nonrelativistic and relativistic behaviour by perturbation theory has vanished. Conversely, it can still be useful to enquire what effect space charge will have when it has been neglected in a preliminary calculation and here perturbation theory is convenient. The final group of aberrations, associated with small mechanical or electrical imperfections in the lenses, is of extreme practical importance; these *mechanical* or *parasitic aberrations* must always be kept small if lens performance is not to be impaired.

As we saw in Part III, it is necessary to distinguish between real and asymptotic coupling between object and image space and this remains true of aberration calculations. To illustrate this, consider a real object immersed in a magnetic field; the image will be affected by the aberrations introduced by the part of the field downstream from the object. If, however, the same magnetic field is used to magnify an intermediate image, the entire field will contribute to the aberrations and we then characterize the field by the coupling between incident and emergent asymptotes. In the first case, we speak of real aberrations, in the second, of asymptotic aberrations.

The study of electron lens aberrations is almost as old as the electron microscope itself, for the first calculations of aberration coefficients were made in the early 1930s by Scherzer, who preferred the trajectory method, and by Glaser, who introduced the eikonal method into electron optics (Glaser, 1933a–c, 1935, 1936a,b, 1937, 1938, 1949; Scherzer, 1933, 1936b, 1937; see also Funk, 1936, 1937; Gratsiatos, 1936; Rogowski, 1937; Ramberg, 1939; Busch and Brüche 1937). During that decade, the real aberrations of round lenses were thoroughly studied and several equivalent formulae for the various coefficients were derived. In particular, Scherzer (1936b) demonstrated that the spherical aberration, which is of particular concern since it limits the resolution of microscopes, always has the same sign, whatever the lens design. This finding is so important that it is known as Scherzer's theorem, the only named theorem in the subject.

During the 1940s, the aberration coefficients of quadrupoles were obtained by the trajectory method (Melkich, 1947) and in a celebrated paper, again by Scherzer (1947), a number of possible ways of cancelling spherical aberration were adumbrated. Also during this period, Grinberg (1942, 1943a,b, republished 1948) derived highly general ray equations, assuming no particular symmetry, and these, supplemented by the work of Vandakurov (1956a,b, 1957) and Kas'yankov (1956, 1957, 1958a), have been widely used by Russian authors, notably Strashkevich and Pilat (1951, 1952) and Strashkevich and Gluzman (1954), who made the first attempts to follow Grinberg's approach in the study of aberrations.

The late 1940s and early 1950s saw several substantial contributions. Tretner returned to Scherzer's proof that the spherical aberration coefficient cannot be made to vanish by ingenious lens design, without infringing any of the conditions required by the proof, and established minimum values for this coefficient and for that of chromatic aberration (for which a similar rule holds true), subject to reasonable practical constraints on lens dimensions and excitation (Tretner, 1950, 1954, 1955, 1956 and especially 1959). Sturrock (1951a,b, 1952, 1955) re-examined the eikonal theory of electron lens aberrations and succeeded in placing it on a firm theoretical foundation; he showed how higher order aberrations can be calculated and how the theory can be used to establish asymptotic aberration coefficients. The credit for recognizing the importance of distinguishing between real and asymptotic aberrations goes to Lenz (1956, 1957), however, closely followed by Seman (1958a). Sturrock also examined in detail the effect of small imperfections in magnetic lens construction—small departures from circularity of the bore, for example—expressing his results in terms of *parasitic* aberration coefficients. These last studies were inspired by the work of Bertein on electrostatic lenses, and simpler analysis had already led Rang (1949a) and Hillier and Ramberg (1947) to introduce the stigmator, a weak quadrupole designed to cancel the principal parasitic aberration, the axial astigmatism (Sturrock, 1949, 1951b; Bertein, 1947d,e, 1948a).

We have mentioned that the formulae for the aberration coefficients can be cast into many equivalent forms. The technique originally employed for this required partial integration and substitution from the paraxial equations to eliminate second derivatives of $x(z)$ and $y(z)$, using (15.12) for round lenses, for example. This is not only laborious but has the severe disadvantage that the form of the result cannot be predicted: we cannot know, until all the possible equivalent forms of a given coefficient have been established, which terms can be eliminated simultaneously and whether there are any that can never be removed. A procedure that permits us to write down a general expression containing all possible forms of every coefficient was introduced by Seman (1951, 1954, 1955a,b, 1958b) for round lenses and has been extensively used by Hawkes (1966/7b, 1967b) for quadrupoles and round lenses.

The methods of Sturrock, whose familiarity with the work of Hamilton and of Herzberger (1931) had enabled him to consolidate the fundamental studies of Glaser, were later applied to quadrupole lenses by Hawkes (1965a,b,c), who derived formulae for all their geometrical aberration coefficients, more compact than those of Melkich (1947). The chromatic aberrations of such lenses were analysed by Kel'man and Yavor (1961), who derived the condition that must be satisfied if a mixed electrostatic–magnetic quadrupole is to be achromatic, subsequently rediscovered by Septier (1963) and generalized by Hawkes (1965d).

In 1963, Verster noticed in the course of his work on 'gauze lenses' (see Chapter 41) that some aberration coefficients can be written as polynomials in reciprocal magnification, the coefficients being determined by lens geometry and excitation. Such a representation had been known to the members of the van Heel school (see van Heel, 1949, 1964) but had not hitherto been recognized in electron optics. This observation led Hawkes (1968, 1970b–f) to examine the structure of the asymptotic aberration coefficients and of the real coefficients of Newtonian fields in some detail; formulae were established for the coefficients occurring in the various polynomials and, by expressing the results in matrix form (cf. Brouwer, 1957, 1964), expressions for the aberrations of multiplets could be obtained explicitly. The aberration polynomials were also obtained by Ade (1973, 1982), by a rather different route.

During the 1960s, the basic aberration theory was again reformulated by Rose (1968, 1968/69; Rose and Petri, 1971), who devised a systematic way of handling higher order aberrations, suitable for studying the complicated combinations of electron optical elements needed for practical aberration correction in an electron microscope.

The most far-reaching innovation of the 1960s was, however, a consequence of the increasing availability of powerful computers. With these, it became possible to calculate optical properties easily and accurately and to apply sophisticated minimization techniques to the task of finding lens combinations possessing some particular property (Moses, 1970, 1971a–c, 1972, 1973, 1974; Rose and Moses, 1973). The 1970s saw the introduction of the finite-element and boundary-element methods for field calculation (Munro, 1970, 1971, 1972, 1973; Adams and Read, 1972a,b; Harrington, 1968; Read et al., 1971; Rauh, 1971), and the extremely difficult problem of calculating electron gun behaviour accurately was gradually solved (see Lauer, 1982; Kasper, 1982; and Part IX). Systems of great complexity can now be analysed in detail and the subject is in a stage of rapid development.

Computers have been used not only for numerical solution of field and trajectory equations but also to derive expressions for aberration coefficients. Here, we require them to perform not arithmetic but algebra and a number of special languages have been devised for this purpose, of which REDUCE is among the most widespread (Hearn, 1985; Fitch, 1985) though CAMAL is very easy to use (Barton and Fitch, 1972; Fitch, 1979; see Ng, 1979 for an overall view of the subject). CAMAL has been employed to calculate aberration coefficients by Hawkes (1977a,b, 1980a, 1983a) and REDUCE by Goto and Soma (1977) and Soma (1977), see Chapter 34.

Computer algebra has also been found useful in the latest addition to the family of methods used to study aberrations, in which the properties of Lie algebra and the associated groups are exploited. These techniques were introduced into particle and later light optics by Dragt (Dragt, 1979, 1982; Dragt and Forest, 1983; Dragt and Douglas, 1983); full accounts are to be found in Dragt and Forest (1986), Dragt et al. (1986) and Mondragón and Wolf (1986).

The foregoing survey is restricted to some of the highlights in the history of electron lens aberration theory. Many contributions of arguably comparable importance have not been mentioned and this account should therefore be regarded as no more than a preliminary glance at the field, charted more fully in the remainder of this Part and in the bibliography, where more detailed attributions are to be found. In particular, we have not mentioned the use of lens models and the calculation of explicit forms of the coefficients for the most important of these; references to these are to be found in the Part on instrumental optics (especially Chapters 35, 36 and 39), but the rise of fast and accurate numerical methods is accompanied by the fall, not to say the demise, of many of these models, which were introduced precisely because accurate calculation was unthinkably laborious.

22
Perturbation Theory: General Formalism

We set out from the variational principle obtained from (4.32) and (4.34):

$$\int_{P_1}^{P_2} M \, dz \to \text{extr.} \tag{22.1}$$

in which we have written

$$\begin{aligned} M :&= \overline{M}(2m_0 e)^{-\frac{1}{2}} \\ &= \hat{\Phi}^{\frac{1}{2}}(1 + x_1'^2 + x_2'^2) - \eta(A_1 x_1' + A_2 x_2' + A_z) \end{aligned} \tag{22.2}$$

for electrons, using (2.19). We adopt the suffix notation here because the coordinates (x_1, x_2, z) may be curvilinear, the optic axis coinciding with the z-axis. At this stage, this simply means that the z-axis is the path of a possible ray. Although much of the text is devoted to systems with a straight axis, whereupon (x_1, x_2, z) become the familiar cartesian coordinates, we shall meet a very important family of curved-axis systems in Part X. We recall that (22.1) asserts that the first-order variation of the integral vanishes when the integration is taken along a ray, provided that the endpoints remain fixed. We now consider the value of this integral for an arbitrary path of integration, and to prevent any possible confusion, we write

$$S^* := \int_{P_1}^{P_2} M \, dz \tag{22.3}$$

reserving S for the value of the integral when the integration is taken over a ray. We find that if the path of integration is altered from $x_j(z)$ to $x_j(z) + \Delta x_j(z)$, then S^* increases by an amount ΔS^*, where

$$\Delta S^* = \int_{z_1}^{z_2} \left(\Delta x_j \frac{\partial M}{\partial x_j} + \Delta x_j' \frac{\partial M}{\partial x_j'} \right) dz \tag{22.4a}$$

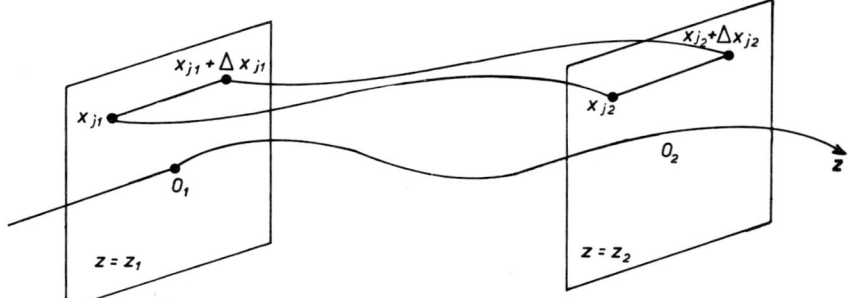

Fig. 22.1: Perturbation from $x_j(z)$ to $x_j(z) + \Delta x_j(z)$.

or integrating by parts,

$$\Delta S^* = -\int_{z_1}^{z_2} \Delta x_j \left(\frac{d}{dz} \frac{\partial M}{\partial x'_j} - \frac{\partial M}{\partial x_j} \right) dz \\ + \left[\frac{\partial M}{\partial x'_j} \right]_{z_2} \Delta x_j(z_2) - \left[\frac{\partial M}{\partial x'_j} \right]_{z_1} \Delta x_j(z_1) \quad (22.4b)$$

Summation over $j = 1, 2$ is implicit here. Furthermore, the variation is restricted to the surfaces $z = z_1$ and $z = z_2$, so that the z-coordinates of the endpoints of the path of integration are held constant (Fig. 22.1). We know that ΔS^* must vanish (22.1) when $\Delta x_{j1} = \Delta x_{j2} = 0$ $(j = 1, 2)$ if the path of integration follows a ray; we have written $\Delta x_j(z_k) =: \Delta x_{jk}$. The ray equations

$$\frac{d}{dz} \frac{\partial M}{\partial x'_j} = \frac{\partial M}{\partial x_j} \quad (22.5)$$

are then satisfied. For this situation, we denote the value of S^* by S and from (22.4), we have

$$\Delta S(z_1, z_2) = \left(\frac{\partial M}{\partial x'_j} \right)_2 \Delta x_{j2} - \left(\frac{\partial M}{\partial x'_j} \right)_1 \Delta x_{j1} \quad (22.6)$$

Introducing the ray vector

$$p_j = \frac{\partial M}{\partial x'_j} \quad (22.7)$$

(22.6) becomes

$$\Delta S(z_1, z_2) = p_{j2} \Delta x_{j2} - p_{j1} \Delta x_{j1} \quad (22.8)$$

It is convenient to write ΔS_{12} instead of $\Delta S(z_1, z_2)$. We recall that, apart from the trivial change of scale relating S and $\overline{S}$, the function S is Hamilton's point characteristic function and its arguments are x_{j1} and x_{j2}, the word 'point' underlining the fact that these are the coordinates of points P_1 and P_2. Characteristic functions with other arguments are sometimes convenient, such as the angle characteristic already encountered in Section 15.5.

Suppose now that the system is perturbed in some way, and that a parameter λ characterizes this perturbation. We need to introduce two perturbation operators, one of which enables us to represent the perturbed form of a function while the other expresses the effect of perturbing its arguments. For the former, we write the perturbed form of a function f as $P_f f$,

$$P_f = 1 + \lambda P_f^I + \lambda^2 P_f^{II} + \ldots \tag{22.9}$$

For M in particular, we write

$$P_f M = M + \lambda M^I + \lambda^2 M^{II} + \ldots \tag{22.10}$$

For the argumental perturbation operator P_a, we use the Taylor expansion of a function in the form $f(x + \epsilon) = \exp(\epsilon \partial/\partial x) f(x)$ so that

$$P_a = \exp(\lambda D^I + \lambda^2 D^{II} + \ldots) \tag{22.11}$$

where

$$\begin{aligned} D^I &= x_j^I \frac{\partial}{\partial x_j} + x_j'^I \frac{\partial}{\partial x_j'} \\ D^{II} &= x_j^{II} \frac{\partial}{\partial x_j} + x_j'^{II} \frac{\partial}{\partial x_j'} \end{aligned} \tag{22.12}$$

The total perturbation is then

$$P = P_a P_f \tag{22.13}$$

We now apply these general remarks to the function S_{12}. Since $S_{12} = \int_{z_1}^{z_2} M \, dz$, we may write

$$PS_{12} = \int_{z_1}^{z_2} PM \, dz \tag{22.14}$$

It is permissible to take the operator P under the integral since the argument z is not varied by P_a. From (22.8), we have

$$\Delta P S_{12} = P p_{j2} \cdot \Delta P x_{j2} - P p_{j1} \cdot \Delta P x_{j1} \tag{22.15}$$

Combining (22.10, 22.11 and 22.13), we find

$$
\begin{aligned}
PS_{12} &= \int_{z_1}^{z_2} \left\{ 1 + \lambda D^I + \lambda^2 D^{II} + \frac{1}{2}\lambda^2 (D^I)^2 + \ldots \right\} \\
&\quad \times \left(M + \lambda M^I + \lambda^2 M^{II} + \ldots \right) dz \\
&= \int_{z_1}^{z_2} \Big[M + \lambda (D^I M + M^I) \\
&\quad + \lambda^2 \left\{ M^{II} + D^I M^I + D^{II} M + \frac{1}{2}(D^I)^2 M \right\} + \ldots \Big] dz
\end{aligned}
\tag{22.16}
$$

In order to keep the notation reasonably compact, we write

$$PS_{12} = S_{12}^{(0)} + \lambda S_{12}^{(1)} + \lambda^2 S_{12}^{(2)} + \ldots \tag{22.17}$$

so that arabic indices (1), (2),... indicate the order of *total* perturbation whereas roman numerals I, II,... signify the order of *functional* perturbation. Then, comparing (22.16) and (22.17), it is immediately clear that

$$S_{12}^{(1)} = \int_{z_1}^{z_2} (M^I + D^I M) \, dz \tag{22.18}$$

From (22.8), we see that for any pair ξ_j,

$$\int_{z_1}^{z_2} \left(\xi_j \frac{\partial M}{\partial x_j} + \xi_j' \frac{\partial M}{\partial x_j'} \right) dz = p_{j2} \xi_{j2} - p_{j1} \xi_{j1} \tag{22.19}$$

so that for $\xi_j = x_j^{(1)}$, we have

$$\int_{z_1}^{z_2} D^I M \, dz = p_{j2} x_{j2}^{(1)} - p_{j1} x_{j1}^{(1)}$$

and hence

$$S_{12}^{(1)} = p_{j2} x_{j2}^{(1)} - p_{j1} x_{j1}^{(1)} + S_{12}^I \tag{22.20}$$

where

$$S_{12}^I := \int_{z_1}^{z_2} M^I \, dz \tag{22.21}$$

The final function, S_{12}^I, is known as the *first-order perturbation characteristic function* and we shall use it frequently. It is in fact the functional contribution to $S_{12}^{(1)}$, the argumental part coming from $\int_{z_1}^{z_2} D^I M \, dz$.

Returning to (22.15), we have

$$\Delta S_{12}^{(1)} = p_{j2}^{(1)} \cdot \Delta x_{j2} + p_{j2} \cdot \Delta x_{j2}^{(1)}$$
$$- p_{j1}^{(1)} \cdot \Delta x_{j1} - p_{j1} \cdot \Delta x_{j1}^{(1)} \tag{22.22}$$

so that from (22.20) and (22.22),

$$\Delta S_{12}^I = (p_{j2}^{(1)} \cdot \Delta x_{j2} - x_{j2}^{(1)} \cdot \Delta p_{j2})$$
$$- (p_{j1}^{(1)} \cdot \Delta x_{j1} - x_{j1}^{(1)} \cdot \Delta p_{j1}) \tag{22.23}$$

This is a *first-order perturbation relation* but it is much more general in form than we commonly need. We may perturb the rays subject to various constraints, which correspond to particular physical situations. Two choices of constraints are of especial relevance. Consider first the implications of setting

$$x_{j1}^{(1)} = p_{j1}^{(1)} = 0 \tag{22.24}$$

and, for simplicity, suppose that the plane $z = z_1$ lies in field-free space. Then $p_j \propto x_j'$ and the constraints (22.24) imply that neither the position nor the gradient of the ray in $z = z_1$ is allowed to change. Thus in Fig. 22.2a, if R is the unperturbed ray, the perturbed ray $\bar{R}$ sets out from the same point with the same gradient. By using this set of constraints, aberration coefficients expressed in terms of position and gradient in the object plane are obtained, always provided that the latter lies in field-free space. If it does not, the relation between $p_j^{(1)}(z_1)$ and $x_j'^{(1)}(z_1)$ becomes more complicated and it is easier to derive the real aberrations expressed in terms of position and gradient in the object plane via those based on position in two separate planes (object and aperture). In the case of asymptotic aberrations, which describe the coupling between (field-free) object space and image space, this problem does not arise and the appropriate aberration coefficients may be obtained without difficulty.

The other physically significant choice of constraints is

$$x_{j1}^{(1)} = x_{j2}^{(1)} = 0 \tag{22.25}$$

so that the unperturbed and perturbed rays must pass through the same point in two different planes, z_1 and z_2. This situation is shown in Fig. 22.2b; the ray $\bar{R}$ is clearly not the same as that in Fig. 22.2a.

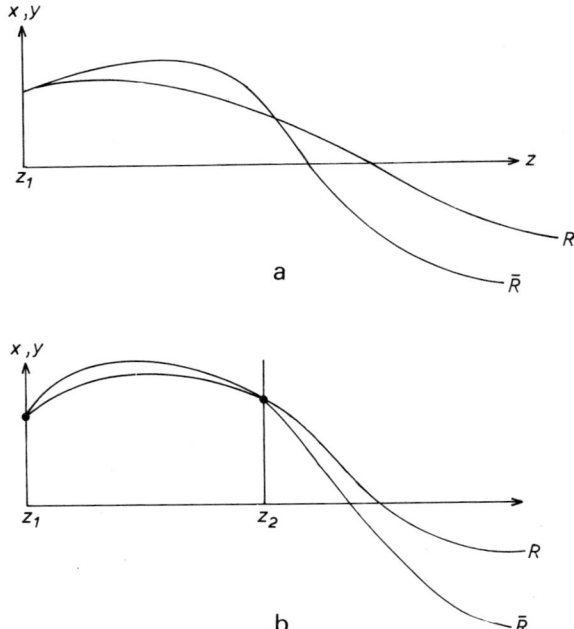

Fig. 22.2: The constraints described by (a) (22.24) and (b) by (22.25).

Some caution is needed when selecting the constraints, since not all combinations are permissible on every occasion. Thus (22.25) cannot be used when the planes $z = z_1$ and $z = z_2$ are conjugate, since all the rays through a point P_2' in the neighbourhood of P_2 will intersect in a point close to P_1; apart from this point, the neighbourhood of P_1 will not be connected by rays to P_2'.

We shall therefore study (22.23) in two forms. For constraints (22.24), we have

$$\Delta S_{12}^I = p_{j2}^{(1)} \cdot \Delta x_{j2} - x_{j2}^{(1)} \cdot \Delta p_{j2} \qquad (22.26)$$

and for (22.25),

$$\Delta S_{12}^I = p_{j2}^{(1)} \Delta x_{j2} - p_{j1}^{(1)} \Delta x_{j1} \qquad (22.27)$$

From (22.26), we obtain the very important pair of equations

$$\begin{aligned}\frac{\partial S_{12}^I}{\partial x_{k1}} &= p_{j2}^{(1)} \frac{\partial x_{j2}}{\partial x_{k1}} - x_{j2}^{(1)} \frac{\partial p_{j2}}{\partial x_{k1}} \\ \frac{\partial S_{12}^I}{\partial p_{k1}} &= p_{j2}^{(1)} \frac{\partial x_{j2}}{\partial p_{k1}} - x_{j2}^{(1)} \frac{\partial p_{j2}}{\partial p_{k1}}\end{aligned} \qquad (22.28)$$

Equation (22.27) we shall handle slightly differently. We shall use it to express aberrations in some arbitrary plane, frequently the image plane, in terms of position coordinates in the object plane (suffix o) and aperture plane (suffix a) in a system. Returning to (22.23), we first set $z_1 = z_o$, $x_{jo}^{(1)} = 0$, giving

$$\Delta S_{o2}^I = p_{j2}^{(1)} \cdot \Delta x_{j2} - x_{j2}^{(1)} \cdot \Delta p_{j2} - p_{jo}^{(1)} \cdot \Delta x_{jo} \qquad (22.29a)$$

and then set $z_1 = z_a$, $x_{ja}^{(1)} = 0$, giving

$$\Delta S_{a2}^I = p_{j2}^{(1)} \cdot \Delta x_{j2} - x_{j2}^{(1)} \cdot \Delta p_{j2} - p_{ja}^{(1)} \cdot \Delta x_{ja} \qquad (22.29b)$$

Hence

$$\begin{aligned}\frac{\partial S_{o2}^I}{\partial x_{ka}} &= p_{j2}^{(1)} \frac{\partial x_{j2}}{\partial x_{ka}} - x_{j2}^{(1)} \frac{\partial p_{j2}}{\partial x_{ka}} \\ \frac{\partial S_{a2}^I}{\partial x_{ko}} &= p_{j2}^{(1)} \frac{\partial x_{j2}}{\partial x_{ko}} - x_{j2}^{(1)} \frac{\partial p_{j2}}{\partial x_{ko}}\end{aligned} \qquad (22.30)$$

These may be solved for the unknown quantities $p_{j2}^{(1)}$ and $x_{j2}^{(1)}$. The determinant of (22.30) is given by

$$\Delta = -\frac{\partial(p_{j2}, x_{j2})}{\partial(x_{jo}, x_{ja})} \qquad (22.31)$$

and may be expanded as a product of Lagrange brackets,

$$\Delta = -[x_{1o}, x_{2o}][x_{1a}, x_{2a}] + [x_{1o}, x_{1a}][x_{2o}, x_{2a}] \\ - [x_{1o}, x_{2a}][x_{2o}, x_{1a}] \qquad (22.32)$$

where $[\lambda, \mu]$ denotes

$$\frac{\partial(p_{12}, x_{12})}{\partial(\mu, \lambda)} + \frac{\partial(p_{22}, x_{22})}{\partial(\mu, \lambda)} \qquad (22.33)$$

We have seen that the Lagrange brackets are invariant (5.34) so that Δ is likewise independent of z. Hence

$$\Delta = \frac{\partial(p_{1o}, p_{2o})}{\partial(x_{1a}, x_{2a})} = \frac{\partial(p_{1a}, p_{2a})}{\partial(x_{1o}, x_{2o})} \qquad (22.34)$$

If there exists an image plane $z = z_i$ in which the object plane $z = z_o$ is imaged stigmatically, all rays from a point in z_o will be reunited in z_i irrespective of their point of intersection with the plane $z = z_a$:

$$\frac{\partial x_{ki}}{\partial x_{ja}} = 0 \qquad (22.35)$$

and, solving (22.30) when (22.35) is satisfied, we find

$$x^{(1)}_{1i} = \frac{1}{\Delta}\left(\frac{\partial p_{2i}}{\partial x_{1a}}\frac{\partial S^I_{oi}}{\partial x_{2a}} - \frac{\partial p_{2i}}{\partial x_{2a}}\frac{\partial S^I_{oi}}{\partial x_{1a}}\right)$$
$$x^{(1)}_{2i} = \frac{1}{\Delta}\left(\frac{\partial p_{1i}}{\partial x_{2a}}\frac{\partial S^I_{oi}}{\partial x_{1a}} - \frac{\partial p_{1i}}{\partial x_{1a}}\frac{\partial S^I_{oi}}{\partial x_{2a}}\right)$$
(22.36)

with

$$\Delta = \frac{\partial(p_{1i}, p_{2i})}{\partial(x_{1a}, x_{2a})}$$
(22.37)

We note that only the perturbation characteristic S^I_{oi} appears in (22.36).

Second-order perturbations

Returning to (22.16) and (22.17), we see that

$$S^{(2)}_{12} = \int_{z_1}^{z_2}\left\{M^{II} + D^I M^I + D^{II} M + \frac{1}{2}(D^I)^2 M\right\} dz$$
(22.38)

and using (22.19) with $\xi_j = x_j^{(2)}$, which tells us that

$$\int_{z_1}^{z_2} D^{II} M \, dz = p_{j2} x^{(2)}_{j2} - p_{j1} x^{(2)}_{j1}$$
(22.39)

we may write

$$S^{(2)}_{12} = \int_{z_1}^{z_2}\left\{M^{II} + D^I M^I + \frac{1}{2}(D^I)^2 M\right\} dz$$
$$+ p_{j2} \cdot x^{(2)}_{j2} - p_{j1} \cdot x^{(2)}_{j1}$$
(22.40)

The perturbed form of (22.19) enables us to simplify this further. Applying the operator P to (22.19), we find

$$\int_{z_1}^{z_2}\left(\xi_j \frac{\partial(PM)}{\partial x_j} + \xi'_j \frac{\partial(PM)}{\partial x'_j}\right) dz$$
$$= \xi_{j2} P p_{j2} - \xi_{j1} P p_{j1}$$
(22.41)

so that for $\xi_j = x_j^{(1)}$,

$$\int_{z_1}^{z_2} D^I(M^I + D^I M) \, dz = p^{(1)}_{j2} x^{(1)}_{j2} - p^{(1)}_{j1} x^{(1)}_{j1}$$
(22.42)

This enables us to eliminate either $D^I M^I$ or $\frac{1}{2}(D^I)^2 M$ from (22.40); we obtain

$$\begin{aligned} S_{12}^{(2)} =&\, p_{j2} x_{j2}^{(2)} + p_{j2}^{(1)} x_{j2}^{(1)} - (p_{j1} x_{j1}^{(2)} + p_{j1}^{(1)} x_{j1}^{(1)}) \\ &+ \int \{M^{II} - \frac{1}{2}(D^I)^2 M\}\, dz \end{aligned} \tag{22.43a}$$

or

$$\begin{aligned} S_{12}^{(2)} =&\, p_{j2} x_{j2}^{(2)} + \frac{1}{2} p_{j2}^{(1)} x_{j2}^{(1)} - (p_{j1} x_{j1}^{(2)} + \frac{1}{2} p_{j1}^{(1)} x_{j1}^{(1)}) \\ &+ \int \{M^{II} + \frac{1}{2} D^I M^I\}\, dz \end{aligned} \tag{22.43b}$$

and writing

$$\begin{aligned} S_{12}^{II} &= \int \left(M^{II} - \frac{1}{2}(D^I)^2 M \right) dz \\ \tilde{S}_{12}^{II} &= \int \left(M^{II} + \frac{1}{2} D^I M^I \right) dz \end{aligned} \tag{22.44}$$

it is easily shown that

$$\begin{aligned} \Delta S_{12}^{II} =&\, (p_{j2}^{(2)} \cdot \Delta x_{j2} - x_{j2}^{(2)} \cdot \Delta p_{j2}) \\ &- (p_{j1}^{(2)} \cdot \Delta x_{j1} - x_{j1}^{(2)} \cdot \Delta p_{j1}) \\ &- (x_{j2}^{(1)} \cdot \Delta p_{j2}^{(1)} - x_{j1}^{(1)} \cdot \Delta p_{j1}^{(1)}) \\ \Delta \tilde{S}_{12}^{II} =&\, (p_{j2}^{(2)} \cdot \Delta x_{j2} - x_{j2}^{(2)} \cdot \Delta p_{j2}) \\ &- (p_{j1}^{(2)} \cdot \Delta x_{j1} - x_{j1}^{(2)} \cdot \Delta p_{j1}) \\ &- \frac{1}{2}(x_{j2}^{(1)} \cdot \Delta p_{j2}^{(1)} - p_{j2}^{(1)} \cdot \Delta x_{j2}^{(1)}) \\ &+ \frac{1}{2}(x_{j1}^{(1)} \cdot \Delta p_{j1}^{(1)} - p_{j1}^{(1)} \cdot \Delta x_{J1}^{(1)}) \end{aligned} \tag{22.45}$$

Hence

$$\begin{aligned} p_{j2}^{(2)} \frac{\partial x_{j2}}{\partial x_{ka}} - x_{j2}^{(2)} \frac{\partial p_{j2}}{\partial x_{ka}} &= \frac{\partial S_{o2}^{II}}{\partial x_{ka}} + x_{j2}^{(1)} \frac{\partial p_{j2}^{(1)}}{\partial x_{ka}} \\ p_{j2}^{(2)} \frac{\partial x_{j2}}{\partial x_{ko}} - x_{j2}^{(2)} \frac{\partial p_{j2}}{\partial x_{ko}} &= \frac{\partial S_{o2}^{II}}{\partial x_{ko}} + x_{j2}^{(1)} \frac{\partial p_{j2}^{(1)}}{\partial x_{ko}} \end{aligned} \tag{22.46}$$

which may be solved for $x_j^{(2)}$ and $p_j^{(2)}$. In an *image* plane, $z = z_i$, the solution collapses to the simpler form

$$
\begin{aligned}
x_{1i}^{(2)} &= \frac{1}{\Delta} \left\{ \frac{\partial p_{2i}}{\partial x_{1a}} \left(\frac{\partial S_{oi}^{II}}{\partial x_{2a}} + x_{ji}^{(1)} \frac{\partial p_{ji}^{(1)}}{\partial x_{2a}} \right) \right. \\
&\quad \left. - \frac{\partial p_{2i}}{\partial x_{2a}} \left(\frac{\partial S_{oi}^{II}}{\partial x_{1a}} + x_{ji}^{(1)} \frac{\partial p_{ji}^{(1)}}{\partial x_{1a}} \right) \right\} \\
x_{2i}^{(2)} &= \frac{1}{\Delta} \left\{ \frac{\partial p_{1i}}{\partial x_{2a}} \left(\frac{\partial S_{oi}^{II}}{\partial x_{1a}} + x_{ji}^{(1)} \frac{\partial p_{ji}^{(1)}}{\partial x_{1a}} \right) \right. \\
&\quad \left. - \frac{\partial p_{1i}}{\partial x_{1a}} \left(\frac{\partial S_{oi}^{II}}{\partial x_{2a}} + x_{ji}^{(1)} \frac{\partial p_{ji}^{(1)}}{\partial x_{2a}} \right) \right\}
\end{aligned}
\qquad (22.47)
$$

where Δ is the corresponding determinant. (Note that the summation is implicit over $j, j = 1, 2$, whereas i is simply a label signifying "image".)

Equations (22.47) are complicated in appearance but a physical meaning may be associated with the various groups of terms. When we analyse a particular system, we first examine its paraxial properties, then its primary aberrations, characterized essentially by M^I and hence S^I. If we then proceed to the secondary aberrations, a contribution will come from M^{II} and hence S^{II}. A further contribution arises because it is the paraxial solutions that we substitute in M^{II} and these are not the best approximation available since we already know the primary aberrations. This further contribution is the secondary aberration term arising from the primary aberrations.

Before applying these general formulae to particular types of system, we must mention one further degree of complication that can arise. In the case of primary aberrations, those obtained by first-order perturbation theory, the results are additive, in the sense that if M^I consists of two parts, $M^I = M_1^I + M_2^I$, the calculations may be performed separately for M_1^I and M_2^I and the results added. This is no longer true for second-order perturbations, and we may need to introduce a new parameter, μ, in addition to λ. This typically arises when geometrical and chromatic aberrations are being considered; for primary aberrations, we have merely to calculate M_1^I for the geometrical aberrations and M_2^I for the chromatic aberrations and add the resulting perturbation terms $x_j^{(1)}$. For the secondary aberrations, there will be mixed terms, arising from the interplay between the two contributions. These are incorporated in the theory in the following way. We denote functional perturbations with respect to μ by the superscript J and $P_f M$ now becomes

$$
\begin{aligned}
P_f M = M &+ \lambda M^I + \mu M^J + \lambda^2 M^{II} \\
&+ \lambda \mu M^{IJ} + \mu^2 M^{JJ} + \ldots
\end{aligned}
\qquad (22.48)
$$

22. PERTURBATION THEORY

The order in which the perturbations are considered is of no importance, as we can readily see on physical grounds. If the contribution to the secondary geometrical aberrations is gradually increased from a negligibly small value, the chromatic contribution remaining fixed, the resulting effects must be the same as those observed if the geometrical aberrations are fixed at their final value and the chromatic aberration gradually increased from a small value.

The formulae for a single perturbation parameter λ given above enable us to manipulate all the terms of an expansion of the form (22.48) except the mixed term $\lambda \mu M^{IJ}$. We merely state the results. The single-parameter reasoning may be followed provided that the following replacements are made:

$$M^{II} \to M^I M^J$$
$$(D^I)^2 M \to 2 D^I D^J \qquad (22.49)$$
$$D^I M^I \to D^I M^J + D^J M^I$$

We define

$$S_{12}^{IJ} := \int_{z_1}^{z_2} \left(M^{IJ} - D^I D^J M \right) dz$$

$$\tilde{S}_{12}^{IJ} := \int_{z_1}^{z_2} \left\{ M^I M^J + \frac{1}{2} \left(D^I M^J + D^J M^I \right) \right\} dz \qquad (22.50)$$

for which

$$\Delta S_{12}^{IJ} = -(x_{j2}^I \cdot \Delta p_{j2}^J + x_{j2}^J \cdot \Delta p_{j2}^I)$$
$$+ (x_{j1}^I \cdot \Delta p_{j1}^J + x_{j1}^J \cdot \Delta p_{j1}^I)$$
$$+ (p_{j2}^{IJ} \cdot \Delta x_{j2} - x_{j2}^{IJ} \cdot \Delta p_{j2})$$
$$- (p_{j1}^{IJ} \cdot \Delta x_{j1} - x_{j1}^{IJ} \cdot \Delta p_{j1})$$

$$\Delta \tilde{S}_{12}^{IJ} = -\frac{1}{2}(x_{j2}^I \cdot \Delta p_{j2}^J + x_{j2}^J \cdot \Delta p_{j2}^I)$$
$$+ \frac{1}{2}(x_{j1}^I \cdot \Delta p_{j1}^J + x_{j1}^J \cdot \Delta p_{j1}^I) \qquad (22.51)$$
$$+ \frac{1}{2}(p_{j2}^I \cdot \Delta x_{j2}^J + p_{j2}^J \cdot \Delta x_{j2}^I)$$
$$- \frac{1}{2}(p_{j1}^I \cdot \Delta x_{j1}^J + p_{j1}^J \cdot \Delta x_{j1}^I)$$
$$+ (p_{j2}^{IJ} \cdot \Delta x_{j2} - x_{j2}^{IJ} \cdot \Delta p_{j2})$$
$$- (p_{j1}^{IJ} \cdot \Delta x_{j1} - x_{j1}^{IJ} \cdot \Delta p_{j1})$$

22. PERTURBATION THEORY : GENERAL FORMALISM

in which we have been obliged to label x_j and p_j with I and J to distinguish the different contributions. Thus x_j^I might represent the primary geometrical aberration, for example, and x_j^J the chromatic aberration.

The mixed forms of (22.46) are as follows:

$$p_{j2}^{(2)} \frac{\partial x_{j2}}{\partial x_{ka}} - x_{j2}^{(2)} \frac{\partial p_{j2}}{\partial x_{ka}} = \frac{\partial S_{o2}^{IJ}}{\partial x_{ka}} + x_{j2}^I \frac{\partial p_{j2}^J}{\partial x_{ka}} + x_{j2}^J \frac{\partial p_{j2}^I}{\partial x_{ka}}$$
$$p_{j2}^{(2)} \frac{\partial x_{j2}}{\partial x_{ko}} - x_{j2}^{(2)} \frac{\partial p_{j2}}{\partial x_{ko}} = \frac{\partial S_{o2}^{IJ}}{\partial x_{ko}} + x_{j2}^I \frac{\partial p_{j2}^J}{\partial x_{ko}} + x_{j2}^J \frac{\partial p_{j2}^I}{\partial x_{ko}} \quad (22.52)$$

23
The Relation Between Permitted Types of Aberration and System Symmetry

23.1 Introduction

The nature of the aberrations of any given system is entirely determined by its symmetry. A classification of the various system types was proposed by Sturrock (1951a) and a detailed exploration of the relation between aberrations and symmetry is to be found in Hawkes (1965a).

Consider a ray passing through an entirely general system, intersecting three planes (or even surfaces) normal to the optic axis, itself in general curved. The point of intersection of the ray with an arbitrary "current" plane, $z = z_c$, is a function of the point of intersection with the other two planes, which we refer to as the object plane $z = z_o$ and the aperture plane $z = z_a$ (Fig. 23.1). Thus

$$w_c = X_c + iY_c = f(w_o, \overline{w}_o, w_a, \overline{w}_a) \qquad (23.1)$$

In this chapter, a bar over a symbol denotes its complex conjugate (thus $\overline{w} = X - iY$). Since we are interested in the *optical* behaviour of the system, and hence in rays that remain in the vicinity of the optic axis, we expand w_c as a power series:

$$w_c = \sum_{\alpha,\beta,\gamma,\delta \geq 0} (\alpha\beta\gamma\delta) w_o^\alpha \overline{w}_o^\beta w_a^\gamma \overline{w}_a^\delta \qquad (23.2)$$

in which the complex coefficient $(\alpha\beta\gamma\delta)$ is a function of z. Suppose now that the system is such that, on simultaneously moving the object point w_o to $w_o \exp(2\pi i/N)$ and the aperture point w_a to $w_a \exp(2\pi i/N)$, the point w_c moves to $w_c \exp(2\pi i/N)$. We then say that the system is N-fold symmetrical. Clearly, not all values of N and shapes of axis are compatible, the higher symmetries only being found with a straight optic axis, for example. From (23.2), it is readily seen that in an N-fold symmetrical system, the indices α, β, γ and δ can take only those non-negative integral values for which

$$\alpha - \beta + \gamma - \delta = 1 + kN \qquad (23.3)$$

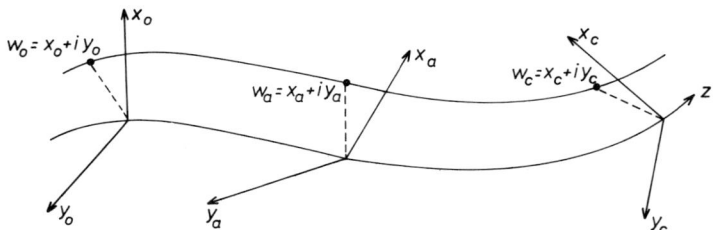

Fig. 23.1: General ray intersecting the object, aperture and current surfaces at the points w_0, w_a and w_c respectively.

where k is a positive or negative integer or zero. For rotationally symmetric systems, only the values that satisfy

$$\alpha - \beta + \gamma - \delta = 1 \tag{23.4}$$

are acceptable. The permitted values of α, β, γ and δ for values of N from 1 to 6 are shown in Table 23.1 (Hawkes and Cosslett, 1962; Hawkes, 1965a).

The condition that restricts the values of α, β, γ and δ contains only the combinations $(\alpha + \gamma)$ and $(\beta + \delta)$, and this suggests a simpler method of obtaining the permissible members in the expansion for w_c. We need consider only one of two particular pencils of rays, either the pencil through the axial object point $(0, 0, z_o)$ or the pencil through the axial aperture point $(0, 0, z_a)$. The point w_c can now be written

$$w_c = \sum_{\xi,\eta \geq 0} (\xi\eta) w_o^\xi \overline{w}_o^\eta \quad \text{or} \quad w_c = \sum_{\xi,\eta \geq 0} (\xi\eta) w_a^\xi \overline{w}_a^\eta \tag{23.5}$$

and hence

$$\xi - \eta = 1 + kN$$

We now have only to replace w_o or w_a by $w_o + w_a$ for the condition $\alpha - \beta + \gamma - \delta = 1 + kN$ to be automatically satisfied, since the binomial theorem ensures that $\alpha + \gamma = \xi$ and $\beta + \delta = \eta$. In optical terms, we have considered first two particular pencils, of which one could be affected only by distortions, and could not produce an indistinct image, while the other could be perturbed only by wholly aperture-dependent aberrations. To obtain the remaining aberrations we replaced these pencils by off-axial pencils, passing through neither the object nor the aperture origin.

As an example of this, we might consider the secondary aberrations of $N = 2$ systems, which are fifth order. From Table 23.1, we see that there are 56 of these. The condition that $\xi - \eta = 1 + 2k$ subject to $\xi + \eta = 5$ allows ξ and η to take the values (50), (41), (32), (23), (14) and

23.1 INTRODUCTION

Table 23.1 The Values of α, β, γ and δ for $N = 1$ to $N = 6$ and Rotational Symmetry

Each row contains terms for which $\alpha + \beta + \gamma + \delta$ is the same (roman numeral), and each column all the terms with some particular value of $\gamma - \delta$ (arabic numeral). The column headed 'D' contains distortions, for which $\gamma = \delta = 0$.

	D	Rotational symmetry : $(\alpha + \gamma) - (\beta + \delta) = 1$					
		-2	-1	0	1	2	3
I	1000	.	.	.	0010	.	.
III	2100	.	2001	1011	0021	0120	.
	.	.	.	.	1110	.	.
V	3200	3002	3101	2111	0032	1220	0230
	.	.	2012	1022	2210	1031	.
	.	.	.	.	1121	.	.

In all the subsequent lists, the rotationally symmetric terms just tabulated all recur; they are therefore not repeated in every list.

	D	$N = 6.$ $(\alpha + \gamma) - (\beta + \delta) = -5, 1$			
		-5	-4	-3	-2	-1
V	0500	0005	0104	0203	0302	0401

	D	$N = 5.$ $(\alpha + \gamma) - (\beta + \delta) = -4, 1$			
		-4	-3	-2	-1
IV	0400	0004	0103	0202	0301

	D	$N = 4.$ $(\alpha + \gamma) - (\beta + \delta) = -3, 1, 5$			
		-4	-3	-2	-1
III	0300	.	0003	0102	0201
V	1400	1004	1103	1202	1301
	5000	.	0014	0113	0212
	0	5	4	3	2	1
V	0311	0050	1040	2030	3020	0410
	.	.	.	.	.	4010

	D	$N = 3.$ $(\alpha + \gamma) - (\beta + \delta) = -5, -2, 1, 4$			
		-5	-4	-3	-2	-1
II	0200	.	.	.	0002	0101
IV	1300	.	.	1003	1102	1201
	4000	.	.	.	0013	0112
V	0500	0005	0104	0203	0302	0401
	0	4	3	2	1
IV	0211	0040	1030	2020	0310
	.	.	.	.	3010

318 23. RELATION BETWEEN ABERRATION AND SYSTEM SYMMETRY

$$N = 2. \quad (\alpha + \gamma) - (\beta + \delta) = \pm 1, \pm 3, \pm 5$$

	D	−5	−4	−3	−2	−1		0	5	4	3	2	1
I	0100	.	.	.	.	0001		.	.	.	.	.	.
III	3000	.	.	0003	1002	0012	III	0111	.	.	0030	1020	0210
	0300	.	.	.	0102	0201		.	.	.	.	.	2010
	1200	.	.	.	.	1101							
V	5000	0005	1004	2003	2102	2201	V	1211	0050	1040	1130	0320	1310
	4100	.	0104	0203	1202	1301		0311	.	0140	0041	3020	3110
	2300	.	.	1103	1013	4001		3011	.	.	2030	2120	0221
	1400	.	.	0014	0113	1112		0122	.	.	.	1031	2021
	0500	.	.	.	0302	0401		.	.	.	.	.	4010
	.	.	.	.	.	0023		.	.	.	.	.	0410
	.	.	.	.	.	0212		.	.	.	.	.	.

$$N = 1. \quad (\alpha + \gamma) - (\beta + \delta) = 0, \pm 1, \pm 2, \pm 3, \pm 4, \pm 5$$
+1 is included in this list

	D	−5	−4	−3	−2	−1		0	5	4	3	2	1
0	0000	.	.	.	.	.		.	.	.	.	.	.
I	1000	.	.	.	.	0001	I	.	.	.	.	.	0010
	0100	.	.	.	.	.							
II	2000	.	.	.	0002	0101	II	0011	.	.	.	0020	1010
	1100	.	.	.	.	1001		.	.	.	.	.	0110
	0200	.	.	.	.	.							
III	3000	.	.	0003	0102	2001	III	1011	.	.	0030	1020	2010
	2100	.	.	.	1002	0201		0111	.	.	.	0120	0210
	1200	.	.	.	.	1101		.	.	.	.	.	1110
	0030	.	.	.	.	0012		.	.	.	.	.	0021
IV	4000	.	0004	0103	2002	3001	IV	2011	.	0040	1030	2020	3010
	3100	.	.	1003	0202	2101		1111	.	.	0130	1120	2110
	2200	.	.	.	1102	1201		0211	.	.	.	0220	1210
	1300	.	.	.	0013	0301		0022	.	.	.	0031	0310
	0400	.	.	.	.	1012		.	.	.	.	.	1021
	.	.	.	.	.	0112		.	.	.	.	.	0121
V	5000	0005	0104	2003	3002	4001	V	3011	0050	1040	2030	3020	4010
	4100	.	1004	0203	2102	3101		2111	.	0140	1130	2120	3110
	3200	.	.	1103	1202	2201		1211	.	.	0230	1220	2210
	2300	.	.	0014	0302	1301		0311	.	.	0041	0320	1310
	1400	.	.	.	1013	0401		1022	.	.	.	1031	0410
	0500	.	.	.	0113	2012		0122	.	.	.	0131	2021
	.	.	.	.	.	0212		.	.	.	.	.	1121
	.	.	.	.	.	1112		.	.	.	.	.	0221
	.	.	.	.	.	0023		.	.	.	.	.	0032

23.1 INTRODUCTION

Table 23.2 The Values of $(\xi\eta)$ Corresponding to each Type of Symmetry

The bracketed symbols indicate the symmetry class of which the corresponding aberrations are typical. R signifies rotationally symmetric.

	$\xi+\eta=1$	$\xi+\eta=2$	$\xi+\eta=3$	$\xi+\eta=4$	$\xi+\eta=5$
$N=1$	10(R)	20	30(2)	40(3)	50(4)
	01(2)	11	21(R)	31	41(2)
	.	02(3)	12(2)	22	32(R)
	.	.	03(4)	13(3)	23(2)
	.	.	.	04(5)	14(4)
	.	.	.	.	05(6)
$N=2$	10(R)	.	30	.	50(4)
	01	.	21(R)	.	41
	.	.	12	.	32(R)
	.	.	03(4)	.	23
	.	.	.	.	14(4)
	.	.	.	.	05(6)
$N=3$	10(R)	02	21(R)	40	32(R)
	.	.	.	13	05(6)
$N=4$	10(R)	.	21(R)	.	50
	.	.	03	.	32(R)
	.	.	.	.	14
$N=5$	10(R)	.	.	04	32(R)
$N=6$	10(R)	.	.	.	05
rotational symmetry	10	.	21	.	32

(05). Of these, $(\xi\eta) = (50)$ generates six aberrations and $(\xi\eta) = (14)$ generates ten aberrations; these together are characteristic of $N = 4$. The term $(\xi\eta) = (32)$ generates the 12 aberrations characteristic of rotationally symmetrical systems. The six aberrations characteristic of $N = 6$ (which also afflict $N = 3$ systems, therefore) are generated by $(\xi\eta) = (05)$. Finally, $(\xi\eta) = (41)$ and $(\xi\eta) = (23)$ generate the 22 aberrations peculiar to $N = 2$ systems. The full results are set out in Table 23.2.

A particularly simple way of discussing the effects of each aberration term, $(\alpha\beta\gamma\delta)w_o^\alpha \overline{w}_o^\beta w_a^\gamma \overline{w}_a^\delta$, or group of aberration terms, has been explored in detail by Chako (1957), who employed only cartesian coordinates and considered only rotationally symmetric systems. Amboss (1959) used the same method with complex coordinates to examine slightly imperfect systems. If we wish to interpret geometrically an expression of the form

$$w_c = \sum (\alpha\beta\gamma\delta) w_o^\alpha \overline{w}_o^\beta w_a^\gamma \overline{w}_a^\delta \tag{23.6}$$

we can obtain a curve that is typical of each type of aberration on a general plane in image space by considering a pencil of rays that emerges from an object point w_o=constant and is restricted by an annular stop in the aperture plane $\mid w_a \mid =$ constant $= r_a$. These are the curves which Chako calls 'characteristic curves'. We can then write

$$w_c = \sum r_o^{\alpha+\beta} r_a^{\gamma+\delta} \mid \alpha\beta\gamma\delta \mid \exp \mathrm{i}\{\widetilde{\alpha\beta\gamma\delta} + (\alpha-\beta)\theta_o + (\gamma-\delta)\theta_a\} \quad (23.7)$$

in which $(\alpha\beta\gamma\delta) =: \mid \alpha\beta\gamma\delta \mid e^{\mathrm{i}\widetilde{\alpha\beta\gamma\delta}}$. For each term, therefore, we have

$$w_c = (\mathbf{A}\mathbf{B}\mathbf{\Gamma}\mathbf{\Delta})e^{\mathrm{i}(\gamma-\delta)\Theta_a} \begin{cases} (\mathbf{A}\mathbf{B}\mathbf{\Gamma}\mathbf{\Delta}) = \mid \alpha\beta\gamma\delta \mid r_o^{\alpha+\beta} r_a^{\gamma+\delta} \\ \Theta_a = \theta_a + \dfrac{\widetilde{\alpha\beta\gamma\delta} + (\alpha-\beta)\theta_o}{\gamma-\delta} \end{cases} \quad (23.8)$$

If $\gamma = \delta = 0$, we can consider the pencil that passes through a fixed point in the aperture plane and intersects the object plane in an annulus, $\mid w_o \mid = r_o =$ constant.

Not all the coefficients $(\alpha\beta\gamma\delta)$ are independent, as we can see by considering the point characteristic function S, which we expand as a power series in X, X', Y and Y':

$$S = S^{(0)} + S^{(1)} + S^{(2)} + \ldots + S^{(r)} + \ldots \quad (23.9)$$

in which each group of terms, $S^{(r)}$, contains only products of degree r in the off-axial coordinates. Of these groups, $S^{(0)}$ is a constant and will not concern us further; $S^{(1)}$ is zero since the axis must also be a possible ray; the Gaussian or first-order (primordial) properties are given by $S^{(2)}$ provided that $N \leq 2$ and the primary, secondary and further aberrations are given by successive nonzero terms of S.

The function S is a definite integral of a physical quantity along a real path between two points; it must therefore be real and must also satisfy the same symmetry conditions as the whole system. This suggests that we might write each of the even components of S as a quadratic or Hermitian form

$$S^{(r)} = Q_r^T R Q_r \quad \text{or} \quad S^{(r)} = H_r^* r H_r \quad (23.10)$$

Q is a column matrix and Q^T its transpose; H is also a column matrix and H^* is the conjugate complex matrix of H^T; R and r are square matrices, which we shall refer to as the coefficient matrices.

We have already used a four-symbol notation $(\alpha\beta\gamma\delta)$ to enable the coefficients of the terms in the series expansion of w_c to be written down directly. We shall employ a similar notation for the elements of the coefficient matrices, R and r. Each element will be denoted by a four-figure symbol $(pqrs)$ that is automatically associated with the term $w_o^p \overline{w}_o^q w_a^r \overline{w}_a^s$.

23.1 INTRODUCTION

Whereas $\alpha+\beta+\gamma+\delta$ is equal to the order of the corresponding aberration, $p+q+r+s$ is equal to the order of the terms in the characteristic function that correspond to the same aberration, namely $\alpha+\beta+\gamma+\delta+1$. Since S must be real, we can immediately conclude that

$$(pqrs) = \overline{(qpsr)} \tag{23.11}$$

All the relations that connect the aberration coefficients stem from this property. In the simplest case, that of rotational symmetry, the relation which we encounter below between the two parts of the coma coefficient is a consequence of

$$(1012) = \overline{(0121)} \tag{23.12}$$

Similarly, the fact that under certain conditions, quadrupole lens systems are fully characterized by three, and not four, aperture aberration coefficients can be deduced from

$$(0013) = \overline{(0013)} \tag{23.13}$$

The condition used by de Broglie (1950, pp.131–2) to obtain the interrelations for axially symmetric systems can be straightforwardly derived from (23.11). This relation between the elements of the characteristic function does, however, bring out the fact that de Broglie's condition is a consequence of more fundamental relations which remain true, even though the actual aberration coefficients may no longer be so simply connected.

If the N-fold symmetry of a system is reinforced by a symmetry plane, the coefficients $(pqrs)$ are either all real or all purely imaginary. For, if the system possesses a plane of symmetry, S will be invariant under a change from right-handed to left-handed axes, and hence $(-1)^{p+q+r+s}(pqrs) = (qpsr)$. Since we have already shown that $(pqrs) = \overline{(qpsr)}$, $(pqrs)$ must be real if $p+q+r+s$ is even, and imaginary if $p+q+r+s$ is odd.

For simplicity, we study the aberrations in field-free image space, taking $z=z_a$ as the exit pupil of the system. Then in image space, we have

$$w = w_a + (z-z_a)\left(\frac{\partial w}{\partial z}\right)_{z=z_a} \tag{23.14}$$

or

$$w(z) = \{1 + \mathrm{i}B(z-z_a)/A\}w_a + 2\frac{z-z_a}{A}\frac{\partial S_{oa}}{\partial \overline{w}_a}$$

in which we have used the fact that the terms in $S^{(2)}$ containing X' and Y' can always be cast into the form $\frac{1}{2}A(X'^2+Y'^2) + B(X'Y-XY')$. Thus $w(z)$ has the form

$$w(z) = Cw_a + D\frac{\partial S_{oa}}{\partial \overline{w}_a} \tag{23.15}$$

in which C may be complex and D is real.

It is to be noted that the analysis set out here enables us to identify all aberrations permitted by symmetry but some of these may be forbidden for other reasons. General laws have not yet been established but some rules of practical importance have been found by Shao (1987a,b) and Shao and Crewe (1987).

We now consider the image formation and aberrations for each type of symmetry, as characterized by N.

23.2 $N = 1$

This is the most general case: the system is wholly arbitrary, the axis may be straight or curved, and few generalizations can be made about the image formation. The function $S_{oa}^{(2)}$ is of the form

$$\begin{pmatrix} w_o \\ \overline{w}_o \\ w_a \\ \overline{w}_a \end{pmatrix}^T \begin{pmatrix} 2000 & 1100(r) & 1010 & 1001 \\ 0 & 2000 & 1001 & 1010 \\ 0 & 0 & 0020 & 0011(r) \\ 0 & 0 & 0 & 0020 \end{pmatrix} \begin{pmatrix} w_o \\ \overline{w}_o \\ w_a \\ \overline{w}_a \end{pmatrix} \quad (23.16)$$

and using (23.15), we find that these coefficients are related to the $(\alpha\beta\gamma\delta)$ of (23.6) thus:

$$(1000) = D(1001); \quad (0100) = D(\overline{1010}); \quad (0001) = 2D(\overline{0020})$$

$$\Re(0010) = D(0011) + \Re(C); \quad \Im(0010) = \Im(C)$$

To understand what imagery such a system would produce, we consider a pencil of rays emerging from a fixed object point and restricted by an annular stop, $|w_a| = r_a = $ constant. We denote $(1000)w_o + (0100)\overline{w}_o$ by U, and we write $(\alpha\beta\gamma\delta) = |\alpha\beta\gamma\delta| \exp i\widetilde{\alpha\beta\gamma\delta}$ and $w = re^{i\theta}$. We find

$$w_c = U + r_a \mid 0010 \mid e^{i(\theta_a + \widetilde{0010})} + r_a \mid 0001 \mid e^{-i(\theta_a - \widetilde{0001})}$$

or

$$(w_c - U)e^{-\frac{1}{2}i(\widetilde{0010} + \widetilde{0001})} = r_a \mid 0001 \mid e^{i\Theta_a} + r_a \mid 0010 \mid e^{-i\Theta_a} \quad (23.17)$$

in which

$$\Theta_a = \theta_a + \frac{1}{2}(\widetilde{0010} - \widetilde{0001})$$

---

* (r) indicates that the adjoining element is real

In general, therefore, the pencil will emerge from the system as an astigmatic bundle of rays, which collapses to two mutually perpendicular focal lines in two particular current planes where $|\ 0010\ | + |\ 0001\ | = 0$ and $|\ 0010\ | - |\ 0001\ | = 0$; these focal lines are inclined to the axes w_c at an angle which depends upon the position of the object point. If a point object approaches the system from infinity, the focal lines may execute any one of three possible manoeuvres. They may swivel round the axis of the system, turning always in the same sense; in this case $(\widetilde{0010 + 0001})$ changes monotonically. Systems in which the focal lines behave in this way were called '*gedrehte*' or '*tordierte Systeme*' by Gullstrand, who first discussed them (1915, p.20). Alternatively, the lines may swivel round the axis until the object point reaches one of the two 'orthogonal points', beyond which they turn in the opposite sense; when the object point reaches the second orthogonal point, the focal lines revert to their original sense of rotation. These are Gullstrand's '*zurückgedrehte*' or '*retordierte Systeme*'. Finally, these two orthogonal points may coincide. Between the two orthogonal points of the preceding type of system, the focal lines unwind through a quarter-turn; in these '*halbgedrehte*' or '*semitordierte Systeme*', the focal lines turn through the same angle instantaneously, with the result that as the object point crosses the coincident orthogonal point, the focal lines appear abruptly to change places, but continue to rotate in the same sense.

In a stigmatic system, the image of a pair of straight lines $x_o = \text{const}$ and $y_o = \text{const}$ will be

$$y_i = -\frac{\Re(1000 - 0100)}{\Im(1000 - 0100)}x_i$$
$$+ \frac{\Re(1000 + 0100)\Re(1000 - 0100) + \Im(1000 + 0100)\Im(1000 - 0100)}{\Im(1000 - 0100)}x_o$$

(23.18a)

and

$$y_i = \frac{\Im(1000 + 0100)}{\Re(1000 + 0100)}x_i$$
$$+ \frac{\Re(1000 + 0100)\Re(1000 - 0100) + \Im(1000 + 0100)\Im(1000 - 0100)}{\Im(1000 + 0100)}y_o$$

(23.18b)

which represents a pair of straight lines, no longer at right angles; a rectangle is therefore imaged as a parallelogram.

The primary aberrations are second order, and correspond to $S^{(3)}$; using $(pqrs) = (\overline{qpsr})$, we can write the latter in the form

$$S^{(3)} = \begin{pmatrix} w_o \\ \overline{w}_o \\ w_a \\ \overline{w}_a \end{pmatrix}^T \begin{pmatrix} 3000 & \overline{1200} & 0 & \overline{1020} & \overline{1002} & \overline{1011} \\ 1200 & 3000 & 0 & 1002 & 1020 & 1011 \\ 2010 & \overline{2001} & 1110 & 0030 & \overline{0021} & 0 \\ 2001 & \overline{2010} & \overline{1110} & 0021 & \overline{0030} & 0 \end{pmatrix} \begin{pmatrix} w_o^2 \\ \overline{w}_o^2 \\ w_o\overline{w}_o \\ w_a^2 \\ \overline{w}_a^2 \\ w_a\overline{w}_a \end{pmatrix}$$

(23.19)

No element is necessarily real.

In an arbitrary image plane, the primary aberrations w_c^I are given by (22.30); we substitute

$$w_c = (1000)w_o + (0100)\overline{w}_o + (0010)w_a + (0001)\overline{w}_a$$

and

$$s_c = Aw_c' - iBw_c$$

and obtain

$$Dw_c^I = 2A \begin{vmatrix} 0010 & 0001 & 0001 & \partial S_{oc}^{(3)}/\partial \overline{w}_a \\ 1000 & 0100 & 0100 & \partial S_{ac}^{(3)}/\partial \overline{w}_o \\ 0001 & 0010 & 0010 & \partial S_{oc}^{(3)}/\partial w_a \\ 0100 & 1000 & 1000 & \partial S_{ac}^{(3)}/\partial w_o \end{vmatrix} \quad (23.20)$$

Each of the elements in the fourth column is composed of ten members. For further details, see Hawkes (1965a).

$N = 1$. Systems with a plane of symmetry

The axis of the system is now a curve lying in a plane, and we shall suppose this to be the plane which also contains the y-axis. Certain deflection systems and β-spectrometers fall within this class. The primary aberrations are still second order, but the function $S^{(3)}$ must be independent of the signs of x_o and x_a; it is therefore of the form

$$\begin{pmatrix} x_o^2 \\ x_o x_a \\ x_a^2 \\ y_o^2 \\ y_a^2 \end{pmatrix}^T \begin{pmatrix} 2100 & 2001 \\ 1110 & 1011 \\ 0120 & 0021 \\ 0300 & 0201 \\ 0102 & 0003 \end{pmatrix} \begin{pmatrix} y_o \\ y_a \end{pmatrix} \quad (23.21)$$

and so

$$\begin{aligned}
k_x x_c^I =& h_{xc}\{2[2100]x_oy_o + 2[2001]x_oy_a + [1110]y_ox_a + [1011]x_ay_a\} \\
& - g_{xc}\{(1110)x_oy_o + (1011)x_oy_a + 2(0120)y_ox_a + 2(0021)x_ay_a\} \\
k_y y_c^I =& h_{yc}\{[2100]x_o^2 + [1110]x_ox_a + [0120]x_a^2 \\
& + 3[0300]y_o^2 + 2[0201]y_oy_a + [0120]y_a^2\} \\
& - g_{yc}\{(2001)x_o^2 + (1011)x_ox_a + (0021)x_a^2 \\
& + (0201)y_o^2 + 2(0102)y_oy_a + 3(0003)y_a^2\}
\end{aligned}$$
(23.22)

In a general plane, therefore, ten coefficients suffice to characterize the imagery, and in the image plane of a stigmatic system, eight.

The aperture aberrations are of the form

$$x_c^I = (0011)x_ay_a, \qquad y_c^I = (0020)x_a^2 + (0002)y_a^2 \qquad (23.23)$$

which represents an ellipse, centred on the point $(0, \frac{1}{2}\{(0020) + (0002)\}r_a^2)$ with axes $\frac{1}{2}r_a^2(0011), \frac{1}{2}r_a^2\{(0020) - (0002)\}$. For a circular aperture, the envelope of this family of ellipses is a pair of straight lines, inclined to the y-axis at an angle arccot $2\sqrt{(0020)(0002)}/(0011)$. In a stigmatic orthomorphic system, $(0011) = 2(0020)$, and the angle becomes arccot $\sqrt{(0002/(0020)}$ or arccot $\sqrt{3(0003/(0021)}$.

The distortions are similar in nature; the remaining aberrations are types of astigmatism, and the aberration curve for a fixed object point and an annular aperture is a tilted centred ellipse.

The secondary aberrations are third order; the contribution from $S^{(4)}$ is analysed in Hawkes (1965a).

23.3 $N = 2$

Systems for which $N = 2$ and there is no symmetry plane may consist of any combination of round lenses and quadrupole lenses in any orientation; the lenses may be either electrostatic or magnetic and the electrostatic lenses need not be excited symmetrically nor need they be geometrically symmetric. The aberrations of quadrupole systems in which the azimuthal alignment of the individual members is imperfect thus fall within this class. The Gaussian imagery is no simpler than that of general systems, but the primary aberrations are now third order. Both the primary and secondary (third-order) aberrations are analysed in detail in Hawkes (1965a).

Electron optical systems with straight axes may be orthogonal without possessing a plane of symmetry. These are the systems which comprise

Dušek's '*erster Hauptfall*' (1959) and since they involve a complicated and delicate balance of electric and magnetic forces, they have as yet found no practical employment. The surfaces in which electrons experience no expulsive force are not planes but curved surfaces, which twist about the axis within the lens fields.

Since such systems are orthogonal, it is simpler to use cartesian coordinates, and $S^{(4)}$ can therefore be written

$$S^{(4)} = \begin{pmatrix} x_o^2 \\ y_o^2 \\ x_a^2 \\ y_a^2 \\ x_o x_a \end{pmatrix}^T A \begin{pmatrix} x_o^2 \\ y_o^2 \\ x_a^2 \\ y_a^2 \\ x_o x_a \\ y_o y_a \\ x_o y_a \\ x_a y_o \\ x_o y_o \\ x_a y_a \end{pmatrix} \tag{23.24}$$

where

$$A = \begin{pmatrix} 4000 & 2200 & 2020 & 2002 & 3010 & 2101 & 3001 & 2110 & 3100 & 2011 \\ 0 & 0400 & 0220 & 0202 & 1210 & 0301 & 1201 & 0310 & 1300 & 0211 \\ 0 & 0 & 0040 & 0022 & 1030 & 0121 & 1021 & 0130 & 1120 & 0031 \\ 0 & 0 & 0 & 0004 & 1012 & 0103 & 1003 & 0112 & 1102 & 0013 \\ 0 & 0 & 0 & 0 & 0 & 1111 & 0 & 0 & 0 & 0 \end{pmatrix}$$

The aberrations associated with perturbation characteristic functions of this form resemble the secondary aberrations of $N = 1$ systems.

The aperture aberrations are of the form

$$\begin{aligned}
x_c^I &= \frac{h_{xc}}{k_x}\{[1030]x_a^3 + [1012]x_a^2 y_a + [1012]x_a y_a^2 + [1003]y_a^3\} \\
&\quad - \frac{g_{xc}}{k_x}\{4(0040)x_a^3 + 3(0031)x_a^2 y_a + 2(0022)x_a y_a^2 + (0013)y_a^3\}, \\
y_c^I &= \frac{h_{yc}}{k_y}\{[0130]x_a^3 + [0121]x_a^2 y_a + [0112]x_a y_a^2 + [0103]y_a^3\} \\
&\quad - \frac{g_{yc}}{k_y}\{(0031)x_a^3 + 2(0022)x_a^2 y_a + 3(0013)x_a y_a^2 + 4(0004)y_a^4\}
\end{aligned} \tag{23.25}$$

When the imagery is stigmatic, the aberration curve in the stigmatic image plane simplifies to the form

$$-Ah'_{xi}x_i^I = \alpha_1 x_a^3 + 3\beta x_a^2 y_a + \gamma x_a y_a^2 + \delta y_a^3$$

$$-Ah'_{yi}y^I_i = \beta x_a^3 + \gamma x_a^2 y_a + 3\delta x_a y_a^2 + \alpha_2 y_a^3$$

The secondary aberrations involve $S^{(6)}$.

$N = 2.$ *Systems possessing a plane of symmetry*

Systems belonging to this class are automatically orthogonal; they represent the special case of the preceding class for which the curved orthogonal surfaces collapse into a pair of mutually perpendicular planes (Dušek's *'verdrehungsfreie Orthogonalsysteme'*). The elements of such systems may be electrostatic and magnetic quadrupole lenses, and round electrostatic lenses; all the electrodes of the electrostatic quadrupoles must lie in the same pair of (mutually perpendicular) azimuthal planes, however, and all the polepieces of the magnetic quadrupoles must lie in the pair of azimuthal planes that are inclined to the electrode planes at 45°. The electrodes need be symmetrical in neither excitation nor geometry, provided of course that any asymmetry is compatible with the symmetry plane.

In complex notation, the presence of the symmetry plane implies that all the elements of the coefficient matrices are real; in the cartesian notation, $S^{(4)}$ and $S^{(6)}$ contain only even powers of x and even powers of y. Examining equation (23.24), we can see that $S^{(4)}$ is obtained by selecting the first six columns of the coefficient matrix, and retaining the first six elements of the ten-element column matrix, thus:

$$S^{(4)} = \begin{pmatrix} x_o^2 \\ y_o^2 \\ x_a^2 \\ y_a^2 \\ x_o x_a \end{pmatrix}^T \begin{pmatrix} 4000 & 2200 & 2020 & 2002 & 3010 & 2101 \\ 0 & 0400 & 0220 & 0202 & 1210 & 0301 \\ 0 & 0 & 0040 & 0022 & 1030 & 0121 \\ 0 & 0 & 0 & 0004 & 1012 & 0103 \\ 0 & 0 & 0 & 0 & 0 & 1111 \end{pmatrix} \begin{pmatrix} x_o^2 \\ y_o^2 \\ x_a^2 \\ y_a^2 \\ x_o x_a \\ y_o y_a \end{pmatrix}$$

(23.26)

The primary aberrations in a general plane are therefore given by

$$\begin{aligned}k_x x_c^I =& h_{xc}\{4[4000]x_o^3 + 2[2200]x_o y_o^2 + [1030]x_a^3 + [1012]x_a y_a^2 \\ & + x_a(3[3010]x_o^2 + [1210]y_o^2) + 2[2101]x_o y_o y_a \\ & + 2[2020]x_o x_a^2 + 2[2002]x_o y_a^2 + [1111]y_o x_a y_a\} \\ & - g_{xc}\{(3010)x_o^3 + (1210)x_o y_o^2 + 4(0040)x_a^3 + 2(0022)x_a y_a^2 \\ & + x_a(2(2020)x_o^2 + 2(0220)y_o^2) + (1111)x_o y_o y_a \\ & + 3(1030)x_o x_a^2 + (1012)x_o y_a^2 + 2(0121)y_o x_a y_a\} \\ k_y y_c^I =& h_{yc}\{2[2200]x_o^2 y_o + 4[0400]y_o^3 + [0121]x_a^2 y_a + [0103]y_a^3\end{aligned}$$

$$+ y_a([2101]x_o^2 + 3[0301]y_o^2) + 2[1210]x_oy_ox_a$$
$$+ 2[0220]y_ox_a^2 + 2[0202]y_oy_a^2 + [1111]x_ox_ay_a\}$$
$$- g_{yc}\{(2101)x_o^2y_o + (0301)y_o^3 + 2(0022)x_a^2y_a + 4(0004)y_a^3 +$$
$$+ y_a(2(2002)x_o^2 + 2(0202)y_o^2) + (1111)x_oy_ox_a$$
$$+ (0121)y_ox_a^2 + 3(0103)y_oy_a^2 + 2(1012)x_ox_ay_a\}$$
(23.27)

In a stigmatic system, there exists an image plane in which $h_x(z_i) = h_y(z_i) = 0$; as we have remarked earlier, the aperture aberrations are then described by only three coefficients:
$$Ah'_{x_i}x_i^I = -4(0040)x_a^3 - 2(0022)x_ay_a^2$$
$$Ah'_{y_i}y_i^I = -2(0022)x_a^2y_a - 4(0004)y_a^3$$
provided the slopes of $h_x(z)$ and $h_y(z)$ are equal at the image plane. $S^{(6)}$ is given by

$$S^{(6)} = \begin{pmatrix} x_o^3 & y_o^3 & x_a^3 & y_a^3 & x_oy_ox_a & x_oy_oy_a & x_ox_ay_a \\ y_ox_ay_a \end{pmatrix} VI \begin{pmatrix} x_o^3 & y_o^3 & x_a^3 & y_a^3 & x_o^2y_o & x_o^2x_a & x_o^2y_a \\ y_o^2x_o & y_o^2x_a & y_o^2y_a & x_a^2x_o & x_a^2y_o & x_a^2y_a & y_a^2x_o \\ y_a^2y_o & y_a^2x_a & x_oy_ox_a & x_oy_oy_a & x_ox_ay_a & y_ox_ay_a \end{pmatrix}^T$$ (23.28)

and

$$VI^T = \begin{pmatrix} 6000 & 0 & 0 & 0 & 0 & 0 & 0 & 0 \\ 0 & 0600 & 0 & 0 & 0 & 0 & 0 & 0 \\ 3030 & 0 & 0060 & 0 & 0 & 0 & 0 & 0 \\ 0 & 0303 & 0 & 0006 & 0 & 0 & 0 & 0 \\ 0 & 2400 & 0 & 2103 & 0 & 0 & 0 & 0 \\ 5010 & 0 & 2040 & 0 & 0 & 0 & 0 & 0 \\ 0 & 2301 & 0 & 2004 & 0 & 0 & 0 & 0 \\ 4200 & 0 & 1230 & 0 & 0 & 0 & 0 & 0 \\ 3210 & 0 & 0240 & 0 & 0 & 0 & 0 & 0 \\ 0 & 0501 & 0 & 0204 & 0 & 0 & 0 & 0 \\ 4020 & 0 & 1050 & 0 & 0 & 0 & 0 & 0 \\ 0 & 0420 & 0 & 0123 & 0 & 0 & 0 & 0 \\ 0 & 0321 & 0 & 0024 & 0 & 0 & 0 & 0 \\ 4002 & 0 & 1032 & 0 & 0 & 0 & 0 & 0 \\ 0 & 0402 & 0 & 0105 & 0 & 0 & 0 & 0 \\ 3012 & 0 & 0042 & 0 & 0 & 0 & 0 & 0 \\ 0 & 1410 & 0 & 1113 & 2220 & 0 & 0 & 0 \\ 4101 & 0 & 1131 & 0 & 0 & 2202 & 0 & 0 \\ 0 & 1311 & 0 & 1014 & 2121 & 0 & 2022 & 0 \\ 3111 & 0 & 0141 & 0 & 0 & 1212 & 0 & 0222 \end{pmatrix}$$ (23.29)

23.4 $N = 3$

Systems for which $N = 3$ may contain any number of round lenses, together with elements of a new kind; these latter may consist of a diaphragm with a triangular opening, for example, or of a symmetric sextupole. Such a device has been employed by Amboss (1959) in an attempt to combat 'anticoma'. If the system does contain round lenses, then the first-order properties will be those of an ordinary rotationally symmetric system and the primary aberrations will be due to the three-fold symmetric element alone. If the system consists only of sextupolar elements, the primordial properties will be due to $S^{(3)}$ and the primary aberrations to $S^{(4)}$.

When both sextupole elements and round lenses are present, the image-forming properties of the system are described by

$$w_c = (1000)w_o + (0010)w_a \tag{23.30}$$

in which (1000) is real, and equal to $g(z)$, and (0010), also real, is equal to $h(z)$.

The component $S^{(3)}$ of S is of the form

$$S^{(3)} = \begin{pmatrix} w_o^2 \\ \overline{w}_o^2 \\ w_a^2 \\ \overline{w}_a^2 \end{pmatrix}^T \begin{pmatrix} 3000 & 0 & 2010 & 0 \\ 0 & \overline{3000} & 0 & \overline{2010} \\ 1020 & 0 & 0030 & 0 \\ 0 & \overline{1020} & 0 & \overline{0030} \end{pmatrix} \begin{pmatrix} w_o \\ \overline{w}_o \\ w_a \\ \overline{w}_a \end{pmatrix} \tag{23.31}$$

so that

$$\begin{aligned} Dw_c^I &= h\frac{\partial S_{ac}^{(3)}}{\partial \overline{w}_o} - g\frac{\partial S_{oc}^{(3)}}{\partial \overline{w}_a} \\ &= h\{3[\overline{3000}]\overline{w}_o^2 + 2[\overline{2010}]\overline{w}_o\overline{w}_a + [\overline{1020}]\overline{w}_a^2\} \\ &\quad - g\{(\overline{2010})\overline{w}_o^2 + 2(\overline{1020})\overline{w}_o\overline{w}_a + (\overline{0030})\overline{w}_a^2\} \end{aligned} \tag{23.32}$$

In the Gaussian image plane, $h(z)$ vanishes. In any plane, however, there are three primary aberrations: a distortion, $(0200)\overline{w}_o^2$, where

$$(0200) = \frac{h}{D}3[\overline{3000}] - \frac{g}{D}(\overline{2010}) \tag{23.33a}$$

an aperture aberration, $(0002)\overline{w}_a^2$,

$$(0002) = \frac{h}{D}[\overline{1020}] - \frac{g}{D}3(\overline{0030}) \tag{23.33b}$$

and an astigmatism, $(0101)\overline{w}_o\overline{w}_a$,

$$(1010) = \frac{h}{D}2[\overline{2010}] - \frac{g}{D}2(\overline{1020}) \qquad (23.33c)$$

We have so far made no assumptions about the alignment of the triangular elements, and the aberration coefficients can therefore all be complex. If there is only one such element, however, or if corresponding points of different elements all lie in the same meridian plane, then all the coefficients will be imaginary, provided the y-axes (say) lie in this plane.

The secondary aberrations will be the same as the primary aberrations of the round lenses, in nature at least; their values will be modified, however, by the presence of the second-order (primary) aberrations. This property has been exploited to correct the spherical aberration of round lenses (see Chapter 41).

If no round lenses are present, we obtain a system from which all the familiar characteristics of a lens system have vanished. The primordial effect is no longer a combination of anisotropic magnification, defocusing and astigmatism, as in the most general cases of $N = 1$ and $N = 2$ systems. Instead we find

$$w_c = (0200)\overline{w}_o^2 + (0002)\overline{w}_a^2 + (0101)\overline{w}_o\overline{w}_a \qquad (23.34)$$

so that even if (0002) and (0101) can be reduced simultaneously to zero, the magnification is not linear.

The primary aberrations will be third order and the secondary aberrations fourth.

23.5 $N = 4$

Apart from rotationally symmetric lenses, the system may contain any number of electrostatic or magnetic octopoles, in any orientation. The Gaussian imagery is identical with that of ordinary round systems, and the primary aberrations are now third order. The corresponding component of S, namely $S^{(4)}$, is most compactly written as the Hermitian form

$$S^{(4)} = \begin{pmatrix} w_o^2 \\ \overline{w}_o^2 \\ w_a^2 \\ \overline{w}_a^2 \\ w_o w_a \end{pmatrix}^T \begin{pmatrix} 2200 & 4000 & 2002 & 2020 & 2101 \\ \overline{4000} & 0 & 0 & 0 & 3010 \\ \overline{2002} & 0 & 0022 & 0040 & \overline{1012} \\ \overline{2020} & 0 & \overline{0040} & 0 & 1030 \\ \overline{2101} & \overline{3010} & 1012 & \overline{1030} & 1111 \end{pmatrix} \begin{pmatrix} \overline{w}_o^2 \\ w_o^2 \\ \overline{w}_a^2 \\ w_a^2 \\ \overline{w}_o \overline{w}_a \end{pmatrix}$$

$$(23.35)$$

The aberrations in a general plane are therefore given by

$$Dw_c^I = h_c \frac{\partial S_{ac}^{(4)}}{\partial w_o} - g_c \frac{\partial S_{oc}^{(4)}}{\partial \overline{w}_a}$$

$$= h \begin{pmatrix} w_o^2 \\ \overline{w}_o^2 \\ w_a^2 \\ \overline{w}_a^2 \\ w_o w_a \end{pmatrix}' \begin{pmatrix} 2[2200] & [2101] \\ 4[\overline{4000}] & 3[\overline{3010}] \\ 2[\overline{2002}] & [\overline{1012}] \\ 2[\overline{2020}] & [\overline{1030}] \\ 2[\overline{2101}] & [\overline{1111}] \end{pmatrix} \begin{pmatrix} \overline{w}_o \\ \overline{w}_a \end{pmatrix}$$

$$- g \begin{pmatrix} w_o^2 \\ \overline{w}_o^2 \\ w_a^2 \\ \overline{w}_a^2 \\ w_o w_a \end{pmatrix}' \begin{pmatrix} (2101) & 2(2002) \\ (\overline{3010}) & 2(\overline{2020}) \\ (\overline{1012}) & 2(0022) \\ 3(\overline{1030}) & 4(\overline{0040}) \\ (\overline{1111}) & 2(\overline{1012}) \end{pmatrix} \begin{pmatrix} \overline{w}_o \\ \overline{w}_a \end{pmatrix} \tag{23.36}$$

There are therefore two aperture aberrations: $(0003)\overline{w}_a^3$ and the ordinary spherical aberration of round lenses, $(0021)w_a^2\overline{w}_a$; two distortions: $(0300)\overline{w}_o^3$ and the round lens distortion, $(2100)w_o^2\overline{w}_o$; two astigmatisms: $(0201)\overline{w}_o^2\overline{w}_a$ and the round lens astigmatism, $(2001)w_o^2\overline{w}_a$; the round lens coma terms $(1011)w_ow_a\overline{w}_a$ and $(0120)\overline{w}_o^2 w_a$ together with an 'anticoma' term $(0102)\overline{w}_o\overline{w}_a^2$; and finally, the round lens field curvature, $(1110)w_o\overline{w}_ow_a$.

If all the elements are aligned in such a way that the system possesses a plane of symmetry—this implies that one electrode of each electrostatic element lies in a single azimuthal plane, and a polepiece of each magnetic element in a plane which is inclined to the electrode plane at $22\frac{1}{2}°$—all the elements of the coefficient matrix will be real.

The secondary aberrations are fifth order. $S^{(6)}$ can be written as a Hermitian form:

$$S^{(6)} = \begin{pmatrix} w_o^3 \\ w_a^3 \\ w_o\overline{w}_o^2 \\ w_o\overline{w}_a^2 \\ \overline{w}_o^2 w_a \\ w_a\overline{w}_a^2 \\ w_o\overline{w}_o\overline{w}_a \\ \overline{w}_ow_a\overline{w}_a \end{pmatrix}^T A \begin{pmatrix} \overline{w}_o^3 \\ \overline{w}_a^3 \\ w_o^2\overline{w}_o \\ \overline{w}_ow_a^2 \\ w_o^2\overline{w}_a \\ w_a^2\overline{w}_a \\ w_o\overline{w}_ow_a \\ w_ow_a\overline{w}_a \end{pmatrix} \tag{23.37}$$

23. RELATION BETWEEN ABERRATION AND SYSTEM SYMMETRY

where

$$A = \begin{pmatrix} 3300(r) & 3003 & 5100 & 3120 & 5001 & 3021 & 4110 & 4001 \\ \overline{3003} & 0033(r) & 2130 & \overline{1005} & 2031 & 0051 & 1140 & 1041 \\ \overline{5100} & \overline{2130} & 0 & \overline{3102} & 3201 & \overline{2112} & 0 & 0 \\ \overline{3120} & 1005 & 3102 & 1122(r) & 0 & 1023 & 0 & 2013 \\ 5001 & 2031 & 3201 & 0 & 2211(r) & 0 & 0 & 0 \\ 3021 & 0051 & 2112 & 1023 & 0 & 0 & 0 & 0 \\ 4110 & 1140 & 0 & 0 & 0 & 0 & 0 & 0 \\ \overline{4011} & \overline{1041} & 0 & \overline{2013} & 0 & 0 & 0 & 0 \end{pmatrix}$$

The derivatives of $S^{(6)}$ with respect to $\overline{w}_o$ and $\overline{w}_a$ can each be written in the form

$$\begin{pmatrix} w_o^3 \\ w_a^3 \\ w_o \overline{w}_o^2 \\ w_o \overline{w}_a^2 \\ \overline{w}_o^2 w_a \\ w_a \overline{w}_a^2 \\ w_o \overline{w}_o \overline{w}_a \\ \overline{w}_o w_a \overline{w}_a \end{pmatrix}^T \Upsilon \begin{pmatrix} w_o^2 \\ \overline{w}_o^2 \\ w_a^2 \\ \overline{w}_a^2 \\ w_o w_a \end{pmatrix} \qquad (23.38)$$

For $\partial S_{ac}^{(6)}/\partial \overline{w}_o$,

$$\Upsilon = \begin{pmatrix} [5100] & 3[3300] & [3120] & [3102] & [4110] \\ [2130] & 3[3003] & \overline{[1005]} & \overline{[1023]} & [1140] \\ 0 & 5[5100] & 3[3102] & 3[3120] & 3[3201] \\ 0 & 0 & [1122] & \overline{[1140]} & [2112] \\ 0 & 5[5001] & 0 & 3[3021] & 0 \\ 0 & 0 & 0 & \overline{[1041]} & 0 \\ 2[3201] & 4[4110] & 2[2112] & 2[2130] & 2[2211] \\ 0 & 4\overline{[4011]} & 2\overline{[2013]} & 2[2031] & 0 \end{pmatrix}$$

and for $\partial S_{oc}^{(6)}/\partial \overline{w}_a$,

$$\Upsilon = \begin{pmatrix} (5001) & (3201) & (3021) & 3(3003) & (4011) \\ (2031) & \overline{(2013)} & (0051) & 3(0033) & (1041) \\ 0 & \overline{(4110)} & \overline{(2112)} & 3\overline{(2130)} & (2211) \\ 0 & 0 & 3(1023) & 5(1005) & 3(2013) \\ 0 & \overline{(4011)} & 0 & 3\overline{(2031)} & 0 \\ 0 & 0 & 0 & 5\overline{(0051)} & 0 \\ 2(3102) & 2\overline{(3120)} & 2(1122) & 4\overline{(1140)} & 2(2112) \\ 0 & 2\overline{(3021)} & 2\overline{(1023)} & 4\overline{(1041)} & 0 \end{pmatrix} \qquad (23.39)$$

These aberrations are all members of the class $N = 2$, and we shall therefore discuss them no further.

23.6 $N = 5$ and 6

(a) $N = 5$. A system possessing this symmetry would consist of 'decapole' elements: these could be produced in the form of ordinary lenses with pentagonal openings instead of round ones, or as elements with ten poles (or electrodes) symmetrically disposed about the axis. Round lenses could of course be present also, and since the function of the decapole element would probably be to correct or diminish aberrations, the primordial properties and primary aberrations of any practical $N = 5$ system would be most likely to be those of a round lens system. The secondary aberrations would then be fourth order, due to the decapole unit, and just as the (primordial) astigmatism of a stigmator is used to annul the third-order astigmatism of round lens systems, it might be possible to use these fourth-order aberrations to combat either the fifth-order aberrations of an axially symmetric system, or the appropriate mechanical aberrations due to constructional shortcomings of the system.

We shall not discuss this case in any detail; we simply state that the component $S^{(5)}$ which leads to these fourth-order aberrations is of the form

$$S^{(5)} = \begin{pmatrix} w_o^4 \\ w_o^2 \overline{w}_a^2 \\ w_a^4 \\ \overline{w}_o^4 \\ \overline{w}_o \overline{w}_a^2 \\ \overline{w}_a^4 \end{pmatrix}^T \begin{pmatrix} 5000 & 4010 & 0 & 0 \\ 3020 & 2030 & 0 & 0 \\ 1040 & 0050 & 0 & 0 \\ 0 & 0 & 5000 & 4010 \\ 0 & 0 & 3020 & 2030 \\ 0 & 0 & 1040 & 0050 \end{pmatrix} \begin{pmatrix} w_o \\ w_a \\ \overline{w}_o \\ \overline{w}_a \end{pmatrix} \qquad (23.40)$$

so that

$$\partial S^{(5)}_{ac}/\partial \overline{w}_o = 5[\overline{5000}]w_o^4 + 4[\overline{4010}]w_o^3 w_a + 3[\overline{3020}]w_o^2 w_a^2$$
$$+ 2[\overline{2030}]w_o w_a^3 + [\overline{1040}]w_a^4$$

and
$$\qquad (23.41)$$

$$\partial S^{(5)}_{oc}/\partial \overline{w}_a = (\overline{4010})w_o^4 + 2(\overline{3020})w_o^3 w_a + 3(\overline{2030})w_o^2 w_a^2$$
$$+ 4(\overline{1040})w_o w_a^3 + 5(\overline{0050})w_a^4$$

(b) $N = 6$. The same general remarks also apply to these systems; the aberrations are the same as certain of those which afflict systems for which $N = 3$.

23.7 Systems with an axis of rotational symmetry

The primordial properties and primary aberrations of these systems have been very thoroughly studied. The primary aberrations, which are third order, are most simply derived from the Hermitian form

$$S^{(4)} = \begin{pmatrix} \overline{w}_o^2 \\ \overline{w}_o \overline{w}_a \\ \overline{w}_a^2 \end{pmatrix}^T \begin{pmatrix} 2200(r) & \overline{2101} & \overline{2002} \\ 2101 & 1111(r) & \overline{1012} \\ 2002 & 1012 & 0022(r) \end{pmatrix} \begin{pmatrix} w_o^2 \\ w_o w_a \\ w_a^2 \end{pmatrix} \quad (23.42)$$

The secondary aberrations are fifth order; the function $S^{(6)}$ can be written

$$S^{(6)} = \begin{pmatrix} \overline{w}_o^3 \\ \overline{w}_o^2 \overline{w}_a \\ \overline{w}_o \overline{w}_a^2 \\ \overline{w}_a^3 \end{pmatrix}^T \begin{pmatrix} 3300(r) & \overline{3201} & \overline{3102} & \overline{3003} \\ 3201 & 2211(r) & \overline{2112} & \overline{2013} \\ 3102 & 2112 & 1122(r) & \overline{1023} \\ 3003 & 2013 & 1023 & 0033(r) \end{pmatrix} \begin{pmatrix} w_o^3 \\ w_o^2 w_a \\ w_o w_a^2 \\ w_a^3 \end{pmatrix}$$

$$(23.43)$$

so that

$$\frac{\partial S_{ac}^{(6)}}{\partial \overline{w}_o} = \begin{pmatrix} \overline{w}_o^2 \\ \overline{w}_o \overline{w}_a \\ \overline{w}_a^2 \end{pmatrix}^T \begin{pmatrix} 3[3300] & 3[\overline{3201}] & 3[\overline{3102}] & 3[\overline{3003}] \\ 2[3201] & 2[2211] & 2[\overline{2112}] & 2[\overline{2013}] \\ [3102] & [2112] & [1122] & [\overline{1023}] \end{pmatrix} \begin{pmatrix} w_o^3 \\ w_o^2 w_a \\ w_o w_a^2 \\ w_a^3 \end{pmatrix}$$

$$(23.44a)$$

and

$$\frac{\partial S_{oc}^{(6)}}{\partial \overline{w}_a} = \begin{pmatrix} \overline{w}_o^2 \\ \overline{w}_o \overline{w}_a \\ \overline{w}_a^2 \end{pmatrix}^T \begin{pmatrix} (3201) & (2211) & (\overline{2112}) & (\overline{2013}) \\ 2(3102) & 2(2112) & 2(1122) & 2(\overline{1023}) \\ 3(3003) & 3(2013) & 3(1023) & 3(0033) \end{pmatrix} \begin{pmatrix} w_o^3 \\ w_o^2 w_a \\ w_o w_a^2 \\ w_a^3 \end{pmatrix}$$

$$(23.44b)$$

In the stigmatic image plane, the contribution to the secondary aberrations that arises from $S^{(6)}$ contains the following terms:

$$\text{distortion: } (3200) = -\frac{(3201)}{Ah_i'}$$

$$\text{aperture aberration: } (0032) = -3\frac{(0033)}{Ah_i'}, \text{ real}$$

23.7 SYSTEMS WITH AN AXIS OF ROTATIONAL SYMMETRY

astigmatism and field curvature: $(2210) = -\dfrac{(2211)}{Ah'_i}$, real

$$(3101) = -2\dfrac{(3102)}{Ah'_i}$$

comas: $(1220) = -\dfrac{\overline{(2112)}}{Ah'_i}$

$$(2111) = -2\dfrac{(2112)}{Ah'_i}$$

$$(3002) = -3\dfrac{(3003)}{Ah'_i}$$

terms in r_a^3: $(0230) = -\dfrac{\overline{(2013)}}{Ah'_i}$

$(1121) = -2\dfrac{(1122)}{Ah'_i}$, real

$$(2012) = -3\dfrac{(2013)}{Ah'_i}$$

terms in r_a^4: $(0131) = -2\dfrac{\overline{(1023)}}{Ah'_i}$

$$(1022) = -3\dfrac{(1023)}{Ah'_i} \qquad (23.45)$$

In an electrostatic system, all the coefficients are real, and the aberrations resemble those of glass lenses. If, however, the system is magnetic, 'anisotropic aberrations' also appear as we should expect, by which we mean that certain of the aberration coefficients are complex numbers.

Consider, for example, the terms in the first power of the aperture coordinates, (2210) and (3101), the astigmatisms:

$$w_c^{II} = (2210)w_o^2\overline{w}_o^2 w_a + (3101)w_o^3\overline{w}_o\overline{w}_a$$

or

$$w_c^{II}e^{i\phi} = r_o^4 r_a\{(2210)e^{i\Theta_a} + |3101|e^{-i\Theta_a}\}$$

in which

$$\phi = -(\tfrac{1}{2}\widetilde{3101} + \theta_o) \text{ and } \Theta_a = \theta_a - (\tfrac{1}{2}\widetilde{3101} + \theta_o)$$

which represents a tilted ellipse, with semi-axes $r_o^4 r_a\{(2210)\pm |\ 3101\ |\}$.

Likewise, the comas produce an overall effect

$$w_c^{II} = Aw_o\overline{w}_o^2 w_a^2 + 2\overline{A}w_o^2\overline{w}_o w_a\overline{w}_a + Bw_o^3\overline{w}_a^2$$

or
$$w_c^{II} e^{i\phi}/r_o^3 r_a^3 = 2\overline{A}e^{i(\theta_o+\phi)} + |A|e^{2i\Theta_a} + |B|e^{-i\Theta_a}$$

with
$$\phi = -\theta_o - \frac{1}{2}(\tilde{A}+\tilde{B}) \text{ and } \Theta_a = \theta_a - \theta_o + \frac{1}{4}(\tilde{A}-\tilde{B})$$

which represents a family of ellipses for different values of r_a, centred on the line
$$w_c^{II} = 2r_o^3 r_a^2 \overline{A} e^{i\theta_o}$$

with semi-axes $r_o^3 r_a^2(|A|\pm|B|)$.

The terms in r_a^4 give
$$w_c^{II} = 2A\overline{w}_o w_a^3 \overline{w}_a + 3\overline{A} w_o w_a^2 \overline{w}_a^2$$

or
$$w_c^{II}/r_o r_a^4 = 2|A|e^{2i\Theta_a} + 3\overline{A}e^{i\theta_o}$$

in which $\Theta_a = \theta_a + \frac{1}{2}(\tilde{A}-\theta_o)$, which represents a family of circles, centred on the line $w_c^{II} = 3r_o r_a^4 \overline{A} e^{i\theta_o}$, radii $2r_o r_a^4 \,|\,A\,|$.

Finally, the terms in r_a^3 lead to
$$w_c^{II} = 3Aw_o^2 w_a \overline{w}_a^2 + \overline{A}\overline{w}_o^2 w_a^3 + Bw_o \overline{w}_o w_a^2 \overline{w}_a$$

or
$$w_c^{II}/r_o^2 r_a^3 = 3|A|e^{i(\tilde{A}+2\theta_o-2\theta_a)} + |A|e^{i(-\tilde{A}-2\theta_o+3\theta_a)} + Be^{i\theta_a}$$

The terms in $|\,A\,|$ can be written as
$$e^{-i\phi}\left\{3|A|e^{-i\Theta_a} + |A|e^{3i\Theta_a}\right\}$$

in which
$$\phi = -\frac{1}{2}\tilde{A} - \theta_o \text{ and } \Theta_a = \theta_a - \theta_o - \frac{1}{2}\tilde{A}$$

and if we write $v = w_c^{II} e^{i\phi}/r_o^2 r_a^3\,|\,A\,|= x' + iy'$, we have
$$x' = 4\cos^3\theta, \qquad y' = -4\sin^3\theta$$

which represents an astroid.

23.8 Note on the classification of aberrations

We see from the earlier parts of this chapter that although the number of aberration coefficients needed to characterize complex systems may be large, the aberrations form families, the importance of which is often very different in different components. In electron microscopes, for example, the objective lens suffers from the same aberrations as the final projector but for the former only the aperture aberrations are important, whereas for the latter all but the distortions are negligible. We therefore introduce a convenient nomenclature, which brings together members of the same family of aberrations for any optical element. Each aberration is associated with a term of the form $x_o^p y_o^q x_a^r y_a^s$ and in the case of stigmatic imagery and primary aberrations, these can be related back to particular terms of S^I.

(i) Terms independent of x_o, $y_o (p = q = 0)$: aperture aberrations

For these aberrations, the aberration figure is independent of the choice of object point and, in particular, their effects do not vanish or become small if the latter lies on the optic axis. They are known generically as *aperture defects* or *aperture aberrations* and include as important special cases the spherical aberration of round lenses and the aperture aberrations of quadrupoles. The defocus also belongs to this group, since an image formed on a plane that is not conjugate to the object plane is blurred in an aperture-dependent way. The axial astigmatism associated, for example, with imperfect roundness of a lens intended to be rotationally symmetric and the primary (axial) chromatic aberration are also aperture defects.

(ii) Terms independent of x_a, $y_a (r = s = 0)$: distortions

These are at the other extreme from aperture aberrations, for they depend only on the position of the object point. Hence all rays from a given object point will be displaced by the same amount from their paraxial points of arrival and, if the latter lie in a stigmatic image plane, such terms will shift the paraxial image point but not blur it. These aberrations are known as distortions and include the isotropic and anisotropic distortions of round lenses and the chromatic aberrations other than the axial term.

(iii) Intermediate terms

The intermediate terms may be classified into those linear in x_o or y_o; those linear in x_a and y_a; and any others. The former may be regarded as comas, by analogy with the primary coma of round lenses. Terms linear in x_a and y_a form the group of field curvatures and astigmatism. The others, which occur when we consider fourth and higher order aberrations, have no particular name.

(iv) Phase shifts

If we are considering a pair of conjugate planes in an imaging system, any terms in S^I that are independent of x_a and y_a will not contribute to the aberrations, which are determined by $\partial S^I/\partial x_a$ and $\partial S^I/\partial y_a$. Such terms are important when we consider the dependence of the aberration coefficients on object and aperture position in Chapter 25 and will reappear in Vol. 3 in connection with the wave theory of aberrations.

24
The Geometrical Aberrations of Round Lenses

24.1 Introduction

We now obtain explicit formulae for the aberration coefficients of systems of round magnetic and electrostatic lenses in which the fields may overlap. In order to avoid repetition, we deal with the real aberrations expressed in terms of object and aperture coordinates in detail, then describe briefly the changes necessary to convert these coefficients when position and gradient in the object plane are used. We then show (in Chapter 25) how a similar calculation yields the asymptotic aberrations. These calculations will be based on the eikonal method but the main stages in the reasoning when the trajectory method is employed are also given.

Within this overall plan, certain decisions remain to be taken: the most important concerns the use of reduced coordinates (15.40). If these are employed, as they are by Sturrock (1955), the expressions for the various coefficients are very different in form from those obtained in conventional coordinates, so different indeed that Glaser (1952) believed them to be wrong. (In fact, they can be shown to be equivalent but the demonstration is undeniably laborious.) We have chosen to use conventional coordinates but we list one form of the coefficients in reduced coordinates for reference purposes (Section 24.8).

The other decision concerns the use of complex coordinates. There is no doubt that these give the analysis a more compact appearance, but expressions involving the third and fourth powers of complex coordinates are much less easy to picture than those involving first and second powers, which we have met in the paraxial domain. We therefore retain a certain flexibility, giving some important formulae in both forms; we avoid repeating very similar equations for x and y by the use of suffix notation, as in Chapter 22.

24.2 Derivation of the real aberration coefficients

The paraxial ray equations for round lenses were obtained in Section 15.2 by writing down the Euler equations corresponding to the variational relation

24. THE GEOMETRICAL ABERRATIONS OF ROUND LENSES

$\delta \int M^{(2)} dz = 0$. If we retain the next higher order terms, $M^{(4)}$, in the series expansion for M and again write down the Euler equations, we obtain the inhomogeneous second-order equations from which the aberrations are obtained in the trajectory method; alternatively, we may set $M^{(p)} = M^{(4)}$ and use the perturbation theory of Chapter 22. In both cases, we need $M^{(4)}$. Substituting terms of higher order for Φ and A_x, A_y into (15.23), we find

$$\begin{aligned} M^{(4)} = &\frac{1}{128\hat{\phi}^{\frac{1}{2}}}\left(\gamma\phi^{(4)} - \frac{\phi''^2}{\hat{\phi}}\right)(X^2 + Y^2)^2 \\ &- \frac{\gamma\phi''}{16\hat{\phi}^{\frac{1}{2}}}(X^2 + Y^2)(X'^2 + Y'^2) \\ &- \frac{\hat{\phi}^{\frac{1}{2}}}{8}(X'^2 + Y'^2) \\ &+ \frac{\eta B''}{16}(X^2 + Y^2)(XY' - X'Y) \end{aligned} \quad (24.1)$$

and introducing the rotating coordinate system (x, y, z) with the aid of (15.26)

$$\begin{aligned} M^{(4)} = &-\frac{1}{4}L_1(x^2 + y^2)^2 - \frac{1}{2}L_2(x^2 + y^2)(x'^2 + y'^2) \\ &- \frac{1}{4}L_3(x'^2 + y'^2)^2 - R(xy' - x'y)^2 \\ &- P\hat{\phi}^{\frac{1}{2}}(x^2 + y^2)(xy' - x'y) \\ &- Q\hat{\phi}^{\frac{1}{2}}(x'^2 + y'^2)(xy' - x'y) \end{aligned} \quad (24.2)$$

in which

$$\begin{aligned} L_1 &= \frac{1}{32\hat{\phi}^{\frac{1}{2}}}\left(\frac{\phi''^2}{\hat{\phi}} - \gamma\phi^{(4)} + \frac{2\gamma\phi''\eta^2 B^2}{\hat{\phi}}\right. \\ &\left.\quad + \frac{\eta^4 B^4}{\hat{\phi}} - 4\eta^2 BB''\right) \\ L_2 &= \frac{1}{8\hat{\phi}^{\frac{1}{2}}}(\gamma\phi'' + \eta^2 B^2) \\ L_3 &= \frac{1}{2}\hat{\phi}^{\frac{1}{2}} \\ P &= \frac{\eta}{16\hat{\phi}^{\frac{1}{2}}}\left(\frac{\gamma\phi'' B}{\hat{\phi}} - B'' + \frac{\eta^2 B^3}{\hat{\phi}}\right) \\ Q &= \frac{\eta B}{4\hat{\phi}^{\frac{1}{2}}} \\ R &= \frac{\eta^2 B^2}{8\hat{\phi}^{\frac{1}{2}}} \end{aligned} \quad (24.3)$$

24.2 DERIVATION OF THE REAL ABERRATION COEFFICIENTS

Note that in purely electrostatic lenses, P, Q and R vanish, and that in purely magnetic lenses, $L_2 = R$; Q is half the rate of rotation (15.9) of the rotating frame (x, y, z).

The trajectory method

We now write

$$\delta \int (M^{(2)} + M^{(4)}) \, dz = 0 \qquad (24.4)$$

or

$$\frac{d}{dz}\left(\frac{\partial M^{(2)}}{\partial x'_j}\right) - \frac{\partial M^{(2)}}{\partial x_j} = -\frac{d}{dz}\left(\frac{\partial M^{(4)}}{\partial x'_j}\right) + \frac{\partial M^{(4)}}{\partial x_j} \qquad (24.5)$$

Here and elsewhere, $j = 1, 2$; $x_1 = x$ and $x_2 = y$. Substituting for $M^{(2)}$ and $M^{(4)}$, we find

$$\frac{d}{dz}(\hat{\phi}^{\frac{1}{2}} x'_j) + \frac{\gamma \phi'' + \eta^2 B^2}{4 \hat{\phi}^{\frac{1}{2}}} x_j = \Lambda_j \qquad (24.6)$$

where

$$\Lambda_1 := (x^2 + y^2)\{-L_1 x + L'_2 x' + L_2 x'' - (P\phi^{\frac{1}{2}})' y - 2P\phi^{\frac{1}{2}} y'\}$$
$$+ (x'^2 + y'^2)\{-L_2 x + L'_3 x' + L_3 x'' - (Q\phi^{\frac{1}{2}})' y - 2Q\phi^{\frac{1}{2}} y'\}$$
$$+ 2(xy' - x'y)\{-P\phi^{\frac{1}{2}} x + (Q\phi^{\frac{1}{2}})' x' + Q\phi^{\frac{1}{2}} x'' - R'y - 2Ry'\}$$
$$+ 2(xx' + yy')(L_2 x' - P\phi^{\frac{1}{2}} y)$$
$$+ 2(x'x'' + y'y'')(L_3 x' - Q\phi^{\frac{1}{2}} y)$$
$$+ 2(xy'' - x''y)(Q\phi^{\frac{1}{2}} x' - 2Ry)$$

$$(24.7)$$

and Λ_2 is obtained by writing $x \to y$, $y \to -x$. Equation (24.6) is solved by replacing x and y by their paraxial expressions on the right-hand side, which thereby becomes a known function of z, and employing the method of variation of parameters. Suppose that $a(z)$ and $b(z)$ are two linearly independent solutions of the homogeneous equation, which is also the paraxial equation. We seek a solution of the inhomogeneous equation (24.6) of the form

$$x_j(z) = A_j(z) a(z) + B_j(z) b(z) \qquad (24.8)$$

Selecting $A_j(z)$ and $B_j(z)$ in such a way that

$$A'_j a + B'_j b = 0 \qquad (24.9)$$

and substituting (24.8) into (24.6), we obtain

$$\hat{\phi}^{\frac{1}{2}}(A'_j a' + B'_j b') = \Lambda_j \tag{24.10}$$

Solving (24.9, 24.10) for A'_j and B'_j yields

$$A'_j = -\frac{\Lambda_j b}{\hat{\phi}^{\frac{1}{2}}(ab' - a'b)} = -\frac{\Lambda_j b}{W}$$
$$B'_j = \frac{\Lambda_j a}{\hat{\phi}^{\frac{1}{2}}(ab' - a'b)} = \frac{\Lambda_j a}{W} \tag{24.11}$$

where W denotes the constant Wronskian (15.51)

$$W = \hat{\phi}^{\frac{1}{2}}(ab' - a'b) \tag{24.12}$$

Hence

$$A_j(z) = A_j(z_1) - \frac{1}{W}\int_{z_1}^{z} \Lambda_j b\, d\zeta$$
$$B_j(z) = B_j(z_2) + \frac{1}{W}\int_{z_2}^{z} \Lambda_j a\, d\zeta \tag{24.13}$$

We have retained the possibility of using different lower limits in the integrals appearing in (24.13) because we need to study two sets of boundary conditions; when these govern ray position in two different planes, this extra flexibility is indispensable.

If the aberrations are expressed in terms of position coordinates in two planes, the object and image planes $z = z_o$ and $z = z_a$, then the aberration terms vanish there. Introducing the paraxial solutions $s(z)$, $t(z)$,

$$\begin{array}{ll} a(z) \Rightarrow s(z), & s(z_o) = t(z_a) = 1 \\ b(z) \Rightarrow t(z), & s(z_a) = t(z_o) = 0 \end{array} \tag{24.14}$$

we find

$$x_j(z) = A_j(z_1)s(z) + B_j(z_2)t(z)$$
$$- \frac{s(z)}{W_s}\int_{z_1}^{z} \Lambda_j t\, d\zeta + \frac{t(z)}{W_s}\int_{z_2}^{z} \Lambda_j s\, d\zeta \tag{24.15}$$

where $W_s := \hat{\phi}^{\frac{1}{2}}(st' - s't) = \hat{\phi}_o^{\frac{1}{2}} t'_o = -\hat{\phi}_a^{\frac{1}{2}} s'_a$ is the Wronskian for the choice (24.14). In $z = z_o$, $x_j = x_{jo}$ so that $A_j(z_1) = x_{jo}$, $z_1 = z_o$. In $z = z_a$,

24.2 DERIVATION OF THE REAL ABERRATION COEFFICIENTS

$x_j = x_{ja}$ and hence $B_j(z_2) = x_{ja}$, $z_2 = z_a$. Finally,

$$x_j(z) = x_{jo} s(z) + x_{ja} t(z)$$
$$+ \frac{1}{W_s} \left\{ t(z) \int_{z_a}^{z} \Lambda_j s \, d\zeta - s(z) \int_{z_o}^{z} \Lambda_j t \, d\zeta \right\} \quad (24.16)$$

In the image plane, $z = z_i$, $t(z)$ is again zero and $s(z_i) = M$. It is usual to refer the aberrations back to the object plane by considering the quantity $\{ x_j(z_i) - M x_{jo} \} / M$:

$$\frac{x_j(z_i) - M x_{jo}}{M} =: \Delta x_{ji} = -\frac{1}{W_s} \int_{z_o}^{z_i} \Lambda_j t \, dz \quad (24.17)$$

The individual aberration coefficients are then extracted by substituting $x_j(z) = x_{jo} s(z) + x_{ja} t(z)$ in the Λ_j and collecting terms of various degree in x_o, y_o, x_a and y_a. We shall perform this step in detail below when using the eikonal method.

If the aberrations are expressed in terms of position and gradient in the object plane, we use the pair of paraxial solutions denoted by $g(z)$ and $h(z)$ (15.56):

$$\begin{aligned} a(z) &\Rightarrow g(z), & g(z_o) &= h'(z_o) = 1 \\ b(z) &\Rightarrow h(z), & g'(z_o) &= h(z_o) = 0 \end{aligned} \quad (24.18)$$

and now

$$x_j(z) = A_j(z_1) g(z) + B_j(z_2) h(z)$$
$$+ \frac{1}{W_g} \left\{ -g(z) \int_{z_1}^{z} \Lambda_j h \, d\zeta + h(z) \int_{z_2}^{z} \Lambda_j g \, d\zeta \right\} \quad (24.19)$$

where $W_g := \hat{\phi}^{\frac{1}{2}}(gh' - g'h) = \hat{\phi}_o^{\frac{1}{2}}$ is the Wronskian for (24.18). In $z = z_o$, $x_j(z_o) = x_{jo}$ and $x'_j(z_o) = x'_{jo}$ so that as before $A_j(z_1) = x_{jo}$, $z_1 = z_o$; now, however, z_2 is also equal to z_o and $B_j(z_2) = x'_{jo}$:

$$x_j(z) = x_{jo} g(z) + x'_{jo} h(z)$$
$$+ \frac{1}{W_g} \left\{ h(z) \int_{z_o}^{z} \Lambda_j g \, d\zeta - g(z) \int_{z_o}^{z} \Lambda_j h \, d\zeta \right\} \quad (24.20)$$

or in the image plane, $h(z_i) = 0$,

$$\frac{x_{ji} - Mx_{jo}}{M} = \Delta x_{ji} = -\frac{1}{W_g}\int_{z_o}^{z_i} \Lambda_j h\, dz \qquad (24.21)$$

Despite their formal resemblance, there is an important difference between (24.17) and (24.21) or (24.16) and (24.19). Since both $t(z)$ and $h(z)$ vanish in the object and image planes, these paraxial solutions are proportional, $t(z) \propto h(z)$. The rays $s(z)$ and $g(z)$, though of course linearly related, are not proportional:

$$\begin{aligned} g(z) &= s(z) + g(z_a)t(z) = s(z) - \frac{s'(z_o)}{t'(z_o)}t(z) \\ s(z) &= g(z) + s'(z_o)h(z) = g(z) - \frac{g(z_a)}{h(z_a)}h(z) \end{aligned} \qquad (24.22)$$

With the aid of (24.17), therefore, the effect of aperture position on the aberration coefficients can be explored.

The eikonal method

Instead of using $M^{(4)}$ to derive (24.6), we can equally well use perturbation theory, regarding $M^{(2)}$ as the unperturbed refractive index and setting $M^{(4)} = M^{(p)}$. For round lenses, the formulae of Chapter 22 reduce to a much simpler form since $M^{(2)}$ contains terms in (x^2+y^2) and $(x'^2+y'^2)$ only (15.29) so that paraxially,

$$p_j = \hat{\phi}^{\frac{1}{2}} x'_j \qquad (24.23)$$

For the boundary conditions (22.24), we have $x_j = x_{jo}g + x'_{jo}h$ and hence

$$\begin{aligned} \frac{\partial x_j(z)}{\partial x_{ko}} &= g(z)\delta_{jk} & \frac{\partial x_j(z)}{\partial p_{ko}} &= \frac{h(z)}{\hat{\phi}_o^{\frac{1}{2}}}\delta_{jk} \\ \frac{\partial p_j(z)}{\partial x_{ko}} &= \hat{\phi}^{\frac{1}{2}}(z)g'(z)\delta_{jk} & \frac{\partial p_j(z)}{\partial p_{ko}} &= h'(z)\delta_{jk} \end{aligned} \qquad (24.24)$$

and (22.28) become

$$\begin{aligned} \frac{\partial S_{12}^I}{\partial x_{jo}} &= p_{j2}^{(1)} g(z_2) - x_{j2}^{(1)} \hat{\phi}^{\frac{1}{2}}(z_2) g'(z_2) \\ \hat{\phi}_o^{\frac{1}{2}} \frac{\partial S_{12}^I}{\partial p_{jo}} &= p_{j2}^{(1)} h(z_2) - x_{j2}^{(1)} \hat{\phi}^{\frac{1}{2}}(z_2) h'(z_2) \end{aligned} \qquad (24.25)$$

24.2 DERIVATION OF THE REAL ABERRATION COEFFICIENTS

Solving for $x_j^{(1)}(z_2)$ and $p_j^{(1)}(z_2)$, we find

$$W_g x_j^{(1)}(z_2) = h(z_2)\frac{\partial S_{12}^I}{\partial x_{jo}} - \hat{\phi}_o^{\frac{1}{2}} g(z_2)\frac{\partial S_{12}^I}{\partial p_{jo}}$$

$$W_g p_j^{(1)}(z_2) = \hat{\phi}^{\frac{1}{2}}(z_2)\left\{h'(z_2)\frac{\partial S_{12}^I}{\partial x_{jo}} - \hat{\phi}_o^{\frac{1}{2}} g'(z_2)\frac{\partial S_{12}^I}{\partial p_{jo}}\right\}$$

(24.26)

or using (24.23),

$$W_g x_j^{(1)}(z_2) = h(z_2)\frac{\partial S_{12}^I}{\partial x_{jo}} - g(z_2)\frac{\partial S_{12}^I}{\partial x'_{jo}} \qquad (24.27a)$$

$$\frac{W_g p_j^{(1)}(z_2)}{\hat{\phi}^{\frac{1}{2}}(z_2)} = h'(z_2)\frac{\partial S_{12}^I}{\partial x_{jo}} - g'(z_2)\frac{\partial S_{12}^I}{\partial x'_{jo}} \qquad (24.27b)$$

We have already mentioned that, provided that the plane $z = z_o$ lies in field-free space, the boundary condition $p_{jo}^{(1)} = 0$ is equivalent to $x_{jo}^{\prime(1)} = 0$; if $z = z_2$ likewise lies in field-free space, so that $p^{(1)}(z_2)/\hat{\phi}^{\frac{1}{2}}(z_2) = x'^{(1)}(z_2)$, then (24.27) give the aberrations of position and gradient in $z = z_2$ by writing the left-hand side of (24.27b) as $W_g x_j^{\prime(1)}(z_2)$.

If the aberrations are to be expressed in terms of position in the object and aperture planes, we use (22.30) with $x_j(z) = x_{jo}s(z) + x_{ja}t(z)$ so that

$$\frac{\partial S_{o2}^I}{\partial x_{ja}} = p_{j2}^{(1)} t(z_2) - x_{j2}^{(1)} \hat{\phi}_2^{\frac{1}{2}} t'(z_2)$$

$$\frac{\partial S_{a2}^I}{\partial x_{jo}} = p_{j2}^{(1)} s(z_2) - x_{j2}^{(1)} \hat{\phi}_2^{\frac{1}{2}} s'(z_2)$$

(24.28)

with solution

$$W_s x_j^{(1)}(z_2) = t(z_2)\frac{\partial S_{a2}^I}{\partial x_{jo}} - s(z_2)\frac{\partial S_{o2}^I}{\partial x_{ja}}$$

$$W_s p_j^{(1)}(z_2) = \hat{\phi}_2^{\frac{1}{2}}\left\{t'(z_2)\frac{\partial S_{a2}^I}{\partial x_{jo}} - s'(z_2)\frac{\partial S_{o2}^I}{\partial x_{ja}}\right\}$$

(24.29)

The final stage of the calculation involves substituting the paraxial solutions for $x_j(z)$ into $M^{(4)}$ and reorganizing the result into convenient groups of terms. We reintroduce x and y, setting

$$x_1(z) = x_o s(z) + x_a(z)$$
$$x_2(z) = y_o s(z) + y_a(z)$$

(24.30)

whereupon $M^{(4)}$ (24.2) takes the form

$$\begin{aligned}
M^{(4)} = &-\frac{1}{4}(L_1 s^4 + 2L_2 s^2 s'^2 + L_3 s'^4)(x_o^2 + y_o^2)^2 \\
&-\frac{1}{4}(L_1 t^4 + 2L_2 t^2 t'^2 + L_3 t'^4)(x_a^2 + y_a^2)^2 \\
&-\{L_1 s^2 t^2 + 2L_2 ss' tt' + L_3 s'^2 t'^2 - R(st' - s't)^2\}(x_o x_a + y_o y_a)^2 \\
&-\frac{1}{2}\{L_1 s^2 t^2 + L_2(s^2 t'^2 + s'^2 t^2) + L_3 s'^2 t'^2 \\
&\quad + 2R(st' - s't)^2\}(x_o^2 + y_o^2)(x_a^2 + y_a^2) \\
&-\{L_1 s^3 t + L_2 ss'(st)' + L_3 s'^3 t'\}(x_o^2 + y_o^2)(x_o x_a + y_o y_a) \\
&-\{L_1 st^3 + L_2(st)' tt' + L_3 s' t'^3\}(x_a^2 + y_a^2)(x_o x_a + y_o y_a) \\
&-W_s(Ps^2 + Qs'^2)(x_o^2 + y_o^2)(x_o y_a - x_a y_o) \\
&-2W_s(Pst + Qs't')(x_o x_a + y_o y_a)(x_o y_a - x_a y_o) \\
&-W_s(Pt^2 + Qt'^2)(x_a^2 + y_a^2)(x_o y_a - x_a y_o) \quad (24.31)
\end{aligned}$$

We write

$$A = \frac{1}{W}\int\left\{L_1 s^2 t^2 + 2L_2 ss' tt' + L_3 s'^2 t'^2 - R(st' - s't)^2\right\} dz$$

$$C = \frac{1}{W}\int\left(L_1 t^4 + 2L_2 t^2 t'^2 + L_3 t'^4\right) dz$$

$$D = \frac{1}{W}\int\left(L_1 s^3 t + L_2 ss'(st)' + L_3 s'^3 t'\right) dz$$

$$F = \frac{1}{W}\int\left\{2L_1 s^2 t^2 + L_2(st' + s't)^2 + 2L_3 s'^2 t'^2 + R(st' - s't)^2\right\} dz$$

$$E = \frac{1}{W}\int\left(L_1 s^4 + 2L_2 s^2 s'^2 + L_3 s'^4\right) dz$$

$$K = \frac{1}{W}\int\left\{L_1 st^3 + L_2(st)' tt' + L_3 s' t'^3\right\} dz$$

$$a = 2\int\left(Pst + Qs't'\right) dz$$

$$d = \int\left(Ps^2 + Qs'^2\right) dz$$

$$k = \int\left(Pt^2 + Qt'^2\right) dz \quad (24.32)$$

---

* The notation adopted here is not that widely encountered in the literature, notably in Glaser (GdE, HdP), but is mnemonically superior: we shall see that A and a are associated with astigmatism, C with spherical aberration, D and d with distortion, F with field curvature and K and k with coma.

24.2 DERIVATION OF THE REAL ABERRATION COEFFICIENTS

It is clear both on general symmetry grounds and from (24.31) that the quantities x_o, y_o, x_a and y_a can only occur in S^I as the rotationally invariant groups r_o, r_a, V and v, where

$$r_o^2 := x_o^2 + y_o^2 = u_o u_o^*$$
$$r_a^2 := x_a^2 + y_a^2 = u_a u_a^*$$
$$V := x_o x_a + y_o y_a = r_o r_a \cos(\varphi_a - \varphi_o) = \frac{1}{2}(u_o u_a^* + u_o^* u_a) \qquad (24.33a)$$
$$v := x_o y_a - x_a y_o = r_o r_a \sin(\varphi_a - \varphi_o) = \frac{i}{2}(u_o u_a^* - u_o^* u_a)$$

with $x_o =: r_o \cos\varphi_o$, $y_o =: r_o \sin\varphi_o$ and similarly for x_a, y_a; the complex numbers $u_o = x_o + iy_o$ and $u_a = x_a + iy_a$ are those defined in (15.11). It will also be convenient to write

$$\varphi := \varphi_a - \varphi_o \qquad (24.33b)$$

The quantities S_{a2}^I and S_{o2}^I have the following generic form, only the lower limit of integration changing:

$$-\frac{S^I}{W} = \frac{E}{4}r_o^4 + \frac{C}{4}r_a^4 + \frac{A}{2}(V^2 - v^2)$$
$$+ \frac{F}{2}r_o^2 r_a^2 + Dr_o^2 V + Kr_a^2 V$$
$$+ v(dr_o^2 + kr_a^2 + aV)$$
$$= \frac{1}{4}Cr_a^4 + r_a^3 r_o(K\cos\varphi + k\sin\varphi)$$
$$+ \frac{1}{2}r_a^2 r_o^2(A\cos 2\varphi + a\sin 2\varphi) + \frac{1}{2}Fr_a^2 r_o^2 \qquad (24.34)$$
$$+ r_a r_o^3(D\cos\varphi + d\sin\varphi) + \frac{1}{4}Er_o^2$$
$$= \begin{pmatrix} u_o^{*2} \\ u_o^* u_a^* \\ u_a^{*2} \end{pmatrix}^T \begin{pmatrix} E/4 & (D-id)/2 & (A-ia)/4 \\ (D+id)/2 & F/2 & (K-ik)/2 \\ (A+ia)/4 & (K+ik)/2 & C/4 \end{pmatrix} \begin{pmatrix} u_o^2 \\ u_o u_a \\ u_a^2 \end{pmatrix}$$

We add the suffix o to the coefficients A, C, ..., d, k when the integration runs from z_o to an arbitrary plane z_2; we add a when the lower limit is z_a; when no suffix is added, the integrals run from the object plane z_o to the image plane $z = z_i$.

In a general plane

$$\begin{aligned}x^{(1)}(z) = {} & r_o^2\{x_o(D_o s - E_a t) + x_a(F_o s - A_o s - D_a t) - y_o d_o s\} \\ & + r_a^2\{x_o(K_o s - F_a t + A_a t) + x_a(C_o s - K_a t) \\ & \quad - y_o(k_o s + a_a t) - y_a k_a t\} \\ & + V\{2x_o(A_o s - D_a t) + 2x_a(K_o s - A_a t) \\ & \quad - y_o(a_o s + d_a t)\} \\ & + v\{x_o(a_o s - 3d_a t) + 2x_a(k_o s - a_a t)\} \end{aligned} \qquad (24.35a)$$

$$\begin{aligned}y^{(1)}(z) = {} & r_o^2\{y_o(D_o s - E_a t) + y_a(F_o s - A_o s - D_a t) + x_o d_o s\} \\ & + r_a^2\{y_o(K_o s - F_a t + A_a t) + y_a(C_o s - K_a t) \\ & \quad + x_o(k_o s + a_a t) + x_a k_a t\} \\ & + V\{2y_o(A_o s - D_a t) + 2y_a(K_o s - A_a t) \\ & \quad + x_o(a_o s + d_a t)\} \\ & + v\{y_o(a_o s - 3d_a t) + 2y_a(k_o s - a_a t)\} \end{aligned} \qquad (24.35b)$$

or

$$u^{(1)}(z) =$$

$$\begin{pmatrix} r_o^2 \\ r_a^2 \\ V \\ v \end{pmatrix}^T \begin{pmatrix} (D_o + id_o)s - E_a t & F_o s - A_o s - D_a t \\ (K_o + ik_o)s - (F_a - A_a - ia_a)t & C_o s - (K_a - ik_a)t \\ (2A_o + ia_o)s - (2D_a - id_a)t & 2(K_o s - A_a t) \\ a_o s - 3d_a t & 2(k_o s - a_a t) \end{pmatrix} \begin{pmatrix} u_o \\ u_a \end{pmatrix} \qquad (24.35c)$$

In the image plane, where $t(z)$ vanishes and $s(z_i) = M$,

$$\begin{aligned}\frac{x^{(1)}(z_i)}{M} =: \Delta x_i = {} & x_a(Cr_a^2 + 2KV + 2kv + (F-A)r_o^2) \\ & + x_o(Kr_a^2 + 2AV + av + Dr_o^2) \\ & - y_o(kr_a^2 + aV + dr_o^2) \\ \frac{y^{(1)}(z_i)}{M} =: \Delta y_i = {} & y_a(Cr_a^2 + 2KV + 2kv + (F-A)r_o^2) \\ & + y_o(Kr_a^2 + 2AV + av + Dr_o^2) \\ & + x_o(kr_a^2 + aV + dr_o^2) \end{aligned} \qquad (24.36)$$

Setting
$$\Delta u_i := \Delta x_i + i\Delta y_i \qquad (24.37)$$
the pair of equations (24.36) can be combined into the compact expression

$$\Delta u_i = \begin{pmatrix} u_o^* \\ 2u_a^* \end{pmatrix} \begin{pmatrix} D + id & F & K - ik \\ \tfrac{1}{2}(A + ia) & K + ik & \tfrac{1}{2}C \end{pmatrix} \begin{pmatrix} u_o^2 \\ u_o u_a \\ u_a^2 \end{pmatrix} \qquad (24.38a)$$

24.2 DERIVATION OF THE REAL ABERRATION COEFFICIENTS

which could have been derived directly from (24.34) since $\partial/\partial x + i\partial/\partial y = 2\partial/\partial u^*$. Equation (24.38a) may also be written in the form

$$\Delta u_i = Cr_a^2 u_a \quad \text{(spherical aberration)}$$
$$+ 2(K + ik)r_a^2 u_o + (K - ik)u_a^2 u_o^* \quad \text{(coma)}$$
$$+ (A + ia)u_o^2 u_a^* \quad \text{(astigmatism)}$$
$$+ Fr_o^2 u_a \quad \text{(field curvature)}$$
$$+ (D + id)r_o^2 u_o \quad \text{(distortion)} \quad (24.38b)$$
$$= Cr_a^2 u_a$$
$$+ 2(K + ik)r_a^2 r_o e^{i\varphi_o} + (K - ik)r_a^2 r_o e^{i(2\varphi_a - \varphi_o)}$$
$$+ (A + ia)r_o^2 r_a e^{i(2\varphi_o - \varphi_a)} + Fr_o^2 r_a e^{i\varphi_a}$$
$$+ (D + id)r_o^3 e^{i\varphi_o} \quad (24.38c)$$

The numerous terms of which Δx_i and Δy_i are composed depend in different ways on the object and aperture coordinates and are in practice of unequal importance. We have indicated in (24.38b) the names by which they are commonly known and in Sections 24.3–24.6, we consider each in turn, pointing out general properties and other features of interest. Formulae for particular field or potential models and information about the dependence of the various coefficients on lens geometry and excitation will be found in Part VII.

The perturbation characteristic function S^I has hitherto appeared as an accessory that enables us to obtain the aberrations Δx_i and Δy_i by differentiation. We saw in Section 5.3, however, that pencils of rays propagate in such a way that they are always orthogonal to the surfaces $S = \text{const}$, $\boldsymbol{p}(\boldsymbol{r}) = \text{grad } \overline{S}(\boldsymbol{r})$ (5.12) or in the absence of magnetic field, $\boldsymbol{g}(\boldsymbol{r}) = \text{grad } \overline{S}(\boldsymbol{r})$. This relationship enables us to interpret (24.34) in physical terms and sheds additional light on the origin of the various types of aberration.

For simplicity, we assume that object and image are in field-free space (or virtual) and we introduce the entrance and exit pupils at $z = z_{ao}$ and $z = z_{ai}$. These are the images of the true aperture by the parts of the lens that precede and follow it, respectively. Figure 24.1 shows the two pairs of conjugate planes, object and image and the two pupils; in the paraxial approximation, the point P_o is conjugate to P_i and the latter is shifted by aberrations to $\overline{P}_i$. Consider now a ray from P_o that intersects the entrance pupil at P'_o and the image pupil at P'_i. In the absence of aberrations, all rays from P_o are orthogonal to a set of spheres centred on P_o; the action of the lens converts the incident sphere into a sphere centred on P_i, and the rays in image space are all orthogonal to a set of spheres likewise centred on P_i. In the presence of aberrations, these spheres in image space are replaced

24. THE GEOMETRICAL ABERRATIONS OF ROUND LENSES

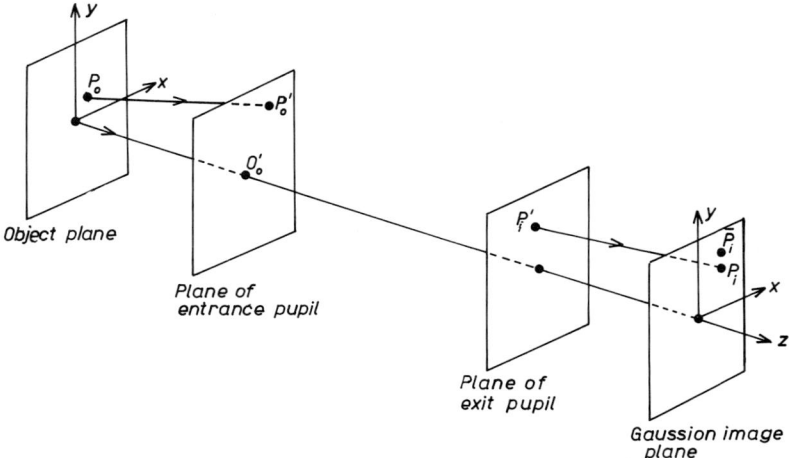

Fig. 24.1: The object and image planes are conjugate as are the pupil planes. In the paraxial approximation, P_i is the image of P_o; aberrations shift P_i to $\overline{P}_i$.

by aspheric surfaces, and the distance between the paraxial spheres and the aberration surfaces is measured by S^I. Figure 24.2 shows the spherical surface S_o centred on P_i that intersects the axis at the exit pupil and the aspheric surface S to which the electron trajectories are orthogonal in the presence of aberrations; we have again chosen the member of the set of such surfaces that intersects the axis at the exit pupil. The ray connecting P'_i to $\overline{P}_i$ cuts the reference sphere S_o at Q_o and the aspheric surface S at $\overline{Q}$. The distance $S(\overline{Q}, Q_o)$ is given by S^I (24.34) and we therefore associate each term of the latter with a particular distortion of the surface S. We return to this interpretation of S briefly in each of the following sections.

24.3 Spherical aberration (terms in x_a and y_a only)

We first examine the *spherical* or *aperture aberration*, of the greatest importance in the first (objective) lens of magnifying systems and in the final lens of probe-forming systems. From (24.36), we see that a pencil of rays from some object point (x_o, y_o) intersects the image plane not at the Gaussian

---

* Readers familiar with similar treatments of the aberrations in light optics, Sections 5.1–5.3 of Born and Wolf (1959) for example, may note that the role of the radius of the Gaussian reference sphere (R in Born and Wolf), which occurs in the aberration formulae, is absorbed into the Wronskian in electron optics. Thus (5.1.10) of Born and Wolf is exactly equivalent to (24.29a) for $t(z_2) = 0$.

24.3 SPHERICAL ABERRATION

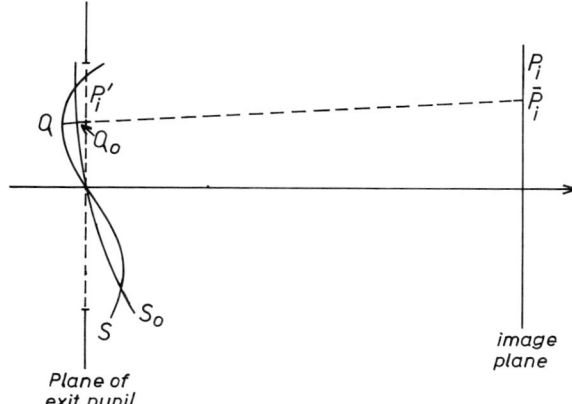

Fig. 24.2: The reference sphere S_o is centred on P_i. Aberrations distort S_o into the aspheric surface S.

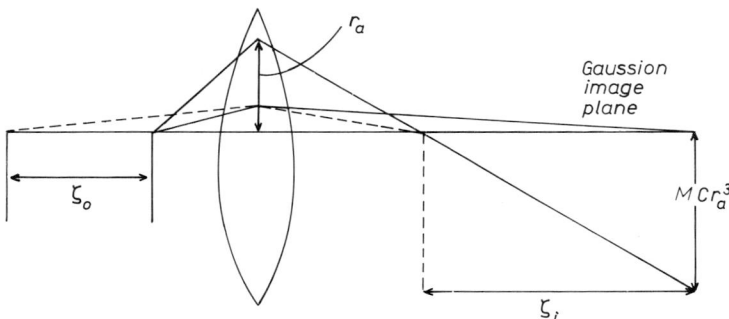

Fig. 24.3: Spherical aberration. In the Gaussian image plane, the image of a point object is a circular disc of radius MCr_a^3. The outermost ray intersects the axis at a distance ζ_i from the Gaussian image plane. The axial object point of which this point of intersection is the Gaussian image is distant ζ_o from the original object plane.

image point (Mx_o, My_o) but at points (x_i, y_i) such that

$$x_i = M\{x_o + Cx_a(x_a^2 + y_a^2)\}$$
$$y_i = M\{y_o + Cy_a(x_a^2 + y_a^2)\}$$
(24.39)

Hence $\Delta x_{ji} = Cx_{ja}r_a^2$ $(j = 1, 2)$. Each Gaussian image point is replaced by an aberration figure, as explained in Chapter 23. If the beam is confined within a circular aperture, radius r_A, so that $x_a^2 + y_a^2 \leq r_A^2$, the aberration figure is circular

$$\{(\Delta x_i)^2 + (\Delta y_i)^2\}^{\frac{1}{2}} \leq C r_A^3 \qquad (24.40)$$

Thus all rays from a fixed object point that intersect the aperture around a circle of radius r_a intersect the image plane around a circle, centred on the Gaussian image point, of radius MCr_a^3. These rays intersect in a plane distant ζ from the Gaussian image plane. For simplicity, we consider an axial object point (Fig. 24.3); a ray in the neighbourhood of the image plane may be written

$$\begin{aligned} x(z_i + \zeta) &= x_a t(z_i + \zeta) + MC x_a r_a^2 \\ &\approx x_a t'(z_i)\zeta + MC x_a r_a^2 \end{aligned} \qquad (24.41)$$

and likewise for $y(z_i + \zeta)$. This ray crosses the axis in the plane for which

$$\zeta = -\frac{MC}{t_i'} r_a^2 \qquad (24.42)$$

and clearly all rays for which r_a is constant intersect the axis at this point. The coefficient MC/t_i' may be written $W_s C/\hat{\phi}_i^{\frac{1}{2}} t_i'^2$; it therefore has the same sign as CW_s and hence as the integral in the expression for C (24.32b). We shall see that this integral is never negative, and so ζ is always negative (24.42). *Rays affected by spherical aberration intersect before reaching the image plane* (Fig. 24.3). The value of ζ corresponding to r_A is known as the longitudinal spherical aberration:

$$\zeta_i := -\frac{MC}{t_i'} r_A^2 = -\frac{M^2 C}{t_o'} \left(\frac{\hat{\phi}_i}{\hat{\phi}_o}\right)^{\frac{1}{2}} r_A^2 \qquad (24.43a)$$

or referred back to object space,

$$\zeta_o := \frac{\zeta_i}{M^2} = -\frac{C}{t_o'} \left(\frac{\hat{\phi}_i}{\hat{\phi}_o}\right)^{\frac{1}{2}} r_A^2 \qquad (24.43b)$$

Another way of picturing the fact that ζ_i is always negative is to say that rays far from the axis are focused more strongly than those closer to it.

It is frequently more convenient to measure the size of the spherical aberration disc in terms of the range of ray gradients at the object instead of the radius of the aperture. Rather than return to the formula for aberration coefficients expressed in terms of position and gradient at the object, we can simply transform (24.39) by recalling that the rays of the pencil from

24.3 SPHERICAL ABERRATION

an object point (x_o, y_o) are described by $x = x_o s + x_a t$, $y = y_o s + y_a t$ so that $x'_o = x_o s'_o + x_a t'_o$, $y'_o = y_o s'_o + y_a t'_o$. Thus

$$\Delta x_i = C x_a (x_a^2 + y_a^2) = \frac{C}{t_o'^3} x'_o (x_o'^2 + y_o'^2) + \text{mixed terms}$$
$$\Delta y_i = C y_a (x_a^2 + y_a^2) = \frac{C}{t_o'^3} y'_o (x_o'^2 + y_o'^2) + \text{mixed terms} \quad (24.44)$$

in which the mixed terms contribute to other aberrations but none of the latter contributes to spherical aberration. The coefficient $C/t_o'^3$ is always denoted by C_s (the suffix recalling the "spherical" origin of this defect in the case of glass lenses):

$$\Delta x_i =: C_s x'_o (x_o'^2 + y_o'^2)$$
$$\Delta y_i =: C_s y'_o (x_o'^2 + y_o'^2) \quad (24.45)$$

and from (24.32b), we see that

$$C_s = \frac{1}{\hat{\phi}^{\frac{1}{2}}} \int_{z_o}^{z_i} (L_1 h^4 + 2L_2 h^2 h'^2 + L_3 h'^4) \, dz \quad (24.46)$$

in which we have used the fact that $h(z) = t(z)/t'_o$.

The radius of the spherical aberration disc in the Gaussian image plane is MCr_A^3. Is it smaller in any neighbouring plane? We return to (24.41), which we now write

$$x(z_i - \zeta) = -x_a t'_i \zeta + MC x_a r_a^2 \quad (24.47)$$

The ray that is most distant from the axis in the image plane intersects the aperture plane around the circle $x_a^2 + y_a^2 = r_A^2$; for a given magnification, this ray and a general ray are equidistant from the axis in some plane $z = z_i - \zeta$ if $x(z_i - \zeta) = -x_A t'_i \zeta + MC x_A r_A^2$ or

$$\zeta t'_i (x_A - x_a) = MC(x_A^3 - x_a^3)$$

or again

$$\zeta = MC(x_A^2 + x_A x_a + x_a^2)/t'_i \quad (24.48)$$

(in which we have set $y_a = 0$ to simplify the calculation—no generality is lost). In this plane

$$x(z_i - \zeta) = -MC(x_a + x_A) x_a x_A \quad (24.49)$$

and this is smallest in the plane for which $dx(z_i - \zeta)/dx_a = 0$, namely, that in which $x_a = -x_A/2$. For this value, (24.48) tells us that

$$\zeta = \frac{3}{4}\frac{MC}{t'_i}x_A^2 = \frac{3}{4}|\zeta_i| \qquad (24.50)$$

and the beam radius is

$$x(z_i - \zeta) = \frac{1}{4}MCx_A^3 \qquad (24.51)$$

Thus the radius of this *disc of least confusion* is only one quarter of that of the spherical aberration disc in the Gaussian image plane; the disc is formed in a plane three-quarters of the way from the Gaussian image plane to the plane of the marginal focus.

In practice, the spherical aberration coefficient is tabulated for the case of greatest interest, objective lenses, used either as the first lens beyond the specimen in a magnifying system or as the final lens just before the target in a demagnifying (probe-forming) system. How are these values related? We shall discuss such relations in detail in the case of asymptotic aberrations in Chapter 25 but a simple argument enables us to relate the values of the real spherical aberration coefficient in the two situations. When the aberration is expressed in terms of aperture coordinates, the question is almost trivial since the only ray occurring in the integral in C is $t(z)$, which is unaffected by interchanging z_o and z_i. The size of the aberration disc, referred back to the particular object plane, is governed by W and hence by t'_o and $\hat{\phi}_o^{\frac{1}{2}}$ only. When the aberration is expressed in terms of gradient (x'_o, y'_o), however, confusion may arise. Consider the situation illustrated in Fig. 24.4a, which shows three rays; ray 1 is a paraxial ray connecting the axial object and image points P_o and P_i. Ray 2 sets out from P_o but strikes the image plane at $\overline{P}_i(\overline{x}_i, \overline{y}_i)$, owing to the spherical aberration; ray 3, also affected by spherical aberration, passes through P_i and must hence have set out from some point $\overline{P}_o$ off the axis. Rays 1 and 2 have the same gradient at P_o, and rays 1 and 3 at P_i. We have $P_i\overline{P}_i = MC_s^{(oi)}\theta_o^3$ and $P_o\overline{P}_o = C_s^{(io)}\theta_i^3/M$, where $C_s^{(oi)}$ and $C_s^{(io)}$ are the values of C_s for the cases in which the electron travel from P_o towards P_i and the reverse, respectively. From (24.39, 24.45), however, we see that

$$0 = MP_o\overline{P}_o + MC_s^{(oi)}\theta_o^3$$

or

$$P_o\overline{P}_o = -C_s^{(oi)}\theta_o^3 \qquad (24.52)$$

24.3 SPHERICAL ABERRATION

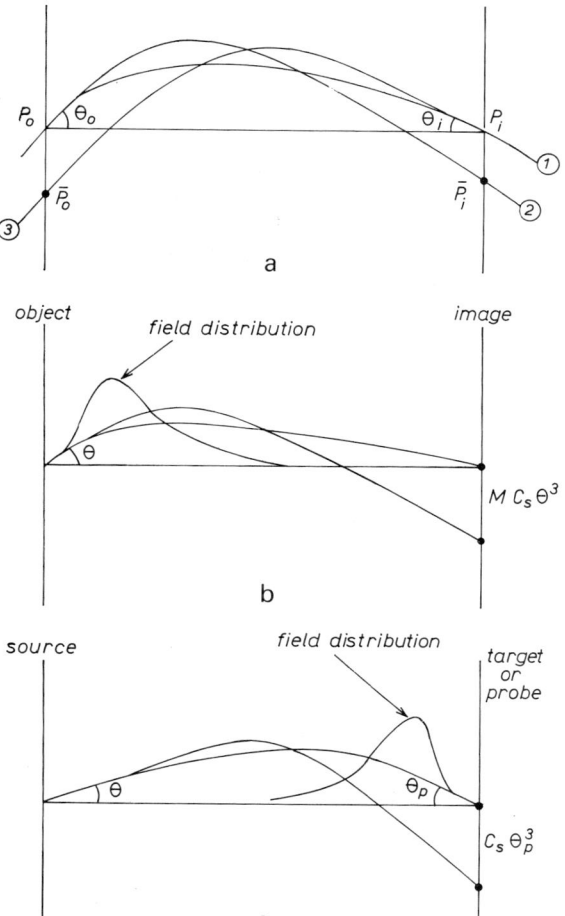

Fig. 24.4: Forward and backward aberration coefficients. (a) Calculation of the relation between the coefficients. (b) Aberration disc ($|M| \gg 1$). (c) Probe size ($|M| \ll 1$).

Hence
$$\frac{1}{M} C_s^{(io)} \theta_i^3 = -C_s^{(oi)} \theta_o^3$$

or
$$C_s^{(io)} = -M \left(\frac{\theta_o}{\theta_i}\right)^3 C_s^{(oi)} = M^4 \left(\frac{\hat{\phi}_i}{\hat{\phi}_o}\right)^{\frac{3}{2}} C_s^{(oi)} \tag{24.53}$$

or for magnetic lenses
$$C_s^{(io)} = M^4 C_s^{(oi)} \tag{24.54}$$

The forward and backward spherical aberration coefficients are thus very different. Equation (24.54) is an obvious consequence of the relation between C_s and C (24.44). Figures 24.4b and c show a field distribution used as a magnifying and demagnifying system. Equation (24.52) shows that if the spherical aberration coefficient for the magnifying situation is C_s, the radius of the corresponding probe will be given by $C_s \theta_p^3$, where θ_p denotes the angular aperture at the probe. That this is consistent with (24.53) is easily seen from the fact that the probe radius r_p is given by

$$\begin{aligned} r_p &= M^{(io)} C_s^{(io)} \theta^3 \\ &= \frac{1}{M} M^4 \left(\frac{\hat{\phi}_i}{\hat{\phi}_o}\right)^{\frac{3}{2}} C_s^{(oi)} \frac{1}{M^3} \left(\frac{\hat{\phi}_o}{\hat{\phi}_i}\right)^{\frac{3}{2}} \theta_p^3 \\ &= C_s^{(oi)} \theta_p^3 \end{aligned} \tag{24.55}$$

In practice, these relations are almost always required for magnetic lenses alone, and the important result is that expressed by (24.54).

We mentioned at the end of Section 24.2 that each aberration can be associated with a characteristic distortion of the surfaces S=const. For the spherical aberration, we have $S^I/W = -\frac{1}{4} C r_a^4$ and so, as shown in Fig. 24.5, the true surface S (including aberrations) is everywhere closer to the image plane than the reference sphere S_o, except at the axis where the two touch. Since the curvature of S is greater than that of S_o, it is immediately obvious that outer rays will be more strongly focused than those close to the axis and hence the marginal focus retreats towards the lens as the aperture is opened more widely.

We now turn to the coefficient itself, C or C_s. The formulae for these can be rewritten in numerous ways, very different in appearance but otherwise equivalent. These expressions can be obtained either by partial integration using the paraxial equations to replace second derivatives of the paraxial rays s and t or g and h whenever they occur, or by an ingenious differential technique introduced by Seman (1951, 1954, 1955a,b, 1958b,c) and exploited extensively by Hawkes (1966/7b, 1967b). Seman in fact applied his method not to individual aberration coefficients but to the characteristic function S^I, as we explain in Section 24.9. We introduce it now, however, as it provides an extremely convenient way of analysing individual coefficients for which a formula is already available, without returning to the characteristic function. We set out from (24.46),

$$C_s = \frac{1}{\hat{\phi}_o^{\frac{1}{2}}} \int (L_1 h^4 + 2L_2 h^2 h'^2 + L_3 h'^4) \, dz$$

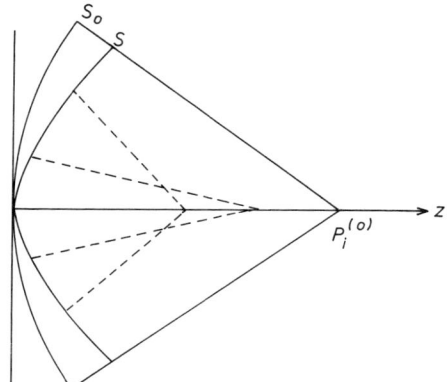

Fig. 24.5: Spherical wave surface S_o centred on the Gaussian image point $P_i^{(0)}$ and aspherical wave surface S distorted by spherical aberration. As the aperture is widened, the image point corresponding to the outermost rays, or wave-normals, retreats towards the lens.

in which L_1, L_2 and L_3 are given by (24.3). Retention of the relativistic terms in the electrostatic case always renders the calculation extremely heavy, however, and we shall therefore apply the method to the nonrelativistic form of C_s, obtained by setting $\epsilon = 0$, $\hat{\phi} = \phi$ in (24.3). The relativistic results are merely listed.

We begin by noticing that the quantities ϕ, B and h and the operator (d/dz) occur in only a few combinations in the integrand of C_s, namely $(d/dz)^2(B^2/\phi^{1/2})h^4$ and $(d/dz)^4\phi^{1/2}h^4$. A set of terms with comparable structure is obtained by differentiating once all the terms generated by $(d/dz)(B^2/\phi^{1/2})h^4$ and $(d/dz)^3\phi^{1/2}h^4$, eliminating h'' as necessary with the aid of the paraxial equation, $h'' = -(\phi'/2\phi)h' - (\phi'' + \eta^2 B^2)h/4\phi$. The first term generates

$$\frac{BB'}{\phi^{\frac{1}{2}}}h^4; \quad \frac{B^2\phi'}{\phi^{\frac{3}{2}}}h^4; \quad \text{and} \quad \frac{B^2}{\phi^{\frac{1}{2}}}h^3h'$$

and the other,

$$\frac{\phi'''}{\phi^{\frac{1}{2}}}h^4; \quad \frac{\phi'\phi''}{\phi^{\frac{3}{2}}}h^4; \quad \frac{\phi'^3}{\phi^{\frac{5}{2}}}h^4;$$

$$\frac{\phi''}{\phi^{\frac{1}{2}}}h^3h'; \quad \frac{\phi'^2}{\phi^{\frac{3}{2}}}h^3h'; \quad (24.56)$$

$$\frac{\phi'}{\phi^{\frac{1}{2}}}h^2h'^2; \quad \text{and} \quad \phi^{\frac{1}{2}}hh'^3$$

24. THE GEOMETRICAL ABERRATIONS OF ROUND LENSES

Each of these is differentiated and then formally integrated; the term in $\phi^{1/2}hh'^3$, for example, yields

$$\frac{d}{dz}(\phi^{\frac{1}{2}}hh'^3) = \frac{\phi'}{2\phi^{\frac{1}{2}}}hh'^3 + \phi^{\frac{1}{2}}h'^4 \qquad (24.57a)$$
$$- 3\phi^{\frac{1}{2}}hh'^2\left(\frac{\phi'}{2\phi}h' + \frac{\phi'' + \eta^2 B^2}{4\phi}h\right)$$

so that

$$\left[\phi^{\frac{1}{2}}hh'^3\right]_{z_o}^{z_i} + \int_{z_o}^{z_i}\left(\frac{\phi'}{\phi^{\frac{1}{2}}}hh'^3 + \frac{3}{4}\frac{\phi'' + \eta^2 B^2}{\phi^{\frac{1}{2}}}h^2h'^2 - \phi^{\frac{1}{2}}h'^4\right) dz \equiv 0 \quad (24.57b)$$

In all we have ten such identities. After first multiplying each by an arbitrary constant p_j, $j = 1$–10, these are all added to the integral in C_s, giving

$$32\phi_o^{\frac{1}{2}}C_s = \int_{z_o}^{z_i} \phi^{\frac{1}{2}}\left[h'^4(16 + p_1) + h'^3h\frac{\phi'}{\phi}(-p_1 + 2p_2)\right.$$
$$+ h'^2h^2\left\{\frac{\phi''}{\phi}\left(8 - \frac{3}{4}p_1 + p_2 + 3p_3\right) + \left(\frac{\phi'}{\phi}\right)^2\left(-\frac{3}{2}p_2 + 3p_4\right)\right.$$
$$\left.+ \frac{\eta^2 B^2}{\phi}\left(8 - \frac{3}{4}p_1 + 3p_5\right)\right\}$$
$$+ h'h^3\left\{\frac{\phi'''}{\phi}(p_3 + 4p_6) + \frac{\phi''\phi'}{\phi^2}(-\frac{1}{2}p_2 - p_3 + 2p_4 + 4p_7)\right.$$
$$+ \left(\frac{\phi'}{\phi}\right)^3(-2p_4 + 4p_8) + \frac{\eta^2 B^2\phi'}{\phi^2}(-\frac{1}{2}p_2 - p_5 + 4p_9)$$
$$\left.+ \frac{\eta^2 BB'}{\phi}(2p_5 + 4p_{10})\right\}$$
$$+ h^4\left\{\frac{\phi^{(4)}}{\phi}(p_6 - 1) + \frac{\phi'''\phi'}{\phi^2}(-\frac{1}{2}p_6 + p_7)\right.$$
$$+ \left(\frac{\phi''}{\phi}\right)^2(1 - \frac{1}{4}p_3 + p_7) - \frac{5}{2}\left(\frac{\phi'}{\phi}\right)^4 p_8$$
$$+ \frac{\phi''\phi'^2}{\phi^3}(-\frac{1}{4}p_4 - \frac{3}{2}p_7 + 3p_8) + \frac{\eta^2 B^2\phi''}{\phi^2}(2 - \frac{p_3}{4} - \frac{p_5}{4} + p_9)$$
$$\left.+ \frac{\eta^2 B^2\phi'^2}{\phi^3}(-\frac{1}{4}p_4 - \frac{3}{2}p_9) + \frac{\eta^2 BB'\phi'}{\phi^2}(2p_9 - \frac{1}{2}p_{10})\right.$$

24.3 SPHERICAL ABERRATION

$$+ \frac{\eta^4 B^4}{\phi^2}\left(1 - \frac{1}{4}p_5\right) + \frac{\eta^2 BB''}{\phi}(-4 + p_{10}) + \frac{\eta^2 B'^2}{\phi}p_{10}\Bigg\} \Bigg] dz$$

$$- \Bigg[\phi^{\frac{1}{2}}(p_1 hh'^3 + p_2 \frac{\phi'}{\phi}h^2 h'^2 + p_3 \frac{\phi''}{\phi}h^3 h'$$

$$+ p_4 \frac{\phi'^2}{\phi^2}h^3 h' + p_5 \frac{\eta^2 B^2}{\phi}h^3 h' + p_6 \frac{\phi'''}{\phi}h^4$$

$$+ p_7 \frac{\phi''\phi'}{\phi^2}h^4 + p_8 \frac{\phi'^3}{\phi^3}h^4 + p_9 \frac{\eta^2 B^2 \phi'}{\phi^2}h^4$$

$$+ p_{10} \frac{\eta^2 BB'}{\phi}h^4\Bigg]_{z_o}^{z_i} \tag{24.58}$$

Since the multipliers p_i are arbitrary, we may choose them in any self-consistent way to eliminate terms that are for some reason undesirable. For numerical work, for example, we usually eliminate high derivatives of ϕ and B; if a field model that permits the integrals to be evaluated in closed form is being studied, we may well prefer to have as few terms to integrate as possible. Again, we might wish to establish whether C_s can change sign, in which case we enquire whether or not the integrand can be written as a sum of squared terms. The integrated terms all vanish since $t(z)$ appears undifferentiated in each of them. They must not, however, be completely forgotten, for they can affect the thin-lens approximation, examined below.

In practice, the simultaneous presence of electrostatic and magnetic fields is rare and we therefore list a number of forms of C_s for the two separate cases, $B = 0$ and $\phi =$ const. The mixed forms may be obtained by manipulating (24.58).

Electrostatic case ($B = 0$, $\phi \neq$ const)
General relativistic expression

$$32\hat{\phi}_o^{\frac{1}{2}} C_s = \int (A_0 h^4 + 2A_1 h^3 h' + 2A_2 h^2 h'^2 \tag{24.59}$$
$$+ 2A_3 hh'^3 + A_4 h'^4) dz$$

in which

$$A_0 = \frac{\gamma \phi^{(4)}}{\hat{\phi}^{\frac{1}{2}}}(p_6 - 1) + \frac{\phi''^2}{\hat{\phi}^{\frac{3}{2}}}\left\{\left(-\frac{p_3}{4} + 1\right)\gamma^2 + p_7\right\}$$

$$- 4\epsilon\frac{\phi''^2}{\hat{\phi}^{\frac{1}{2}}}(1 + p'_7) + \frac{\phi'\phi'''}{\hat{\phi}^{\frac{3}{2}}}(-\frac{1}{2}\gamma^2 p_6 + p_7)$$

$$- 4\epsilon\frac{\phi'\phi'''}{\hat{\phi}^{\frac{1}{2}}}(p'_7 - \frac{1}{2}p_6) - \gamma\frac{\phi'^2\phi''}{\hat{\phi}^{\frac{5}{2}}}(-3p_8 + \frac{3}{2}p_7 + \frac{1}{4}p_4)$$

$$- 2\epsilon \frac{\phi'^2 \phi''}{\hat{\phi}^{\frac{3}{2}}} \{p'_{10} + \epsilon\phi p'_{11} - \gamma(p'_7 + p'_5)\}$$

$$- \frac{5}{2} \frac{\gamma^2 \phi'^4}{\hat{\phi}^{\frac{7}{2}}} p_8 + \frac{\epsilon \phi'^4}{\hat{\phi}^{\frac{5}{2}}} (\gamma p'_{10} + 2p_8 + \gamma\epsilon\phi p'_{11})$$

$$- \frac{2}{3} \frac{\epsilon^2 \phi'^4}{\hat{\phi}^{\frac{3}{2}}} p'_{11} \qquad (24.60a)$$

$$A_1 = \frac{\gamma \phi'''}{\hat{\phi}^{\frac{1}{2}}} (\frac{1}{2} p_3 + 2p_6) - \frac{\phi'' \phi'}{\hat{\phi}^{\frac{3}{2}}} \{(\frac{1}{2} p_3 + \frac{1}{4} p_2) \gamma^2$$

$$- (2p_7 + p_4)\} - \frac{\epsilon \phi' \phi''}{\hat{\phi}^{\frac{1}{2}}} (8p'_7 + 8p'_5 - p_3)$$

$$+ \frac{\gamma \phi'^3}{\hat{\phi}^{\frac{5}{2}}} (2p_8 - p_4) - \frac{4\epsilon \phi'^3}{\hat{\phi}^{\frac{3}{2}}} (\frac{p'_{10}}{3} - p'_5)$$

$$- \frac{8\epsilon^2 \phi \phi'^3}{\hat{\phi}^{\frac{3}{2}}} (\frac{1}{6} p'_{11} - p'_5) \qquad (24.60b)$$

$$A_2 = \frac{\gamma \phi''}{\hat{\phi}^{\frac{1}{2}}} (4 + \frac{3}{2} p_3 + \frac{1}{2} p_2 - \frac{3}{8} p_1)$$

$$+ 3 \frac{\phi'^2}{\hat{\phi}^{\frac{3}{2}}} (\frac{1}{2} p_4 - \gamma^2 \frac{p_2}{4})$$

$$- \frac{4\epsilon \phi'^2}{\hat{\phi}^{\frac{1}{2}}} (3p'_5 - \frac{1}{4} p_2) \qquad (24.60c)$$

$$A_3 = \frac{\gamma \phi'}{\hat{\phi}^{\frac{1}{2}}} (p_2 - \frac{1}{2} p_1) \qquad (24.60d)$$

$$A_4 = \hat{\phi}^{\frac{1}{2}} (p_1 + 16) \qquad (24.60e)$$

The nonrelativistic forms of these are as follows:

$$A'_0 := A_0(\epsilon \to 0) = \frac{\phi^{(4)}}{\hat{\phi}^{\frac{1}{2}}} (p_6 - 1) + \frac{\phi''^2}{\hat{\phi}^{\frac{3}{2}}} (1 - \frac{p_3}{4} + p_7)$$

$$+ \frac{\phi' \phi'''}{\hat{\phi}^{\frac{3}{2}}} (p_7 - \frac{1}{2} p_6) - \frac{\phi'^2 \phi''}{\hat{\phi}^{\frac{5}{2}}} (\frac{1}{4} p_4 + \frac{3}{2} p_7 - 3p_8)$$

$$- \frac{5}{2} \frac{\phi'^4}{\hat{\phi}^{\frac{7}{2}}} p_8$$

$$A'_1 := A_1(\epsilon \to 0) = \frac{\phi'''}{2\hat{\phi}^{\frac{1}{2}}} (p_3 + 4p_6) - \frac{\phi'' \phi'}{\hat{\phi}^{\frac{3}{2}}} (\frac{1}{2} p_3 + \frac{1}{4} p_2 - 2p_7 - p_4)$$

$$+ \frac{\phi'^3}{\hat{\phi}^{\frac{5}{2}}} (2p_8 - p_4)$$

24.3 SPHERICAL ABERRATION

$$A'_2 := A_2(\epsilon \to 0) = \frac{\phi''}{\phi^{\frac{1}{2}}}(4 + \frac{3}{2}p_3 + \frac{1}{2}p_2 - \frac{3}{8}p_1) + 3\frac{\phi'^2}{\phi^{\frac{3}{2}}}(\frac{1}{2}p_4 - \frac{p_2}{4})$$

$$A'_3 := A_3(\epsilon \to 0) = \frac{\phi'}{\phi^{\frac{1}{2}}}(p_2 - \frac{1}{2}p_1)$$

$$A'_4 := A_4(\epsilon \to 0) = \phi^{\frac{1}{2}}(p_1 + 16)$$

General nonrelativistic expression

$$32\phi_o^{\frac{1}{2}} C_s = \int \phi^{\frac{1}{2}} \Big[h'^4(16 + p_1) + h'^3 h \frac{\phi'}{\phi}(-p_1 + 2p_2)$$
$$+ h'^2 h^2 \Big\{ \frac{\phi''}{\phi}(8 - \frac{3}{4}p_1 + p_2 + 3p_3) + (\frac{\phi'}{\phi})^2(-\frac{3}{2}p_2 + 3p_4) \Big\}$$
$$+ h' h^3 \Big\{ \frac{\phi'''}{\phi}(p_3 + 4p_6) + \frac{\phi''\phi'}{\phi^2}(-\frac{1}{2}p_2 - p_3 + 2p_4 + 4p_7)$$
$$+ \Big(\frac{\phi'}{\phi}\Big)^3 (-2p_4 + 4p_8) \Big\}$$
$$+ h^4 \Big\{ \frac{\phi^{(4)}}{\phi}(p_6 - 1) + \frac{\phi'''\phi'}{\phi^2}(-\frac{1}{2}p_6 + p_7)$$
$$+ \Big(\frac{\phi''}{\phi}\Big)^2 (1 - \frac{1}{4}p_3 + p_7) - \frac{5}{2}\Big(\frac{\phi'}{\phi}\Big)^4 p_8$$
$$+ \frac{\phi''\phi'^2}{\phi^3}(-\frac{1}{4}p_4 - \frac{3}{2}p_7 + 3p_8) \Big\} \Big] dz$$
$$- \Big[\phi^{\frac{1}{2}}(p_1 hh'^3 + p_2 \frac{\phi'}{\phi} h^2 h'^2 + p_3 \frac{\phi''}{\phi} h^3 h'$$
$$+ p_4 \frac{\phi'^2}{\phi^2} h^3 h' + p_6 \frac{\phi'''}{\phi} h^4 + p_7 \frac{\phi''\phi'}{\phi^2} h^4$$
$$+ p_5 (\frac{\phi'}{\phi})^3 h^4 \Big]_{z_o}^{z_i} \quad (24.61)$$

Numerous forms are to be found in the literature, all of which correspond to various choice of the coefficients p_i. For example, eliminating all terms involving $h'(z)$ from (24.61), we find

$$32\phi_o^{\frac{1}{2}} C_s = \int \phi^{\frac{1}{2}} \Big(-\frac{\phi'''\phi'}{2\phi^2} + 2\frac{\phi''^2}{\phi^2} + 5\Big(\frac{\phi'}{\phi}\Big)^4 - 5\frac{\phi''\phi'^2}{\phi^3} \Big) h^4 \, dz \quad (24.62)$$

which may also be written in terms of

$$\psi(z) := \phi'(z)/\phi(z) \quad (24.63)$$

as
$$64\phi_0^{\frac{1}{2}} C_s = \int \phi^{\frac{1}{2}}(4\psi'^2 + 3\psi^4 - 5\psi^2\psi' - \psi\psi'')h^4\, dz \tag{24.64}$$

Magnetic case ($\phi = \text{const}, B \neq 0$)
General relativistic case

$$\begin{aligned}
32 C_s = \int &\left\{ h^{14}(16+p_1) + h^2 h'^2 \frac{\eta^2 B^2}{\hat{\phi}}\left(8 - \frac{3}{4}p_1 + 3p_5\right) \right.\\
&+ h^3 h' \frac{\eta^2 BB'}{\hat{\phi}}(2p_5 + 4p_{10}) + h^4 \frac{\eta^4 B^4}{\hat{\phi}^2}\left(1 - \frac{1}{4}p_5\right)\\
&+ h^4 \frac{\eta^2 BB''}{\hat{\phi}}(-4 + p_{10}) + h^4 \frac{\eta^2 B'^2}{\hat{\phi}} p_{10}\bigg\}\, dz\\
&- \left[p_1 h h'^3 + p_5 \frac{\eta^2 B^2}{\hat{\phi}} h^3 h' + p_{10}\frac{\eta^2 BB'}{\hat{\phi}}h^4\right]_{z_o}^{z_i}
\end{aligned} \tag{24.65}$$

All derivatives of h can be eliminated by writing $p_1 = -16, p_5 = -20/3$ and $p_{10} = 10/3$, giving

$$C_s = \frac{1}{48}\int\left(5\frac{\eta^2 B'^2}{\hat{\phi}} - \frac{\eta^2 BB''}{\hat{\phi}} + 4\frac{\eta^4 B^4}{\hat{\phi}^2}\right)h^4\, dz \tag{24.66}$$

Other useful forms are as follows:

$$\begin{aligned}
C_s &= \int\left\{\left(\frac{\eta^4 B^4}{32\hat{\phi}^2} - \frac{\eta^2 BB''}{8\hat{\phi}}\right)h^4 + \frac{\eta^2 B^2}{4\hat{\phi}}h^2 h'^2 + \frac{1}{2}h'^4\right\}\, dz & (a)\\
&= \int\left\{\left(\frac{\eta^4 B^4}{32\hat{\phi}^2} - \frac{\eta^2 BB''}{8\hat{\phi}}\right)h^4 + \frac{5\eta^2 B^2}{8\hat{\phi}}h^2 h'^2\right\}\, dz & (b)\\
&= \int\left\{\left(\frac{3\eta^4 B^4}{32\hat{\phi}^2} + \frac{\eta^2 B'^2}{8\hat{\phi}}\right)h^4 - \frac{\eta^2 B^2}{8\hat{\phi}}h^2 h'^2\right\}\, dz & (c)\\
&= \int\left\{\frac{\eta^4 B^4}{16\hat{\phi}^2}h^4 + (hB' + h'B)^2\frac{\eta^2 h^2}{8\hat{\phi}} + \frac{\eta^2 B^2}{8\hat{\phi}}h^2 h'^2\right\}\, dz & (d)
\end{aligned}$$
$$\tag{24.67}$$

Scherzer's theorem

One of the aims of the designers of the first electron lenses was to find combinations of lens geometry and excitation for which the aberration coefficients, especially C_s, were small, preferably zero. It was not, however, long before Otto Scherzer (1936b) demonstrated that the formula for C_s can be written as the integral of a sum of squared terms, so that unless all these vanish, we can at best find the lens that corresponds to a minimum value of C_s. Scherzer's expression was nonrelativistic but Rose (1967/8) has since derived the relativistic expression:

$$C_s = \frac{1}{32\hat{\phi}_o^{\frac{1}{2}}} \int_{z_0}^{z_i} \frac{\phi^2}{\hat{\phi}^{\frac{3}{2}}}(C_1 + C_2)h^4 \, dz$$

$$C_1 = \left\{\gamma\left(\frac{\phi''}{\phi} + \frac{\phi' h'}{\phi h}\right) + \frac{\eta^2 B^2}{\phi} - \frac{1}{4}\left(\frac{5+8\epsilon\hat{\phi}}{1+\epsilon\phi}\right)\frac{\phi'^2}{\phi^2}\right\}^2$$

$$+ \frac{3}{2}\left(\frac{\phi''}{\phi} + \frac{\phi' h'}{\phi h} - \frac{\gamma\phi'^2}{\phi\hat{\phi}}\right)^2$$

$$+ \frac{\phi'^2}{\phi^2}\left(\frac{h'}{h} + \frac{5\gamma\phi'}{6\hat{\phi}}\right)^2$$

$$C_2 = \frac{4\eta^2 B^2 (1+\epsilon\phi)}{\phi}\left\{\left(\frac{h'}{h} + \frac{B'}{B} - \frac{3}{4}\frac{\gamma\phi'}{\hat{\phi}}\right)^2\right.$$

$$\left. + \left(\frac{h'}{h} + \frac{\gamma\phi'}{4\hat{\phi}}\right)^2\right\}$$

$$+ \frac{\phi'^2}{\phi^2}\left(\frac{\gamma h'}{h} + \frac{2+3\gamma^2}{6\hat{\phi}}\phi'\right)^2$$

$$+ \frac{\eta^4 B^4}{\phi^2} + \frac{1+8\epsilon\hat{\phi}}{36\hat{\phi}^2}\frac{\phi'^4}{\phi^2} + \frac{29}{8}\frac{\eta\phi'^2 B^2}{\phi^2\hat{\phi}} \quad (24.68)$$

Scherzer's original formula is given by (24.58) if we set

$$p_1 = -16; \quad p_2 = -8; \quad p_3 = -4; \quad p_4 = -5/2; \quad p_5 = -4;$$
$$p_6 = 1; \quad p_7 = 1/2; \quad p_8 = -13/8; \quad p_9 = -4; \quad p_{10} = 4 \quad (24.69)$$

The electrostatic part of the relativistic expression given by Rose corresponds to the following values of the p_i given in (24.59, 24.60):

$$p_1 = -16; \quad p_2 = -8; \quad p_3 = -4; \quad p_4 = -5/2; \quad p_5' = 1;$$
$$p_6 = 1; \quad p_7 = 1/2; \quad p_8 = -43/24; \quad p_7' = 0; \quad p_{10}' = 3; \quad (24.70)$$
$$p_{11}' = 6$$

For magnetic lenses, the appropriate formula is (24.67d).

This result has had an immense influence on electron optical studies. It is true provided that the derivations of the various expressions are correct, which requires that the lens be round, static and free of space charge; furthermore, ϕ and its derivatives must be continuous and the object and image must be real. Scherzer himself (1947) proposed methods of correcting spherical aberration by relaxing one or other of these conditions: the use of components with lower symmetry, such as quadrupoles and octopoles; the introduction of space charge or a potential discontinuity; excitation of the electrodes of an "electrostatic" lens at high frequency. We shall return to these various possibilities in Chapter 41. Meanwhile, Glaser (who constantly sought loopholes in Scherzer's proof, see HdP p.227 footnote and GdE, p.677 n.163) had the ingenious idea of seeking a magnetic lens for which $C_s = 0$ by solving the differential equation obtained by setting the integrand in C_s equal to zero (Glaser, 1940a; recalled in Lenz, 1982b and Hawkes, 1986); this failed because the resulting field distribution proved to be incapable of producing a real image of a real object (Rebsch, 1940) but was found useful in β-ray spectroscopy (Siegbahn, 1946). Recknagel (1941) analysed the electrostatic lens integrand in the same way. Rather later, Tretner established the minimum value of C_s as a function of the various lens parameters. His work and the complementary analysis of Moses is examined in Part VII.

Thin-lens approximation

If the lens is assumed to be thin, in the sense discussed in Section 17.4, each of the aberration coefficients collapses to a simpler form. For the spherical aberration, we set out from (24.66) in the magnetic case. An expression may likewise be derived for electrostatic lenses (Riedl, 1937) but has been little used in practice, no doubt because the potential distribution in the latter is rarely as narrow as the field distribution in many magnetic lenses; see however, the discussion in Hawkes (1987) concerning Renau and Heddle (1986) and their reply (1987). Equation (24.66) gives straightforwardly

$$C_s = z_o^4 \left\{ \frac{\eta^4}{12\hat{\phi}^2} \int_{-\infty}^{\infty} B^4(z)\, dz + \frac{\eta^2}{8\hat{\phi}} \int \left(\frac{dB}{dz}\right)^2 dz \right\} \qquad (24.71)$$

in which z_o is the distance of the object from the lens centre but it is usual to rewrite this in terms of dimensionless quantities by scaling distances with respect to some characteristic length of the lens, such as the gap between the polepieces, S. Writing

$$z/S =: \zeta \qquad (24.72)$$

and setting $B(z) = B_0 b(z)$, where B_0 is the maximum value of $B(z)$, (24.71) becomes

$$\frac{C_s}{S} = S^4 \zeta_o^4 \left(\frac{\eta^4 B_0^4}{12\hat{\phi}^2} \tau_4 + \frac{\eta^2 B_0^2}{8 S^2 \hat{\phi}} \tau_0 \right) \qquad (24.73)$$

in which

$$\zeta_o = \frac{z_o}{S}$$

$$\tau_4 := \int_{-\infty}^{\infty} b^4(\zeta)d\zeta \qquad (24.74)$$

$$\tau_0 := \int_{-\infty}^{\infty} \left(\frac{db}{d\zeta}\right)^2 d\zeta$$

in agreement with the expression given by Glaser (1952, eq.124.7) for high magnification, $z_o = f$. Approximate expressions for (24.73) as a function of a parameter characterizing the lens geometry have been derived by Der-Shvarts (1970) and Der-Shvarts and Makarova (1972, 1973); these are analysed in Hawkes (1980b). Several early papers discuss these thin and/or weak lens approximations; see, for example, Scherzer (1936a), Rebsch and Schneider (1937), Gratsiatos (1937), Voit (1939) and Marschall (1939).

24.4 Coma (terms linear in x_o, y_o)

For lenses such as objectives and probe-forming elements, in which the rays are inclined to the axis at a relatively steep angle, the coma is the next most important aberration after the spherical aberration. If we consider a pencil of rays intersecting the aperture around a circle of radius r_a, the corresponding aberration figure will be given by

$$\Delta u_i - 2(K+ik)r_a^2 r_o e^{i\varphi_o} = (K-ik)r_a^2 r_o e^{i(2\varphi_a - \varphi_o)} \qquad (24.75)$$

in which we have transferred the term independent of φ_a to the left-hand side since it merely shifts the image point by a distance $2\sqrt{K^2+k^2}r_a^2 r_o$ along a line inclined at an angle $\arctan(k/K)$ to the radius vector of the Gaussian image point. The term on the right-hand side causes the image point to rotate around a circle, centred on the shifted image point already discussed, of radius $\sqrt{K^2+k^2}r_a^2 r_o$. The circle is described twice as φ_a varies from zero to 2π. This behaviour may also be seen by multiplying each side of (24.75) by its complex conjugate, which gives

$$\begin{aligned}\{\Delta x - 2r_a^2(Kx_o - ky_o)\}^2 + \{\Delta y - 2r_a^2(Ky_o + kx_o)\}^2 \\ = (K^2 + k^2)r_a^4 r_o^2 \end{aligned} \qquad (24.76)$$

This clearly represents a circle of radius $(K^2+k^2)^{\frac{1}{2}}r_a^2 r_o$ centred on the point $(2r_a^2(Kx_o - ky_o), 2r_a^2(Ky_o + kx_o))$. The tangents to this circle from

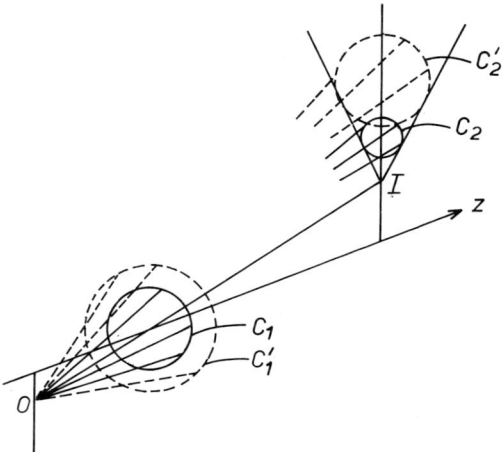

Fig. 24.6: Coma. Rays from an object point O intersect the aperture plane around concentric circles C_1, C_1' and subsequently intersect the image plane around non-concentric circles C_2, C_2'.

the origin are inclined to the line joining the origin to the centre at an angle ψ and $\sin \psi = \frac{1}{2}$; hence $\psi = 30°$. For different values of r_a, we obtain a family of circles, the centres of which lie on a straight line, all tangent to the pair of straight lines inclined to the line of centres at $\pm 30°$ (Fig. 24.6). A beam filling the aperture will therefore generate a comet-shaped aberration figure, from which the aberration takes its name.

Coma is characterized by two coefficients, K and k; if $k = 0$, the line of centres passes through the Gaussian image point and the origin in the image plane: the coma is *radial* and is exactly analogous to that of glass lenses; K is thus known as the *isotropic coma coefficient*. If, on the other hand, $K = 0$ but $k \neq 0$, then the line of centres is perpendicular to the line joining the origin to the Gaussian image point: the coma is *sagittal* (or tangential) and has no analogue in light optics; k is known as the *anisotropic coma coefficient*, and is peculiar to magnetic lenses.

Like every aberration, the coma may be interpreted in terms of the corresponding distortion of the surface $S = $ const; we have $-S^I/W = r_a^3 r_o (K \cos\varphi + k \sin\varphi) = r_a^3 r_o \sqrt{K^2 + k^2} \cos(\varphi - \varphi_k)$, $\tan\varphi_k := k/K$ and the surface S is therefore partly on the image side of the reference sphere S_o, partly on the object side. It is not so easy to relate the distortion of S to the aberration figure as in the case of spherical aberration; Fig. 24.7 shows the difference between S and S_o.

The coefficients K and k may be written in numerous forms. The

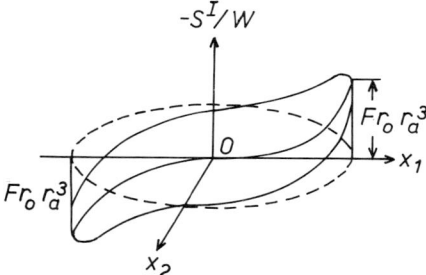

Fig. 24.7: Coma. The distance between S and S_o as measured by $-S^I/W$. The axes are such that x_1 corresponds to $\varphi - \varphi_k = 0$ and x_2 to $\varphi - \varphi_k = \pi/2$.

relativistic expressions for mixed lenses (24.32 with 24.3) in terms of object and aperture coordinates are

$$K = \frac{1}{W_s} \int_{z_o}^{z_i} \left(L_1 st^3 + L_2(st)'tt' + L_3 s't'^3 \right) dz$$

$$k = \int_{z_o}^{z_i} \left(Pt^2 + Qt'^2 \right) dz \tag{24.77}$$

By substituting $x_a = x'_o/t'_o - x_o s'_o/t'_o$ and likewise for y_a in the complete expressions for $\Delta x_i, \Delta y_i$ and using the relations $g = s - s'_o t/t'_o$, $h = t/t'_o$, it is readily shown that the coma coefficients have the same form as (24.77), namely

$$K = \frac{1}{W_g} \int \left(L_1 g h^3 + L_2(gh)'hh' + L_3 g'h'^3 \right) dz$$

$$k = \int \left(Ph^2 + Qh'^2 \right) dz \tag{24.78}$$

in terms of position and gradient in the object plane; note that both the spherical aberration and the coma contribute when making the substitutions. In the following expressions, therefore, s and t may be replaced by g and h at will.

For *electrostatic lenses*, the general relativistic expression for K is

$$32\hat{\phi}_o^{\frac{1}{2}} K = \frac{1}{t'_o} \int_{z_o}^{z_i} \left\{ A_0 st^3 + \frac{1}{2} A_1(3s't^2 + st'^2)t \right.$$

$$+ A_2(st + s't')tt' + \frac{1}{2}A_3(s't^2 + 3st'^2)t'$$
$$+ A_4 s't'^3 \bigg\} dz$$
$$- \frac{p_1}{4}\hat{\phi}_o^{\frac{1}{2}} t'_o(t'^2_i - t'^2_o) \tag{24.79}$$

where the A_i are given by (24.60). The corresponding nonrelativistic form is obtained by setting A'_i (24.60') in place of A_i.

For *magnetic lenses*, we find

$$K = \frac{1}{t'_o}\int_{z_o}^{z_i}\bigg\{st^3\bigg(\frac{\eta^4 B^4}{32\hat{\phi}^2} - \frac{\eta^2 BB''}{8\hat{\phi}} + p_1\frac{\eta^2 B'^2}{\hat{\phi}} + p_1\frac{\eta^2 BB''}{\hat{\phi}}$$
$$- p_2\frac{\eta^4 B^4}{4\hat{\phi}^2} - p_3\frac{\eta^4 B^4}{4\hat{\phi}^2}\bigg)$$
$$+ \frac{\eta^2 BB'}{\hat{\phi}}s't^3(p_1 + 2p_3)$$
$$+ \frac{\eta^2 BB'}{\hat{\phi}}st^2 t'(3p_1 + 2p_2) \tag{24.80}$$
$$+ \frac{\eta^2 B^2}{\hat{\phi}}s't^2 t'\bigg(\frac{1}{8} + p_2 - \frac{1}{2}p_4 + 3p_3\bigg)$$
$$+ \frac{\eta^2 B^2}{\hat{\phi}}stt'^2\bigg(\frac{1}{8} + 2p_2 - \frac{1}{4}p_4 - \frac{3}{4}p_5\bigg)$$
$$+ s't'^3\bigg(\frac{1}{2} + p_4 + p_5\bigg)\bigg\} dz - \frac{p_5}{t'_o}\bigg[st^3\bigg]_{z_o}^{z_i} - \frac{p_4}{t'_o}\bigg[s'tt'^2\bigg]_{z_o}^{z_i}$$

Eliminating the term in t'^2 from k gives

$$k = \frac{1}{16}\int\bigg(\frac{\eta B''}{\hat{\phi}^{\frac{1}{2}}} + \frac{2\eta^3 B^3}{\hat{\phi}^{\frac{3}{2}}}\bigg)t^2\, dz \tag{24.81}$$

(The general form of k, set out in Hawkes (1980b), is of less interest.)

Thin-lens formulae

Some care is needed in deriving the thin-lens formula for K. Choosing the p_i in such a way that terms involving the discontinuous function t' vanish in the integrand, we apparently find that $K = 0$; the presence of this

24.5 ASTIGMATISM AND FIELD CURVATURE

discontinuity at z_a means, however, that the integrated term in $[s'tt'^2]_{z_o}^{z_i}$ is not zero, even though $s'tt'^2$ vanishes at the endpoints. We now have

$$\frac{1}{t'_o}\left[s'tt'^2\right]_{z_o}^{z_i} = \frac{1}{t'_o}\left[s'tt'^2\right]_{z_o}^{z_{a-}} + \frac{1}{t'_o}\left[s'tt'^2\right]_{z_{a+}}^{z_i}$$

$$= -t'^2_o + t'^2_i \qquad (24.82)$$

$$= \frac{1}{z_o^2}(\frac{1}{M^2} - 1)$$

and with $p_4 = -\frac{1}{2}$

$$K \to \frac{1}{2}\frac{1}{z_o^2}(\frac{1}{M^2} - 1) \qquad (24.83)$$

or if the aberrations are expressed in terms of position and gradient at the object plane

$$K \to \frac{1}{2}(\frac{1}{M^2} - 1) \qquad (24.84)$$

This integrated term is neglected by Glaser (1952, eq.124.30), who thus finds $K \to 0$.

For the anisotropic coefficient, we find

$$k \to \frac{\eta^3 B_0^3 S}{16\hat{\phi}^{\frac{3}{2}}}\tau_3 + \frac{\eta B_0 S}{8\hat{\phi}^{\frac{1}{2}}}\tau_1(t'^2_o + t'^2_i) \qquad (24.85)$$

in which

$$\tau_1 := \int_{-\infty}^{\infty} b(\zeta)\, d\zeta$$

$$\tau_3 := \int_{-\infty}^{\infty} b^3(\zeta)\, d\zeta \qquad (24.86)$$

(cf. 24.74). Here too, we do not agree with Glaser (1952, e.q.124.31); for further discussion, see Hawkes (1980b).

24.5 Astigmatism and field curvature (terms linear in x_a, y_a)

These aberrations are usually the least important of the third-order defects of electron lenses. The corresponding aberration figure is obtained as usual by considering rays from an object point that intersect the aperture plane around a circle of radius r_a. We have

$$\Delta u_i = (A + ia)r_o^2 r_a e^{i(2\varphi_o - \varphi_a)} + F r_o^2 r_a e^{i\varphi_a} \qquad (24.87)$$

24. THE GEOMETRICAL ABERRATIONS OF ROUND LENSES

The term in F resembles a defocus term; returning to the paraxial approximation, $u = u_o s + u_a t$, we see that in a plane close to the image plane z_i, $z = z_i + \zeta$, say, we have $u(z_i + \zeta) \approx u(z_i) + \zeta u_a t'_i$. Hence $u(z_i + \zeta) - u(z_i) \approx \zeta u_a t'_i$. The term in F therefore resembles the blur caused by a defocus $\zeta, \zeta \propto F r_o^2$. After reading Chapter 31, it will be apparent that the term in A and a is the third-order analogue of the paraxial astigmatism of imperfectly round lenses.

Equation (24.87) is the parametric representation of a tilted ellipse, as we can see by suitably rotating the coordinate frame. With an angle of rotation χ, we have

$$\Delta u_i e^{-i\chi} = (A + ia) r_o^2 r_a e^{i(2\varphi_o - \varphi_a - \chi)} + F r_o^2 r_a e^{i(\varphi_a - \chi)}$$
$$= \sqrt{A^2 + a^2}\, r_o^2 r_a e^{i(2\varphi_o - \varphi_a + \varphi_A - \chi)} + F r_o^2 r_a e^{i(\varphi_a - \chi)} \quad (24.88)$$

where $A + ia =: \sqrt{A^2 + a^2} \exp(i\varphi_A)$. On choosing χ so that the exponents are equal and opposite, for which

$$\chi = \varphi_o + \frac{1}{2}\varphi_A \quad (24.89)$$

we have

$$\Delta \overline{x}_i + i\Delta \overline{y}_i := \Delta \overline{u}_i := \Delta u_i e^{-i\chi}$$
$$= F r_o^2 r_a e^{i(\varphi_a - \varphi_o - \varphi_A/2)} + \sqrt{A^2 + a^2} r_o^2 r_a e^{-i(\varphi_a - \varphi_o - \varphi_A/2)}$$
$$= (F + \sqrt{A^2 + a^2}) r_o^2 r_a \cos(\varphi_a - \varphi_o - \varphi_A/2)$$
$$+ i(F - \sqrt{A^2 + a^2}) r_o^2 r_a \sin(\varphi_a - \varphi_o - \varphi_A/2) \quad (24.90)$$

Hence

$$\left(\frac{\Delta \overline{x}_i}{a_1}\right)^2 + \left(\frac{\Delta \overline{y}_i}{a_2}\right)^2 = 1 \quad, \quad a_{1,2} = r_o^2 r_a (F \pm \sqrt{A^2 + a^2}) \quad (24.91)$$

For a fixed object point, therefore, the beam that fills the aperture $(x_a, y_a | x_a^2 + y_a^2 \leq r_A^2)$ occupies an ellipse in the image plane, centred on the Gaussian image point, with semi-axes a_1, a_2. The major and minor axes are inclined to the line joining the origin to the image point. Some typical aberration figures are shown in Fig. 24.8.

The aberration characterized by F is known as the *field curvature* while the coefficients A and a describe the *isotropic* and *anisotropic astigmatism*. The last vanishes in the absence of any magnetic field. The third-order astigmatism is often known as the Seidel astigmatism (as in light optics),

24.5 ASTIGMATISM AND FIELD CURVATURE

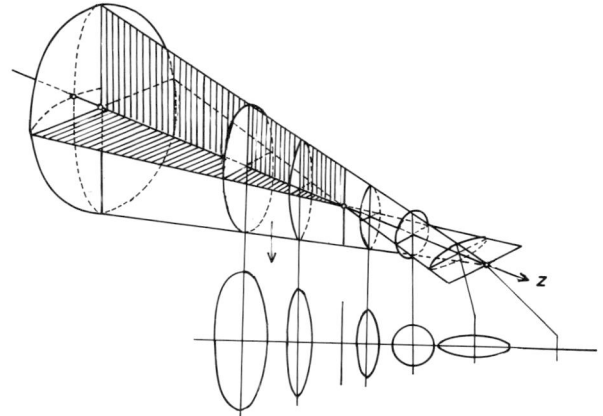

Fig. 24.8: Astigmatism. Cross-section of an initially circular beam at various planes in the neighbourhood of the Gaussian image plane. To simplify the drawing, the ellipses are shown vertical or horizontal relative to the principal ray.

to distinguish it from the paraxial astigmatism, which we shall meet in Chapter 31.

The field curvature is so called from the observation that it vanishes if the image is studied not in the Gaussian image plane but on a curved surface tangent to the latter. In order to see why this is so, we now enquire how the foregoing reasoning would be modified if the object and image points lay not on planes but on curved surfaces, tangent to the object and image planes at the axis. We examine only the case of spherically curved surfaces, S_o and S_i in Fig. 24.9.

We consider a ray from the object point $P_o(x_o, y_o)$, intersecting S_o at $\overline{P}_o(\overline{x}_o, \overline{y}_o)$, S_i at $\overline{P}_i(\overline{x}_i, \overline{y}_i)$ and the Gaussian plane at $P_i(x_i, y_i)$. The distances ζ_o and ζ_i are the sagittae of the chords through $\overline{P}_o$ and $\overline{P}_i$ respectively, normal to the optic axes, and are hence given by

$$\zeta_o = \frac{\overline{x}_o^2 + \overline{y}_o^2}{2\varrho_o} \approx \frac{x_o^2 + y_o^2}{2\varrho_o} = \frac{r_o^2}{2\varrho_o}$$
$$\zeta_i = -\frac{\overline{x}_i^2 + \overline{y}_i^2}{2\varrho_i} \approx -\frac{x_i^2 + y_i^2}{2\varrho_i} = -\frac{M^2}{2\varrho_i}(x_o^2 + y_o^2) = -\frac{M^2 r_o^2}{2\varrho_i} \quad (24.92)$$

The sign convention is such that ϱ_o and ϱ_i are regarded as positive if $z(C_o) > z_o$ and $z(C_i) < z_i$, as is the case in Fig. 24.9; ζ_o clearly has the same sign as ϱ_o, and ζ_i the opposite sign to ϱ_i.

24. THE GEOMETRICAL ABERRATIONS OF ROUND LENSES

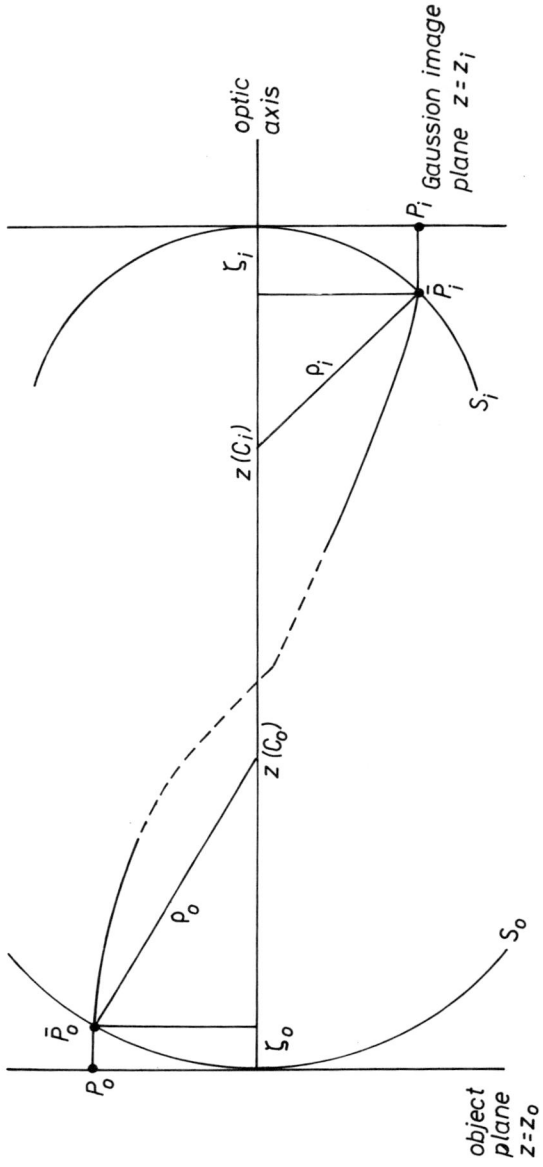

Fig. 24.9: Field curvature. Notation used when the object and image surfaces may be curved.

24.5 ASTIGMATISM AND FIELD CURVATURE

In the Gaussian approximation, we have

$$\bar{x}_o = x_o s(z_o + \zeta) + x_a t(z_o + \zeta)$$
$$= x_o + \frac{s'_o}{2\varrho_o} x_o r_o^2 + \frac{t'_o}{2\varrho_o} x_a r_o^2 \quad (24.93)$$
$$\bar{y}_o = y_o + \frac{s'_o}{2\varrho_o} y_o r_o^2 + \frac{t'_o}{2\varrho_o} y_a r_o^2$$

and similarly

$$\bar{x}_i = x_o s(z_i + \zeta_i) + x_a t(z_i + \zeta_i)$$
$$\approx x_i + x_o s'_i \zeta_i + x_a t'_i \zeta_i$$
$$= x_i - \frac{M^2 s'_i}{2\varrho_i} x_o r_o^2 - \frac{M^2 t'_i}{2\varrho_i} x_a r_o^2 \quad (24.94)$$
$$\bar{y}_i = y_i - \frac{M^2 s'_i}{2\varrho_i} y_o r_o^2 - \frac{M^2 t'_i}{2\varrho_i} y_a r_o^2$$

Thus

$$\Delta \bar{x}_i := \frac{\bar{x}_i}{M} - \bar{x}_o = \frac{x_i}{M} - x_o - \left(\frac{M s'_i}{2\varrho_i} + \frac{s'_o}{2\varrho_o}\right) x_o r_o^2$$
$$- \left(\frac{M t'_i}{2\varrho_i} + \frac{t'_o}{2\varrho_o}\right) x_a r_o^2 \quad (24.95)$$
$$\Delta \bar{y}_i = \Delta y_i - \left(\frac{M s'_i}{2\varrho_i} + \frac{s'_o}{2\varrho_o}\right) y_o r_o^2 - \left(\frac{M t'_i}{2\varrho_i} + \frac{t'_o}{2\varrho_o}\right) y_a r_o^2$$

Returning to (24.87), and considering first the terms in F only, we have

$$\Delta \bar{x}_i = -\left(\frac{M s'_i}{2\varrho_i} + \frac{s'_o}{2\varrho_o}\right) x_o r_o^2 + \left(F - \frac{M t'_i}{2\varrho_i} - \frac{t'_o}{2\varrho_o}\right) x_a r_o^2 \quad (24.96)$$

with a similar expression for $\Delta \bar{y}_i$. The term in $x_o r_o^2$ is a distortion and will be dealt with in the next section. For a plane image surface, $\varrho_i \to \infty$, the last term vanishes if ϱ_o is chosen so that

$$\varrho_o = \frac{t'_o}{2F} \quad (24.97)$$

If now we retain all the terms of (24.87), again postponing discussion of the term in $x_o r_o^2$, we obtain expressions for $\Delta \bar{x}_i$ and $\Delta \bar{y}_i$ identical with those given by (24.90) except that $F - Mt'_i/2\varrho_i - t'_o/2\varrho_o$ replaces F. The axes of the ellipse corresponding to particular values of x_o, y_o and r_A are

$$a_{1,2} = r_A r_o^2 \left\{ F - \frac{M t'_i}{2\varrho_i} - \frac{t'_o}{2\varrho_o} \pm (A^2 + a^2)^{\frac{1}{2}} \right\} \quad (24.98)$$

If ϱ_i takes the value $\varrho_i^{(1)}$ such that

$$\frac{Mt'_i}{2\varrho_i^{(1)}} = F - \frac{t'_o}{2\varrho_o} + (A^2 + a^2)^{\frac{1}{2}} \qquad (24.99)$$

the semi-axis a_1 is zero and the ellipse then collapses to a line of length $4(A^2 + a^2)^{1/2} r_o^2 r_A$; likewise, for $\varrho_i = \varrho_i^{(2)}$ such that

$$\frac{Mt'_i}{2\varrho_i^{(2)}} = F - \frac{t'_o}{2\varrho_o} - (A^2 + a^2)^{\frac{1}{2}} \qquad (24.100)$$

a_2 is zero and the ellipse collapses to a line of the same length. These lines are known as the tangential or meridional and sagittal line foci and the corresponding values of ϱ_i as the tangential and sagittal field curvatures. For an intermediate value of ϱ_i, $\bar{\varrho}_i$, the ellipse becomes a circle; here, $|a_1| = |a_2|$ or

$$\frac{Mt'_i}{2\bar{\varrho}_i} = F - \frac{t'_o}{2\varrho_o}$$

or

$$\frac{1}{\bar{\varrho}_i} = \frac{1}{2}\left(\frac{1}{\varrho_i^{(1)}} + \frac{1}{\varrho_i^{(2)}}\right) \qquad (24.101)$$

This is sometimes known as the mean field curvature and the corresponding circle is again called a 'circle of least confusion'; it is essential to mention that the confusion caused by A, a and F is intended.

We have not mentioned ϱ_o in the foregoing reasoning since microscope objects and microprobe targets are usually (assumed to be) flat; it is, however, necessary to consider it in some devices, especially those involving electron emission from surfaces.

Finally, we note that to each ϱ_i corresponds a value of ζ_i. Writing (24.92)

$$\zeta_i^{(1,2)} := -\frac{M^2 r_o^2}{2\varrho_i^{(1,2)}} \qquad (24.102)$$

we have

$$\Delta\zeta := \zeta_i^{(2)} - \zeta_i^{(1)} = -\frac{M^2 r_o^2}{2}\left(\frac{1}{\varrho_i^{(2)}} - \frac{1}{\varrho_i^{(1)}}\right)$$
$$= \frac{2Mr_o^2}{t'_i}\sqrt{A^2 + a^2} \qquad (24.103)$$

Similarly,

$$\bar{\zeta} := -\frac{M^2 r_o^2}{2\bar{\varrho}_i} = \frac{Mr_o^2}{t'_i}(F - \frac{t'_o}{\varrho_i}) \qquad (24.104)$$

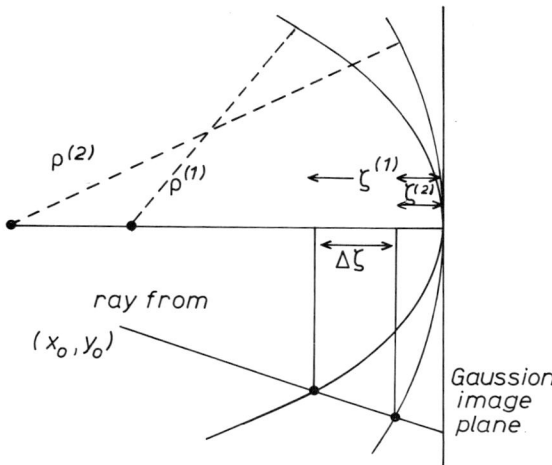

Fig. 24.10: Notation used in connection with a curved image surface.

which vanishes if $F = 0$ and $\varrho_i \to \infty$: the circle of least confusion due to astigmatism alone lies in the Gaussian image plane. The quantity $\Delta \zeta$ is known as the astigmatic difference. Figure 24.10 illustrates the geometrical meanings of $\zeta_i^{(1,2)}$ and $\Delta \zeta$.

In terms of the distortion of the surfaces S=constant, the astigmatism and field curvature have different effects. We have $-S^I/W = \frac{1}{2} r_a^2 r_o^2 (A \cos 2\varphi + a \sin 2\varphi)$ and $-S^I/W = \frac{1}{2} F r_a^2 r_o^2$. Thus field curvature, like spherical aberration, shifts the surface S uniformly away from the reference sphere S_o towards the image plane ($F > 0$) by an amount that now varies with the distance of the object point from the axis.

For the astigmatism, we write $-S^I/W = \frac{1}{2} r_a^2 r_o^2 \sqrt{A^2 + a^2} \cos(2\varphi - \varphi_A)$, $\tan \varphi_A := a/A$ and we see that the distance from S_o to S is now described by a saddle-shaped surface (Fig. 24.11), S being shifted towards the image around $\varphi - \varphi_A/2 = 0, \pi$ and away from it in the neighbourhood of $\varphi - \varphi_A/2 = \pi/2, 3\pi/2$. The formation of line foci is readily understood.

The coefficients F, A and a are given by the following formulae:

$$F = \frac{1}{W_s} \int_{z_o}^{z_i} \{2L_1 s^2 t^2 + L_2(st' + s't)^2 + 2L_3 s'^2 t'^2 \\ + R(st' - s't)^2\} dz \qquad (24.105)$$

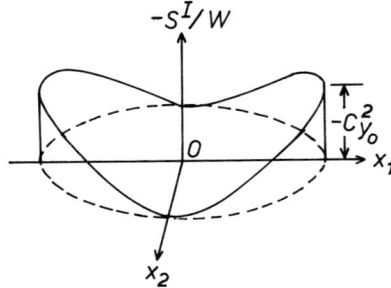

Fig. 24.11: The effect of astigmatism on the wave surface.

$$A = \frac{1}{W_s} \int_{z_o}^{z_i} \{L_1 s^2 t^2 + 2L_2 s s' t t' + L_3 s'^2 t'^2 \qquad (24.106)$$

$$- R(st' - s't)^2\} \, dz$$

$$a = 2 \int (Pst + Qs't') \, dz \qquad (24.107)$$

Replacing s by g and t by h, we obtain the expressions in terms of position and gradient in the object plane; spherical aberration, coma and the expressions for F and A now all contribute.

The structures of F and A are very similar; we see that

$$F - 2A = W_s \int_{z_o}^{z_i} \frac{L_2 + 3R}{\hat{\phi}} \, dz$$

$$= W_s \int_{z_o}^{z_i} \frac{\gamma \phi'' + 4\eta^2 B^2}{8\hat{\phi}^{\frac{3}{2}}} \, dz \qquad (24.108)$$

which is a function of $\phi(z)$ and $B(z)$ only and does not contain $s(z)$ or $t(z)$ except in the Wronskian. This quantity is known as the *Petzval coefficient*; it can be shown that it is the only linear combination of the third-order aberration coefficients that possesses this property.

For *electrostatic lenses* the general relativistic expressions for F and A are

$$32\hat{\phi}_o^{\frac{1}{2}} A = \frac{1}{t'_o} \int \{A_0 s^2 t^2 + A_1 st(st)' + A_3 s't'(st)'$$

$$+ A_4 s'^2 t'^2 + \frac{1}{2} A_5 (s^2 t'^2 + s'^2 t^2) + A_6 ss' tt'\} \, dz$$

24.5 ASTIGMATISM AND FIELD CURVATURE

$$-\frac{1}{2}p_1[\hat{\phi}^{\frac{1}{2}}ss't'^2] + \frac{p_2' - p_2}{4}\left[\frac{\gamma\phi'}{\hat{\phi}^{\frac{1}{2}}}s^2t'^2\right] \qquad (24.109)$$

$$32\hat{\phi}_o^{\frac{1}{2}}F = \frac{2}{t'_o}\int\{A_0 s^2 t^2 + A_1 st(st)' + A_3 s't'(st)'$$

$$+ A_4 s'^2 t'^2 + \frac{1}{4}A_6(st' + s't)^2 + A_5 ss'tt'\}\,dz$$

$$- p_1[\hat{\phi}^{\frac{1}{2}}ss't'^2] - \frac{p_2' + p_2}{4}\left[\frac{\gamma\phi'}{\hat{\phi}^{\frac{1}{2}}}s^2t'^2\right] \qquad (24.110)$$

in which

$$A_5 := \frac{\gamma\phi''}{\hat{\phi}^{\frac{1}{2}}}\{p_3 + \frac{1}{2}(p_2 - p_2') - \frac{1}{4}p_1\}$$

$$+ \frac{\phi'^2}{\hat{\phi}^{\frac{3}{2}}}\{p_4 - \frac{3}{4}\gamma^2(p_2 - p_2')\} \qquad (24.111)$$

$$- \frac{\epsilon\phi'^2}{\hat{\phi}^{\frac{1}{2}}}\{8p_5' - (p_2 - p_2')\}$$

$$A_6 := \frac{\gamma\phi''}{\hat{\phi}^{\frac{1}{2}}}\{8 + 2p_3 + \frac{1}{2}(p_2 + p_2') - \frac{1}{2}p_1\}$$

$$+ \frac{\phi'^2}{\hat{\phi}^{\frac{3}{2}}}\{2p_4 - \frac{3}{4}\gamma^2(p_2 + p_2')\} \qquad (24.112)$$

$$- \frac{\epsilon\phi'^2}{\hat{\phi}^{\frac{1}{2}}}\{16p_5' - (p_2 + p_2')\}$$

For *magnetic lenses*, we have

$$A = \frac{1}{t'_o}\int_{z_o}^{z_i}\{s^2 t^2\{\frac{\eta^4 B^4}{\hat{\phi}^2}(\frac{1}{32} - \frac{1}{4}p_2 - \frac{1}{4}p_5)$$

$$+ \frac{\eta^2 BB''}{\hat{\phi}}(p_1 - \frac{1}{8}) + \frac{\eta^2 B'^2}{\hat{\phi}}p_1\}$$

$$+ \frac{2\eta^2 BB'}{\hat{\phi}}st\{s't(p_1 + p_5) + st'(p_1 + p_2)\}$$

$$+ \frac{\eta^2 B^2}{\hat{\phi}}\{s'^2 t^2(p_5 - \frac{1}{4}p_3) + s^2 t'^2(p_2 - \frac{1}{4}p_4)$$

$$+ 2ss'tt'(\frac{1}{8} + p_5 + p_2 - \frac{1}{4}p_3 - \frac{1}{4}p_4)\}$$

$$+ s'^2 t'^2(\frac{1}{2} + p_3 + p_4)\}\,dz$$

$$-\frac{\eta^2 t'_o}{8\hat{\phi}} \int B^2\, dz$$

$$-\frac{1}{t'_o}\left[p_4 s s' t'^2 + p_3 s'^2 t t'\right]_{z_o}^{z_i} \tag{24.113}$$

$$F = 2A + \frac{\eta^2 t'_o}{2\hat{\phi}} \int B^2\, dz \tag{24.114}$$

using (24.108), and

$$a = \frac{1}{8}\int_{z_o}^{z_i}\left(2\frac{\eta^3 B^3}{\hat{\phi}^{\frac{3}{2}}} + \frac{\eta B''}{\hat{\phi}^{\frac{1}{2}}}\right)st\, dz + \frac{1}{8}\left[\frac{\eta B}{\hat{\phi}^{\frac{1}{2}}}st'\right]_{z_o}^{z_i} \tag{24.115}$$

Thin-lens formulae

For magnetic lenses, the isotropic and anisotropic astigmatism both vanish,

$$A \to 0, \quad a \to 0 \tag{24.116}$$

and the field curvature is hence equal to the Petzval coefficient:

$$F \to \frac{\eta^2 t'_o}{2\hat{\phi}}\int B^2\, dz = \frac{\eta^2 B_0^2 \tau_2}{2\hat{\phi}\zeta_o} \tag{24.117}$$

Glaser (1952, eq.124.29,31) also finds $a \to 0$ but his expression for A, and hence that for F, does not agree with ours. Der-Shvarts and Makarova (1973) find the same thin-lens expression for F but do not agree that $A \to 0$. (Note that these authors do not adopt the definitions of field curvature and astigmatism employed here.)

24.6 Distortion (terms in x_o and y_o only)

These aberrations are of most importance in lenses in which the rays are comparatively far from the axis at the object plane. They are usually negligible in microscope objectives, therefore, but are the dominant geometrical defects of projectors. They can be kept acceptably small either by careful design of the column of a conventional instrument or by employing a very compact type of lens, which we describe in Part VII. As their name indicates, distortions do not blur the image but merely destroy the proportionality between image and object coordinates.

From (24.36), we see that

$$\Delta x_i = r_o^2(Dx_o - dy_o)$$
$$\Delta y_i = r_o^2(Dy_o + dx_o)$$ (24.118)

and to avoid repetition, we modify these expressions to include the terms that arise if the object and image surfaces are curved (spherical), as in Section 24.5; this gives

$$\Delta \bar{x}_i = \left\{ \left(D - \frac{Ms'_i}{2\varrho_i} - \frac{s'_o}{2\varrho_o} \right) x_o - dy_o \right\} r_o^2$$ (24.119a)

$$\Delta \bar{y}_i = \left\{ \left(D - \frac{Ms'_i}{2\varrho_i} - \frac{s'_o}{2\varrho_o} \right) y_o + dx_o \right\} r_o^2$$ (24.119b)

or

$$\Delta \bar{u}_i = (D^* + id) r_o^3 e^{i\varphi_o}$$

with $D^* = D - Ms'_i/2\varrho_i - s'_o/2\varrho_o$. Writing

$$\Delta \bar{u}_i =: e^{i\varphi_o}(\Delta r_o + i r_o \Delta \varphi_o)$$ (24.120)

we see that

$$\Delta r_o = D^* r_o^3 \qquad \Delta \varphi_o = d r_o^2$$ (24.121)

so that D^* shifts the image point radially relative to its Gaussian position whereas d shifts it azimuthally. Consider now a square grid in the object plane, $x_o = m\tilde{x}$ for all y_o and $y_o = n\tilde{y}$ for all x_o, $m,n = 0,1,...$ (Fig. 24.12a). For $d = 0$, the grid will be shrunk, as shown in Fig. 24.12b if $D^* < 0$; this is known as *barrel distortion*. If $D^* > 0$, the grid is distended as in Fig. 24.12c; we then speak of *pincushion distortion*.

The coefficient D is the *isotropic distortion coefficient*.

If $D = 0$ but $d \neq 0$, the grid is warped as shown in Fig. 24.12d. The coefficient d is known as the *anisotropic distortion coefficient* and vanishes in the absence of magnetic fields. It is often referred to as the *spiral distortion* or occasionally (Sturrock, 1955) as the *pocket-handkerchief distortion*.

In terms of the shape of the surfaces S=const, the distortion has a very simple effect. We have

$$-S^I/W = r_a r_o^3 (D \cos \varphi + d \sin \varphi) = r_a r_o^3 \sqrt{D^2 + d^2} \cos(\varphi - \varphi_D)$$

with $\tan \varphi_D = d/D$ and the distance between S_o and S is therefore represented by a plane. The sphere S_o may thus be pictured as being shifted

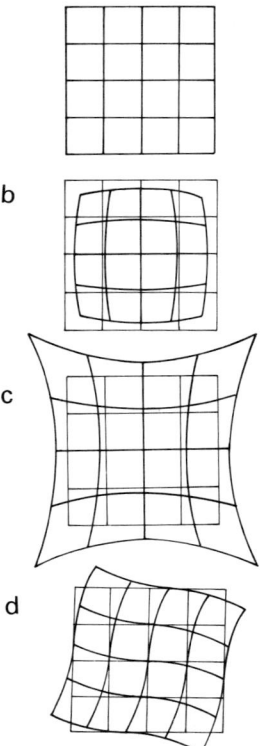

Fig. 24.12: Distortion. (a) Square grid in the object plane. (b) Image with barrel distortion. (c) Image with pincushion distortion. (d) Image with anisotropic or "spiral" distortion.

bodily so that the Gaussian image point is shifted by an amount that is constant for each object point but different ($\propto r_o^3$) for object points at varying distances from the axis. This is exactly the behaviour we have already described as characteristic of distortion. The coefficients D and d are given in general by

$$D = \frac{1}{W_s} \int_{z_o}^{z_i} \{L_1 s^3 t + L_2 ss'(st)' + L_3 s'^3 t'\}\, dz$$

$$d = \int_{z_o}^{z_i} (Ps^2 + Qs'^2)\, dz \tag{24.122}$$

For *electrostatic lenses*, the general expression is

$$32\hat{\phi}_o^{\frac{1}{2}} D = \frac{1}{t'_o} \int_{z_o}^{z_i} \Big\{ A_0 s^3 t + \frac{1}{2} A_1 s (3s^2 t' + s'^2 t) + A_2 ss'(st + s't')$$

$$+ \frac{1}{2} A_3 s'(s^2 t' + 3s'^2 t) + A_4 s'^3 t' \Big\} dz$$

$$- \Big[\frac{3}{4} p_1 \hat{\phi}^{\frac{1}{2}} ss'^2 t' + \frac{1}{2} p_2 \frac{\gamma \phi'}{\hat{\phi}^{\frac{1}{2}}} s^2 s' t'$$

$$+ \Big(\frac{1}{4} p_3 \frac{\gamma \phi''}{\hat{\phi}^{\frac{1}{2}}} + \frac{1}{4} p_4 \frac{\phi'^2}{\hat{\phi}^{\frac{3}{2}}} - 2 p'_5 \frac{\epsilon \phi'^2}{\hat{\phi}^{\frac{1}{2}}} \Big) s^3 t' \Big]_{z_o}^{z_i} \quad (24.123)$$

For *magnetic lenses*,

$$D = \frac{1}{t'_o} \int \Big\{ s^3 t \Big(\frac{\eta^4 B^4}{32 \hat{\phi}^2} - \frac{\eta^2 BB''}{8\hat{\phi}} + p_1 \frac{\eta^2 B'^2}{\hat{\phi}}$$

$$+ p_1 \frac{\eta^2 BB''}{\hat{\phi}} - (p_2 + p_3) \frac{\eta^4 B^4}{4\hat{\phi}^2} \Big)$$

$$+ \frac{\eta^2 BB'}{\hat{\phi}} s^3 t' (p_1 + 2p_3)$$

$$+ \frac{\eta^2 BB'}{\hat{\phi}} s^2 s' t (3p_1 + 2p_2)$$

$$+ \frac{\eta^2 B^2}{\hat{\phi}} s^2 s' t' \Big(\frac{1}{8} + p_2 - \frac{1}{2} p_4 + 3p_3 \Big)$$

$$+ \frac{\eta^2 B^2}{\hat{\phi}} ss'^2 t \Big(\frac{1}{8} + 2p_2 - \frac{1}{4} p_4 - \frac{3}{4} p_5 \Big)$$

$$+ s'^3 t' \Big(\frac{1}{2} + p_4 + p_5 \Big) \Big\} dz$$

$$- \frac{1}{t'_o} \Big[p_5 s'^3 t + p_4 ss'^2 t' + p_3 \frac{\eta^2 B^2}{\hat{\phi}} s^3 t' \Big]_{z_o}^{z_i} \quad (24.124)$$

$$d = \frac{1}{16 \hat{\phi}^{\frac{1}{2}}} \int_{z_o}^{z_i} \Big(\eta B'' + \frac{2\eta^3 B^3}{\hat{\phi}} \Big) s^2 \, dz$$

$$+ \Big[\frac{\eta B}{4 \hat{\phi}^{\frac{1}{2}}} ss' - \frac{\eta B'}{8 \hat{\phi}^{\frac{1}{2}}} s^2 \Big]_{z_o}^{z_i} \quad (24.125)$$

Thin-lens formulae
For magnetic lenses, we find

$$D = 0$$
$$d = \frac{\eta B_0 \tau_1}{4\hat{\phi}^{\frac{1}{2}} S \zeta_o^2} \qquad (24.126)$$

Glaser (1952, 124.30) likewise finds that the isotropic distortion coefficient vanishes but, unlike us, he also finds $d = 0$ (123.31); the origin of this disagreement lies in his use of a partially integrated expression.

24.7 The variation of the aberration coefficients with aperture position

The aberration coefficients clearly vary with the position of the object and aperture. In the case of the *real* aberration coefficients, with which we are concerned here, it is principally the aperture dependence that is important, since objective and probe-forming lenses are mostly operated with the specimen close to a focus. Furthermore, it is not possible to enunciate general laws about the dependence of the aberration coefficients on *object* position when the latter lies within the field owing to the fact that z_o is the lower limit of integration in S_{oi}^I. Their dependence on the *aperture* position can, however, be studied as we now show, and it will transpire that some aberration coefficients can be made to vanish by choosing z_a suitably.

Each of the integrands appearing in the aberration coefficients consists of one or more terms of the form $f(z) s^p s'^q t^r t'^s$, $p + q = m$, $r + s = 3 - m$, in which, we recall, $s(z)$ and $t(z)$ correspond to some particular aperture position. For this choice, we have (24.36)

$$\Delta x_i = \begin{pmatrix} x_o \\ x_a \\ y_o \end{pmatrix}^T \begin{pmatrix} D & K & 2A & a \\ F-A & C & 2K & 2k \\ -d & -k & -a & 0 \end{pmatrix} \begin{pmatrix} r_o^2 \\ r_a^2 \\ V \\ v \end{pmatrix} \qquad (24.127)$$

with a similar expression for Δy_i $(x \to y, y \to -x)$. For another choice of aperture position, we should have

$$\Delta \overline{x}_i = \begin{pmatrix} x_o \\ x_a \\ y_o \end{pmatrix}^T \begin{pmatrix} \overline{D} & \overline{K} & 2\overline{A} & \overline{a} \\ \overline{F}-\overline{A} & \overline{C} & 2\overline{K} & 2\overline{k} \\ -\overline{d} & -\overline{k} & -\overline{a} & 0 \end{pmatrix} \begin{pmatrix} r_o^2 \\ r_a^2 \\ V \\ \overline{v} \end{pmatrix} \qquad (24.128)$$

24.7 THE VARIATION OF THE ABERRATION COEFFICIENTS

with

$$\bar{r}_a^2 := \bar{x}_a^2 + \bar{y}_a^2$$
$$\bar{V} := x_o\bar{x}_a + y_o\bar{y}_a \qquad (24.129)$$
$$\bar{v} := x_o\bar{y}_a - \bar{x}_a y_o$$

In (24.128), the coefficients must have the same overall structure as those of Δx_i but contain $\bar{s}$ and $\bar{t}$ satisfying $\bar{s}(z_o) = \bar{t}(\bar{z}_a) = 1$, $\bar{s}(\bar{z}_a) = \bar{t}(z_o) = 0$. Setting

$$\bar{s}(z) = s(z) + \sigma t(z)$$
$$\bar{t}(z) = \tau t(z) \qquad (24.130)$$

we see that

$$\sigma = -\frac{s(\bar{z}_a)}{t(\bar{z}_a)}$$
$$\tau = \frac{1}{t(\bar{z}_a)} \qquad (24.131)$$

and

$$\bar{s}\bar{t}' - \bar{s}'\bar{t} = \tau(st' - s't)$$
$$W_{\bar{s}} = \tau W_s \qquad (24.132)$$

From the expressions for $A, C, \ldots, k$, it is a simple matter to show that

$$\begin{pmatrix} \overline{C} \\ \overline{K} \\ \overline{A} \\ \overline{F} \\ \overline{D} \\ \overline{k} \\ \overline{a} \\ \overline{d} \end{pmatrix} = \begin{pmatrix} \tau^3 & 0 & 0 & 0 & 0 & & & \\ \tau^2\sigma & \tau^2 & 0 & 0 & 0 & & & \\ \tau\sigma^2 & 2\tau\sigma & \tau & 0 & 0 & & 0 & \\ 2\tau\sigma^2 & 4\tau\sigma & 0 & \tau & 0 & & & \\ \sigma^3 & 3\sigma^2 & 2\sigma & \sigma & 1 & & & \\ & & & & & \tau^2 & 0 & 0 \\ & & 0 & & & 2\sigma\tau & \tau & 0 \\ & & & & & \sigma^2 & \sigma & 1 \end{pmatrix} \begin{pmatrix} C \\ K \\ A \\ F \\ D \\ k \\ a \\ d \end{pmatrix} \qquad (24.133)$$

A number of general conclusions about the various aberrations can be drawn from (24.133). Disappointingly, there is no privileged aperture position so far as the spherical aberration is concerned: (24.133) merely confirms the obvious result that since the aberration is expressed in terms of aperture coordinates, the closer the aperture is to the image, the smaller will the coefficient be—but x_a and y_a will be large!

The expression for $\overline{K}$ shows that if σ can be chosen so that

$$K + \sigma C = 0 \qquad (24.134)$$

the coma will vanish. The existence of a coma-free point has long been known in optics: see Herzberger (1931, 1958) for extensive discussion and Czapski and Eppenstein (1924) for the earlier history.

The isotropic astigmatism vanishes if

$$A + 2\sigma K + \sigma^2 C = 0 \tag{24.135}$$

There will be two astigmatism-free aperture positions if $K^2 > AC$, one if it so happens that $K^2 = AC$ (the Finsterwalder condition, see Finsterwalder 1892), and otherwise none. In the case $K^2 = AC$, we have $\sigma = -K/C$, so that the coma too vanishes. Similar reasoning can be applied, not very profitably, to the field curvature and distortion.

Turning to the anisotropic aberrations, we see that the coma (k) cannot be eliminated by choosing the aperture position suitably; the astigmatism vanishes if $\sigma = -a/2k$, and the distortion if $\sigma = \{-a \pm (a^2 - 4dk)\}/2k$. If $a^2 = 4dk$, astigmatism and coma vanish simultaneously for $\sigma = -a/2k$.

Of these results, only the existence of a coma-free aperture position has found any practical application.

We note that relations identical with (24.133) are obtained if, instead of replacing $\bar{s}$ and $\bar{t}$ by s and t, as we have done, we write $\bar{x}_a := x(\bar{z}_a) = x_o s(\bar{z}_a) + x_a t(\bar{z}_a)$ and use this to replace x_a by $\bar{x}_a$ and likewise y_a by $\bar{y}_a$ in the expressions for Δx_i and Δy_i. Although this leads to the same result with less effort, it is not immediately obvious that the boundary conditions, $\Delta x = \Delta y = 0$ in the object and aperture planes, are satisfied. In fact, both techniques are legitimate, as the agreement between the results confirms.

24.8 Reduced coordinates

For reference purposes, we reproduce the expression for each aberration coefficient in terms of reduced coordinates (Section 15.3.1), as derived by Sturrock (1951c, 1955). Note that not only is $v(z)$ related to $u(z)$ by (15.40) but that the paraxial solutions satisfy the reduced equation (15.38). This has been a pitfall in the past (Glaser, 1952, p.676; Kuyatt, 1978).

The forms of the coefficients $A, C, \ldots k$ listed in Sturrock (1955) are as follows; these are a generalization of the earlier expressions, for electrostatic and magnetic lenses separately, to be found in Sturrock (1951c). The functions $\sigma(z)$ and $\tau(z)$ are the solutions of the reduced paraxial equations

---

* For those unfamiliar with Gaussian units, a c.g.s. system employing both electromagnetic units (e.m.u.) and electrostatic units (e.s.u.), we note that Sturrock's units may be converted to SI by writing $\Phi[\text{Sturrock}] \rightarrow 2\epsilon\phi[\text{SI}]$; $p[\text{St}] \rightarrow 2\sqrt{\epsilon\phi(1+\epsilon\phi)} = 2(\epsilon\hat{\phi})^{1/2}[\text{SI}]$; $H[\text{St}] \rightarrow (1/100)(e/m_0 c)B[\text{SI}]$. Note that when converting Sturrock's formulae, centimetres must be replaced by metres, particularly in derivatives. The presence of the factor $(1/100)$ above is an instance of this.

24.8 REDUCED COORDINATES

satisfying the boundary conditions $\sigma(z_o) = \tau(z_a) = 1$, $\sigma(z_a) = \tau(z_o) = 0$, and W_σ is the corresponding Wronskian, $W_\sigma = \tau'(z_o) = -\sigma'(z_a)$.

$$A + ia = \frac{1}{16}\left[\left(\frac{\hat{\phi}_o\hat{\phi}_a}{\hat{\phi}^2}\right)^{\frac{1}{4}}\left(4\sigma'\tau' - \gamma W_\sigma\frac{\phi'}{\hat{\phi}} + 2\mathrm{i}W_\sigma\frac{\eta B}{\hat{\phi}^{\frac{1}{2}}}\right)\right]_o^i$$

$$- \frac{4}{W_\sigma}\int_o^i \left(\frac{\hat{\phi}_o\hat{\phi}_a}{\hat{\phi}^2}\right)^{\frac{1}{4}}\{\sigma\tau(P\sigma'\tau' + Q\sigma\tau) - W_\sigma^2 T$$

$$+ \mathrm{i}W_\sigma(R\sigma'\tau' + S\sigma\tau)\}\,dz$$

$$C = -\frac{4}{W}\int_o^i \left(\frac{\hat{\phi}_a^3}{\hat{\phi}_o\hat{\phi}^2}\right)^{\frac{1}{4}} \tau^2(P\tau'^2 + Q\tau^2)\,dz$$

$$D + id = -\left[\left(\frac{\hat{\phi}_o}{\hat{\phi}}\right)^{\frac{1}{2}}\left(-\frac{3}{8}\sigma'^2 + \frac{\gamma}{16}\frac{\phi'}{\hat{\phi}}\sigma\sigma'\right)\right.$$

$$- \frac{1}{16}\left(\frac{\gamma^2}{4}\frac{\phi'^2}{\hat{\phi}^2} + \frac{\gamma}{2}\frac{\phi''}{\hat{\phi}} + \frac{\eta^2 B^2}{\hat{\phi}}\right)\sigma^2$$

$$\left. - \frac{\mathrm{i}}{8}\frac{\eta B}{\hat{\phi}^{\frac{1}{2}}}\sigma\sigma' + \frac{\mathrm{i}}{16}\frac{\eta B'}{\hat{\phi}^{\frac{1}{2}}}\sigma^2\right]_o^i$$

$$- \frac{2}{W_\sigma}\int_o^i \left(\frac{\hat{\phi}_o}{\hat{\phi}}\right)^{\frac{1}{2}}\left\{2(P\sigma'\tau' + Q\sigma\tau)\sigma^2 - W_\sigma P\sigma\sigma'\right.$$

$$\left. + \mathrm{i}W_\sigma(R\sigma'^2 + S\sigma^2)\right\}dz$$

$$F = \frac{1}{2}\left[\left(\frac{\hat{\phi}_o\hat{\phi}_a}{\hat{\phi}^2}\right)^{\frac{1}{4}}\sigma't'\right]_o^i$$

$$- \frac{8}{W_\sigma}\int_o^i \left(\frac{\hat{\phi}_o\hat{\phi}_a}{\hat{\phi}^2}\right)^{\frac{1}{4}}\sigma\tau(P\sigma'\tau' + Q\sigma\tau)\,dz$$

$$- 2W_\sigma\int_o^i \left(\frac{\hat{\phi}_o\hat{\phi}_a}{\hat{\phi}^2}\right)^{\frac{1}{4}}(P + 2T)\,dz$$

$$K + \mathrm{i}k = \frac{1}{8}\left[\left(\frac{\hat{\phi}_a}{\hat{\phi}}\right)^{\frac{1}{2}}\tau'^2\right]_o^i$$

$$-\frac{2}{W_\sigma}\int_o^i \left(\frac{\hat{\phi}_a}{\hat{\phi}}\right)^{\frac{1}{2}}\left\{2(P\tau'^2 + Q\tau^2)\sigma\tau - W_\sigma P\tau\tau'\right\}dz$$

$$-2\mathrm{i}\int_o^i \left(\frac{\hat{\phi}_a}{\hat{\phi}}\right)^{\frac{1}{2}}(R\tau'^2 + S\tau^2)\,dz \qquad (24.136)$$

in which

$$P(z) := \frac{3}{128}\left(\frac{\phi'}{\hat{\phi}}\right)^2\left(1 + \frac{4}{3}\epsilon\hat{\phi}\right) + \frac{\eta^2 B^2}{32\hat{\phi}}$$

$$Q(z) := -\frac{59}{1024}\left(1 + \frac{234}{59}\epsilon\hat{\phi} + \frac{120}{59}\epsilon^2\hat{\phi}^2\right)\left(\frac{\phi'}{\hat{\phi}}\right)^4$$

$$+ \frac{65}{1024}\left(1 + \frac{12}{13}\epsilon\hat{\phi}\right)\frac{\gamma\phi'^2\phi''}{\hat{\phi}^3}$$

$$- \frac{5}{256}\left(1 + \frac{8}{5}\epsilon\hat{\phi}\right)\left(\frac{\phi''}{\hat{\phi}}\right)^2$$

$$- \frac{27}{512}\left(1 + \frac{8}{3}\epsilon\hat{\phi}\right)\frac{\phi'^2\eta^2 B^2}{\hat{\phi}^3}$$

$$- \frac{3\gamma}{256}\frac{\eta^2 B^2\phi''}{\hat{\phi}^2} + \frac{9\gamma}{128}\frac{\eta^2 BB'\phi'}{\hat{\phi}^2}$$

$$- \frac{3}{128}\frac{\eta^4 B^4}{\hat{\phi}^2} - \frac{1}{32}\frac{\eta^2 B'^2}{\hat{\phi}}$$

$$R(z) := -\frac{\eta B}{16\hat{\phi}^{\frac{1}{2}}}$$

$$S(z) := -\frac{5}{256}\left(1 + \frac{12\epsilon\hat{\phi}}{5}\right)\frac{\phi'^2\eta B}{\hat{\phi}^{\frac{5}{2}}}$$

$$- \frac{\gamma}{32}\frac{\phi''\eta B}{\hat{\phi}^{\frac{3}{2}}} + \frac{\gamma}{32}\frac{\phi'\eta B'}{\hat{\phi}^{\frac{3}{2}}} - \frac{3}{64}\frac{\eta^3 B^3}{\hat{\phi}^{\frac{3}{2}}}$$

$$T(z) := -\frac{5}{256}\left(1 + \frac{12}{5}\epsilon\hat{\phi}\right)\left(\frac{\phi'}{\hat{\phi}}\right)^2 - \frac{3}{64}\frac{\eta^2 B^2}{\hat{\phi}} \qquad (24.137)$$

24.9 Seman's transformation of the characteristic function

We have seen how useful is the technique introduced by Seman for performing partial integration on aberration coefficients in a systematic fashion. Originally, however, it was not the individual coefficients but the perturbation characteristic S^I to which the technique was applied, thus generating a set of coefficients all possessing similar characteristics, such as absence of high order derivatives of $\phi(z)$ or $B(z)$. Seman considered electrostatic and magnetic round lenses. The idea behind his method is to be found in two early notes (Seman, 1951, 1954) and a full account appeared shortly afterwards (Seman, 1955a,b, 1958b). We illustrate the power and simplicity of the technique in the case of orthogonal systems consisting of suitably orientated magnetic and electrostatic quadrupoles and round electrostatic lenses. Only the main steps in the reasoning are presented here; for additional details, see Hawkes (1966/7b). This discussion is limited to the nonrelativistic approximation but the relativistic case has been explored in detail for round electrostatic lenses (see Hawkes, 1977a and Chapter 34).

It is convenient to introduce the vectors $\boldsymbol{x} := (x, y)$, $\tilde{\boldsymbol{x}} := (x, -y)$, whereupon $M^{(2)}$ and $M^{(4)}$ may be written

$$M^{(2)} = -\frac{\phi''}{8\phi^{\frac{1}{2}}}\boldsymbol{x}^2 + \frac{1}{4}Q(\boldsymbol{x}\cdot\tilde{\boldsymbol{x}})\frac{1}{2}\phi^{\frac{1}{2}}\boldsymbol{x}'^2$$

$$M^{(4)} = \left(\frac{\phi^{(4)}}{128\phi^{\frac{1}{2}}} - \frac{\phi''^2}{128\phi^{\frac{3}{2}}} - \frac{1}{2}O\right)\boldsymbol{x}^4$$
$$+ \left(-\frac{p_2^2}{32\phi^{\frac{3}{2}}} + O\right)(\boldsymbol{x}\cdot\tilde{\boldsymbol{x}})^2$$
$$+ \left(\frac{p_2\phi''}{32\phi^{\frac{3}{2}}} - \frac{p_2''}{48\phi^{\frac{1}{2}}} + \frac{\eta Q_2''}{48}\right)\boldsymbol{x}^2(\boldsymbol{x}\cdot\tilde{\boldsymbol{x}})$$
$$- \frac{\phi''}{16\phi^{\frac{1}{2}}}\boldsymbol{x}^2\boldsymbol{x}'^2 + \frac{p_2}{8\phi^{\frac{1}{2}}}(\boldsymbol{x}\cdot\tilde{\boldsymbol{x}})\boldsymbol{x}'^2 - \frac{\phi^{\frac{1}{2}}}{8}\boldsymbol{x}'^4$$
$$+ \frac{\eta Q_2'}{16}\left\{(\boldsymbol{x}\cdot\tilde{\boldsymbol{x}})\boldsymbol{x}^{2\prime} - \boldsymbol{x}^2(\boldsymbol{x}\cdot\tilde{\boldsymbol{x}})'\right\} \qquad (24.138)$$

in which

$$Q := \frac{p_2}{\phi^{\frac{1}{2}}} - 2\eta Q_2$$
$$O := \frac{1}{24}\left(\frac{p_4}{\phi^{\frac{1}{2}}} - 2\eta Q_4\right) \qquad (24.139)$$

and $\boldsymbol{x}^2$ and $\boldsymbol{x}^4$ denote $(\boldsymbol{x}\cdot\boldsymbol{x})$ and $(\boldsymbol{x}\cdot\boldsymbol{x})^2$ respectively.

The individual terms of $M^{(4)}$ fall into groups, which have the following 'dimensions':

$$\left[\phi^{\frac{1}{2}}\right]\left[\boldsymbol{x}^2\right]^2\left[d/dz\right]^4$$

$$\left[p_4/\phi^{\frac{1}{2}} \text{ or } Q_4\right]\left[\boldsymbol{x}^2\right]^2 \text{ and } \left[p_4/\phi^{\frac{1}{2}} \text{ or } Q_4\right]\left[\boldsymbol{x}\cdot\tilde{\boldsymbol{x}}\right]^2$$

$$\left[p_2/\phi^{\frac{1}{2}} \text{ or } Q_2\right]\left[\boldsymbol{x}^2\right]\left[\boldsymbol{x}\cdot\tilde{\boldsymbol{x}}\right]\left[d/dz\right]^2$$

$$\left[\phi^{-\frac{1}{2}}\right]\left[p_2/\phi^{\frac{1}{2}} \text{ or } Q_2\right]\left[\boldsymbol{x}\cdot\tilde{\boldsymbol{x}}\right]^2 \tag{24.140}$$

Since $S^I = \int M^{(4)} dz$, the terms of S^I are generated by quantities with one lower power of d/dz than (24.140); the only terms in which d/dz survives are therefore

$$\left[\phi^{\frac{1}{2}}\right]\left[\boldsymbol{x}^2\right]^2\left[d/dz\right]^3$$

$$\left[p_2/\phi^{\frac{1}{2}} \text{ or } Q_2\right]\left[\boldsymbol{x}^2\right]\left[\boldsymbol{x}\cdot\tilde{\boldsymbol{x}}\right]\left[d/dz\right] \tag{24.141}$$

These generate 16 terms possessing the appropriate dimensions, eight involving ϕ alone ($r_1 - r_8$), three each involving p_2 alone ($e_1 - e_3$) and Q_2 alone ($m_1 - m_3$) and two mixed terms (r_9 and r_{10}):

$$r_1 = \phi^{\frac{1}{2}}\boldsymbol{x}^{2\prime}\boldsymbol{x}^{\prime 2}$$

$$r_2 = \frac{\phi'}{\phi^{\frac{1}{2}}}(\boldsymbol{x}^{2\prime})^2$$

$$r_3 = \frac{\phi'}{\phi^{\frac{1}{2}}}\boldsymbol{x}^2\boldsymbol{x}^{\prime 2}$$

$$r_4 = \frac{\phi''}{\phi^{\frac{1}{2}}}\boldsymbol{x}^2\boldsymbol{x}^{2\prime}$$

$$r_5 = \frac{\phi'^2}{\phi^{\frac{3}{2}}}\boldsymbol{x}^2\boldsymbol{x}^{2\prime}$$

$$r_6 = \frac{\phi'''}{\phi^{\frac{1}{2}}}\boldsymbol{x}^4$$

$$r_7 = \frac{\phi''\phi'}{\phi^{\frac{3}{2}}}\boldsymbol{x}^4$$

$$r_8 = \frac{\phi'^3}{\phi^{\frac{5}{2}}}\boldsymbol{x}^4$$

$$r_9 = \frac{p_2\phi'}{\phi^{\frac{3}{2}}}\boldsymbol{x}^2(\boldsymbol{x}\cdot\tilde{\boldsymbol{x}})$$

$$r_{10} = \frac{\eta Q_2\phi'}{\phi}\boldsymbol{x}^2(\boldsymbol{x}\cdot\tilde{\boldsymbol{x}})$$

$$e_1 = \frac{p_2}{\phi^{\frac{1}{2}}}\boldsymbol{x}^{2\prime}(\boldsymbol{x}\cdot\tilde{\boldsymbol{x}})$$

$$e_2 = \frac{p_2}{\phi^{\frac{1}{2}}}\boldsymbol{x}^2(\boldsymbol{x}\cdot\tilde{\boldsymbol{x}})'$$

$$e_3 = \frac{p_2'}{\phi^{\frac{1}{2}}}\boldsymbol{x}^2(\boldsymbol{x}\cdot\tilde{\boldsymbol{x}})$$

$$m_1 = \eta Q_2\boldsymbol{x}^{2\prime}(\boldsymbol{x}\cdot\tilde{\boldsymbol{x}})$$

$$m_2 = \eta Q_2\boldsymbol{x}^2(\boldsymbol{x}\cdot\tilde{\boldsymbol{x}})' \tag{24.142}$$

$$m_3 = \eta Q'\boldsymbol{x}^2(\boldsymbol{x}\cdot\tilde{\boldsymbol{x}})$$

24.9 SEMAN'S TRANSFORMATION

The terms $r_1, r_1, \ldots r_8$ are identical with eight of Seman's terms i_j, those not containing the magnetic field.

The refractive index can be altered in form without affecting the eikonal by adding to it expressions that vanish; such expressions are obtained by differentiation of each of the quantities m_j, e_j and r_j and elimination of $\boldsymbol{x}''$ with the aid of the paraxial equations. We find

$$r_1' + \frac{1}{2}R_{17} + \frac{1}{4}R_{22} + R_{23} - 2R_{24} - E_5 - \frac{1}{2}E_6 + 2M_5 + M_6 = 0$$

$$r_2' + R_{13} - 2R_{15} + 4R_{19} + \frac{3}{2}R_{21} - R_{22} - 4R_{23} = 0$$

$$r_3' + \frac{1}{4}R_{13} - \frac{1}{2}R_{16} - R_{17} + \frac{3}{2}R_{18} + R_{20} - R_{23} = 0$$

$$r_4' + \frac{1}{2}R_3 - R_6 + 2R_9 + R_{13} - R_{14} - 2R_{17} - R_{22} = 0$$

$$r_5' + \frac{1}{2}R_2 - R_7 + 2R_{10} + 2R_{12} - 2R_{13} - 2R_{18} - R_{21} = 0$$

$$r_6' + \frac{1}{2}R_4 - R_5 - 2R_{14} = 0$$

$$r_7' + \frac{3}{2}R_2 - R_3 - R_4 - 2R_{13} = 0$$

$$r_8' + \frac{5}{2}R_1 - 3R_2 - 2R_{12} = 0$$

$$r_9' - 2R_6 + 3R_7 - 2R_8 - 2R_{15} - 2R_{16} = 0$$

$$r_{10}' - R_9 + R_{10} - R_{11} - R_{19} - R_{20} = 0$$

$$e_1' + R_6 + 2R_{15} - 2E_1 - 2E_3 - 4E_5 - 2E_6 + 4E_8 = 0$$

$$e_2' + R_6 + 2R_{16} - 2E_2 - 2E_4 + 4E_5 - 4E_6 + 4M_8 = 0$$

$$e_3' + R_8 - 2E_1 - 2E_2 - 2E_7 = 0$$

$$m_1' + \frac{1}{2}R_9 + \frac{1}{2}R_{19} - E_8 - M_1 + 2M_3 - 2M_5 - M_6 = 0$$

$$m_2' + \frac{1}{2}r_9 + \frac{1}{2}R_{20} - M_2 + 2M_4 + 2M_5 - 2M_6 - M_8 = 0$$

$$m_3' - M_1 - M_2 - M_7 = 0 \qquad (24.143)$$

24. THE GEOMETRICAL ABERRATIONS OF ROUND LENSES

in which

$$R_1 = \frac{\phi'^4}{\phi^{7/2}} x^4$$

$$R_2 = \frac{\phi'^2 \phi''}{\phi^{5/2}} x^4$$

$$R_3 = \frac{\phi''^2}{\phi^{3/2}} x^4$$

$$R_4 = \frac{\phi''' \phi'}{\phi^{3/2}} x^4$$

$$R_5 = \frac{\phi^{(4)}}{\phi^{1/2}} x^4$$

$$R_6 = \frac{p_2 \phi''}{\phi^{3/2}} x^2 (\boldsymbol{x} \cdot \tilde{\boldsymbol{x}})$$

$$R_7 = \frac{p_2 \phi'^2}{\phi^{5/2}} x^2 (\boldsymbol{x} \cdot \tilde{\boldsymbol{x}})$$

$$R_8 = \frac{p_2' \phi'}{\phi^{3/2}} x^2 (\boldsymbol{x} \cdot \tilde{\boldsymbol{x}})$$

$$R_9 = \eta \frac{Q_2 \phi''}{\phi} x^2 (\boldsymbol{x} \cdot \tilde{\boldsymbol{x}})$$

$$R_{10} = \eta \frac{Q_2 \phi'^2}{\phi^2} x^2 (\boldsymbol{x} \cdot \tilde{\boldsymbol{x}})$$

$$R_{11} = \eta \frac{Q_2' \phi'}{\phi} x^2 (\boldsymbol{x} \cdot \tilde{\boldsymbol{x}})$$

$$R_{12} = \frac{\phi'^3}{\phi^{5/2}} x^2 x^{2\prime}$$

$$E_1 = \frac{p_2'}{\phi^{1/2}} x^{2\prime} (\boldsymbol{x} \cdot \tilde{\boldsymbol{x}})$$

$$E_2 = \frac{p_2'}{\phi^{1/2}} x^2 (\boldsymbol{x} \cdot \tilde{\boldsymbol{x}})'$$

$$E_3 = \frac{p_2^2}{\phi^{3/2}} (\boldsymbol{x} \cdot \tilde{\boldsymbol{x}})^2$$

$$E_4 = \frac{p_2^2}{\phi^{3/2}} x^4$$

$$E_5 = \frac{p_2}{\phi^{1/2}} x'^2 (\boldsymbol{x} \cdot \tilde{\boldsymbol{x}})$$

$$E_6 = \frac{p_2}{\phi^{1/2}} x^{2\prime} (\boldsymbol{x} \cdot \tilde{\boldsymbol{x}})$$

$$E_7 = \frac{p_2''}{\phi^{1/2}} x^2 (\boldsymbol{x} \cdot \tilde{\boldsymbol{x}})$$

$$E_8 = \eta \frac{p_2 Q_2}{\phi} (\boldsymbol{x} \cdot \tilde{\boldsymbol{x}})^2$$

$$R_{13} = \frac{\phi'' \phi'}{\phi^{3/2}} x^2 x^{2\prime}$$

$$R_{14} = \frac{\phi'''}{\phi^{1/2}} x^2 x^{2\prime}$$

$$R_{15} = \frac{p_2 \phi'}{\phi^{3/2}} x^{2\prime} (\boldsymbol{x} \cdot \tilde{\boldsymbol{x}})$$

$$R_{16} = \frac{p_2 \phi'}{\phi^{3/2}} (\boldsymbol{x} \cdot \tilde{\boldsymbol{x}})'$$

$$R_{17} = \frac{\phi''}{\phi^{1/2}} x^2 x'^2$$

$$R_{18} = \frac{\phi'^2}{\phi^{3/2}} x^2 x'^2$$

$$R_{19} = \eta \frac{Q_2 \phi'}{\phi} x^{2\prime} x^2 (\boldsymbol{x} \cdot \tilde{\boldsymbol{x}})$$

$$R_{20} = \eta \frac{Q_2 \phi'}{\phi} x^2 (\boldsymbol{x} \cdot \tilde{\boldsymbol{x}})'$$

$$R_{21} = \frac{\phi'^2}{\phi^{3/2}} (x^{2\prime})^2$$

$$R_{22} = \frac{\phi''}{\phi^{1/2}} (x^{2\prime})^2$$

$$R_{23} = \frac{\phi'}{\phi^{1/2}} x'^2 x^{2\prime}$$

$$R_{24} = \phi^{\frac{1}{2}} x'^4$$

$$M_1 = \eta Q_2' x^{2\prime} (\boldsymbol{x} \cdot \tilde{\boldsymbol{x}})$$

$$M_2 = \eta Q_2' x^2 (\boldsymbol{x} \cdot \tilde{\boldsymbol{x}})'$$

$$M_3 = \eta^2 \frac{Q_2^2}{\phi^{1/2}} (\boldsymbol{x} \cdot \tilde{\boldsymbol{x}})^2$$

$$M_4 = \eta^2 \frac{Q_2^2}{\phi^{1/2}} x^4$$

$$M_5 = \eta Q_2 x'^2 (\boldsymbol{x} \cdot \tilde{\boldsymbol{x}})$$

$$M_6 = \eta Q_2 x^{2\prime} (\boldsymbol{x} \cdot \tilde{\boldsymbol{x}})'$$

$$M_7 = \eta Q_2' x^2 (\boldsymbol{x} \cdot \tilde{\boldsymbol{x}})$$

$$M_8 = \eta \frac{p_2 Q_2}{\phi} x^4$$

(24.144)

24.9 SEMAN'S TRANSFORMATION

Each of the identities (24.143) is now weighted by an arbitrary multiplier, p_j, ϵ_j and μ_j, and added to the expression for $\int M^{(4)}\,dz$ which has the primitive form

$$\begin{aligned}S^I &= \int M^{(4)}\,dz \\ &= \frac{1}{2}\int O\{-\boldsymbol{x}^4 + 2(\boldsymbol{x}\cdot\tilde{\boldsymbol{x}})^2\}\,dz \\ &\quad + \frac{1}{128}\int\bigg(-R_3 + R_5 + 4R_6 - 8R_{17} - 16R_{24} \\ &\qquad\qquad - 4E_3 + 16E_5 - \frac{8}{3}E_7 + 8M_1 - 8M_2 + \frac{8}{3}M_7\bigg)\,dz\end{aligned} \qquad (24.145)$$

The general form of S^I is thus as follows:

$$\begin{aligned}128 S^I &- 64\int O\{-\boldsymbol{x}^4 + 2(\boldsymbol{x}\cdot\tilde{\boldsymbol{x}})^2\}\,dz \\ &= \Bigg[\sum_{j=1}^{10}\varrho_j r_j + \sum_{j=1}^{3}\epsilon_j e_j + \sum_{j=1}^{3}\mu_j m_j\Bigg] \\ &\quad + \int\bigg\{\frac{5}{2}\varrho_8 R_1 + \bigg(\frac{1}{2}\varrho_5 + \frac{3}{2}\varrho_7 - 3\varrho_8\bigg)R_2 + \bigg(-1 + \frac{1}{2}\varrho_4 - \varrho_7\bigg)R_3 \\ &\quad + \bigg(\frac{1}{2}\varrho_6 - \varrho_7\bigg)R_4 + (1 - \varrho_6)R_5 + (4 - \varrho_4 - 2\varrho_9 + \epsilon_1 + \epsilon_2)R_6 \\ &\quad + (-\varrho_5 + 3\varrho_9)R_7 + (-2\varrho_9 + \epsilon_3)R_8 \\ &\quad + \bigg(2\varrho_4 - \varrho_{10} + \frac{1}{2}\mu_1 + \frac{1}{2}\mu_2\bigg)R_9 + (2\varrho_5 + \varrho_{10})R_{10} - \varrho_{10}R_{11} \\ &\quad + 2(\varrho_5 - \varrho_8)R_{12} \\ &\quad + \bigg(\varrho_2 + \frac{1}{4}\varrho_3 + \varrho_4 - 2\varrho_5 - 2\varrho_7\bigg)R_{13} + (-\varrho_4 - 2\varrho_6)R_{14} \\ &\quad + 2(-\varrho_2 - \varrho_9 + \epsilon_1)R_{15} + \bigg(-\frac{1}{2}\varrho_3 - 2\varrho_9 + 2\epsilon_2\bigg)R_{16} \\ &\quad + \bigg(-8 + \frac{1}{2}\varrho_1 - \varrho_3 - 2\varrho_4\bigg)R_{17} + \bigg(\frac{3}{2}\varrho_3 - 2\varrho_5\bigg)R_{18} \\ &\quad + \bigg(4\varrho_2 - \varrho_{10} + \frac{1}{2}\mu_1\bigg)R_{19} + \bigg(\varrho_3 - \varrho_{10} + \frac{1}{2}\mu_2\bigg)R_{20} + \bigg(\frac{3}{2}\varrho_2 - \varrho_5\bigg)R_{21} \\ &\quad + \bigg(\frac{1}{4}\varrho_1 - \varrho_2 - \varrho_4\bigg)R_{22} + (\varrho_1 - 4\varrho_2 - \varrho_3)R_{23} + (-16 - 2\varrho_1)R_{24} \\ &\quad + 2(-\epsilon_1 - \epsilon_3)E_1 + 2(-\epsilon_2 - \epsilon_3)E_2 + (-4 - 2\epsilon_1)E_3\end{aligned}$$

$$\begin{aligned}
&- 2\epsilon_2 E_4 + (16 - \varrho_1 - 4\epsilon_1 + 4\epsilon_2)E_5 + \left(-\frac{1}{2}\varrho_1 - 2\epsilon_1 - 4\epsilon_2\right)E_6 \\
&+ \left(-\frac{8}{3} - 2\epsilon_3\right)E_7 + (4\epsilon_1 - \mu_1)E_8 \\
&+ (8 - \mu_1 - \mu_3)M_1 + (-8 - \mu_2 - \mu_3)M_2 + 2\mu_1 M_3 \\
&+ 2\mu_2 M_4 + (2\varrho_1 - 2\mu_1 + 2\mu_2)M_5 \\
&+ (\varrho_1 - \mu_1 - 2\mu_2)M_6 + \left(\frac{8}{3} - \mu_3\right)M_7 \\
&+ (4\epsilon_2 - \mu_2)M_8 \Bigg\} dz
\end{aligned} \qquad (24.146)$$

From this very general expression, many useful forms of S^I and hence of the aberration coefficients can be derived. These are exhaustively analysed in Hawkes (1966/7b) and we say no more about them here.

Before leaving this topic, we should, however, mention that not quite all possible forms of the eikonal or aberration coefficients emerge from the foregoing expression. Further transformations, which are in practice very often simplifications, can be made by recalling that the Wronskian is a constant. Thus $\phi^{\frac{1}{2}}(st' - s't)$ or its square may be differentiated, weighted and added to the aberration integrals; this frequently enables us to eliminate otherwise persistent but unwanted terms.

25
Asymptotic Aberration Coefficients

For most lenses, projectors and condensers in particular, the "object" is in fact an intermediate image of the specimen or the source created by the lenses upstream, so that the entire lens field participates in the image formation. We are thus interested in the coordination between incident and emergent *asymptotes*, as explained in Chapter 16, and the corresponding aberrations are hence said to be *asymptotic*. They were first thoroughly studied by Lenz (1956, 1957), using the trajectory method; the correct form of the characteristic function is to be found in Sturrock (1955) and this was later employed by Hawkes (1968, 1970b,c) to explore them in more detail. As for the real aberrations we must distinguish between aberrations expressed in terms of position and gradient at the (asymptotic) object and in terms of position in the (asymptotic) object plane and *real* aperture plane (or the entrance pupil). The former are generally more useful and we deal with them first.

In order to introduce the boundary conditions $x_o^{(1)} = y_o^{(1)} = x_o'^{(1)} = y_o'^{(1)} = 0$ directly, we need the relation between $p^{(1)}, q^{(1)}, x'^{(1)}$ and $y'^{(1)}$; to the paraxial approximation, we have

$$p := p^{(p)} = \hat{\phi}^{\frac{1}{2}} x' \qquad q := q^{(p)} = \hat{\phi}^{\frac{1}{2}} y' \tag{25.1}$$

while to the third-order approximation,

$$\begin{aligned} p := p^{(p)} + p^{(1)} &= \hat{\phi}^{\frac{1}{2}} x' \left(1 - \frac{x'^2 + y'^2}{2} \right) \\ q := q^{(p)} + q^{(1)} &= \hat{\phi}^{\frac{1}{2}} y' \left(1 - \frac{x'^2 + y'^2}{2} \right) \end{aligned} \tag{25.2}$$

Setting $x' = x'^{(p)} + x'^{(1)}$ and $y' = y'^{(p)} + y'^{(1)}$ we obtain

$$\begin{aligned} p^{(1)} &= \hat{\phi}^{\frac{1}{2}} \left\{ x'^{(1)} - \frac{1}{2} x'^{(p)} (x'^{(p)2} + y'^{(p)2}) \right\} \\ q^{(1)} &= \hat{\phi}^{\frac{1}{2}} \left\{ y'^{(1)} - \frac{1}{2} y'^{(p)} (x'^{(p)2} + y'^{(p)2}) \right\} \end{aligned} \tag{25.3}$$

Thus if we impose the condition $x'^{(1)} = y'^{(1)} = 0$ in the object plane, the quantities $p^{(1)}$ and $q^{(1)}$ must take the values

$$p_o^{(1)} = -\frac{1}{2}\hat{\phi}_o^{\frac{1}{2}} x'_o(x'^2_o + y'^2_o)$$
$$q_o^{(1)} = -\frac{1}{2}\hat{\phi}_o^{\frac{1}{2}} y'_o(x'^2_o + y'^2_o) \qquad (25.4)$$

in that plane, where we have dropped the index (p). Substituting this into the expression for S_{o2}^I (22.23), we find

$$\Delta S_{o2}^I = p_2^{(1)} \cdot \Delta x_2 + q_2^{(1)} \cdot \Delta y_2 - (x_2^{(1)} \cdot \Delta p_2 + y_2^{(1)} \cdot \Delta q_2)$$
$$+ \frac{1}{2}\hat{\phi}_o^{\frac{1}{2}}(x'_o \cdot \Delta x_o + y'_o \cdot \Delta y_o)(x'^2_o + y'^2_o) \qquad (25.5)$$

Substituting $x(z) = x_o G(z) + x'_o H(z)$, $y(z) = y_o G(z) + y'_o H(z)$ from (16.1), with

$$\lim_{z \to -\infty} G(z) = 1 \qquad \lim_{z \to -\infty} H(z) = z - z_o \qquad (25.6)$$

we find

$$x^{(1)}(z_2) = \frac{1}{\hat{\phi}_o^{\frac{1}{2}}}\left(H_2^* \frac{\partial S_{o2}^I}{\partial x_o} - G_2^* \frac{\partial S_{o2}^I}{\partial x'_o}\right) - \frac{1}{2}H_2^* x'_o \theta_o^2$$
$$y^{(1)}(z_2) = \frac{1}{\hat{\phi}_o^{\frac{1}{2}}}\left(H_2^* \frac{\partial S_{o2}^I}{\partial y_o} - G_2^* \frac{\partial S_{o2}^I}{\partial y'_o}\right) - \frac{1}{2}H_2^* y'_o \theta_o^2 \qquad (25.7a)$$

and

$$x'^{(1)}(z_2) = \frac{1}{\hat{\phi}_o^{\frac{1}{2}}}\left(H_2^{*\prime} \frac{\partial S_{o2}^I}{\partial x_o} - G_2^{*\prime} \frac{\partial S_{o2}^I}{\partial x'_o}\right)$$
$$- \frac{1}{2}H_2^{*\prime} x'_o \theta_o^2 + \frac{1}{2}x'_2 \theta_2^2$$
$$y'^{(1)}(z_2) = \frac{1}{\hat{\phi}_o^{\frac{1}{2}}}\left(H_2^{*\prime} \frac{\partial S_{o2}^I}{\partial y_o} - G_2^{*\prime} \frac{\partial S_{o2}^I}{\partial y'_o}\right) \qquad (25.7b)$$
$$- \frac{1}{2}H_2^{*\prime} y'_o \theta_o^2 + \frac{1}{2}y'_2 \theta_2^2$$

in which G_2^* and H_2^* denote the asymptotes to $G(z)$ and $H(z)$ in image space and we extend (24.33a) by writing

$$\theta_o^2 := x'^2_o + y'^2_o \qquad V_o = x_o x'_o + y_o y'_o$$
$$v_o = x_o y'_o - x'_o y_o$$

25. ASYMPTOTIC ABERRATION COEFFICIENTS

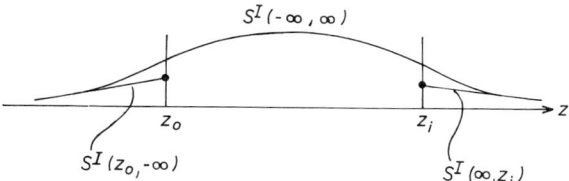

Fig. 25.1: Branches of S_{oi}^I. The asymptotic branches $S^I(z_o, -\infty)$ and $S^I(\infty, z_i)$ are straight lines, the asymptotes to $S^I(-\infty, \infty)$.

These formulae are exactly analogous to (24.27) and by following the same reasoning as we employed for the real aberrations, aberration integrals involving B, ϕ, G and H similar in appearance to those listed in Chapter 24 are obtained. Before doing this, it is advantageous to replace the ray $H(z)$, which varies with object position, by the ray $\overline{G}(z)$ introduced in Chapter 16, which does not. In this way, we obtain aberration integrals that are characteristic of the distribution $B(z)$ or $\phi(z)$, irrespective of the working conditions.

The ray $\overline{G}(z)$ is, we recall, defined by the boundary condition (16.1) $\lim_{z \to \infty} \overline{G}(z) = 1$ and hence

$$H(z) = -\frac{f_o}{M} G(z) + f_o \overline{G}(z)$$
$$= -f_o(mG - \overline{G}) \qquad (25.9)$$

where

$$m := \frac{1}{M} \qquad (25.10)$$

is the *reciprocal magnification* (not to be confused with the demagnification, the term commonly employed to describe a magnification $|M|$ less than unity).

The characteristic function S_{oi}^I now consists of three branches (Fig. 25.1). We write

$$S_{oi}^I =: S^I(z_o, -\infty) + S^I(-\infty, \infty) + S^I(\infty, z_i) \qquad (25.11a)$$

The contribution $S^I(-\infty, \infty)$ is given by

$$S^I(-\infty, \infty) = \int_{-\infty}^{\infty} M^{(4)} \, dz \qquad (25.11b)$$

into which we must substitute $x(z) = x_o G(z) + x_o' H(z) = (x_o - f_o x_o' m) G(z) + f_o x_o' \overline{G}(z)$ and similarly for $y(z)$. The other two contributions are "virtual". In them, $\phi(z) = $ const and $B(z) = 0$ and the rays $G(z)$ and $H(z)$ or

$\overline{G}(z)$ are replaced by their asymptotes. Thus

$$S^I(z_o, -\infty) = -\frac{\hat{\phi}_o^{\frac{1}{2}}}{8}\theta_o^4 \lim_{z \to -\infty}(z - z_o)$$

$$= -\frac{\hat{\phi}_o^{\frac{1}{2}}}{8}\theta_o^4 \left\{ \lim_{z \to -\infty}(z - z_{Fo}) - f_o m \right\} \quad (25.11c)$$

$$S^I(\infty, z_i) = -\frac{\hat{\phi}_i^{\frac{1}{2}}}{8}\left\{ \frac{r_o^2}{f_i^2} - 2\frac{V_o}{f_i M}\left(\frac{\hat{\phi}_o}{\hat{\phi}_i}\right)^{\frac{1}{2}} \right.$$

$$\left. + \frac{\theta_o^2}{M^2}\frac{\hat{\phi}_o}{\hat{\phi}_i} \right\}^2 \left\{ \lim_{z \to \infty}(z_{Fi} - z) - f_i M \right\} \quad (25.11d)$$

The reason for introducing z_{Fo} and z_{Fi} will become clearer below but we can anticipate that the convergence of integrals involving G'^4 and $\overline{G}'^4$ will be easier to understand in combination with the limit terms in (25.11c,d).

Substituting for $x(z)$ and $y(z)$ in $M^{(4)}$, we find

$$\frac{S^I_{oi}}{\hat{\phi}_o^{\frac{1}{2}}} = -\begin{pmatrix} u_o^{*2} \\ u_o^* u_o^{*\prime} \\ u_o^{*\prime 2} \end{pmatrix}^T \begin{pmatrix} E/4 & (D-id)/2 & (A-ia)/4 \\ (D+id)/2 & F/2 & (K-ik)/2 \\ (A+ia)/4 & (K+ik)/2 & C/4 \end{pmatrix} \begin{pmatrix} u_o^2 \\ u_o u_o' \\ u_o'^2 \end{pmatrix}$$
(25.12)

as in (24.33) or (24.34). The nine magnification-dependent quantities $A, C, ...k$ can now be expressed in terms of nine integrals, six for the isotropic aberration coefficients ($i_1 - i_6$) and three for the anisotropic coefficients ($i_7 - i_9$), that are *independent of the magnification*: they are properties of the lens, determined by its geometry and excitation but unaffected by the working conditions. The dependence of the coefficients $A, C, ...k$ on magnification is expressed by a series of polynomials in reciprocal magnification m in which simple multiples of the integrals $i_j (j = 1 - 9)$ are the coefficients of the various powers of m.

Spherical aberration

$$C = C_4 m^4 + C_3 m^3 + C_2 m^2 + C_1 m + C_0 \quad (25.13)$$

25. ASYMPTOTIC ABERRATION COEFFICIENTS

$$\left.\begin{aligned} C_4 &= i_1 f_o^4 \\ C_3 &= -4i_2 f_o^4 - \frac{1}{2} r^2 f_o \\ C_2 &= 2(i_3 + i_4) f_o^4 \\ C_1 &= -4i_5 f_o^4 - \frac{1}{2} f_o \\ C_0 &= i_6 f_o^4 \end{aligned}\right\} \quad (25.14)$$

Coma

$$K = K_3 m^3 + K_2 m^2 + K_1 m + K_0 \quad (25.15)$$

$$\left.\begin{aligned} K_3 &= -i_1 f_o^3 \\ K_2 &= 3i_2 f_o^3 + \frac{1}{2} r^2 \\ K_1 &= -(i_3 + i_4) f_o^3 \\ K_0 &= i_5 f_o^3 \end{aligned}\right\} \quad (25.16)$$

$$k = k_2 m^2 + k_1 m + k_0 \quad (25.17)$$

$$\left.\begin{aligned} k_2 &= i_7 f_o^2 \\ k_1 &= -i_8 f_o^2 \\ k_0 &= i_9 f_o^2 \end{aligned}\right\} \quad (25.18)$$

Astigmatism and field curvature

$$A = A_2 m^2 + A_1 m + A_0 \quad (25.19)$$

$$F = F_2 m^2 + F_1 m + F_0 \quad (25.20)$$

$$\left.\begin{aligned} A_2 &= \frac{1}{2} F_2 = i_1 f_o^2 \\ A_1 &= \frac{1}{2} F_1 = -2i_2 f_o^2 - \frac{r^2}{2 f_o} \\ A_0 &= i_3 f_o^2 \qquad F_0 = i_4 f_o^2 \end{aligned}\right\} \quad (25.21)$$

$$a = a_1 m + a_0 \quad (25.22)$$

$$\left.\begin{aligned} a_1 &= -2i_7 f_o \\ a_0 &= i_8 f_o \end{aligned}\right\} \quad (25.23)$$

Distortion

$$D = D_1 m + D_0 \quad (25.24)$$

$$D_1 = -i_1 f_o$$
$$D_0 = i_2 f_o + \frac{1}{2f_i^2}$$
(24.25)

$$d = d_0 = i_7 \qquad (25.26)$$

where
$$r := (\frac{\hat{\phi}_o}{\hat{\phi}_i})^{\frac{1}{2}} = \frac{f_o}{f_i} \qquad (25.27)$$

More compactly

$$\begin{pmatrix} C \\ K \\ A \\ F \\ D \end{pmatrix} = Q \begin{pmatrix} m^4 \\ m^3 \\ m^2 \\ m \\ 1 \end{pmatrix} \qquad \begin{pmatrix} k \\ a \\ d \end{pmatrix} = q \begin{pmatrix} m^2 \\ m \\ 1 \end{pmatrix} \qquad (25.28)$$

with

$$Q = \begin{pmatrix} i_1 f_o^4 & -4i_2 f_o^4 - \frac{1}{2} r^2 f_o & 2(i_3 + i_4) f_o^4 & -4i_5 f_o^4 - \frac{1}{2} f_o & i_6 f_o^4 \\ 0 & -i_1 f_o^3 & 3i_2 f_o^3 + \frac{1}{2} r^2 & -(i_3 + i_4) f_o^3 & i_5 f_o^3 \\ 0 & 0 & i_1 f_o^2 & -2i_2 f_o^2 - r/2 f_o & i_3 f_o^2 \\ 0 & 0 & 2i_1 f_o^2 & -4i_2 f_o^2 - r/f_o & i_4 f_o^2 \\ 0 & 0 & 0 & -i_1 f_o & i_2 f_o + 1/2 f_i^2 \end{pmatrix}$$

(25.29)

and

$$q = \begin{pmatrix} i_7 f_o & -i_8 f_o^2 & i_9 f_o^2 \\ 0 & -2i_7 f_o & i_8 f_o \\ 0 & 0 & i_7 \end{pmatrix} \qquad (25.30)$$

The integrals $i_1 - i_9$ can as usual be cast into a host of different forms, which can be established without difficulty by the methods explained in Section 24.3. We list here the primitive form, on which no partial integration has been performed, and two particularly simple forms, one for magnetic lenses ($\phi =$ const) and the other for electrostatic lenses ($B = 0$).

25. ASYMPTOTIC ABERRATION COEFFICIENTS

Another set of (nonrelativistic) integrals for electrostatic lenses, well suited for numerical work, is to be found in Kuyatt *et al.* (1974).

$$i_1 = \frac{1}{\hat{\phi}_o^{\frac{1}{2}}} \int_{-\infty}^{\infty} (L_1 G^4 + 2L_2 G^2 G'^2 + L_3 G'^4) \, dz$$

$$+ \frac{1}{2 f_i^3 f_o} \lim_{z \to \infty} (z_{Fi} - z)$$

$$\xrightarrow[\phi=\text{const}]{} \int_{-\infty}^{\infty} \Lambda_m G^4 \, dz$$

$$\xrightarrow[B=0]{} \int_{-\infty}^{\infty} \Lambda_e G^4 \, dz$$

$$i_2 = \frac{1}{\hat{\phi}_o^{\frac{1}{2}}} \int_{-\infty}^{\infty} \{L_1 G^3 \overline{G} + L_2 GG'(G\overline{G})' + L_3 G'^3 \overline{G}'\} \, dz$$

$$\xrightarrow[\phi=\text{const}]{} \int_{-\infty}^{\infty} \Lambda_m G^3 \overline{G} \, dz - \frac{1}{8 f^3}$$

$$\xrightarrow[B=0]{} \int_{-\infty}^{\infty} \Lambda_e G^3 \overline{G} \, dz - \frac{1}{8 f_o f_i^2}$$

$$i_3 = \frac{1}{\hat{\phi}_o^{\frac{1}{2}}} \int_{-\infty}^{\infty} (L_1 G^2 \overline{G}^2 + 2L_2 GG' \overline{G}\overline{G}' + L_3 G'^2 \overline{G}'^2) \, dz$$

$$- \frac{1}{8 f_o^2} \int \left(\frac{\hat{\phi}_o}{\hat{\phi}}\right)^{\frac{1}{2}} \frac{\eta^2 B^2}{\hat{\phi}} \, dz$$

$$\xrightarrow[\phi=\text{const}]{} \int_{-\infty}^{\infty} \Lambda_m G^2 \overline{G}^2 - \frac{1}{6 f^2} \int \frac{\eta^2 B^2}{\hat{\phi}} \, dz$$

$$\xrightarrow[B=0]{} \int_{-\infty}^{\infty} \Lambda_e G^2 \overline{G}^2 - \frac{\hat{\phi}_o^{\frac{1}{2}}}{24 f_o^2} \int \frac{\gamma \phi''}{\hat{\phi}^{\frac{3}{2}}} \, dz$$

25. ASYMPTOTIC ABERRATION COEFFICIENTS

$$i_4 = \frac{1}{\hat{\phi}_o^{\frac{1}{2}}} \int_{-\infty}^{\infty} \{2L_1 G^2 \overline{G}^2 + L_2(G\overline{G})'^2 + 2L_3 G'^2 \overline{G}'^2\} \, dz$$

$$+ \frac{1}{8f_o^2} \int_{-\infty}^{\infty} \left(\frac{\hat{\phi}_o}{\hat{\phi}}\right)^{\frac{1}{2}} \frac{\eta^2 B^2}{\hat{\phi}} \, dz$$

$$\xrightarrow[\phi=\text{const}]{} 2 \int_{-\infty}^{\infty} \Lambda_m G^2 \overline{G}^2 + \frac{1}{6f^2} \int_{-\infty}^{\infty} \frac{\eta^2 B^2}{\hat{\phi}} \, dz$$

$$\xrightarrow[B=0]{} 2 \int_{-\infty}^{\infty} \Lambda_e G^2 \overline{G}^2 + \frac{\hat{\phi}_o^{\frac{1}{2}}}{24 f_o^2} \int_{-\infty}^{\infty} \frac{\gamma \phi''}{\hat{\phi}^{\frac{3}{2}}} \, dz$$

$$i_5 = \frac{1}{\hat{\phi}_o^{\frac{1}{2}}} \int_{-\infty}^{\infty} \{L_1 G \overline{G}^3 + L_2(G\overline{G})' \overline{G} G' + L_3 G' \overline{G}'^3\} \, dz$$

$$\xrightarrow[\phi=\text{const}]{} \int_{-\infty}^{\infty} \Lambda_m G \overline{G}^3 \, dz - \frac{1}{8f^3}$$

$$\xrightarrow[B=0]{} \int_{-\infty}^{\infty} \Lambda_e G \overline{G}^3 \, dz - \frac{1}{8f_o^3}$$

$$i_6 = \frac{1}{\hat{\phi}_o^{\frac{1}{2}}} \int_{-\infty}^{\infty} (L_1 \overline{G}^4 + 2L_2 \overline{G}^2 \overline{G}'^2 + L_3 \overline{G}'^4) \, dz$$

$$+ \frac{1}{2 f_o^4} \lim_{z \to -\infty} (z - z_{Fo})$$

$$\xrightarrow[\phi=\text{const}]{} \int_{-\infty}^{\infty} \Lambda_m \overline{G}^4 \, dz$$

$$\xrightarrow[B=0]{} \int_{-\infty}^{\infty} \Lambda_e \overline{G}^4 \, dz$$

$$i_7 = \int_{-\infty}^{\infty} (PG^2 + QG'^2) \, dz$$

$$\xrightarrow[\phi=\text{const}]{} \int_{-\infty}^{\infty} \Lambda_a G^2 \, dz$$

25. ASYMPTOTIC ABERRATION COEFFICIENTS

$$i_8 = 2 \int_{-\infty}^{\infty} (P G \overline{G} + Q G' \overline{G}') \, dz$$

$$\xrightarrow[\phi=\text{const}]{} 2 \int_{-\infty}^{\infty} \Lambda_a G \overline{G} \, dz$$

$$i_9 = \int_{-\infty}^{\infty} (P \overline{G}^2 + Q \overline{G'}^2) \, dz$$

$$\xrightarrow[\phi=\text{const}]{} \int_{-\infty}^{\infty} \Lambda_a \overline{G}^2 \, dz \qquad (25.31)$$

In these integrals, the functions L_1, L_2, L_3, P and Q are given by (24.3) and Λ_m, Λ_e and Λ_a by

$$\Lambda_m := \frac{1}{48} \left(\frac{4\eta^4 B^4}{\hat{\phi}^2} + \frac{5\eta^2 B'^2}{\hat{\phi}} - \frac{\eta^2 B B''}{\hat{\phi}} \right)$$

$$\Lambda_e := \frac{1}{192 \hat{\phi}_o^{\frac{1}{2}}} \{ 4(3 + 5\epsilon\hat{\phi}) \frac{\phi''^2}{\hat{\phi}^{\frac{3}{2}}} - (3 + 4\epsilon\hat{\phi}) \frac{\phi' \phi'''}{\hat{\phi}^{\frac{3}{2}}}$$

$$- 6(5 + 17\epsilon\hat{\phi}) \frac{\gamma \phi'^2 \phi''}{\hat{\phi}^{\frac{5}{2}}} + (22\gamma^4 + 7\gamma^2 + 1) \frac{\phi'^4}{\hat{\phi}^{\frac{7}{2}}} \} \qquad (25.32)$$

$$\Lambda_a := \frac{1}{16} \left(\frac{\eta B''}{\hat{\phi}^{\frac{1}{2}}} + \frac{2\eta^3 B^3}{\hat{\phi}^{\frac{3}{2}}} \right)$$

(For Λ_m, see Jandeleit and Lenz (1959) and for Λ_e, see Hawkes (1985b).)

For symmetric lenses (or lens combinations), $G(z) = \overline{G}(z)$ if the origin is set in the symmetry plane, and hence

$$i_1 = i_6 \quad i_2 = i_5 \quad i_7 = i_9 \qquad (25.33)$$

The asymptotic Petzval coefficient is given by $F - 2A$, which is simply $F_o - 2A_o$ (25.21) or $f_o^2(i_4 - 2i_3)$. Thus

$$F - 2A = \int_{-\infty}^{\infty} \left(\frac{\hat{\phi}_o}{\hat{\phi}} \right)^{\frac{1}{2}} \frac{\gamma \phi'' + 4\eta^2 B^2}{8\hat{\phi}} \, dz$$

$$\xrightarrow[\phi=\text{const}]{} \int_{-\infty}^{\infty} \frac{\eta^2 B^2}{2\hat{\phi}} \, dz \qquad (25.34)$$

$$\xrightarrow[B=0]{} \int_{-\infty}^{\infty} \left(\frac{\hat{\phi}_o}{\hat{\phi}} \right)^{\frac{1}{2}} \frac{\gamma \phi''}{8\hat{\phi}} \, dz$$

Expressions (25.13–25.28) for the aberration coefficients as polynomials in reciprocal magnification m are convenient in that any symmetries in the lens are immediately reflected in the coefficients. For design purposes, it may be convenient to replace m by the object (or image) position, which is easily done with the aid of (16.25). Again, we might wish to know the magnitudes of the coefficients for a specific value of the magnification (or object position), given those for some other value. The appropriate formulae have been derived from first principles by Ade (1973, 1982). They can also be read off from the polynomial expressions (25.28), see Hawkes (1984a). Thus suppose that C, K, A, F and D are known for some inverse magnification m and that the latter is altered to $m + \mu$, so that for the spherical aberration, for example,

$$C(m + \mu) = \sum_{j=0}^{4} C_j (m+\mu)^j$$

and likewise for the other coefficients. Clearly,

$$C(m + \mu) = \sum_{j=0}^{4} C'_j \mu^j \qquad (25.35)$$

where

$$\begin{aligned}
C'_4 &= C_4 \\
C'_3 &= 4C_4 m + C_3 \\
C'_2 &= 6C_4 m^2 + 3C_3 m + C_2 \\
C'_1 &= 4C_4 m^3 + 3C_3 m^2 + 2C_2 m + C_1 \\
C'_0 &= C(m)
\end{aligned} \qquad (25.36)$$

But

$$\begin{aligned}
4C_4 m + C_3 &= 4i_1 f_o^4 m - 4i_2 f_o^4 - \frac{1}{2} r^2 f_o \\
&= 4f_o^4 (i_2 - i_1 m) - \frac{1}{2} r^2 f_o \\
&= -4f_o^3 \{D(m) - \frac{1}{2f_i^2}\} - \frac{1}{2} r^2 f_o \\
&= -4f_o^3 D(m) - \frac{3}{2} \frac{f_o^3}{f_i^2}
\end{aligned} \qquad (25.37)$$

The remaining coefficients C'_i and their counterparts for all the other aberrations can likewise be expressed in terms of the known coefficients $A(m)...k(m)$ plus one extra quantity, i_1, with the following results. Recalling that the change in inverse magnification μ corresponds to a shift

$\Delta z_o = -f_o\mu$ in object position with the sign convention employed by Ade (1982), these expressions are identical with those derived by him.

$$\begin{pmatrix} C(m+\mu) \\ K(m+\mu) \\ A(m+\mu) \\ F(m+\mu) \\ D(m+\mu) \end{pmatrix} = Q_\mu \begin{pmatrix} \mu^4 \\ \mu^3 \\ \mu^2 \\ \mu \\ 1 \end{pmatrix} \tag{25.38a}$$

$$\begin{pmatrix} k(m+\mu) \\ a(m+\mu) \\ d(m+\mu) \end{pmatrix} = q_\mu \begin{pmatrix} \mu^2 \\ \mu \\ 1 \end{pmatrix} \tag{25.38b}$$

where the elements of Q_μ are as follows:

$Q_{\mu 11} = f_o^4 i_1$, $Q_{\mu 12} = -4f_o^3 D - 3r^2 f_o/2$
$Q_{\mu 13} = 2f_o^2(A+F) + 3r^2 f_o m/2$, $Q_{\mu 14} = -4f_o K + f_o(r^2 m^2 - 1)/2$
$Q_{\mu 15} = C$, $Q_{\mu 22} = -f_o^3 i_1$, $Q_{\mu 23} = 3f_o^2 D - r^2$
$Q_{\mu 24} = -f_o(A+F) - r^2 m/2$, $Q_{\mu 25} = K$, $Q_{\mu 33} = f_o^2 i_1$
$Q_{\mu 34} = -2f_o D + r^2/2f_o$, $Q_{\mu 35} = A$, $Q_{\mu 43} = 2f_o^2 i_1$
$Q_{\mu 44} = -4f_o D + r^2/f_o$, $Q_{\mu 45} = F$, $Q_{\mu 54} = -f_o i_1$, $Q_{\mu 55} = D$
$Q_{\mu 21} = Q_{\mu 31} = Q_{\mu 32} = Q_{\mu 41} = Q_{\mu 42}$
$= Q_{\mu 51} = Q_{\mu 52} = Q_{\mu 53} = 0$ (25.39a)

and

$$q_\mu = \begin{pmatrix} f_o^2 d & -f_o a & k \\ 0 & -2f_o d & a \\ 0 & 0 & d \end{pmatrix} \tag{25.39b}$$

One other situation remains to be examined: hitherto the aberrations have been expressed in terms of position and gradient in the asymptotic object plane but, in practice, the effect of the real aperture will often be important; in projectors, for example, where the dominant geometrical aberration is distortion, the sign of the coefficient can be changed by varying the aperture position, since the aperture selects different pencils of rays from the beam at the asymptotic object as its position is altered. We therefore consider the situation illustrated in Fig. 25.2, which shows that we shall be concerned with the aberrations of the asymptotic image of an asymptotic object with boundary conditions requiring aberrations of position to

---

* The following account is confined to magnetic lenses; it can be extended straightforwardly to electrostatic lenses if needed.

25. ASYMPTOTIC ABERRATION COEFFICIENTS

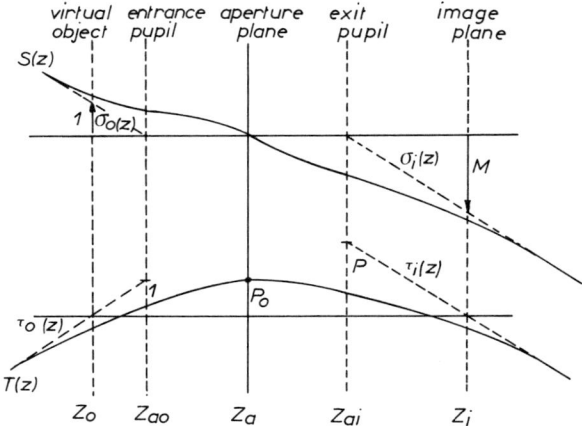

Fig. 25.2: The rays $S(z)$ and $T(z)$ have asymptotes $\sigma_o(z)$ and $\tau_o(z)$ in object space and $\sigma_i(z)$ and $\tau(z_i)$ in image space. The object–image magnification M and the pupil magnification N are defined in terms of these asymptotes.

vanish in two planes (cf. 24.14, 24.18): the asymptotic aberrations must vanish in the asymptotic object plane and the real aberrations in the real aperture plane. We shall, however, find it useful to introduce the *entrance* and *exit pupils* associated with the real aperture; these are the asymptotic images of the latter formed by the part of the lens field before the real aperture (entrance pupil) and the part beyond it (exit pupil). This in turn introduces the important notion of pupil magnification, the magnification between these two pupil planes, which are (asymptotically) conjugate.

Figure 25.2 shows two paraxial solutions, analogous to $s(z)$ and $t(z)$, which we here denote $S(z)$ and $T(z)$; these satisfy the conditions

$$\lim_{z \to -\infty} S(z) =: \sigma_o(z) = \sigma_o'(z - z_{ao}) \qquad (25.40a)$$

$$\lim_{z \to \infty} S(z) =: \sigma_i(z) = \sigma_i'(z - z_{ai}) \qquad (25.40b)$$

$$\lim_{z \to -\infty} T(z) =: \tau_o(z) = \tau_o'(z - z_o) \qquad (25.40c)$$

$$\lim_{z \to \infty} T(z) =: \tau_i(z) = \tau_i'(z - z_i) \qquad (25.40d)$$

and

$$\begin{aligned} S(z_a) &= 0 \\ \sigma(z_o) &= 1 \end{aligned} \qquad (25.41)$$

Clearly, $z = z_{ao}$ and $z = z_{ai}$ are the entrance and exit pupil planes, respectively. From the Wronskian, we see that

$$\tau_o' = -S_a' T_a = M\tau_i' = -\sigma_o' \tau_{ao} \qquad (25.42)$$

25. ASYMPTOTIC ABERRATION COEFFICIENTS

It is convenient to denote the pupil magnification by N, and we write

$$N = N_o N_i \tag{25.43}$$

where N_o is the magnification between the entrance pupil and the real aperture and N_i that between the latter and the exit pupil:

$$\frac{S'_a}{\sigma'_o} = \frac{1}{N_o} \qquad \frac{\sigma'_i}{S'_a} = \frac{1}{N_i} \tag{25.44}$$

We have not yet fully specified $T(z)$, for we see from (25.40c) that the object asymptote τ_o intersects the entrance pupil at $\tau_o(z_{ao}) = \tau'_o(z_{ao} - z_o)$ so that from (25.40–25.42)

$$\tau_{ao} = -\frac{\tau'_o}{\sigma'_o} = \frac{S'_a T_a}{\sigma'_o} \tag{25.45}$$

or

$$\tau_{ao} = \frac{T_a}{N_o} \tag{25.46}$$

By choosing

$$T(z_a) = N_o \tag{25.47}$$

so that

$$\tau_{ao} = 1 \tag{25.48}$$

we express the aberrations in terms of asymptotic object coordinates (in $z = z_o$) and entrance pupil coordinates (in $z = z_{ao}$). (If we had set $T(z_a) = 1, \tau_{ao} = 1/N_o$, the aberrations would have been expressed in terms not of entrance pupil coordinates but of real aperture coordinates, which is not so convenient; the important point is that the aberrations do actually vanish in $z = z_a$.) These relations and some simple consequences of them are listed in Table 25.1

The general paraxial solution is now

$$\begin{aligned} x(z) &= x_o S(z) + x_a T(z) \\ y(z) &= y_o S(z) + y_a T(z) \end{aligned} \tag{25.49}$$

in which x_a, y_a are position coordinates in the entrance pupil plane. (They should strictly speaking be denoted x_{ao}, y_{ao}, but we drop the second suffix since no confusion should arise.) With the boundary conditions already explained, the asymptotic aberrations at the image plane z_i are given by

$$\begin{aligned} \Delta x_i &= -\frac{1}{\tau'_o} \frac{\partial S^I_{oi}}{\partial x_a} \\ \Delta y_i &= -\frac{1}{\tau'_o} \frac{\partial S^I_{oi}}{\partial y_a} \end{aligned} \tag{25.50}$$

Table 25.1 Relations between the asymptotes to $S(z)$ and $T(z)$ and various paraxial properties: pupil magnification, N; magnification, M; focal length, f; and others.

$$\sigma_o(z) = \sigma'_o(z - z_o) + 1 = \sigma'_o(z - z_{ao}) \qquad \sigma_i(z) = \sigma'_i(z - z_i) + M = \sigma'_i(z - z_{ai})$$

$$\tau_o(z) = \tau'_o(z - z_o) = \tau'_o(z - z_{ao}) + 1 \qquad \tau_i(z) = \tau'_i(z - z_i) = \tau'_i(z - z_{ai}) + N$$

$$\sigma'_o = \frac{1}{z_o - z_{ao}} = \frac{1}{f(m - m_a)} \qquad \sigma'_i = \frac{M}{z_i - z_{ai}} = \frac{\sigma'_o}{N}$$

$$\tau'_o = \frac{1}{z_{ao} - z_o} = -\frac{1}{f(m - m_a)} \qquad \tau'_i = \frac{\tau'_o}{M} = \frac{N}{z_{ai} - z_i}$$

$$T_a = N_o = \frac{\sigma'_o}{S'_a} \qquad N_i = \frac{S'_a}{\sigma'_i}$$

$$z_i - z_{Fi} = -fM \qquad z_o - z_{Fo} = \frac{f}{M} = fm$$

$$N = N_o N_i$$

and the various coefficients can hence be written down immediately. Our purpose here is to enquire how they vary when the (real) aperture and (asymptotic) object are shifted, and how this variation compares with the simple polynomial forms obtained earlier.

Let us suppose that the image magnification is initially M_1 and the pupil magnification N_1; we add the label 1 to the other relevant quantities, S_1, T_1 in particular. After moving object and aperture, the magnifications become M_2 and N_2 and the new fundamental paraxial solutions, S_2 and T_2, are linear combinations of their predecessors,

$$\begin{aligned} S_2 &= \lambda_s S_1 + \lambda_t T_1 \\ T_2 &= \mu_s S_1 + \mu_t T_1 \end{aligned} \tag{25.51}$$

By considering the relations between the asymptotes to these rays and the object and aperture positions (Table 25.1), it is easily seen that

$$\begin{aligned} \lambda_s &= \frac{z_{o1} - z_{ao2}}{z_{o2} - z_{ao2}} = \frac{m_1 - n_2}{m_2 - n_2} \\ \lambda_t &= \frac{z_{ao1} - z_{ao2}}{z_{o2} - z_{ao2}} = \frac{n_1 - n_2}{m_2 - n_2} \\ \mu_s &= \frac{z_{o2} - z_{o1}}{z_{o2} - z_{ao2}} = \frac{m_2 - m_1}{m_2 - n_2} \\ \mu_t &= \frac{z_{o2} - z_{ao1}}{z_{o2} - z_{ao2}} = \frac{m_2 - n_1}{m_2 - n_2} \end{aligned} \tag{25.52}$$

The individual aberration coefficients can be expressed in terms of the integrals

$$I_{ij} := \frac{1}{\tau'_o} \int_{-\infty}^{\infty} \Lambda_m(z) S^i T^j \, dz \qquad i+j=4, \quad i,j=0-4$$

$$I_{00} := \frac{\eta^2}{\hat{\phi}} \int_{-\infty}^{\infty} B^2(z) \, dz \qquad (25.53)$$

in which (25.32)

$$\Lambda_m(z) := \frac{\eta^4 B^4}{12\hat{\phi}^2} + \frac{5}{48}\frac{\eta^2 B'^2}{\hat{\phi}} - \frac{\eta^2 B B''}{48\hat{\phi}} \qquad (25.54)$$

and

$$J_{ij} := \int_{-\infty}^{\infty} \Lambda_a(z) S^i T^j \, dz \qquad i+j=2 \quad i,j=0-2 \qquad (25.55)$$

in which (25.32)

$$\Lambda_a(z) := \frac{1}{16}\left(\frac{\eta B''}{\hat{\phi}^{\frac{1}{2}}} + \frac{2\eta^3 B^3}{\hat{\phi}^{\frac{3}{2}}}\right) \qquad (25.56)$$

For $m = m_1$ and $n = n_1$, we have

$$\begin{aligned}
C_1 &= I_{04}^{(1)} \\
K_1 &= I_{13}^{(1)} - (1-m_1^2)\Big/8f^2(m_1-n_1)^2 & k_1 &= J_{02} \\
A_1 &= I_{22}^{(1)} - \tfrac{1}{6}\tau'_{10}I_{00} + (1-m_1 n_1)\Big/4f^2(m_1-n_1)^2 & a_1 &= 2J_{11} \\
F_1 &= 2I_{22}^{(1)} + \tfrac{1}{6}\tau'_{10}I_{00} + (1-m_1 n_1)\Big/2f^2(m_1-n_1)^2 \\
D_1 &= I_{31}^{(1)} - 3(1-n_1^2)\Big/8f^2(m_1-n_1)^2 & d_1 &= J_{20}
\end{aligned} \qquad (25.57)$$

For any other pair of reciprocal magnifications m_2, n_2, similar expressions relate $C_2, K_2, ...d_2$ to $I_{ij}^{(2)}$ and $J_{ij}^{(2)}$. Using (25.51), it is easy to express the integrals $I_{ij}^{(2)}$ and $J_{ij}^{(2)}$ in terms of $I_{ij}^{(1)}$ and $J_{ij}^{(1)}$ and hence $C_2...d_2$ in terms of $C_1,...d_1$. The results are as follows.

$$\begin{pmatrix} I_{40}^{(2)} \\ I_{31}^{(2)} \\ I_{22}^{(2)} \\ I_{13}^{(2)} \\ I_{04}^{(2)} \end{pmatrix} = I \begin{pmatrix} I_{40}^{(1)} \\ I_{31}^{(1)} \\ I_{22}^{(1)} \\ I_{13}^{(1)} \\ I_{04}^{(1)} \end{pmatrix} \qquad \begin{pmatrix} J_{20}^{(2)} \\ J_{11}^{(2)} \\ J_{02}^{(2)} \end{pmatrix} = J \begin{pmatrix} J_{20}^{(1)} \\ J_{11}^{(1)} \\ J_{02}^{(1)} \end{pmatrix} \qquad (25.58)$$

where
$$I = \frac{m_1 - n_1}{m_2 - n_2} I^*$$
and

$$I_{11}^* = \lambda_s^4 \quad , \quad I_{12}^* = 4\lambda_s^3 \lambda_t \quad , \quad I_{13}^* = 6\lambda_s^2 \lambda_t^2 \quad , \quad I_{14}^* = 4\lambda_s \lambda_t^3 \quad , \quad I_{15}^* = \lambda_t^4$$
$$I_{21}^* = \lambda_s^3 \mu_s \quad , \quad I_{22}^* = 3\lambda_s^2 \lambda_t \mu_s + \lambda_s^3 \mu_t \quad , \quad I_{23}^* = 3\lambda_s \lambda_t^2 \mu_s + 3\lambda_s^2 \lambda_t \mu_t$$
$$I_{24}^* = \lambda_t^3 \mu_s + 3\lambda_s \lambda_t^2 \mu_t \quad , \quad I_{25}^* = \lambda_t^3 \mu_t$$
$$I_{31}^* = \lambda_s^2 \mu_s^2 \quad , \quad I_{32}^* = 2(\lambda_s^2 \mu_s \mu_t + \lambda_s \lambda_t \mu_s^2) \quad , \quad I_{33}^* = \lambda_s^2 \mu_t^2 + \mu_s^2 \lambda_t^2 + 4\lambda_s \lambda_t \mu_s \mu_t$$
$$I_{34}^* = 2(\lambda_s \lambda_t \mu_t^2 + \mu_s \mu_t \lambda_t^2) \quad , \quad I_{35}^* = \lambda_t^2 \mu_t^2$$
$$I_{41}^* = \lambda_s \mu_s^3 \quad , \quad I_{42}^* = 3\lambda_s \mu_s^2 \mu_t + \lambda_t \mu_s^3 \quad , \quad I_{43}^* = 3(\lambda_t \mu_s^2 \mu_t + \lambda_s \mu_s \mu_t^2)$$
$$I_{44}^* = \lambda_s \mu_t^3 + 3\lambda_t \mu_s \mu_t^2 \quad , \quad I_{45}^* = \lambda_t \mu_t^3$$
$$I_{51}^* = \mu_s^4 \quad , \quad I_{52}^* = 4\mu_s^3 \mu_t \quad , \quad I_{53}^* = 6\mu_s^2 \mu_t^2 \quad , \quad I_{54}^* = 4\mu_s \mu_t^3 \quad , \quad I_{55}^* = \mu_t^4$$

(25.59)

$$J = \begin{pmatrix} \lambda_s^2 & 2\lambda_s \lambda_t & \lambda_t^2 \\ \lambda_s \mu_s & \lambda_s \mu_t + \lambda_t \mu_s & \lambda_t \mu_t \\ \mu_s^2 & 2\mu_s \mu_t & \mu_t^2 \end{pmatrix} \quad (25.60)$$

This question, the variation of the aberrations when object and aperture (or stop) are shifted, has a long history in light optics, and we draw attention to the work that is closest to the contents of this chapter. The earliest thorough exploration, using the eikonal or characteristic function, is that of T. Smith (1921/22) but owing to the extreme generality of the results, and to the fact that the latter were not expressed in terms of familiar quantities, his findings were frequently referred to but little used. The problem was solved in language of direct use to the lens designer by a series of members of van Heel's laboratory in Delft (Korringa, 1942; Stephan, 1947; Brouwer, 1957). The full matrix treatment developed by Brouwer renders his work easy to apply and we can derive (25.59–25.60) straightforwardly by his methods. This requires the notion of aberration matrices, however, and we therefore postpone further discussion to Chapter 27, where these are introduced. For a very scholarly account of the various formulations of aberration theory with many more references, see Focke (1965) and for a recent development, Velzel (1987) and Velzel and de Meijere (1988).

26

Chromatic Aberrations

26.1 Real chromatic aberrations

Electron guns do not furnish rigorously monoenergetic beams of particles and the voltage and current supplies needed to excite electron lenses are never perfectly stable. The spread of energies in a beam is usually further broadened by its passage through a specimen, for inelastically scattered electrons will emerge with diminished energy. The initial energy spread is typically a few electronvolts for thermionic emitters; power supplies can be stabilized to within a few parts in a million (better than 1 V at 100 kV) and although the specimen energy losses may reach tens or hundreds of electronvolts for moderately thick specimens, the mean loss is often much smaller.

In a column of lenses, the image plane conjugate to some fixed plane will therefore inevitably fluctuate, retreating towards the fixed plane if the electron energy falls or the lens strength increases and *vice versa*. Since fluctuations of lens strength may be rapid and the energy of the beam will be spread over a finite if narrow range, these phenomena must be expected to blur the image. The resulting effects are known as the *chromatic aberrations* and are characterized by a set of *chromatic aberration coefficients*. The latter are conveniently calculated by perturbation theory, taking $M^{(2)}$ as the unperturbed refractive index and writing

$$M^{(P)} = \frac{\partial M^{(2)}}{\partial \phi}\Delta\phi + \frac{\partial M^{(2)}}{\partial B}\Delta B \qquad (26.1)$$

From (15.24), we see that

$$\begin{aligned} \frac{\partial M^{(2)}}{\partial \phi} &= \frac{\phi''}{16\hat{\phi}^{\frac{3}{2}}}(X^2+Y^2) + \frac{\gamma}{4\hat{\phi}^{\frac{1}{2}}}(X'^2+Y'^2) \\ \frac{\partial M^{(2)}}{\partial B} &= -\frac{1}{2}\eta(XY'-X'Y) \end{aligned} \qquad (26.2)$$

so that introducing rotating coordinates (15.9),

$$M^{(c)} := M^{(P)} = A_1(x^2 + y^2) + A_2(x'^2 + y'^2) + A_3(xy' - x'y)$$

$$A_1 = \frac{\phi'' + \gamma\eta^2 B^2}{16\hat{\phi}^{\frac{3}{2}}}\Delta\phi - \frac{\eta^2 B}{4\hat{\phi}^{\frac{1}{2}}}\Delta B$$

$$A_2 = \frac{\gamma}{4\hat{\phi}^{\frac{1}{2}}}\Delta\phi \qquad (26.3)$$

$$A_3 = \frac{1}{4}\eta B(\frac{\gamma}{\phi}\Delta\phi - 2\frac{\Delta B}{B})$$

With $x = x_o s + x_a t$, $y = y_o s + y_a t$, we find

$$M^{(c)} = a_1(x_o^2 + y_o^2) + a_2(x_a^2 + y_a^2) + a_3(x_o x_a + y_o y_a)$$
$$\qquad + a_4(x_o y_a - x_a y_o)$$
$$a_1 = A_1 s^2 + A_2 s'^2$$
$$a_2 = A_1 t^2 + A_2 t'^2 \qquad (26.4)$$
$$a_3 = 2(A_1 st + A_2 s't')$$
$$a_4 = A_3(st' - s't)$$

Using (24.29), we obtain the following expressions in a general plane:

$$x^{(c)} = x_o s + x_a t + \frac{t}{W_s}\left(2x_o\int_{z_a} a_1\,dz + x_a\int_{z_a} a_3\,dz + y_a\int_{z_a} a_4\,dz\right)$$
$$\qquad - \frac{s}{W_s}\left(2x_a\int_{z_o} a_2\,dz + x_o\int_{z_o} a_3\,dz - y_o\int_{z_o} a_4\,dz\right)$$
$$y^{(c)} = y_o s + y_a t + \frac{t}{W_s}\left(2y_o\int_{z_a} a_1\,dz + y_a\int_{z_a} a_3\,dz - x_a\int_{z_a} a_4\,dz\right) \qquad (26.5)$$
$$\qquad - \frac{s}{W_s}\left(2y_a\int_{z_o} a_2\,dz + y_o\int_{z_o} a_3\,dz + x_o\int_{z_o} a_4\,dz\right)$$

26.1 REAL CHROMATIC ABERRATIONS

and in the image plane

$$\Delta x^{(c)} := \frac{x_i^{(c)} - M x_o}{M} = -\frac{1}{W_s}\left(2x_a \int_{z_o}^{z_i} a_2\, dz + x_o \int_{z_o}^{z_i} a_3\, dz - y_o \int_{z_o}^{z_i} a_4\, dz\right)$$

$$\Delta y^{(c)} := \frac{y_i^{(c)} - M y_o}{M} = -\frac{1}{W_s}\left(2y_a \int_{z_o}^{z_i} a_2\, dz + y_o \int_{z_o}^{z_i} a_3\, dz + x_o \int_{z_o}^{z_i} a_4\, dz\right)$$

$$\Delta u^{(c)} := \Delta x(c) + i\Delta y^{(c)} = -\frac{1}{W_s}\left(2u_a \int_{z_o}^{z_i} a_2\, dz + u_o \int_{z_o}^{z_i} a_3\, dz + iu_o \int_{z_o}^{z_i} a_4\, dz\right)$$

(26.6)

The integrals may be rewritten in the convenient forms

$$\int_{z_o}^{z_i} a_2\, dz = \int_{z_o}^{z_i} \Delta\phi\left\{\frac{\gamma\eta^2 B^2}{8\hat{\phi}^{\frac{3}{2}}} + \frac{\gamma\phi'^2(3+2\epsilon\hat{\phi})}{16\hat{\phi}^{\frac{5}{2}}}\right\}t^2\, dz$$

$$- \int_{z_o}^{z_i} \Delta B \frac{\eta^2 B}{4\hat{\phi}^{\frac{1}{2}}} t^2\, dz$$

$$\int_{z_o}^{z_i} a_3\, dz = \int_{z_o}^{z_i} \gamma\Delta\phi\left\{\frac{\eta^2 B^2}{4\hat{\phi}^{\frac{3}{2}}} + \frac{\phi'^2(3+2\epsilon\hat{\phi})}{8\hat{\phi}^{\frac{5}{2}}}\right\}st\, dz$$

$$- \int_{z_o}^{z_i} \Delta B \frac{\eta^2 B}{2\hat{\phi}^{\frac{1}{2}}} st\, dz + \Delta\phi\left(\frac{\gamma_i s_i t'_i}{\hat{\phi}_i^{\frac{1}{2}}} - \frac{\gamma_o t'_o}{\hat{\phi}_o^{\frac{1}{2}}}\right)$$

$$\int_{z_o}^{z_i} a_4\, dz = \hat{\phi}^{\frac{1}{2}}(st' - s't) \int_{z_o}^{z_i} \frac{\eta B}{4\hat{\phi}^{\frac{1}{2}}}\left(\frac{\gamma}{\hat{\phi}}\Delta\phi - 2\frac{\Delta B}{B}\right) dz \qquad (26.7)$$

In practice, we take $\Delta\phi$ to be a variation of the accelerating voltage ϕ_0, and ΔB a variation of the maximum magnetic field B_0. Thus if $B(z) = B_0 b(z)$, $\Delta B = b(z)\Delta B_0 = B\Delta B_0/B_0$, provided that $b(z)$ does not alter; this is true if the metal of the lens is not saturated. We can then apply Ampère's circuital theorem (Fig. 36.2), so that for a lens of traditional design, with a winding of N turns carrying a current I,

$$\int_{-\infty}^{\infty} B(z)\, dz = B_0 \int_{-\infty}^{\infty} b(z)\, dz = \mu_0 N I \qquad (26.8)$$

and so
$$\frac{\Delta B_0}{B_0} = \frac{\Delta(NI)}{NI} \qquad (26.9)$$

In the magnetic case ($\phi \equiv \phi_0$), we write

$$\Delta x^{(c)} := -(C_c^* x_a + C_D^* x_o - C_\theta^* y_o)\left(\frac{\gamma\Delta\phi_0}{\hat{\phi}_0} - 2\frac{\Delta B_0}{B_0}\right)$$

$$\Delta y^{(c)} := -(C_c^* y_a + C_D^* y_o + C_\theta^* x_o)\left(\frac{\gamma\Delta\phi_0}{\hat{\phi}_0} - 2\frac{\Delta B_0}{B_0}\right) \qquad (26.10)$$

$$\Delta u^{(c)} = -\{C_c^* u_a + (C_D^* + iC_\theta^*)u_o\}\left(\frac{\gamma\Delta\phi_0}{\hat{\phi}_0} - 2\frac{\Delta B_0}{B_0}\right)$$

where

$$C_c^* = \frac{1}{t_o'}\int_{z_o}^{z_i} \frac{\eta^2 B^2 t^2}{4\hat{\phi}_0}\,dz$$

$$C_D^* = \frac{1}{t_o'}\int_{z_o}^{z_i} \frac{\eta^2 B^2 st}{4\hat{\phi}_0}\,dz \qquad (26.11)$$

$$C_\theta^* = \int_{z_o}^{z_i} \frac{\eta B}{4\hat{\phi}_0^{\frac{1}{2}}}\,dz$$

In the electrostatic case, $\Delta B \equiv 0$ and we write

$$\Delta u^{(c)} = \Delta x^{(c)} + i\Delta y^{(c)} = -(C_c^* u_a + C_D^* u_o)\left(\frac{\Delta\phi_0}{\hat{\phi}_0}\right) \qquad (26.12)$$

where

$$C_c^* = \frac{\hat{\phi}_0^{\frac{1}{2}}}{t_o'}\int_{z_o}^{z_i} \frac{\gamma\phi'^2(3+2\epsilon\hat{\phi})}{8\hat{\phi}^{\frac{5}{2}}}t^2\,dz$$

$$C_D^* = \frac{\hat{\phi}_0^{\frac{1}{2}}}{t_o'}\int_{z_o}^{z_i} \frac{\gamma\phi'^2(3+2\epsilon\hat{\phi})}{8\hat{\phi}^{\frac{5}{2}}}st\,dz + \frac{\hat{\phi}_0}{4}\left(\frac{\gamma_i}{\hat{\phi}_i} - \frac{\gamma_o}{\hat{\phi}_0}\right) \qquad (26.13)$$

On measuring the variations in accelerating voltage or energy spread in terms not of ϕ_0 but of $\hat{\phi}_0$ in the magnetic case, so that $\Delta\hat{\phi}_0 = \gamma\Delta\phi_0$, and replacing x_a, y_a by x_o', y_o', we find the following expressions, which are the most common form of the chromatic aberration coefficients:

$$\Delta u^{(c)} = -\{C_c u_o' + (C_D + iC_\theta)u_o\}\left(\frac{\Delta\hat{\phi}_0}{\hat{\phi}_0} - 2\frac{\Delta B_0}{B_0}\right) \qquad (26.14)$$

26.1 REAL CHROMATIC ABERRATIONS

with

$$C_c = \int_{z_o}^{z_i} \frac{\eta^2 B^2}{4\hat{\phi}_0} h^2 \, dz$$

$$C_D = \int_{z_o}^{z_i} \frac{\eta^2 B^2}{4\hat{\phi}_0} gh \, dz \qquad (26.15)$$

$$C_\theta = \int_{z_o}^{z_i} \frac{\eta B}{4\hat{\phi}_0^{\frac{1}{2}}} \, dz$$

It is not convenient to use $\Delta\hat{\phi}_0/\hat{\phi}_0$ in the electrostatic case, and the counterparts of (26.14–26.15) are then

$$\Delta u^{(c)} = -(C_c u'_o + C_D u_o)\frac{\Delta\phi_0}{\hat{\phi}_0} \qquad (26.16)$$

with

$$C_c = \hat{\phi}_0^{\frac{1}{2}} \int_{z_o}^{z_i} \frac{\gamma\phi'^2(3+2\epsilon\hat{\phi})h^2}{8\hat{\phi}^{\frac{5}{2}}} \, dz$$

$$C_D = \hat{\phi}_0^{\frac{1}{2}} \int_{z_o}^{z_i} \frac{\gamma\phi'^2(3+2\epsilon\hat{\phi})gh}{8\hat{\phi}^{\frac{5}{2}}} \, dz + \frac{\hat{\phi}_0}{4}\left(\frac{\gamma_i}{\hat{\phi}_i} - \frac{\gamma_o}{\hat{\phi}_o}\right) \qquad (26.17)$$

The chromatic aberrations are therefore of two kinds: an "aperture" aberration, characterized by C_c, which is invariably known simply as the *chromatic aberration coefficient*, and a distortion, characterized by $C_D + iC_\theta$. The term in C_c, like the spherical aberration, creates an aberration disc, the radius of which in the image plane is the same for all points in the object plane. This radius, $M\Delta r^{(c)}$, is given by

$$M\Delta r^{(c)} = M(\Delta x^{(c)2} + \Delta y^{(c)2})^{1/2}$$

or referred back to object space,

$$\Delta r^{(c)} = C_c(x'^2_o + y'^2_o)^{1/2}\Delta_c = C_c\theta_o\Delta_c$$

or

$$\Delta r^{(c)} = C_c^*(x_a^2 + y_a^2)^{1/2}\Delta_c = C_c^* r_a \Delta_c \qquad (26.18)$$

where $\theta_o := (x'^2_o + y'^2_o)^{1/2}$, $r_a := (x_a^2 + y_a^2)^{1/2}$ and Δ_c is the appropriate term in $\Delta\phi_0$ or $\Delta\hat{\phi}_0$ and ΔB_0. (Note that the radius of the aberration disc is the same for a variation of the form $\phi_0 \to \phi_0 + \Delta\phi_0$, and likewise for B_0, as it is for a variation $\phi_0 \to \phi_0 - \Delta\phi_0$. Some thought must therefore be given to the meaning of $\Delta\phi_0$ when calculating the size of the disc. Cf. Orloff, 1983b.)

From (26.15, 26.17), we see that, again like C_s, C_c is always positive with the sign convention we have adopted (minus signs have been inserted in (26.14, 26.16) purely for convenience, since $\Delta\phi_0$ and ΔB_0 are usually oscillatory fluctuations about ϕ_0 and B_0 rather than steady drifts, though the latter are not of course excluded if present). This result too was first obtained by Scherzer (1936b). Methods of correcting or eliminating this chromatic aberration will be discussed at the same time as those for combating spherical aberration.

The chromatic aberration C_c is of most concern in objective lenses, where the angles are comparatively large, while the distortions characterized by C_D and C_θ mainly affect projectors, where the angles are small but the rays are farther from the axis. The isotropic chromatic distortion coefficient, C_D, is commonly known as the *chromatic aberration of magnification*, since we may write

$$u_i = M\left\{1 - C_D\left(\frac{\Delta\hat{\phi}_0}{\hat{\phi}_0} - 2\frac{\Delta B_0}{B_0}\right)\right\} u_o \qquad (26.19)$$

in the magnetic case, and similarly for electrostatic lenses.

The *anisotropic chromatic distortion coefficient*, C_θ, is often known as the chromatic aberration of rotation, since it causes a small change in the image rotation in magnetic lenses. We note that C_θ is equal to half the image rotation (15.27).

A slightly different way of considering the chromatic aberrations yields forms for the coefficients that can save a lot of calculation. We now denote the paraxial solution corresponding to accelerating voltage $\hat{\phi}_0$ by $w(z, \hat{\phi}_0)$, so that

$$w(z, \hat{\phi}_0) = w_o g(z, \hat{\phi}_0) + w'_o h(z, \hat{\phi}_0) \qquad (26.20)$$

In the image plane z_i for $\hat{\phi}_0$, we have $h(z_i, \hat{\phi}_0) = 0$. Hence

$$w(z_i, \hat{\phi}_0 + \Delta\hat{\phi}_0) = w_o g(z_i, \hat{\phi}_0 + \Delta\hat{\phi}_0) + w'_o h(z_i, \hat{\phi}_0 + \Delta\hat{\phi}_0)$$

$$\approx w_o\left(M + \frac{\partial g_i}{\partial \hat{\phi}_0}\Delta\hat{\phi}_0\right) + \Delta\hat{\phi}_0 w'_o \frac{\partial h_i}{\partial \hat{\phi}_0}$$

In the fixed coordinate system, this is equivalent to

$$u(z_i, \hat{\phi}_0 + \Delta\hat{\phi}_0) = w(z_i, \hat{\phi}_0 + \Delta\hat{\phi}_0)\exp\left\{\frac{i\eta}{2(\hat{\phi}_0 + \Delta\hat{\phi}_0)^{\frac{1}{2}}}\int B\,dz\right\}$$

$$= \exp(i\theta_i)\left[Mw_o + \Delta\hat{\phi}_0\left\{w_o\left(\frac{\partial g_i}{\partial \hat{\phi}_0} - \frac{Mi\theta_i}{2\hat{\phi}_0}\right) + w'_o\frac{\partial h_i}{\partial \hat{\phi}_0}\right\}\right]$$

26.2 ASYMPTOTIC CHROMATIC ABERRATIONS

which must have the same form as

$$u(z_i, \hat{\phi}_0 + \Delta\hat{\phi}_0) = \exp(i\theta_i)\left[Mw_o - M\frac{\Delta\hat{\phi}_0}{\hat{\phi}_0}\{C_c w'_o + (C_D + iC_\theta)w_o\}\right]$$

so that

$$C_c = -\frac{\hat{\phi}_0}{M}\frac{\partial h_i}{\partial\hat{\phi}_0}$$

$$C_D = -\frac{\hat{\phi}_0}{M}\frac{\partial g_i}{\partial\hat{\phi}_0} \qquad (26.21)$$

$$C_\theta = \frac{1}{2}\theta_i$$

This form is most useful when the lens is operating at high magnification, $z_o \to z_{F_o}$. Since in general

$$x_i = Mx_o - MC_c x'_o\frac{\Delta\hat{\phi}_0}{\hat{\phi}_0} = Mx_o + \hat{\phi}_0 x'_o\frac{\partial h_i}{\partial\hat{\phi}_0}\frac{\Delta\hat{\phi}_0}{\hat{\phi}_0}$$

a ray intersecting the axis at $z = z_i$ leaves the object plane at $x_o = (-1/M)x'_o\Delta\hat{\phi}_0(\partial h_i/\partial\hat{\phi}_0)$. But $x_o = -x'_o\Delta\hat{\phi}_0(\partial z_{F_o}/\partial\hat{\phi}_0)$ so that

$$\frac{\partial z_{F_o}}{\partial\hat{\phi}_0} = \frac{1}{M}\frac{\partial h_i}{\partial\hat{\phi}_0}$$

and hence

$$C_c = -\hat{\phi}_0\frac{\partial z_{F_o}}{\partial\hat{\phi}_0} \qquad (26.22a)$$

Likewise,

$$C_D = \frac{\hat{\phi}_0}{f_o}\frac{\partial f_o}{\partial\hat{\phi}_0} \qquad (26.22b)$$

These expressions are particularly convenient when field models that yield explicit formulae for the cardinal elements are being studied.

26.2 Asymptotic chromatic aberrations

These aberration coefficients can be written down immediately as there is no term analogous to $L_3(x'^2 + y'^2)^2$ (24.2) needing special attention. We find

$$\Delta u^{(c)} = -\{C_c u'_o + (C_D + iC_\theta)u_o\}(\frac{\gamma\Delta\phi_0}{\hat{\phi}_0} - 2\frac{\Delta B_0}{B_0}) \qquad (26.23a)$$

or

$$\Delta u^{(c)} = -(C_c u'_o + C_D u_o)\frac{\Delta \phi_0}{\hat{\phi}_0} \qquad (24.23b)$$

for magnetic and electrostatic lenses, respectively, where

$$C_c = \int_{-\infty}^{\infty} \frac{\eta^2 B^2 H^2}{4\hat{\phi}_0} dz \quad \text{or} \quad C_c = \hat{\phi}_0^{\frac{1}{2}} \int_{-\infty}^{\infty} \frac{\gamma(3+2\epsilon\hat{\phi})\phi'^2 H^2}{8\hat{\phi}^{\frac{5}{2}}} dz$$

$$C_D = \int_{-\infty}^{\infty} \frac{\eta^2 B^2 GH}{4\hat{\phi}_0} dz \quad \text{or} \quad C_D = \hat{\phi}_0^{\frac{1}{2}} \int_{-\infty}^{\infty} \frac{\gamma(3+2\epsilon\hat{\phi})\phi'^2 GH}{8\hat{\phi}^{\frac{5}{2}}} dz$$

$$C_\theta = \int_{-\infty}^{\infty} \frac{\eta B}{4\hat{\phi}_0^{\frac{1}{2}}} dz \qquad\qquad\qquad + \frac{\hat{\phi}_0^{\frac{1}{2}}}{4}\left(\frac{\gamma_i}{\hat{\phi}_i} - \frac{\gamma_o}{\hat{\phi}_0}\right)$$

$$(26.24)$$

These can of course be cast into polynomial form with the aid of (25.9) (Shimoyama, 1982; Hawkes, 1980a):

$$\begin{pmatrix} C_c \\ C_D \\ C_\theta \end{pmatrix} = \begin{pmatrix} f_o^2 C_2 & -2f_o^2 C_1 & f_o^2 C_0 \\ 0 & -f_o C_2 & f_o C_1 + C_e \\ 0 & 0 & C_\theta \end{pmatrix} \begin{pmatrix} m^2 \\ m \\ 1 \end{pmatrix} \qquad (26.25)$$

in which for magnetic lenses,

$$C_2 = \frac{\eta^2}{4\hat{\phi}_0} \int_{-\infty}^{\infty} B^2 G^2 \, dz \qquad C_1 = \frac{\eta^2}{4\hat{\phi}_0} \int_{-\infty}^{\infty} B^2 G\overline{G} \, dz$$

$$C_0 = \frac{\eta^2}{4\hat{\phi}_0^{\frac{1}{2}}} \int_{-\infty}^{\infty} B^2 \overline{G}^2 \, dz \qquad C_\theta = \frac{\eta^2}{4\hat{\phi}_0} \int_{-\infty}^{\infty} B^2 \, dz \qquad (26.26)$$

$$C_e = 0$$

26.2 ASYMPTOTIC CHROMATIC ABERRATIONS

and for electrostatic lenses

$$C_2 = \hat{\phi}_0^{\frac{1}{2}} \int_{-\infty}^{\infty} \frac{\gamma(3 + 2\epsilon\hat{\phi})\phi'^2 G^2}{8\hat{\phi}^{\frac{5}{2}}} \, dz$$

$$C_1 = \hat{\phi}_0^{\frac{1}{2}} \int_{-\infty}^{\infty} \frac{\gamma(3 + 2\epsilon\hat{\phi})\phi'^2 G\overline{G}}{8\hat{\phi}^{\frac{5}{2}}} \, dz$$

$$C_0 = \hat{\phi}_0^{\frac{1}{2}} \int_{-\infty}^{\infty} \frac{\gamma(3 + 2\epsilon\hat{\phi})\phi'^2 \overline{G}^2}{8\hat{\phi}^{\frac{5}{2}}} \, dz$$

$$C_e = \frac{\hat{\phi}_0^{\frac{1}{2}}}{4}(\frac{\gamma_i}{\hat{\phi}_i} - \frac{\gamma_o}{\hat{\phi}_o})$$

(26.27)

27
Aberration Matrices and the Aberrations of Lens Combinations

We have seen in Section 16.2 that the cardinal elements of doublets and hence multiplets can be obtained straightforwardly by multiplying the appropriate transfer matrices. A similar procedure can be devised to express the aberration coefficients of a doublet (or multiplet) in terms of those of the individual components. From this, the coefficients of the aberration polynomials of the multiplet can likewise be expressed in terms of those of the separate lenses.

For this, we introduce column vectors in an arbitrary pair of conjugate planes; we write

$$\boldsymbol{u}_o = \begin{pmatrix} u_o \\ u'_o \\ u_o r_o^2 \\ u'_o r_o^2 \\ u_o V_o \\ u'_o V_o \\ u_o \theta_o^2 \\ u'_o \theta_o^2 \\ u_o v_o \\ u'_o v_o \end{pmatrix} \qquad (27.1)$$

in the object plane, with a similar expression for $\boldsymbol{u}_m$, the corresponding vector in the image plane, magnification M. Then we may write

$$\boldsymbol{u}_m = \boldsymbol{M} \boldsymbol{u}_o \qquad (27.2)$$

where $\boldsymbol{M}$ is a 10×10 matrix, which divides naturally into four block matrices:

$$\boldsymbol{M} = \begin{pmatrix} \boldsymbol{M}_1 & \boldsymbol{M}_2 \\ \boldsymbol{M}_3 & \boldsymbol{M}_4 \end{pmatrix} \qquad (27.3)$$

$\boldsymbol{M}_1$ is the 2×2 paraxial matrix (16.20),

$$\boldsymbol{M}_1 = \begin{pmatrix} M & 0 \\ c & rm \end{pmatrix} \qquad (27.4)$$

27. ABERRATION MATRICES

in which we have introduced the *convergence*, c and the relative refractive index r (25.27):

$$c := -\frac{1}{f_i} \qquad r := \frac{f_o}{f_i} = \left(\frac{\hat{\phi}_o}{\hat{\phi}_i}\right)^{\frac{1}{2}} \qquad (27.5)$$

The matrix M_2 has two rows and eight columns, the upper row containing the coefficients describing the aberrations of position and the lower row those of gradient:

$$M_2 =: \begin{pmatrix} Mm_{11} & Mm_{12} & Mm_{13} & Mm_{14} & Mm_{15} & Mm_{16} & Mm_{17} & Mm_{18} \\ m_{21} & m_{22} & m_{23} & m_{24} & m_{25} & m_{26} & m_{27} & m_{28} \end{pmatrix} \qquad (27.6)$$

and

$$\begin{aligned} m_{11} &= D + \mathrm{i}d & m_{15} &= K + \mathrm{i}k \\ m_{12} &= F - A & m_{16} &= C \\ m_{13} &= 2A + \mathrm{i}a & m_{17} &= a \\ m_{14} &= 2K & m_{18} &= 2k \end{aligned} \qquad (27.7)$$

$$m_{21} = c(D + \mathrm{i}d) - r i_1 m$$

$$m_{22} = c(F - A) - (D - \mathrm{i}d)rm + \frac{1}{2}c^2 rm$$

$$m_{23} = c(2A + \mathrm{i}a) - 2Drm + c^2 rm$$

$$m_{24} = 2cK - (2A - \mathrm{i}a)rm + cr^2 m^2$$

$$m_{25} = c(K + \mathrm{i}k) - (F - A)rm + \frac{1}{2}cr^2 m^2 \qquad (27.8)$$

$$m_{26} = cC - (K - \mathrm{i}k)rm - \frac{1}{2}rm(1 - r^2 m^2)$$

$$m_{27} = ca - 2drm$$

$$m_{28} = 2ck - arm$$

in which i_1 is defined in (25.31). The matrix M_3 is null and M_4, generated by M_1, expresses the rules needed for adding the aberration coefficients. It is easily seen that

$$M_4 = \begin{pmatrix} M^3 & 0 & 0 & 0 & 0 & 0 & 0 & 0 \\ cM^2 & rM & 0 & 0 & 0 & 0 & 0 & 0 \\ cM^2 & 0 & rM & 0 & 0 & 0 & 0 & 0 \\ c^2M & rc & rc & r^2m & 0 & 0 & 0 & 0 \\ c^2M & 0 & 2rc & 0 & r^2m & 0 & 0 & 0 \\ c^3 & rc^2m & 2rc^2m & 2r^2cm^2 & r^2cm^2 & r^3m^3 & 0 & 0 \\ 0 & 0 & 0 & 0 & 0 & 0 & rM & 0 \\ 0 & 0 & 0 & 0 & 0 & 0 & rc & r^2m \end{pmatrix} \qquad (27.9)$$

Suppose now that a second lens (or lens system) is characterized by a similar 10×10 matrix, $\boldsymbol{M}'$, and that just as $\boldsymbol{u}_m = \boldsymbol{M}\boldsymbol{u}_o$, so $\boldsymbol{u}'_m = \boldsymbol{M}'\boldsymbol{u}_m$ or

$$\boldsymbol{u}'_m = \boldsymbol{M}'\boldsymbol{M}\boldsymbol{u}_o \qquad (27.10)$$

It is convenient to write

$$\boldsymbol{P} := \boldsymbol{M}'\boldsymbol{M} \qquad P := M'M \qquad p := P^{-1} \qquad (27.11)$$

and to write $\boldsymbol{u}_p$ instead of $\boldsymbol{u}'_m$, so that (27.10) becomes

$$\boldsymbol{u}_p = \boldsymbol{P}\boldsymbol{u}_o \qquad (27.12)$$

Clearly $\boldsymbol{P}$ must have the same block structure as $\boldsymbol{M}$ and $\boldsymbol{M}'$ and indeed $\boldsymbol{P}_1$ is just the paraxial matrix of the combination, which we have already met in Chapter 16 (16.29–16.34):

$$\boldsymbol{P}_1 = \boldsymbol{M}'_1 \boldsymbol{M}_1 = \begin{pmatrix} P & 0 \\ c_p & r_p p \end{pmatrix} = \begin{pmatrix} P & 0 \\ D_p cc' & r_p p \end{pmatrix} \qquad (27.13)$$

with $D_p := z'_{F_o} - z_{Fi}$ and $r_p := rr'$; $\boldsymbol{P}_2$ is of course null and $\boldsymbol{P}_4$ has the same structure as $\boldsymbol{M}_4$ (27.9). It is $\boldsymbol{P}_2$ that is of interest, for it gives us the recipe for adding aberrations, long known but intermittently rediscovered (e.g. Orloff, 1983a). Explicitly, we find

$$p_{11} = m_{11} + M^2 m'_{11} + Mc(m'_{12} + m'_{13}) + c^2(m'_{14} + m'_{15}) + c^3 mm'_{16}$$

or

$$D_p + id_p = D + id + M^2(D' + id') + Mc(F' + A' + ia)$$
$$+ c^3(3K' + ik') + c^3 C'$$
$$p_{12} = m_{12} + rm'_{12} + rcmm'_{14} + rc^2 mm'_{16}$$

or

$$F_p = F + rF' + 4rcmK' + 2rc^2 mC'$$
$$p_{13} = m_{13} + rm'_{13} + rcm(m'_{14} + 2m'_{15}) + 2rc^2 mm'_{16}$$

or

$$2A_p + ia_p = 2A + ia + r(2A' + ia') + rcm(4k' + 2ik')$$
$$+ 2rc^2 mC'$$
$$p_{14} = m_{14} + r^2 m^2 m'_{14} + 2r^2 cm^3 m'_{16}$$

or

$$K_p = K + r^2 m^2 K' + r^2 cm^3 C'$$
$$p_{15} = m_{15} + r^2 m^2 m'_{15} + r^2 cm^3 m'_{16}$$

or

27. ABERRATION MATRICES

$$K_p + ik_p = K + ik + r^2 m^2 (K' + ik') + r^2 cm^3 C'$$
$$p_{16} = m_{16} + r^3 m^4 m'_{16}$$

or

$$C_p = C + r^3 m^4 C'$$
$$p_{17} = m_{17} + rm'_{17} + rcmm'_{18}$$

or

$$a_p = a + ra' + 2rcmk'$$
$$p_{18} = m_{18} + r^2 m^2 m'_{18}$$

or

$$k_p = k + r^2 m^2 k' \tag{27.14}$$

(The last two relations, for p_{17} and p_{18}, merely repeat information already provided by p_{13} and p_{15}.)

Just as P must have the same overall structure as M, so must the elements of the aberration submatrix P_2 have the polynomial dependence on p that the elements of M_2 have on m. A somewhat lengthy calculation reveals that the relation between the polynomial coefficients appearing in P_2 and those in M_2 and M'_2 is as follows, in which i_j corresponds to M, i_j' to M' and $i_j^{(p)}$ to P ($j = 1 - 9$), cf. (25.29–25.30).

$$i_1^{(p)} = i_1 + \frac{c^4}{r}\left\{i_1' D_p^4 - 4i_2' D_p^3 \frac{r'}{c} + 2(i_3' + i_4') D_p^2 \left(\frac{r'}{c'}\right)^2 \right.$$
$$\left. - 4i_5' D_p \left(\frac{r'}{c'}\right)^3 + i_6' \left(\frac{r'}{c'}\right)^4 + \frac{1}{2} D_p^3 \left(c'^2 + \frac{1}{D_p^2}\right)\right\}$$

$$i_2^{(p)} = \frac{c'}{r'}\left(i_2 D_p - \frac{i_1}{c}\right)$$
$$+ \frac{c^3}{r}\left\{i_2' D_p^3 - (i_3' + i_4') D_p^2 \frac{r'}{c'} + 3i_5' D_p \left(\frac{r'}{c'}\right)^2 \right.$$
$$\left. - i_6' \left(\frac{r'}{c'}\right)^3 - \frac{1}{2} D_p \frac{c'}{r'}\right\}$$

$$i_3^{(p)} = \left(\frac{c'}{r'}\right)^2 \left(i_3 D_p^2 - 2i_2 \frac{D_p}{c} + \frac{i_1}{c^2}\right)$$
$$+ \frac{c^2}{r}\left\{i_3' D_p^2 - 2i_5' D_p \frac{r'}{c'} + i_6' \left(\frac{r'}{c'}\right)^2 + \frac{1}{2} D_p \left(\frac{c'}{r'}\right)^2\right\}$$

$$i_4^{(p)} = \left(\frac{c'}{r'}\right)^2 \left(i_4 D_p^2 - 4i_2 \frac{D_p}{c} + 2\frac{i_1}{c^2}\right)$$

$$+ \frac{c^2}{r}\left\{ i'_4 D_p^2 - 4i'_5 D_p \frac{r'}{c'} + 2i'_6\left(\frac{r'}{c'}\right)^2 + D_p\left(\frac{c'}{r'}\right)^2 \right\}$$

$$i_5^{(p)} = \left(\frac{c'}{r'}\right)^3 \left\{ i_5 D_p^3 - (i_3 + i_4)\frac{D_p^2}{c} + 3i_2 \frac{D_p}{c^2} \right) - \frac{i_1}{c^3} \right\}$$

$$+ \frac{c}{r}\left\{ i'_5 D_p - i'_6 \frac{r'}{c'} - \frac{1}{2} D_p\left(\frac{c'}{r'}\right)^3 \right\}$$

$$i_6^{(p)} = \left(\frac{c'}{r'}\right)^4 \left\{ i_6 D_p^4 - 4i_5 \frac{D_p^3}{c} + 2(i_3 + i_4)\frac{D_p^2}{c^2} \right.$$

$$\left. - 4i_2 \frac{D_p}{c^3} + \frac{i_1}{c^4} \right\}$$

$$+ \frac{1}{r}\left\{ i'_6 + \frac{1}{2} D_p\left(D_p^2 \frac{c^2}{r^2} + 1 \right)\left(\frac{c'}{r'}\right)^4 \right\}$$

$$i_7^{(p)} = i_7 + c^2\left\{ i'_7 D_p^2 - i'_8 D_p \frac{r'}{c'} + i'_9\left(\frac{r'}{c'}\right)^2 \right\}$$

$$i_8^{(p)} = \frac{c'}{r'}\left(-\frac{2i_7}{c} + i_8 D_p \right) + c\left(i'_8 D_p - 2i'_9 \frac{r'}{c'} \right)$$

$$i_9^{(p)} = \left(\frac{c'}{r'}\right)\left(\frac{i_7}{c^2} - \frac{i_8 D_p}{c} + i_9 D_p^2\right) + i'_9 \qquad (27.15)$$

Another problem that is more easily solved with the aid of these aberration matrices than in any other way is that of determining the aberration coefficients for some magnification $\overline{M}$ given those for another magnification M. We have provided a solution in Chapter 25 (25.58–25.60) but this required a detailed knowledge of the aberration structure. The same result can be obtained with somewhat less effort by straightforward matrix multiplication. Suppose that the elements of M_2 are known and that

$$\boldsymbol{u}_i = M\boldsymbol{u}_o \qquad (27.16)$$

If the object and image are shifted so that the magnification is now $\overline{M}$, we seek shift matrices $S^{(o)}$ and $S^{(i)}$ that take us from $\boldsymbol{u}_i$ to $\overline{\boldsymbol{u}}_i$ and from $\boldsymbol{u}_o$ to $\overline{\boldsymbol{u}}_o$, where

$$\overline{\boldsymbol{u}}_i = \overline{M}\overline{\boldsymbol{u}}_o \qquad (27.17)$$

Explicitly, we seek matrices $S^{(o)}$ and $S^{(i)}$ such that

$$\overline{\boldsymbol{u}}_i =: S^{(i)}\boldsymbol{u}_i \quad \text{and} \quad \boldsymbol{u}_o =: S^{(o)}\overline{\boldsymbol{u}}_o \qquad (27.18)$$

and hence

$$\overline{\boldsymbol{u}}_i = S^{(i)}\boldsymbol{u}_i = S^{(i)}M\boldsymbol{u}_o = S^{(i)}MS^{(o)}\overline{\boldsymbol{u}}_o \qquad (27.19)$$

27. ABERRATION MATRICES

The unknown matrix $\overline{M}$ will then be obtained from (27.17) and (27.19):

$$\overline{M} = S^{(i)} M S^{(o)} \tag{27.20}$$

Since u_o and $\overline{u}_o$ are connected by the incident asymptote and u_i and $\overline{u}_i$ by the emergent asymptote we have

$$\begin{aligned} \overline{u}_i &= u_i + (\overline{z}_i - z_i)u'_i & \overline{u}'_i &= u'_i \\ u_o &= \overline{u}_o - (\overline{z}_o - z_o)\overline{u}'_o & u'_o &= \overline{u}'_o \end{aligned} \tag{27.21}$$

and so $S^{(i)}$ and $S^{(o)}$ will have block structures similar to that of M:

$$S^{(o)} = \begin{pmatrix} S_1^{(o)} & S_2^{(o)} \\ S_3^{(o)} & S_4^{(o)} \end{pmatrix} \qquad S^{(i)} = \begin{pmatrix} S_1^{(i)} & S_2^{(i)} \\ S_3^{(i)} & S_4^{(i)} \end{pmatrix} \tag{27.22}$$

where now $S_2^{(o)} = S_3^{(o)} = S_2^{(i)} = S_3^{(i)} = 0$,

$$S_1^{(o)} = \begin{pmatrix} 1 & -\zeta_o \\ 0 & 1 \end{pmatrix} \qquad S_1^{(i)} = \begin{pmatrix} 1 & \zeta_i \\ 0 & 1 \end{pmatrix} \tag{27.23}$$

and

$$S_4^{(o)} = \begin{pmatrix} 1 & -\zeta_o & -2\zeta_o & 2\zeta_o^2 & \zeta_o^2 & -\zeta_o^3 & 0 & 0 \\ 0 & 1 & 0 & -2\zeta_o & 0 & \zeta_o^2 & 0 & 0 \\ 0 & 0 & 1 & -\zeta_o & -\zeta_o & \zeta_o^2 & 0 & 0 \\ 0 & 0 & 0 & 1 & 0 & -\zeta_o & 0 & 0 \\ 0 & 0 & 0 & 0 & 1 & -\zeta_o & 0 & 0 \\ 0 & 0 & 0 & 0 & 0 & 1 & 0 & 0 \\ 0 & 0 & 0 & 0 & 0 & 0 & 1 & -\zeta_o \\ 0 & 0 & 0 & 0 & 0 & 0 & 0 & 1 \end{pmatrix} \tag{27.24}$$

$S_4^{(i)}$ has the same appearance as $S_4^{(o)}$ but ζ_i replaces $-\zeta_o$ everywhere. We have written

$$\zeta_o := \overline{z}_o - z_o \qquad \zeta_i := \overline{z}_i - z_i \tag{27.25}$$

We are interested only in $\overline{M}_2$ and from the common block structure of $S^{(o)}$, $S^{(i)}$ and M, it is obvious that

$$\overline{M}_2 = S_1^{(i)} M_2 S_4^{(o)} \tag{27.26}$$

Since $S_1^{(i)}$ is only 2×2 and $S_4^{(o)}$ is so sparse, the elements of $\overline{M}_2$ may be read off immediately and of course prove to be identical with those already listed (25.39). The expressions (25.59) for the change in aberrations when object and aperture are shifted can likewise be obtained with less effort by deriving the appropriate shift matrices, as discussed in detail by Brouwer (1957) (cf. Hawkes, 1985a). The appropriate starting point is the matrix equation $u_i = M u_o$ analogous to (27.16) except that u'_o is replaced by u_{ao} in u_o and u'_i by u_{ai} in u_i, where u_{ao} is the complex cartesian coordinate in the entrance pupil and u_{ai} is that in the exit pupil, between which the magnification is N. The paraxial block matrix M_1 therefore has the form

$$M_1 = \begin{pmatrix} M & 0 \\ 0 & N \end{pmatrix} \qquad (27.27)$$

and M_4 is diagonal. M_3 is of course still null and M_2 is the new aberration matrix. The upper row of M_2 has the usual form but the lower row, representing the aberrations between the pupil planes, must be derived from expressions analogous to (25.7a) or from (25.57–25.60), setting $m_2 = n_1$ and $n_2 = m_1$. The results are as follows, in which the new coefficient C^* is equal to I_{04} (25.53):

$$\begin{aligned}
m_{21} &= C^* \\
m_{22} &= -D + \mathrm{i}d + \frac{n^2 - 1}{2f^2(n-m)^2} \\
m_{23} &= -2D + \frac{n^2 - 1}{f^2(n-m)^2} \\
m_{24} &= -2A + \mathrm{i}a + \frac{1 - mn}{f^2(n-m)^2} \\
m_{25} &= -(F - A) + \frac{1 - nm}{2f^2(n-m)^2} \\
m_{26} &= -K + \mathrm{i}k + \frac{m^2 - 1}{2f^2(n-m)^2} \\
m_{27} &= -2d \\
m_{28} &= -a
\end{aligned} \qquad (27.28)$$

The shift matrices may be established without difficulty and the shifted aberration coefficients are obtained as the elements of $S_1^{(i)} M_2 S_4^{(o)}$ (27.26). The results agree with those listed earlier.

28
The Aberrations of Mirrors and Cathode Lenses

In this chapter, we extend the theories outlined in Chapter 18 beyond the paraxial domain, using the notation introduced there. There are again two possible representations of the theory, the parametric form and the cartesian form. We go into details only for the former.

28.1 The parametric form of the theory

We again set out from (3.12), as in Section 18.2. Introducing the fields corresponding to (28.10) into the nonrelativistic form of (3.12), and setting

$$W(\sigma) := X(\sigma) + iY(\sigma) \quad , \quad |W| =: r(\sigma) \tag{28.1}$$

we obtain the coupled differential equations

$$\ddot{W} = -\frac{1}{4U}\left\{\phi''(Z) - \frac{r^2}{8}\phi^{(4)}(Z)\right\}W(\sigma)$$
$$+ i\eta U^{-1/2}\left\{\left(B'(Z) - \frac{r^2}{8}B'''\right)\frac{\dot{Z}}{2}W(\sigma) + \left(B - \frac{r^2}{4}B''\right)\dot{W}\right\} + O(5) \tag{28.2a}$$

$$\ddot{Z} = \frac{1}{2U}\left\{\phi'(Z) - \frac{r^2}{4}\phi'''\right\} - \frac{\eta}{2}U^{-1/2}B'(Z)\Im(W^*\dot{W}) + O(4) \tag{28.2b}$$

where primes denote differentiation with respect to Z, while, as before, dots denote differentiation with respect to the curve parameter σ.

As in Section 18.2, we introduce the rotating frame. With (28.1) and the new complex variable $w := x + iy$, the transform (18.5) is now concisely rewritten as

$$W(\sigma) = w(\sigma)\exp\left\{i\theta(Z(\sigma))\right\} \quad , \quad r = |W| \tag{28.3}$$

with (18.6) for $\dot{\theta}$. In the lowest order, the coupling terms cancel out and we find

$$\ddot{w} + \frac{1}{4U}\left\{\phi''(Z) + \eta^2 B^2(Z)\right\}w(\sigma)$$
$$= \frac{r^2 w}{32U}\left\{\phi^{(4)} + 4\eta^2 BB''\right\} - \frac{i\eta r^2}{16U^{1/2}}\left\{4B''\dot{w} + B'''\dot{Z}w\right\} \tag{28.4a}$$

$$\ddot{Z} = \frac{1}{2U}\phi'(Z) - \frac{r^2}{8U}\left\{\phi''' + 2\eta^2 BB'\right\} - \frac{\eta}{2}U^{-1/2}B'(Z)\Im(w^*\dot{w}) \quad (28.4b)$$

These differential equations are valid for all values of the total energy $e\Phi_T$. They do not contain any singularities and we can therefore carry out the necessary series expansions.

In the *paraxial approximation* we recover (18.3c) and (18.8a,b), which are still valid for all total energies $e\Phi_T$. As in Section 18.2, we now specify a nominal value $\Phi_T = 0$ and a corresponding solution $z(\sigma)$ with $\phi(0) = 0$. The coordinate Z is not identical with z, and so we must introduce an axial deviation $s(\sigma)$:

$$Z(\sigma) =: z(\sigma) + s(\sigma) \quad (28.5)$$

This shift $s(\sigma)$ is usually very small but may be larger in the vicinity of the mirror surface. It is now sufficient to extend the series expansions only to the terms of lowest order and to insert the paraxial solutions $w_p(\sigma)$, $z(\sigma)$ in the aberration terms. The paraxial equations are then

$$\ddot{w}_p(\sigma) + F(\sigma)w_p(\sigma) = 0 \quad (28.6)$$

with (18.11) and

$$\ddot{z}(\sigma) = \phi'(z)/2U \quad (28.7)$$

The total lateral coordinate $w(\sigma)$ is rewritten as

$$w(\sigma) = w_p(\sigma) + w_a(\sigma)$$

the last term denoting the lateral aberration. This quantity then satisfies an inhomogeneous linear differential equation

$$\ddot{w}_a(\sigma) + F(\sigma)w_a(\sigma) = R(\sigma) \quad (28.8)$$

with the *transverse* aberration term

$$R(\sigma) = \frac{r_p^2 w_p}{32U}\left[\phi^{(4)} + 4\eta^2 BB''\right]_{(z)} - \frac{i\eta r_p^2}{16U^{1/2}}\left[4B''\dot{w}_p + B'''\dot{z}w_p\right]_{(z)}$$
$$- \frac{1}{4U}\left\{\phi'''(z) + 2\eta^2 B(z)B'(z)\right\}s(\sigma)w_p(\sigma)$$
$$(28.9)$$

The final term arises from the series expansion of the coefficient on the left-hand side of (28.4a). In this term the contributions linear in s are sufficient, since sw_p is always very small. In (28.4b) the approximation $\phi'(Z) = \phi'(z) + s\phi''(z)$ is sufficient. Considering (28.7), we find:

$$\ddot{s}(\sigma) + \frac{1}{2U}\phi''(z(\sigma))s(\sigma) = S(\sigma) \quad (28.10)$$

28.1 THE PARAMETRIC FORM OF THE THEORY

with the *longitudinal* aberration term

$$S(\sigma) = -\frac{r_p^2}{8U}\left[\phi''' + 2\eta^2 BB'\right]_{(z)} - \frac{\eta}{2}U^{-1/2}B'(z)\Im(w_p^* \dot{w}_p) \tag{28.11}$$

We now have to solve these differential equations with the appropriate initial conditions. The first step is the solution of the paraxial ray equations, as outlined in Section 18.2. This gives us the appropriate paraxial solutions $z = z(\sigma)$ and

$$w_p(\sigma) = w_o g(\sigma) + \dot{w}_o h(\sigma) \quad , \quad r_p = |w_p| \tag{28.12}$$

with

$$w_o = x_o + iy_o = X_o + iY_o \tag{28.13}$$

and

$$\dot{w}_o = \dot{x}_o + i\dot{y}_o = \dot{X}_o + i\dot{Y}_o - \frac{i\eta B(z_o)}{2\sqrt{U}}\left(X_o + iY_o\right) \tag{28.13}$$

The solution (28.12) is now introduced into (28.11); from the constancy of the Wronskians we have

$$D := \Im(w_p^* \dot{w}_p) = x_p \dot{y}_p - \dot{x}_p y_p = x_o \dot{y}_o - \dot{x}_o y_o \tag{28.14}$$

which expresses the conservation of axial angular momentum, and this leads to some simplification.

With (28.12) the function $S(\sigma)$ is now well-specified and we can solve (28.10), for instance by the method of variation of parameters. The appropriate initial value for $\dot{s}(\sigma_o) = \dot{s}_o$ is determined by the conservation of energy:

$$\dot{X}_o^2 + \dot{Y}_o^2 + \dot{Z}_o^2 = U^{-1}\left(\Phi_T + \Phi(r_o)\right) \tag{28.15}$$

Here the approximation $\Phi(r_o) = \phi(z_o) - r_o^2 \phi''(z_o)/4$ is sufficient. With $s_o = s(z_o) = 0$ and $\dot{z}_o^2 = \phi(z_o)/U$, we find

$$2\dot{z}_o \dot{s}_o + \dot{s}_o^2 = U^{-1}\left\{\Phi_T - \frac{1}{4}r_o^2 \phi''(z_o)\right\} - \left(\dot{X}_o^2 + \dot{Y}_o^2\right) \tag{28.16}$$

Once (28.10) has been solved, the function $R(\sigma)$, given by (28.9), is well-defined and we can solve (28.8) again by the method of variation of parameters; the appropriate initial conditions are here simply $w_a(\sigma_o) = 0$ and $\dot{w}_a(\sigma_o) = 0$.

The solution of (28.8) yields the value of the quantity $w_a(\sigma_b)$, where σ_b is the value of the parameter σ at the image (the subscript b is chosen to avoid confusion); $w_a(\sigma_b)$ does *not*, however, represent the required lateral aberration. There are additional terms arising from aberrations of the image rotation and from the longitudinal shift s. The angle of image rotation is given by integration of (18.6). Here a linearized series expansion

$$\dot\theta(Z) = \dot\theta(z+s) = \frac{\eta}{2} U^{-1/2} B(z+s)$$
$$= \frac{\eta}{2} U^{-1/2} \left(B(z) + sB'(z) + O(s^2) \right)$$

is necessary. After calculation of the functions $z(\sigma)$ and $s(\sigma)$ we can perform the necessary integration whereupon we find that $\theta_b = \theta_p + \Delta\theta$, where

$$\theta_p = \frac{\eta}{2} U^{-1/2} \int_{\sigma_o}^{\sigma_b} B(z(\sigma))\, d\sigma \qquad (28.17a)$$

is the paraxial contribution and

$$\Delta\theta = \frac{\eta}{2} U^{-1/2} \left\{ \int_{\sigma_o}^{\sigma_b} s(\sigma) B'(z(\sigma))\, d\sigma - s(\sigma_b) B(z_b)/\dot z(\sigma_b) \right\} \qquad (28.17b)$$

the aberration. The second term under the integral is the contribution from the longitudinal shift.

The complex coordinate W_b in the image plane is now given by

$$W_b = \exp\left\{ i\left(\theta_p + \Delta\theta\right) \right\} \left\{ w_p(\sigma_b) + w_a(\sigma_b) \right.$$
$$\left. - s(\sigma_b) \dot w_p(\sigma_p)/\dot z(\sigma_b) \right\}$$

where the last contribution is again the contribution from the longitudinal shift. With $\exp(i\Delta\theta) \approx 1 + i\Delta\theta$ we obtain finally:

$$\Delta W_b := W_b - e^{i\theta_p} w_p(\sigma_b) =: e^{i\theta_p} \Delta w_b$$
$$= e^{i\theta_p} \left[w_a + i\Delta\theta w_p - \dot w_p s/\dot z \right]_{(\sigma_b)} \qquad (28.18)$$

where all products of small aberrations have been neglected. The expression in brackets, to be evaluated at $\sigma = \sigma_b$, represents the lateral aberration

28.2 SYSTEMS WITH CURVED CATHODES

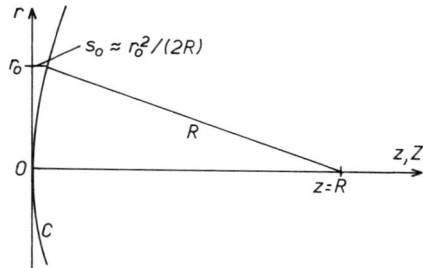

Fig. 28.1: Parameters characterizing a spherical cathode surface C.

in the rotating frame. The division by $\dot{z}(\sigma_b)$ never creates any problem because in practice the image plane is never located in the mirror surface or on the cathode.

This theory is not very familiar, principally because the total aberration is not found in a direct manner but only emerges gradually. An analogous theory, which is marginally more straightforward, can be found in the book of Zworykin et al. (1945). The version presented here has the advantage that at each stage of the calculation only *regular* expressions are involved, so that the procedure can be easily put into a computer program.

28.2 Systems with curved cathodes

So far we have tacitly assumed the object to be planar but this excludes cathode lenses with curved cathode surfaces. Concave spherical cathodes are, however, quite common in image converters. Usually, the radius R of curvature is fairly large, so that $2Z = r^2/R$ is a sufficiently good approximation for the equation of the cathode surface (Fig. 28.1). The theory can easily be generalized to include this case. The cathode surface must be an equipotential $\Phi(r) = 0$. From the series expansion (7.37), we see that

$$\Phi(Z, r) = Z\phi'_o + \frac{Z^2}{2}\phi''_o - \frac{r^2}{4}\phi''_o + \ldots \qquad (28.19)$$

is valid in the vicinity of the cathode. Introducing $2Z = r^2/R$ for the surface $\Phi = 0$ and retaining only terms of order r^2, we find that

$$\phi''_o = 2\phi'_o/R \qquad (28.20)$$

It only remains to define the initial conditions for the longitudinal aberration $s(\sigma)$. We now have

$$\sigma_o = 0 \quad , \quad z_o = 0 \quad , \quad Z_o \equiv s_o = r_o^2/2R \qquad (28.21)$$

Recalling that $\Phi(\mathbf{r}_o)$ is zero on the cathode surface, we obtain immediately from (28.15) the condition

$$\dot{Z}_o^2 \equiv \dot{s}_o^2 = \Phi_T/U \quad , \quad \dot{s}_o \geq 0 \tag{28.22}$$

Apart from these special initial conditions, no other aspects of systems with a curved cathode need be considered here.

28.3 Structure of the aberrations

We have not yet expressed the complex lateral aberration in terms of the initial conditions; we shall do this in the rotating frame. Since we know that the aberration must be invariant with respect to any constant rotation of the coordinate system, the only possible invariants of second order in the transverse initial values are

$$\left.\begin{array}{l} p_1 = r_o^2 = x_o^2 + y_o^2 = w_o^* w_o \\ p_2 = r_o \dot{r}_o = x_o \dot{x}_o + y_o \dot{y}_o = (w_o^* \dot{w}_o + \dot{w}_o^* w_o)/2 \\ p_3 = D = x_o \dot{y}_o - y_o \dot{x}_o = (w_o^* \dot{w}_o - \dot{w}_o^* w_o)/2\mathrm{i} \\ p_4 = \dot{x}_o^2 + \dot{y}_o^2 = \dot{w}_o^* \dot{w}_o \end{array}\right\} \tag{28.23}$$

From the paraxial solution (28.12), we find

$$r_p^2 = |w_p|^2 = p_1 g^2(\sigma) + 2p_2 g(\sigma) h(\sigma) + p_4 h^2(\sigma) \tag{28.24}$$

From the solution of (28.10) with (28.11), it is now obvious that the longitudinal aberration must have the basic structure

$$s(\sigma) = \sum_{j=1}^{5} p_j a_j(\sigma) \quad \text{with} \quad p_5 := \dot{s}_o \tag{28.25}$$

This holds even in the case (28.21), since $s_o = p_1/2R$. Up to third order, the only possible complex invariants are

$$p_j w_o \quad \text{and} \quad p_j \dot{w}_o \quad , \quad j = 1\ldots 5$$

Introducing (28.24) and (28.25) into (28.9), we find that $R(\sigma)$ has the form

$$R(\sigma) = \sum_{j=1}^{5} p_j \left\{ b_j^{(1)}(\sigma) w_o + b_j^{(2)}(\sigma) \dot{w}_o \right\}$$

28.4 THE CARTESIAN FORM OF THE ABERRATION THEORY

The solution of (28.8) must have an analogous structure. Introducing (28.25) into (28.17b), we find that the angular aberration $\Delta\theta$ must be linear in $p_1 \ldots p_5$. Finally, (28.18) shows that the lateral aberration must have the basic form

$$\Delta w_b = \sum_{j=1}^{5} p_j \left\{ c_j^{(1)} w_o + c_j^{(2)} \dot{w}_o \right\} \tag{28.26}$$

This all holds under the assumption that the longitudinal shift $s(\sigma)$ is very small, which is usually justified. It is, of course, possible to extend the theory further to include all terms in $\dot{s}_o^2$; it then becomes more realistic. We can rewrite the general result very concisely in the form

$$\Delta w_b = \sum_{j=0}^{6} p_j \left\{ c_j^{(1)} w_o + c_j^{(2)} \dot{w}_o \right\} \tag{28.27a}$$

with (28.23) and

$$p_o = 1 \quad , \quad p_5 = \dot{s}_o \quad , \quad p_6 = \dot{s}_o^2 \tag{28.27b}$$

For completeness, we have included here the contributions of any defocus ($j = 0$). It is possible to represent the complex coefficients in (28.27a) in the form of aberration integrals, but this is very laborious. It is much easier to determine them by a least-squares-fit method, as outlined in Chapter 33. It is then not even necessary to consider the rotating frame, since an analogous structure is also found in the laboratory frame.

The value of the present theory is that it explains the origin of the series expansion (28.27a). This must *not* be expressed in terms of slopes, since the latter can be arbitrarily large. Moreover (28.16) must not be used with a prescribed value of Φ_T; instead, $\dot{s}_o$ must be chosen as an *independent* parameter, since otherwise the series expansion contains complicated square-root expressions. Equation (28.15) can be used afterwards to determine the energy $e\Phi_T$ of the particular ray. Since the determination of the coefficients in (28.27a) by means of a least-squares-fit method entails using many different sets of parameters, the resulting set of calculated rays is necessarily polyenergetic.

Although we have suggested how systems with a curved cathode surface may be included, the present theory must not be applied to electron guns with sharply pointed cathodes. In the latter, the curvature of the cathode is so great and the associated aberrations are so large that a series expansion in the form (28.27) makes no sense. A completely different approach is then necessary, which is the subject of Part IX.

28.4 The cartesian form of the aberration theory

Cartesian theories of the aberrations of cathode lenses and, by extension, of electron mirrors have been presented intermittently for several decades and their legimitacy and accuracy have been explored in some detail. These theories fall into two families, those that remain very close to the familiar aberration theory of round lenses and those that introduce a new independent variable as explained in Chapter 18. The first group begins with the paper of Ximen (1957), to be followed by Bonshtedt (1964) (who does not cite Ximen). The Chinese work was repeated and completed by Zhou et al. (1983) and by Ximen et al. (1983) but, meanwhile, numerous Russian publications had been published on the subject, in particular Kulikov (1971, 1972, 1973, 1975), Monastyrskii and Kulikov (1976, 1978), Monastyrskii (1978, 1980), Kulikov et al. (1978) and Smirnov et al. (1979).

The second group is associated initially with the work of Kel'man et al. (1971, 1972, 1973) who examined rotationally and cylindrically symmetric systems; their analysis was recast into a more general form by Daumenov et al. (1978).

It was clearly important to establish whether one of these approaches was superior and Dodin and Nesvizhskii (1981) in particular have examined the approximations involved. They come to the conclusion that the easier method associated with Bonshtedt, Kulikov et al. and Ximen is reliable.

The most detailed analyses of the aberrations by the earlier method are to be found in the papers of Kulikov et al. (1978), Monastyrskii and Kulikov (1978) and Ximen et al. (1983), in which aberration integrals for the usual geometrical aberration coefficients and for chromatic aberration coefficients defined in terms of the axial component of the initial electron velocity are listed. The behaviour of these expressions close to the cathode is analysed by Monastyrskii (1978). We draw particular attention to a more recent paper by Nesvizhskii (1986), who shows how the aberration coefficients of complex cathode systems can be derived with the aid of the traditional methods of aberration calculation by an ingenious change of variable (see also the references for Chapter 38).

Similar lists of coefficients are to be found in the papers of Kel'man et al. already cited but the amount of computation needed seems very heavy when the chromatic aberrations are being calculated. This can be seen by returning to (18.23), which describes any polyenergetic paraxial beam, for which the constant Φ_T in (18.22) will have a spectrum of values. For each of these, the corresponding turning point $Z_o(\Phi_T)$ has to be found by solving $V(Z_o) = \phi(Z_o) + \Phi_T = 0$. Series expansions about each of these points Z_o must now be made: the origin is not the appropriate singularity

28.4 THE CARTESIAN FORM OF THE ABERRATION THEORY

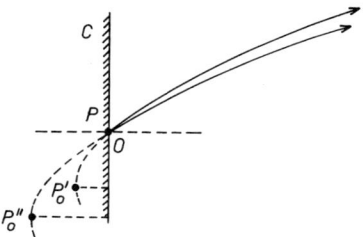

Fig. 28.2: Points of inflexion P'_o and P''_o of two rays starting from a common point P of a cathode surface with different initial velocities.

when $\Phi_T \neq 0$, as is shown in Fig. 28.2.

29

The Aberrations of Quadrupole Lenses and Octopoles

29.1 Introduction

The aberration coefficients of systems characterized by a pair of symmetry planes are of less interest at moderate energies than those of rotationally symmetric components but they are needed in two practical situations: instruments in which a line focus is required and quadrupole–octopole systems intended to reduce or cancel the spherical aberration of round lenses. In the latter case, a sequence of quadrupole lenses, typically four, may be suitably combined with octopoles either as a corrector, the combination providing little or no paraxial focusing, or as a corrected lens, in which case both focusing and correction are required. Spherical aberration compensation by means of quadrupoles and octopoles was one of the ways of circumventing Scherzer's theorem proposed by Scherzer himself (1947); the idea has attracted much attention over the years, both experimental and theoretical, but the final result is disappointing. There is no doubt that the principle is sound and a measure of correction has been demonstrated experimentally (see Chapter 41), but the mechanical and electrical complexity of suitable quadrupole–octopole systems is such that they have never yet been capable of surpassing a high-quality round lens in performance, so far as the spherical aberration is concerned.

29.2 Geometrical aberration coefficients

Although it is in theory necessary to distinguish between real and asymptotic coefficients, we concentrate on the latter for it is unlikely that a real specimen or target would be immersed within the field of a quadrupole. Real aberration coefficients will therefore be mentioned only in passing; they are accorded much more space in earlier surveys (Hawkes, 1966, 1970a) and in the textbooks of Strashkevich (1959, 1966), Yavor (1968) and Baranova and Yavor (1986).

We set out from the perturbation characteristic function denoted by

29.2 GEOMETRICAL ABERRATION COEFFICIENTS

S_{12}^I in Chapter 22 (22.21); for two planes $z = z_o$ and $z = z_c$, we may write

$$S_{oc}^I =$$

$$\begin{pmatrix} x_o^2 \\ y_o^2 \\ x_o'^2 \\ y_o'^2 \\ x_o x_o' \end{pmatrix}^T \begin{pmatrix} \overline{(4000)} & \overline{(2200)} & \overline{(2020)} & \overline{(2002)} & \overline{(3010)} & \overline{(2101)} \\ 0 & \overline{(0400)} & \overline{(0220)} & \overline{(0202)} & \overline{(1210)} & \overline{(0301)} \\ 0 & 0 & \overline{(0040)} & \overline{(0022)} & \overline{(1030)} & \overline{(0121)} \\ 0 & 0 & 0 & \overline{(0004)} & \overline{(1012)} & \overline{(0103)} \\ 0 & 0 & 0 & 0 & 0 & \overline{(1111)} \end{pmatrix} \begin{pmatrix} x_o^2 \\ y_o^2 \\ x_o'^2 \\ y_o'^2 \\ x_o x_o' \\ y_o y_o' \end{pmatrix}$$

(29.1)

The notation for the matrix elements has been chosen so that

$$S_{oc}^I = \sum_{0 \leq p,q,r,s \leq 4} \overline{(pqrs)} x_o^p y_o^q x_o'^r y_o'^s \quad , \quad p+q+r+s = 4 \tag{29.2}$$

Introducing the dimensionless coordinates

$$\xi := \frac{x_o}{f_{xi}} \quad , \quad \eta := \frac{y_o}{f_{yi}} \tag{29.3}$$

S_{oc}^I becomes

$$S_{oc}^I =$$

$$\begin{pmatrix} \xi_o^2 \\ \eta_o^2 \\ x_o'^2 \\ y_o'^2 \\ \xi_o x_o' \end{pmatrix}^T \begin{pmatrix} (4000) & (2200) & (2020) & (2002) & (3010) & (2101) \\ 0 & (0400) & (0220) & (0202) & (1210) & (0301) \\ 0 & 0 & (0040) & (0022) & (1030) & (0121) \\ 0 & 0 & 0 & (0004) & (1012) & (0103) \\ 0 & 0 & 0 & 0 & 0 & (1111) \end{pmatrix} \begin{pmatrix} \xi_o^2 \\ \eta_o^2 \\ x_o'^2 \\ y_o'^2 \\ \xi_o x_o' \\ \eta_o y_o' \end{pmatrix}$$

(29.4)

with

$$(4000) = \overline{(4000)} f_{xi}^4, \quad (0400) = \overline{(0400)} f_{yi}^4, \quad (0022) = \overline{(0022)}$$
$$(2200) = \overline{(2200)} f_{xi}^2 f_{yi}^2, \quad (0220) = \overline{(0220)} f_{yi}^2, \quad (1030) = \overline{(1030)} f_{xi}$$
$$(2020) = \overline{(2020)} f_{xi}^2, \quad (0202) = \overline{(0202)} f_{yi}^2, \quad (0121) = \overline{(0121)} f_{yi}$$
$$(2002) = \overline{(2002)} f_{xi}^2, \quad (1210) = \overline{(1210)} f_{xi} f_{yi}^2, \quad (0004) = \overline{(0004)}$$
$$(3010) = \overline{(3010)} f_{xi}^3, \quad (0301) = \overline{(0301)} f_{yi}^3, \quad (1012) = \overline{(1012)} f_{xi}$$
$$(2101) = \overline{(2101)} f_{xi}^2 f_{yi}, \quad (0040) = \overline{(0040)}, \quad (0103) = \overline{(0103)} f_{yi}$$
$$(1111) = \overline{(1111)} f_{xi} f_{yi}$$

(29.5)

or

$$(pqrs) = f_{xi}^p f_{yi}^q \overline{(pqrs)}$$

If the lens or system of lenses is *symmetric*, in the sense that all the field or potential functions are symmetric about some mid-plane, the object and image focal lengths are equal and several of the $(pqrs)$ are likewise equal. To see this, we write $x_o \to -f_x x'_c$, $y_o \to -f_y y'_c$, $x'_o \to x_c/f_x =: \xi_c$ and $y'_o \to y_c/f_y =: \eta_c$ in (29.4), giving

$$S^I_{oc} = \begin{pmatrix} \xi_c^2 \\ \eta_c^2 \\ x_c'^2 \\ y_c'^2 \\ \xi_c x'_c \end{pmatrix}^T \begin{pmatrix} (0040) & (0022) & (2020) & (0220) & (1030) & (0121) \\ 0 & (0004) & (2002) & (0202) & (1012) & (0103) \\ 0 & 0 & (4000) & (2200) & (3010) & (2101) \\ 0 & 0 & 0 & (0400) & (1210) & (0301) \\ 0 & 0 & 0 & 0 & 0 & (1111) \end{pmatrix} \begin{pmatrix} \xi_c^2 \\ \eta_c^2 \\ x_c'^2 \\ y_c'^2 \\ \xi_c x'_c \\ \eta_c y'_c \end{pmatrix}$$
(29.6)

so that

$$\begin{array}{ll}
(4000) = (0040) & (1030) = (3010) \\
(0400) = (0004) & (0103) = (0301) \\
(2200) = (0022) & (0121) = (2101) \\
(2002) = (0220) & (1210) = (1012)
\end{array} \quad \text{(symmetric system)} \tag{29.7}$$

or

$$(pqrs) = (rspq)$$

In another common combination, the system is *antisymmetric*: the centre plane is a plane of geometrical symmetry and electrical antisymmetry (Fig. 29.1). Writing (29.4) in the form

$$S^I_{oc} = \begin{pmatrix} \eta_c^2 \\ \xi_c^2 \\ y_c'^2 \\ x_c'^2 \\ \eta_c y'_c \end{pmatrix}^T \begin{pmatrix} (0004) & (0022) & (0202) & (2002) & (0103) & (1012) \\ 0 & (0040) & (0220) & (2020) & (0121) & (1030) \\ 0 & 0 & (0400) & (2200) & (0301) & (1210) \\ 0 & 0 & 0 & (4000) & (2101) & (3010) \\ 0 & 0 & 0 & 0 & 0 & (1111) \end{pmatrix} \begin{pmatrix} \eta_c^2 \\ \xi_c^2 \\ y_c'^2 \\ x_c'^2 \\ \eta_c y'_c \\ \xi_c x'_c \end{pmatrix}$$
(29.8)

we see that

$$\begin{array}{ll}
(4000) = (0004) & (2101) = (1012) \\
(0022) = (2200) & (0400) = (0040) \\
(2020) = (0202) & (1210) = (0121) \\
(3010) = (0103) & (0301) = (1030)
\end{array} \tag{29.9}$$

29.2 GEOMETRICAL ABERRATION COEFFICIENTS

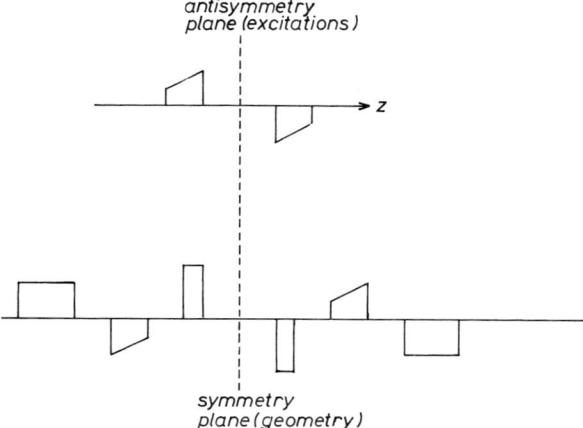

Fig. 29.1: Antisymmetric multiplets. The simplest case is the doublet (above) but the quadruplet and the sextuplet (below), having more variable parameters, are more useful.

or

$$(pqrs) = (srqp)$$

Proceeding as in Chapters 22 and 24, we obtain expressions for the aberrations but with an additional degree of complexity, already mentioned in Chapter 19. A quadrupole lens focuses a point object into a pair of real or virtual line foci, and since quadrupoles are mostly used in combination, the "object" for an intermediate member of a multiplet will almost always be astigmatic. We therefore list the aberrations for the general case of astigmatic object and image. If z_{ox} and z_{ix} are conjugate in the plane x-z and z_{oy} and z_{iy} in the plane y-z, with magnifications M_x and M_y respectively, we write

$$\Delta x_i := \frac{x(z_{ix}) - M_x x_o}{M_x}$$
$$\Delta y_i := \frac{y(z_{iy}) - M_y y_o}{M_y} \quad (29.10)$$

and

$$\begin{aligned}
\Delta x_i =: &\{(3000)\xi_o^2 + (1200)\eta_o^2\}\xi_o & &\text{distortions} \\
&+\{(2010)\xi_o^2 + (0210)\eta_o^2\}x'_o + (1101)\xi_o\eta_o y'_o & &\begin{cases}\text{field curvature} \\ \text{and astigmatism}\end{cases} \\
&+\{(1020)x'^2_o + (1002)y'^2_o\}\xi_o + (0111)\eta_o x'_o y'_o & &\text{comas} \\
&+\{(0030)x'^2_o + (0012)y'^2_o\}x'_o & &\begin{cases}\text{aperture} \\ \text{aberrations}\end{cases} \\
\Delta y_i =: &\{(0300)\eta_o^2 + (2100)\xi_o^2\}\eta_o & &\text{distortions} \\
&+\{(2001)\xi_o^2 + (0201)\eta_o^2\}y'_o + (1110)\xi_o\eta_o x'_o & &\begin{cases}\text{field curvature} \\ \text{and astigmatism}\end{cases} \\
&+\{(0120)x'^2_o + (0102)y'^2_o\}\eta_o + (1011)\xi_o x'_o y'_o & &\text{comas} \\
&+\{(0003)y'^2_o + (0021)x'^2_o\}y'_o & &\begin{cases}\text{aperture} \\ \text{aberrations}\end{cases}
\end{aligned} \qquad (29.11)$$

In the case of real aberrations and a stigmatic object, $x_c^{(1)}$ and $y_c^{(1)}$ are given by (22.30), into which we substitute

$$\begin{aligned} x(z) &= x_o s_x(z) + x_a t_x(z) \\ y(z) &= y_o s_y(z) + y_a t_y(z) \end{aligned} \qquad (29.12)$$

giving

$$\begin{aligned} x_c^{(1)}(z) &= \frac{t_x}{W_x}\frac{\partial S^I_{ac}}{\partial x_o} - \frac{g_x}{W_x}\frac{\partial S^I_{oc}}{\partial x_a} \\ y_c^{(1)}(z) &= \frac{t_y}{W_y}\frac{\partial S^I_{ac}}{\partial y_o} - \frac{g_y}{W_y}\frac{\partial S^I_{oc}}{\partial y_a} \end{aligned} \qquad (29.13)$$

where W_x, W_y are the appropriate Wronskians. In the line-image plane where $t_x = 0$, $s_x = M_x$ for example, we have

$$\Delta x_i = -\frac{1}{W_x}\frac{\partial S^I_{oc}}{\partial x_a} \qquad (29.14a)$$

while in the other line-image plane, where $t_y = 0$ and $g_y = M_y$,

$$\Delta y_i = -\frac{1}{W_y}\frac{\partial S^I_{oc}}{\partial y_a} \qquad (29.14b)$$

In the case of asymptotic aberrations, we generalize (25.9) to include astigmatic imagery:

$$\begin{aligned} S^I_x &= S^I(z_{ox}, -\infty) + S^I(-\infty, \infty) + S^I(\infty, z_{ix}) \\ S^I_y &= S^I(z_{oy}, -\infty) + S^I(-\infty, \infty) + S^I(\infty, z_{iy}) \end{aligned} \qquad (29.15)$$

29.2 GEOMETRICAL ABERRATION COEFFICIENTS

and the aberration coefficients are then obtained from equations analogous to (25.7):

$$x^{(1)}(z_c) = \frac{1}{\hat{\phi}_o^{\frac{1}{2}}}\left(H_{xc}^*\frac{\partial S_x^I}{\partial x_o} - G_{xc}^*\frac{\partial S_x^I}{\partial x_o'}\right) - \frac{1}{2}H_{xc}^*x_o'(x_o'^2 + y_o'^2)$$
$$y^{(1)}(z_c) = \frac{1}{\hat{\phi}_o^{\frac{1}{2}}}\left(H_{yc}^*\frac{\partial S_y^I}{\partial y_o} - G_{yc}^*\frac{\partial S_y^I}{\partial y_o'}\right) - \frac{1}{2}H_{yc}^*y_o'(x_o'^2 + y_o'^2)$$
(29.16)

in which the asterisk indicates that the asymptote to $G_x(z)$, $G_y(z)$, $H_x(z)$, $H_y(z)$ is to be understood.

Each of the numerous coefficients appearing in (29.11) has a polynomial structure in inverse magnification, but now the two inverse magnifications, $m_x := 1/M_x$ and $m_y := 1/M_y$, appear. Explicitly, we find

Aperture aberrations

$$(0030) = \sum_{i=0}^{4}(0030)_i m_x^i \qquad (0012) = \sum_{i,j=0}^{2}(0012)_{ij}m_x^i m_y^j$$
$$(0003) = \sum_{i=0}^{4}(0003)_i m_y^i \qquad (0021) = \sum_{i,j=0}^{2}(0021)_{ij}m_x^i m_y^j$$
(29.17a)

Comas

$$(1020) = \sum_{i=0}^{3}(1020)_i m_x^i \qquad (0102) = \sum_{i=0}^{3}(0102)_i m_y^i$$
$$(1002) = \sum_{i=0}^{1}\sum_{j=0}^{2}(1002)_{ij}m_x^i m_y^j \qquad (0120) = \sum_{i=0}^{2}\sum_{j=0}^{1}(0120)_{ij}m_x^i m_y^j$$
$$(0111) = \sum_{i=0}^{2}\sum_{j=0}^{1}(0111)_{ij}m_x^i m_y^j \qquad (1011) = \sum_{i=0}^{1}\sum_{j=0}^{2}(1011)_{ij}m_x^i m_y^j$$
(29.17b)

Astigmatisms

$$(2010) = \sum_{i=0}^{2}(2010)_i m_x^i \qquad (0201) = \sum_{i=0}^{2}(0201)_i m_y^i$$
$$(0210) = \sum_{i=0}^{2}(0210)_i m_x^i \qquad (2001) = \sum_{i=0}^{2}(2001)_i m_y^i$$
$$(1101) = \sum_{i,j=0}^{1}(1101)_{ij}m_x^i m_y^j \qquad (1110) = \sum_{i,j=0}^{1}(1110)_{ij}m_x^i m_y^j$$
(29.17c)

Distortions

$$(3000) = \sum_{i=0}^{1}(3000)_i m_x^i \qquad (1200) = \sum_{i=0}^{1}(1200)_i m_x^i$$
$$(0300) = \sum_{i=0}^{1}(0300)_i m_y^i \qquad (2100) = \sum_{i=0}^{1}(2100)_i m_y^i$$

(29.17d)

The coefficients $(pqrs)_i$ and $(pqrs)_{ij}$ are themselves simply related to a set of integrals. We write (25.27)

$$r = \frac{f_{xo}}{f_{xi}} = \frac{f_{yo}}{f_{yi}} = \left(\frac{\hat{\phi}_o}{\hat{\phi}_i}\right)^{\frac{1}{2}} \qquad (29.18)$$

Aperture aberrations

$$(0030)_4 = -4(4000)r^3 \qquad (0003)_4 = -4(0400)r^3$$
$$(0030)_3 = \left\{4(3010) - \frac{f_{xo}}{2}\right\}r^2 \qquad (0003)_3 = \left\{4(0301) - \frac{f_{yo}}{2}\right\}r^2$$
$$(0030)_2 = -4(2020)r \qquad (0003)_2 = -4(0202)r$$
$$(0030)_1 = 4(1030) - \frac{f_{xo}}{2} \qquad (0003)_1 = 4(0103) - \frac{f_{yo}}{2}$$
$$(0030)_0 = -4(0040)r^{-1} \qquad (0003)_0 = -4(0004)r^{-1}$$
$$(0012)_{22} = -2(2200)_F r^3 \qquad (0021)_{22} = -2(2200)_G r^3$$
$$(0012)_{21} = 2(2101)r^2 \qquad (0021)_{21} = \left\{2(2101) - \frac{f_{yo}}{2}\right\}r^2$$
$$(0012)_{12} = \left\{(2110) - \frac{f_{xo}}{2}\right\}r^2 \qquad (0021)_{12} = 2(1210)r^2$$
$$(0012)_{20} = -2(2002)r \qquad (0021)_{20} = -2(2002)r$$
$$(0012)_{11} = -2(1111)r \qquad (0021)_{11} = -2(1111)r$$
$$(0012)_{02} = -2(0220)r \qquad (0021)_{02} = -2(0220)r$$
$$(0012)_{10} = 2(1012) - \frac{f_{xo}}{2} \qquad (0021)_{10} = 2(1012)$$
$$(0012)_{01} = 2(0121) \qquad (0021)_{01} = 2(0121) - \frac{f_{yo}}{2}$$
$$(0012)_{00} = -2(0022)_F r^{-1} \qquad (0021)_{00} = -2(0022)_G r^{-1}$$

(29.19a)

29.2 GEOMETRICAL ABERRATION COEFFICIENTS

Comas

$(1020)_3 = 12(4000)r^2$ $\qquad (0102)_3 = 12(0400)r^2$

$(1020)_2 = \left\{-9(3010) + 3\dfrac{f_{xo}}{2}\right\}r$ $\qquad (0102)_2 = \left\{-9(0301) + 3\dfrac{f_{yo}}{2}\right\}r$

$(1020)_1 = 6(2020)$ $\qquad (0102)_1 = 6(0202)$

$(1020)_0 = -3(1030)r^{-1}$ $\qquad (0102)_0 = -3(0103)r^{-1}$

$(1002)_{12} = 2(2200)_F r^2$ $\qquad (0120)_{21} = 2(2200)_G r^2$

$(1002)_{11} = -2(2101)r$ $\qquad (0120)_{11} = -2(1210)r$

$(1002)_{02} = \left\{-(1210) + \dfrac{f_{xo}}{2}\right\}r$ $\qquad (0120)_{20} = \left\{-(2101) + \dfrac{f_{yo}}{2}\right\}r$

$(1002)_{10} = 2(2002)$ $\qquad (0120)_{10} = (1111)$

$(1002)_{01} = (1111)$ $\qquad (0120)_{01} = 2(0220)$

$(1002)_{00} = -(1012)r^{-1}$ $\qquad (0120)_{00} = -(0121)r^{-1}$

$(0111)_{21} = 4(2200)_F r^2$ $\qquad (1011)_{12} = 4(2200)_G r^2$

$(0111)_{20} = -2(2101)r$ $\qquad (1011)_{02} = -2(1210)r$

$(0111)_{11} = \left\{-4(1210) + f_{xo}\right\}r$ $\qquad (1011)_{11} = \left\{-4(2101) + f_{yo}\right\}r$

$(0111)_{10} = 2(1111)$ $\qquad (1011)_{10} = 4(2002)$

$(0111)_{01} = 4(0220)$ $\qquad (1011)_{01} = 2(1111)$

$(0111)_{00} = -2(0121)r^{-1}$ $\qquad (1011)_{00} = -2(1012)r^{-1}$

$(29.19b)$

Astigmatism

$(2010)_2 = -12(4000)r$ $\qquad (0201)_2 = -12(0400)r$

$(2010)_1 = 6(3010) - 3\dfrac{f_{xo}}{2}$ $\qquad (0201)_1 = 6(0301) - 3\dfrac{f_{yo}}{2}$

$(2010)_0 = -2(2020)r^{-1}$ $\qquad (0201)_0 = -2(0202)r^{-1}$

$(1101)_{11} = -4(2200)_F r$ $\qquad (1110)_{11} = -4(2200)_G r$

$(1101)_{10} = 2(2101)$ $\qquad (1110)_{10} = 2(2101) - f_{yo}$

$(1101)_{01} = 2(1210) - f_{xo}$ $\qquad (1110)_{01} = 2(1210)$ $\qquad (29.19c)$

$(1101)_{00} = -(1111)r^{-1}$ $\qquad (1110)_{00} = -(1111)r^{-1}$

$(0210)_2 = -2(2200)_F r$ $\qquad (2001)_2 = -2(2200)_G r$

$(0210)_1 = 2(1210) - \dfrac{f_{xo}}{2}$ $\qquad (2001)_1 = 2(2101) - \dfrac{f_{yo}}{2}$

$(0210)_0 = -2(0220)r^{-1}$ $\qquad (2001)_0 = -2(2002)r^{-1}$

Distortions

$$(3000)_1 = 4(4000) \qquad\qquad (0300)_1 = 4(0400)$$

$$(3000)_0 = \left\{-(3010) + \frac{f_{xo}}{2}\right\}r^{-1} \qquad (0300)_0 = \left\{-(0301) + \frac{f_{yo}}{2}\right\}r^{-1}$$

$$(1200)_1 = 2(2200)_F \qquad\qquad (2100)_1 = 2(2200)_G$$

$$(1200)_0 = \left\{-(1210) + \frac{f_{xo}}{2}\right\}r^{-1} \qquad (2100)_0 = \left\{-(2101) + \frac{f_{yo}}{2}\right\}r^{-1}$$

$$(29.19d)$$

The coefficients characterizing the aberrations of gradient, defined by

$$\begin{aligned}
\Delta x'_i &= f_x\left\{x'(z_{ix}) + x_o - f_x \frac{x'_o}{M_x}\right\} \\
&= \xi\{[3000]\xi^2 + [1200]\eta^2\} + x'_o\{[0030]x'^2_o + [0012]y'^2_o\} \\
&\quad + x'_o\{[2010]\xi^2 + [0210]\eta^2\} + \xi\{[1020]x'^2_o + [1002]y'^2_o\} \\
&\quad + \eta y'_o\{[1101]\xi + [0111]x'_o\} \\
\Delta y'_i &= f_y\left\{y'(z_{iy}) + y_o - f_y \frac{y'_o}{M_y}\right\} \\
&= \eta\{[0300]\eta^2 + [2100]\xi^2\} + y'_o\{[0003]y'^2_o + [0021]x'^2_o\} \\
&\quad + y'_o\{[2001]\xi^2 + [0201]\eta^2\} + \eta\{[0120]x'^2_o + [0102]y'^2_o\} \\
&\quad + \xi x'_o\{[1110]\eta + [1011]y'_o\}
\end{aligned} \qquad (29.20)$$

are also polynomials in inverse magnification. These have the following form:

$$[3000] = (3010) - \frac{1}{2}f_{xo}$$

$$[1200] = (1210) - \frac{1}{2}f_{xo}$$

$$[2010] = -3\{(3010) - \frac{1}{2}f_{xo}\}rm_x + 2(2020)$$

$$[0210] = -\{(1210) - \frac{1}{2}f_{xo}\}rm_x + 2(0220)$$

$$[1101] = -2\{(1210) - \frac{1}{2}f_{xo}\}rm_y + (1111)$$

$$[1020] = 3\{(3010) - \frac{1}{2}f_{xo}\}r^2m_x - 4(2020)rm_x + 3(1030)$$

$$[1002] = \{(1210) - \frac{1}{2}f_{xo}\}r^2m_y^2 - (1111)rm_y + (1012)$$

29.2 GEOMETRICAL ABERRATION COEFFICIENTS

$$[0111] = 2\{(1210) - \frac{1}{2}f_{xo}\}r^2 m_x m_y - 4(0220)rm_y$$
$$- (1111)m_x + 2(0121)$$

$$[0030] = -\{(3010) - \frac{1}{2}f_{xo}\}r^3 m_x^3 + 2(2020)r^2 m_x^2$$
$$- 3(1030)rm_x + 4(0040)$$

$$[0012] = -\{(1210) - \frac{1}{2}f_{xo}\}r^3 m_x m_y^2 + 2(0220)r^2 m_y^2$$
$$+ (1111)r^2 m_x m_y - (1012)rm_x - 2(0121)rm_y$$
$$+ 2(0022)_F$$

$$[0300] = (0301) - \frac{1}{2}f_{yo}$$

$$[2100] = (2101) - \frac{1}{2}f_{yo}$$

$$[2001] = -\{(2101) - \frac{1}{2}f_{yo}\}rm_y + 2(2002)$$

$$[0201] = -3\{(0301) - \frac{1}{2}f_{yo}\}rm_y + 2(0202)$$

$$[1110] = -2\{(2101) - \frac{1}{2}f_{yo}\}rm_x + (1111)$$

$$[0120] = \{(2101) - \frac{1}{2}f_{yo}\}r^2 m_x^2 - (1111)rm_x + (0121)$$

$$[0102] = 3\{(0301) - \frac{1}{2}f_{yo}\}r^2 m_y^2 - (0202)rm_y + 3(0103)$$

$$[1011] = 2\{(2101) - \frac{1}{2}f_{yo}\}r^2 m_x m_y - 4(2002)rm_x - (1111)rm_y$$
$$+ 2(1012)$$

$$[0003] = -\{(0301) - \frac{1}{2}f_{yo}\}r^3 m_y^3 + 2(0202)r^2 m_y^2$$
$$- 3(0103)rm_y + 4(0004)$$

$$[0021] = -\{(2101) - \frac{1}{2}f_{yo}\}r^3 m_x^2 m_y + 2(2002)r^2 m_x^2$$
$$+ (1111)r^2 m_x m_y - 2(0102)rm_x - (0121)rm_y$$
$$+ 2(0022)_G \tag{29.21}$$

The coefficients $(pqrs)$, $p+q+r+s = 4$, denote the following integrals:

$$(4000) = f_{xi}^4 \int_{-\infty}^{\infty} (A_x G_x^4 + B_x G_x^2 G_x'^2 + N G_x'^2)\, dz$$
$$+ \frac{1}{8} \lim_{z \to \infty} (z - z_{Fi}^{(x)})$$

$$(0400) = f_{yi}^4 \int_{-\infty}^{\infty} (A_y G_y^4 + B_y G_y^2 G_y'^2 + N G_y'^2)\, dz$$

$$+ \frac{1}{8} \lim_{z \to \infty} (z - z_{Fi}^{(y)})$$

$$(0040) = \int_{-\infty}^{\infty} (A_x \overline{\Gamma}_x^4 + B_x \overline{\Gamma}_x^2 \overline{\Gamma}_x'^2 + N \overline{\Gamma}_x'^2)\, dz$$

$$- \frac{1}{8} r \lim_{z \to -\infty} (z - z_{Fo}^{(x)})$$

$$(0004) = \int_{-\infty}^{\infty} (A_y \overline{\Gamma}_y^4 + B_y \overline{\Gamma}_y^2 \overline{\Gamma}_y'^2 + N \overline{\Gamma}_y'^4)\, dz$$

$$- \frac{1}{8} r \lim_{z \to -\infty} (z - z_{Fo}^{(y)})$$

$$(3010) = 4 f_{xi}^3 \int_{-\infty}^{\infty} \{A_x G_x^3 \overline{\Gamma}_x + \frac{1}{4} B_x (G_x^2)'(G_x \overline{\Gamma}_x)' + N G_x'^3 \overline{\Gamma}_x'\}\, dz$$

$$(0301) = 4 f_{yi}^3 \int_{-\infty}^{\infty} \{A_y G_y^3 \overline{\Gamma}_y + \frac{1}{4} B_y (G_y^2)'(G_y \overline{\Gamma}_y)' + N G_y'^3 \overline{\Gamma}_y'\}\, dz$$

$$(1030) = 4 f_{xi} \int_{-\infty}^{\infty} \{A_x G_x \overline{\Gamma}_x^3 + \frac{1}{4} B_x (G_x \overline{\Gamma}_x)'(\overline{\Gamma}_x^2)' + N G_x' \overline{\Gamma}_x'^3\}\, dz$$

$$(0103) = 4 f_{yi} \int_{-\infty}^{\infty} \{A_y G_y \overline{\Gamma}_y^3 + \frac{1}{4} B_y (G_y \overline{\Gamma}_y)'(\overline{\Gamma}_y^2)' + N G_y' \overline{\Gamma}_y'^3\}\, dz$$

$$(2020) = 6 f_{xi}^2 \int_{-\infty}^{\infty} \{A_x G_x^2 \overline{\Gamma}_x^2 + \frac{1}{6} B_x (G_x^2 \overline{\Gamma}_x'^2 + G_x'^2 \overline{\Gamma}_x^2$$

$$+ (G_x^2)'(\overline{\Gamma}_x^2)') + N G_x'^2 \overline{\Gamma}_x'^2\}\, dz$$

$$(0202) = 6 f_{yi}^2 \int_{-\infty}^{\infty} \{A_y G_y^2 \overline{\Gamma}_y^2 + \frac{1}{6} B_y (G_y^2 \overline{\Gamma}_y'^2 + G_y'^2 \overline{\Gamma}_y^2$$

$$+ (G_y^2)'(\overline{\Gamma}_y^2)') + N G_y'^2 \overline{\Gamma}_y'^2\}\, dz$$

$$(2002) = f_{xi}^2 \int_{-\infty}^{\infty} \{C G_y^2 \overline{\Gamma}_y^2 + B_x G_x^2 \overline{\Gamma}_y'^2 + B_y G_x'^2 \overline{\Gamma}_y^2 + 2 N G_x'^2 \overline{\Gamma}_y'^2$$

$$+ RG_x\overline{\Gamma}_y(G_x\overline{\Gamma}'_y - G'_x\overline{\Gamma}_y)\} \, dz$$

$$(0220) = f_{yi}^2 \int_{-\infty}^{\infty} \{C\overline{\Gamma}_x^2 G_y^2 + B_x\overline{\Gamma}_x^2 G_y'^2 + B_y\overline{\Gamma}_x'^2 G_y^2 + 2N\overline{\Gamma}_x'^2 G_y'^2$$

$$+ R\overline{\Gamma}_x G_y(\overline{\Gamma}_x G'_y - \overline{\Gamma}'_x G_y)\} \, dz$$

$$(2200) = f_{xi}^2 f_{yi}^2 \int_{-\infty}^{\infty} \{CG_x^2 G_y^2 + B_x G_x^2 G_y'^2 + B_y G_x'^2 G_y^2 + 2NG_x'^2 G_y'^2$$

$$+ RG_x G_y(G_x G'_y - G'_x G_y)\} \, dz$$

$$(2200)_F = (2200) + \frac{1}{4} \lim_{z \to \infty} (z - z_{Fi}^{(x)})$$

$$(2200)_G = (2200) + \frac{1}{4} \lim_{z \to \infty} (z - z_{Fi}^{(y)})$$

$$(0022) = \int_{-\infty}^{\infty} \{C\overline{\Gamma}_x^2 \overline{\Gamma}_y^2 + B_x\overline{\Gamma}_x^2 \overline{\Gamma}_y'^2 + B_y\overline{\Gamma}_x'^2 \overline{\Gamma}_y^2 + 2N\overline{\Gamma}_x'^2 \overline{\Gamma}_y'^2$$

$$+ R\overline{\Gamma}_x \overline{\Gamma}_y(\overline{\Gamma}_x \overline{\Gamma}'_y - \overline{\Gamma}'_x \overline{\Gamma}_y)\} \, dz$$

$$(0022)_F = (0022) - \frac{1}{4} r \lim_{z \to -\infty} (z - z_{Fo}^{(x)})$$

$$(0022)_G = (0022) - \frac{1}{4} r \lim_{z \to -\infty} (z - z_{Fo}^{(y)})$$

$$(2101) = 2f_{xi}^2 f_{yi} \int_{-\infty}^{\infty} [CG_x^2 G_y \overline{\Gamma}_y + B_x G_x^2 G'_y \overline{\Gamma}'_y + B_y G_x'^2 G_y \overline{\Gamma}_y + 2NG_x'^2 G'_y \overline{\Gamma}'_y$$

$$+ \frac{1}{2} R\{G_x^2(G_y \overline{\Gamma}_y)' - (G_x^2)' G_y \overline{\Gamma}_y\}] \, dz$$

$$(1210) = 2f_{xi} f_{yi}^2 \int_{-\infty}^{\infty} [CG_x \overline{\Gamma}_x G_y^2 + B_x G_x \overline{\Gamma}_x G_y'^2 + B_y G'_x \overline{\Gamma}'_x G_y^2 + 2NG'_x \overline{\Gamma}'_x G_y'^2$$

$$+ \frac{1}{2} R\{G_x \overline{\Gamma}_x (G_y^2)' - (G_x \overline{\Gamma}_x)' G_y^2\}] \, dz$$

$$(1012) = 2f_{xi} \int_{-\infty}^{\infty} [CG_x \overline{\Gamma}_x \overline{\Gamma}_y^2 + B_x G_x \overline{\Gamma}_x \overline{\Gamma}_y'^2 + B_y G'_x \overline{\Gamma}'_x \overline{\Gamma}_y^2 + 2NG'_x \overline{\Gamma}'_x \overline{\Gamma}_y'^2$$

$$+ \frac{1}{2} R\{G_x \overline{\Gamma}_x (\overline{\Gamma}_y^2)' - (G_x \overline{\Gamma}_x)' \overline{\Gamma}_y^2\}] \, dz$$

$$(0121) = 2f_{yi} \int_{-\infty}^{\infty} [C\overline{\Gamma}_x^2 G_y \overline{\Gamma}_y + B_x \overline{\Gamma}_x^2 G'_y \overline{\Gamma}'_y + B_y \overline{\Gamma}_x'^2 G_y \overline{\Gamma}_y + 2N\overline{\Gamma}_x'^2 G'_y \overline{\Gamma}'_y$$
$$+ \frac{1}{2} R\{\overline{\Gamma}_x^2 (G_y \overline{\Gamma}_y)' - (\overline{\Gamma}_x^2)' G_y \overline{\Gamma}_y\}] \, dz$$

$$(1111) = 4f_{xi}f_{yi} \int_{-\infty}^{\infty} [CG_x \overline{\Gamma}_x G_y \overline{\Gamma}_y + B_x G_x \overline{\Gamma}_x G'_y \overline{\Gamma}'_y + B_y G'_x \overline{\Gamma}'_x G_y \overline{\Gamma}_y$$
$$+ 2NG'_x \overline{\Gamma}'_x G'_y \overline{\Gamma}'_y + \frac{1}{2} R\{G_x \overline{\Gamma}_x (G_y \overline{\Gamma}_y)' - (G_x \overline{\Gamma}_x)' G_y \overline{\Gamma}_y\}] \, dz$$
$$(29.22)$$

in which

$$A_x = \frac{1}{128} \left(\frac{\hat{\phi}}{\hat{\phi}_i}\right)^{\frac{1}{2}} \left(\frac{\gamma \phi^{(4)}}{\hat{\phi}} - \frac{\phi''^2 + 4p_2^2}{\hat{\phi}^2} - \frac{8\gamma p_2''}{3\hat{\phi}}\right.$$
$$\left. + \frac{4p_2 \phi''}{\hat{\phi}^2} + \frac{8\eta Q_2''}{3\hat{\phi}^{\frac{1}{2}}}\right) + \Xi$$

$$A_y = \frac{1}{128} \left(\frac{\hat{\phi}}{\hat{\phi}_i}\right)^{\frac{1}{2}} \left(\frac{\gamma \phi^{(4)}}{\hat{\phi}} - \frac{\phi''^2 + 4p_2^2}{\hat{\phi}^2} + \frac{8\gamma p_2''}{3\hat{\phi}}\right.$$
$$\left. - \frac{4p_2 \phi''}{\hat{\phi}^2} - \frac{8\eta Q_2''}{3\hat{\phi}^{\frac{1}{2}}}\right) + \Xi$$

$$B_x = -\frac{\gamma}{16} \left(\frac{\hat{\phi}}{\hat{\phi}_i}\right)^{\frac{1}{2}} \frac{\phi'' - 2p_2}{\hat{\phi}}$$

$$B_y = -\frac{\gamma}{16} \left(\frac{\hat{\phi}}{\hat{\phi}_i}\right)^{\frac{1}{2}} \frac{\phi'' + 2p_2}{\hat{\phi}}$$

$$C = \frac{1}{64} \left(\frac{\hat{\phi}}{\hat{\phi}_i}\right)^{\frac{1}{2}} \left(\frac{\gamma \phi^{(4)}}{\hat{\phi}} - \frac{\phi''^2 - 4p_2^2}{\hat{\phi}^2}\right) - 6\Xi$$

$$N = -\frac{1}{8} \left(\frac{\hat{\phi}}{\hat{\phi}_i}\right)^{\frac{1}{2}}$$

$$R = \frac{\eta Q_2'}{4\hat{\phi}_i^{\frac{1}{2}}} \tag{29.23}$$

and $\Xi(z)$ characterizes the octopole distribution:

$$\Xi(z) := \frac{1}{48} \left(\frac{\hat{\phi}}{\hat{\phi}_i}\right)^{\frac{1}{2}} \left(\frac{\gamma p_4}{\hat{\phi}} - \frac{2\eta Q_4}{\hat{\phi}^{\frac{1}{2}}}\right) \tag{29.24}$$

29.2 GEOMETRICAL ABERRATION COEFFICIENTS

In the foregoing formulae, we have imposed no restrictions on $\phi(z)$, $p_2(z)$ and $Q_2(z)$, which may all be present simultaneously. Commonly, however, ϕ is constant and any electrostatic and magnetic quadrupole fields do not overlap. In these conditions, the formulae can be recast in a much simpler form by writing

$$p_2(z) =: p_{20}\ q(z) \qquad Q_2(z) =: Q_{20}\ q(z)$$
$$p_4(z) =: p_{40}\ \omega(z) \qquad Q_4(z) =: Q_{40}\ \omega(z) \tag{29.25}$$

where $q(z)$ and $\omega(z)$ are functions that reach a maximum value of unity. We define an excitation parameter β^2 and a label n as follows:

$$\frac{\gamma p_{20}}{2\hat{\phi}} =: \beta_E^2 \qquad \frac{\eta Q_{20}}{\hat{\phi}^{\frac{1}{2}}} =: \beta_M^2 \qquad \beta^2 := \beta_M^2 - \beta_E^2 \qquad n := \frac{\beta_E^2}{\beta_M^2 - \beta_E^2} \tag{29.26}$$

Thus for purely electrostatic quadrupoles, $n = -1$, for magnetic quadrupoles, $n = 0$, and as we shall see in Section 29.4, for achromatic quadrupoles, $n = \gamma^2 \approx 1$. From (29.25–29.26), we see that

$$\frac{\gamma p_2(z)}{\hat{\phi}} = 2n\beta^2 q(z) \quad \text{and} \quad \frac{\eta Q_2(z)}{\hat{\phi}^{\frac{1}{2}}} = (n+1)\beta^2 q(z) \tag{29.27}$$

For octopoles, we write

$$\frac{\gamma p_{40}}{48\hat{\phi}} =: -\tau_E \qquad \frac{\eta Q_{40}}{24\hat{\phi}^{\frac{1}{2}}} =: -\tau_M \qquad \tau_M - \tau_E := \tau \tag{29.28}$$

so that

$$\Xi(z) = \tau\omega(z) \tag{29.29}$$

The functions $A_x, A_y \ldots R$ now become

$$\begin{aligned}
A_x &= -\frac{1}{8}\beta^2(\beta^2 q^2 \overline{n}^2 + \frac{n-1}{6}q'') + \tau\omega \\
A_y &= -\frac{1}{8}\beta^2(\beta^2 q^2 \overline{n}^2 - \frac{n-1}{6}q'') + \tau\omega \\
B_x &= -B_y = \frac{1}{4}\beta^2 nq \\
C &= \frac{1}{4}\beta^4 \overline{n}^2 q^2 - 6\tau\omega \\
N &= -\frac{1}{8} \\
R &= \frac{1}{4}\beta^2(n+1)q'
\end{aligned} \tag{29.30}$$

29. THE ABERRATION OF QUADRUPOLES

where $\bar{n} := n/\gamma$. The following set of coefficients $(pqrs)$ is now more convenient:

$$(4000) = \int_{-\infty}^{\infty} a_x \Gamma_x^4 \, dz$$

$$= \int_{-\infty}^{\infty} b\Gamma_x^4 \, dz - \frac{1}{24} \int_{-\infty}^{\infty} \Gamma_x'^4 \, dz + \frac{1}{24} \lim_{z \to \infty} (z - z_{Fi}^{(x)})$$

$$(0400) = \int_{-\infty}^{\infty} a_y \Gamma_y^4 \, dz$$

$$= \int_{-\infty}^{\infty} b\Gamma_y^4 \, dz - \frac{1}{24} \int_{-\infty}^{\infty} \Gamma_y'^4 \, dz + \frac{1}{24} \lim_{z \to \infty} (z - z_{Fi}^{(y)})$$

$$(0040) = \int_{-\infty}^{\infty} a_x \overline{\Gamma}_x^4 \, dz$$

$$= \int_{-\infty}^{\infty} b\overline{\Gamma}_x^4 \, dz - \frac{1}{24} \int_{-\infty}^{\infty} \overline{\Gamma}_x'^4 \, dz - \frac{1}{24} \lim_{z \to -\infty} (z - z_{Fo}^{(x)})$$

$$(0004) = \int_{-\infty}^{\infty} a_y \overline{\Gamma}_y^4 \, dz$$

$$= \int_{-\infty}^{\infty} b\overline{\Gamma}_y^4 \, dz - \frac{1}{24} \int_{-\infty}^{\infty} \overline{\Gamma}_y'^4 \, dz - \frac{1}{24} \lim_{z \to -\infty} (z - z_{Fo}^{(y)})$$

$$(3010) = 4 \int_{-\infty}^{\infty} a_x \Gamma_x^3 \overline{\Gamma}_x \, dz + \frac{f_x}{8}$$

$$= 4 \int_{-\infty}^{\infty} b\Gamma_x^3 \overline{\Gamma}_x \, dz - \frac{1}{6} \int_{-\infty}^{\infty} \Gamma_x'^3 \overline{\Gamma}_x' \, dz + \frac{f_x}{12}$$

$$(0301) = 4 \int_{-\infty}^{\infty} a_y \Gamma_y^3 \overline{\Gamma}_y \, dz + \frac{f_y}{8}$$

$$= 4 \int_{-\infty}^{\infty} b\Gamma_y^3 \overline{\Gamma}_y \, dz - \frac{1}{6} \int_{-\infty}^{\infty} \Gamma_y'^3 \overline{\Gamma}_y' \, dz + \frac{f_y}{12}$$

29.2 GEOMETRICAL ABERRATION COEFFICIENTS

$$(1030) = 4\int_{-\infty}^{\infty} a_x \Gamma_x \overline{\Gamma}_x^3 \, dz + \frac{f_x}{8}$$

$$= 4\int_{-\infty}^{\infty} b\Gamma_x \overline{\Gamma}_x^3 \, dz - \frac{1}{6}\int_{-\infty}^{\infty} \Gamma'_x \overline{\Gamma}_x'^3 \, dz + \frac{f_x}{12}$$

$$(0103) = 4\int_{-\infty}^{\infty} a_y \Gamma_y \overline{\Gamma}_y^3 \, dz + \frac{f_y}{8}$$

$$= 4\int_{-\infty}^{\infty} b\Gamma_y \overline{\Gamma}_y^3 \, dz - \frac{1}{6}\int_{-\infty}^{\infty} \Gamma'_y \overline{\Gamma}_y'^3 \, dz + \frac{f_y}{12}$$

$$(2020) = 6\int_{-\infty}^{\infty} a_x \Gamma_x^2 \overline{\Gamma}_x^2 \, dz$$

$$= 6\int_{-\infty}^{\infty} b\Gamma_x^2 \overline{\Gamma}_x^2 \, dz - \frac{1}{4}\int_{-\infty}^{\infty} \Gamma'^2_x \overline{\Gamma}'^2_x \, dz$$

$$(0202) = 6\int_{-\infty}^{\infty} a_y \Gamma_y^2 \overline{\Gamma}_y^2 \, dz$$

$$= 6\int_{-\infty}^{\infty} b\Gamma_y^2 \overline{\Gamma}_y^2 \, dz - \frac{1}{4}\int_{-\infty}^{\infty} \Gamma'^2_y \overline{\Gamma}'^2_y \, dz$$

$$(2002) = \int_{-\infty}^{\infty} \{c\Gamma_x^2 \overline{\Gamma}_y^2 + \frac{3}{8}\beta^2 q' \Gamma_x \overline{\Gamma}_y (\Gamma_x \overline{\Gamma}'_y - \Gamma'_x \overline{\Gamma}_y)\} \, dz$$

$$= \int_{-\infty}^{\infty} \{\bar{c}\Gamma_x^2 \overline{\Gamma}_y^2 - \frac{3}{8}\beta^2 q (\Gamma_x^2 \overline{\Gamma}_y'^2 - \Gamma'^2_x \overline{\Gamma}_y^2)\} \, dz$$

$$= \int_{-\infty}^{\infty} \bar{c}\Gamma_x^2 \overline{\Gamma}_y^2 \, dz - \frac{3}{4}\int_{-\infty}^{\infty} \Gamma'^2_x \overline{\Gamma}_y'^2 \, dz$$

$$(0220) = \int_{-\infty}^{\infty} \{c\overline{\Gamma}_x^2 \Gamma_y^2 + \frac{3}{8}\beta^2 q' \overline{\Gamma}_x \Gamma_y (\overline{\Gamma}_x \Gamma'_y - \overline{\Gamma}'_x \Gamma_y)\} \, dz$$

$$= \int_{-\infty}^{\infty} \{\bar{c}\bar{\Gamma}_x^2\Gamma_y^2 - \frac{3}{8}\beta^2 q(\bar{\Gamma}_x^2\Gamma_y'^2 - \bar{\Gamma}_x'^2\Gamma_y^2)\}\, dz$$

$$= \int_{-\infty}^{\infty} \bar{c}\bar{\Gamma}_x^2\Gamma_y^2\, dz - \frac{3}{4}\int_{-\infty}^{\infty} \bar{\Gamma}_x'^2\Gamma_y'^2\, dz$$

$$(2200) = \int_{-\infty}^{\infty} \{c\Gamma_x^2\Gamma_y^2 + \frac{3}{8}\beta^2 q'\Gamma_x\Gamma_y(\Gamma_x\Gamma_y' - \Gamma_x'\Gamma_y)\}\, dz$$
$$- \frac{1}{8}\lim_{z\to\infty}(2z - z_{Fi}^{(x)} - z_{Fi}^{(y)})$$

$$= \int_{-\infty}^{\infty} \{\bar{c}\Gamma_x^2\Gamma_y^2 - \frac{3}{8}\beta^2 q(\Gamma_x^2\Gamma_y'^2 - \Gamma_x'^2\Gamma_y^2)\}\, dz - \frac{1}{8}\lim_{z\to\infty}(2z - z_{Fi}^{(x)} - z_{Fi}^{(y)})$$

$$= \int_{-\infty}^{\infty} \bar{c}\Gamma_x^2\Gamma_y^2\, dz - \frac{3}{4}\int_{-\infty}^{\infty} \Gamma_x'^2\Gamma_y'^2\, dz + \frac{1}{4}\lim_{z\to\infty}(2z - z_{Fi}^{(x)} - z_{Fi}^{(y)})$$

$$(2200)_F = \int_{-\infty}^{\infty} \{c\Gamma_x^2\Gamma_y^2 + \frac{3}{8}\beta^2 q'\Gamma_x\Gamma_y(\Gamma_x\Gamma_y' - \Gamma_x'\Gamma_y)\}\, dz - \frac{z_{Fi}^{(x)} - z_{Fi}^{(y)}}{8}$$

$$= \int_{-\infty}^{\infty} \{\bar{c}\Gamma_x^2\Gamma_y^2 - \frac{3}{8}\beta^2 q(\Gamma_x^2\Gamma_y'^2 - \Gamma_x'^2\Gamma_y^2)\}\, dz - \frac{z_{Fi}^{(x)} - z_{Fi}^{(y)}}{8}$$

$$= \int_{-\infty}^{\infty} \bar{c}\Gamma_x^2\Gamma_y^2\, dz - \frac{3}{4}\int_{-\infty}^{\infty} \Gamma_x'^2\Gamma_y'^2\, dz + \frac{1}{4}\lim_{z\to\infty}(3z - 2z_{Fi}^{(x)} - z_{Fi}^{(y)})$$

$$(2200)_G = \int_{-\infty}^{\infty} \{c\Gamma_x^2\Gamma_y^2 + \frac{3}{8}\beta^2 q'\Gamma_x\Gamma_y(\Gamma_x\Gamma_y' - \Gamma_x'\Gamma_y)\}\, dz + \frac{z_{Fi}^{(x)} - z_{Fi}^{(y)}}{8}$$

$$= \int_{-\infty}^{\infty} \{\bar{c}\Gamma_x^2\Gamma_y^2 - \frac{3}{8}\beta^2 q(\Gamma_x^2\Gamma_y'^2 - \Gamma_x'^2\Gamma_y^2)\}\, dz + \frac{z_{Fi}^{(x)} - z_{Fi}^{(y)}}{8}$$

$$= \int_{-\infty}^{\infty} \bar{c}\Gamma_x^2\Gamma_y^2\, dz - \frac{3}{4}\int_{-\infty}^{\infty} \Gamma_x'^2\Gamma_y'^2\, dz + \frac{1}{4}\lim_{z\to\infty}(3z - 2z_{Fi}^{(y)} - z_{Fi}^{(x)})$$

$$(0022) = \int_{-\infty}^{\infty} \{c\bar{\Gamma}_x^2\bar{\Gamma}_y^2 + \frac{3}{8}\beta^2 q'\bar{\Gamma}_x\bar{\Gamma}_y(\bar{\Gamma}_x\bar{\Gamma}_y' - \bar{\Gamma}_x'\bar{\Gamma}_y)\}\, dz$$

29.2 GEOMETRICAL ABERRATION COEFFICIENTS

$$+ \frac{1}{8} \lim_{z \to -\infty} (2z - z_{Fo}^{(x)} - z_{Fo}^{(y)})$$

$$= \int_{-\infty}^{\infty} \{\bar{c}\bar{\Gamma}_x^2 \bar{\Gamma}_y^2 - \frac{3}{8}\beta^2 q(\bar{\Gamma}_x^2 \bar{\Gamma}_y'^2 - \bar{\Gamma}_x'^2 \bar{\Gamma}_y^2)\}\, dz + \frac{1}{8} \lim_{z \to -\infty} (2z - z_{Fo}^{(x)} - z_{Fo}^{(y)})$$

$$= \int_{-\infty}^{\infty} \bar{c}\bar{\Gamma}_x^2 \bar{\Gamma}_y^2\, dz - \frac{3}{4} \int_{-\infty}^{\infty} \bar{\Gamma}_x'^2 \bar{\Gamma}_y^2\, dz - \frac{1}{4} \lim_{z \to -\infty} (2z - z_{Fo}^{(x)} - z_{Fo}^{(y)})$$

$$(0022)_F = \int_{-\infty}^{\infty} \{c\bar{\Gamma}_x^2 \bar{\Gamma}_y^2 + \frac{3}{8}\beta^2 q' \bar{\Gamma}_x \bar{\Gamma}_y (\bar{\Gamma}_x \bar{\Gamma}_y' - \bar{\Gamma}_x' \bar{\Gamma}_y)\}\, dz + \frac{z_{Fo}^{(x)} - z_{Fo}^{(y)}}{8}$$

$$= \int_{-\infty}^{\infty} \{\bar{c}\bar{\Gamma}_x^2 \bar{\Gamma}_y^2 - \frac{3}{8}\beta^2 q(\bar{\Gamma}_x^2 \bar{\Gamma}_y'^2 - \bar{\Gamma}_x'^2 \bar{\Gamma}_y^2)\}\, dz + \frac{z_{Fo}^{(x)} - z_{Fo}^{(y)}}{8}$$

$$= \int_{-\infty}^{\infty} \bar{c}\bar{\Gamma}_x^2 \bar{\Gamma}_y^2\, dz - \frac{3}{4} \int_{-\infty}^{\infty} \bar{\Gamma}_x'^2 \bar{\Gamma}_y^2\, dz - \frac{1}{4} \lim_{z \to -\infty} (3z - 2z_{Fo}^{(x)} - z_{Fo}^{(y)})$$

$$(0022)_G = \int_{-\infty}^{\infty} \{c\bar{\Gamma}_x^2 \bar{\Gamma}_y^2 + \frac{3}{8}\beta^2 q' \bar{\Gamma}_x \bar{\Gamma}_y (\bar{\Gamma}_x \bar{\Gamma}_y' - \bar{\Gamma}_x' \bar{\Gamma}_y)\}\, dz - \frac{z_{Fo}^{(x)} - z_{Fo}^{(y)}}{8}$$

$$= \int_{-\infty}^{\infty} \{\bar{c}\bar{\Gamma}_x^2 \bar{\Gamma}_y^2 - \frac{3}{8}\beta^2 q(\bar{\Gamma}_x^2 \bar{\Gamma}_y'^2 - \bar{\Gamma}_x'^2 \bar{\Gamma}_y^2)\}\, dz - \frac{z_{Fo}^{(x)} - z_{Fo}^{(y)}}{8}$$

$$= \int_{-\infty}^{\infty} \bar{c}\bar{\Gamma}_x^2 \bar{\Gamma}_y^2\, dz - \frac{3}{4} \int_{-\infty}^{\infty} \bar{\Gamma}_x'^2 \bar{\Gamma}_y'^2\, dz - \frac{1}{4} \lim_{z \to -\infty} (3z - 2z_{Fo}^{(y)} - z_{Fo}^{(x)})$$

$$(2101) = \int_{-\infty}^{\infty} [2c\bar{\Gamma}_x^2 \Gamma_y \bar{\Gamma}_y + \frac{3}{8}\beta^2 q' \{\Gamma_x^2 (\Gamma_y \bar{\Gamma}_y)' - (\Gamma_x^2)' \Gamma_y \bar{\Gamma}_y\}]\, dz + \frac{f_y}{8}$$

$$= 2 \int_{-\infty}^{\infty} \{\bar{c}\Gamma_x^2 \Gamma_y \bar{\Gamma}_y - \frac{3}{8}\beta^2 q(\Gamma_x^2 \Gamma_y' \bar{\Gamma}_y' - \Gamma_x'^2 \Gamma_y \bar{\Gamma}_y)\}\, dz + \frac{f_y}{8}$$

$$= 2 \int_{-\infty}^{\infty} \bar{c}\Gamma_x^2 \Gamma_y \bar{\Gamma}_y\, dz - \frac{3}{2} \int_{-\infty}^{\infty} \Gamma_x'^2 \Gamma_y' \bar{\Gamma}_y'\, dz - \frac{f_y}{4}$$

$$(1210) = \int_{-\infty}^{\infty} [2c\Gamma_x \bar{\Gamma}_x \Gamma_y^2 + \frac{3}{8}\beta^2 q' \{\Gamma_x \bar{\Gamma}_x (\Gamma_y^2)' - (\Gamma_x \bar{\Gamma}_x)' \Gamma_y^2\}]\, dz + \frac{f_x}{8}$$

29. THE ABERRATION OF QUADRUPOLES

$$= 2\int_{-\infty}^{\infty} \{\bar{c}\Gamma_x\bar{\Gamma}_x\Gamma_y^2 - \frac{3}{8}\beta^2 q(\Gamma_x\bar{\Gamma}_x\Gamma_y'^2 - \Gamma_x'\bar{\Gamma}_x'\Gamma_y'^2)\}\, dz + \frac{f_x}{8}$$

$$= 2\int_{-\infty}^{\infty} \bar{c}\Gamma_x\bar{\Gamma}_x\Gamma_y^2\, dz - \frac{3}{2}\int_{-\infty}^{\infty} \Gamma_x'\bar{\Gamma}_x'\Gamma_y'^2\, dz - \frac{f_x}{4}$$

$$(1012) = \int_{-\infty}^{\infty} [2c\Gamma_x\bar{\Gamma}_x\bar{\Gamma}_y^2 + \frac{3}{8}\beta^2 q'\{\Gamma_x\bar{\Gamma}_x(\bar{\Gamma}_y^2)' - (\Gamma_x\bar{\Gamma}_x)'\bar{\Gamma}_y^2\}]\, dz + \frac{f_x}{8}$$

$$= 2\int_{-\infty}^{\infty} \{\bar{c}\Gamma_x\bar{\Gamma}_x\bar{\Gamma}_y^2 - \frac{3}{8}\beta^2 q(\Gamma_x\bar{\Gamma}_x\bar{\Gamma}_y'^2 - \Gamma_x'\bar{\Gamma}_x'\bar{\Gamma}_y^2)\}\, dz + \frac{f_x}{8}$$

$$= 2\int_{-\infty}^{\infty} \bar{c}\Gamma_x\bar{\Gamma}_x\bar{\Gamma}_y^2\, dz - \frac{3}{2}\int_{-\infty}^{\infty} \Gamma_x'\bar{\Gamma}_x'\bar{\Gamma}_y'^2\, dz - \frac{f_x}{4}$$

$$(0121) = \int_{-\infty}^{\infty} [2c\bar{\Gamma}_x^2\Gamma_y\bar{\Gamma}_y + \frac{3}{8}\beta^2 q'\{\bar{\Gamma}_x^2(\Gamma_y\bar{\Gamma}_y)' - (\bar{\Gamma}_x^2)'\Gamma_y\bar{\Gamma}_y\}]\, dz + \frac{f_y}{8}$$

$$= 2\int_{-\infty}^{\infty} \{\bar{c}\bar{\Gamma}_x^2\Gamma_y\bar{\Gamma}_y - \frac{3}{8}\beta^2 q(H_x^2\Gamma_y'\bar{\Gamma}_y' - \bar{\Gamma}_x'^2\Gamma_y\bar{\Gamma}_y)\}\, dz + \frac{f_y}{8}$$

$$= 2\int_{-\infty}^{\infty} \bar{c}\bar{\Gamma}_x^2\Gamma_y\bar{\Gamma}_y\, dz - \frac{3}{2}\int_{-\infty}^{\infty} \bar{\Gamma}_x'^2\Gamma_y'\bar{\Gamma}_y'\, dz - \frac{f_y}{4}$$

$$(1111) = 2\int_{-\infty}^{\infty} [2c\Gamma_x\bar{\Gamma}_x\Gamma_y\bar{\Gamma}_y + \frac{3}{8}\beta^2 q'\{\Gamma_x\bar{\Gamma}_x(\Gamma_y\bar{\Gamma}_y)' - (\Gamma_x\bar{\Gamma}_x)'\Gamma_y\bar{\Gamma}_y\}]\, dz$$

$$= 4\int_{-\infty}^{\infty} \{\bar{c}\Gamma_x\bar{\Gamma}_x\Gamma_y\bar{\Gamma}_y - \frac{3}{8}\beta^2 q(\Gamma_x\bar{\Gamma}_x\Gamma_y'\bar{\Gamma}_y' - \Gamma_x'\bar{\Gamma}_x'\Gamma_y\bar{\Gamma}_y)\}\, dz$$

$$= 4\int_{-\infty}^{\infty} \bar{c}\Gamma_x\bar{\Gamma}_x\Gamma_y\bar{\Gamma}_y\, dz - 3\int_{-\infty}^{\infty} \Gamma_x'\bar{\Gamma}_x'\Gamma_y'\bar{\Gamma}_y'\, dz \tag{29.31}$$

In these expressions, the functions a_x, a_y, b, c and $\bar{c}$ are as follows:

$$a_x := -\frac{1}{24}\beta^4(3\bar{n}^2 - 2n + 3)q^2 - \frac{1}{96}\beta^2 q'' + \tau\omega$$

$$a_y := -\frac{1}{24}\beta^4(3\overline{n}^2 - 2n + 3)q^2 + \frac{1}{96}\beta^2 q'' + \tau\omega$$

$$b := -\frac{1}{24}\beta^4(3\overline{n}^2 - 2n + 2)q^2 + \tau\omega$$

$$c := \frac{1}{4}\beta^4(\overline{n}^2 - 2n + 1)q^2 - 6\tau\omega$$

$$\overline{c} := \frac{1}{4}\beta^4(\overline{n}^2 - 2n - 2)q^2 - 6\tau\omega \qquad 29.32)$$

We have lightened the notation by setting

$$\Gamma_x = f_x G_x \qquad \overline{\Gamma}_x = f_x \overline{G}_x \qquad \Gamma_y = f_y G_y \qquad \overline{\Gamma}_y = f_y \overline{G}_y \qquad (29.33)$$

so that Γ_x and Γ_y are the rays incident from object space parallel to the axis at heights f_x and f_y in the two principal sections. If the function $q(z)$ is even in z, $\Gamma_x(z) = \overline{\Gamma}_x(-z)$ and $\Gamma_y(z) = \overline{\Gamma}_y(-z)$; if $q(z)$ is odd, then $\Gamma_x(z) = \overline{\Gamma}_y(-z)$ and $\Gamma_y(z) = \overline{\Gamma}_x(-z)$. The interrelations between the coefficients, derived earlier by symmetry arguments (29.7, 29.9), become obvious.

29.3 Aperture aberrations

Each of the aberration coefficients of quadrupoles is associated with an aberration figure. We shall examine only the aperture aberrations, however, and we likewise confine the detailed analysis of the aberration formulae to this family of defects.

In a general plane z_c, the real aperture aberrations expressed in terms of x_o, y_o, x_a and y_a are given by

$$\begin{aligned} x_c - s_{xc}x_o &= x_a\{t_{xc} + (30)x_a^2 + (12)y_a^2\} \\ y_c - s_{yc}y_o &= y_a\{t_{yc} + (03)y_a^2 + (21)x_a^2\} \end{aligned} \qquad (29.34)$$

in which we have shortened $(00rs)$ to (rs). As in Chapter 24, we consider a family of rays intersecting the aperture plane around a circle, $x_a = r_a \cos\theta$, $y_a = r_a \sin\theta$. We find

$$\begin{aligned} x_c - s_{xc}x_o &= r_a\{t_{xc} + (30)r_a^2\}\cos\theta + r_a^3\{(12) - (30)\}\sin^2\theta\cos\theta \\ y_c - s_{yc}y_o &= r_a\{t_{yc} + (03)r_a^2\}\sin\theta + r_a^3\{(21) - (03)\}\cos^2\theta\sin\theta \end{aligned} \qquad (29.35)$$

or writing

$$\begin{aligned} \delta x &:= x_c - s_{xc}x_o =: (\kappa + \lambda\sin^2\theta)\cos\theta \\ \delta y &:= y_c - s_{yc}y_o =: (\mu + \nu\cos^2\theta)\sin\theta \end{aligned} \qquad (29.36)$$

we obtain
$$\delta X = (1 + \epsilon + \alpha \sin^2 \theta) \cos \theta$$
$$\delta Y = (1 - \epsilon + \alpha \cos^2 \theta) \sin \theta \qquad (29.37)$$

with
$$\delta X = \frac{2\nu}{\kappa\nu + \lambda\mu} \delta x \qquad \delta Y = \frac{2\lambda}{\kappa\nu + \lambda\mu} \delta y$$
$$\epsilon = \frac{\kappa\nu - \lambda\mu}{\kappa\nu + \lambda\mu} \qquad \alpha = \frac{2\lambda\nu}{\kappa\nu + \lambda\mu} \qquad (29.38)$$

Equations (29.37) are the parametric representation of the aberration figure in a general plane. The resulting curve is always symmetric about the planes $\theta = 0$, $\theta = \pi/2$ and has an oval, star- or rosette-shaped appearance according to the relative values of α and ϵ. Figure 29.2 shows a division of the $\alpha - \epsilon$ plane into regions associated with various aberration figures, proposed by Meads (1963) (and extensively analysed in Hawkes (1966)). A classification of the aberrations for quadrupole systems producing a stigmatic but not necessarily orthomorphic image has also been proposed by Burfoot (1954a,b), who writes

$$x_i - s_{xi}x_o = x_a\{\alpha_x r_a^2 + \beta x_a^2 + \gamma(x_a^2 - y_a^2)\}$$
$$y_i - s_{yi}y_o = y_a\{\alpha_y r_a^2 - \beta y_a^2 - \gamma(x_a^2 - y_a^2)\} \qquad (29.39)$$

so that
$$\alpha_x = \frac{1}{4}\{(30) + (03) + 3(12) - (21)\}$$
$$\alpha_y = \frac{1}{4}\{(30) + (03) - (12) + 3(21)\}$$
$$\beta = \frac{1}{2}\{(30) - (03) - (12) + (21)\} \qquad (29.40)$$
$$\gamma = \frac{1}{4}\{(30) + (03) - (12) - (21)\}$$

The pair of coefficients α_x and α_y convert an image point into an ellipse which reduces to a circle, the familiar spherical aberration disc of round lenses, if $M_x = M_y$, so that $(12) = (21)$. The coefficient β measures the so-called "star" aberration, and γ the the "rosette".

The aperture aberration coefficients can be written in many different ways. We now list a few of the formulae that have been found convenient in practice.

(i) *Real aberrations*, $\phi(z)$, $p_2(z)$ and $Q_2(z)$ may all be present. For this case only, we list distortions as well as aperture aberrations, to show that

29.3 APERTURE ABERRATIONS

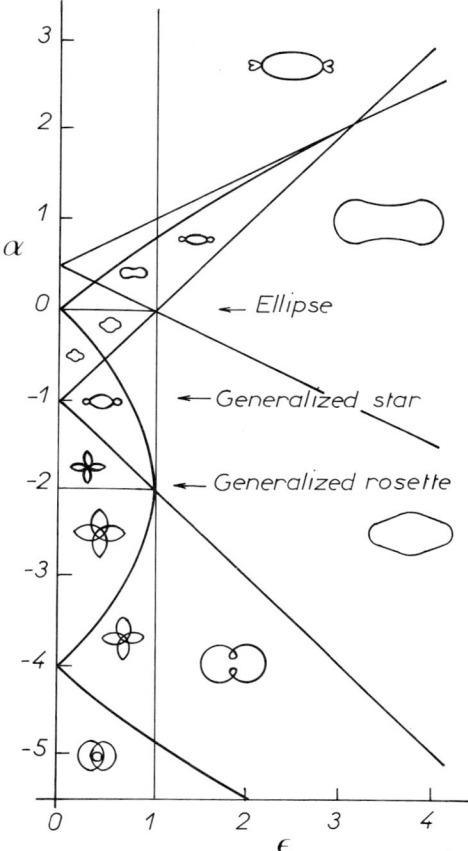

Fig. 29.2: The various forms of the aberration figure associated with the aperture aberrations.

they can be derived from one another by symmetry arguments.

$$W_x x^{(1)}(z_c) = (t_{xc} \int_a^c \sigma_{x4}\, dz - s_{xc} \int_0^c \sigma_{x3}\, dz) x_o^3$$
$$+ (t_{xc} \int_a^c \sigma_{xy}\, dz - s_{xc} \int_0^c \sigma_{12}\, dz) x_o y_o^2$$

$$+ (t_{xc} \int_a^c \tau_{x3}\, dz - s_{xc} \int_0^c \tau_{x4}\, dz) x_a^3$$

$$+ (t_{xc} \int_a^c \tau_{12}\, dz - s_{xc} \int_0^c \tau_{xy}\, dz) x_a y_a^2$$

$$W_y y^{(1)}(z_c) = (t_{yc} \int_a^c \sigma_{y4}\, dz - s_{yc} \int_0^c \sigma_{y3}\, dz) y_o^3$$

$$+ (t_{yc} \int_a^c \sigma_{xy}\, dz - s_{yc} \int_0^c \sigma_{21}\, dz) x_o^2 y_o$$

$$+ (t_{yc} \int_a^c \tau_{y3}\, dz - s_{yc} \int_0^c \tau_{y4}\, dz) y_a^3$$

$$+ (t_{yc} \int_a^c \tau_{21}\, dz - s_{yc} \int_0^c \tau_{xy}\, dz) x_a^2 y_a \qquad (29.41)$$

with $c := z_c$ and

$$\sigma_{x4} = 4(A_x s_x^4 + B_x s_x^2 s_x'^2 + N s_x'^4)\hat{\phi}_i^{\frac{1}{2}}$$
$$\sigma_{y4} = 4(A_y s_y^4 + B_y s_y^2 s_y'^2 + N s_y'^4)\hat{\phi}_i^{\frac{1}{2}}$$
$$\tau_{x4} = 4(A_x t_x^4 + B_x t_x^2 t_x'^2 + N t_x'^4)\hat{\phi}_i^{\frac{1}{2}}$$
$$\tau_{y4} = 4(A_y t_y^4 + B_y t_y^2 t_y'^2 + N t_y'^4)\hat{\phi}_i^{\frac{1}{2}}$$
$$\sigma_{x3} = 2\hat{\phi}_i^{\frac{1}{2}}(2A_x s_x^3 t_x + B_x s_x s_x'(s_x t_x)' + 2N s_x'^3 t_x')$$
$$\sigma_{y3} = 2\hat{\phi}_i^{\frac{1}{2}}(2A_y s_y^3 t_y + B_y s_y s_y'(s_y t_y)' + 2N s_y'^3 t_y')$$
$$\tau_{x3} = 2\hat{\phi}_i^{\frac{1}{2}}(2A_x s_x t_x^3 + B_x (s_x t_x)' t_x t_x' + 2N s_x' t_x'^3)$$
$$\tau_{y3} = 2\hat{\phi}_i^{\frac{1}{2}}(2A_y s_y t_y^3 + B_y (s_y t_y)' t_y t_y' + 2N s_y' t_y'^3)$$
$$\sigma_{xy} = \hat{\phi}_i^{\frac{1}{2}}[2C s_x^2 s_y^2 + R\{s_x^2 (s_y^2)' - (s_x^2)' s_y^2\}$$
$$\qquad + 2B_x s_x^2 s_y'^2 + 2B_y s_x'^2 s_y^2 + 4N s_x'^2 s_y'^2]$$
$$\tau_{xy} = \hat{\phi}_i^{\frac{1}{2}}[2C t_x^2 t_y^2 + R\{t_x^2 (t_y^2)' - (t_x^2)' t_y^2\}$$
$$\qquad + 2B_x t_x^2 t_y'^2 + 2B_y t_x'^2 t_y^2 + 4N t_x'^2 t_y'^2]$$

29.3 APERTURE ABERRATIONS

$$\sigma_{21} = \hat{\phi}_i^{\frac{1}{2}}[2Cs_x^2 s_y t_y + R\{s_x^2(s_y t_y)' - (s_x^2)'s_y t_y\}$$
$$+ 2B_x s_x^2 s_y' t_y' + 2B_y s_x'^2 s_y t_y + 4N s_x'^2 s_y' t_y']$$

$$\tau_{21} = \hat{\phi}_i^{\frac{1}{2}}[2Ct_x^2 s_y t_y + R\{t_x^2(s_y t_y)' - (t_x^2)'s_y t_y\}$$
$$+ 2B_x t_x^2 s_y' t_y' + 2B_y t_x'^2 s_y t_y + 4N t_x'^2 s_y' t_y']$$

$$\sigma_{12} = \hat{\phi}_i^{\frac{1}{2}}[2C s_x t_x s_y^2 + R\{s_x t_x (s_y^2)' - (s_x t_x)'s_y^2\}$$
$$+ 2B_x s_x t_x s_y'^2 + 2B_y s_x' t_x' s_y^2 + 4N s_x' t_x' s_y'^2]$$

$$\tau_{12} = \hat{\phi}_i^{\frac{1}{2}}[2C s_x t_x t_y^2 + R\{s_x t_x (t_y^2)' - (s_x t_x)'t_y^2\}$$
$$+ 2B_x s_x t_x t_y'^2 + 2B_y s_x' t_x' t_y^2 + 4N s_x' t_x' t_y'^2] \qquad (29.42)$$

(ii) *Real aberrations, quadrupoles only, formulae indicating the sign of the various coefficients.*

At the real line image at which t_x vanishes, $z = z_i =: i$, we have

$$(0030) = \frac{1}{6}\int_o^i t_x'^4 \, dz + \frac{1}{6}\int_o^i \left\{ 4\left(\frac{3\gamma p_2}{\hat{\phi}} - \frac{4\eta Q_2}{\hat{\phi}^{\frac{1}{2}}}\right)^2 \right.$$
$$\left. + \frac{4p_2^2}{\hat{\phi}^2}(6-\gamma^2)\right\} t_x^4 \, dz$$

$$= \frac{1}{6}\int_o^i t_x'^4 \, dz + \frac{1}{6}(3\overline{n}^2 - 2n + 2)\beta^4 \int_o^i q^2 t_x^4 \, dz$$

$$(0012) = -\frac{1}{2}t_{xi}'^2 t_{yi} t_{yi}' + \frac{3}{2}\int_o^i t_x'^2 t_y'^2 \, dz \qquad (29.43)$$

$$- \frac{1}{8}\int_o^i \left(\frac{p_2^2}{\hat{\phi}^2} + \frac{4\eta\gamma p_2 Q_2}{\hat{\phi}^{\frac{3}{2}}} - \frac{8\eta^2 Q_2^2}{\hat{\phi}}\right) t_x^2 t_y^2 \, dz$$

$$= -\frac{1}{2}t_{xi}'^2 t_{yi} t_{yi}' + \frac{3}{2}\int_o^i t_x'^2 t_y'^2 \, dz$$

$$- \frac{1}{2}(\overline{n}^2 - 2n - 2)\beta^4 \int_o^i q^2 t_x^2 t_y^2 \, dz$$

At the real line image at which t_y vanishes, $z = z_j$, (0003) and (0021) are obtained from (29.43) by interchanging t_x and t_y. If the image is stigmatic, $z_i = z_j$, then (0012) = (0021). As before, $\overline{n} = n/\gamma$.

It is clear from (29.43) that (0030) and (0003) cannot change sign in the nonrelativistic approximation ($\gamma = 1$) and that the mixed coefficients cannot change sign for stigmatic magnetic quadrupole systems (Ovsyannikova and Yavor, 1965; Moses, 1966; Hawkes, 1966/67b). Still in the nonrelativistic case, (0012)=(0021) is always positive in stigmatic systems provided that $1 - \sqrt{3} \le n \le 1 + \sqrt{3}$; this range includes magnetic ($n = 0$) and achromatic ($n = 1$) systems. When the relativistic correction becomes large, however, (0030) and (0003) can vanish as a negative term appears in the integrand when $6 - \gamma^2 \le 0$, that is, when the accelerating voltage exceeds about 0.7 MV (Rose, 1967).

(iii) *Quadrupoles and octopoles only* ($\phi = $ const). Asymptotic aberrations expressed in terms of x_o and x'_o in $z = z_{ox}$, y_o and y'_o in $z = z_{oy}$ and evaluated at the asymptotic image planes $z = z_{ix}$ and $z = z_{iy}$ conjugate to $z = z_{ox}$ and $z = z_{oy}$ respectively.

$$(0030) = \int_{-\infty}^{\infty} (F - \overline{F}) H_x^4 \, dz$$

$$(0003) = \int_{-\infty}^{\infty} (F + \overline{F}) H_y^4 \, dz$$

$$(0012) = \int_{-\infty}^{\infty} \{K H_x^2 H_y^2 + L H_x H_y (H_x H'_y - H'_x H_y)\} \, dz$$
$$+ \frac{1}{4}(d_i m_x^2 m_y^2 - d_o)$$

$$(0021) = \int_{-\infty}^{\infty} \{K H_x^2 H_y^2 + L H_x H_y (H_x H'_y - H'_x H_y)\} \, dz$$
$$- \frac{1}{4}(d_i m_x^2 m_y^2 - d_o) \tag{29.44}$$

where

$$F = \frac{1}{6\hat{\phi}^2}(2p_2^2 - 4\eta\gamma p_2 Q_2 \hat{\phi}^{\frac{1}{2}} + 3\eta^2 Q_2^2 \hat{\phi}) + \frac{5\epsilon p_2^2}{6\hat{\phi}} - 2\Xi$$

$$= \frac{\beta^4 q^2}{6}(3\bar{n}^2 - 2n + 3) - 2\tau\omega$$

$$\overline{F} = \frac{1}{48\hat{\phi}}(\gamma p_2'' - 2\eta Q_2'' \hat{\phi}^{\frac{1}{2}}) = -\frac{1}{24}\beta^2 q''$$

$$K = -\frac{1}{2\hat{\phi}^2}(\gamma p_2 - \eta Q_2 \hat{\phi}^{\frac{1}{2}})^2 + \frac{\epsilon p_2^2}{2\hat{\phi}} + 6\Xi$$

29.3 APERTURE ABERRATIONS

$$= -\frac{1}{2}\beta^4 q^2(\overline{n}^2 - 2n + 1) + 6\tau\omega$$

$$L = \frac{3}{8}\frac{\gamma p_2' - 2\eta Q_2' \hat{\phi}^{\frac{1}{2}}}{\hat{\phi}} = -\frac{3}{4}\beta^2 q_2'$$

$$d_i = z_{ix} - z_{iy}$$

$$d_o = z_{ox} - z_{oy} \tag{29.45}$$

(iv) *Quadrupoles and octopoles only* ($\phi = const$). As (ii) but free of derivatives of p_2 and Q_2.

$$(0030) = \int_{-\infty}^{\infty} \left\{ \frac{1}{24}\left(\frac{7 + 16\epsilon\hat{\phi}}{\hat{\phi}}p_2^2 - 12\eta\gamma\frac{p_2 Q_2}{\hat{\phi}^{\frac{3}{2}}} + \frac{8\eta^2 Q_2^2}{\hat{\phi}}\right)H_x^4 \right.$$

$$\left. - \frac{1}{4}\left(\frac{\gamma p_2}{\hat{\phi}} - \frac{2\eta Q_2}{\hat{\phi}^{\frac{1}{2}}}\right)H_x^2 H_x'^2 \right\} dz - 2\int_{-\infty}^{\infty} \Xi H_x^4 \, dz$$

$$= \int_{-\infty}^{\infty} \{\frac{1}{6}(3\overline{n}^2 - 2n + 2)\beta^4 q^2 H_x^4 \, dz + \frac{1}{2}\beta^2 q H_x^2 H_x'^2\} \, dz$$

$$- 2\tau \int_{-\infty}^{\infty} \omega H_x^4 \, dz$$

$$(0003) = \int_{-\infty}^{\infty} \left\{ \frac{1}{24}\left(\frac{7 + 16\epsilon\hat{\phi}}{\hat{\phi}}p_2^2 - 12\eta\gamma\frac{p_2 Q_2}{\hat{\phi}^{\frac{3}{2}}} + \frac{8\eta^2 Q_2^2}{\hat{\phi}}\right)H_y^4 \right.$$

$$\left. + \frac{1}{4}\left(\frac{\gamma p_2}{\hat{\phi}} - \frac{2\eta Q_2}{\hat{\phi}^{\frac{1}{2}}}\right)H_y^2 H_y'^2 \right\} dz - 2\int_{-\infty}^{\infty} \Xi H_y^4 \, dz$$

$$= \int_{-\infty}^{\infty} \{\frac{1}{6}(3\overline{n}^2 - 2n + 2)\beta^4 q^2 H_y^4 \, dz - \frac{1}{2}\beta^2 q H_y^2 H_y'^2\} \, dz$$

$$- 2\tau \int_{-\infty}^{\infty} \omega H_y^4 \, dz$$

$$\begin{matrix}(0012)\\(0021)\end{matrix} = -\frac{1}{8}\int_{-\infty}^{\infty} \left\{ \left(\frac{p_2^2}{\hat{\phi}^2} + \frac{4\eta\gamma p_2 Q_2}{\hat{\phi}^{\frac{3}{2}}} - \frac{8\eta^2 Q_2^2}{\hat{\phi}}\right)H_x^2 H_y^2 \right.$$

$$\left. + 3\left(\frac{\gamma p_2}{\hat{\phi}} - \frac{2\eta Q_2}{\hat{\phi}^{\frac{1}{2}}}\right)(H_x^2 H_y'^2 - H_x'^2 H_y^2) \right\} dz + 6\int_{-\infty}^{\infty} \Xi H_x^2 H_y^2 \, dz$$

$$\pm \frac{1}{4}(d_i m_x^2 m_y^2 - d_o)$$

$$= -\int_{-\infty}^{\infty} \left\{ \left(\frac{\overline{n}}{2} - n - 1\right) \beta^4 q^2 H_x^2 H_y^2 + \frac{3}{4} \beta^2 q (H_x^2 H_y'^2 - H_x'^2 H_y^2) \right\} dz$$

$$+ 6\tau \int_{-\infty}^{\infty} \omega H_x^2 H_y^2 \, dz \pm \frac{1}{4}(d_i m_x^2 m_y^2 - d_o) \tag{29.46}$$

29.4 Chromatic aberrations

In the general case, in which $\phi(z)$, $p_2(z)$ and $Q_2(z)$ may all be present, the asymptotic chromatic aberration is given in the planes z_{ix} and z_{iy} conjugate to z_{ox} and z_{oy} respectively by the following expressions:

$$\begin{aligned}\Delta x_i &= (C_{cx} x'_o + C_{Mx} x_o) \frac{\Delta \phi}{\hat{\phi}_0} \\ \Delta y_i &= (C_{cy} y'_o + C_{My} y_o) \frac{\Delta \phi}{\hat{\phi}_0}\end{aligned} \tag{29.47}$$

where

$$\begin{aligned} C_{cx} &= \int_{-\infty}^{\infty} (k_\phi + k_q) H_x^2 \, dz \\ C_{cy} &= \int_{-\infty}^{\infty} (k_\phi - k_q) H_y^2 \, dz \\ C_{Mx} &= \int_{-\infty}^{\infty} (k_\phi + k_q) G_x H_x \, dz \\ C_{My} &= \int_{-\infty}^{\infty} (k_\phi - k_q) G_y H_y \, dz \end{aligned} \tag{29.48}$$

and

$$\begin{aligned} k_\phi &= -\left(\frac{\hat{\phi}_0}{\hat{\phi}}\right)^{\frac{1}{2}} \frac{\gamma(5+\gamma^2)\phi'^2}{16 \hat{\phi}^2} \\ k_q &= \left(\frac{\hat{\phi}_0}{\hat{\phi}}\right)^{\frac{1}{2}} \frac{(1+\gamma^2)p_2 - 2\eta\gamma Q_2 \hat{\phi}^{\frac{1}{2}}}{4\hat{\phi}} \end{aligned} \tag{29.49}$$

With the notation of (29.25–29.29), the coefficients (29.48) take the following form in the pure quadrupole case:

$$C_{cx} = \frac{1}{2}\nu\beta^2 \int_{-\infty}^{\infty} q(z)H_x^2 \, dz \, ; \qquad C_{cy} = -\frac{1}{2}\nu\beta^2 \int_{-\infty}^{\infty} q(z)H_y^2 \, dz \, ;$$

$$C_{Mx} = \frac{1}{2}\nu\beta^2 \int_{-\infty}^{\infty} q(z)G_x H_x \, dz \, ; \qquad C_{My} = -\frac{1}{2}\nu\beta^2 \int_{-\infty}^{\infty} q(z)G_y H_y \, dz \, ;$$

(29.50)

in which

$$\nu = \frac{n - \gamma^2}{\gamma} \approx n - 1 \qquad (29.51)$$

Thus if $n = \gamma^2 \approx 1$, which implies $\beta_M^2 = 2\beta_E^2$, a combined electrostatic and magnetic quadrupole will be *achromatic*, provided that the field functions $p_2(z)$ and $Q_2(z)$ are identical. Figure 29.3 shows a quadrupole in which this condition is satisfied to a very good approximation.

This achromatic condition was first derived by Kel'man and Yavor (1961) and rediscovered independently by Septier (1963). The relativistic formulae were first given by Hawkes (1964, 1965c). Extensive experimental work on mixed magnetic and electromagnetic quadrupoles was subsequently performed by Hardy (1967). The achromatic condition was originally established for real aberration coefficients but these are identical with the asymptotic ones apart from the presence of different pairs of paraxial solutions. The condition is thus applicable in both cases.

In a practical design, it is more than likely that $p_2(z)$ and $Q_2(z)$ will not have quite the same shape. The effect of small differences has been analysed with the aid of a particular field model, the rectangular distribution (see Chapter 39) by Shpak and Yavor (1964, 1965) (cf. Yavor et al., 1964).

29.5 Quadrupole multiplets

Since quadrupole lenses have a diverging action in one plane and a converging effect in the other, they are commonly combined into multiplets. Formulae from which the total aberration coefficients can be calculated, given those of the individual components, are therefore required. These have a strong family resemblance to their counterparts for round lenses, listed in Chapter 27; we give only the principal expressions here, referring to Hawkes (1970a) for further details.

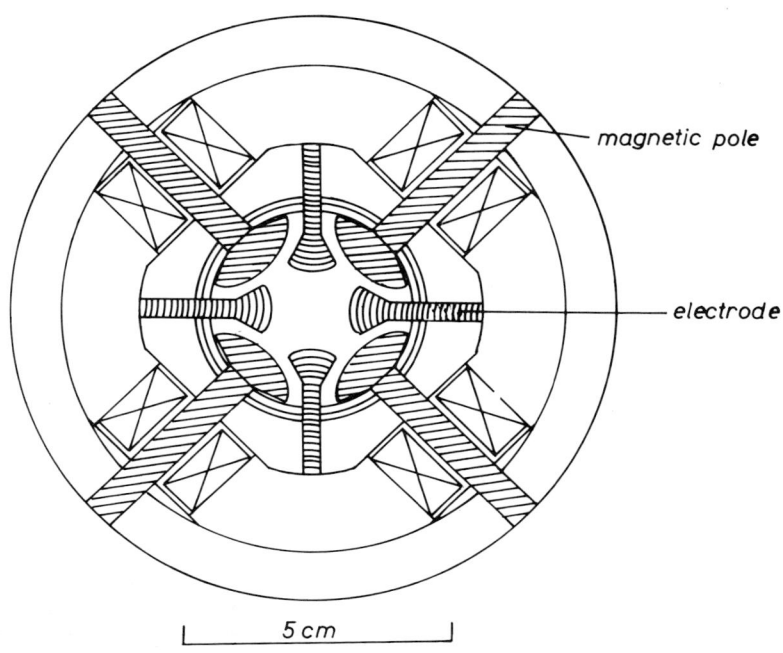

Fig. 29.3a: A combined electrostatic and magnetic quadrupole, with which the achromatic condition can be satisfied to a very good approximation. (a) Cross-section and (b) view of such a lens. Courtesy D.F. Hardy.

As in Chapter 27, we introduce vectors,

$$\boldsymbol{x} = (x \ x' \ x^3 \ xy^2 \ x^2x' \ y^2x' \ xyy' \ xx'^2 \ xy'^2 \ yx'y' \ x'^3 \ x'y'^2)^T$$
$$\boldsymbol{y} = (y \ y' \ y^3 \ x^2y \ y^2y' \ x^2y' \ xyx' \ yy'^2 \ yx'^2 \ xx'y' \ y'^3 \ x'^2y')^T$$
(29.52)

(in which T denotes transpose) with the convention that suffix o attached to x, x' indicates quantities in the object plane z_{ox}, but attached to y, y', quantities in $z = z_{oy}$ are meant. We write

$$\boldsymbol{x}_m = \boldsymbol{M}\boldsymbol{x}_o \qquad \boldsymbol{y}_n = \boldsymbol{N}\boldsymbol{y}_o \qquad (29.53)$$

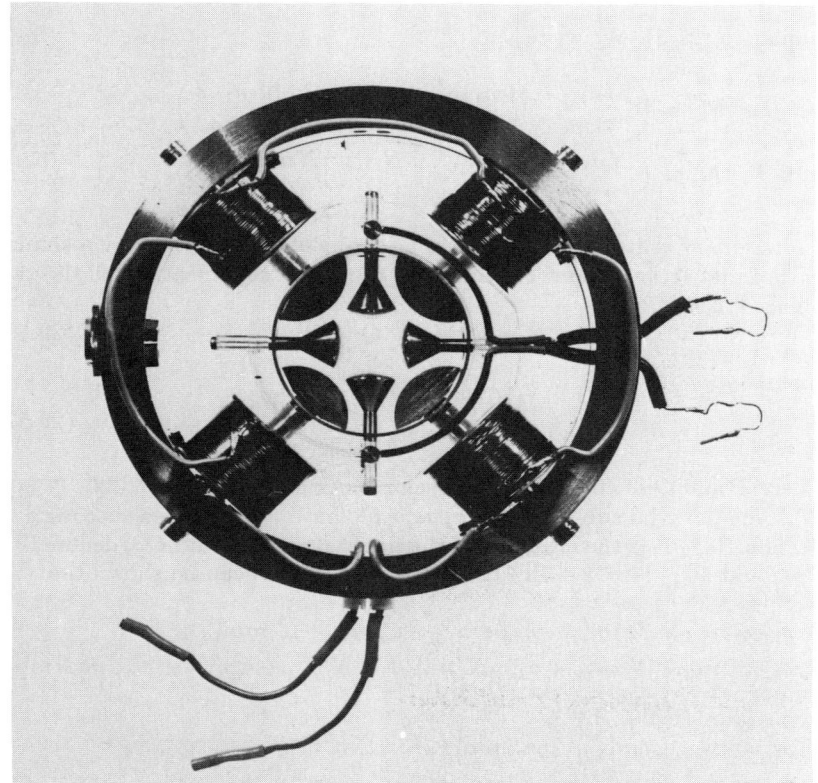

Fig. 29.3b: View of a combined electrostatic and magnetic quadrupole.

in which M_x is now denoted by M and M_y by N, to prevent undue proliferation of the suffixes. The transfer matrices M and N relating z_{ix} and z_{ox}, z_{iy} and z_{oy} respectively, have the block structure of (27.3):

$$M = \begin{pmatrix} M_1 & M_2 \\ M_3 & M_4 \end{pmatrix} \qquad N = \begin{pmatrix} N_1 & N_2 \\ N_3 & N_4 \end{pmatrix} \qquad (29.54)$$

Here M_1 and N_1 are the 2×2 paraxial transfer matrices in the $x - z$ and $y - z$ planes; M_3 and N_3 are null; M_4 and N_4 encode the addition rules for the aberration coefficients; and M_2, N_2 are the 2×10 aberration matrices similar to (27.6). We write

$$M_2 = \begin{pmatrix} Mm_{11} & Mm_{12} & \ldots & Mm_{19} & Mm_{1,10} \\ m_{21} & m_{22} & \ldots & m_{29} & m_{2,10} \end{pmatrix}$$
$$N_2 = \begin{pmatrix} Nn_{11} & Nn_{12} & \ldots & Nn_{19} & Nn_{1,10} \\ n_{21} & n_{22} & \ldots & n_{29} & n_{2,10} \end{pmatrix} \quad (29.55)$$

Suppose now that we have a second quadrupole characterized by a similar pair of matrices, M' and N', and that the total magnifications in the two planes are P, Q:

$$P = M'M \qquad Q = N'N \quad (29.56)$$

Then
$$\begin{aligned} \boldsymbol{x}_p &= \boldsymbol{M}'\boldsymbol{x}_m = \boldsymbol{M}'\boldsymbol{M}\,\boldsymbol{x}_o =: \boldsymbol{P}\,\boldsymbol{x}_o \\ \boldsymbol{y}_p &= \boldsymbol{N}'\boldsymbol{y}_n = \boldsymbol{N}'\boldsymbol{N}\,\boldsymbol{y}_o =: \boldsymbol{Q}\,\boldsymbol{y}_o \end{aligned} \quad (29.57)$$

where $\boldsymbol{P}$ and $\boldsymbol{Q}$ of course have the same block structure as $\boldsymbol{M}$, $\boldsymbol{M}'$, $\boldsymbol{N}$ and $\boldsymbol{N}'$. The paraxial sub-matrices $\boldsymbol{P}_1$ and $\boldsymbol{Q}_1$ have already been discussed in Section 19.1. It is the elements of the upper row of $\boldsymbol{P}_2$ and $\boldsymbol{Q}_2$, denoted by Pp_{1j} and Qq_{1j} ($j = 1-10$), that are of interest. It can be shown that

$$\begin{aligned}
p_{11} &= m_{11} + M^2 m'_{11} + c_m M m'_{13} + c_m^2 m'_{16} + c_m^3 m m'_{19} \\
p_{12} &= m_{12} + N^2 m'_{12} + c_m m N^2 m'_{14} + c_n N m'_{15} + c_n^2 m'_{17} + c_m c_n m N m'_{18} \\
&\quad + c_m c_n^2 m m'_{1,10} \\
p_{13} &= m_{13} + r m'_{13} + 2 r c_m m m'_{16} + 3 r c_m^2 m^2 m'_{19} \\
p_{14} &= m_{14} + r m^2 N^2 m'_{14} + r c_n m^2 N m'_{18} + r c_n^2 m^2 m'_{1,10} \\
p_{15} &= m_{15} + r m'_{15} + 2 r c_n n m'_{17} + r c_m m m'_{18} + 2 r c_m c_n m n m'_{1,10} \\
p_{16} &= m_{16} + r^2 m^2 m'_{16} + 3 r^2 c_m m^3 m'_{19} \\
p_{17} &= m_{17} + r^2 n^2 m'_{17} + r^2 c_m m n^2 m'_{1,10} \\
p_{18} &= m_{18} + r^2 m^2 m'_{18} + 2 r^2 c_n m^2 n m'_{1,10} \\
p_{19} &= m_{19} + r^3 m^4 m'_{19} \\
p_{1,10} &= m_{1,10} + r^3 m^2 n^2 m'_{1,10}
\end{aligned}$$

$$\begin{aligned}
q_{11} &= n_{11} + N^2 n'_{11} + c_n N n'_{13} + c_n^2 n'_{16} + c_n^3 n n'_{19} \\
q_{12} &= n_{12} + M^2 n'_{12} + c_n M^2 n n'_{14} + c_m M n'_{15} + c_m^2 n'_{17} + c_m c_n M n n'_{18} \\
&\quad + c_n c_m^2 n n'_{1,10} \\
q_{13} &= n_{13} + r n'_{13} + 2 r c_n n n'_{16} + 3 r c_n^2 n^2 n'_{19} \\
q_{14} &= n_{14} + r M^2 n^2 n'_{14} + r c_m M n^2 n'_{18} + r c_m^2 n^2 n'_{1,10}
\end{aligned}$$

$$q_{15} = n_{15} + rn'_{15} + 2rc_m mn'_{17} + rc_n nn'_{18} + 2rc_m c_n mnn'_{1,10}$$
$$q_{16} = n_{16} + r^2 n^2 n'_{16} + 3r^2 c_n n^3 n'_{19}$$
$$q_{17} = n_{17} + r^2 m^2 n'_{17} + r^2 c_n m^2 nn'_{1,10}$$
$$q_{18} = n_{18} + r^2 n^2 n'_{18} + 2r^2 c_m n^2 mn'_{1,10}$$
$$q_{19} = n_{19} + r^3 n^4 n'_{19}$$
$$q_{1,10} = n_{1,10} + r^3 m^2 n^2 n'_{1,10} \tag{29.58}$$

in which $c_m := -1/f_{xi}$ and $c_n = -1/f_{yi}$ as in (27.5).

The coefficients m_{1j}, n_{1j} m'_{1j}, n'_{1j}, and hence p_{1j} and q_{1j} also, may be written as polynomials in the corresponding inverse magnification m and n (for m_{1j} and n_{1j}), m' and n' (for m'_{1j} and n'_{1j}) and p and q (for p_{1j} and q_{1j}). The formulae giving the numerous coefficients that occur in these polynomials for the complete system in terms of those of the two individual lenses (or partial systems) are set out explicitly in Hawkes (1970e) and are not reproduced here.

30
The Aberrations of Cylindrical Lenses

The aberrations of cylindrical lenses, the paraxial properties of which were outlined in Chapter 20, have been analysed by several authors. For electrostatic lenses, they were first calculated by Leitner (1942), who used the method of variation of parameters to obtain formulae in several different forms for all the aberration coefficients. Strashkevich (1965) likewise listed formulae for the latter, some of which had appeared earlier in Strashkevich and Pilat (1951, 1952) and Strashkevich and Gluzman (1954). Full expressions for the primary aberrations of combined electrostatic and magnetic lenses were derived by Shtepa (1952), Laudet (1955a,b) and Rheinfurth (1955). Shi (1956) used Seman's formulation of the method of characteristic functions to analyse electrostatic lenses, employing the reduced equations of motion and eliminating derivatives of the Gaussian solutions. The coefficients, recalculated using characteristic functions, are listed in Hawkes (1966/67a), where a comparison with all the earlier formulae is to be found. For details of the calculation, the reader is referred to that paper; here we merely indicate the form of the results in the general case and list the full formulae only for electrostatic cylindrical lenses. We express the aberrations in terms of object and aperture coordinates; for details of the alternative choice, again see the paper cited above.

The paraxial solution expressed in terms of position coordinates in the object and aperture planes was not derived explicitly in Chapter 20. With $s(z_o) = t(z_a) = 1$ and $t(z_o) = s(z_a) = 0$, a lengthy but straightforward calculation yields

$$x(z) = -r(z)(y_a - y_o) + x_o \left\{ s(z) + \eta r(z) \int_o^a \frac{Bs}{\phi^{\frac{1}{2}}} \, dz \right\}$$

$$+ x_a \left\{ t(z) + \eta r(z) \int_o^a \frac{Bt}{\phi^{\frac{1}{2}}} \, dz \right\} \qquad (30.1)$$

$$y(z) = \{1 - R(z)\} y_o + R y_a + S x_o + T x_a$$

with

$$r(z) := -\frac{\eta}{FW} \left(s \int_o^z \frac{Bt}{\phi^{\frac{1}{2}}} \, d\zeta - t \int_o^z \frac{Bs}{\phi^{\frac{1}{2}}} \, d\zeta \right)$$

30. THE ABERRATIONS OF CYLINDRICAL LENSES

$$R(z) := \frac{1}{F} \int_0^z \frac{d\zeta}{\phi^{\frac{1}{2}}}$$

$$S(z) := \eta \int_0^z \frac{Bs}{\phi^{\frac{1}{2}}} d\zeta - \frac{\eta}{F} \int_0^a \frac{Bs}{\phi^{\frac{1}{2}}} dz \int_0^z \frac{d\zeta}{\phi^{\frac{1}{2}}}$$

$$+ \eta^2 \int_0^a \frac{Bs}{\phi^{\frac{1}{2}}} dz \int_0^z \frac{Br}{\phi^{\frac{1}{2}}} d\zeta$$

$$T(z) := \eta \int_0^z \frac{Bt}{\phi^{\frac{1}{2}}} d\zeta - \frac{\eta}{F} \int_0^a \frac{Bt}{\phi^{\frac{1}{2}}} dz \int_0^z \frac{d\zeta}{\phi^{\frac{1}{2}}}$$

$$+ \eta^2 \int_0^a \frac{Bt}{\phi^{\frac{1}{2}}} dz \int_0^z \frac{Br}{\phi^{\frac{1}{2}}} d\zeta$$

$$F := \int_0^a \frac{dz}{\phi^{\frac{1}{2}}} + \frac{\eta^2}{W} \int_0^a \frac{Bs}{\phi^{\frac{1}{2}}} \left(\int_0^z \frac{Bt}{\phi^{\frac{1}{2}}} d\zeta \right) dz$$

$$- \frac{\eta^2}{W} \int_0^a \frac{Bt}{\phi^{\frac{1}{2}}} \left(\int_0^z \frac{Bs}{\phi^{\frac{1}{2}}} d\zeta \right) dz$$

$$W := \phi^{\frac{1}{2}}(st' - s't) \tag{30.2}$$

The expressions (30.1) are substituted into the perturbation $M^{(4)}$ (M^I in Chapter 22), and aberration coefficients are extracted by partial differentiation (22.30). In the general case, the geometrical aberrations in an arbitrary plane take the form

$$\begin{aligned}x^{(1)}(z) = & (300)_x x_o^3 + (210)_x x_o^2 x_a + (201)_x x_o^2 Y \\ & + (120)_x x_o x_a^2 + (102)_x x_o Y^2 + (111)_x x_o x_a Y \\ & + (030)_x x_a^3 + (021)_x x_a^2 Y + (012)_x x_a Y^2 + (003)_x Y^3\end{aligned} \tag{30.3}$$

with a similar expression for $y^{(1)}(z)$. We have introduced $Y := y_a - y_o$. The formulae for the coefficients are extremely complicated in the mixed case, in which electrostatic and magnetic fields are both present; they are listed in full in Hawkes (1966/67a) and are not reproduced here. In the simpler case of electrostatic cylindrical lenses, there are fewer terms. The results are then as follows.

The function $M^{(4)}$ is given by

$$M^{(4)} = \left(\frac{\phi^{(4)}}{48\phi^{\frac{1}{2}}} - \frac{\phi''^2}{32\phi^{\frac{3}{2}}} \right) x^4 - \frac{\phi''}{8\phi^{\frac{1}{2}}} x^2(x'^2 + y'^2) - \frac{1}{8} \phi^{\frac{1}{2}}(x'^2 + y'^2)^2 \tag{30.4}$$

30. THE ABERRATIONS OF CYLINDRICAL LENSES

in the nonrelativistic approximation. Into this, we substitute

$$x = sx_o + tx_a$$
$$y = (1 - R)y_o + Ry_a \tag{30.5}$$

where

$$R = \frac{R_z}{F} \qquad R_z := \int_{z_o}^{z} \frac{d\zeta}{\phi^{\frac{1}{2}}}, \qquad F := \int_{z_o}^{z_a} \frac{dz}{\phi^{\frac{1}{2}}} \tag{30.6}$$

giving

$$x^{(1)}(z_c) = (300)x_o^3 + (030)x_a^3 + (120)x_o x_a^2 + (210)x_o x_a$$
$$\quad + (102)x_o Y^2 + (012)x_a Y^2 \tag{30.7}$$
$$y^{(1)}(z_c) = (201)x_o^2 Y + (111)x_o x_a Y + (021)x_a^2 Y + (003)Y^3$$

in some arbitrary plane $z = z_c$. The coefficients are given by

$$(300) = \frac{t}{2W} \int_a^c \left\{ \left(\frac{\phi^{(4)}}{6\phi^{\frac{1}{2}}} - \frac{\phi''^2}{4\phi^{\frac{3}{2}}} \right) s^4 - \frac{\phi''}{\phi^{\frac{1}{2}}} s^2 s'^2 - \phi^{\frac{1}{2}} s'^4 \right\} dz$$

$$- \frac{s}{2W} \int_o^c \left\{ \left(\frac{\phi^{(4)}}{6\phi^{\frac{1}{2}}} - \frac{\phi''^2}{4\phi^{\frac{3}{2}}} \right) s^3 t - \frac{\phi''}{\phi^{\frac{1}{2}}} (s^2)'(st)' - \phi^{\frac{1}{2}} s'^3 t' \right\} dz$$

and $(300) \to (030)$ when $s \to t$, $t \to -s$, $o \leftrightarrow a$;

$$(102) = -\frac{t}{4F^2} \int_a^c \left(\frac{\phi''}{\phi^{\frac{3}{2}}} s^2 + 2\phi^{-\frac{1}{2}} s'^2 \right) dz$$

$$+ \frac{s}{4F^2} \int_o^c \left(\frac{\phi''}{\phi^{\frac{3}{2}}} st + 2\phi^{-\frac{1}{2}} s't' \right) dz$$

and $(102) \to (012)$ when $s \to t$, $t \to -s$, $o \leftrightarrow a$;

$$(120) = \frac{3t}{4W} \int_a^c \left\{ \left(\frac{\phi^{(4)}}{3\phi^{\frac{1}{2}}} - \frac{\phi''^2}{2\phi^{\frac{3}{2}}} \right) s^2 t^2 \right.$$
$$\left. - \frac{\phi''}{3\phi^{\frac{1}{2}}} (s^2 t'^2 + 4ss'tt' + s'^2 t^2) - 2\phi^{\frac{1}{2}} s'^2 t'^2 \right\} dz$$

$$- \frac{3s}{4W} \int_o^c \left\{ \left(\frac{\phi^{(4)}}{3\phi^{\frac{1}{2}}} - \frac{\phi''^2}{2\phi^{\frac{3}{2}}} \right) st^3 - \frac{\phi''}{\phi^{\frac{1}{2}}} (st)'tt' - 2\phi^{\frac{1}{2}} s't'^3 \right\} dz$$

30. THE ABERRATIONS OF CYLINDRICAL LENSES

and $(120) \to (210)$ when $s \to t$, $t \to -s$, $o \leftrightarrow a$

$$(003) = \frac{1}{2F^4}\left(F\int_0^c \frac{dz}{\phi^{\frac{3}{2}}} - R_z \int_0^c \frac{dz}{\phi^{\frac{3}{2}}}\right)$$

$$(201) = \frac{1}{4F}\int_0^c \left(\frac{\phi''}{\phi^{\frac{3}{2}}}s^2 + 2\phi^{-\frac{1}{2}}s'^2\right) dz$$

$$-\frac{R_z}{4F}\int_0^c \left(\frac{\phi''}{\phi^{\frac{3}{2}}}s^2 + 2\phi^{-\frac{1}{2}}s'^2\right) dz$$

and $(201) \to (111)$ when $s^2 \to 2st$, $s'^2 \to 2s't'$ while $(201) \to (021)$ when $s^2 \to t^2$ and $s'^2 \to t'^2$.

31
Parasitic Aberrations

We have so far been assuming that the electron optical system under consideration is perfect, in the sense that the various symmetries are exactly respected. This is certainly unrealistic, for every real system is imperfect: there is a limit to the precision attainable during the manufacture and assembly of its components. For instance, the electrodes and polepieces of a lens system designed to have rotational symmetry are never truly round but slightly elliptical, and after assembly their individual axes are shifted and tilted with respect to one another. Moreover, in magnetic lenses, the polepiece material will be slightly inhomogeneous, and in electrostatic lenses, an asymmetric layer of insulating but charged contamination may be deposited on the electrode surfaces, again impairing the original symmetry. It is clear that the number of degrees of freedom in the imperfections increases drastically with the number of individual components in the device, so that complex systems are particularly prone to problems of this kind.

The electron optical defects caused by the various forms of imperfection are known as *parasitic aberrations*. Unlike the systematic aberrations of perfect systems, they cannot be determined accurately since the mechanical defects that cause them are never known exactly. The aim of the theory is to furnish an understanding of the possible types of parasitic aberrations and to establish tolerance limits on the precision of machining and alignment of the various components of a system.

31.1 Small deviations from rotational symmetry; axial astigmatism

This situation is the most important, for in the vast majority of practical devices, the lens systems are intended to be round. Any deviations of the boundaries from the ideal rotational symmetry generate weak multipole fields of various orders, as outlined in Section 9.4.6 where magnetic lenses were considered. From the properties of multipole fields presented in Chapter 7, it is obvious that the fields with the lowest multipole orders, $m = 1, 2$ and 3, will be the most important in the paraxial domain, as their potentials increase as r^m with the radial coordinate r.

Eccentricities, in the sense of shifts and tilts of the local optic axes, produce very weak dipole fields ($m = 1$), which merely cause a very small

31.1 SMALL DEVIATIONS FROM ROTATIONAL SYMMETRY;

lateral deflection of the whole electron beam. The image is hence displaced bodily but remains stigmatic, with the result that these defects are not serious as long as they remain small.

Quadrupole fields ($m = 2$) are caused by ellipticity of the electrodes or polepieces and produce an astigmatism; the latter does not vanish even for the axial object point and is therefore often called the *axial* astigmatism. Owing to its great practical importance, this aberration was studied comprehensively in earlier decades. Substantial contributions were made by Glaser (1942/3), Bertein (1947e,1948a), Bertein *et al.* (1947), Hillier and Ramberg (1947), Rang (1949a,b), Sturrock (1951b,1955), which should be read in conjunction with Archard (1953), Glaser (1952), Glaser and Schiske (1953), Der-Shvarts (1954) and Stoyanov (1955). With the advance of modern computational methods, these early investigations have lost much of their significance and we discuss them only briefly. It should, however, not be forgotten that the idea of analysing departures from rotational symmetry into Fourier components, thereby revealing the character of the aberrations to be expected, first appeared in these studies, notably those of Bertein and Sturrock; this in turn led to the idea of the *stigmator* and permitted Archard to relate individual types of imperfection of magnetic lenses (corrugation of the polepieces, eccentricity and misorientation of the axes of the latter) to particular image defects.

It is customary to introduce dimensionless ellipticity parameters $\epsilon_1(z)$, $\epsilon_2(z)$; the electrostatic potential Φ in the paraxial domain then takes the form

$$\Phi(z,r,\varphi) = \phi(z) - \frac{r}{4}\phi''(z)\{1 + \epsilon_1(z)\cos 2\varphi + \epsilon_2(z)\sin 2\varphi\} + O(r^4) \quad (31.1)$$

This represents the superposition of a dominant round focusing field and two weak quadrupole fields with variable strengths and different orientations; we must assume that $|\epsilon_1| \ll 1$, $|\epsilon_2| \ll 1$. In the case of *constant* ellipticity, $\epsilon_1 = \text{const}$, $\epsilon_2 = \text{const}'$, it is possible to relate these parameters to the semiaxes $a > b$ of the bores in the electrodes. Without loss of generality, we introduce a constant rotation of the coordinate system about the optic axis, so that $\epsilon_1 =: \bar{\epsilon} > 0$, $\epsilon_2 = 0$. We then have

$$\bar{\epsilon} = \frac{a-b}{a+b} \quad (31.2)$$

The motions of the electrons in the two symmetry planes, $X_1 = 0$ and $X_2 = 0$, are decoupled and each coordinate satisfies a paraxial ray equation of the type associated with quadrupoles:

$$\phi X''_{1,2} + \frac{1}{2}\phi' X'_{1,2} + \frac{1}{4}\phi''(1 \pm \bar{\epsilon})X_{1,2} = 0 \quad (31.3)$$

With the Picht transformation (15.40), $X_j = v_j \phi^{-1/4}$, this becomes

$$v''_{1,2} + \left\{\frac{3}{16}\left(\frac{\phi'}{\phi}\right)^2 \pm \frac{\bar{\epsilon}\phi''}{4\phi}\right\} v_{1,2} = 0 \tag{31.4}$$

In the Busch approximation for weak lenses, we first find the average focal length $\bar{f}$,

$$1/\bar{f} = (f_i f_o)^{-\frac{1}{2}} = \frac{3}{16} \int_{-\infty}^{\infty} \left(\frac{\phi'}{\phi}\right)^2 dz \tag{31.5}$$

The ellipticity causes small deviations from $\bar{f}$, given by

$$\Delta(1/\bar{f}) = \pm\frac{\bar{\epsilon}}{4} \int_{-\infty}^{\infty} \frac{\phi''}{\phi} dz$$

Partial integration yields an expression proportional to formula (31.5) for $\bar{f}^{-1}$ and as this deviation is very small, we can simplify the relation to

$$\Delta \bar{f} = \mp \frac{4\bar{\epsilon}}{3} \bar{f} \tag{31.6}$$

This expression is certainly an oversimplification since in most practical applications, the lenses are not weak, but it gives an estimate of the axial astigmatism and this is sufficient for rough calculations.

For the scalar potential in magnetic lenses an expansion similar to (31.1) holds and, in the case of constant ellipticity of the bores a relation of the form (31.2) can be derived, but thereafter the situation becomes far more complicated. Owing to the Larmor rotation of the electron rays in the lens, there is no coordinate system in which the ray equations decouple like eqs (31.1). We now have to employ one of the general methods for calculating aberrations. This theory is comprehensively presented by Glaser (1952); we shall not repeat it here, since more informative techniques are now available.

31.2 Classification of the parasitic aberrations

Quite generally, there must exist an eikonal or characteristic function, from which the lateral aberrations can be obtained by partial differentiation. For rotationally symmetric systems these relations can be cast into the convenient form

$$\Delta x_o = \frac{\partial}{\partial x'_o} S(x_o, y_o, x'_o, y'_o, z_o, z_i), \qquad \Delta y_o = \frac{\partial S}{\partial y'_o} \tag{31.7}$$

where S is a characteristic function, often called the 'wave aberration', and Δx_o, Δy_o are the lateral aberrations referred back to the object plane in the usual way. These relations remain valid even if only the paraxial unperturbed contributions are rotationally symmetric; small asymmetric aberrations are then superimposed on the intrinsic aberrations of the round lens.

The assumption that the asymmetries are small allows us to neglect their fourth-order contributions to S in comparison with those from the rotationally symmetric part. We can therefore terminate the series expansion of the eikonal after the *third* order if only the asymmetric aberrations are being studied. Even so, the possible aberrations are quite numerous since S is now a polynomial consisting of all terms in the variables x_o, y_o, x'_o and y'_o up to third order. We first examine the *second-order* terms. In the expansion, we omit any terms that depend on x_o and y_o only, since they do not contribute to the derivatives (31.7); we then find

$$S = x'_o(c_1 + a_{11}x_o + a_{12}y_o) + y'_o(c_2 + a_{21}x_o + a_{22}y_o)$$
$$+ \frac{1}{2}(b_{11}x'^2_o + 2b_{12}x'_oy'_o + b_{22}y'^2_o) \quad (31.8)$$

Here a_{12} is not necessarily the same as a_{21}. It is now convenient to introduce new coefficients, writing

$$a_{11} =: a'_1 + a_1, \quad a_{22} =: a'_1 - a_1, \quad a_{21} =: a_2 + a'_2, \quad a_{12} =: a_2 - a'_2,$$
$$b_{11} =: b'_1 + b_1, \quad b_{22} =: b'_1 - b_1, \quad b_{12} =: b_2$$

We can then rewrite (31.8) as the sum of a rotationally symmetric term

$$S_s = a'_1(x_ox'_o + y_oy'_o) + a'_2(x_oy'_o - x'_oy_o) + \frac{1}{2}b'^2_1(x'^2_o + y'^2_o)$$

and an essentially asymmetric contribution

$$S_a = a_1(x_ox'_o - y_oy'_o) + a_2(x_oy'_o + y_ox'_o)$$
$$+ \frac{1}{2}b_1(x'^2_o - y'^2_o) + b_2x'_oy'_o + c_1x'_o + c_2y'_o \quad (31.9)$$

with $S = S_a + S_s$. The expression for S_s, being composed of rotation-invariants, yields the aberrations to be expected in perfectly round lenses. These are an isotropic and an anisotropic alteration of the magnification (a'_1 and a'_2, respectively) and a defocus (b'_1). Since these are of no interest in the present context, we disregard them here.

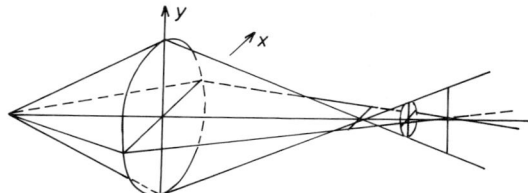

Fig. 31.1: Axial astigmatism.

The lateral aberrations, arising from (31.9), are now given by

$$\Delta x_o = c_1 + a_1 x_o + a_2 y_o + b_1 x'_o + b_2 y'_o \tag{31.10a}$$
$$\Delta y_o = c_2 + a_2 x_o - a_1 y_o + b_2 x'_o - b_1 y'_o \tag{31.10b}$$

The constants c_1 and c_2 describe a *lateral shift* of the image: $\Delta x_i = Mc_1$, $\Delta y_i = Mc_2$, M being the magnification. Such a shift results from tilts or lateral displacements of the lens parts relative to each other, which produce a weak deflection. Since the image remains stigmatic, this is not serious. The terms that are proportional to x_o and y_o describe a *paraxial distortion*. From the affine imaging relation,

$$\begin{aligned} x_i &= M(x_o + \Delta x_o) = M\{(1+a_1)x_o + a_2 y_o\} \\ y_i &= M(y_o + \Delta y_o) = M\{a_2 x_o + (1-a_1)y_o\} \end{aligned} \tag{31.11}$$

it can be seen that there are two principal directions in the image, which are orthogonal to each other and in which the magnification takes the extreme values

$$M_{1,2} = M\{1 \pm (a_1^2 + a_2^2)^{1/2}\} \tag{31.12}$$

Since $|M_1 - M_2| \ll |M|$, the deviations from uniform magnification are so small that this error is usually of very little practical concern.

The terms in x'_o and y'_o in (31.10) are the most important. By an appropriate rotation of the coordinate system about the optic axis, we can bring them into an uncoupled form:

$$\Delta \tilde{x}_o = b\tilde{x}'_o, \qquad \Delta \tilde{y}_o = -b\tilde{y}'_o, \qquad b = \sqrt{b_1^2 + b_2^2} \tag{31.13}$$

These describe the *axial astigmatism*, which is presented schematically in Fig. 31.1. Equations (31.13) represent the lateral aberration in the circle of least confusion: with $\tilde{x}'_o = \alpha \cos\varphi$ and $\tilde{y}'_o = \alpha \sin\varphi$, α being the semi-aperture angle, we find $\Delta \tilde{x}_o = b\alpha \cos\varphi$, $\Delta \tilde{y}_o = -b\alpha \sin\varphi$. Obviously the sense of rotation on the circle of least confusion is opposite to that on the aperture cone. Referred back to object space, the distance between the two line foci is $2b$.

Among the lateral aberrations of *second* order, the corresponding pure aperture term is the most important. This is described by the wave aberration term

$$S_3 = A_{3,0} x_o'^3 + A_{2,1} x_o'^2 y_o' + A_{1,2} x_o' y_o'^2 + A_{0,3} y_o'^3 \qquad (31.14)$$

with the corresponding lateral aberrations

$$\Delta x_o = 3A_{3,0} x_o'^2 + 2A_{2,1} x_o' y_o' + A_{1,2} y_o^2 \qquad (31.15a)$$
$$\Delta y_o = A_{2,1} x_o'^2 + 2A_{1,2} x_o' y_o' + 3A_{0,3} y_o^2 \qquad (31.15b)$$

This aberration consists of two components with different azimuthal symmetry, as can be seen by writing (31.14) in polar coordinates,

$$x_o' = r_o' \cos\theta, \qquad y_o' = r_o' \sin\theta$$

Expressing all powers of x_o' and y_o' in terms of trigonometric functions of multiple arguments, we find $S_3 = S_3^{(1)} + S_3^{(1)}$ with

$$S_3^{(1)} = \frac{1}{4} r_o'^3 \{(A_{3,0} + A_{1,2})\cos\theta + (A_{2,1} + 3A_{0,3})\sin\theta\} \qquad (31.16a)$$

$$S_3^{(3)} = \frac{1}{4} r_o'^3 \{(A_{3,0} - A_{1,2})\cos 3\theta + (A_{2,1} - A_{0,3})\sin 3\theta\} \qquad (31.16b)$$

The different multiplicities with respect to the azimuth θ show that the aberrations due to (31.16a) are caused by weak *deflection* fields, produced by misalignments, while the aberrations corresponding to (31.16b) are caused by *threefold* deviations from the rotational symmetry. The latter are produced by corresponding threefold corrugations of the electrodes or polepieces. We shall call these two different aberrations the deflection astigmatism and the threefold astigmatism, respectively. They are hardly ever seen in their pure forms, since they are practically always superimposed on the numerous other lens aberrations.

31.3 Numerical determination of parasitic aberrations

Determination of tolerance limits for the machining and alignment of electron optical systems requires calculation of the parasitic aberrations. Apart from a few oversimplified cases, this can only be accomplished numerically.

The main problem is the computation of the electromagnetic fields in the imperfect system. A rigorous calculation would proceed as follows:

1. First, the fields in the perfect system are calculated by means of the methods outlined in Part II.

2. From the solution of the appropriate boundary-value problem, the boundary values of the field strengths $E(r)$ and $H(r)$ are determined. This is often a very tedious task, since the numerical differentiation of potentials in the vicinity of their surface sources is rarely straightforward.

3. It is now necessary to adopt realistic values of the shifts, tilts, ellipticities and any other deformations and to determine from these a surface deformation $s(r)$, defined as the shift from a point r at the ideal surface to the point $r' = r + s$ at the real surface. With the necessary assumption that $|s|$ is very small, such a deformation s corresponds to an alteration $\delta\Phi = E \cdot s$ or $\delta\chi = H \cdot s$ of the potential on the *ideal* boundary.

4. With the boundary values $\delta\Phi(r)$ or $\delta\chi(r)$, the boundary-value problem in the *undeformed* domain is solved and the resulting solution is added to the unperturbed one. This procedure is usually much easier than a direct field calculation in the deformed domain, since the explicit treatment of irregularly deformed surfaces is extremely tedious.

5. A number of electron trajectories with appropriate initial conditions are then traced through the total fields. From the sets of final coordinates and slopes in a terminal (image) plane $z = z_i$, the aberration figures and—if needed—the aberration coefficients are determined. The corresponding methods are outlined in Part V. Very often the least-squares-fit methods explained there are much simpler than the evaluation of aberration integrals.

This whole procedure requires a major effort and is therefore often much simplified. The greatest simplification is obtained if the axial harmonics, the axial amplitudes of the various multipole components in the field, are determined not from the solution of a boundary-value problem but from simple analytic models. An approximate field calculation is then possible by evaluation of the corresponding radial series expansions given in Chapter 7. This is certainly advisable whenever some parts of the boundaries are not round even in the ideal case, as in electric deflectors or in quadrupole systems.

For systems with circular boundaries, such as those composed of ordinary lenses and magnetic deflectors with a round ferrite core, the more rigorous method is not too complicated. The Fourier-series method, outlined in Chapter 7, can then be applied to the potential variations $\delta\Phi(r)$ and $\delta\chi(r)$, and there remains the task of repeatedly solving the boundary-value problems for differential equations having the structure of eq. (7.10). Calculations of this kind were first carried out, with a view to determining tolerance limits for the shift $s(r)$, by Janse (1971), who demonstrated their practical applicability. For a recent numerical study starting from Sturrock's theory, see Munro (1988).

31.4 The isoplanatic approximation

The isoplanatic approximation is invoked whenever we study the aberrations that affect the imaging of very small objects. This situation arises in practice in electron microscopes operating at very high magnification, say $|M| \gtrsim 10^5$, where only a very small part of the object is visible on the viewing screen. In a perfectly round system, the best image quality would be obtained for an object in the vicinity of the straight optic axis, but since no unique axis exists in a real and imperfect instrument, the question of the best alignment is not at all trivial.

We confine the following discussion to the parasitic aberrations of first order, together with the intrinsic aberrations of a round system. We express all lateral aberrations in complex form in terms of a complex object coordinate $u_o := x_o + iy_o$ $u_A := x_A + iy_A$. The resulting lateral image aberration δu_i can then be written

$$\begin{aligned}\Delta u_i = &S + Pu_o + P'u_o^* + Qy_A + Q'u_A^* \\ &+ Cu_A^2 u_A^* + 2(K+ik)u_o u_A u_A^* + (K-ik)u_o^* u_A^2 \\ &+ (A+ia)u_o^2 u_A^* + Fu_o u_o^* u_A + (D+id)u_o^2 u_o^*\end{aligned} \quad (31.17)$$

where the aberrations of third order are those dealt with comprehensively in Chapter 24. The constant S is a small shift due to weak parasitic deflection fields. The coefficient P is the deviation of the magnification from its nominal value due to a defocus, an error in the image rotation, and to lateral chromatic aberrations

$$P = \delta M + iM\delta\theta + \hat{\phi}_i^{-1}\Delta\hat{\phi}_i(C_M + iC_\theta). \quad (31.18)$$

The term $P'u_o^*$ represents the paraxial distortion. The factor Q expresses the effect of the defocus Δ and the axial chromatic aberration and is always real:

$$Q = \Delta + \hat{\phi}_i^{-1}\Delta\hat{\phi}_i \cdot C_c \quad (31.19)$$

Finally the term $Q'u_A^*$ represents the axial astigmatism due to ellipticities.

It is a crucial feature of imperfect systems that there is no way of determining whether the aperture is aligned perfectly relative to the optic axis because there is no unique axis. It hence makes sense to assume that the centre of the aperture is *shifted laterally* relative to the z-axis of the coordinate system, $w^{(0)} := x_A^{(0)} + iy_A^{(0)}$, and to express the aberrations in terms of

$$w_A := u_A - w^{(0)} = x_A - x_A^{(0)} + i(y_A - y_A^{(0)}) \tag{31.20}$$

On doing this, we notice that there is one particular shift

$$w^{(0)} = -(K + ik)\frac{u_o}{C} \tag{31.21}$$

for which the coma vanishes. Experimentally, this value is always chosen approximately, since one always tries to get the centre of the image as sharp as possible by an adjustment of the aperture. With (31.20) and (31.21) we now establish a series expansion in terms of u_o and w_A instead of (31.17), the new coefficients being

$$\begin{aligned}
C_0 &= C, \qquad K_0 + ik_0 = 0 \\
A_0 + ia_0 &= A + ia - (K + ik)^2/C \\
F_0 &= F - 2(K^2 + k^2)/C \\
D_0 + id_0 &= D + id - F(K + ik)/C + (A + ia)(K - ik)/C \\
&\quad + 2(K + ik)^2(K - ik)/C^2 \\
S_0 &= S, \qquad Q_0 = Q, \qquad Q'_0 = Q' \\
P_0 &= P - (K + ik)Q/C \\
P'_0 &= P' - (K - ik)Q'/C
\end{aligned} \tag{31.22}$$

In the isoplanatic approximation the object field in question is assumed to be so narrow that the variable u_o can be replaced by the complex coordinate of the *centre* of the observed field. This replacement is to be made in all the aberration terms. We then arrive at the representation:

$$\Delta u_i = s + qw_A + q'w_A^* + Cw_A^2 w_A^* \tag{31.23}$$

with the coefficients

$$\begin{aligned}
s &= S_0 + P_0 u_o + P'_0 u_o^* + (D_0 + id_0)u_o^2 u_o^* \\
q &= Q_0 + F_0 u_o u_o^*, \qquad q' = Q'_0 + (A_0 + ia_0)u_o^2
\end{aligned} \tag{31.24}$$

The resulting lateral shift s is of very little practical importance, since the microscope is always adjusted in such a way that the centre of the object field of interest is imaged in the centre of the screen; this is achieved by means of weak deflection fields. We can thus ignore the shifts. The remaining aberrations are now the defocus, the axial chromatic aberration,

31.4 THE ISOPLANATIC APPROXIMATION

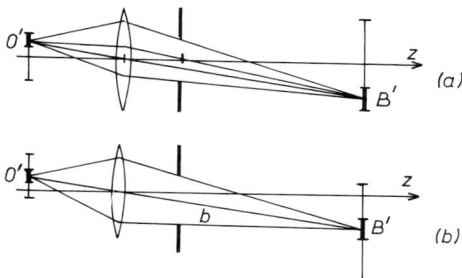

Fig. 31.2: Lateral shift of an aperture.
(a) Aperture centred on the lens axis leading to asymmetric confinement of the beam with coma.
(b) Aperture shifted in such a way that the beam is confined symmetrically; the beam axis b is now straight and the coma vanishes.

(both combined in the term qw_A), the astigmatism $q'w_A^*$ and the spherical aberration $Cw_A^2 w_A^*$. As will be shown in Chapter 32, the astigmatism, here the combined effect of the axial and off-axis contributions, can be compensated with a stigmator; the resolution of the instrument is thus finally determined by the spherical and the axial chromatic aberration.

Elimination of the coma by a shift of the aperture is important; it is shown schematically in Fig. 31.2. It can be understood as a straightening of the beam axis passing simultaneously through the centres of the object, the lens, the aperture and the image. Apart from the axial aberrations of perfectly round lenses there remains only an *astigmatism*, generated by the ellipticity and the inclination of the beam axis relative to the lens.

Part V

Deflection Systems

32
Deflection Systems and their Aberrations

32.1 Introduction

In virtually every kind of electron beam device, some provision is made for deflection, that is, for lateral shift of the electrons with as little disturbance of the beam structure as possible. In fixed-beam instruments, essentially in conventional transmission electron microscopes, deflection plays a minor role and is provided only to permit nonmechanical alignment of the column. In scanning devices, however, the role of the optics of the deflection system is to move a focused spot (often called the electron probe) in a raster pattern over a prescribed area of a specimen or a viewing screen, or to sweep a two-dimensional image over a small detector, and its design is at least as important as that of the lenses. The obvious examples here are cathode-ray oscillographs and television tubes, together with the various types of scanning electron and ion microscopes. In the more recent generations of electron microscopes, provision is made for both fixed-beam (conventional) transmission imaging and scanning transmission microscopy, so that the deflection system becomes just one member of the complex sequence of optical components that make up these hybrid instruments. In the past decade, deflection systems, mainly of the magnetic type, have acquired new importance, for it is their properties that determine the optical performance of electron lithography devices. Here, not only is the raster technique employed but—far more advantageously—the shaped-beam technique: instead of a narrow beam, focused into a small probe, a comparatively broad beam, shaped by masks, is deflected. The accuracy required in such devices is very high. More details will be found in Chapter 40.

In scanning electron microscopes, the magnification is altered by varying the area scanned with the result that high resolution is associated with small deflection angles. The probe size is then determined by the spherical and chromatic aberrations of the probe-forming lens and it is important to place the specimen as close as possible to the lens to keep the corresponding coefficients small. The scan coils are hence placed upstream from the lens. In lithography devices, on the other hand, the area scanned is large and the deflection aberrations become important; there is therefore no need to keep the working distance (the distance between the lens and the target) small and the scan coils can be inserted into this space, which has long been known to be advantageous (e.g. Owen and Nixon, 1973). In the most recent designs, however, the fields of the probe-forming lens and of the scan coils overlap.

Deflection systems may be either magnetic or electric and each type finds practical application. In television tubes, for example, the power consumption is higher for magnetic deflectors, but these are nevertheless preferred, for it is easier to provide the required deflection current than the voltage needed in an electric system; moreover, the distortions are lower. In the range of frequencies used in television systems, inductance effects do not yet play a role.

In oscilloscopes, on the other hand, electric fields have been widely used since the deflection plates draw little or no current and only at very high frequency does the displacement cease to follow the signal. In magnetic systems, particularly in those with ferrite shielding, the inductance of the device begins to cause problems at much lower frequencies. In spite of this, magnetic systems are almost exclusively used in scanning microscopes since their deflection aberrations can be kept small; electric deflection is often included for beam blanking, that is, switching off the illuminating beam by abruptly deflecting it sideways so that it no longer passes through some centred aperture. Magnetic systems are also usually adopted in lithography devices, though electrostatic and hybrid arrangements are of interest for particular purposes.

In the electrostatic case, deflectors usually consist of pairs of plates, symmetrical about a plane through the optical axis. They may have a wide variety of shapes, ranging from simple rectangles parallel to each other to tilted or curved surfaces. Some typical designs are illustrated in Fig. 32.1. A more advanced design consists of a family of eight or 12 plates, arranged symmetrically around the axis and forming sectors of a cylindrical surface (Fig. 32.2). By choosing the electrode potentials appropriately, not only can deflection at an arbitrary azimuth be achieved but the aberrations can also be partially corrected.

In the magnetic case, two geometries are common: saddle coils and toroidal coils. *Saddle coils* (Fig. 32.3) are usually enclosed in a ferrite sheath, thereby reducing the wastage of flux. The shielding is omitted only in devices designed to function at high deflection frequencies in order to decrease the inductance. In *toroidal structures* (Fig. 32.3b,c) the turns of the two individual coils are wound meridionally, or in other words, lengthwise around a ferrite yoke, which may have a more complicated shape than a simple cylinder or cone (Fig. 32.3d). This is especially true of television tubes, where distortions must be kept very small. In these, hybrid schemes are often used; Fig. 32.4 illustrates a deflection system in which the vertical deflection is achieved by means of toroidal coils while the horizontal

32.1 INTRODUCTION

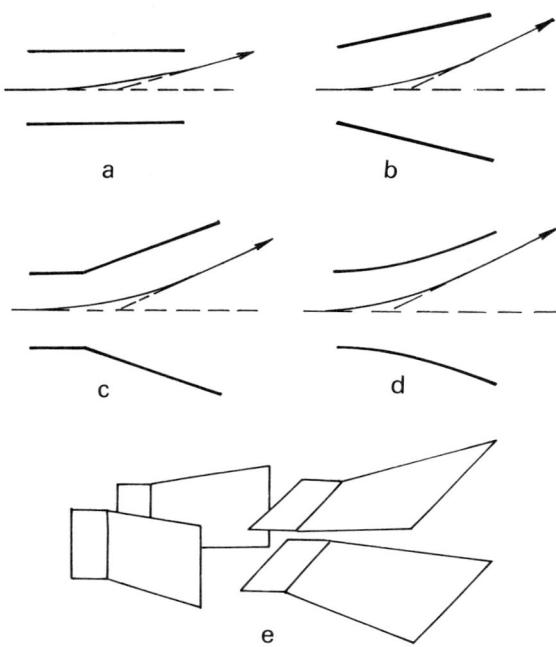

Fig. 32.1: (a)–(d) Cross-sections through electric deflectors of various kinds; the flared design (c) is frequently employed. (e) Perspective view of an electric deflector with flared plates.

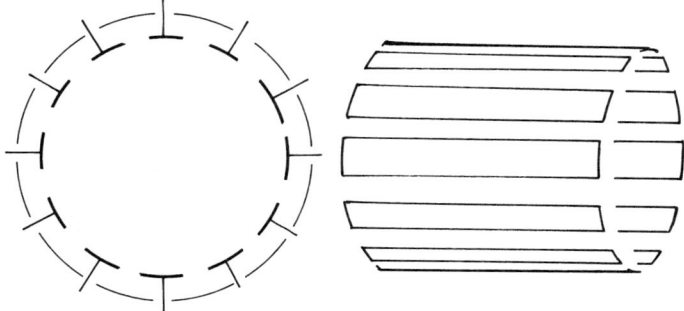

Fig. 32.2: Simplified drawing of an electric 12-pole deflector. The outer screening cylinder is omitted in the perspective view (right).

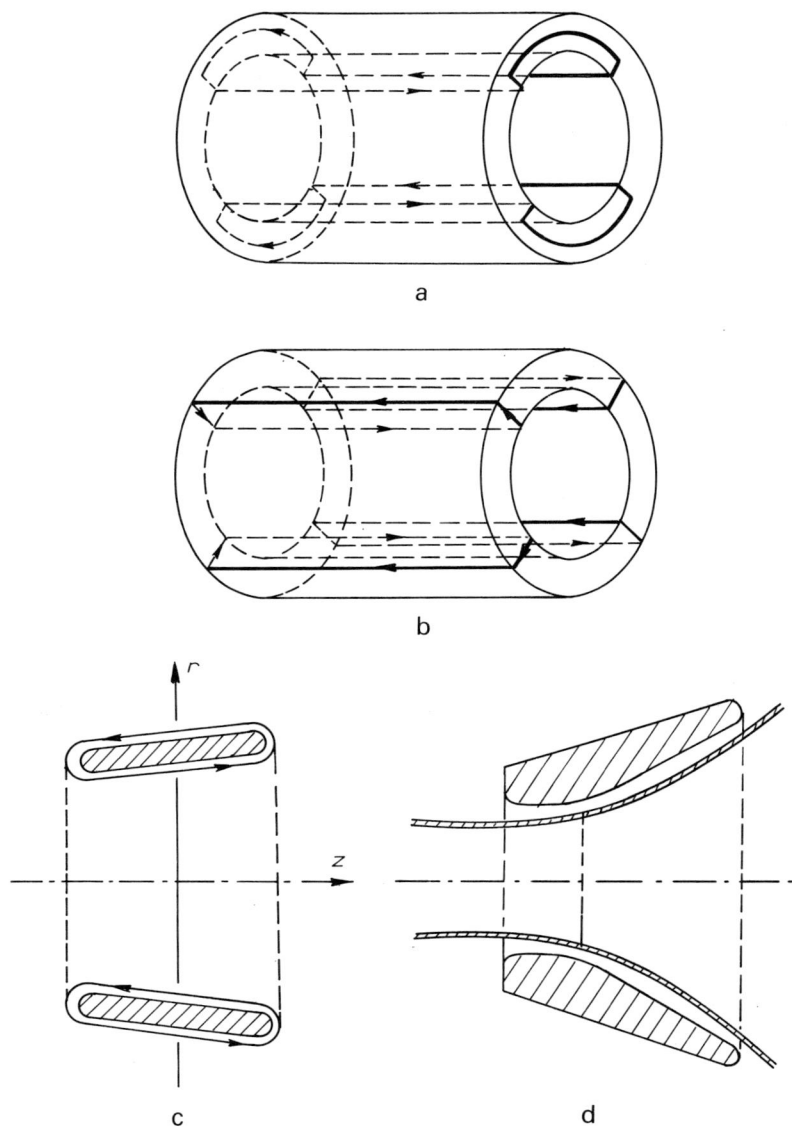

Fig. 32.3: Different types of magnetic deflector. (a) Windings in a pair of saddle coils. (b) Windings in a pair of toroidal coils. (c) Axial section through a pair of toroidal coils wound round a conical yoke. (d) Axial section through a ferrite cone in a television tube; part of the tube wall is indicated.

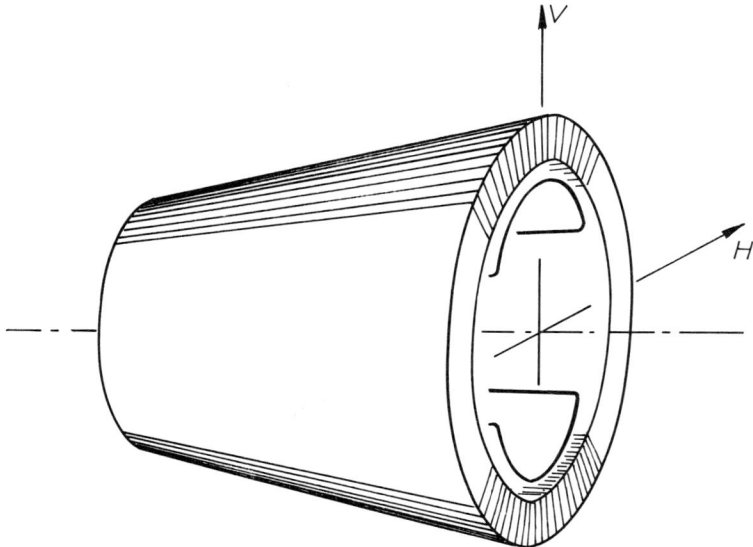

Fig. 32.4: Hybrid system consisting of saddle coils for the horizontal deflection (H) and toroidal coils for the vertical deflection (V). On the optic axis, the magnetic field generated by the saddle coils is in the vertical direction while that created by the toroidal coils is horizontal.

deflection is produced by the shielded saddle coils. The former have smaller aberrations, the latter lower inductance.

These introductory remarks can give only a preliminary impression of the host of deflection systems in practical use. More technical details will be found in Chapter 40. We now develop the corresponding theory of paraxial properties and aberrations.

32.2 The paraxial optics of deflection systems

The study of the optics of deflection systems passes through essentially the same stages as those already encountered for round lenses or quadrupoles; the novel aspects arise from the new symmetry conditions. Hitherto, we have considered systems in which an electron travelling along the straight optic axis—the axis of rotational symmetry of round lenses and the line of intersection of the symmetry or antisymmetry planes in quadrupole systems—experienced no transverse force. Terms of odd degree in x and y were thus excluded from the expansions (7.36) for $\Phi(\boldsymbol{r})$ and (7.45) for A_z, as were terms of even degree from A_x, A_y in (7.43–7.44).

In this section, we first derive the most general form of the paraxial equations and then specialize these to the cases of practical importance. Just as for round lenses and quadrupole systems, these equations can be obtained in two equivalent ways, namely, by using the trajectory method or the eikonal method. Here we use the trajectory method, which is convenient for the paraxial theory; the eikonal method is presented in detail by Glaser (1952, 1956), Kaashoek (1968) and Ximen (1986).

32.2.1 The general paraxial equations

The paraxial equations for an arbitrary system with a straight optic axis are easily obtained by linearization of (3.22). Reduction to linear form implies that $\varrho = 1$ from (3.16), and $B_t = B_z$ from (3.19). A nontrivial simplification arises from the fact that there is a linear relation between the transverse deflection field strengths at the optic axis and the coordinates x_i, y_i of the trajectory in a given recording plane $z = z_i$, as will become obvious later. The axial values of $E_x = -\Phi_{|x}$, $E_y = -\Phi_{|y}$, B_x and B_y are thus to be considered as *linear* quantities, though this is not explicitly obvious from the corresponding series expansions. We may therefore replace $\hat{\Phi}(\mathbf{r})$ in the denominators of (3.22) by its axial value $\hat{\phi}(z)$ and $\nabla\hat{\Phi}$ by $(1 + 2\epsilon\phi)\nabla\Phi(\mathbf{r})$. Bringing all this together, we obtain

$$\left. \begin{aligned} x'' &= \frac{\gamma}{2\hat{\phi}}(\Phi_{|x} - x'\Phi_{|z}) + \eta\hat{\phi}^{-\frac{1}{2}}(B_y - y'B_z) \\ y'' &= \frac{\gamma}{2\hat{\phi}}(\Phi_{|y} - y'\Phi_{|z}) + \eta\hat{\phi}^{-\frac{1}{2}}(-B_x + x'B_z) \end{aligned} \right\} \quad (32.1)$$

Differentiating (7.36) and retaining only terms that are of linear order when substituted into (32.1), we find $\Phi_{|z} = \phi'(z)$ and

$$\begin{aligned} \Phi_{|x} &= -F_1 - \frac{\phi''}{2}x + xp_2 + yq_2 \\ \Phi_{|y} &= -F_2 - \frac{\phi''}{2}y + xq_2 - yp_2 \end{aligned} \quad (32.2a)$$

The linear terms of $\mathbf{B}$ (7.46–7.48) are given by $B_z = B$ and

$$\begin{aligned} B_x &= B_1 - \frac{x}{2}B' - P_2 x - Q_2 y \\ B_y &= B_2 - \frac{y}{2}B' - Q_2 x + P_2 y \end{aligned} \quad (32.2b)$$

All this is now to be introduced into (32.1). The result of these elementary calculations takes a very concise form if the following complex quantities

32.2 THE PARAXIAL OPTICS OF DEFLECTION SYSTEMS

are introduced:

$$w := x + iy \tag{32.3a}$$
$$F_T(z) := F_1 + iF_2 \tag{32.3b}$$
$$B_T(z) := B_1 + iB_2 \tag{32.3c}$$
$$Q_E(z) := p_2 + iq_2 \tag{32.3d}$$
$$Q_M(z) := P_2 + iQ_2 \tag{32.3e}$$

These have the following physical meaning: w is the transverse coordinate, F_T and B_T are the complex transverse axial field strengths, while Q_E and Q_M are quadrupole coefficients. Quadrupole terms are not introduced deliberately in deflection systems but may arise from constructional imperfections. Using (32.3), an elementary calculation yields

$$w''(z) = A_1 w^* + A_2 w + A_3 w' + A_4 \tag{32.4}$$

the coefficients being given by

$$A_1 = -\frac{\gamma}{2}\frac{Q_E}{\hat{\phi}} + i\eta\hat{\phi}^{-\frac{1}{2}} Q_M \tag{32.5a}$$

$$A_2 = -\frac{\gamma}{4}\frac{\phi''}{\hat{\phi}} + \frac{i}{2}\eta\hat{\phi}^{-\frac{1}{2}} B' \tag{32.5b}$$

$$A_3 = -\frac{\gamma}{2}\frac{\phi'}{\hat{\phi}} + i\eta\hat{\phi}^{-\frac{1}{2}} B \tag{32.5c}$$

$$A_4 = -\frac{\gamma}{2}\frac{F_T}{\hat{\phi}} - i\eta\hat{\phi}^{-\frac{1}{2}} B_T \tag{32.5d}$$

The complex linear differential equation (32.4) is seldom solved in full generality, though its numerical solution provides absolutely no problem. The reason is that an electron optical system of so general a type is very rarely encountered in practice. We now turn to the important special cases.

First of all, a round electrostatic lens is never combined with electrostatic elements of lower symmetry in such a way that the fields overlap, since the boundary conditions governing the electric field on the electrode surfaces do not allow this (although round lens components can of course be generated by elements such as quadrupoles, suitably excited). Round electrostatic lenses well separated from any other elements are better treated independently by the methods outlined earlier and we hence ignore them here.

A second simplification concerns the term $A_1 w^*$ in (32.4), which may be neglected. Such a term would cause axial astigmatism, which can be compensated by means of stigmators; in practice, therefore, this term is always negligible. Stigmators are dealt with in Section 32.4; we say no more about them here and assume forthwith that $A_1(z) \equiv 0$.

The combination of round magnetic lenses and deflection units is common in electron lithography systems, but advantageous only under very special conditions, as will become obvious later. If these conditions are not satisfied, it is better to remove the round lens from the deflection field region since its presence destroys the electron optical symmetry properties of the deflection unit. In order to establish these properties, we omit the magnetic lens in the first step.

32.2.2 Ideal deflection

With all these simplifying assumptions, (32.4) and (32.5) reduce to

$$w'' = D(z) \tag{32.6}$$

where

$$D(z) \equiv A_4(z) = -\frac{\gamma F_T(z)}{2\hat{\phi}} - i\eta \hat{\phi}^{-\frac{1}{2}} B_T(z) \tag{32.7}$$

in which $\hat{\phi}$ is now a constant. Equation (32.6) is readily integrated, the solution with the general initial conditions

$$w(z_o) = w_o, \quad w'(z_o) = w'_o$$

being given by

$$w(z) = w_o + (z - z_o)w'_o + w_d(z) \tag{32.8}$$

with the particular solution

$$w_d(z) = \int_{z_o}^{z} (z - z')D(z')dz' \tag{32.9}$$

This is easily verified. The solution (32.9) is of particular interest since it describes the trajectory of a particle incident along the optic axis, so that (32.9) is the equation of a curved axis representing the central ray of a deflected beam. The most important property of $w_d(z)$ is its asymptotic behaviour for large values of z beyond the domain of the deflecting field. This is given by

$$w_d(z) = z \int_{-\infty}^{\infty} D(z')\,dz' - \int_{-\infty}^{\infty} z'D(z')\,dz' \tag{32.10}$$

32.2 THE PARAXIAL OPTICS OF DEFLECTION SYSTEMS

The limits of integration have been extended to infinity, which is justified if $D(z')$ has appreciable values only in a finite interval and the coordinates z_o and z are located outside the latter. In practice this simplification is always permissible.

Equation (32.10) describes a straight line, the asymptote of the deflected principal ray. In general this does *not* intersect the optic axis: it is a skew ray. This is easily seen by writing (32.10) in real form:

$$\left.\begin{aligned} x_d(z) &= z \int_{-\infty}^{\infty} D_x(z')\,dz' - \int_{-\infty}^{\infty} z' D_x(z')\,dz' \\ y_d(z) &= z \int_{-\infty}^{\infty} D_y(z')\,dz' - \int_{-\infty}^{\infty} z' D_y(z')\,dz' \end{aligned}\right\} \quad (32.11)$$

This can be cast into the form

$$\left.\begin{aligned} x_d(z) &= x'_d(\infty)(z - z_1) \\ y_d(z) &= y'_d(\infty)(z - z_2) \end{aligned}\right\} \quad (32.12)$$

with the asymptotic slopes

$$x'_d(\infty) = \int_{-\infty}^{\infty} D_x(z)\,dz, \qquad y'_d(\infty) = \int_{-\infty}^{\infty} D_y(z)\,dz \quad (32.13)$$

and pivot-point coordinates

$$z_1 = \frac{\int_{-\infty}^{\infty} z D_x(z)\,dz}{\int_{-\infty}^{\infty} D_x(z)\,dz}, \qquad z_2 = \frac{\int_{-\infty}^{\infty} z D_y(z)\,dz}{\int_{-\infty}^{\infty} D_y(z)\,dz} \quad (32.14)$$

The latter depend only on the geometric forms of the function $D_x(z)$ and $D_y(z)$, not on their absolute values. In the general case, these forms are different and, hence $z_1 \neq z_2$. Figures 32.5a,b illustrate this for an electrostatic deflection system, of the type employed in oscilloscopes.

The general solution (32.8) represents the linear superposition of the principal ray solution $w_d(z)$ and a free-space motion $w_o + (z - z_o)w'_o$. A consequence of (32.8) is that if there is stigmatic focusing in the absence of deflection ($w_d \equiv 0$), this remains true with deflection; there is only a lateral shift of the focus. Let us assume that with no deflection, the image

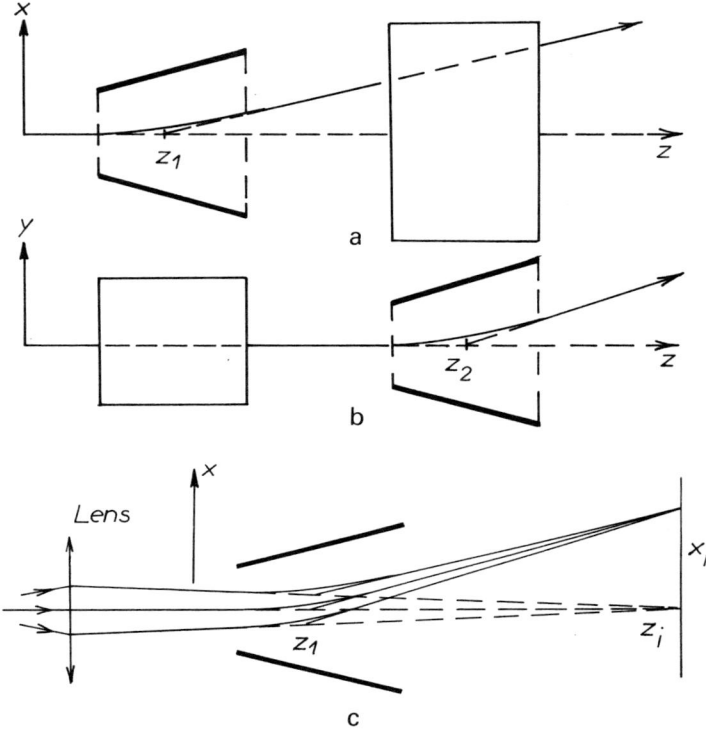

Fig. 32.5: (a)–(b) Principal sections through an electric deflection system. (a) x–z plane; (b) y–z plane. (c) Schematic description of the 'ideal' deflection in the x–z plane; the situation in the y–z plane is analogous.

is a point $(\bar{x}_i, \bar{y}_i)$ in the plane $z = z_i$: $w_d(z) \equiv 0$, $w(z_i) = \bar{w}_i$ irrespective of w'_o, as is sketched in Fig. 32.5c. The initial values in the plane $z = z_o$ in front of the deflection unit must satisfy the condition

$$w_o + (z_i - z_o)w'_o = \bar{w}_i$$

Substituting this into (32.8) we find

$$w(z) = (z - z_i)w'_o + w_d(z) + \bar{w}_i \tag{32.15}$$

In the plane of the focus, $z = z_i$, we then have

$$w_i := w(z_i) = w_d(z_i) + \bar{w}_i \tag{32.16}$$

32.2 THE PARAXIAL OPTICS OF DEFLECTION SYSTEMS

or in real form, with the abbreviations $x'_i = x'_d(\infty)$, $y'_i = y'_d(\infty)$,

$$\left.\begin{array}{l} x_i = x'_i(z_i - z_1) + \overline{x}_i \\ y_i = y'_i(z_i - z_2) + \overline{y}_i \end{array}\right\} \quad (32.17)$$

irrespective of the initial slopes x'_o, y'_o. This is characteristic of ideal deflection and illustrated in Fig. 32.5c.

32.2.3 The dependence on the electrical input signals

The input signals for deflection systems are time-dependent capacitor voltages or coil currents, which generate the necessary deflection fields. Production of the appropriate time dependence of the signals may be a major technical problem, but this does not concern us here. The inertia of the electron beam and the contributions of the Maxwell-displacement terms to the field equations can always be neglected. The time is thus a mere parameter, and a quasi-stationary approximation is quite sufficient. There is no need to study any time-dependent effects here.

The axial electric field strengths F_1, F_2 are proportional to the corresponding deflection voltages U_1, U_2 at the capacitor plates:

$$F_j(z) = U_j a_j(z) \quad , \quad j = 1, 2 \quad (32.18)$$

The functions $a_j(z)$ depend only on the geometric forms of the axial field distributions and have the dimension of a reciprocal length. The axial magnetic field strengths B_1, B_2 are likewise proportional to the corresponding coil currents I_1, I_2:

$$B_j(z) = \mu_0 I_j b_j(z) \quad , \quad j = 1, 2 \quad (32.19)$$

The permeability μ_0 has been introduced for dimensional reasons: b_1 and b_2 also have the dimensions of a reciprocal length.

Let us first consider systems with no round magnetic lens term. We rewrite (32.7) in real form and introduce (32.18) and (32.19), giving

$$\left.\begin{array}{l} D_x = -\dfrac{\gamma a_1(z)}{2\hat{\phi}} U_1 + \dfrac{\mu_0 \eta\, b_2(z)}{\hat{\phi}^{1/2}} I_2 \\[2mm] D_y = -\dfrac{\gamma a_2(z)}{2\hat{\phi}} U_2 - \dfrac{\mu_0 \eta\, b_1(z)}{\hat{\phi}^{1/2}} I_1 \end{array}\right\} \quad (32.20)$$

We see that there is indeed a linear relation between the input signals and the deflection functions D_x, D_y, but not a simple one. If we introduce

(32.20) into (32.13) and (32.14), and then substitute the resulting expressions into (32.17), we certainly obtain linear relations between the input signals U_1, U_2, I_1, I_2 and the corresponding deflections x_i, y_i, as we assumed at the outset, but these linear relations are often too complicated for practical applications. We recall that the functions $a_1(z)$, $a_2(z)$, $b_1(z)$, $b_2(z)$ are in general different from one other, and the same may also be true of integrals over these functions.

In practice, apart from very special applications like Wien filters, combined electric-magnetic deflection systems are of rare occurrence though the possibility of using magnetic deflection for large shifts and electrostatic fields for small deflections has been explored in lithography. We first specialize to the pure cases.

In purely *electric* deflection systems, (32.14) become

$$z_j = \frac{\int_{-\infty}^{\infty} z a_j(z)\,dz}{\int_{-\infty}^{\infty} a_j(z)\,dz}, \qquad j = 1, 2 \tag{32.21}$$

and (32.17) then take the form

$$x_i = d_1^E U_1, \qquad y_i = d_2^E U_2 \tag{32.22}$$

with the proportionality factors

$$d_j^E = -\frac{\gamma}{2\hat{\phi}}(z_i - z_j) \int_{-\infty}^{\infty} a_j(z)\,dz, \qquad j = 1, 2 \tag{32.23}$$

These factors are called the *sensitivities* and have the dimensions of a reciprocal field strength. Even though $z_1 \neq z_2$ (see Fig. 32.5), the condition $d_1^E = d_2^E$ can be satisfied by appropriate choice of the functions $a_j(z)$, and in practice this can be achieved by suitable design of the deflection plates. The response is then *isotropic*. Whether this is necessary depends on the particular purpose of the device in question.

In the case of purely *magnetic* deflection systems, the formulae are analogous; it is only necessary to modify certain factors, signs and subscripts. Equation (32.14) may be written

$$z_j = \frac{\int_{-\infty}^{\infty} z b_j(z)\,dz}{\int_{-\infty}^{\infty} b_j\,dz}, \qquad j = 1, 2 \tag{32.24}$$

32.2 THE PARAXIAL OPTICS OF DEFLECTION SYSTEMS

The change of sign in (32.20) here cancels out. The response relations take the form

$$x_i = d_2^M I_2, \qquad y_i = -d_1^M I_1 \qquad (32.25)$$

with the sensitivities

$$d_j^M = \mu_0 \eta \hat{\phi}^{-\frac{1}{2}}(z_i - z_j) \int_{-\infty}^{\infty} b_j(z)\, dz, \qquad j = 1, 2 \qquad (32.26)$$

Once again, the sensitivities can be made equal by appropriate design of the deflection coils, even if the functions $b_1(z)$ and $b_2(z)$ are different. The case $b_1(z) \neq b_2(z)$ certainly occurs in television tubes equipped with hybrid deflection systems (Fig. 32.4).

32.2.4 Rotation-invariant systems

We now examine the effect on the electron optical properties of superimposing a round magnetic field on a purely magnetic deflection system, a situation that occurs in lithography devices. In this situation, (32.4) and (32.5), in combination with (32.19), become

$$w'' = i\eta\hat{\phi}^{-\frac{1}{2}}(\frac{1}{2}B'w + Bw') + D(z) \qquad (32.27)$$

in which the inhomogeneous term $D(z)$ takes the form

$$D(z) = -i\mu_0\eta\hat{\phi}^{-\frac{1}{2}}\{I_1 b_1(z) + iI_2 b_2(z)\} \qquad (32.28)$$

In the theory of round magnetic lenses, it is advantageous to introduce the familiar coordinate frame rotating with the Larmor frequency. In the present case, however, it is no longer obvious that this is advantageous since the natural symmetries of the series expansion for the magnetic deflection field would be destroyed. For this reason, we shall continue to use the fixed (cartesian) frame (X, Y, z), which we here denote (x, y, z). This causes no particular complication. We note that this choice is not mandatory, and some authors have pursued the calculation in the rotating frame.

We must now solve (32.27) for arbitrary initial conditions $w(z_o) = w_o$, $w'(z_o) = w'_o$, in some given object plane. The fundamental solutions $\sigma(z)$ and $\tau(z)$ of the homogeneous differential equation with the initial conditions

$$\sigma(z_o) = \tau'(z_o) = 1, \qquad \sigma'(z_o) = \tau(z_o) = 0 \qquad (32.29)$$

and the Wronskian

$$W_0 = e^{-2i\theta(z)}(\tau'\sigma - \sigma'\tau) = \tau'\sigma^* - \sigma'\tau^* \equiv 1$$

are now appropriate, $\theta(z)$ being the angle of Larmor rotation (2.39, 15.9); the functions $\sigma(z)$, $\tau(z)$ are now complex. Solution of (32.27) by variation of parameters is straightforward and we find

$$w(z) = w_o \sigma(z) + w'_o \tau(z) + w_d(z) \tag{32.30}$$

with the particular integral

$$\begin{aligned} w_d(z) &= \int_{z_o}^{z} e^{-2i\theta(\zeta)} D(\zeta) \{\sigma(\zeta)\tau(z) - \tau(\zeta)\sigma(z)\} \, d\zeta \\ &= \int_{z_o}^{z} D(\zeta) \{\sigma^*(\zeta)\tau(z) - \tau^*(\zeta)\sigma(z)\} \, d\zeta \end{aligned} \tag{32.31}$$

The value of this solution in the image plane $z = z_i$ is the most important. There we have $\tau(z_i) = 0$ and $\sigma(z_i) = M\exp(i\theta_i)$, $\theta_i = \theta(z_i)$ being the angle of image rotation relative to the object. Hence,

$$w_d(z_i) = -Me^{i\theta_i} \int_{z_o}^{z_i} D(z)\tau^*(z)\, dz =: w_{di}$$

It will prove convenient to introduce the real fundamental solution $h(z)$ in the rotating frame, with $h(z_o) = 0$, $h'(z_o) = 1$ as usual. From (32.29) with $\theta(z_o) = 0$, we see that

$$\tau(z) = h(z) e^{i\theta(z)}$$

and so

$$w_{di} = -Me^{i\theta_i} \int_{z_o}^{z_i} e^{-i\theta(z)} D(z) h(z)\, dz \tag{32.32}$$

We notice that, since $\tau(z_i) = 0$, the paraxial deflection is again *ideal*, in the sense that the general solution $w(z_i) = w_o\sigma(z_i) + w_{di}$ is independent of w'_o. Equation (32.28) shows that it also depends linearly on the currents I_1 and I_2. In the general case, however, it is *not rotation-invariant*. This means that if we apply the coil currents

$$I_1 = I_0 \cos\alpha \quad , \quad I_2 = I_0 \sin\alpha \tag{32.33}$$

with a time-dependent phase α, intending to trace out a circle, the deflection signal (with $w_o = 0$) does not in fact respect this but instead describes an ellipse. This effect is called the *paraxial distortion*. In order to obtain

32.3 THE ABERRATIONS OF DEFLECTION SYSTEMS

a circle, the deflection signal w_{di}, given by (32.32), must be proportional to the complex current variable I_c,

$$I_c := I_1 + iI_2 = I_0 e^{i\alpha} \tag{32.34}$$

This proportionality can only be achieved if

$$\int_{z_o}^{z_i} e^{-i\theta} b_1(z) h(z)\, dz = \int_{z_o}^{z_i} e^{-i\theta} b_2(z) h(z)\, dz \tag{32.35}$$

When different types of coil systems are combined, in hybrid toroidal- and saddle-coil systems for instance, this condition is in general not satisfied. Even if (32.35) is true for a particular lens excitation, a slight variation of the latter to adjust the focusing renders the two integrals unequal. The only reasonable way of satisfying (32.35) in general is hence to impose the condition $b_1(z) \equiv b_2(z)$, which means that the coils, rotated at 90° to each other, must have the same geometry. Systems of this type are *rotation-invariant*. They are the only reasonable choice when a deflection field is to be combined with a magnetic round lens field and, in our discussion of aberration theory, we shall consider only this case.

We can then drop the suffices of $b_1(z)$ and $b_2(z)$; combining (32.32) with (32.28), we find

$$w_{di} = d_c I_c \tag{32.36}$$

in which the complex sensitivity is given by

$$d_c = i\mu_0 \eta \hat{\phi}^{-\frac{1}{2}} M e^{i\theta_i} \int_{z_o}^{z_i} e^{-i\theta(z)} b(z) h(z)\, dz \tag{32.37}$$

The modulus of this complex quantity is of great technical importance, whereas the phase is not important since it merely describes a *constant* rotation between the input and output signals.

32.3 The aberrations of deflection systems

In this section, we again distinguish between systems with pure deflection fields and hybrid systems containing a round magnetic lens because the symmetry properties in these two situations are essentially different. We concentrate on the nature of the aberrations, referring to the original publications for the full lists of aberration integrals for the following reasons:

in the more general case, the terms are extremely numerous whereas in the simpler cases of pure deflection fields, only a few of the aberration integrals have found practical use. We do, however, reproduce one set of aberration integrals, which is particularly convenient in that the expressions required in many practical situations can be obtained as special cases.

32.3.1 Pure deflection systems

The most important technical devices employing this kind of deflector are oscilloscopes and television tubes. These do of course also contain electrostatic lenses to focus the beam on the viewing screen but these lenses are well separated from the deflectors, and have little effect so far as the deflection aberrations are concerned. The situation in scanning electron microscopes is more complicated. For medium and high resolution operation, the deflection structure most commonly used, the double deflector, can be situated far enough before the lens for the field overlap to be negligible. If, however, the microscope is designed to produce a good image at very low magnification where the angles involved become large, it will be necessary to place the the deflector within the lens field and the situation is that discussed in Section 32.3.2, always recalling that the deflector precedes the probe-forming lens. The so-called Grigson coils that are used to scan a diffraction pattern over a small detector do, on the other hand, belong to the present section.

For simplicity, we shall assume that the undeflected beam is directed along the optic axis and produces a round spot, blurred by spherical and chromatic aberrations. When the beam is deflected, the aberration coordinates Δx_i, Δy_i of a particular ray in the undeflected beam can simply be added to the corresponding deflection aberration components.

Two different symmetries

Devices that fall into this category are always constructed so as to have two orthogonal planes of symmetry, intersecting along the optic axis within the limits of experimental accuracy. We have already tacitly assumed this in Sections 32.2.2 and 32.2.3, where these symmetry planes were the coordinate planes $x = 0$ and $y = 0$.

In magnifying systems, it is usually the object plane that is of interest and the aberrations are therefore most usefully expressed in terms of position and gradient in that plane or in terms of position in the object and entrance pupil planes. In probe-forming and other demagnifying systems, on the other hand, the plane of interest is the target or image plane and the most suitable variables for characterizing deflection aberrations are therefore the position coordinates x_i and y_i corresponding to ideal deflection, introduced earlier, and the paraxial difference of gradient between that of the ray in question (gradient x'_i, y'_i) and the deflected central ray (gradient

32.3 THE ABERRATIONS OF DEFLECTION SYSTEMS

$x_i^{(0)\prime}$, $y_i^{(0)\prime}$); we write $\alpha := x_i' - x_i^{(0)\prime}$, $\beta := y_i' - y_i^{(0)\prime}$.

The lateral aberrations Δx_i and Δy_i can be derived from a perturbation eikonal, which we denote as in Part IV by S. The symmetry properties impose the following conditions on $S(x_i, y_i, \alpha, \beta)$:

$$S(-x_i, y_i, -\alpha, \beta) = S(x_i, y_i, \alpha, \beta)$$
$$S(x_i, -y_i, \alpha, -\beta) = S(x_i, y_i, \alpha, \beta) \qquad (32.38)$$

These are the same symmetry properties as those we have encountered for well-aligned systems of quadrupole lenses. At first sight, this is surprising, since in quadrupoles the potentials are positively symmetric with respect to the symmetry planes, while in deflection systems they are antisymmetric. We must, however, remember that the deflection input signals are not fixed constants, but proportional to x_i and y_i respectively.

From (32.38) it can be concluded that S will contain aberration terms of *even* order only, the lowest order being characterized by terms of fourth degree. We can simply adopt the classification investigated in Chapter 29 without further ado; the cartesian representation of the transverse aberrations is given by (29.12) if we exchange the variables as follows:

$$\xi_o \to x_i, \qquad \eta_o \to y_i, \qquad x_o' \to \alpha, \qquad y_o' \to \beta$$

One additional simplification arises from the condition that, when the deflection fields are switched off ($x_i = y_i = 0$), the aberration must collapse to the spherical aberration of a round lens and hence $C := (0030) = (0012) = (0003) = (0021)$ in (29.12).

We now introduce a simplified notation, in terms of which the perturbation eikonal has the basic form

$$\begin{aligned} S = &\frac{1}{4}(E_{11}x_i^4 + 2E_{12}x_i^2 y_i^2 + E_{22}y_i^4) \\ &+ \alpha x_i(D_{11}x_i^2 + D_{12}y_i^2) + \beta y_i(D_{21}x_i^2 + D_{22}y_i^2) \\ &+ \frac{\alpha^2}{2}(A_{11}x_i^2 + A_{12}y_i^2) + \frac{\beta^2}{2}(A_{21}x_i^2 + A_{22}y_i^2) \\ &+ A_3 \alpha\beta x_i y_i + \frac{C}{4}(\alpha^2 + \beta^2)^2 \\ &+ \frac{x_i \alpha}{3}(K_{11}\alpha^2 + 3K_{12}\beta^2) + \frac{y_i \beta}{3}(3K_{21}\alpha^2 + K_{22}\beta^2) \end{aligned} \qquad (32.39)$$

There are three phase shifts (E_{jk}), four distortions (D_{jk}), five astigmatisms (A_{jk}), disregarding differences between these and the field curvature, four comas (K_{jk}) and the spherical aberration (C). Since the phase shifts (E_{jk})

do not contribute to the lateral aberrations, 14 relevant real independent terms remain.

The function S can be normalized in such a way that the lateral aberrations are simply given by

$$\Delta x_i = \frac{\partial S}{\partial \alpha}, \qquad \Delta y_i = \frac{\partial S}{\partial \beta} \qquad (32.40)$$

Differentiation of (32.39) then gives

$$\begin{aligned}\Delta x_i = &\; x_i(D_{11}x_i^2 + D_{12}y_i^2) \\ &+ \alpha(A_{11}x_i^2 + A_{12}y_i^2) + A_3\beta x_i y_i \\ &+ x_i(K_{11}\alpha^2 + K_{12}\beta^2) + 2K_{21}y_i\alpha\beta \\ &+ C\alpha(\alpha^2 + \beta^2)\end{aligned} \qquad (32.41a)$$

$$\begin{aligned}\Delta y_i = &\; y_i(D_{21}x_i^2 + D_{22}y_i^2) \\ &+ \beta(A_{21}x_i^2 + A_{22}y_i^2) + A_3\alpha x_i y_i \\ &+ y_i(K_{21}\alpha^2 + K_{22}\beta^2) + 2K_{12}x_i\alpha\beta \\ &+ C\beta(\alpha^2 + \beta^2)\end{aligned} \qquad (32.41b)$$

Fourfold symmetry

A further simplification is obtained if the two symmetry planes $x = 0$ and $y = 0$ are equivalent, in the sense that the system is invariant under rotation through 90° about the optic axis. This means that interchanging x and y ($x \leftrightarrow y$) and α and β ($\alpha \leftrightarrow \beta$) must leave the formulae unaffected, which in turn implies that for all sets of coefficients,

$$\lambda_{11} = \lambda_{22}, \qquad \lambda_{12} = \lambda_{21} \qquad (32.42)$$

where λ stands for each of E, D, A and K.

In order to reveal the physical meaning of the surviving coefficients, it is helpful to introduce complex and polar position and angular coordinates thus:

$$w_i := x_i + iy_i = r_i e^{i\varphi_i} \qquad (32.43a)$$
$$s := \alpha + i\beta = \sigma e^{i\varphi_a} \qquad (32.43b)$$

On expanding (32.39) as a Fourier series with respect to the azimuths φ_i

32.3 THE ABERRATIONS OF DEFLECTION SYSTEMS

and φ_a, we find

$$\begin{aligned}
S = &\frac{1}{4}r_i^4(E_0 + E_4\cos 4\varphi_i) \\
&+ \sigma r_i^3\{D_0\cos(\varphi_a - \varphi_i) + D_4\cos(3\varphi_i + \varphi_a)\} \\
&+ \frac{1}{2}\sigma^2 r_i^2\{A_0\cos(2\varphi_a - 2\varphi_i) + A_4\cos(2\varphi_i + 2\varphi_a)\} \\
&+ \sigma^3 r_i\{K_0\cos(\varphi_a - \varphi_i) + \frac{1}{3}K_4\cos(\varphi_i + 3\varphi_a)\} \\
&+ \frac{1}{2}F_0\sigma^2 r_i^2 + \frac{1}{4}C_0\sigma^4
\end{aligned} \qquad (32.44)$$

in which

$$\begin{aligned}
E_0 &= \frac{1}{4}(3E_{11} + E_{12}), & E_4 &= \frac{1}{4}(E_{11} - E_{12}) \\
D_0 &= \frac{1}{4}(3D_{11} + D_{12}), & D_4 &= \frac{1}{4}(D_{11} - D_{12}) \\
A_0 &= \frac{1}{4}(A_{11} - A_{12} + A_3), & A_4 &= \frac{1}{4}(A_{11} - A_{12} - A_3) \qquad (32.45) \\
K_0 &= \frac{1}{4}(K_{11} + K_{12}), & K_4 &= \frac{1}{4}(K_{11} - 3K_{12}) \\
F_0 &= \frac{1}{2}(A_{11} + A_{12}), & C_0 &\equiv C
\end{aligned}$$

The transverse aberrations are now obtained by forming the gradient in complex polar coordinates:

$$\Delta w_i = \frac{\partial S}{\partial\alpha} + i\frac{\partial S}{\partial\beta} = e^{i\varphi_a}\left(\frac{\partial}{\partial\sigma} + \frac{i}{\sigma}\frac{\partial}{\partial\varphi_a}\right)S \qquad (32.46)$$

the result being

$$\begin{aligned}
\Delta w_i = &r_i^3(D_0 e^{i\varphi_i} + D_4 e^{-3i\varphi_i}) \\
&+ r_i^2\sigma(A_0 e^{i(2\varphi_i - \varphi_a)} + A_4 e^{-i(2\varphi_i + \varphi_a)}) \\
&+ r_i^2\sigma F_0 e^{i\varphi_a} + 2\sigma^2 r_i K_0 e^{i\varphi_i} \qquad (32.47) \\
&+ r_i\sigma^2(K_0 e^{i(2\varphi_a - \varphi_i)} + K_4 e^{-i(2\varphi_a + \varphi_i)}) \\
&+ \sigma^3 C_0 e^{i\varphi_a}
\end{aligned}$$

or in the general complex form with (32.45):

$$\begin{aligned}
\Delta w_i = &D_0 w_i^2 w_i^* + D_4 w_i^{*3} \\
&+ A_0 w_i^2 s^* + A_4 w_i^{*2} s^* \\
&+ F_0 w_i w_i^* s + 2K_0 s s^* w_i \qquad (32.48) \\
&+ K_0 s^2 w_i^* + K_4 s^{*2} w_i^* \\
&+ C_0 s^2 s^*
\end{aligned}$$

Fig. 32.6: Various combinations of fourfold aberrations and defocus values.

Apart from the choice of notation and the limitation to systems without superimposed magnetic lens fields, this is the representation which has long been known and has been explored in great detail in connection with electron lithography devices (Munro, 1974; Goto and Soma, 1977). In comparison with the aberrations of round lenses (see 24.36–24.37), there are three new types of error, those indicated by the subscript 4; these are known as *fourfold errors* and in particular as fourfold distortion (D_4), astigmatism (A_4) and coma (K_4). This nomenclature reflects the behaviour of these aberrations when the entire system is rotated bodily about the optic axis. Let us substitute $\varphi_i = \varphi'_i + \psi$ and $\varphi_a = \varphi'_a + \psi$ in (32.47). It quickly becomes clear that all the *isotropic* aberration terms, labelled with the subscript zero and familiar from the theory of round lenses, are modified by a common factor $\exp(i\psi)$, which means that these lateral aberrations rotate in synchronism with the system, just as in round systems. The *fourfold* error terms, on the other hand, are modified by a common factor $\exp(-3i\psi)$, and their rotation relative to the system is therefore described by a phase factor $\exp(-4i\psi)$.

The pure fourfold errors have fourfold symmetry; some typical examples are shown in Fig. 32.6. In practice, however, such symmetric error figures are never obtained. The complex superposition of all the terms appearing in (32.47) or (32.48) can result in quite complicated asymmetric patterns. The possibility of designing electrostatic deflection electrodes free of fourfold aberrations is examined by Dodin (1983).

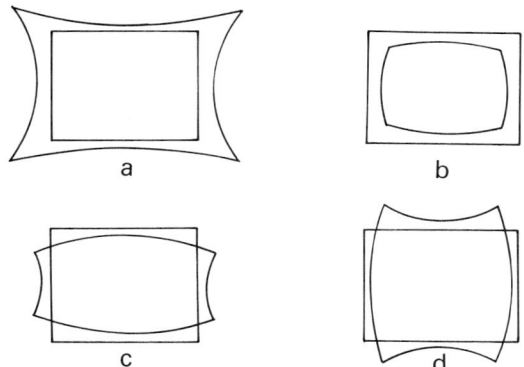

Fig. 32.7: Deflection distortions. (a) Cushion; (b) barrel; (c)–(d) hammock.

General considerations

Although the classification of the possible error types in deflection systems is an intellectually pleasing and somehow beautiful task, the determination of the corresponding aberration coefficients is very laborious. In principle, the calculation is straightforward, whether the trajectory method or the eikonal method is employed, and formulae for the various coefficients have been derived by Glaser (1938, 1941, 1949, see 1952), Wendt (1939, 1942a,b, 1947, 1953), Picht and Himpan (1941), Picht (1943) and Hutter (1947, 1948). They were subsequently reconsidered by Haantjes and Lubben (1957, 1959), by Kanaya (Kanaya and Kawakatsu, 1961a,b, 1962; Kawakatsu and Kanaya, 1961; Kanaya et al., 1961, 1963, 1964) and especially by Kaashoek (1968) and re-examined by Ding (1982). We shall not, however list them here as they are not much used in practice. For numerical values, see Wang (1966,1967a–c, 1971) and Amboss and Wolf (1971) as well as the papers cited above.

In television devices and oscilloscopes the deflection angles can become quite large, since the viewing screen must have a given minimum size and the tube-length must be kept reasonably small. For such large angles a perturbation theory of third order is clearly insufficient, but the practical evaluation of error terms of higher than third order by means of the standard perturbation calculus becomes extremely laborious. The fifth-order aberrations are examined by Kaashoek (1968) but it is arguable that if a knowledge of the third-order aberrations is insufficient, it is better to abandon perturbation theory and adopt the exact ray-tracing methods outlined in Chapter 33.

In electron lithography devices, the off-axis distances and deflection an-

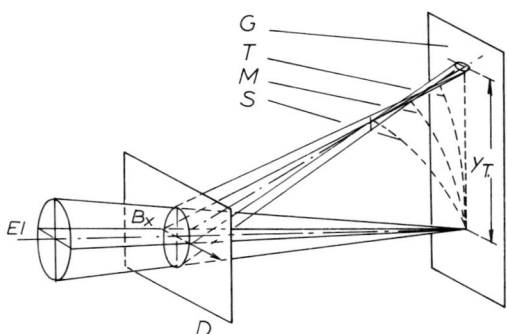

Fig. 32.8: Deflection astigmatism (the deflection is shown in only one symmetry plane); G: Gaussian image plane; T, M, S: tangential, mean and sagittal image curvature; D: principal plane of deflection.

gles remain small enough for the third-order approximation to be adequate but, in such systems, a round magnetic lens is usually strongly coupled to the deflection. The classification given above is then incomplete. This is the subject of the next section.

Before leaving the present topic, we briefly discuss how systems with large deflection angles can be investigated. It is unreasonable to analyse all the aberrations of fifth and even higher order, since the number of terms becomes unmanageable. The discussion must be confined to the technically most important defects, which are clearly the distortions and, with lower priority, the astigmatisms (see Figs 32.7, 32.8). Since fourfold symmetry is not always present, a more general form of (32.41) is required. For the distortions of third and fifth order, we may write

$$\Delta x_i = x_i(D_{11}x_i^2 + D_{12}y_i^2 + D_{13}x_i^4 + 2D_{14}x_i^2 y_i^2 + D_{15}y_i^4)$$
$$\Delta y_i = y_i(D_{21}x_i^2 + D_{22}y_i^2 + D_{23}x_i^4 + 2D_{24}x_i^2 y_i^2 + D_{25}y_i^4)$$
(32.49)

The coefficients in these expansions can be determined by exact numerical calculation of a sufficiently large number of electron trajectories followed by a least-squares fit of (32.49) to the set of numerically determined aberrations (Δx_i, Δy_i).

32.3.2 Deflection systems with magnetic lenses
We now study the influence of a round magnetic lens field on the deflection aberrations. As we have already mentioned, this is of great importance

32.3 THE ABERRATIONS OF DEFLECTION SYSTEMS

in the theory of electron lithography devices. The practical necessity for superimposing deflection and lens fields will be discussed in Chapter 40.

Geometric aberrations

We again begin by classifying the permitted types of aberrations. As we have seen in Section 32.2.4, only systems with a fourfold geometric symmetry of the coil system are advantageous, since the device will otherwise exhibit a paraxial distortion. In contrast to the case treated in the preceding section, the planes $x = 0$ and $y = 0$ are no longer planes of mirror symmetry, since the round magnetic lens forces the beam to rotate about the optic axis. There thus remains only an invariance of the eikonal with respect to rotations through 90° about the optic axis. Instead of (32.38) we have

$$S(y_i, -x_i, \beta, -\alpha) = S(x_i, y_i, \alpha, \beta)$$
$$S(-x_i, -y_i, -\alpha, -\beta) = S(x_i, y_i, \alpha, \beta) \tag{32.50}$$

From the second property it can be concluded that the power series expansion of S can have only terms of *even* total order, the interesting ones being again those of fourth order. The first condition expresses the fourfold symmetry.

The mathematical form of the perturbation eikonal S is most easily obtained by a Fourier series expansion similar to (32.44). Only terms that are invariant with respect to the simultaneous transforms $\varphi_i = \varphi'_i + \pi/2$ and $\varphi_a = \varphi'_a + \pi/2$ can occur in this expression. Furthermore, terms higher than fourth order need not be retained. The only permissible trigonometric dependences are then

$$\cos n(\varphi_a - \varphi_i), \qquad \sin n(\varphi_a - \varphi_i), \qquad n = 0, 1, 2$$

together with the cosine and sine of the arguments

$$4\varphi_i, \qquad 3\varphi_i + \varphi_a, \qquad 2\varphi_i + 2\varphi_a, \qquad \varphi_i + 3\varphi_a, \qquad 4\varphi_a$$

the sum of the factors of φ_i and φ_a being always 4. In (32.44) only the cosine terms were obtained, since only these are compatible with (32.38) and (32.50). If (32.38) is given up, the corresponding sine terms are also permitted. A fourfold spherical aberration, given by the terms in $4\varphi_a$, is compatible with the symmetries, but is excluded for technical reasons, since the ordinary spherical aberration of the round lens must be obtained when the deflection currents are switched off.

We now have to complete the Fourier series expansion (32.44) by including the appropriate sine terms. The resulting expression can be cast into a more concise form by introducing a complex notation. For this purpose we define the following complex aberration coefficients:

$$\begin{aligned}
& & \tilde{E}_4 &= E_4 + ie_4 \\
\tilde{D}_0 &= D_0 + id_0, & \tilde{D}_4 &= D_4 + id_4 \\
\tilde{A}_0 &= A_0 + ia_0, & \tilde{A}_4 &= A_4 + ia_4 \\
\tilde{K}_0 &= K_0 + ik_0, & \tilde{K}_4 &= K_4 + ik_4
\end{aligned} \quad (32.51)$$

The real parts denote isotropic coefficients, while the imaginary parts represent anisotropic ones. Then, instead of (32.44), we have

$$\begin{aligned}
S = \Re \Bigg[&\frac{1}{4} r_i^4 (E_0 + \tilde{E}_4 \exp(-4i\varphi_i)) \\
&+ \sigma r_i^3 \left\{ \tilde{D}_0 \exp(i\varphi_i - i\varphi_a) + \tilde{D}_4 \exp(-3i\varphi_i - i\varphi_a) \right\} \\
&+ \frac{1}{2}\sigma^2 r_i^2 \left\{ \tilde{A}_0 \exp(2i\varphi_i - 2i\varphi_a) + \tilde{A}_4 \exp(-2i\varphi_i - 2i\varphi_a) \right\} \quad (32.52)\\
&+ \sigma^3 r_i \left\{ \tilde{K}_0 \exp(i\varphi_i - i\varphi_a) + \frac{1}{3}\tilde{K}_4 \exp(-i\varphi_i - 3i\varphi_a) \right\} \Bigg] \\
&+ \frac{1}{2} F_0 \sigma^2 r_i^2 + \frac{1}{4} C_0 \sigma^4
\end{aligned}$$

After some elementary calculations, differentiation of (32.52) using (32.46) gives the geometric aberrations (suffix g)

$$\begin{aligned}
\Delta w_g &= C_0 s^2 s^* + \tilde{D}_0 w_i^2 w_i^* + \tilde{D}_4 w_i^{*3} \\
&\quad + F_0 w_i w_i^* s + \tilde{A}_0 w_i^2 s^* + \tilde{A}_4 w_i^{*2} s^* \\
&\quad + 2\tilde{K}_0 s s^* w_i + \tilde{K}_0^* s^2 w_i^* + \tilde{K}_4 s^{*2} w_i^* \\
&= (w_i^* \ s^*) \begin{pmatrix} \tilde{D}_0 & F_0 & \tilde{K}_0^* & \tilde{D}_4 & \tilde{K}_4 \\ \tilde{A}_0 & 2\tilde{K}_0 & C_0 & \tilde{A}_4 & 0 \end{pmatrix} \begin{pmatrix} w_i^2 \\ w_i s \\ s^2 \\ w_i^{*2} \\ s^{*2} \end{pmatrix} \\
&= (w_i^* \ s^*) \begin{pmatrix} \tilde{D}_0 & F_0 & \tilde{K}_0^* \\ \tilde{A}_0 & 2\tilde{K}_0 & C_0 \end{pmatrix} \begin{pmatrix} w_i^2 \\ w_i s \\ s^2 \end{pmatrix} \\
&\quad + (w_i^* \ s^*) \begin{pmatrix} \tilde{D}_4 & \tilde{K}_4 \\ \tilde{A}_4 & 0 \end{pmatrix} \begin{pmatrix} w_i^{*2} \\ s^{*2} \end{pmatrix} \quad (32.53)
\end{aligned}$$

Apart from the novel fact that, with the exception of F_0 and C_0, all the coefficients have become complex, (32.53) has the same structure as (32.48). The formula clearly exhibits an interesting pattern: fourfold error terms contain only complex conjugate variables, w_i^* and s^*; moreover, the field curvature and the astigmatisms have the same functional structure as the comas, if we exchange the variables $w_i \leftrightarrow s$, $w_i^* \leftrightarrow s^*$.

Chromatic aberrations

The accuracy requirements in electron lithography devices are so demanding that chromatic aberrations must also be taken into account. As in the corresponding theory of round lenses, a *linear* approximation is sufficient. The chromatic effects can be easily understood in the following way. In (32.30) the principal solutions $\sigma(z)$, $\tau(z)$ and $w_d(z)$ depend on the acceleration voltage $\hat{\phi}$ and the lens excitation current I_L as parameters, which may fluctuate around their mean values. For sufficiently small fluctuations, $\Delta\hat{\phi}$ and ΔI_L, an expansion of the form

$$\Delta w_c = w_o \left(\frac{\partial \sigma}{\partial \hat{\phi}} \Delta\hat{\phi} + \frac{\partial \sigma}{\partial I_L} \Delta I_L \right) + w'_o \left(\frac{\partial \tau}{\partial \hat{\phi}} \Delta\hat{\phi} + \frac{\partial \tau}{\partial I_L} \Delta I_L \right) + \frac{\partial w_d}{\partial \hat{\phi}} \Delta\hat{\phi} + \frac{\partial w_d}{\partial I_L} \Delta I_L$$

is justified. Usually, $|w_o|$ is so small that the corresponding term can be neglected; w'_o is proportional to the final slope s. Owing to the rotation invariance in the paraxial domain, the whole solution $w_d(z)$ is proportional to the complex deflection current I_c and also, from (32.36), to w_{di}. Since this is true for all values of $\hat{\phi}$ and I_L, this proportionality must also hold for the derivatives $\partial w_d/\partial\hat{\phi}$ and $\partial w_d/\partial I_L$. Moreover, $\tau(z)$, being a paraxial solution for an ordinary magnetic lens, depends only on $I_L^2/\hat{\phi}$ and hence $\hat{\phi}\partial\tau/\partial\hat{\phi} = -2I_L\partial\tau/\partial I_c$. Putting all this together and referring to the image plane $z = z_i$, we obtain the general form

$$\Delta w_c = C_c s \left(\frac{\Delta\hat{\phi}}{\hat{\phi}} - 2\frac{\Delta I_L}{I_L} \right) + \left(C_{T1} \frac{\Delta\hat{\phi}}{\hat{\phi}} + C_{T2} \frac{\Delta I_L}{I_L} \right) w_{di} \qquad (32.54)$$

with $\Delta w_{di} = \Delta w_c + \Delta w_g$ as the total lateral aberration. The first term in (32.54) is the familiar axial chromatic aberration of a round magnetic lens, referred to the image plane. The second term, being proportional to the deflection w_i, is called the *transverse* chromatic error and represents a new type of aberration. Between the corresponding coefficients C_{T1} and C_{T2} no simple relation can be expected, since any variation of $\theta(z)$ in (32.32) causes complications.

Detailed aberration analyses

Apart from the choice of notation, the classifications (32.53) and (32.54) are the same as those first derived by Ohiwa (1970) and Munro (1974), who used the eikonal and trajectory methods respectively. Munro lists the geometric and chromatic aberration integrals for purely magnetic deflection and round lens systems. Goto and Soma (1977) pointed out that (32.53) is too limited for practical purposes, since dynamic correction of the aberrations is not represented whereas such correction has long been an important feature of deflection units. Additional weak multipole fields are applied, generated by currents that are usually nonlinear functions of the principal currents. Owen and Nixon (1973) had already pointed out that deflection field curvature and both isotropic and anisotropic distortion can be eliminated in this way, and Goto and Soma demonstrated that this is also true of both types of astigmatism. Moreover, linear correction currents are capable of eliminating all the pure deflection aberrations. Their paper concludes with a list of aberration integrals, which are a generalization of those to be found in Munro (1974). Later, Soma (1977) generalized the formulae still further to include all possible focusing and deflection effects of both magnetic and electrostatic type for arbitrary superposition of the fields; even relativistic effects were taken into account. A correction to these expressions was made by Li (1983).

Meanwhile, the possibility of eliminating or compensating the aberrations in demagnifying systems had been examined by Koops (1972, 1973; Koops and Bernhard, 1975) and ways of correcting all the aberrations in scanning microscopes by Crewe and Parker (1976). A systematic analysis of aberration minimization for post-lens deflection, double deflection before the lens and the 'moving objective lens' (MOL) of Ohiwa (1970, 1978, 1979; Ohiwa *et al.*, 1971), is to be found in Kern (1979). We shall return to the MOL in Chapter 40 but we mention that Ohiwa (1979) relates this concept to the coma-free condition and states explicitly that the introduction of pre-lens deflection is equivalent to replacement of the position and angle variables by linear combinations of these and the deflection current (in the magnetic case). A systematic procedure for eliminating aberrations from magnetic scanning and focusing combinations is set out in Hosokawa (1980), and the aberration coefficients of a double-deflection unit for which the second deflector coincides with a round magnetic lens field are listed in Kuroda (1980). More general formulae for adding deflection aberrations are to be found in Lencová (1981); these are applied very tellingly to several practical situations by Lencová (1988), using a model field (Lenc and Lencová, 1988).

We postpone discussion of a series of papers by Munro and Chu while we examine the work of Ximen and Li. In the first contribution (Ximen,

1981), which extends this author's earlier work (1977, 1978), expressions for the terms of second and fourth order in the eikonal function are given explicitly in the rotating coordinate system; the fact that deflection has the effect of replacing position and angle (in the object or target plane) by a linear combination of these quantities and the deflecting current or voltage, mentioned above, is used to introduce a generating function, from which the various aberration coefficients are extracted by differentiation. This ingenious technique enables Ximen to formulate the aberration theory very compactly. Explicit aberration integrals are given in later papers by Li and Ximen (1982a,b), once again in the rotating coordinate system, together with formulae for adding the aberration coefficients of sets of deflectors. Finally, Ximen and Li (1982) show how dynamic correction can be used to eliminate various aberrations. Further details are given in Li (1981). The asymptotic aberration coefficients of both position and gradient are listed in Hawkes (1988), where the coefficients of the corresponding polynomials are also to be found. A less full list is given by Tang (1986).

One important family of aberrations is not considered in any of these papers, namely, the parasitic aberrations; this gap has been filled by Plies (1982), who examined the aberrations that arise from misalignments of these complex hybrid systems. The purpose of his studies was to find practical tolerance limits by exploring the influence of various kinds of misalignment on the aberrations.

We now return to the series of papers by Munro and Chu (1982a,b; Chu and Munro, 1982a,b) devoted to the numerical analysis of electron beam lithography systems. The first two papers are concerned with field calculation and the fourth with computer optimization of complex systems. It is the third part that particularly concerns us here, for in it Chu and Munro give a list of aberration integrals, which can be used to study systems consisting of any combination of magnetic and electrostatic lenses and deflectors; these expressions are of immediate practical use and we reproduce them here, in the notation defined earlier.

The paraxial equation for a system consisting of any combination of round lenses and deflectors takes the (non-relativistic) form

$$w'' + \frac{\phi'}{2\phi}w' + \frac{\phi''}{4\phi}w - \frac{i\eta}{\phi^{\frac{1}{2}}}(Bw' + \frac{1}{2}B'w)$$
$$= -\frac{F_T}{2\phi} - i\frac{\eta}{\phi^{\frac{1}{2}}}B_T = -\frac{U_c}{2\phi}a(z) - \frac{i\eta\mu_0 I_c}{\phi^{\frac{1}{2}}}b(z) \tag{32.55}$$

which is a generalization of (32.27) using the notation of (32.3b,c) and (32.18-32.29). The solution of (32.55) is written

$$w(z) = w_o\sigma(z) + w_o'\tau(z) + I_c m(z) + U_c e(z) \tag{32.56}$$

which replaces (32.30), I_c being the complex current of (32.34) and U_c its electrical analogue; $m(z)$ and $e(z)$ are the magnetic and electrical contributions to $w_d(z)$ (32.31), after separating the current I_c and the voltage U_c as factors:

$$m(z) = -\frac{i\eta\mu_0}{\phi_o^{\frac{1}{2}}} \int_{z_o}^{z} b(\zeta)\{\tau(z)\sigma^*(\zeta) - \sigma(z)\tau^*(\zeta)\}\,d\zeta$$

$$e(z) = -\frac{1}{2\phi_o^{\frac{1}{2}}} \int_{z_o}^{z} \frac{a(\zeta)}{\phi^{\frac{1}{2}}(\zeta)}\{\tau(z)\sigma^*(\zeta) - \sigma(z)\tau^*(\zeta)\}\,d\zeta$$

(32.57)

These expressions are obtained by means of the method of variation of parameters, applied to (32.55), as in (32.31); examination of the Wronskian now tells us that

$$\phi^{\frac{1}{2}}(\sigma\tau' - \sigma'\tau) = e^{2i\theta}\phi_o^{\frac{1}{2}} \qquad (32.58a)$$

or

$$\phi^{\frac{1}{2}}(\sigma^*\tau' - \sigma'\tau^*) = \phi_o^{\frac{1}{2}} \qquad (32.58b)$$

(cf. 32.39b).

The aberrations are obtained by retaining higher order terms in the potential and field expansions and again using the method of variation of parameters—the standard trajectory method. Into the inhomogeneous terms is substituted a slightly modified form of (32.56),

$$w(z) = w_i \frac{\sigma(z)}{\sigma_i} + s_i \frac{\tau(z)}{\tau_i'} + w_i^{(m)} \frac{m(z)}{m_i} + w_i^{(e)} \frac{e(z)}{e_i} \qquad (32.59)$$

in which the image parameters in terms of which the aberrations are expressed have been introduced; $\sigma_i := \sigma(z_i)$ and likewise for τ_i, m_i and e_i. The shifts $w_i^{(m)}$ and $w_i^{(e)}$ are given by equations analogous to (32.36): $w_i^{(m)} = d_c^{(m)} I_c$ and $w_i^{(e)} = d_c^{(e)} U_c$.

In the worst case, in which no term is negligible, 59 complex geometric aberration coefficients emerge from this calculation. It is fortunately extremely rare that all these are required and indeed, we return below to the utility of an analysis such as this when so large a number of coefficients are relevant. At the other extreme, for a system in which the Gaussian spot size is negligible compared with the magnetic deflection $w_i^{(m)}$ and there is no electrostatic contribution, only nine coefficients survive (Munro, 1974, for example). In practice, the most general situation that is liable to be encountered is the *dual-channel* system, in which the Gaussian spot size is typically negligible, and both magnetic and electric deflection are employed (Fig. 40.20): the magnetic deflection provides coverage of a large field while the electrostatic deflection offers fast response.

Substituting (32.59) into the aberration formula, 27 aberration coefficients are obtained, as shown in the Table 32.1. The coefficients are as follows:

$$C_0 = F(\tau, \tau, \tau^*)$$
$$\tilde{K}_0^{*(m)} = F(\tau, \tau, m^*) \quad , \quad \tilde{K}_0^{*(e)} = F(\tau, \tau, e^*)$$
$$2\tilde{K}_0^{(m)} = F(\tau, m, \tau^*) + F(m, \tau, \tau^*)$$
$$2\tilde{K}_0^{(e)} = F(\tau, e, \tau^*) + F(e, \tau, \tau^*)$$
$$\tilde{A}_0^{(m)} = F(m, m, \tau^*) \quad , \quad \tilde{A}_0^{(e)} = F(e, e, \tau^*)$$
$$\tilde{A}_0^{(me)} = F(m, e, \tau^*) + F(e, m, \tau^*)$$
$$F_0^{(m)} = F(\tau, m, m^*) + F(m, \tau, m^*)$$
$$F_0^{(e)} = F(\tau, e, e^*) + F(e, \tau, e^*)$$
$$F_0^{(me)} = F(\tau, m, e^*) + F(m, \tau, e^*)$$
$$F_0^{(em)} = F(\tau, e, m^*) + F(e, \tau, m^*)$$
$$\tilde{D}_0^{(m)} = F(m, m, m^*) \quad , \quad \tilde{D}_0^{(e)} = F(e, e, e^*)$$
$$\tilde{D}_0^{(mme^*)} = F(m, m, e^*) \quad , \quad \tilde{D}_0^{(m^*ee)} = F(e, e, m^*)$$
$$\tilde{D}_0^{(mm^*e)} = F(m, e, m^*) + F(e, m, m^*)$$
$$\tilde{D}_0^{(mee^*)} = F(e, m, e^*) + F(m, e, e^*)$$
$$\tilde{K}_4^{(m)} = G(\tau^*, \tau^*, m^*) \quad , \quad \tilde{K}_4^{(e)} = G(\tau^*, \tau^*, e^*)$$
$$\tilde{A}_4^{(m)} = 2G(\tau^*, m^*, m^*) \quad , \quad \tilde{A}_4^{(e)} = 2G(\tau^*, e^*, e^*)$$
$$\tilde{A}_4^{(me)} = 2G(\tau^*, m^*, e^*) + 2G(\tau^*, e^*, m^*)$$
$$\tilde{D}_4^{(m)} = G(m^*, m^*, m^*) \quad , \quad \tilde{D}_4^{(e)} = G(e^*, e^*, e^*)$$
$$\tilde{D}_4^{(me)} = G(m^*, m^*, e^*) + 2G(m^*, e^*, m^*)$$
$$\tilde{D}_4^{(em)} = G(e^*, e^*, m^*) + 2G(e^*, m^*, e^*) \tag{32.60}$$

(These follow Chu and Munro and are a striking example of the advantage of the eikonal method over the trajectory method: the fact that the for-

	Axial aberration	Magnetic deflection aberration	Electrostatic deflection aberration	Mixed deflection aberration
Spherical aberration	$C_0 s^2 s^*$			
Coma	—	$\tilde{K}_0^{(m)*} s^2 w^{(m)*}$	$\tilde{K}_0^{(e)*} s^2 w^{(e)*}$	
Astigmatism and	—	$2\tilde{K}_0^{(m)} ss^* w^{(m)}$	$2\tilde{K}_0^{(e)} ss^* w^{(e)}$	$\tilde{F}_0^{(em)} sw^{(m)*} w^{(e)}$
Field curvature	—	$\tilde{F}_0^{(m)} sw^{(m)} w^{(m)*}$	$\tilde{F}_0^{(e)} sw^{(e)} w^{(e)*}$	$\tilde{F}_0^{(me)} sw^{(m)} w^{(e)*}$
	—			$\tilde{A}_0^{(me)} s^* w^{(m)} w^{(e)}$
Distortion	—	$\tilde{A}_0^{(m)} s^* w^{(m)2}$	$\tilde{A}_0^{(e)} s^* w^{(e)2}$	$\tilde{D}_0^{(mme^*)} w^{(m)2} w^{(e)*}$
	—	$\tilde{D}_0^{(m)} w^{(m)2} w^{(m)*}$	$\tilde{D}_0^{(e)} w^{(e)2} w^{(e)*}$	$\tilde{D}_0^{(mme^*)} w^{(m)} w^{(m)*} w^{(e)}$
				$\tilde{D}_0^{(mee^*)} w^{(m)} w^{(e)} w^{(e)*}$
Fourfold coma	—	$\tilde{K}_4^{(m)*} s^{*2} w^{(m)*}$	$\tilde{K}_4^{(e)*} s^{*2} w^{(e)*}$	— — —
Fourfold astigmatism	—	$\tilde{A}_4^{(m)} s^* w^{(m)*2}$	$\tilde{A}_4^{(e)} s^* w^{(e)*2}$	$\tilde{A}_4^{(me)} s^* w^{(m)*} w^{(e)*}$
Fourfold distortion	—	$\tilde{D}_4^{(m)} w^{(m)*3}$	$\tilde{D}_4^{(e)} w^{(e)*3}$	$\tilde{D}_4^{(me)} w^{(m)*2} w^{(e)*}$
				$\tilde{D}_4^{(em)} w^{(m)*} w^{(e)*2}$

Table 32.1

Third-order geometrical aberrations of a combined focusing and dual-channel deflection system with magnetic main-field deflection and electrostatic secondary deflection

32.3 THE ABERRATIONS OF DEFLECTION SYSTEMS

mulae for $\tilde{K}_0^{(e)}$ and $\tilde{K}_0^{(m)}$ and their complex conjugates, given separately above, are the same is far from obvious.)

The functions F and G are as follows:

$$F(x_1, x_2, x_3) = F_1 + \frac{1}{\phi_i^{\frac{1}{2}} \tau_i'^* v_{1i} v_{2i} v_{3i}^*} \int_{z_o}^{z_i} \phi^{\frac{1}{2}} (F_2 + F_3) \, dz \quad (32.61a)$$

where

$$F_1 = \frac{n_1 n_2 n_3}{\phi_i^{\frac{1}{2}} \tau_i'^* v_{1i} v_{2i} v_{3i}^*} \left[\phi^{\frac{1}{2}} \tau'^* v_2 \left\{ v_1 v_3^* \left(\frac{\phi''}{32\phi} - \frac{i\eta B'}{16\phi^{\frac{1}{2}}} \right) \right. \right.$$
$$\left. \left. + v_1 \left(\frac{s_3 a}{16\phi} + \frac{i\eta m_3 b}{8\phi^{\frac{1}{2}}} \right) + v_3^* \left(\frac{s_1 a}{8\phi} + \frac{i\eta m_1 b}{4\phi^{\frac{1}{2}}} \right) \right\} \right]_{z_o}^{z_i} \quad (32.61b)$$

$$F_2 = \frac{1}{2} \tau'^* x_1' x_2' x_3'^* + \frac{3}{64} \left(\frac{\phi''}{\phi} \right)^2 \tau^* x_1 x_2 x_3''^*$$
$$- \frac{\phi''}{32\phi} (\tau''^* x_1 x_2 x_3^* + 2\tau^* x_1'' x_2 x_3^* + \tau^* x_1 x_2 x_3''^*$$
$$+ 2\tau'^* x_1 x_2 x_3'^* + 2\tau^* x_1' x_2' x_3^*)$$
$$+ \frac{\phi'' \phi'}{32\phi^2} (\tau'^* x_1 x_2 x_3^* + 2\tau x_1' x_2 x_3^* + \tau x_1 x_2 x_3'^*)$$
$$- \frac{3\phi'' \phi'^2}{128\phi^3} \tau^* x_1 x_2 x_3^*$$
$$+ \frac{i\eta B'}{16\phi^{\frac{1}{2}}} (\tau''^* x_1 x_2 x_3^* - 2\tau^* x_1'' x_2 x_3^* + \tau^* x_1 x_2 x_3''^*$$
$$+ 2\tau'^* x_1 x_2 x_3'^* - 2\tau^* x_1' x_2' x_3^*) \quad (32.61c)$$

$$F_3 = \frac{s_1 a}{8\phi} \left\{ \tau^* \left(\frac{\phi''}{\phi} x_2 x_3^* + \frac{s_2 a}{\phi} x_3^* + \frac{s_3 a}{\phi} x_2 \right) \right.$$
$$- (\tau''^* x_2 x_3^* + \tau^* x_2'' x_3^* + \tau^* x_2 x_3''^* + 2\tau'^* (x_2 x_3^*)')$$
$$\left. + \frac{\phi'}{\phi} (\tau^* x_2 x_3^*)' - \frac{3\phi'^2}{4\phi^2} \tau^* x_2 x_3^* \right\} + \frac{s_2 a}{16\phi} x_3^* \left(4\tau'^* x_1' + \frac{\phi''}{\phi} \tau^* x_1 \right)$$
$$+ \frac{s_3 a}{16\phi} \left\{ \frac{3\phi''}{2\phi} \tau^* x_1 x_2 - \tau''^* x_1 x_2 - 2\tau x_1'' x_2 - 2\tau^* x_1' x_2' \right.$$
$$\left. + \frac{\phi'}{\phi} x_2 (\tau'^* x_1 + 2\tau^* x_1') - \frac{3\phi'^2}{4\phi^2} \tau^* x_1 x_2 \right\}$$
$$- \frac{i\eta b m_1}{4\phi^{\frac{1}{2}}} (\tau''^* x_1 x_3^* - \tau x_2'' x_3^* + \tau x_2 x_3''^* + 2\tau'^* x_2 x_3'^*)$$

$$-\frac{i\eta b m_3}{8\phi^{\frac{1}{2}}}(\tau''^* x_1 x_2 - 2\tau^* x_1'' x_2 - 2\tau^* x_1' x_2') \tag{32.61d}$$

and

$$G(x_1^*, x_2^*, x_3^*) := \frac{1}{2\phi_i^{\frac{1}{2}} \tau'^* x_{1i}^* x_{2i}^* x_{3i}^*} \int_{z_o}^{z_i} \phi^{\frac{1}{2}} \left(\frac{s_3 p_e}{2\phi} + \frac{\eta m_3 P_m}{\phi^{\frac{1}{2}}} \right) \tau^* x_1^* x_2^* \, dz \tag{32.62}$$

The quantities v_j, n_j, m_j and s_j that figure in these definitions take the following values:

w_j	v_j	n_j	m_j	s_j
τ	τ'	0	0	0
σ	σ	1	0	0
m	m	1	1	0
e	e	1	0	1

(32.63)

The field functions p_e and P_m that appear in $G(x_1^*, x_2^*, x_3^*)$ denote the contributions from the field components with threefold symmetry:

$$p_e := \frac{p_3 - iq_3}{U_c} \qquad P_m := -i\frac{P_3 - iQ_3}{I_c} \tag{32.64}$$

(*Note*: the above formulae are slightly less general than those of Chu and Munro, who retain the possibility that the different deflection fields are not aligned; we have set the angles between them equal to zero.)

These expressions can be generalized to include the other principal situations, namely, dual-channel systems in which each channel can be electrostatic or magnetic and shaped-beam systems (Gaussian spot large) with electrostatic or magnetic deflection (but not both). For such shaped-beam systems, with magnetic deflection say, (32.59) becomes

$$w(z) = w_i \frac{\sigma(z)}{\sigma_i} + s_i \frac{\tau(z)}{\tau_i'} + w_i^{(m)} \frac{m(z)}{m_i} \tag{32.65}$$

so that $w_i^{(e)}$ must be replaced by w_i throughout Table 32.1 and column four now contains 'shaped-beam aberrations'; (32.60–32.62) give the aberration integrands on substituting $\sigma(z)$ for $e(z)$. A similar set of substitutions yields the corresponding expressions for a shaped-beam system with electrostatic deflection.

32.3 THE ABERRATIONS OF DEFLECTION SYSTEMS

The general dual-channel system coefficients are again given by (32.60) but the functions F_1, F_3 and G are more complicated:

$$F_1(x_1, x_2, x_3) = \frac{n_1 n_2 n_3}{\phi^{\frac{1}{2}} \tau'^* v_{1i} v_{2i} v_{3i}^*} \left[\phi^{\frac{1}{2}} \tau'^* v_2 \left\{ v_1 v_3^* \left(\frac{\phi''}{32\phi} - \frac{i\eta B'}{16\phi^{\frac{1}{2}}} \right) \right. \right.$$

$$+ v_1 \left(\frac{m_3 a_1 + s_3 a_2}{16\phi} - \frac{i\eta}{8\phi^{\frac{1}{2}}} (m_3 b_1 + s_3 b_2) \right)$$

$$\left. \left. + v_3^* \left(\frac{m_1 a_1 + s_1 a_2}{8\phi} + \frac{i\eta}{4\phi^{\frac{1}{2}}} (m_1 b_1 + s_1 b_2) \right) \right\} \right]_{z_o}^{z_i}$$

$$F_3(x_1, x_2, x_3) = -\frac{m_1 a_1 + s_1 a_2}{8\phi} \left\{ \tau^* \left(\frac{\phi''}{\phi} x_2 x_3^* \right. \right.$$

$$+ \frac{m_2 a_1 + s_2 a_2}{\phi} x_3^* + \frac{m_3 a_1 + s_3 a_2}{\phi} x_2 \right)$$

$$- (\tau''^* x_2 x_3^* + \tau^* x_2'' x_3^* + \tau^* x_2 x_3''^* + 2\tau'^* (x_2 x_3^*)')$$

$$\left. + \frac{\phi'}{\phi} (\tau^* x_2 x_3^*)' - \frac{3\phi'^2}{4\phi^2} \tau^* x_2 x_3^* \right\}$$

$$+ \frac{m_2 a_1 + s_2 a_2}{16\phi} x_3^* \left(4\tau'^* x_1' - \frac{\phi''}{\phi} \tau^* x_1 \right)$$

$$+ \frac{m_3 a_1 + s_3 a_2}{16\phi} \left\{ \frac{3\phi''}{2\phi} \tau x_1 x_2 - \tau''^* x_1 x_2 - 2\tau^* x_1'' x_2 - 2\tau^* x_1' x_2' \right.$$

$$\left. + \frac{\phi'}{\phi} x_2 (\tau'^* x_1 + 2\tau^* x_1') - \frac{3\phi'^2}{4\phi^2} \tau x_1 x_2 \right\}$$

$$- \frac{i\eta}{4\phi^{\frac{1}{2}}} (m_1 b_1 + s_1 b_2)(\tau'''^* x_2 x_3^* - \tau^* x_2'' x_3^* + \tau^* x_2 x_3''^* + 2\tau^* x_2 x_3'^*)$$

$$- \frac{i\eta}{8\phi^{\frac{1}{2}}} (m_3 b_1 + s_3 b_2)(\tau'''^* x_1 x_2 - 2\tau^* x_1'' x_2 - 2\tau^* x_1' x_2') \tag{32.66}$$

$$G(x_1^*, x_2^*, x_3^*) = \frac{1}{2\phi_i^{\frac{1}{2}} \tau_i'^* v_{1i}^* v_{2i}^* v_{3i}^*} \int_{z_o}^{z_i} \phi^{\frac{1}{2}} \left\{ \frac{m_3 p_{e1} + s_3 p_{e2}}{2\phi} \right.$$

$$\left. + \frac{\eta}{\phi^{\frac{1}{2}}} (m_3 P_{m1} + s_3 P_{m2}) \right\} \tau^* x_1^* x_2^* \, dz \tag{32.67}$$

The suffixes 1 and 2 added to $a(z)$, $b(z)$ and $p_e(z)$ and $P_m(z)$ indicate the role of the deflection field in the dual-channel system; thus a_1 and p_{e1} are the field functions for the principal electrostatic deflection field while b_1 and P_{m1} are those for the principal magnetic deflection field; these are of course mutually exclusive. The functions a_2, p_{e2}, b_2 and P_{m2} are the corresponding functions for the subfield.

Chu and Munro also give compact formulae for the chromatic aberration coefficients denoted by C_c and C_{T1} in (32.54). For the most important situation, magnetic main-field deflection and electrostatic sub-field deflection, C_T is divided into two parts, contributing $C_T^{(m)} w_i^{(m)} \Delta\phi/\phi_i$ and $C_T^{(e)} w_i^{(e)} \Delta\phi/\phi_i$ to Δw_c. The coefficients are given by

$$C_c = H(\tau) \qquad C_T^{(m)} = H(m) \qquad C_T^{(e)} = H(e) \qquad (32.68)$$

in which

$$H(x) = \frac{\phi_i^{\frac{1}{2}}}{\tau'^* v_{1i}} \int_{z_o}^{z_i} \phi^{-\frac{1}{2}} \left\{ -\frac{1}{2}\tau'^* x + \frac{\phi'}{\phi}(\tau'^* x + \tau^* x') \right.$$
$$\left. - \frac{3}{16}\left(\frac{\phi'}{\phi}\right)^2 \tau^* x - \frac{s_1 a}{4\phi}\tau^* \right\} dz \qquad (32.69)$$

A shaped-beam system may be incorporated as explained above. In the more general case of the dual-channel system, the function $H(x)$ becomes

$$H(x) = \frac{\phi_i^{\frac{1}{2}}}{\tau'^* v_{1i}} \int_{z_o}^{z_i} \phi^{-\frac{1}{2}} \left\{ -\frac{1}{2}\tau'^* x_1' + \frac{\phi'}{8\phi}(\tau'^* x_1 + \tau^* x_1') \right.$$
$$\left. - \frac{3}{16}\left(\frac{\phi'}{\phi}\right)^2 \tau^* x_1 - \frac{\tau^*}{4\phi}(m_1 a_1 + s_1 a_2) \right\} dz \qquad (32.70)$$

For a further generalization, see Smith and Munro (1986, 1987).

The formulae for the various deflection aberration coefficients, of which these are a representative sample, are manifestly extremely complicated in appearance and, in cases that are becoming common in practice, very numerous. Programming them with no errors is not a light task though it is now possible to output the results of a computer algebra calculation (Chapter 34) in Fortran directly, which reduces the chance of human error. Moreover, when the list of aberration coefficients is long, it is difficult to estimate their relative importance and one is driven to wonder whether this kind of aberration theory is the most appropriate. An alternative approach may prove to be more favourable.

32.4 Stigmators

Stigmators are systems of weak electric or magnetic multipole fields, suitably located within an electron optical device, the role of which is to provide

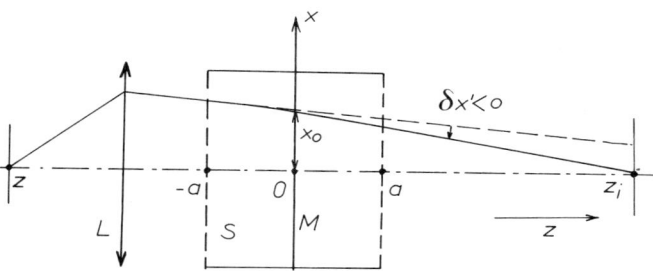

Fig. 32.9: Action of a stigmator. L: lens, S: stigmator, M: midplane. The stigmator field is confined to the domain $-a \leq z \leq a$.

minor correction of the beam. Such corrections are necessary to eliminate some of the aberrations caused by lenses and deflectors and, still more important, those caused by misalignments and similar imperfections (parasitic aberrations). The corrections may be both static and dynamic. Static correction of the geometric imperfections is always necessary; in this case the electrode voltages or coil currents are constant. Dynamic correction is mainly employed in electron lithography devices; the correction currents are then functions of the time-dependent deflection currents.

32.4.1 Necessary simplifications

In this section we shall outline the operating principles of a stigmator. Since an exact treatment requires extensive numerical calculations, we make a number of simplifying assumptions that permit analytic relations to be found. These assumptions are approximately valid in practice. First of all, we assume that the necessary correction fields are *weak* and not superimposed on other strong fields. This means that the stigmator has to be located outside the lens or deflector field region. In the lowest order approximation, the trajectories are then straight lines within the stigmator, and a first-order perturbation yields the small deviations from these straight lines.

The second assumption is illustrated in Fig. 32.9. The stigmator fields must be essentially confined between two planes $z = -a$ and $z = +a$ and symmetric with respect to the midplane $z = 0$. For asymmetric fields, the theory will become more complicated but the conclusions made below will be essentially the same. Moreover, there is no need to produce asymmetric correction fields, which would complicate the construction with no real gain; the limitation to symmetric fields is thus not a serious restriction. This assumption implies that the functions Φ, E_x, E_y, A_z, B_x, B_y are positively symmetric in z, while E_z, A_x, A_y, B_z are antisymmetric. These symmetry requirements can still be satisfied by certain round lens fields, but we shall exclude these, since it is not the purpose of a stigmator to alter the focusing. This is better done by alteration of the lens excitations.

Finally we assume that $x'^2 + y'^2 \ll 1$ for the slopes of the trajectories. Since some correction will result even for $x' = 0$ or $y' = 0$, it is permissible to neglect all terms of second and higher order in x' and y'. The latter would already represent aberrations caused by the stigmator itself. Since the stigmator is assumed to be weak, such aberrations will be extremely small and can hence be ignored. With all these simplifying assumptions, the theory of the stigmator can be expressed in closed form.

32.4.2 The wave aberration

We start from (4.34) and (4.35), and with $x'^2 + y'^2 \ll 1$, $Q = -e$, $g = \sqrt{2m_0 e \hat{\Phi}}$, $z_o = -a$, $z_1 = a$ we obtain

$$S \approx \int_{-a}^{a} \sqrt{2m_0 e} \left\{ \hat{\Phi}^{\frac{1}{2}}(\mathbf{r}) - \eta(x' A_x + y' A_y + A_z) \right\} dz$$

The electric potential has the form

$$\Phi(\mathbf{r}) = \phi + \Phi_s(\mathbf{r}), \qquad |\Phi_s| \ll \phi$$

ϕ being the accelerating voltage and Φ_s the very small stigmator potential. The approximation

$$\sqrt{\hat{\Phi}} = \sqrt{\hat{\phi}} + \frac{\gamma}{2\hat{\phi}^{1/2}} \Phi_s$$

is then justified. We now drop the constant term in S and normalize with respect to the factor $(2m_0 e \hat{\phi})^{1/2}$; this yields a convenient form of the characteristic function, called now the *wave aberration*:

$$W := 2a - \frac{S}{(2m_0 e \hat{\phi})^{1/2}} = \int_{-a}^{a} \left\{ -\frac{\gamma}{2\hat{\phi}} \Phi_s(\mathbf{r}) + \frac{\eta}{\sqrt{\hat{\phi}}} (x' A_x + y' A_y + A_z) \right\} dz \tag{32.71}$$

This integral is to be evaluated for an arbitrary *straight* trajectory with small slopes. The corresponding cartesian representation is

$$x(z) = x_0 + z x', \qquad y(z) = y_0 + z y' \tag{32.72}$$

x_0 and y_0 being the coordinates of the point of intersection with the symmetry plane. Introducing (32.72) into (32.71) and recalling the assumptions

32.4 STIGMATORS

made above, we find that the terms in A_x and A_y cancel out. In order to evaluate the remaining terms, we introduce the Taylor series expansions of Φ_s and A_z in x' and y' and retain only linear terms:

$$\Phi_s(\mathbf{r}) = \Phi_s(z, x_0, y_0) + z\left(x'\Phi_{s|x} + y'\Phi_{s|y}\right)_{(z,x_0,y_0)}$$

$$A_z(\mathbf{r}) = A_z(z, x_0, y_0) + z\left(x'A_{z|x} + y'A_{z|y}\right)_{(z,x_0,y_0)}$$

Introducing this into (32.71) and bearing in mind the symmetry properties, we see that the terms in x' and y' again cancel out. We hence obtain a very simple formula for W:

$$W(x_0, y_0) = \int_{-a}^{a} \left(-\frac{\gamma \Phi_s(z, x_0, y_0)}{2\hat{\phi}} + \frac{\eta}{\sqrt{\hat{\phi}}} A_z(z, x_0, y_0)\right) dz \qquad (32.73)$$

The corrections introduced by the stigmator are therefore determined by the coordinates (x_0, y_0) of the trajectory in the midplane.

An interesting rule is obtained by forming the two-dimensional Laplacian of W. Differentiation under the integral results in

$$\nabla_2^2 W(x_0, y_0) \equiv \frac{\partial^2 W}{\partial x_0^2} + \frac{\partial^2 W}{\partial y_0^2}$$

$$= -\frac{\gamma}{2\hat{\phi}} \int_{-a}^{a} \left(\frac{\partial^2 \Phi_s}{\partial x_0^2} + \frac{\partial^2 \Phi_s}{\partial y_0^2}\right) dz + \frac{\eta}{\sqrt{\hat{\phi}}} \int_{-a}^{a} \left(\frac{\partial^2 A_z}{\partial x_0^2} + \frac{\partial^2 A_z}{\partial y_0^2}\right) dz$$

Now $\Phi_s(\mathbf{r})$ and $A_z(\mathbf{r})$ must satisfy the Laplace equation, since the space charge of the beam can be neglected in the vast majority of practical applications and other sources are always located outside the beam. For Φ_s,

$$\int_{-a}^{a} \left(\frac{\partial^2 \Phi_s}{\partial x_0^2} + \frac{\partial^2 \Phi_s}{\partial y_0^2}\right) dz = -\int_{-a}^{a} \frac{\partial^2 \Phi_s}{\partial z^2} dz = \left[-\Phi_{s|z}(z, x_0, y_0)\right]_{-a}^{a} = 0$$

since at and beyond the boundaries $z = \pm a$, all the corrector functions considered vanish in view of our assumptions about the confinement. A similar argument shows that the magnetic term also vanishes and hence

$$\nabla_2^2 W \equiv \frac{\partial^2 W}{x_0^2} + \frac{\partial^2 W}{\partial y_0^2} = 0 \qquad (32.74)$$

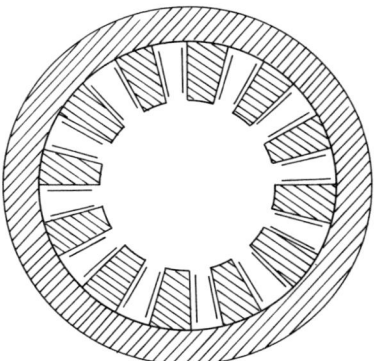

Fig. 32.10: Cross-section through a magnetic 12-pole stigmator.

The operation of a stigmator can now be understood in the following way. In the midplane, the system of lenses and deflectors located upstream produces a wave aberration $W_L(x_0, y_0)$. The stigmator adds the contribution $W(x_0, y_0)$ so that for the part of the beam downstream the wave aberration is $W_c = W_L + W$. The average value of $|W_c|$ over the cross-section of the beam at $z = 0$ should be made as small as possible. It is obvious that only geometric aberrations whose wave aberrations already satisfy $\nabla_2^2 W_L = 0$ can be eliminated completely. These are paraxial deflections and *astigmatisms* of various multiplicity. Their wave aberrations are best given in polar form

$$W_L(r_0, \varphi_0) = \sum_{n=1}^{N} r_0^n (a_n \cos n\varphi_0 + b_n \sin n\varphi_0) \qquad (32.75)$$

$4N$ being the number of electrodes or poles of the stigmator. Figure 32.10 shows a magnetic 12-pole stigmator which is suitable for the correction of astigmatisms up to $N = 3$. It is, however, never possible to correct spherical aberration, since the latter produces a wave aberration

$$W_L = \mathrm{const}(x_0^2 + y_0^2)^2 = \mathrm{const} \times r_0^4, \qquad \nabla_2^2 W_L \neq 0$$

32.4.3 The deflection of trajectories

The influence of a stigmator on the electron beam can also be understood in another useful manner. The trajectories are weakly bent by the stigmator fields, the alteration of direction being simply given by

$$\Delta x' = -\frac{\partial W}{\partial x_0}, \qquad \Delta y' = -\frac{\partial W}{\partial y_0} \qquad (32.76)$$

32.4 STIGMATORS

Carrying out this differentiation in the integrand of (32.73) and recalling that $\boldsymbol{E} = -\nabla\Phi$, we obtain

$$\Delta x'_j = -\frac{\gamma}{2\hat{\phi}} \int_{-a}^{a} E_j(z, x_0, y_0)\, dz - \frac{\eta}{\sqrt{\hat{\phi}}} \int_{-a}^{a} A_{z|j}\, dz, \qquad j = 1, 2$$

In the second term we substitute $A_{z|x} = A_{x|z} - B_y$ and $A_{z|y} = A_{y|z} + B_x$. The derivatives with respect to z do not contribute to the integral, since $\int_{-a}^{a} A_{x|z}\, dz = [A_x]_{-a}^{a} = 0$, for example. Finally, then, we obtain

$$\Delta x' = -\frac{\gamma}{2\hat{\phi}} \int_{-a}^{a} E_x(z, x_0, y_0)\, dz + \frac{\eta}{\sqrt{\hat{\phi}}} \int_{-a}^{a} B_y(z, x_0, y_0)\, dz$$

$$\Delta y' = -\frac{\gamma}{2\hat{\phi}} \int_{-a}^{a} E_y(z, x_0, y_0)\, dz - \frac{\eta}{\sqrt{\hat{\phi}}} \int_{-a}^{a} B_x(z, x_0, y_0)\, dz \tag{32.77}$$

This is a Busch approximation of first order, since the true trajectory over which the integration should strictly be made is replaced by a simple straight line parallel to the optic axis. Apart from correction terms of second order in x' and y', this gives essentially the same results as integration over the straight line given by (32.72).

The stigmator can now be considered as an optically thin element, which bends the trajectories sharply in the midplane, the change of gradient being given by (32.77). This result is very useful for practical applications.

Part VI

Computer-Aided Electron Optics

33
Numerical Calculation of Trajectories, Paraxial Properties and Aberrations

33.1 Introduction

In the preceding section, and also in many later paragraphs, various forms of trajectory equations, paraxial quantities and aberration coefficients are presented but their practical evaluation has not yet been discussed, important though this topic is. Attitudes to this question have changed dramatically with the rapid advance of computer technology in the past few decades. When the basic trajectory equations were derived (Störmer, 1907; Busch, 1927) and the aberration theory developed (Glaser, 1933, 1935, 1952; Scherzer, 1936), the resulting differential equations and aberration integrals had to be solved by laborious hand calculations. It is clear that under these circumstances the amount of numerical calculation necessary had to be reduced to the absolute minimum and that analytical calculations were preferred, however sophisticated, whenever this was possible. There is no doubt that these constraints strongly influenced the early development of electron optics.

Fortunately, the accurate solution of systems of ordinary differential equations is nowadays no longer an obstacle, even for comparatively complicated mathematical structures; in Section 33.2 we shall present a numerical procedure that has proved very useful. In combination with advanced techniques for the differentiation of axial fields (see Chapter 13) and for integration, all the paraxial properties and aberration coefficients of electron optical systems can be calculated with sufficient accuracy.

For highly complex systems like the arrangements of lenses and deflectors in lithography devices or the sequences of multipole lenses needed for aberration correction, the derivation of aberration coefficients and the writing of a computer program to evaluate them are themselves major tasks. The labour involved can be alleviated with the aid of computer algebra systems (Chapter 34) but without such facilities, when only hand calculation remains, the organization of the corresponding programs really does seem hopeless. A brief account of the types of task that can be performed by computer algebra languages is given in Chapter 34.

Another way of determining the effects of aberrations is to plot the endpoints of a large number of accurately calculated electron trajectories, the spot diagrams of light optics. This will be discussed in some detail in Sections 33.5 and 33.6. Although this is not yet very common in electron optics, it does offer a means of assessing the performance of a complex system.

33.2 Numerical solution of ordinary differential equations

We have seen that the equations describing the motion of electrons can be cast into various different forms. We shall not discuss these again, but simply state that finally, after certain mathematical transformations, they can always be cast into the general form

$$y'_i(x) = f_i(x, y_1(x), \ldots y_N(x)) \qquad i = 1 \ldots N \qquad (33.1)$$

or in more concise vector notation

$$\boldsymbol{y}'(x) = \boldsymbol{f}(x, \boldsymbol{y}) \qquad (33.2)$$

We assume that this transformation has been made before embarking on any attempt to obtain a numerical solution. The variables x and $\boldsymbol{y}$ may have various physical meanings or even none. For instance, the variable x may be the time, the arc-length, the axial coordinate or, indeed, none of these. In the following discussion, the particular meaning is unimportant. For physical reasons, the vector function $\boldsymbol{f}(x, \boldsymbol{y})$ must be smooth in all its arguments and we exclude forthwith any cases in which exceptions such as singularities or discontinuities appear.

The solution of (33.2) for given initial values

$$\boldsymbol{y}(x_0) = \boldsymbol{y}_0(x_0) \quad \Rightarrow \quad \boldsymbol{y}'(x_0) = \boldsymbol{f}(x_0, \boldsymbol{y}_0) =: \boldsymbol{y}'_0 \qquad (33.3)$$

is a standard problem of numerical analysis. In practically every major textbook on this subject, it is dealt with in detail and a subprogram for solving it numerically is available in every computer routine library. The methods employed are, however, of unequal suitability for electron optical applications. We therefore outline the various standard methods very briefly with a few remarks concerning their advantages and disadvantages. A good general reference is the collection of surveys edited by Jacobs (1977).

33.2.1 The Fox–Goodwin–Numerov method
This method, proposed by Numerov (1923), Manning and Millman (1938)

SOLUTION OF ORDINARY DIFFERENTIAL EQUATIONS 527

and Fox and Goodwin (1949) and introduced into electron optics by Burfoot (1952) and Jennings and Pratt (1955), is designed for the solution of differential equations of the form

$$y''(x) + f(x)y(x) = g(x) \tag{33.4}$$

which can be solved directly without conversion into the standard form (33.2). The method is a two-step procedure, which requires *constant* step-width h in x. With the abbreviation $y_n := y(x_0 + nh)$ and likewise for f and g, the corresponding discretization formula is given by

$$(1 + af_{n-1})y_{n-1} - (2 - 10af_n)y_n + (1 + af_{n+1})y_{n+1}$$
$$= a(g_{n-1} + 10g_n + g_{n+1}) + O(h^6) \tag{33.5}$$

with $a := h^2/12$.

This formula can be used in two ways. For *initial-value problems*, two starting values y_0 and y_1 are required. In each integration step with $n \geq 1$, (33.5) is then to be solved for y_{n+1}. In *boundary-value problems*, two boundary values y_0 and y_{N+1} are given and the solution is required within the boundaries $x_0 \leq x \leq x_{N+1}$. Then (33.5) represents a *tridiagonal* system of equations for $y_1 \ldots y_N$. This can be solved by well-known standard procedures, for instance the Gauss-elimination algorithm.

The main advantage of (33.5) is its simplicity. A suitable program can be implemented even on a small computer. In electron optics, the typical examples for the application of this method are the solution of the paraxial trajectory equations in their forms (15.32) and (15.38) and the inhomogeneous generalizations of these. The Numerov method does, however, have serious drawbacks. The first is its lack of flexibility: in very many practical cases, the special form (33.4) does not occur. The second important drawback is the restriction to constant step-width together with the lack of any control over the accuracy. In doubtful cases, the whole calculation has to be repeated after halving the step-width until sufficient convergence of the solution is obtained. We emphasize that in this simple form (33.5), the procedure is sensitive to rounding errors when a very small step-width is chosen. Other drawbacks are the need to set up a special routine to obtain the value y_1, if the initial values y_0 and y_0' are given, and the fact that additional formulae for the differentiation must be introduced when $y'(x)$ is needed as well as $y(x)$. In view of all these disadvantages, this method can hardly be recommended if a large computer is available.

33.2.2 The Runge–Kutta method

This method is free of most of the disadvantages of the Numerov method, when applied to the paraxial ray equations; moreover, it is suitable for solving the general problem (33.2) with (33.3). The Runge–Kutta (RK) method is the standard procedure for dealing with such equations and has thus been widely employed in physics and engineering. A complete list of all the publications in which it is used, even for the specific task of electron trajectory calculation, would be much too long to include here.

The RK method may be formulated with different orders of approximation up to the fourth order and for the latter there is more than one version. The classical fourth-order RK procedure is to be found in every textbook on numerical analysis. A more refined version, which rapidly became standard, was derived by Gill (1951; see also Romanelli, 1960). We now present this procedure briefly, without giving the derivation. For conciseness, we introduce the notation

$$y_n = y(x_n), \qquad y'_n = y'(x_n) = f(x_n, y_n), \qquad n \geq 0 \qquad (33.6)$$

and assume that this data set has been computed correctly. We now wish to compute the next data set referring to $x_{n+1} = x_n + h$. In a form in which the symbol ':=' means replacement, in the sense employed by computer languages and which is hence suitable for practical programming, the RK algorithm is given by

$$D := 0, \qquad y' := y'_n, \qquad (q := 0) \qquad (33.7a)$$

$$\left. \begin{array}{l} k := y' \\ D := D + ha_j(k - b_j q) \\ q := q + 3a_j(k - b_j q) - c_j k \\ y' := f(x_n + hd_j, y_n + D) \end{array} \right\} j = 1 \ldots 4 \qquad (33.7b)$$

$$y_{n+1} := y_n + D, \qquad y'_{n+1} := y' \qquad (33.7c)$$

The constants appearing in this procedure are as follows:

j	a_j	b_j	c_j	d_j
1	$\frac{1}{2}$	2	$\frac{1}{2}$	$\frac{1}{2}$
2	$1 - \sqrt{\frac{1}{2}}$	1	$1 - \sqrt{\frac{1}{2}}$	$\frac{1}{2}$
3	$1 + \sqrt{\frac{1}{2}}$	1	$1 + \sqrt{\frac{1}{2}}$	1
4	$\frac{1}{6}$	2	$\frac{1}{2}$	1

The introduction of the increment D is not absolutely necessary but is favourable in connection with the method outlined in the next section,

SOLUTION OF ORDINARY DIFFERENTIAL EQUATIONS   529

and furthermore, y' and y'_{n+1} can occupy the same memory location. The starting values $q = 0$ are only necessary at the first step of the routine ($n = 0$); in the subsequent steps, the accumulation of q helps to minimize rounding errors (Romanelli, 1960).

The step-width h can be altered in the RK procedures. There are, however, no direct criteria for choosing it appropriately and this is the main drawback of all RK methods. The accuracy can be controlled reliably only by repeating each integration step with half the step-width and comparing the two solutions $y^{(1)}_{n+1}$ and $y^{(2)}_{n+1}$ referring to the same abscissa. Their differences, or better that between the corresponding increments D (see 33.7), is a reliable measure of the discretization error and can be used to adjust the step-width appropriately. This will be described in the next section. Moreover, this difference can be used to reduce the error still further by means of an extrapolation:

$$y_{n+1} = y^{(2)}_{n+1} + \frac{1}{15}(y^{(2)}_{n+1} - y^{(1)}_{n+1}) \qquad (33.8)$$

where $y^{(2)}_{n+1}$ denotes the result with twice as many steps of length $h/2$.

In this way, a highly accurate solution can be obtained but at the price of evaluating the function $f(x, y)$ very many times. In electron optics, each call of this function entails a complete calculation of the electromagnetic field strengths, E and B, at the reference point. Since these time-consuming calculations represent the major part of the computation time, their total number should be kept as small as possible. In this respect, the RK method is unfavourable; it is nevertheless necessary to initiate the predictor-corrector method, which is the subject of the next section.

33.2.3 The predictor-corrector method

The predictor-corrector (PC) method is a multistep procedure; this means that the last vector y_{n+1} is calculated by forming an appropriate linear combination of the preceding ones. Since this can be done in different ways, it automatically provides an accuracy control. We have to distinguish between predictor and corrector formulae. The former have the general structure

$$y_{n+1} \approx P_{n+1} = \sum_{\mu=0}^{m} a_\mu y_{n-\mu} + h \sum_{\nu=0}^{m'} a'_\nu y'_{n-\nu} \qquad (33.9a)$$

The corresponding abscissae must be *equidistant* and y'_{n+1} must not appear on the right-hand side. By means of (33.9a) an approximate vector $y'_{n+1} \approx f(x_{n+1}, P_{n+1})$ is obtained. This is then substituted into the corrector formula

$$y_{n+1} \approx D_{n+1} = \sum_{\mu=0}^{s} b_\mu y_{n-\mu} + h \sum_{\nu=-1}^{s'} b'_\nu y'_{n-\nu} \qquad (33.9b)$$

which generally gives a better approximation. The error vector

$$\mathbf{\Delta}_{n+1} := D_{n+1} - P_{n+1} \qquad (33.10)$$

represents the discretization error; its magnitude $\Delta_{n+1} = |\mathbf{\Delta}_{n+1}|$ can be used as a control and for adjustment of the step-width h. If we assume that the discretization error is of fifth order in h and has to be smaller than a given tolerance limit ϵ, then the procedure

$$\left. \begin{array}{l} \Delta_{n+1} \geq \epsilon : h \text{ to be halved,} \\ 0.02\epsilon \leq \Delta_{n+1} < \epsilon : h \text{ maintained} \\ \Delta_{n+1} < 0.02\epsilon : h \text{ to be doubled} \end{array} \right\} \qquad (33.11)$$

is appropriate. In the case $\Delta_{n+1} \geq \epsilon$ the result for y_{n+1} is too inaccurate and is hence rejected. By means of interpolation formulae, the vectors

$$y_{n-\mu/2} := y(x_n - \mu h/2), \qquad \mu = 1, 3, 5, \ldots \qquad (33.12)$$

and the corresponding derivatives are then calculated; the explicit formulae will be given below. The last integration step, yielding y_{n+1}, is now repeated with $h/2$. The other two cases in (33.11) provide no problem.

The procedure outlined above is essentially the same in all versions of the PC method. A review of the various choices for the sets of coefficients and some refinements are presented by Ralston (1960). In Tübingen, one of these versions was employed for the computation of electron trajectories until 1980 (Hoch et al., 1976; Hauke, 1977; Kern, 1978; Niemitz, 1980) but the very high accuracy required for the calculation of Lorentz trajectories in electron guns stimulated further improvements (Kasper, 1982, 1984). In order to minimize rounding errors, the whole procedure is represented in incremental form. This means that the increments

$$D_n := y_n - y_{n-1}$$

are calculated directly and not by subtraction. A suitable procedure, called HPCD, is as follows ($n \geq 3$):

$$P_{n+1} := \frac{h}{3}\{7(y'_n + y'_{n-2}) - 8y'_{n-1}\} - D_{n-2} + \frac{29}{90}h^5 y_n^{(5)}$$

SOLUTION OF ORDINARY DIFFERENTIAL EQUATIONS 531

$$Q_{n+1} := y_n + P_{n+1} + \frac{116}{125}\Delta_n$$
$$x_{n+1} := x_n + h, \qquad Q'_{n+1} = f(x_{n+1}, Q_{n+1}) \approx y'_{n+1}$$
$$D_{n+1} := \frac{1}{8}\{D_n + D_{n-1} + 3h(Q'_{n+1} + 2y'_n - y'_{n-1})\} - \frac{h^5}{40}y_n^{(5)}$$
$$\Delta_{n+1} := D_{n+1} - P_{n+1}, \qquad \text{control (33.11)}$$
$$D_{n+1} := D_{n+1} - \frac{9}{125}\Delta_{n+1}$$
$$y_{n+1} := y_n + D_{n+1}$$
$$y'_{n+1} := f(x_{n+1}, y_{n+1}) \quad \text{(exceptional case)} \tag{33.13}$$

This sequence represents a regular step after the necessary initializations. The vector P_{n+1} is here an incremental predictor; note that this is not the classical formula, which is less suitable for this purpose. Q_{n+1} is a *modified* predictor: the last term approximates the discretization error to a high degree of accuracy and hence Q_{n+1} is already a very good approximation for y_{n+1} and thus $Q'_{n+1} = y'_{n+1}$. Normally, this latter approximation should be quite sufficient and this has been confirmed in practical tests. The vector D_{n+1} is Hamming's corrector in incremental form (for the original form, see Ralston, 1960). The error vector Δ_{n+1} is here less sensitive to rounding errors, since only small increments are to be subtracted and not large y-vectors; D_{n+1} is now the incremental *modified* corrector, the factors 116/125 and $-9/125$ being chosen in such a way that the error terms of fifth order, given explicitly in (33.13), cancel out. The final statement is only necessary in exceptional cases and hence the function f has to be evaluated only once in each regular integration step. The remainder is $h^6 y^{(6)}/57.6$.

The necessary initial values y_1, y_2, y_3, their increments and derivatives can be calculated by means of the RK method, as explained above. The routine (33.7) already has the desired incremental form if we identify D with D_{n+1}. The difference

$$\Delta_{n+1} := y_{n+1}^{(2)} - y_{n+1}^{(1)} \equiv D_{n+1}^{(2)} - D_{n+1}^{(1)} \tag{33.14}$$

is introduced into (33.11) in order to adjust the step-width appropriately and into the extrapolation formula (33.8). The last error vector Δ_3 supplied by the initialization routine is, however, not that needed at the start of (33.13). This problem will be dealt with below.

We now return to the alteration of the step-width h. Doubling it raises no problem apart from the appropriate reorganization of the data sets. This can be carried out, once five acceptable data sets with subscripts $n+1, \ldots n-3$ have been stored. When the step-width needs to be halved, two interpolations must be carried out. The original formulae are given by Ralston (1960). Their incremental forms, needed here, are given by

$$D_{n-\frac{1}{2}} \equiv y_n - y_{n-\frac{1}{2}} = \frac{1}{256}(176D_n + 41D_{n-1} + D_{n-2})$$
$$- \frac{h}{256}(-15y'_n + 90y'_{n-1} + 15y'_{n-2})$$
$$D_{n-\frac{3}{2}} \equiv y_{n-1} - y_{n-\frac{3}{2}} = \frac{1}{256}(-12D_n + 109D_{n-1} + D_{n-2})$$
$$+ \frac{h}{256}(3y'_n + 54y'_{n-1} - 27y'_{n-2})$$

(33.15)

The derivatives $y'_{n-1/2}$ and $y'_{n-3/2}$ are obtained not by interpolation but by evaluation of the function $f(x, y)$. Thereafter the appropriate reorganization of the data is straightforward.

At the start ($n = 3$) of the PC-loop and immediately after each alteration of the step-width, the vector Δ_n appearing in (33.13) does not have the appropriate components. Instead of deriving approximate formulae (Ralston, 1960), it is better to adopt the following procedure: the sequence (33.13) is started with $\Delta_n = 0$ and the exceptional instruction is executed. The resulting derivative y'_{n+1} is identical with Q'_{n+1} and again introduced into D_{n+1}, after which Δ_{n+1}, D_{n+1} and y_{n+1} are recalculated.

The whole integration procedure can be stopped when the abscissa x or one of the components of y or y' satisfies some prescribed final condition. Since this will often not be reached exactly, the integration is halted once a solution has been found within the appropriate interval. An interpolation in x can then be carried out on the basis of a quintic polynomial using the values (x_i, y_i, y'_i) for $i = n + 1, n, n - 1$. When one component $y_m(x)$ of y must satisfy a condition of the form $y_m(x) = y_m^E$, this equation, set up with the corresponding interpolation polynomial, is to be solved numerically for x. An interpolation then gives all correct end-values y^E, $y^{E\prime}$.

Practical experience with this method has been quite satisfactory. Once a corresponding computer program has been written and tested, the solution of (33.2), (33.3) is a mere matter of routine, whenever there are no singularities or discontinuities. The latter are revealed automatically, as the repeated halving of the step-width h is terminated when h reaches some given lower bound. With an error limit of $\epsilon = 10^{-9}$, an accuracy of this order is obtained, for not too long integration intervals at least. The procedure remains numerically stable when ϵ is decreased further and the accuracy increases accordingly.

In this context it is essential that the programs for calculating the electromagnetic fields supply sufficiently smooth results. When the field strengths are calculated by means of interpolation between values stored in a grid, all their components must at least have continuous derivatives of first order at the grid lines, as major discontinuities may cause a breakdown of the algorithm for adjusting the step size. The interpolation techniques described in Chapter 13 satisfy this requirement.

For practical purposes it is helpful to provide an output routine capable of supplying the data $(x_j, \boldsymbol{y}_j, \boldsymbol{y}'_j)$, $j = n, n+1$, immediately after each successful integration step. The integrands of the aberration coefficients can then be summed externally while the differential equations are being solved without any modification of the HPCD program itself. The fullest flexibility in every situation is obtained if the output routine permits the tolerance limits and the final conditions to be altered according to the particular requirements.

We conclude that today, a single standard routine is capable of providing efficiently an accurate solution of any non-singular system of ordinary differential equations and that there is hence no need to simplify these before attempting to find a solution. This is undoubtedly important for the future development of electron optics.

33.3 Standard applications in electron optics

In very many practical cases, the geometrical forms and the excitations of a lens system are given and hence the electromagnetic field is known, either as a result of calculation or from measurement. It is then the asymptotic or real paraxial properties or a mixed form of these and the corresponding aberration coefficients that are usually required.

33.3.1 Initial-value problems

Let us assume that no conditions are imposed at the aperture; discussion of such conditions is deferred to the next section. We are thus confronted with an ordinary initial-value problem. The starting plane may be located in field-free space at a reasonable distance from the lenses or coincide with a real object plane $z = z_o$. We have to choose the appropriate form of the fundamental solutions $v_{1,2}(z)$ of the paraxial ray equations and calculate them for the specific initial values. The integration runs either to a terminal plane in the field-free space on the far side of the lens system or to the conjugate image plane $z = z_i$, which is a zero of the solution $v_2(z)$ for which $v_2(z_o) = 0$. The paraxial ray equations have the general form

$$\frac{d}{dz}\left(P(z)v'_j(z)\right) + Q(z)v_j(z) = 0, \quad j = 1, 2 \qquad (33.16)$$

Identifying y_1 with v_1 and y_2 with v_2, the corresponding system of first order may be written:

$$y_1' = y_3/P, \quad y_2' = y_4/P \\ y_3' = -Qy_1, \quad y_4' = -Qy_2 \quad \} \qquad (33.17)$$

The components $y_3 = Pv_1'$, $y_4 = Pv_2'$ are then the associated momenta. An equivalent form is

$$v_j''(z) + a(z)v_j' + b(z)v_j(z) = 0, \qquad j = 1, 2 \qquad (33.18)$$

Here we set

$$y_1' = y_3, \quad y_2' = y_4 \\ y_3' = -ay_3 - by_1 \quad y_4' = -ay_4 - by_2 \quad \} \qquad (33.19)$$

and $y_3 = v_1'$, $y_4 = v_2'$ are now the derivatives of first order. With HPCD both possibilities can likewise be assimilated. In the latter case, the derivatives of second order, $y_3' \equiv v_1''$, $y_4' = v_2''$, are also calculated automatically. On the other hand, the Numerov method requires the Picht transform that converts $P(z)$ to unity or eliminates $a(z)$ to be applied and does not even give v_1' and v_2'; it is thus distinctly unattractive.

The above transformations remain valid if all the functions involved take complex values, the abscissa z remaining real. A solution could then be found with a program for complex systems of ordinary differential equations but very often such a program is not available. It is not even necessary to write it, as the systems (33.17) and (33.19) can easily be split into their real and imaginary parts. The resulting system of real differential equations then has the rank $N = 8$. Very often the coefficients $P(z)$, $Q(z)$ and $a(z)$, $b(z)$, are *real*. There is then no loss of generality in assuming that the solutions $v_1(z)$ and $v_2(z)$, which must, of course, be linearly independent, are likewise real. Arbitrary complex solutions can then be obtained by appropriate linear combinations of these, the coefficients being complex.

The aberration integrals may be of the quite general form

$$C = \int_{z_o}^{z_i} F(z; f, f', \ldots f^{(4)}; v_1, v_1', v_1''; v_2, v_2', v_2'') \, dz \qquad (33.20)$$

$f(z)$ being any axial lens function such as the axial potential $\phi(z)$ or the flux density $B(z)$ or even any set of such functions. With the techniques for differentiation and interpolation outlined in Chapter 13, even derivatives of high order can be calculated accurately, so that it is not necessary to eliminate them by partial integration, though the latter is certainly favourable.

33.3 STANDARD APPLICATIONS IN ELECTRON OPTICS

At least in the absence of such higher order derivatives, simple integration by means of the trapezoidal rule is quite sufficient.

33.3.2 Boundary-value problems

In electron optics, the imposition of aperture conditions is the classic example of a boundary-value problem. We wish to solve (33.16) or (33.18) for two paraxial rays $s(z)$ and $t(z)$ satisfying

$$s(z_o) = t(z_a) = 1, \qquad s(z_a) = t(z_o) = 0 \tag{33.21}$$

This creates no intrinsic problem. First we calculate two other independent solutions $g(z)$, $h(z)$ satisfying the standard initial conditions

$$g(z_o) = h'(z_o) = 1, \qquad g'(z_o) = h(z_o) = 0 \tag{33.22}$$

These give us the values $g(z_a)$, $h(z_a)$ in the aperture plane; $s(z)$ and $t(z)$ are now obtained as linear combinations of g and h. Explicit determination of these linear combinations after running the integration routines is unfavourable since all the computed data would have to be stored. Instead, it is better to recommence the integration with the initial conditions

$$\begin{aligned} s(z_o) &= 1, & s'(z_o) &= -\frac{g(z_a)}{h(z_a)} \\ t(z_o) &= 0, & t'(z_o) &= \frac{1}{h(z_a)} \end{aligned} \tag{33.23}$$

While this second integration is proceeding, the appropriate integrals for the aperture-dependent aberration coefficients can be evaluated simultaneously.

33.3.3 Eigenvalue problems

Very often, the positions of the object and image planes are fixed by technical requirements and the appropriate lens excitations are not known. This problem can arise for electrostatic lenses as well as for magnetic ones. For clarity, we shall consider these two cases separately.

In *electrostatic lenses* the field is usually produced by two or three electrodes with two different potentials U_1 and U_2. In a lens with three electrodes, the two outer ones are usually at the same potential U_1, and we shall assume this to be the case, although a further generalization provides no problem. The axial potential $\phi(z)$ must now be a linear combination of the form

$$\phi(z) = U_1 \phi_1(z) + U_2 \phi_2(z) \tag{33.24}$$

ϕ_1 and ϕ_2 being the partial potentials for $U_1 = 1$, $U_2 = 0$ and $U_1 = 0$, $U_2 = 1$ respectively. We now introduce a dimensionless lens parameter $\lambda := U_2/U_1$. The nonrelativistic ray equation in reduced coordinates can be cast into the form

$$\tau'' + \frac{3}{16}\left(\frac{\phi_1'(z) + \lambda\phi_2'(z)}{\phi_1(z) + \lambda\phi_2(z)}\right)^2 \tau(z) = 0 \qquad (33.25)$$

with

$$\tau(z_o) = \tau(z_i) = 0, \qquad \tau'(z_o) = 1 \qquad (33.26)$$

The problem is to determine the parameter λ.

In the case of *magnetic lenses* we must assume that there is *no saturation*, so that the field strength is proportional to the coil current I. Otherwise an alteration of I leads to a completely different field distribution whereupon the entire field calculation procedure has to be repeated. We now cast the axial flux density into the form

$$B(z) = \mu_0 J b(z) \qquad (33.27)$$

$J := NI$ being the number of ampère-turns and $b(z)$ a geometric form factor with $\int_{-\infty}^{\infty} b(z)dz = 1$. We also introduce a dimensionless lens strength

$$\lambda^2 := \frac{e(\mu_0 J)^2}{8m_0 \hat{\phi}} \qquad (33.28)$$

The ray equation then takes the form

$$\tau''(z) + \lambda^2 b^2(z) \tau(z) = 0 \qquad (33.29)$$

where (33.26) is again to be satisfied.

Obviously the basic underlying structure in both special cases is of the form

$$\tau''(z, \lambda) + a(z, \lambda)\tau'(z, \lambda) + b(z, \lambda)\tau(z, \lambda) = 0 \qquad (33.30)$$

in combination with (33.26) or even more general boundary conditions. An appropriate technique for the solution of such problems is the well-known *shooting-method*. Here we shall describe only an elementary version of it, which is quite sufficient in electron optics (see Fig. 33.1).

Let us ignore for a moment the condition $\tau(z_i) = 0$, or more completely $\tau(z_i, \lambda_i) = 0$, where λ_i is the appropriate, but unknown, eigenvalue of the parameter λ. The function $\tau(z, \lambda)$ then has a zero at $z = z_f(\lambda)$, which can

33.4 DIFFERENTIAL EQUATIONS FOR THE ABERRATIONS

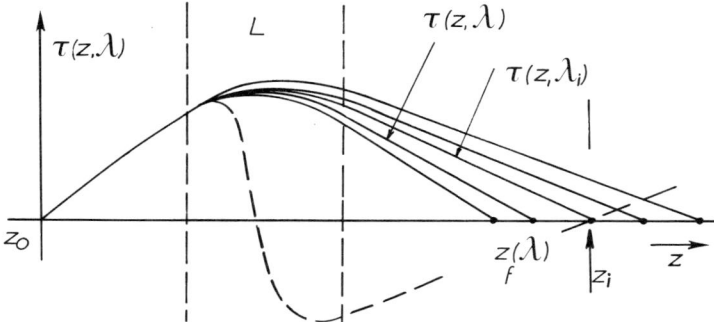

Fig. 33.1: The shooting method in electron optics. The variable strength λ of the lens L is to be determined in such a way that the paraxial ray $\tau(z, \lambda_i)$ has a zero at $z = z_i$. The heavy line indicates the first eigenfunction and the dashed line the second.

always be determined by solving the corresponding initial-value problem. We now define a function

$$F(\lambda) := \frac{(z_i - z_o)}{(z_f(\lambda) - z_o)} - 1 \qquad (33.31)$$

which remains finite for all values of $z_f(\lambda) \neq z_o$, even for $|z_f(\lambda)| \to \infty$. Obviously $F(\lambda_i) = 0$ leads to the required condition $z_f(\lambda_i) = z_i$ and hence we have to determine the zeros of $F(\lambda)$. This can be done iteratively with a suitable program for the computation of zeros of arbitrary continuous functions. Since each evaluation of $F(\lambda)$ requires a complete ray trace, this program for finding zeros must work very efficiently, but this is no problem.

In the case of very high magnification, the task of solving $F(\lambda) = 0$ as $z_i \to \infty$ causes problems. In such a case, it is more reasonable to replace (33.31) by $F(\lambda) = \tau'(z_T) \Rightarrow 0$, z_T being an end-coordinate in the field-free space beyond the lens. In both situations, the function $F(\lambda)$ can have a sequence of zeros, and which of these zeros is found will depend on the initial guess for λ.

Once this problem has been solved, the remaining tasks fall within the province of the earlier sections.

33.4 Differential equations for the aberrations

Hitherto we have been dealing with the familiar form of the aberration theory and with the practical evaluation of the quantities involved. As we have already observed, this approach to aberration studies becomes very burdensome when the integral expressions for the coefficients are complicated. In these circumstances, we notice that there is no real need to use perturbation theory since the fields and the trajectories through them can be computed completely generally. But then we are confronted with another difficulty: in the most important case of very low aberrations, for which the lens designer is always striving, determination of the coefficients becomes inaccurate because we are obliged to calculate the small differences between the lateral image-plane coordinates, which may be large. This numerical instability would be avoided if the ray equations could be transformed in such a way that small aberrations or shifts appeared directly in an incremental form, without subtraction. This proves to be possible, and we now describe the appropriate transformations for two general classes of trajectory equations.

33.4.1 Electrostatic systems with a straight optic axis

For systems with a straight optic axis the axial coordinate z is the best parameter and (3.22) is then a suitable form of the trajectory equation. For conciseness, we specialize to electrostatic systems and introduce $w = x + iy$ as usual, whereupon (3.22) reduces to

$$w''(z) = \hat{\Phi}^{-1}(1 + |w'|^2)\left(\frac{\partial \hat{\Phi}}{\partial w^*} - \frac{1}{2}w'\frac{\partial \hat{\Phi}}{\partial z}\right) \tag{33.32}$$

We now write the electrostatic potential in the form

$$\Phi(\mathbf{r}) = \phi(z) + V(\mathbf{r})$$

$\phi(z)$ being the axial potential and V the remainder. In practice, $|eV| \ll m_0 c^2$ is always satisfied so that all terms in ϵV^2 can be neglected. Equation (33.32) takes the simpler form

$$w'' = \frac{1 + |w'|^2}{\widetilde{\phi}(z) + V}\left(\frac{\partial V}{\partial w^*} - \frac{1}{2}(\phi' + \frac{\partial V}{\partial z})w'\right) \tag{33.33}$$

$\widetilde{\phi} := \phi(1 + \epsilon\phi)/(1 + 2\epsilon\phi)$ being the reduced axial potential. The off-axis potential $V(\mathbf{r})$ consists of paraxial terms $V_p(\mathbf{r})$ and the remainder $V_s(\mathbf{r})$ causing aberrations. The paraxial term certainly contains the expression $-\frac{1}{4}\phi''(x^2 + y^2) = -\frac{1}{4}\phi''ww^*$ if $\phi'' \neq 0$, but may also include deflection and quadrupole terms. Whether such terms are main paraxial terms or aberrations caused by imperfections will depend on the definition of the electron optical system; the separation of V into V_p and V_s is to be made according to the particular situation. The term $w'\partial V/\partial z$ is always an aberration, and so the paraxial ray equations (with subscript p) are here

$$\widetilde{\phi}w_p'' = \frac{\partial V_p}{\partial w_p^*} - \frac{1}{2}\phi'w_p' =: L(z, w_p, w_p') \tag{33.34}$$

33.4 DIFFERENTIAL EQUATIONS FOR THE ABERRATIONS

We now rewrite (33.33) in the form

$$w'' = \frac{1+|w'|^2}{\tilde{\phi}+V}(L+S)$$

and $w = w_p + w_s$. Introducing (33.34) into (33.33) and observing that L is a *linear* operator, we find

$$L + S = \underbrace{L(z, w_p, w'_p) + L(z, w_s, w'_s)}_{L(z,w,w')} + \underbrace{\frac{\partial V_s}{\partial w^*} - \frac{1}{2}w'\frac{\partial V}{\partial z}}_{S(z,w,w')}$$

Note that the separation into w_p and w_s is only made in the linear term! In order to perform the complete separation into paraxial and aberration terms, we use the identity

$$\frac{1+|w'|^2}{\tilde{\phi}+V} = \frac{1}{\tilde{\phi}} + \frac{1}{\tilde{\phi}+V}\left(|w'|^2 - \frac{V}{\tilde{\phi}}\right)$$

After a minor calculation we find

$$w''_s = \frac{1}{\tilde{\phi}}\{L(z, w_s, w'_s) - L(z, 0, 0)\} + \frac{(1+|w'|^2)S(z,w,w')}{\tilde{\phi}+V}$$
$$+ \frac{1}{\tilde{\phi}+V}\left(|w'|^2 - \frac{V}{\tilde{\phi}}\right)L(z,w,w') \quad (33.35)$$

This differential equation contains exactly *all* aberrations that are included in (33.33). The term in $L(z,0,0)$ arises from the fact that the inhomogeneous term representing the axial deflection is already present in (33.34) and must therefore not appear again in the aberrations; the term in braces in (33.35) is hence *homogeneously* linear in the aberrations.

Equation (33.35) has a comparatively complicated structure and can be solved only in combination with the paraxial equation (33.34). By means of the method outlined in Section 33.2.3, an exact numerical solution can be obtained straightforwardly for various initial conditions. The result is numerically stable as $|w_s| \to 0$ and hence an analysis of the aberrations obtained in this way will lead to no numerical problems. A similar reasoning can be applied to magnetic systems, but we shall not present this here for reasons of space (see Kasper, 1987a,b).

33.4.2 Separation in arbitrary systems

A form of the ray equations that has proved very useful in numerical calculation is derived in Section 3.2 where it is given by (3.12). This is the starting point for the following theory, developed by Kasper (1984, 1985).

In order to remove unnecessary constants, we introduce a normalized magnetic field
$$b(r) := \eta \hat{U}^{-1/2} B(r) \qquad (33.36)$$
having the dimension of a reciprocal length, where $\hat{U}$ is a constant accelerating potential. We likewise define a dimensionless electrostatic potential $\varphi(r)$ and its gradient $a(r)$ by
$$\varphi(r) := \frac{1}{2}\frac{\hat{\Phi}(r)}{\hat{U}}, \qquad a = \nabla\varphi \qquad (33.37)$$
Denoting derivatives with respect to the curve parameter σ by dots, we can rewrite (3.12) more concisely as
$$\ddot{r} = a(r) + b(r) \times \dot{r} \qquad (33.38)$$
Let us now consider a neighbouring ray shifted by a distance $s(\sigma)$ relative to the first. This must satisfy the equation
$$\ddot{r} + \ddot{s} = a(r+s) + b(r+s) \times (\dot{r} + \dot{s})$$
By subtraction of (33.38) from this equation we find
$$\ddot{s} = a(r+s) - a(r) + b(r+s) \times \dot{s} + \{b(r+s) - b(r)\} \times \dot{r} \qquad (33.39)$$

This is not very helpful unless we can get rid of the differences between the field strengths. These can, in fact, be eliminated by means of a program that not only furnishes the cartesian components of the field strengths but also the gradients of these components, or in other words, the second-order derivatives of the potentials. For any scalar differentiable function, the relation
$$\Delta F := F(r+s) - F(r) = \int_0^1 (s \cdot \nabla') F(r'|r' = r + st) \, dt$$
is an identity, in which the differentiation refers to a variable argument r', over which the integration is to be performed after the differentiation. This integration is to be carried out by means of Gauss quadrature formulae. With the abbreviation
$$DF := s \cdot \nabla F = (s_x \partial_x + s_y \partial_y + s_z \partial_z) F \qquad (33.40)$$

33.4 DIFFERENTIAL EQUATIONS FOR THE ABERRATIONS

we have

$$\Delta F = \frac{1}{2} DF(\boldsymbol{r} + a_1 \boldsymbol{s}) + \frac{1}{2} DF(\boldsymbol{r} + a_2 \boldsymbol{s}) + O(s^5)$$

with $\quad a_{1,2} = \dfrac{1}{2} \pm \dfrac{\sqrt{2}}{4}$ (33.41a)

$$\Delta F = \frac{1}{18} \left\{ 5DF(\boldsymbol{r} + b_1 \boldsymbol{s}) + 8DF(\boldsymbol{r} + \frac{\boldsymbol{s}}{2}) + 5DF(\boldsymbol{r} + b_2 \boldsymbol{s}) \right\} + O(s^7)$$

with $\quad b_{1,2} = \dfrac{1}{2} \pm \sqrt{\dfrac{3}{20}}$ (33.41b)

In this way we can calculate increments of functions quite accurately in a numerically stable manner. In the case of vector functions these operations are carried out separately for the three cartesian components. With this in mind, we rewrite (33.39) in the form

$$\ddot{\boldsymbol{s}} = \Delta \boldsymbol{a}(\boldsymbol{r}) + \boldsymbol{b}(\boldsymbol{r}) \times \dot{\boldsymbol{s}} + \Delta \boldsymbol{b}(\boldsymbol{r}) \times (\dot{\boldsymbol{r}} + \dot{\boldsymbol{s}}) \qquad (33.42)$$

which is quite generally valid and easily programmable (Kasper, 1985).

So far we have tacitly assumed chromatic aberrations to be absent, since we treated $\boldsymbol{b}(\boldsymbol{r})$ in (33.36) and $\boldsymbol{a}(\boldsymbol{r})$ in (33.37) as unique functions. We now specify explicitly that (33.36), (33.37), (33.38) and (33.42) are true for electrons with the *nominal* value of the kinetic starting energy and for magnetic fields with the nominal values of the coil currents. In accordance with (3.13), the relations

$$\frac{1}{2} \dot{\boldsymbol{r}}^2 = \varphi(\boldsymbol{r}), \qquad \dot{\boldsymbol{r}} \cdot \dot{\boldsymbol{s}} + \frac{1}{2} \dot{\boldsymbol{s}}^2 = \Delta \varphi(\boldsymbol{r}) \qquad (33.43)$$

must then be satisfied at each trajectory point (including the starting point) if the energy has its nominal value.

33.4.3 Chromatic shifts

We now study the effect of altering φ and $\boldsymbol{b}$. If the magnetic field is generated by a coil with current I and the lens is not saturated, an alteration δI changes the magnetic field by $\delta \boldsymbol{b} = I^{-1} \boldsymbol{b} \delta I$. A nonzero kinetic starting energy $e\delta\Phi$ at the surface $\Phi(\boldsymbol{r}) = 0$ alters the function $\varphi(\boldsymbol{r})$ by

$$\delta\varphi(\boldsymbol{r}) = (2\hat{U})^{-1}(1 + 2\epsilon\Phi(\boldsymbol{r}) + \epsilon\delta\Phi)\delta\Phi \qquad (33.44)$$

To prevent any confusion, we denote the shift, caused by chromatic and geometric effects together, by the symbol $\boldsymbol{u}$ instead of $\boldsymbol{s}$. The generalization of (33.43) with $\Delta\varphi := \varphi(\boldsymbol{r} + \boldsymbol{u}) - \varphi(\boldsymbol{r})$ is

$$\dot{\boldsymbol{r}} \cdot \dot{\boldsymbol{u}} + \frac{1}{2} \dot{\boldsymbol{u}}^2 = \Delta\varphi(\boldsymbol{r}) + \delta\varphi(\boldsymbol{r} + \boldsymbol{u}) \qquad (33.45)$$

This condition is to be satisfied only once, at the *starting point*, and will then be valid for the whole trajectory. The practical application of (33.43) and (33.45) proceeds as follows: the starting vectors r_0, s_0 and u_0 and the *directions* of $\dot{r}_0$, $\dot{s}_0$ and $\dot{u}_0$ can be chosen independently, after which (33.43) and (33.45) are used to determine the appropriate *lengths* of the vectors $\dot{r}_0$, $\dot{s}_0$ and $\dot{u}_0$ respectively. The alterations δI and $\delta\Phi$ introduce some additional terms in (33.42). Considering all possible increments more thoroughly, we find

$$\ddot{u} = \Delta a(r) + b(r) \times \dot{u} + \left(\Delta b(r) + \frac{\delta I}{I} b(r+u)\right) \times (\dot{r}+\dot{u})$$
$$+ \ \epsilon\delta\Phi \hat{U}^{-1} \nabla \Phi(r+u) \tag{33.46}$$

This differential equation contains all possible types of geometric and chromatic errors and all allowed combinations of them in full generality. The integration formula (33.41b) is already so accurate that its remainder can be neglected in every practical case. Equation (33.46) shows that the shift between two arbitrary neighbouring trajectories can be computed quite accurately in a numerically stable manner. In practice some simplifications can be made. In (33.46) it is inconvenient to have $\nabla\Phi$ and $\nabla\hat{\Phi}$ (in a) together in the same formula. Since $\epsilon\delta\Phi \sim 10^{-6}$ little error will result from replacing $\nabla\Phi$ by $\nabla\hat{\Phi}$ in the last term of (33.46). Furthermore, it is preferable to avoid explicit evaluation for the argument $r+u$ or $r+s$, since the latter does not appear in (33.41a,b). Thus a more favourable form (see Kasper, 1985) is

$$\ddot{u} = \Delta a(r) + \kappa_1 \Big(a(r) + \Delta a(r)\Big) + b(r) \times \dot{u}$$
$$+ \left\{\Delta b(r) + \kappa_2\Big(b(r) + \Delta b(r)\Big)\right\} \times (\dot{r}+\dot{u}) \tag{33.47}$$

with $\kappa_1 = 2\epsilon\delta\Phi \equiv \dfrac{e\delta\Phi}{m_0 c^2}$, $\kappa_2 = \dfrac{\delta I}{I}$

Comparing the formalism outlined here with that of Section 33.4.1, we notice an important difference. Equation (33.45) is already the differential equation for the lateral geometric aberration: its solution gives the required aberration immediately without further transformations. The price to be paid for this convenience is the necessary specialization. On the other hand, (33.47) is quite *generally* applicable to any electron optical system with stationary fields. This equation describes, however, not the aberrations themselves but the shift between neighbouring trajectories, from which the aberrations must then be determined. This formalism is most useful for systems with a curved optic axis, since this axis can simply be adopted as the reference solution of (33.38).

33.5 Least-squares-fit methods in electron optics

The methods discussed in the preceding sections enable us to compute individual geometric and chromatic aberrations with high accuracy, even in the most complicated cases. The question now arises, how can the corresponding aberration coefficients be calculated from a set of such data? One suitable procedure is the least-squares-fit (LSF) method.

This method is already familiar in physics and numerical mathematics, since it is a general tool for the analysis and approximation of measured or calculated data. In light optics, for instance, the LSF method is in practical use for the determination of aberration coefficients, since this is easier than the numerical evaluation of Seidel's aberration theory. In spite of the close analogy with our present concerns, the LSF method is *not* so familiar in electron optics, for the following reasons. The determination of aberration coefficients by means of the LSF method requires the calculation of the end-points of many rays (up to 100). Unlike the situation in light optics, the tracing of electron trajectories is still a time-consuming procedure. Nevertheless, this method can be very useful and sometimes there is no alternative. This will be demonstrated by examining some telling examples once we have presented the general theory.

33.5.1 General complex formulation

We disregard for the moment the particular purpose and study a more general problem. We assume here that M measurements or computations of some complex function $w(u)$ yield the complex numbers $w_1, w_2 \ldots w_M$, that is $w_\mu = w(u_\mu)$, $\mu = 1 \ldots M$. We now wish to expand the function $w(u)$ as a series of the form

$$w(u) = \sum_{\nu=1}^{N} c_\nu \psi_\nu(u) \tag{33.48}$$

in terms of well-defined functions $\psi_\nu(u)$ $(\nu = 1 \ldots N)$, the *trial functions*, and initially unknown coefficients c_ν; of course $N \leq M$. Since the values $w_1 \ldots w_M$ may be afflicted with small errors or the choice of the trial functions may not be entirely appropriate, it is often impossible to satisfy (33.48) for all w_μ. We hence introduce the less stringent condition

$$\sum_{\mu=1}^{M} G_\mu \left| w_\mu - \sum_{\nu=1}^{N} c_\nu \psi_\nu(u_\mu) \right|^2 =: \epsilon^2 = \min \tag{33.49}$$

$G_1 \ldots G_M$ here being positive *weight factors* normalized to a unit sum.

Minimization of ϵ^2 with respect to the coefficients c_λ,

$$\frac{\partial \epsilon^2}{\partial c_\lambda^*} = 0, \qquad \lambda = 1 \ldots N$$

leads immediately to the *normal equations*:

$$\sum_{\nu=1}^{N} S_{\lambda\nu} c_\nu = T_\lambda, \qquad \lambda = 1 \ldots N \tag{33.50}$$

with the Hermitian matrix

$$S_{\lambda\nu} = \sum_{\mu=1}^{M} G_\mu \psi_\lambda^*(u_\mu) \psi_\nu(u_\mu) \tag{33.51}$$

and the column vector

$$T_\lambda = \sum_{\mu=1}^{M} G_\mu \psi_\lambda^*(u_\mu) w_\mu, \qquad \lambda = 1 \ldots N \tag{33.52}$$

The normal equations have the advantage of being easy to program and the flexibility of including weight factors, which can be chosen according to the significance of the input data $w_1 \ldots w_M$. If no preferences are apparent, then $G_1 = G_2 = \ldots G_M = 1/M$ is appropriate. Substituting the solution back into (33.49) gives us the standard deviation ϵ, which is a good measure of the quality of the approximation.

The main objection to the normal equations is that the matrix S is often ill-conditioned. In order to avoid this, Householder (1964) introduced a new method for the determination of $c_1 \ldots c_\mu$ from (33.49) by means of suitable orthogonalizations. This method requires equal weights $G_1 = \ldots = G_M$. Equation (33.49) can then be interpreted as the norm of a vector in an M-dimensional complex space; this norm is invariant under any *unitary* transformation. We cannot go into details here but simply state the essential procedure. The user of a Householder-transformation program is required to solve

$$\left| \begin{pmatrix} A_{11} & \ldots & A_{1N} & w_1 \\ \ldots & \ldots & \ldots & \ldots \\ A_{M1} & \ldots & A_{MN} & w_M \end{pmatrix} \begin{pmatrix} c_1 \\ \vdots \\ c_N \\ -1 \end{pmatrix} \right|^2 = \min \tag{33.53}$$

33.5 LEAST-SQUARES-FIT METHODS IN ELECTRON OPTICS

with $A_{\mu\nu} = \psi_\nu(u_\mu)$. After setting up this matrix including the column for the w_μ, it is left-multiplied by a succession of suitable unitary matrices of rank M until a triangular structure is obtained: $A'_{\mu\nu} = 0$ for $\mu > \nu$ with $\nu \leq N$. Extraction of the coefficients $c_1, \ldots, c_N$ is then straightforward. For more details, we refer to the standard textbooks on numerical mathematics (e.g. Stoer, 1979; Lawson and Hanson, 1974).

33.5.2 The determination of deflection aberrations

As an example, we now study the third- and fifth-order distortions of deflection systems, defined in (32.49). Since all coordinates refer to the image plane, we drop the subscript i. Inspection of (32.49) shows that in this particular case we have ten real coefficients D_{jk} ($j = 1, 2; k = 1 \ldots 5$) and ten real trial functions

$$\psi_1 \ldots \psi_{10} = x^3, x^2y, xy^2, y^3; x^5, x^4y, x^3y^2, x^2y^3, xy^4, y^5$$

We therefore have to solve a linear system of ten real equations for the unknown coefficients. The real form of the LSF method is a simpler special case of the general complex form.

The system of ten equations can be solved *en bloc*. This is, however, the least efficient way; we use this example to demonstrate how such a system can be profitably split into smaller subsystems. Careful inspection of (32.49) shows that the two equations have no common coefficients and can hence be treated separately. Thus the full system is already partitioned into two uncoupled subsystems of rank 5. We next consider the fact that the distortions are antisymmetric with respect to the coordinate planes; this tells us that it is sufficient to calculate trajectories with end-points in one quarter of the image screen. In order to obtain a true LSF, which will enable us to verify that (32.49) is a valid representation, it is advantageous to use more than the minimum number of trajectories; we therefore choose the pattern of ideal deflections shown in Fig. 33.2. As it is undesirable to raise large numbers to high powers, we *normalize* the equations by introducing dimensionless coordinates:

$$\xi = \frac{x}{a}, \quad \eta = \frac{y}{b}, \quad -\Delta\xi = \frac{\Delta x}{a}, \quad -\Delta\eta = \frac{\Delta y}{b},$$
$$0 \leq \xi \leq 1, \quad 0 \leq \eta \leq 1$$

We then have subsystems with the following aberration components, trial functions and scaled coefficients:

1. $\eta = 0$, $\Delta\xi_1$, $\Delta\xi_2$, $\Delta\xi_3$, $\psi_1 = \xi^3$, $\psi_2 = \xi^5$,
$c_1 = a^2 D_{11}$, $c_2 = a^4 D_{13}$.

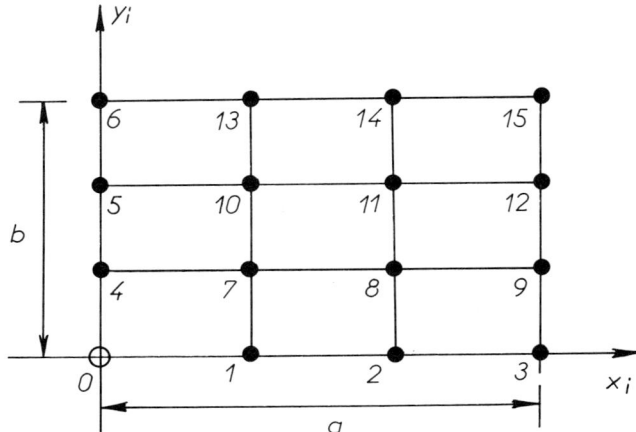

Fig. 33.2: Pattern of ideal deflection in one quarter of a viewing screen. The centre of the screen coincides with the origin, O.

2. $\xi = 0$, $\Delta\eta_4$, $\Delta\eta_5$, $\Delta\eta_6$, $\psi_3 = \eta^3$, $\psi_4 = \eta^5$,
   $c_3 = b^2 D_{22}$, $c_4 = b^4 D_{25}$.

3. $\eta\xi \neq 0$, $\Delta\xi_7 \ldots \Delta\xi_{15}$, $\psi_5 = \xi\eta^2$, $\psi_6 = \xi^3\eta^2$, $\psi_7 = \xi\eta^4$,
   $c_5 = b^2 D_{12}$, $c_6 = a^2 b^2 D_{14}$, $c_7 = b^4 D_{15}$

4. $\xi\eta \neq 0$, $\Delta\eta_7 \ldots \Delta\eta_{15}$, $\psi_8 = \eta\xi^2$, $\psi_9 = \eta^3\xi^2$, $\psi_{10} = \eta\xi^4$,
   $c_8 = a^2 D_{21}$, $c_9 = a^2 b^2 D_{24}$, $c_{10} = a^4 D_{23}$.

The LSF equations for subsystems 1 and 2 are solved first and the results obtained are then introduced into the remaining equations. We shall not go into this elementary procedure. It is clear that such a calculation presents no particular difficulty whereas the second iteration of the perturbation calculation (evaluation of S^{II} in Chapter 22) is extremely complicated, so complicated indeed that this second-order theory has hardly ever been used except in connection with the Darmstadt aberration correction project (Chapter 41). Although a list of aberration integrals for the fifth-order aberrations of round lenses exists, there is no proper study in the literature of any fifth-order aberration coefficient, not even the fifth-order spherical aberration of round lenses.

33.5.3 Some other examples

Another case in which the LSF method is obligatory is the *filter lens*. These are quite ordinary electrostatic einzel lenses (see Chapter 35), but the potential of the central electrode is so low that electrons with an energy of about 5 to 10 eV below the nominal energy cannot pass the central potential wall and are reflected , as in an electrostatic mirror. The electrons that do pass over the wall are slowed down so much that the associated slopes and off-axis distances may be very large; Fig. 33.3 shows a typical example. It is clear that the third-order approximation for the aberrations is then quite insufficient. Niemitz (1980), who investigated such lenses numerically, considered geometric aberrations of at least fifth order and chromatic errors up to the third order in the energy loss. The whole system of LSF equations then becomes so large that it is essential to split it into subsystems and to introduce suitable scaling.

An interesting application of the complex LSF method concerns systems of round magnetic lenses and deflectors. Here the complex trial functions are given by (32.53). In order to simplify the program, it may well be preferable to establish nine independent complex coefficients, treating C_0 and F_0 as though they were complex ($\tilde{C}_0$ and $\tilde{F}_0$) and regarding $\hat{K}_0^*$ as independent of $\hat{K}_0$ (writing $\hat{K}_c$ for $\hat{K}_0^*$ in the defining relation). If the numerical procedure is sound, it will be found that $|\Im \tilde{C}_0| \ll C_0$, $|\Im \tilde{F}_0| \ll F_0$ and $\hat{K}_c = \hat{K}_0^*$ with a high degree of accuracy. This also provides a useful check. Generally, the paraxial properties and the spherical aberration can be separated from the rest of the errors; the distortions also form a separable subsystem. Once again, it is possible to study aberrations of higher than third order.

33.6 Determination and evaluation of aberration discs

A complete determination of all permitted aberrations of a very complex electron optical system is a major task, regardless of the choice of method. We are therefore led to seek simpler special classes of aberrations, which can be determined with a more modest effort. One such class consists of the aberrations associated with a pencil of rays that start from a common fixed object point and pass through an aperture (see Fig. 33.4). Neither the object point nor the centre of the aperture need be situated on the optic axis.

As a result of the lens aberrations, a blurred intensity pattern is formed in the image plane, the aberration disc. This is of extreme interest in many practical respects. For monochromatic electrons, this disc can be interpreted as the shadow projection cast by a grid placed over the aperture. This is shown in Fig. 33.5. If we assume that the grid is illuminated uniformly and that the meshes of the grid in the aperture have equal areas, then the intensities in the distorted areas will also be equal. In this way we obtain some idea of the *intensity distribution* in the aberration disc. The

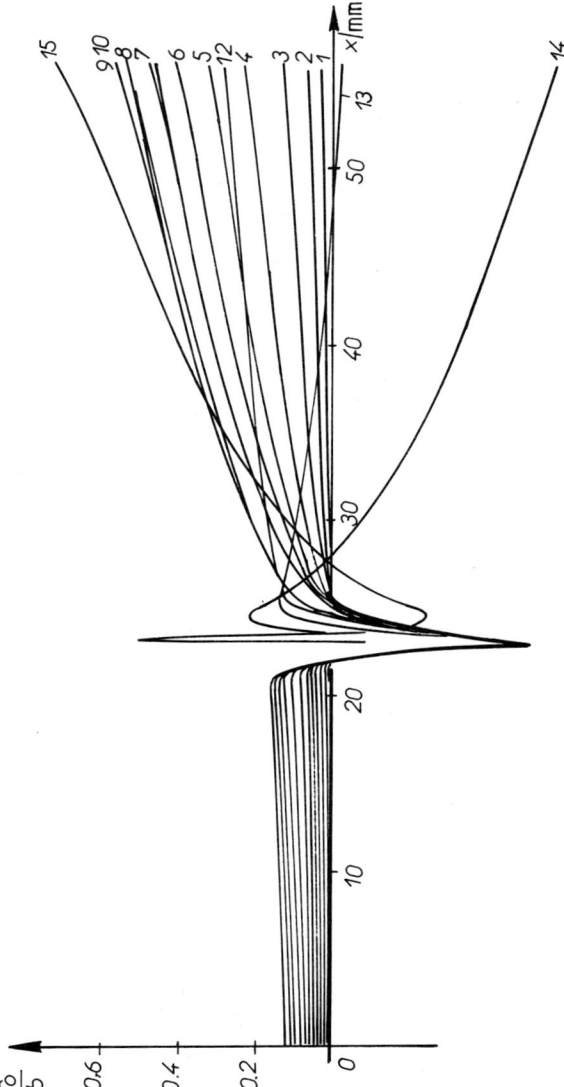

Fig. 33.3: Parallel beam of electrons incident on a filter lens; b denotes the bore radius of the central electrode. Note that the scales on the axes are different. Courtesy of P. Niemitz (1980).

33.6 DETERMINATION AND EVALUATION OF ABERRATION DISCS

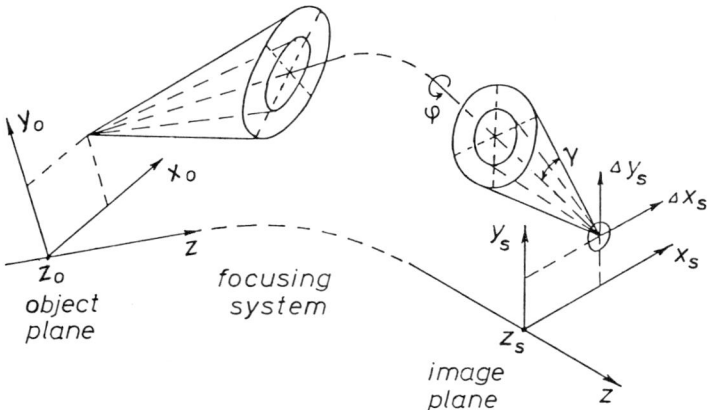

Fig. 33.4: Perspective sketch of the coordinate system and of a focused electron beam. The aperture cone in image space has a circular cross-section. Note that the aperture angles are much exaggerated. After Kasper (1985).

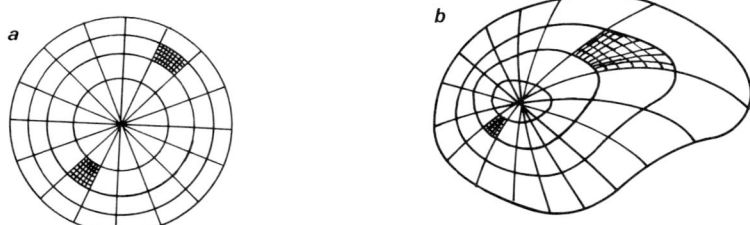

Fig. 33.5: (a) Zone pattern of an aperture. (b) Aberration disc regarded as a shadow projection of the zone structure created by monoenergetic electrons from a single object point.

aberrations may become so large that parts of the disc overlap others; in overlapping areas the corresponding intensities are to be summed.

The determination of patterns such as those shown in Fig. 33.5b is quite easy, if the space between the object and the aperture is field-free, so that the relations between the object coordinates (x_o, y_o, z_o), the initial slopes (x'_o, y'_o) and the coordinates (x_a, y_a, z_a) in the aperture plane are simple. In all other cases, the calculation of the ray that starts from the given object point and strikes a particular aperture point while under the influence of aberrations is already a difficult procedure, since a nonlinear boundary-value problem has to be solved. In order to circumvent such an impractical procedure, we propose that first, the corresponding *paraxial* boundary-value problem should be solved. This tells us the required initial slopes x'_o, y'_o. With these, the ray affected by aberrations is now calculated. The latter will *not*, of course, pass exactly through the aperture point (x_a, y_a, z_a), but if the deviation is small, as it should be in every well-designed system, the disparity will be tolerable.

33.6.1 Fourier analysis of the aberrations

In very many practical devices, the image plane or the viewing screen is located in essentially field-free space. The rays are then orthogonal trajectories of a characteristic function. Since this is already true for the paraxial rays, the same must hold for the aberrations after the paraxial contributions have been separated. If we have an aperture at a distance d from the image plane in the field-free space in front of the latter, then the gradient relations can be cast into the form

$$\Delta x_i = -d\frac{\partial S}{\partial x_a}, \qquad \Delta y_i = -d\frac{\partial S}{\partial y_i}, \qquad (x_i - x_a)^2 + (y_i - y_a)^2 \ll d^2$$

We have $x'_i = (x_i - x_a)/d$, $y'_i = (y_i - y_a)/d$ and hence

$$\Delta x_i = \frac{\partial S}{\partial x'_i}, \qquad \Delta y_i = \frac{\partial S}{\partial y'_i}$$

so that this representation is independent of the particular distance between the aperture and the image plane.

As we have seen above, it is necessary to approximate the slopes x'_i, y'_i by their *paraxial* values, since only for these can the simple prescribed conditions be satisfied. In accordance with the definitions introduced in Section 32.3, we now write

$$\left.\begin{array}{l} x'_i = x_i^{(0)\prime} + \sigma \cos\varphi = x_i^{(0)\prime} + \alpha \\ y'_i = y_i^{(0)\prime} + \sigma \cos\varphi = y_i^{(0)\prime} + \beta \end{array}\right\} \qquad (33.54)$$

for the paraxial slopes, where the subscript a has been dropped. As in (32.40) and (32.46), we have

$$\Delta w = \Delta x_i + \mathrm{i}\Delta y_i = \left(\frac{\partial}{\partial \alpha} + \mathrm{i}\frac{\partial}{\partial \beta}\right) S = e^{\mathrm{i}\varphi}\left(\frac{\partial}{\partial \sigma} + \frac{\mathrm{i}}{\sigma}\frac{\partial}{\partial \varphi}\right) S \qquad (33.55)$$

Since this representation is independent of the distance d to the assumed aperture, we may, in theory at least, consider an arbitrarily close aperture. In the limit $d \to 0$, (33.55) must hence remain valid even if the image plane

33.6 DETERMINATION AND EVALUATION OF ABERRATION DISCS

is located in an electric field, as is the case for the post-acceleration fields employed in cathode-ray tubes.

For conciseness we first define a normalized radial coordinate $r := \sigma/a$, a being the maximum semiaperture angle, hence $0 \leq r \leq 1$. Following Kasper (1985) closely, we now introduce a Fourier series expansion with initially unknown coefficients, which are to be determined from the computed aberrations:

$$S(r,\varphi) = \sum_{m=0}^{\infty} \sum_{k=0}^{\infty} \frac{r^{m+2k}}{2(m+k)} (A_{mk}^* e^{im\varphi} + A_{mk} e^{-im\varphi})$$

$$(m+k > 0, \quad A_{0k} \text{ real}) \tag{33.56}$$

Differentiation gives

$$\Delta w = \sum_{m} \sum_{k} r^{m+2k-1} \left\{ \frac{k}{m+k} A_{mk}^* e^{i(m+1)\varphi} + A_{mk} e^{-i(m-1)\varphi} \right\} \tag{33.57}$$

All the coefficients for $m \leq 4$, $m + 2k \leq 5$ can be determined from a few trajectories. The Fourier series expansion (33.57) then takes the explicit form

$$\begin{aligned}
\Delta w = &\, e^{-3i\varphi} \cdot r^3 A_{40} + e^{-2i\varphi}(r^2 A_{30} + r^4 A_{31}) \\
&+ e^{-i\varphi}(r A_{20} + r^3 A_{12}) + A_{10} + r^2 A_{11} + r^4 A_{12} \\
&+ 2e^{i\varphi}(r A_{01} + r^3 A_{02}) + \frac{1}{2} e^{2i\varphi}(r^2 A_{11}^* + \frac{4}{3} r^4 A_{12}^*) \\
&+ \frac{1}{3} e^{3i\varphi} r^3 A_{21}^* + \frac{1}{4} e^{4i\varphi} r^4 A_{31}^*
\end{aligned} \tag{33.58}$$

This can be rewritten more concisely in the form

$$\Delta w =: u_A(r,\varphi) = \sum_{n=-N+1}^{N} \varrho_n(r) e^{in\varphi}, \quad (N=4) \tag{33.59}$$

We now try to approximate an arbitrary complex function $\Delta w = u(r,\varphi)$, defined on a disc $r \leq 1$, by a series expansion of the form (33.59). Since every part of the disc is *a priori* of equal importance, this approximation is best made by an integral least-squares-fit

$$F := \int_{r=0}^{1} \int_{\varphi=0}^{2\pi} |u(r,\varphi) - u_A(r,\varphi)|^2 r \, dr \, d\varphi = \min \tag{33.60}$$

Quite generally this integration is to be approximated by a suitable discrete summation. With $2N$ equidistant azimuths $\varphi_j = (j-1)\pi/N$, $j = 1\ldots 2N$ and $L \geq 2$ different radii $r_1 \ldots r_L$, such a summation formula has the basic form

$$F := \int_{r=0}^{1}\int_{\varphi=0}^{2\pi} f(r,\varphi) r\, dr\, d\varphi = \frac{\pi}{2N}\sum_{l=1}^{L}\sum_{j=1}^{2N} G_l f_{lj}$$

with discrete function values $f_{lj} := f(r_l, \varphi_j)$ and positive weight factors G_l, normalized to unit sum.

Introducing (33.59) into (33.60) and employing this summation formula, we obtain

$$F = \frac{\pi}{2N}\sum_{l=1}^{L}\sum_{j=1}^{2N} G_l |u_{lj}|^2 + \pi \sum_{l=1}^{L}\sum_{n=1-N}^{N} G_l \{|\rho_{nl}|^2 + 2\Re(\rho_{ln} K_{ln}^*)\} = \min$$

with $u_{lj} := u(r_l, \varphi_j)$ and $\rho_{ln} := \rho_n(r_l)$ and the trigonometric sums

$$K_{ln} := \frac{1}{2N}\sum_{j=1}^{2N} u_{lj} \exp\left\{\frac{-i\pi(j-1)n}{N}\right\}$$

$$(n = -(N-1)\ldots N) \tag{33.61}$$

Since ray tracing in electron optics is still a time-consuming procedure, we choose L and N as small as possible but without significant loss of accuracy. A reasonable compromise is $N = 4$ and $L = 2$, which suffices for the determination of all the coefficients in (33.58). The discretization of the radial integration is then simply the two-point Gauss quadrature for the variable r^2; we hence have $G_1 = G_2 = 1/2$ and $r_{1,2}^2 = 0.5(1 \pm 1/\sqrt{3})$,

$$r_{1,2} = \left(\frac{1}{2} \pm \frac{1}{\sqrt{12}}\right)^{\frac{1}{2}} = 0.4597,\quad 0.8881$$

The corresponding points in the aperture disc are shown in Fig. 33.6. The minimizing condition is now $\partial F/\partial A_{\mu\nu}^* = 0$ for all values of the subscripts μ and ν. With the convenient abbreviation $[X] := \sum X_l$ for the sum of any set of quantities X, we arrive at a complex linear system of equations for the unknowns $A_{\mu\nu}$:

$$\left.\begin{array}{l}[r^2]A_{01} + [r^4]A_{02} = \dfrac{1}{2}[rK_1] \\[6pt] [r^4]A_{01} + [r^6]A_{02} = \dfrac{1}{2}[r^3 K_1]\end{array}\right\} \tag{33.62a}$$

33.6 DETERMINATION AND EVALUATION OF ABERRATION DISCS

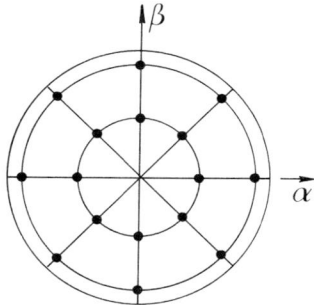

Fig. 33.6: Relative directional angles, α and β, of the selected trajectories near the image plane; the outer circle indicates the full extent of the electron beam.

$$\left. \begin{array}{l} [r^0]A_{10} + [r^2]A_{11} + [r^4]A_{12} = [K_0] \\ [r^2]A_{10} + \dfrac{5}{4}[r^4]A_{11} + \dfrac{4}{3}[r^6]A_{12} = [r^2(K_0 + \dfrac{1}{2}K_2^*)] \\ [r^4]A_{10} + \dfrac{4}{3}[r^6]A_{11} + \dfrac{13}{9}[r^8]A_{12} = [r^4(K_0 + \dfrac{2}{3}K_2^*)] \end{array} \right\} \quad (33.62b)$$

$$\left. \begin{array}{l} [r^2]A_{20} + [r^4]A_{21} = [rK_{-1}] \\ [r^4]A_{20} + \dfrac{10}{9}[r^6]A_{21} = [r^3(K_{-1} + \dfrac{1}{3}K_3^*)] \end{array} \right\} \quad (33.62c)$$

$$\left. \begin{array}{l} [r^4]A_{30} + [r^6]A_{31} = [r^2K_{-2}] \\ [r^6]A_{30} + \dfrac{17}{16}[r^8]A_{31} = [r^4(K_{-2} + \dfrac{1}{4}K_4^*)] \end{array} \right\} \quad (33.62d)$$

$$[r^6]A_{40} = [r^3K_{-3}] \quad (33.62e)$$

These form five uncoupled symmetric subsystems with maximum rank 3. The coefficients on the left-hand side are fixed positive numbers, which need to be calculated only once and are then stored. Only the coefficients on the right-hand side have to be recalculated in repeated applications; the surviving subscript is the last one (n) in K_{ln}.

The procedure is self-consistent in the sense that a function $S(r, \varphi)$, whose gradient contains exactly the terms considered in (33.58) with arbitrary coefficients, is found exactly apart from the unavoidable rounding errors. Moreover, the method provides an intrinsic accuracy control, which is effected by substituting the calculated coefficients into the functional F.

Let the expression

$$\|f\| := \left\{\frac{1}{\pi}\int_0^1\int_0^{2\pi} f(r,\varphi) r\, dr\, d\varphi\right\}^{\frac{1}{2}} = \left(\frac{1}{2LN}\sum_{l=1}^{L}\sum_{0}^{2N}|f_{lj}|^2 G_l\right)^{\frac{1}{2}}$$

denote the norm of any function $f(r,\varphi)$, then $\epsilon = \|\Delta w - u_A\|$ is the absolute error and $\delta = \epsilon/\|\Delta w\|$ the relative error of the approximation; the latter must satisfy $\delta \ll 1$.

The Fourier analysis method not only provides a way of determining aberration coefficients $A_{\mu\nu}$, but is also a quite general mathematical procedure for the reconstruction of a two-dimensional real function from its gradient. In electron optics as well as in light optics, the function $S(r,\varphi)$, the wave aberration, is of particular interest in wave-theoretical considerations.

33.6.2 Some practical aspects

The application of this method requires that the paraxial rays form a round cone in the image space, which implies that the corresponding cone in the object space is elliptic, as is sketched in Fig. 33.4. How then should the initial conditions be chosen? This is very rarely a serious problem. In most cases we know—from the symmetry properties of the system in question— the directions of maximum and minimum lateral magnifications M_x, M_y, respectively. Let us suppose that these magnifications and the refractive index $\nu = (\hat{\Phi}_i/\hat{\Phi}_o)^{1/2}$ of the image relative to the object are known. The appropriate starting slopes (in the rotating frame) are then

$$\left.\begin{array}{l} x'_{lj} = x_o^{(0)\prime} + \nu M_x a r_l \cos\varphi_j \\ y'_{lj} = y_o^{(0)\prime} + \nu M_y a r_l \sin\varphi_j \end{array}\right\} \begin{array}{l} l = 1\ldots L, \\ j = 1\ldots 2N \end{array} \quad (33.63)$$

For equal magnifications, as in all arrangements of round lenses and deflectors with a straight optic axis, it is not really necessary to refer to the image side. We can simply identify the angles appearing in the theory as those at the starting point, thereby representing the aberrations in terms of initial conditions, which is often very convenient.

Until now we have considered only the aberrations in a fixed image plane $z = z_i$. Determination of the variation of the aberration coefficients with the coordinate z is, however, straightforward, requiring no additional ray tracing. The routines for solving systems of ordinary differential equations usually furnish the solution together with its first-order derivative. In the approximation in which each trajectory is replaced by its local tangent, we have

$$\Delta w_j(z_i + \Delta z) = \Delta w_j(z_i) + \Delta z \cdot \Delta w'_j(z_i) + O(\Delta z^2)$$

33.6 DETERMINATION AND EVALUATION OF ABERRATION DISCS 555

It is now easy to repeat the Fourier analysis with these new aberrations and to study in this way the effect of a defocus.

Another necessary extension is the calculation of *chromatic aberrations*. The simplest but also the slowest method is to repeat the whole procedure for a few different values of the kinetic starting energy. In this way we could even determine the chromatic variation of each aberration coefficient. In almost all practical applications, such highly detailed information is hardly ever needed. The following procedure is then more favourable.

For the four rays with aperture conditions

$$r = r_1, \qquad \varphi = (j-1)\frac{\pi}{2}, \qquad j = 1\ldots 4$$

we solve (33.47) with the initial conditions $\boldsymbol{u}(0) = 0$, $\dot{\boldsymbol{u}}(0) \parallel \dot{\boldsymbol{r}}(0)$, the length $|\dot{\boldsymbol{u}}(0)|$ being determined from (33.45). The resulting aberrations are then purely chromatic ones. For the complex lateral aberrations $\Delta w_j, j = 1\ldots 4$, obtained from these solutions, a series expansion similar to (33.57) can be set up, but this is now truncated after the terms with $m \leq 2$, $m + 2k \leq 3$, hence

$$\Delta w = C_{20} r e^{-i\varphi} + C_{10} + C_{11} r^2 + 2 C_{01} r e^{i\varphi} + \frac{1}{2} C_{11}^* r^2 e^{2i\varphi} \qquad (33.64)$$

This approximation is quite sufficient, since the energy shift $e\Phi_0$ is usually very small. Writing down (33.64) for the four rays specified above and then forming the trigonometric interpolation sums (33.61) with $l = 1$, $N = 2$, we find the four simple relations

$$\left.\begin{aligned}
K_{-1} &= \frac{1}{4}(\Delta w_1 + i\Delta w_2 - \Delta w_3 - i\Delta w_4) = C_{20} r_1 \\
K_0 &= \frac{1}{4}(\Delta w_1 + \Delta w_2 + \Delta w_3 + \Delta w_4) = C_{10} + C_{11} r_1^2 \\
K_1 &= \frac{1}{4}(\Delta w_1 - i\Delta w_2 - \Delta w_3 + i\Delta w_4) = 2C_{01} r_1 \\
K_2 &= \frac{1}{4}(\Delta w_1 - \Delta w_2 + \Delta w_3 - \Delta w_4) = \frac{1}{2} C_{11}^* r_1^2
\end{aligned}\right\} \qquad (33.65)$$

(the subscript $l = 1$ being omitted in the K-coefficients). These suffice for the determination of the coefficients in (33.64). An accuracy control is given by the requirement that C_{01} must be real so that the imaginary part of C_{01} represents directly a numerical error. Ignoring $\Im(C_{01})$ we then have the perturbation eikonal, which is analogous to (33.56).

33.6.3 Integral properties of aberration discs

So far we have been exclusively concerned with the *analysis* of aberrations.

A list of coefficients, which may well be long, is, however, not very easy to interpret in terms of the imaging quality of a practical device. A few geometric parameters characterizing the size and shape of the aberration disc would be more helpful. This problem is usually solved in the following way. Depending on the purpose of the device in question, the various types of aberrations have different priorities. The aberrations with low priority are ignored; for each of the remaining ones, a corresponding aberration radius is estimated on the basis of rough conjectural formulae. Finally a root-mean-square radius is calculated, on the unjustified assumption that the aberrations superimpose statistically. This procedure is as unsatisfactory as it is simple. A better proposal has been made by Scherle (1983, 1984), who approximated the cross-section of the beam by a suitably defined ellipse. It is then not even necessary to determine any aberration coefficients. The price of this simplification is, however, that a rather large number of trajectories must be traced and this requires considerable computation time. This disadvantage can be avoided by combining Scherle's method with the Fourier analysis procedure.

First of all we notice that every realistic electron beam has an energy spectrum, characterized by a distribution function $g_s(\epsilon)$ for $\epsilon_1 \le \epsilon \le \epsilon_2$, $\int g_s(\epsilon)\, d\epsilon = 1$. Here $\epsilon = e\delta\Phi$ is the deviation of the energy from its nominal value. We also take into account the fact that the aperture may not be illuminated uniformly; this non-uniform illumination is described by an intensity distribution $g_a(\sigma, \varphi)$, over the aperture of radius $\sigma_{max} = a$

$$\int_{\sigma=0}^{a} \int_{\varphi=0}^{2\pi} g_a(\sigma,\varphi)\sigma\, d\sigma\, d\varphi = 1$$

We now define the expectation value $<f>$ of any real or complex function f in the recording plane $z=$const.

$$<f(z)> = \int_{\sigma=0}^{a} \int_{\varphi=0}^{2\pi} \int_{\epsilon=\epsilon_1}^{\epsilon_2} g_s(\epsilon) g_a(\sigma,\varphi) f(z,\epsilon;\sigma,\varphi) \sigma\, d\sigma\, d\varphi\, d\epsilon \quad (33.66)$$

We recall that (σ, φ) refer to the aperture, while z does not. In order to apply (33.64) to the aberrations, we notice first that

$$\Delta w(z,\epsilon;\sigma,\varphi) = \Delta w(z,0;\sigma,\varphi) + \frac{\epsilon}{\epsilon_m}\Delta u(z,\epsilon_m;\sigma,\varphi) \quad (33.67)$$

is a good approximation for the superposition of geometric and chromatic aberrations Δu, if ϵ_m is the most probable energy. It is exact if only the

33.6 DETERMINATION AND EVALUATION OF ABERRATION DISCS

chromatic and geometric aberrations of lowest order are considered (first-order chromatic and third-order geometric aberrations for round lenses, for example). We can now define an intensity-weighted distortion $<\Delta w(z)>$ by setting $f := \Delta w$ in (33.66). We also introduce a root-mean-square (rms) radius

$$\rho(z) := <|\Delta w - <\Delta w>|^2>^{1/2} \qquad (33.68a)$$

and ellipticity parameters

$$e_1(z) + ie_2(z) := <(\Delta w - <\Delta w>)^2> \qquad (33.68b)$$

The meaning of the latter quantities is as follows: if we shift the origin of the coordinates to the centre $<\Delta w>$ of the ellipse:

$$\xi := \Delta x - <\Delta x>, \qquad \eta := \Delta y - <\Delta y> \qquad (33.69)$$

we obtain

$$2<\xi^2> = \rho^2 + e_1, \qquad 2<\eta^2> = \rho^2 - e_1, \qquad 2<\xi\eta> = e_2 \qquad (33.70)$$

The transformation to the principal axes of the ellipse is given by

$$\bar{\xi} = \xi\cos\theta + \eta\sin\theta, \qquad \bar{\eta} = -\xi\sin\theta + \eta\cos\theta \qquad (33.71a)$$

where the angle θ is to be calculated from

$$\tan 2\theta = \frac{2<\xi\eta>}{<\xi^2> - <\eta^2>} \equiv \frac{e_2(z)}{e_1(z)} \qquad (33.71b)$$

This rotation describes the overall effect of the *anisotropic* errors, if the directions obtained in this way are not either meridional or sagittal. The main axes themselves are now given by

$$2<\bar{\xi}^2> = \rho^2 + \sqrt{e_1^2 + e_2^2}, \qquad 2<\bar{\eta}^2> = \rho^2 - \sqrt{e_1^2 + e_2^2} \qquad (33.72)$$

The axes obtained in this way are smaller than the true dimensions of the aberration figure. Following Scherle's (1983) proposal, we therefore introduce a dilatation factor, which may almost always be set equal to two. Finally, we obtain the semi-axes

$$E_{1,2}(z) = \sqrt{2}\left(\rho^2 \pm \sqrt{e_1^2 + e_2^2}\right)^{1/2} \qquad (33.73)$$

In his thesis, Scherle proposed that the averaging should be performed by summation over many individual aberrations. This is certainly necessary if the initial conditions of the trajectories in the beam are so general that a wave aberration S cannot be used. Such a case arises, for instance, if the electrons start from a cathode surface or from a crossover with finite extent, so that the initial conditions themselves have a statistical distribution.

In very many cases, this method is unnecessarily general and would entail an unreasonably large effort. Instead, we can first determine the aberration coefficients (A_{mk}), either from integral expressions or from a Fourier analysis. These are then introduced into (33.58) or more generally into (33.57) and, for the chromatic effects, into (33.64). These formulae are now regarded as continuous *interpolation* formulae for the aberrations. Numerical evaluation of the necessary integrals is then perfectly practical. This is always a comparatively fast procedure since it does not require any new ray tracing. The approximate ellipse can be determined rapidly for a sequence of image planes, after which the optimal defocus can be found easily.

The advantage of this method lies in the fact that it gives us very clear and simple criteria for optimal focusing. The best approximation to a stigmatic focus is obtained when, after exploring all permissible variations of the system parameters, the largest value of the semi-axis $E_1(z)$ is least for the object point with the worst aberrations. This idea has not yet been much exploited in practice but it seems very likely that the method will prove useful and gain wide acceptance. The conclusions to which it leads are certainly realistic. Figure 33.7 shows, for example, how the size of the spherical aberration disc can be reduced by defocusing. With Scherle's procedure, the plane in which the beam radius is smallest is found to be at $(8/9)\Delta z$, where Δz is the defocus of the theoretical plane of least confusion (24.50); the radius of the disc differs by a factor of only $(8/9)^{1/2} = 0.94$ from the familiar radius of least confusion (24.51). This is well within the experimental confidence limits.

We conclude this account of the various numerical methods of computing aberrations and assessing their importance with the observation that the choice of practical procedures is wide enough to enable us to study virtually any electron optical system, however complicated, with sufficient accuracy for most practical needs.

33.7 Optimization procedures

Once programs for the computation of fields, trajectories, focusing properties and aberrations have been successfully completed, it is very useful to

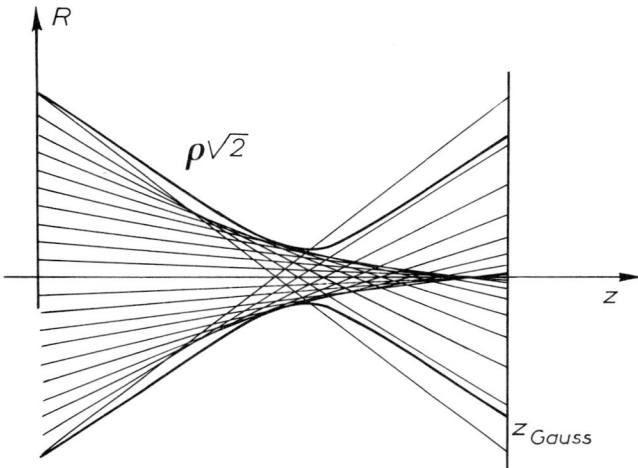

Fig. 33.7: Form of the caustic and of the r.m.s radius as functions of the axial coordinate for a round lens with spherical aberration.

incorporate them into a program for the optimization of electron optical systems. Usually the purpose of optimization is to determine geometric and electromagnetic configurations of electrodes, polepieces and coils that minimize certain electron optical aberrations of the system in question subject to given constraints, typically a fixed focal length, a minimum working distance or an upper bound on a coil current or some other technical limit. It is clear that this problem cannot be solved automatically in a perfectly general manner, since the technical requirements and the constraints will differ widely for the various kinds of electron optical devices. For this reason we cannot go into much detail here.

33.7.1 The defect function

The first step in any optimization procedure must be the definition of an appropriate *defect function* or 'merit function'. This must be a strictly positive quantity, which is to be minimized by varying the system parameters within the allowed limits. This definition already requires a precise formulation of the specific requirements and constraints. The *arguments* $(x_1 \ldots x_n)$ of the defect function are the set of all those system parameters that are allowed to vary, and their domain D of definition is bounded by constraints: $(x_1 \ldots x_n) \in D$. Quite generally the constraints can be represented by inequalities:

$$\theta_k(x_1 \ldots x_n) \geq 0, \qquad k = 1 \ldots r \tag{33.74}$$

which include equations as special cases.

The functional expression ψ of the defect function is usually the square of a total aberration radius. This is defined in the following way. First, for each individual aberration a simple effective defect radius is introduced, for instance $\rho_1 = MC_s\alpha^3/4$ for the spherical aberration of a round lens with magnification M and aperture angle α, or $\rho_2 = MC_c\alpha\Delta\hat{\Phi}/\hat{\Phi}$ for the chromatic aberration, and so on. Next, a set of non-negative weight factors $g_1^2 \ldots g_m^2$ is introduced, which characterize the importance of the m individual aberrations. Large weights mean that the corresponding aberrations are very serious, while low weights are associated with contributions of little importance. Vanishing weights mean that the corresponding aberrations are completely ignored. The defect function is then

$$\psi(x_1 \ldots x_n) = \sum_{i=1}^{m} g_i^2 \rho_i^2 = \sum_{i=1}^{m} f_i^2(x_1 \ldots x_n) \tag{33.75}$$

with $f_i := g_i \rho_i$ for $i = 1 \ldots m$.

It is possible to incorporate the constraints (33.74) into the defect function:

$$\psi(x_1 \ldots x_n) =: \sum_{j=1}^{M} f_j^2 = \sum_{i=1}^{m} g_i^2 \rho_i^2 + \sum_{k=1}^{r} G_k^2 (\theta_k - |\theta_k|)^2 \tag{33.76}$$

with $M = m + r$ being the total number of terms. The second contribution is a 'penalty' function: the defect function increases for $\theta_k < 0$. This form of the defect function may be useful if very small violations of constraints are allowed; such violations can be tolerated if the constraints themselves represent only roughly guessed technical bounds. The magnitudes of these violations will depend on the particular choice of the new weights $G_1^2 \ldots G_r^2$.

As a simple example, we describe the appropriate defect function for axial focusing by an ordinary unsaturated magnetic round lens. Here we have to consider the spherical aberration and the axial chromatic aberration, and we include a rough guess for the blurring caused by diffraction. The defect function is again a sum of squares of aberration radii. This definition is convenient for numerical purposes, though of course the chromatic and geometric aberrations add linearly. Thus

$$\psi = \left(\frac{1}{4}g_1 C_s \alpha^3\right)^2 + \left(g_2 C_c \alpha \frac{\Delta\hat{\phi}}{\hat{\phi}}\right)^2 + \left(\frac{0.61 g_3 h}{\alpha\sqrt{2me\hat{\phi}}}\right)^2 \tag{33.77}$$

α being the aperture angle at the object point, $\hat{\phi}$ the acceleration potential and h Planck's constant. (For further details of the diffraction term, see Volume 3.) Since all three terms are equally important, we set $g_1 = g_2 = g_3 = 1$. In the absence of constraints and parameters to be varied, this function is still incomplete. Reasonable constraints are, for instance, fixed values of the object coordinate z_o and of the image coordinate z_i and the acceptable interval $f_{min} \leq f \leq f_{max}$ for the focal length f. The definition of system parameters requires a decision concerning the class of shapes allowed for the axial field distribution $B(z)$. If, for instance, the simple model with a gap of width S and cylindrical bores with radii R_1 and R_2 is employed, we should use the parameters $x_1 = R_1$, $x_2 = R_2$, $x_3 = S$ and $x_4 = J$, the latter being the number of ampère-turns. If the midplane of the gap is kept fixed at $z_M = 0$, we have a well-specified optimization problem, so that ψ =minimum leads to a unique solution.

This very simple example makes it clear that only with very detailed specification can a reasonable answer to the optimization problem be expected. Moreover, this example is a reminder that rough estimates of the form (33.77) are commonly used. These can be replaced by improved formulae for more realistic intensity distributions and their expectation values, but this is still uncommon.

33.7.2 The optimization of axial distributions

As already pointed out, the focusing properties and the aberrations of electron optical systems are generally determined by certain axial field distributions, such as the electrostatic potential $\phi(z)$ and the magnetic field strength $B(z)$ on the optic axis. It is therefore tempting to try to optimize these functions. This means that, subject to the given constraints, the defect function is minimized by finding the 'best shapes' of these axial distributions.

This procedure has been followed several times and with different approaches. Moses (1973), for instance, employed variational calculus for the minimization of spherical aberration with simultaneously vanishing coma. Crewe et al. (1968) designed a field-emission electron gun under the assumption that the axial potential $\phi(z)$ in the space between the first and second anode can be represented by a cubic polynomial, the coefficients of which were the optimization parameters; this yielded the 'Butler gun'. Later Munro (1973) showed that this design does not represent a true optimum. Szilagyi (1977) introduced dynamic programming (see Section 35.3.2). In this approach the integration interval $z_o \leq z \leq z_i$ between object and image is dissected into a set of small subintervals. Then, starting from the object coordinate z_o, the integrand of an aberration integral, for instance that for C_s, is minimized under given constraints in each of these subintervals. The result is then a piecewise analytic axial field distribution, for example (Szilagyi, 1984, 1987) a cubic spline, which is then assumed to be the best field.

Common to all methods starting from axial field distributions is the severe difficulty of finding reasonably shaped electrodes or polepieces to generate these axial fields. Analytic continuation of the axial potentials into off-axis domains by means of the radial series expansions of Chapter 7 generally leads to equipotentials with a singular character such as sharp edges in the vicinity of the optic axis or even vanishing bore radii. In order to avoid these, the electrode structure has to be altered, but the additional fringe fields thus introduced cast doubt on the whole design procedure.

For these reasons, no really convincing design, based on an optimization of an axial field function, has so far been published and we can hence not recommend this method. There remains only the laborious technique in which the appropriate boundary-value problems are solved repeatedly for different geometric shapes, with the risk that the best solution may not belong to the set of configurations analysed.

33.7.3 The damped least-squares method

The damped least-squares method was first described by Levenberg (1944) and has been successfully applied to light optical design problems. In 1982, it was introduced into electron optics (Chu and Munro, 1982; Munro and Chu, 1982). We now briefly outline their method.

The defect function is represented in the form (33.75). If the arguments x_j are varied by small increments Δx_j ($j = 1 \ldots n$), the functions f_i alter by

$$\hat{f}_i = f_i + \sum_{j=1}^{n} a_{ij} \Delta x_j, \qquad i = 1 \ldots m, \qquad (33.78a)$$

$$a_{ij} = \frac{\partial f_i}{\partial x_j}, \qquad j = 1 \ldots n \qquad (33.78b)$$

This linear approximation is adequate if all the increments Δx_j are sufficiently small. The new defect function is now

$$\hat{\psi} = \sum_{i=1}^{m} \left(f_i + \sum_{j=1}^{n} a_{ij} \Delta x_j \right)^2 \qquad (33.79)$$

The minimization conditions $\partial \hat{\psi} / \partial (\Delta x_j) = 0$ for $j = 1 \ldots n$ lead to n simultaneous linear equations for the increments $\Delta x_1 \ldots \Delta x_n$, which could be solved by straightforward techniques.

33.7 OPTIMIZATION PROCEDURES

Unfortunately this simple undamped least-squares method is unstable and can diverge. In order to avoid this instability, (33.79) is modified to

$$\psi^\dagger = \hat{\psi} + \sum_{j=1}^{n} p_j^2 \Delta x_j^2$$
$$= \sum_{i=1}^{m} \left(f_i + \sum_{j=1}^{m} a_{ij} \Delta x_j \right)^2 + \sum_{j=1}^{n} p_j^2 \Delta x_j^2 \tag{33.80}$$

where the factors p_j are called damping coefficients. Obviously we have $\hat{\psi} < \psi^\dagger$, so that a minimum of $\psi^\dagger$ corresponds to a minimum of $\hat{\psi}$. The minimization conditions $\partial \psi^\dagger / \partial(\Delta x_k) = 0$ now take the form

$$\sum_{i=1}^{m} \left\{ a_{ik} \left(f_i + \sum_{j=1}^{n} a_{ij} \Delta x_j \right) \right\} + p_k^2 \Delta x_k = 0$$
$$(k = 1 \ldots n) \tag{33.81}$$

These form a system of n simultaneous linear equations for the unknowns $\Delta x_1 \ldots \Delta x_n$, the coefficient matrix being symmetric and positive definite.

The new coordinates $x_j^{(n)} = x_j + \Delta x_j$ depend on the choice of the damping coefficients. Two different choices are proposed by Chu and Munro:

$$p_j^2 = p^2 \qquad \text{(additive damping method)} \tag{33.82a}$$

$$p_j^2 = p^2 \sum_{i=1}^{j} a_{ij}^2 \qquad \text{(multiplicative damping method)} \tag{33.82b}$$

p being a constant damping factor. With (33.82a), each diagonal element of the matrix in (33.81) is enlarged by an additive term p^2, while with (33.82b), each diagonal element is multiplied by $(1 + p^2)$. The numerical solution of (33.81) must be repeated for various values of p, until the smallest value of $\psi^\dagger$ is found.

The partial derivatives a_{ij} that appear in the above equations are to be calculated numerically from a simple two-point formula:

$$a_{ij} = \frac{\partial f_i}{\partial x_j} = \frac{f_i(x_j + \delta x_j) - f_i(x_j)}{\delta x_j} \tag{33.83}$$

δx_j being a small increment. According to Chu and Munro, this linear approximation is sufficient, since the whole minimization procedure must in any case be repeated if the starting point is far distant from the final optimal configuration.

Chu and Munro have made extensive investigations concerning the applicability of this method and set up an interactive program. For reasons of space we cannot outline this here and refer to the corresponding publication, where full details are to be found. Concerning the optimization of geometric configurations, Chu and Munro made the following compromise: for each electron optical element, for instance round electrostatic and magnetic lenses and electric or magnetic deflectors, reasonable geometric shapes were assumed and for these the boundary-value problems were solved exactly beforehand. Later, during the optimization process, all operations that can be performed without a new field calculation were allowed: changes of electric or magnetic parameters such as the electrode potentials or coil currents, axial shifts of the object plane, the image plane or of entire deflectors, rotation of deflectors about the optic axis and scale-transformations in independent elements. In principle it is also possible to alter the geometric shape of electrode or polepiece surfaces but then an entirely new field calculation is necessary, which requires considerable computing time.

The great advantage of this method is that each optimization step results in a realistic configuration, which could be constructed. The applicability of this method was demonstrated for focusing and dual-channel deflection systems, which are employed in electron beam lithography. A very detailed knowledge of the technical requirements in electron beam lithography is needed to understand the particular features of the optimization; much more discussion is to be found in Chapter 40 and we merely observe that the results published by Chu and Munro are distinctly encouraging.

We conclude that the damped least-squares method is at present the most effective way of solving optimization problems.

34
The Use of Computer Algebra Languages

34.1 Introduction

The familiar computer languages are designed to perform numerical calculations efficiently but are extremely ill-adapted to any kind of symbolic calculation: it is trivial to list $z := (x+y)^p$ for values of x and y that may contain many digits and any reasonable values of p but much less easy to output the binomial expansion; numerical integration of a function such as $x^p \sin qx$ is again trivial but the closed form of the integral cannot be found; the same is true of differentiation. There are powerful routines for solving differential equations numerically, as we have seen in Chapter 33, but there is no way of knowing whether the equation has a solution in terms of tabulated (or easily computed) functions. It was for reasons such as these that the various members of the family of algebra languages were born, and indeed many were created to perform specific calculations of great complexity in such fields as celestial mechanics, general relativity and quantum electrodynamics.

We shall not describe any particular language here. We simply mention the principal tasks that can be undertaken and indicate their relevance in electron optics. Details are of course to be found in the manuals of each language and a useful survey is given by van Hulzen and Calmet (1982). The languages which are most widespread among physicists at present are probably REDUCE (Hearn, 1985; Fitch, 1985; Rayna, 1987) and MACSYMA (MIT, 1983; Rand, 1984; Pavelle, 1985; Rand and Armbruster, 1987). We know of only two examples of the use of a general language to perform algebraic calculations in electron optics: ALGOL (Dodin and Nesvizhskii, 1981; Nesvizhskii, 1986) and FORTRAN (Berz and Wollnik, 1987). The proceedings of the conferences that are regularly held on progress in computer algebra give a vivid picture of the growth of the subject (e.g. Ng, 1979; Calmet, 1982). For a full list with many other references, see the background text edited by Buchberger *et al.* (1982) and Davenport *et al.* (1988).

34.2 Computer algebra, its role in electron optics

In physics, algebra languages were first used mainly for calculations that involved performing elementary operations on very large operands. The number of terms being enormous, the risk of human error was correspondingly large and the computer was used first to check the hand calculation and later, as confidence in these languages grew, to replace it altogether. A basic operation is therefore multinomial expansion of expressions of the form $(x_1 + x_2 \ldots + x_n)^m$, where the variables x_i may themselves consist of several terms. Since the resulting expression will usually contain a very large number of terms, another important family of operations permits sophisticated sorting of these into groups: all terms $x_i^p x_j^q$ for given p and q, for example, or such that $p + q$ takes a given value, or such that $p \leq q$. If the x_i are circular or hyperbolic functions, or functions of such functions, some kind of reduction or simplification will probably be desirable and facilities for replacing $\sin^p x \sin^q x$ and similar terms by functions of multiples of x are provided or can be incorporated. An important aspect of these languages is that they can be 'taught' results that are not in their regular repertoire. In a program in which Bessel functions and their derivatives appeared, for example, the language could be instructed to use the well-known recurrence formulae that relate J_n, J'_n, J_{n+1} and J_{n-1}; indeed the numerous relations between contiguous hypergeometric functions (Whittaker and Watson, 1927, Section 14.7) could just as easily be included if required. A further elementary operation is substitution: flexible commands permitting substitution of an expression for a variable are provided. The next family of operations is concerned with calculus: these languages are capable of differentiating the everyday functions such as exponentials, logarithms, circular and hyperbolic functions and of course powers of a variable. They can also work correctly with the derivative of a general function so that an expression of the form $\int f'(x) \sin x \, dx$, for example, can be evaluated by partial integration. Depending on the degree of sophistication of the language, indefinite integration of more or less complicated expressions can likewise be performed. Integration of simple expressions is a routine matter but recent work has shown that these algebra languages can be extended to integrate any expression for which a solution in closed form exists, provided that the functions occurring in the integrand belong to a certain set of functions. See Norman and Davenport (1979) for a very readable account of an earlier stage in this development and Norman (1982).

Expansion, substitution, sorting, differentiation and integration: although other operations are available in some languages, it is these that have proved most useful in electron optics, where computer algebra has been used primarily to derive or check integral expressions for the aberration coefficients of various types of electron optical component. Let us consider the steps for a typical aberration coefficient. First we must substitute the expansions for Φ and the components of $\mathbf{A}$ into the refractive index $\{\hat{\Phi}(1+X'^2+Y'^2)\}^{\frac{1}{2}} - \eta(A_X X' + A_Y Y' + A_z)$ (15.23) truncated after terms of a particular degree. If the rotating coordinate system is to be employed, the appropriate transformation from (X, Y, z) to (x, y, z) must be made. The resulting terms must then be sorted according to their degree in x, y, x', y'. It is generally necessary to make a further substitution at this point, replacing the coordinates that appear in the group of primary aberration terms by the paraxial approximation. The result must be sorted according to the powers of the paraxial coordinates (e.g. x_o, y_o, x'_o, y'_o), in order to extract individual aberration coefficients. If higher order perturbation theory is to be used, a substitution involving paraxial terms and the primary aberrations in an arbitrary plane is also needed, followed by sorting of the same kind.

Once the aberration integrals have been derived, further laborious calculations may be needed. If the integrals are to be evaluated numerically, using measured or computed values of the various field functions that occur, it may be desirable to remove terms in which high-order derivatives appear by partial integration. The systematic procedure described in Chapter 24 at some length is the best way of doing this, and once the fundamental structures (e.g. 24.56) have been recognized, the subsequent differentiation, incorporation of the paraxial equation, multiplication by arbitrary constants and addition to the original coefficients or eikonal function can all be performed by the computer.

If, however, the behaviour of some unfamiliar component is being studied as a preliminary to exact numerical computation of ranges of values of the parameters of apparent interest, it is often helpful to adopt a simple but realistic field model. Despite its shortcomings, Glaser's bell-shaped model (Chapter 36) is invaluable for grasping the behaviour of magnetic lenses and is invariably used for teaching purposes; for quadrupoles, the rectangular and bell-shaped models are perfectly adequate for many purposes, and several other types of component are likewise represented rather well by simple models. Evaluation of the aberration integrals for these models is another lengthy task, involving the manipulation of unwieldy expressions frequently involving powers of circular and hyperbolic functions. Computer algebra languages are well-equipped to perform this.

We have mentioned the use of partial integration to cast aberration integrals into a form suitable for numerical evaluation, but how is the latter to be implemented? Computer algebra languages originally provided their answers only as algebraic formulae, which had then to be programmed manually in one of the ordinary high-level languages suited to numerical computation. This was clearly inefficient since one of the attractions of deriving large formulae by computer is to eliminate the risk of human error, but such mistakes are just as likely to creep in when translating the algebraic output into FORTRAN or PASCAL. By 1980, therefore, as a result of a considerable effort, symbolic–numeric interfaces had been developed, notably for REDUCE. The user of this language can now generate a FORTRAN routine automatically from his algebraic output, which is clearly a great improvement.

The calculation of aberration integrals is not of course the only laborious task facing the theoretician in electron optics, and the operations mentioned above are only a selection of those available. Recent versions of REDUCE thus include new packages for integration, for polynomial factorization, for solving ordinary differential equations and integral equations, for arbitrary precision floating-point arithmetic and for network analysis problems, including calculation of the determinants of sparse matrices (van Hulzen and Calmet, 1982).

34.3 Two practical examples

The languages CAMAL (Barton and Fitch, 1972; Fitch, 1985) and REDUCE (Hearn, 1985) have been used in electron optics to obtain relativistically correct formulae for the third-order geometric aberrations of round electrostatic lenses (Hawkes, 1977), to derive aberration coefficients for combined deflection and focusing fields (Soma, 1977, cf. Section 32.3.2) and to study the aberrations of microwave cavities acting as dynamic electron lenses (Hawkes, 1983; cf. Section 41.5). We shall describe very briefly the first two of these; the first illustrates the simplicity of the programming while the second shows that a gargantuan task is well within the capacity of such languages.

34.3.1 Round electrostatic lenses
The language CAMAL, which is extremely easy to learn, represents algebraic expressions in a notation very close to that in common use, apart from the inevitable linearization, that is, replacement of suffixes and indices by symbols in brackets or preceded by a code indicating, for example, exponentiation. It was adopted for the calculation of the relativistic forms of electrostatic lens aberration coefficients, the calculation dividing naturally into a series of short blocks. An extremely detailed account is given in Hawkes (1977) and here we reproduce only two short sections of program which illustrate two important points: first, that the actual calculation re-

quires very few lines of program whereas the sorting may occupy a great many, unless the notation is chosen to facilitate this step, in which case the resemblance between the programmed formulae and their familiar versions will be reduced. The second point is that the output can be organized in such a way that the effort of interpretation is slight.

The first passage represents the expansion of the refractive index M, $F := \hat{\phi}^{\frac{1}{2}}[1 + \{\hat{\Phi}(1 + x'^2 + y'^2) - \hat{\phi}\}/\hat{\phi}]^{\frac{1}{2}}$ (M has been replaced by F since M always denotes an integer). This is accomplished by writing

$$A = u<0> - \frac{1}{4}u<2>b + \frac{1}{64}u<4>b.2$$

$$F = A(1 + eA)(1 + c) - u<0>(1 + eu<0>)$$

$$F = \frac{1}{f} + \frac{1}{2}Ff - \frac{1}{8}F.2f.3$$

together with some simple housekeeping instructions, where $f := \hat{\phi}^{-\frac{1}{2}}$, $u<0> := \phi$, $u<2> := \phi''$, $u<4> := \phi^{(4)}$, $b := (x^2 + y^2)$ and $c := (x'^2 + y'^2)$; e denotes ϵ. Into F we substitute $x = x_o g(z) + x'_o h(z)$ and similarly for y; this entails two definitions and two instructions, each pair of the form

$$V = z<0>g<0>.2 + z<1>h<0>.2 + 2z<2>g<0>h<0>$$
$$F = \text{SUB}(F, V, b)$$

which replace $b = x^2 + y^2$ by V in F, the $z<I>$ denoting $x_o^2 + y_o^2$, $x'^2_o + y'^2_o$, $x_o x'_o + y_o y'_o$, for $I = 0, 1, 2$ respectively. Dots signify powers, in the sense that b.2 denotes b^2.

Some 50 lines are then occupied by sorting instructions, after which we obtain the coefficients $F[I]$ in

$$F[4] = F[0](x_o^2 + y_o^2)^2 + F[1](x'^2_o + y'^2_o)^2$$
$$+ F[2](x_o x'_o + y_o y'_o)^2 + F[3](x_o^2 + y_o^2)(x'^2_o + y'^2_o)$$
$$+ F[4](x_o^2 + y_o^2)(x_o x'_o + y_o y'_o) + F[5](x'^2_o + y'^2_o)(x_o x'_o + y_o y'_o)$$

in terms of the coefficients of the various powers of $g(z) =: g<0>$, $h(z) =: h<0>$ and their derivatives $g'(z) =: g<1>$ and $h'(z) =: h<1>$. The following passage outputs a text to remind the user of the definitions of the $A[I]$ and then lists the latter in the form $A[I] = \ldots$:

TEXT
F[0] = (1/4) (A[0] g<0>.4 + 2 A[1] g<0>.3 g<1>
   + 2 A[2] g<0>.2 g<1>.2
   + 2 A[3] g<0> g<1>.3 + A[4] g<1>.4)
F[1] = (1/4) (A[0] h<0>.4 + 2 A[1] h<0>.3 h<1>
   + 2 A[2] h<0>.2 h<1>.2
   + 2 A[3] h<0> h<1>.3 + A[4] h<1>.4)
$\vdots$

F[5] = A[0] g<0> h<0>.3
   −(1/2) A[1] h<0> (3g<1> h<0>.2 + g<0> h<1>.2)
   + A[2] h<0> h<1> (g<0> h<0> + g<1> h<1>)
   + (1/2) A[3] h<1> (g<1> h<0>.2 + 3 g<0> h<1>.2)
   + A[4] g<1> h<1>.3
NOTE A[5] = A[6] − 4A[7]
$\vdots$

FOR I = 0:1:7
PRINT [A[I]]
REPEAT
PAGE
V = 0
FOR I = 0:1:7
A[I] = SUB(A[I],V ,e)
PRINT [A[I]]
REPEAT

The last few lines list the non-relativistic approximation by setting e(= ϵ) equal to zero with the aid of the instruction SUB.

The other feature of interest in the program is the use of a SET facility to impose side-relations during a calculation. Here these side-relations are needed in the generation of functions A[I] including partial integration. The side-relations supplement the relations known to the language by specifying the derivatives of the various quantities involved and including simplification by means of the paraxial equation. The original refractive index (F) is extended as explained in Chapter 24.

34.3.2 Superimposed focusing and deflection fields
We have seen in Chapter 32 that the number of aberration coefficients becomes very large when the various deflection fields may be magnetic or electrostatic or even both; relativistic terms render the formulae still more complicated as does inclusion of dynamic as well as static potentials. This very general case was studied with the aid of REDUCE by Soma (1977),

who obtained relativistically correct aberration integrals for all the geometric and chromatic aberration coefficients. Soma does not give any details of his REDUCE routines but two important points are worth singling out. First, the manuscript of the published article, including all formulae, Greek letters and mathematical symbols, was typed directly by means of an on-line text-editing system. Such systems are now becoming common, if not yet commonplace, but in 1977 such an original output facility was rightly seen as a considerable asset. Secondly, by using REDUCE, Soma was able to generate a FORTRAN routine directly for the numerical evaluation of his numerous aberration coefficients, an essential step in the search for configurations with low aberrations by the methods described in Chapter 33, for example, and applied to this specific case by Chu and Munro (1982), who employed the damped least-squares procedures very effectively.

Contents of Volume 2

PART VII – INSTRUMENTAL OPTICS

Chapter 35   Electrostatic Lenses
Chapter 36   Magnetic Lenses
Chapter 37   Electron Mirrors
Chapter 38   Cathode Lenses and Field-Emission Microscopy
Chapter 39   Quadrupole Lenses
Chapter 40   Deflection Systems

PART VIII – ABERRATION CORRECTION AND BEAM INTENSITY DISTRIBUTION (CAUSTICS)

Chapter 41   Aberration Correction
Chapter 42   Caustics and their Applications

PART IX – ELECTRON GUNS

Chapter 43   General Features of Electron Guns
Chapter 44   Theory of Electron Emission
Chapter 45   Pointed Cathodes without Space Charge
Chapter 46   Space Charge Effects
Chapter 47   Brightness
Chapter 48   Emittance
Chapter 49   The Boersch Effect
Chapter 50   Complete Electron Guns

PART X – SYSTEMS WITH A CURVED OPTIC AXIS

Chapter 51   General Curvilinear Systems
Chapter 52   Magnetic Sector Fields
Chapter 53   Unified Theories of Ion Optical Systems

Notes and References

Notes and References

The following lists of references follow the main divisions of the book with the exception of those corresponding to Part VII (Instrumental Optics). The lists for Chapters 35 (electrostatic lenses) and 36 (magnetic lenses) are so long that it seemed preferable to give them separately.

In order to avoid repetition, standard abbreviations have been adopted for several series of conference proceedings, namely, the European and International conferences on electron microscopy, which have alternated every two years since 1954 (prior to that date they were not quite so regular); the occasional conferences on high-voltage electron microscopy; the annual meetings of the Electron Microscopy Society of America; and the biennial meetings of the Electron Microscopy and Analysis Group of the British Institute of Physics. The European and International Conferences are identified by date and place, the high-voltage conferences by date followed by HVEM and place, the American meetings by date followed by EMSA and meeting number and the British meetings by date followed by EMAG. Full bibliographic details of all these conference volumes are to be found at the end of this section, after the references for Part X.

Some of the lists that follow contain papers that are not cited in the text. With one exception, however, all these additional references are cited in the notes that precede the lists with some indication of their contents. The exception concerns the Preface and the introductory first chapter. This contains a very full list of books on electron optics; texts that are devoted mainly to electron microscopy are not included, however.

Despite the length of these lists, we make no claim to completeness and indeed, the coverage is deliberately uneven. For some topics, excellent bibliographies have been compiled and we have referred to these rather than merely repeating their contents. There are others, however, for which the literature is very scattered—electron guns are an example—and here we have attempted to give rather thorough coverage. We shall be most grateful to have any errors or serious omissions drawn to our attention.

Preface and Chapter 1

Afanas'ev, V.P. and Yavor, S.Ya. (1978). *Elektrostaticheskie Energoanalizatory dlya Puchkov Zaryazhennykh Chastits* (Nauka, Moscow).

Alyamovskii, I.V. (1966). *Elektronnyi Puchki i Elektronnyi Pushki* (Sovetskoe Radio, Moscow)

Ardenne, M. von (1938). Z. Physik **109**, 553–572 and Z. Tech. Phys. **19**, 407–416.

Ardenne, M. von (1940). *Elektronen-Übermikroskopie* (Springer, Berlin).

Ardenne, M. von (1956). *Tabellen der Elektronenphysik, Ionenphysik und Übermikroskopie*, 2 vols (Deutscher Verlag der Wissenschaften, Berlin).

Ardenne, M. von (1962). *Tabellen zur Angewandten Physik. I. Elektronenphysik, Übermikroskopie, Ionenphysik* (Deutscher Verlag der Wissenschaften, Berlin); 2nd ed., 1973.

Ardenne, M. von (1978). Optik **50**, 177–188 and Bild und Ton **31**, 293–297.

Ardenne, M. von (1985). On the history of scanning electron microscopy, of the electron microprobe, and of early contributions to transmission electron microscopy. In Hawkes (1985), pp. 1–21.

Balladore, J.-L. and Murillo, R. (1977). J. Microsc. Spectrosc. Electron. **2**, 211–222.

Banford, A.P. (1966). *The Transport of Charged Particle Beams* (Spon, London).

Baranova, L.A. and Yavor, S. Ya. (1986). *Elektrostaticheskie Elektronnye Linzy* (Nauka, Moscow); Adv. Electron. Electron Phys.

Boersch, H., Bostanjoglo, O. and Grohmann, K. (1966). Z. Angew. Phys. **20**, 193–194.

Bonshtedt, B.E. and Markovich, M.G. (1967). *Fokusirovka i Otklonenie Puchkov v Elektronno-luchevykh Priborakh* (Sovetskoe Radio, Moscow).

Borries, B. von (1949). *Die Übermikroskopie* (Editio Cantor, Aulendorf).

Borries, B. von and Ruska, E. (1940). Ergebn. Exakt. Naturwiss. **19**, 237–272.

Broglie, L. de (1925). Ann. Physique (Paris) **3**, 22–128.

Broglie, L. de, ed. (1946). *L'Optique Electronique* (Edns Rev. Opt. Theor. Instrum., Paris).

Broglie, L. de (1950). *Optique Electronique et Corpusculaire* (Hermann, Paris).

Brüche, E. and Recknagel, A. (1941). *Elektronengeräte, Prinzipien und Systematik* (Springer, Berlin).

Brüche, E. and Scherzer, O. (1934). *Geometrische Elektronenoptik* (Springer, Berlin).

Busch, H. (1926). Ann. Physik (Leipzig) **81**, 974–993.

Busch, H. (1927). Arch. Elektrotech. **18**, 583–594.

Busch, H. and Brüche, E., eds (1937). *Beiträge zur Elektronenoptik* (Barth, Leipzig).

Carey, D.C. (1987). *The Optics of Charged Particle Beams* (Harwood, Chur, London and New York).

Castaing, R. (1951). *Application des Sondes Electroniques à une Méthode d'Analyse Ponctuelle Chimique et Cristallographique*. Thèse, Paris, 92 pp.

Castaing, R. and Henry, L. (1962). C.R. Acad. Sci. Paris **255**, 76–78.

Chen, W.-x. (1986). *Introduction to Electron Optics* [in Chinese] (Beijing University Press, Beijing).

Cosslett, V.E. (1946). *Introduction to Electron Optics* (Clarendon, Oxford); 2nd ed., 1950.

Cosslett, V.E. (1950). *Bibliography of Electron Microscopy* (Arnold, London).

Cosslett, V.E. (1981). Contemp. Phys. **22**, 3–36 and 147–182.

Cosslett, V.E. and Duncumb, P. (1956). Stockholm, pp. 12–14.

Crewe, A.V. (1965). A scanning transmission microscope utilizing electron energy losses. Unpublished abstract, Conference on Non-conventional Electron Microscopy, Cambridge.

Crewe, A.V. (1970). Q. Rev. Biophys. **3**, 137–175.

Crewe, A.V. (1973). Prog. Opt. **11**, 223–246.

Crewe, A.V. (1983). Science **221**, 325–330.

Crewe, A.V. (1984). J. Ultrastruct. Res. **88**, 94–104.

Crewe, A.V., Wall, J. and Welter, L.M. (1968). J. Appl. Phys. **39**, 5861–5868.

Dahl, P. (1973). *Introduction to Electron and Ion Optics* (Academic Press, New York and London).

Davisson, C.J. and Germer, L.H. (1927). Phys. Rev. **30**, 705–740; Nature **119**, 558–560.

Devyatkov, N.D., ed. (1977). *Metody Rascheta Elektronno-opticheskikh Sistem* (Nauka, Moscow).

Dietrich, I. (1976). *Superconducting Electron-optic Devices* (Plenum, New York and London).

Dietrich, I., Pfisterer, H. and Weyl, R. (1969). Z. Angew. Phys. **28**, 35–39.

Driest, E. and Müller, H.O. (1935). Z. Wiss. Mikrosk. **52**, 53–57.

Duncumb, P. (1958). Berlin, vol. 1, pp. 267–269.

Dupouy, G. (1952). *Eléments d'Optique Electronique* (Armand Colin, Paris); 2nd ed., 1961.

Dupouy, G. (1968). Adv. Opt. Electron Microsc. **2**, 167–250.

Dupouy, G. (1985). In *The Beginnings of Electron Microscopy* (Hawkes, P.W., ed.) pp. 103–165; Adv. Electron. Electron Phys. Suppl. 16.

Dupouy, G., Perrier, F. and Durrieu, L. (1960). C.R. Acad. Sci. Paris **251**, 2836–2841.

El-Kareh, A.B. and El-Kareh, J.C.J. (1970). *Electron Beams, Lenses and Optics*, 2 vols (Academic Press, New York and London).

Felici, N.J. (1965). *Introduction à l'Optique Corpusculaire* (Gauthier-Villars, Paris).

Fernández-Morán, H. (1965). Proc. Natl Acad. Sci. USA **53**, 445–451.

Fujita, H. (1986). *History of Electron Microscopes* (Business Center for Academic Societies Japan, Tokyo).

Gabor, D. (1945). *The Electron Microscope* (Hulton, London).

Gabor, D. (1957). Elektrotech. Z. **A78**, 522–530.

Gettner, M.E. and Ornstein, L. (1956). In *Physical Techniques in Biological Research* (Oster, G. and Pollister, A.W., eds) vol. 3, pp. 627–686 (Academic Press, New York and London).

Glaser, W. (1952). *Grundlagen der Elektronenoptik* (Springer, Vienna).

Glaser, W. (1956). Elektronen- und Ionenoptik. In *Handbuch der Physik* (Flügge, S., ed.) vol. 33, pp. 123–395 (Springer, Berlin).

Good, R.H. and Müller, E.W. (1956). Field emission. In *Handbuch der Physik* (Flügge, S., ed.) vol. 21, pp. 176–231 (Springer, Berlin).

Grime, G.W. and Watt, F. (1984). *Beam Optics of Quadrupole Probe-forming Systems* (Adam Hilger, Bristol).

Grinberg, G.A. (1942). Dokl. Akad. Nauk SSR **37**, 172–178 and 261–268.

Grinberg, G.A. (1943a). Dokl. Akad. Nauk SSR **38**, 78–81.

Grinberg G.A. (1943b). Zh. Tekh. Fiz. **13**, 361–388 (Russian translation of Grinberg 1942, 1943 a).

Grivet, P., Bernard, M.Y. and Septier, A. (1955). *Optique Electronique*, vol. 1, *Lentilles électroniques* (Bordas, Paris).

Grivet, P., Bernard, M.Y., Bertein, F., Castaing, R., Gauzit, M. and Septier, A. (1958). *Optique Electronique*, vol. 2, *Microscopes, diffractographes, spectrographes de masse, oscillographes cathodiques* (Bordas, Paris).

Grivet, P. (1965). *Electron Optics* (Pergamon, London and New York); 2nd ed., 1972.

Grosser, J. (1983). *Einführung in die Teilchenoptik* (Teubner, Stuttgart).

Haine, M.E. (1961). *The Electron Microscope, the Present State of the Art* (Spon, London).

Hall, C.E. (1953). *Introduction to Electron Microscopy* (McGraw-Hill, New York and London); 2nd ed., 1966.

Hardy, D.F. (1973). Adv. Opt. Electron Microsc. **5**, 201–237.

Harting, E. and Read, F.H. (1976). *Electrostatic Lenses* (Elsevier, Amsterdam, Oxford and New York).

Hatschek, P. (1937). *Optik des Unsichtbaren. Eine Einführung in die Welt der Elektronen-Optik* (Franckh'sche Verlagshandlung, Stuttgart); *Electron Optics* (American Photographic Publ. Co., Boston, 1944).

Hawkes, P.W. (1966). *Quadrupole Optics* (Springer, Berlin and New York); Springer Tracts in Modern Physics, vol. 42.

Hawkes, P.W. (1968). Optik **27**, 287–304.

Hawkes, P.W. (1970). *Quadrupoles in Electron Lens Design* (Academic Press, New York and London); Adv. Electron. Electron Phys., Suppl. 7.

Hawkes, P.W. (1972). *Electron Optics and Electron Microscopy* (Taylor and Francis, London).

Hawkes, P.W., ed. (1973). *Image Processing and Computer-aided Design in Electron Optics* (Academic Press, London and New York).

Hawkes, P.W., ed. (1982). *Magnetic Electron Lenses* (Springer, Berlin and New York); Topics in Current Physics, vol. 18.

Hawkes, P.W., ed. (1985). *The Beginnings of Electron Microscopy* (Academic Press, Orlando and London); Adv. Electron. Electron Phys. Suppl. 16.

Hawkes, P.W. and Valdrè, U. (1977). J. Phys. E: Sci. Instrum. **10**, 309–328.

Humphries, S. (1986). *Principles of Charged Particle Acceleration* (Wiley-Interscience, New York and Chichester).

Il'in, V.P. (1974). *Chislennye Metody Resheniya Zadach Elektrooptiki* (Nauka, Novosibirsk).

Il'in, V.P., ed. (1982). *Metody Rascheta Elektronno-opticheskikh Sistem* (Computer Centre, Novosibirsk).

Il'in, V.P., ed. (1983). *Algoritmy i Metody Rascheta Elektronno-opticheskikh Sistem* (Computer Centre, Novosibirsk).

Il'in, V.P., Kateshov, V.A., Kulikov, Yu. V. and Monastyrskii, M.A. (1987). *Chislennye Metody Optimizatsii Emissionnykh Elektronno-opticheskikh Sistem* (Nauka, Novosibirsk); Adv. Electron. Electron Phys.

Jacob, L. (1950). *Introduction to Electron Optics* (Methuen, London).

Kasper, E. (1982). In *Magnetic Electron Lenses* (Hawkes, P.W., ed.) pp. 57–118 (Springer, Berlin and New York).

Kasper, E. (1984). In *Electron Optical Systems for Microscopy, Microanalysis and Microlithography* (Hren, J.J., Lenz, F.A., Munro, E. and Sewell, P.B., eds) pp. 63–73 (Scanning Electron Microscopy, AMF O'Hare).

Kasper, E. and Lenz, F. (1980). The Hague, vol. 1, pp. 10–15.

Kas'yankov, P.P. (1956). *Teoriya Elektromagnitnykh Sistem s krivolineinoi Os'yu* (University Press, Leningrad).

Kel'man, V.M. (1955). *Elektronnaya Optika* (Izd. Akad. Nauk SSSR, Moscow and Leningrad); 2nd ed., 1968 (Nauka, Moscow).

Kel'man, V.M. and Yavor, S. Ya. (1959). *Elektronnaya Optika* (Izd. Akad. Nauk SSSR, Moscow and Leningrad); 2nd ed., 1963; 3rd ed., 1968 (Nauka, Leningrad).

Kel'man, V.M., Karetskaya, S.P., Fedulina, L.V. and Yakushev, E.M. (1979). *Elektronnoopticheskie Elementy Prismennykh Spektrometrov Zaryazhennykh Chastits* (Nauka, Alma-Ata).

Kel'man, V.M., Rodnikova, I.V. and Sekunova, L.M. (1985). *Staticheskie Mass-Spektrometry* (Nauka, Alma-Ata).

Kim, K.-H. and Choi, S.-K. (1963). *Optics of Charged Particles* [in Korean] (Korean Academy of Sciences, Pyong-Yang).

Klemperer, O. (1939). *Electron Optics* (University Press, Cambridge); 2nd ed., 1953; 3rd ed., see Klemperer and Barnett (1971).

Klemperer, O. (1961). *Electron Physics, the Physics of the Free Electron* (Butterworths, London).

Klemperer, O. and Barnett, M.E. (1971). *Electron Optics* (University Press, Cambridge).

Knoll, M. and Eichmeier, J. (1966). *Technische Elektronik*, vol. 2, *Stromsteuernde und elektronenoptische Entladungsgeräte* (Springer, Berlin and New York).

Knoll, M. and Ruska, E. (1932a). Ann. Physik (Leipzig) **12**, 607–640 and 641–661.

Knoll, M. and Ruska, E. (1932b). Z. Physik **78**, 318–339.

Kotov, V.I. and Miller, V.V. (1969). *Fokusirovka i Razdelenie po Massam Chastits Vysokykh Energii* (Atomizdat, Moscow).

Krause, F. (1936). Z. Physik **102**, 417–422.

Laberrigue, A. and Levinson, P. (1964). C.R. Acad. Sci. Paris **259**, 530–532.

Lawson, J.D. (1977). *The Physics of Charged-Particle Beams* (Clarendon, Oxford); 2nd ed., 1988.

Lapostolle, P. and Septier, A., eds (1970). *Linear Accelerators* (North-Holland, Amsterdam).

Lebedev, A.A., ed. (1954). *Elektronnaya Mikroskopiya* (Gostekhizdat, Moscow).

Lefranc, G., Knapek, E. and Dietrich, I. (1982). Ultramicroscopy **10**, 111–124.

Leisegang, S. (1956). Elektronenmikroskope. In *Handbuch der Physik* (Flügge, S., ed.) vol. 33, pp. 396–545.

Lenz, F. (1956). Stockholm, pp. 48–51.

Lenz, F. (1957). Optik **14**, 74–82.

Lenz, F. (1973). In *Image Processing and Computer-aided Design in Electron Optics* (Hawkes, P.W., ed.) pp. 274–282 (Academic Press, London and New York).

Livingood, J.J. (1969). *The Optics of Dipole Magnets* (Academic Press, New York and London).

Lukoshkov, V.S., Zamorozkov, B.M. and Shubin, L.V. (1968). *Elektronnye Puchki i Elektronno-opticheskie Sistemy* (Institut 'Elektronika', Moscow).

Magnan, C., ed. (1961). *Traité de Microscopie Electronique*, 2 vols (Hermann, Paris).

Marchuk, G.I., ed. (1967). *Chislennye Metody Rascheta Elektronno-opticheskikh Sistem* (Nauka, Novosibirsk).

Marchuk, G.I., ed. (1970). *Metody Rascheta Elektronno-Opticheskikh Sistem* (Computer Centre, Novosibirsk).

Marchuk, G.I., ed. (1973). *Metody Rascheta Elektronno-Opticheskikh Sistem*, 2 vols (Computer Centre, Novosibirsk).

Marton, C., Sass, S., Swerdlow, M., Bronkhorst, A. van and Meryman, H. (1950). *Bibliography of Electron Microscopy*, NBS Circular No. 502.

Marton, L. (1934). Nature **133**, 911; Phys. Rev. **46**, 527–528.

Marton, L. (1934-1937). Bull. Cl. Sci. Acad. R. Belg. **20**, 439–446 (1934); **21**, 553–564 and 606–617 (1935); **22**, 1336–1344 (1936); **23**, 672–675 (1937).

Marton, L. (1935). Rev. Opt. **14**, 129–145.

Marton, L. (1968). *Early History of the Electron Microscope* (San Francisco Press, San Francisco).

McMullan, D. (1953). Proc. Inst. Elec. Eng. **100**, 245–259.

Molokovskii, S.I. and Sushkov, A.D. (1965). *Elektronnoopticheskie Systemy Priborov Sverkhvysokikh Chastot* (Moscow and Leningrad).

Mulvey, T. (1962). Brit. J. Appl. Phys. **13**, 197–207.

Mulvey, T. (1967). Proc. R. Microsc. Soc. **2**, 201–227.

Mulvey, T. (1973). Phys. Bull. **24**, 147–154.

Mulvey, T. (1982). In *Magnetic Electron Lenses* (Hawkes, P.W., ed.) pp. 359–412 (Springer, Berlin and New York).

Mulvey, T. (1984). In *Electron Optical Systems for Microscopy, Microanalysis and Microlithography* (Hren, J.J., Lenz, F.A., Munro, E. and Sewell, P.B., eds) pp. 15–27 (Scanning Electron Microscopy, AMF O'Hare).

Mulvey, T. (1985). In *The Beginnings of Electron Microscopy* (Hawkes, P.W., ed.) pp. 417–442 (Academic Press, Orlando and London); Adv. Electron. Electron Phys. Suppl. 16.

Myers, L.M. (1939). *Electron Optics, Theoretical and Practical* (Chapman and Hall, London).

Newman, S.B., Borysko, E. and Swerdlow, M. (1949). J. Res. Natl. Bur. Stand. **A43**, 183–199.

Oatley, C.W. (1972). *The Scanning Electron Microscope. Part I. The Instrument* (University Press, Cambridge).

Oatley, C.W. (1982). J. Appl. Phys. **53**, R1–R13.

Oatley, C.W., Nixon, W.C. and Pease, R.F.W. (1965). Adv. Electron. Electron Phys. **21**, 181–247.

Oatley, C.W., McMullan, D. and Smith, K.C.A. (1985). In *The Beginnings of Electron Microscopy* (Hawkes, P.W., ed.) pp. 443–482 (Academic Press, Orlando and London); Adv. Electron. Electron Phys. Suppl. 16.

Ollendorff, F. (1955). *Technische Elektrodynamik*, vol. II, *Innere Elektronik*, Pt I, *Elektronik des Einzelelektrons* (Springer, Vienna).

Ozasa, S., Katagiri, S., Kimura, H. and Tadano, B. (1966). Kyoto, vol. 1, pp. 149–150.

Paszkowski, B. (1960). *Optyka Elektronowa* (Wydawnictwa Naukowo-Techniczne, Warsaw); 2nd ed., 1965. Translated into English as *Electron Optics* (Iliffe, London and American Elsevier, New York, 1968).

Pease, D.C. and Baker, R.F. (1948). Proc. Soc. Exp. Biol. Med. **67**, 470–474.

Picht, J. (1939). *Einführung in die Theorie der Elektronenoptik* (Barth, Leipzig); 2nd ed., 1957; 3rd ed. 1963.

Picht, J. (1945). *Was Wir über Elektronen Wissen* (Carl Winter, Heidelberg).

Picht, J. and Gain, R. (1955). *Das Elektronenmikroskop* (Fachbuch, Leipzig).

Picht, J. and Heydenreich, J. (1966). *Einführung in die Elektronenmikroskopie* (Verlag Technik, Berlin).

Pierce, J.R. (1949). *Theory and Design of Electron Beams* (Van Nostrand, Princeton and London); 2nd ed., 1954.

Prismennye Beta-Spektrometry i ikh Primenenie (Vilnius, 1971).

Ramsauer, C., ed. (1941). *Zehn Jahre Elektronenmikroskopie; ein Selbstbericht des AEG Forschungs-Instituts* (Springer, Berlin); 2nd ed., 1942; 3rd ed., entitled *Elektronenmikroskopie; Bericht über Arbeiten des AEG Forschungs-Instituts, 1930 bis 1942* (1943).

Riecke, W.D. (1982). In *Magnetic Electron Lenses* (Hawkes, P.W., ed.) pp. 163–357 (Springer, Berlin and New York).

Rose, H. (1968). Optik **27**, 466–474 and 497–514.

Ruska, E. (1957). Elektrotech. Z. **A78**, 531–543.

Ruska, E. (1979). *Die Frühe Entwicklung der Elektronenlinsen und der Elektronenmikroskopie.* Acta Hist. Leopold. No. 12, 136 pp.

Ruska, E. (1980). *The Early Development of Electron Lenses and Electron Microscopy* (Hirzel, Stuttgart).

Ruska, E. and Knoll, M. (1931). Z. Tech. Phys. **12**, 389–400 and 448.

Rusterholz, A.A. (1950). *Elektronenoptik*, vol. 1, *Grundzüge der theoretischen Elektronenoptik* (Birkhäuser, Basle).

Scherzer, O. (1936). Z. Physik **101**, 593–603.

Scherzer, O. (1947). Optik **2**, 114–132.

Seman, O.I. (1951). Dokl. Akad. Nauk SSSR **81**, 775–778.

Seman, O.I. (1954). Dokl. Akad. Nauk SSSR **96**, 1151–1154.

Seman, O.I. (1955). Trudy Inst. Fiz. Astron. Akad. Nauk Eston. SSSR, No. 2, 3–49.

Seman, O.I. (1958a). Uch. Zap. Rostov. Gos. Univ. (Ser. Fiz.) **68**, No. 8, 63–75 and 77–90.

Seman, O.I. (1958b). Radiotekh. Elektron **3**, 283–287; Radio Eng. Electron. **3**, 402–409.

Seman, O.I. (1959). *Theoretical Foundations of Electron Optics* [in Chinese] (Higher Education Press, Peking).

Septier, A., ed. (1967). *Focusing of Charged Particles*, 2 vols (Academic Press, New York and London).

Septier, A., ed. (1980). *Applied Charged Particle Optics*. Parts A and B; Part C, 1983 (Academic Press, New York and London); Supplements 13A, 13B and 13C to Adv. Electron Electron. Phys.

Siegel, B.M., Kitamura, N., Kropfli, R.A. and Schulhof, M.P. (1966). Kyoto, vol. 1, pp. 151–152.

Sjöstrand, F.S. (1953). Experientia **9**, 114–115.

Sjöstrand, F.S. (1967). *Electron Microscopy of Cells and Tissues*, vol. 1, *Instrumentation and Techniques* (Academic Press, New York and London).

Smith, K.C.A., Considine, K. and Cosslett, V.E. (1966). Kyoto, vol. 1, pp. 99–100.

Steffen, K.G. (1965). *High Energy Beam Optics* (Wiley-Interscience, New York and London).

Strashkevich, A.M. (1959). *Elektronnaya Optika Elektrostaticheskikh Polei ne Obladayushchikh Osevoi Simmetriei* (Gos. Izd. Fiz.-Mat. Lit., Moscow).

Strashkevich, A.M. (1966). *Elektronnaya Optika Elektrostaticheskykh Sistem* (Energiya, Moscow and Leningrad).

Sturrock, P.A. (1952). Phil. Trans. Roy. Soc. London **A245**, 155–187.

Sturrock, P.A. (1955). *Static and Dynamic Electron Optics* (University Press, Cambridge).

Sugata, E. (1968). *Electron Microscopes of Japan 1936–1965. A Historical Survey* (Maruzen, Tokyo).

Sushkin, N.G. (1949). *Elektronnyi Mikroskop* (Gos. Izd. Tekh.-Teor. Lit., Moscow).

Szilagyi, M. (1988). *Electron and Ion Optics* (Plenum, New York and London).

Szymański, H., Mulak, A., Duda, A. and Romanowski, A. (1977). *Optyka Elektronowa*; 2nd ed., 1984 (Wydawnictwa Naukowo-Techniczne, Warsaw).

Tang, T.-T. (1987). *Introduction to Charged Particle Optics with Applications* [in Chinese] (Xi'an Jiantong University Press, Xi'an).

Taranenko, V.P. (1964). *Elektronnye Pushki* (Tekhnika, Kiev).

Thomson, G.P. and Reid, A. (1927). Nature **119**, 890.

Tsukkerman, I.I. (1958). *Elektronnaya Optika v Televidenii* (Gosenergoizdat, Moscow and Leningrad); English translation *Electron Optics in Television* (Pergamon, Oxford and New York, 1961).

Vainryb, E.A. and Milyutin, V.I. (1951). *Elektronnaya Optika* (Gos. Energ. Izd., Moscow and Leningrad); German translation by K. Keller as Wainrib, E.A. and Miljutin, W.I., *Elektronenoptik* (Verlag Technik, Berlin, 1954).

Vandakurov, Yu. V. (1955). Zh. Tekh. Fiz. **25**, 2545–2555.

Vandakurov, Yu. V. (1956). Zh. Tekh. Fiz. **26**, 1555–1610 and 2578–2594; Sov. Phys. Tech. Phys. **1**, 1558–1569 and 2491–2507.

Vandakurov, Yu. V. (1957). Zh. Tekh. Fiz. **27**, 1850–1862; Sov. Phys. Tech. Phys. **2**, 1719–1733.

Watt, F. and Grime, G.W., eds (1987). *Principles and Applications of High-energy Ion Microbeams* (Adam Hilger, Bristol).

Weyl, R., Dietrich, I. and Zerbst, H. (1972). Optik **35**, 280–286.

Wollnik, H. (1987). *Optics of Charged Particles* (Academic Press, Orlando and London).

Ximen, J.-y. (1986). *Aberration Theory in Electron and Ion Optics* (Academic Press, Orlando and London); Supplement 17 to Adv. Electron. Electron Phys.

Ximen, J.-y. and Ge, Z.-s. (1979). *Principles of the Electron Microscope and its Design* [in Chinese] (Science Press, Peking).

Yavor, S. Ya. (1968). *Fokusirovka Zaryazhennykh Chastits Kvadrupol'nymi Linzami* (Atomizdat, Moscow).

Zinchenko, N.S. (1961). *Kurs Lektsii po Elektronnoi Optike*, 2nd ed. (Izd. Khar'kovsk. Ordena Trudovogo Krasnogo Znameni Gos. Univ. im A.M. Gor'kogo, Khar'kov).

Zworykin, V.K., Hillier, J. and Snyder, R.L. (1942). ASTM. Bull. No. 117, 15–23.

Zworykin, V.K., Morton, G.A., Ramberg, E.G., Hillier, J. and Vance, A.W. (1945). *Electron Optics and the Electron Microscope* (Wiley, New York and Chapman and Hall, London).

Part I, Chapters 2–5

The examination by Ehrenberg and Siday (1949) of the notion of refractive index provoked a comment by Glaser (1951) and a reply by the original authors (Ehrenberg and Siday, 1951).

Aharonov, Y. and Bohm, D. (1959). Phys. Rev. **115**, 485–491.

Born, M. and Wolf, E. (1959). *Principles of Optics* (Pergamon, London and New York); 6th ed., 1980.

Bruns, H. (1895). Abh. K. Sächs. Ges. Wiss., Math-Phys. Kl. **21**, 321–436.

Ehrenberg, W. and Siday, R.E. (1949). Proc. Phys. Soc. (London) **B62**, 8–21.

Ehrenberg, W. and Siday, R.E. (1951). Proc. Phys. Soc. (London) **B64**, 1088–1089.

Glaser, W. (1951). Proc. Phys. Soc. (London) **B64**, 114–118 and 1089.

Glaser, W. (1952). *Grundlagen der Elektronenoptik* (Springer, Vienna).

Goldstein, H. (1959). *Classical Mechanics* (Addison-Wesley, Reading and London).

Grivet, P. (1965). *Electron Optics* (Pergamon, London and New York).

Kasper, E. (1972). Optik **35**, 83–89.

Kasper, E. (1985). Optik **69**, 117–125.

Kel'man, V.M. and Yavor, S.Ya. (1959). *Elektronnaya Optika* (Izd. Akad. Nauk, Moscow and Leningrad); 2nd ed., 1963; 3rd ed., 1968 (Izd. Nauka, Leningrad).

Kel'man, V.M., Sekunova, L.M. and Yakushev, E.M. (1972). Zh. Tekh. Fiz. **42**, 2279–2287; Sov. Phys. Tech. Phys. **17**, 1786–1791.

Kel'man, V.M., Sekunova, L.M. and Yakushev, E.M. (1973). Zh. Tekh. Fiz. **43**, 1799–1806 and 1807–1817; Sov. Phys. Tech. Phys. **18**, 1142–1146 and 1147–1152.

Picht, J. (1939). *Einführung in die Theorie der Elektronenoptik* (Barth, Leipzig); 2nd ed., 1957; 3rd ed., 1963.

Störmer, C. (1904). Vidensk. Selsk. Skr. Math. Nat. Kl. No. 3, 32 pp.

Störmer, C. (1906). C.R. Acad. Sci. Paris **142**, 1580–1583 and **143**, 140–142, 408–411 and 460–464.

Störmer, C. (1933). Ann. Physik (Leipzig) **16**, 685–696.

Sturrock, P.A. (1955). *Static and Dynamic Electron Optics* (University Press, Cambridge).

Part II, Chapters 6–13

For further details of the work of the Imperial College school, see Munro (1980, 1987a,b) and for accounts of later developments in the Tübingen group, see Kasper (1987a,b,c). The BEM is applied to electrostatic systems of arbitrary shape by Desbruslais and Munro (1987). The BEM is surveyed by Costabel (1987) and the FEM by Morton (1987) and Reid (1987).

Abramowitz, M. and Stegun, I.A. (1964). *Handbook of Mathematical Functions* (NBS, Washington; Dover, New York, 1965).

Adams, A. and Read, F.H. (1972). J. Phys. E: Sci. Instrum. **5**, 150–155 and 156–160.

Ames, W.F. (1969). *Numerical Methods for Partial Differential Equations* (Nelson, London).

Balladore, J.L., Murillo, R., Hawkes, P.W., Trinquier, J. and Jouffrey, B. (1981). Nucl. Instrum. Meth. **187**, 209–215.

Balladore, J.L., Murillo, R. and Trinquier, J. (1984). In *Electron Optical Systems for Microscopy, Microanalysis and Microlithography* (Hren, J.J., Lenz, F.A., Munro, E. and Sewell, P.B., eds) pp. 29–35 (Scanning Electron Microscopy, AMF O'Hare).

Bettess, P. (1977). Int. J. Numer. Meth. Eng. **11**, 53–64.

Boerboom, A.J.H. and Chen, H.-N. (1984). Int. J. Mass Spectrom. Ion Proc. **61**, 1–13.

Bonjour, P. (1980). In *Applied Charged Particle Optics* (Septier, A., ed.) vol. A, pp. 1–44 (Academic Press, New York and London); Adv. Electron. Electron Phys., Suppl. 13A.

Buneman, O. (1971). J. Comput. Phys. **8**, 500–505.

Buneman, O. (1973). J. Comput. Phys. **11**, 307–314 and 447–448.

Carré, B.A. (1961). Comput. J. **4**, 73–78.

Chari, M.V.K. and Silvester, P.P. (1980). *Finite Elements in Electrical and Magnetic Field Problems* (Wiley, Chichester and New York).

Chu, H.C. and Munro, E. (1982). Optik **61**, 121–145 and 213–236.

Colonias, J.S. (1974). *Particle Accelerator Design: Computer Programs* (Academic Press, New York and London).

Costabel, M. (1987). Principles of boundary element methods. Comput. Phys. Repts **6**, 243–274.

Courant, R. (1943). Bull. Am. Math. Soc. **49**, 1–23.

Cruise, D.R. (1963). J. Appl. Phys. **34**, 3477–3479.

Cuthill, E. and McKee, J. (1969). In *Proc. 24th Natl. Conf. ACM* pp. 157–172 (Brandon Systems Press, N.J.).

Desbruslais, S.R. and Munro, E. (1987). In *Proc. Int. Symp. Electron Optics Beijing* (Ximen, J.-y., ed.) pp. 45–48 (Institute of Electronics, Academia Sinica, Beijing).

Dommaschk, W. (1965). Optik **23**, 472–477.

Duff, I.S. (1977). Proc. IEEE **65**, 500–535.

Durand, E. (1966). *Electrostatique* (Masson, Paris).

Eupper, M. (1982). Optik **62**, 299–307.

Eupper, M. (1985). *Eine verbesserte Integralgleichungsmethode zur numerischen Lösung dreidimensionaler Dirichletprobleme und ihre Anwendung in der Elektronenoptik.* Dissertation, Tübingen.

Forsythe, G.E. and Wasow, W.R. (1960). *Finite Difference Methods for Partial Differential Equations* (Wiley, New York).

Franzen, N. (1984). In *Electron Optical Systems for Microscopy, Microanalysis and Microlithography* (Hren, J.J., Lenz, F.A., Munro, E. and Sewell, P.B., eds) pp. 115–126 (Scanning Electron Microscopy, AMF O'Hare).

Gibbs, N.E., Poole, W.G. and Stockmeyer, P.K. (1976). SIAM J. Numer. Anal. **13**, 236–250.

Glaser, W. (1952). *Grundlagen der Elektronenoptik* (Springer, Vienna).

Harrington, R.F. (1967). Proc. Inst. Electr. Electron. Eng. **55**, 136–149.

Harrington, R.F. (1968). *Field Computation by Moment Methods* (Macmillan, New York and Collier-Macmillan, London).

Harrington, R.F., Pontoppidan, K., Abrahamsen P. and Albertsen, N.C. (1969). Proc. Inst. Electr. Eng. **116**, 1715–1720.

Harting, E. and Read, F.H. (1976). *Electrostatic Lenses* (Elsevier, Amsterdam, Oxford and New York).

Hermeline, F. (1982). RAIRO **16**, 211–242.

Hoch, H., Kasper, E. and Kern, D. (1978). Optik **50**, 413–425.

Iselin, C.F. (1981). IEEE Trans. **MAG-17**, 2168–2177.

Jacobs, D.H., ed. (1977). *The State of the Art in Numerical Analysis* (Academic Press, London and New York).

Janse, J. (1971). Optik **33**, 270–281.

Kang, N.K., Orloff, J., Swanson, L.W. and Tuggle, D. (1981). J. Vac. Sci. Technol. **19**, 1077–1081.

Kang, N.K., Tuggle, D. and Swanson, L.W. (1983). Optik **63**, 313–331.

Kasper, E. (1976). Optik **46**, 271–286.

Kasper, E. (1982). Magnetic field calculation and the determination of electron trajectories. In *Magnetic Electron Lenses* (Hawkes, P.W., ed.) pp. 57–118 (Springer, Berlin and New York).

Kasper, E. (1983). Optik **64**, 157–169.

Kasper, E. (1984a). Optik **68**, 341–362.

Kasper, E. (1984b). In *Electron Optical Systems for Microscopy, Microanalysis and Microlithography* (Hren, J.J., Lenz, F.A., Munro, E. and Sewell, P.B. eds) pp. 63–73 (Scanning Electron Microscopy, AMF O'Hare).

Kasper, E. (1987a). Nucl. Instrum. Meth. Phys. Res. **A258**, 466–479.

Kasper, E. (1987b). Optik **77**, 3–12.

Kasper, E. (1987c). In *Proc. Int. Symp. Electron Optics Beijing* (Ximen J.-y., ed.) pp. 115–118 (Institute of Electronics, Academia Sinica, Beijing).

Kasper, E. and Lenz, F. (1980). The Hague, vol. 1, pp. 10–15.

Kasper, E. and Scherle, W. (1982). Optik **60**, 339–352.

Kern, D. (1978). *Theoretische Untersuchungen an rotationssymmetrischen Strahlerzeugungssystemen mit Feldemissionsquelle.* Dissertation, Tübingen.

Killes, P. (1985). Optik **70**, 64–71.

Konrad, A. and Silvester, P. (1973). Comput. Phys. Commun. **5**, 437–455.

Kuroda, K. (1983). Optik **64**, 125–131.

Kuroda, K. and Suzuki, T. (1972). Jpn. J. Appl. Phys. **11**, 1382.

Lencová, B. and Lenc, M. (1982). Hamburg, vol. 1, pp. 317–318.

Lencová, B. and Lenc, M. (1984). Optik **68**, 37–60.

Lencová, B. and Lenc, M. (1986). Scanning Electron Microsc. 897–915.

Lenz, F. (1956). Ann. Physik (Leipzig) **19**, 82–88.

Lenz, F. (1973). In *Image Processing and Computer-aided Design in Electron Optics* (Hawkes, P.W., ed.) pp. 274–282 (Academic Press, London and New York).

Lewis, H.R. (1966). J. Appl. Phys. **37**, 2541–2550.

Lucas, I. (1976). J. Appl. Phys. **47**, 1645–1652.

Mautz, J.R. and Harrington, R.F. (1970). Proc. Inst. Electr. Eng. **117**, 850–852.

McDonald, B.H. and Wexler, A. (1972). IEEE Trans. **MTT-20**, 841–847.

McDonald, B.H., Friedman, M., Decreton, M. and Wexler, A. (1973). Electron. Lett. **9**, 242.

Merwe, J.P. van der (1978). Univ. Zululand Publ., Ser. 3, No. 23, No. 24 and No. 25.

Merwe, J.P. van der (1979). S. Afr. J. Phys. **2**, No. 4, 117–124.

Morton, K.W. (1987). Basic course in finite element methods. Comput. Phys. Repts **6**, 1–72.

Mulvey, T. (1982). Unconventional lens design. In *Magnetic Electron Lenses* (Hawkes, P.W., ed.) pp. 359–412 (Springer, Berlin and New York).

Munro, E. (1971). *Computer-aided Design Methods in Electron Optics.* Dissertation, Cambridge.

Munro, E. (1973). In *Image Processing and Computer-aided Design in Electron Optics* (Hawkes, P.W., ed.) pp. 284–323 (Academic Press, London and New York).

Munro, E. (1980). Electron beam lithography. In *Applied Charged Particle Optics* (Septier, A., ed.) vol. B, pp. 73–131 (Academic Press, New York and London); Supplement 13B to Adv. Electron. Electron Phys.

Munro, E. (1987a). Nucl. Instrum. Meth. Phys. Res. **A258**, 443–461.

Munro, E. (1987b). In *Proc. Int. Symp. Electron Optics Beijing* (Ximen, J.-y., ed.) pp. 177–180 (Institute of Electronics, Academia Sinica, Beijing).

Munro, E. and Chu, H.C. (1982). Optik **60**, 371–390 and **61**, 1–16.

Norrie, D.H. and Vries, G. de (1973). *The Finite Element Method* (Academic Press, New York and London).

Peaceman, D.W. and Rachford, H.H. (1955). J. SIAM **3**, 28–41.

Rauh, H. (1971). Z. Naturforsch. **26a**, 1667–1675.

Read, F.H., Adams, A. and Soto-Montiel, J.R. (1971). J. Phys. E: Sci. Instrum. **4**, 625–632.

Regenstreif, E. (1951). Ann. Radioél. **6**, 51–83 and 114–155.

Reid, J.K. (1987). Algebraic aspects of finite element solutions. Comput. Phys. Repts **6**, 385–414.

Rose, H. (1971). Optik **34**, 285–311.

Rose, H. and Plies, E. (1973). In *Image Processing and Computer-aided Design in Electron Optics* (Hawkes, P.W., ed.) pp. 344–369 (Academic Press, London and New York).

Schaefer, C.H. (1982). *Methoden zur numerischen Lösung der Laplacegleichung bei komplizierten Randwertaufgaben in drei Dimensionen und ihre Anwendung auf Probleme der Elektronenoptik.* Dissertation, Tübingen.

Schaefer, C.H. (1983). Optik **65**, 347–359.

Scherle, W. (1983). Optik **63**, 217–226.

Schwertfeger, W. and Kasper, E. (1974). Optik **41**, 160–173.

Silvester, P. and Konrad, A. (1973). Int. J. Num. Meth. Eng. **5**, 481–497.

Singer, B. and Braun, M. (1970). IEEE Trans. **ED-17**, 926–934.

Stone, H.L. (1968). SIAM J. Numer. Anal. **5**, 530–558.

Ströer, M. (1987). Optik **77**, 15–25.

Sturrock, P.A. (1951). Phil. Trans. Roy. Soc. (London) **A243**, 387–429.

Uchikawa, Y., Ohye, T. and Gotoh, K. (1981). Trans. IEE Japan **101-A**, 263–270; Elec. Eng. Japan **101**, No. 3, 8–13.

Varga, R.S. (1962). *Matrix Iterative Analysis* (Prentice-Hall, Englewood Cliffs).

Wallington, M.J. (1970). J. Phys. E: Sci. Instrum. **3**, 599–604.

Wallington, M.J. (1971). J. Phys. E: Sci. Instrum. **4**, 1–8.

Weber, C. (1967). Philips Res. Repts Suppl. No. 6.

Weysser, R. (1983). *Feldberechnung in rotationssymmetrischen Elektronenstrahlerzeugern mit Spitzkathode unter Berücksichtigung von Raumladungen.* Dissertation, Tübingen.

Whittaker, E.T. and Watson, G.N. (1927). *A Course of Modern Analysis* (University Press, Cambridge).

Winslow, A.M. (1967). J. Comput. Phys. **2**, 149–172.

Zienkiewicz, O.C. (1967). *The Finite Element Method in Structural and Continuum Mechanics* (McGraw-Hill, London and New York).

Zienkiewicz, O.C. (1971). *The Finite Element Method in Engineering Science* (McGraw-Hill, London and New York).

Part III, Chapters 14–20

All books on electron optics devote considerable space to the paraxial properties of round lenses though the distinction between the real and asymptotic cardinal elements is not always emphasized as strongly as it is here. Information about electrostatic and magnetic lenses is to be found in the surveys of de Broglie (1946, 1950), Marton (1946), Bruck and Grivet (1950), Mulvey and Wallington (1973), Hawkes (1982), Riecke (1982, 1984) and Baranova and Yavor (1984, 1986). For some early work not cited in the main text, see Picht (1933a,b, 1939b), Dyachenko (1935), Dyachenko and Sakharov (1935, 1937, 1938), Glaser (1936) and on Newtonian fields, Glaser (1950).

Additional information about electron mirrors (Chapter 18) is to be found in Henneberg and Recknagel (1935), Hottenroth (1936, 1937), Picht (1939a), Regenstreif (1947), Kot (1952), Ivanov and Abalmazova (1966), Kel'man *et al.* (1982) and Yakushev and Sekunova (1986). For further details of the mirror in the Castaing-Henry device, see Castaing and Henry (1964) and Metherell (1971).

Only a small section of the literature on quadrupoles (Chapter 19) is presented here. Several books deal extensively with them (Strashkevich, 1959, 1966; Hawkes, 1966, 1970; and Yavor, 1968) and more are listed in Chapter 39. Many monographs and review articles aimed at higher energies likewise deal with them at length, though here a particular model (see Chapter 39) is usually adopted; see in particular Bernard (1953, 1954), Grivet and Septier (1960), Chamberlain (1960), Septier (1961), Luckey

(1961), King (1964), Steffen (1965, 1985), Banford (1966), Bruck (1966a,b), Brown (1967), Busse and Zelazny (1984), Wollnik (1987), Carey (1987) and Schriber and Taylor (1987). On the origins of the strong-focusing idea, see Thomas (1938), Christofilos (1950), Courant *et al.* (1952) and (for electrostatic strong focusing) Blewett (1952). We also list an interesting early paper of Strashkevich (1954) and a discussion of various quadruplets by Dymnikov *et al.* (1963b).

The optics of cylindrical lenses (Chapter 20) or their potential distributions are further examined by Fry (1932), Henneberg (1935), Glaser and Henneberg (1935), Strashkevich and Glushko (1940, 1941), Glushko and Strashkevich (1940), Strashkevich (1940c, 1952a,b, 1955), Rabin and Strashkevich (1950), Rabin *et al.* (1951), Strashkevich and Yurchenko (1952), Bálta Elías and Gómez Garcia (1950), Archard (1954), Laudet (1953, 1955, 1956), Septier (1954), Baranovskii *et al.* (1955), Gautier and Latour (1959), Yavor and Szilágyi (1960), Yavor *et al.* (1960), Bacquet *et al.* (1961, 1963), Kochanov (1962, 1963), Glikman and Yakushev (1967), Glikman *et al.* (1967a,b), Hibi *et al.* (1967), Ćirić *et al.* (1976) and Vukanić *et al.* (1976).

Archard, G.D. (1954). Brit. J. Appl. Phys. **5**, 179–181 and 395–399.

Bacquet, G., Gautier, P. and Santouil, A. (1961). C.R. Acad. Sci. Paris **252**, 522–524.

Bacquet, G., Gautier, P. and Santouil, A. (1963). J. Microscopie **2**, 393–406.

Bálta Elías, J. and Gómez Garcia, J.A. (1950). An. R. Soc. Esp. Fis. Quim. **A46**, 83–92.

Banford, A.P. (1966). *The Transport of Charged Particle Beams* (Spon, London).

Baranova, L.A. and Yavor, S.Ya. (1984). Zh. Tekh. Fiz. **54**, 1417–1453; Sov. Phys. Tech. Phys. **29**, 827–848.

Baranova, L.A. and Yavor, S.Ya. (1986). *Elektrostaticheskie Elektronnye Linzy* (Nauka, Moscow); English translation to appear in Adv. Electron. Electron Phys.

Baranovskii, S.N. Kaminskii, D.L. and Kel'man, V.M. (1955). Zh. Tekh. Fiz. **25**, 610–624 and 1954–1956.

Bernard, M. (1952). C.R. Acad. Sci. Paris **234**, 606–608.

Bernard, M.-Y. (1953). C.R. Acad. Sci. Paris **236**, 185–187 and 902–904.

Bernard, M.-Y. (1954). Ann. Physique (Paris) **9**, 633–682.

Blewett, J.P. (1952). Phys. Rev. **88**, 1197–1199.

Boerboom, A.J.H. (1959). Z. Naturforsch. **14a**, 809–816.

Boerboom, A.J.H. (1960). Z. Naturforsch. **15a**, 244–252 and 253–259.

Bok, A.B., Le Poole, J.B., Roos, J., Lang, H. de, Bethge, H., Heydenreich, J. and Barnett, M.E. (1971). Mirror electron microscopy. Adv. Opt. Electron Microsc. **4**, 161–261.

Bonjour, P. (1974). J. Microscopie **20**, 219–239.

Bonjour, P. (1975). J. Phys. E: Sci. Instrum. **8**, 764–768.

Born, M. and Wolf, E. (1959). *Principles of Optics* (Pergamon, Oxford).

Brodskii, G.N. and Yavor, S.Ya. (1970). Zh. Tekh. Fiz. **40**, 1310–1313; Sov. Phys. Tech. Phys. **15**, 1011–1012.

Brodskii, G.N. and Yavor, S.Ya. (1971). Zh. Tekh. Fiz. **41**, 1550–1552; Sov. Phys. Tech. Phys. **16**, 1218–1219.

Broglie, L. de, ed. (1946). *L'Optique Electronique* (Editions Revue d'Optique, Paris).

Broglie, L. de (1950). *Optique Electronique et Corpusculaire* (Hermann, Paris).

Brown, K.L. (1967). A first and second-order matrix theory for the design of beam transport systems and charged particle spectrometers, SLAC-75; published in *Advances in Particle Physics* (Cool, R.L. and Marshak, R.E., eds) vol. 1, pp. 71–134 (Wiley-Interscience, New York and London, 1968).

Brown, K.L. and Servranckx, R.V. (1985). First- and second-order charged particle optics. In *Physics of High Energy Particle Accelerators* (Month, M., Dahl, P.F. and Dienes, M., eds) pp. 62–138 (American Institute of Physics, New York).

Bruck, H. (1966a). *Accélérateurs Circulaires de Particules* (Presses Universitaires de France, Paris).

Bruck, H. (1966b). *Théorie et Technique des Accélérateurs de Particules* (Centre d'Etudes Nucléaires de Saclay, Gif-sur-Yvette).

Bruck, H. and Grivet, P. (1950). Rev. Opt. **29**, 164–170.

Busch, H. (1927). Arch. Elektrotech. **18**, 583–594.

Busse, W. and Zelazny, R. (1984). *Computing in Accelerator Design and Operation*, Lecture Notes in Physics, vol. 215 (Springer, Berlin and New York).

Carathéodory, C. (1937). *Geometrische Optik*. Ergeb. Math. Grenzgeb., vol. 5 (Springer, Berlin).

Carey, D.C. (1987). *The Optics of Charged Particle Beams* (Harwood, Chur, London and New York).

Castaing, R. and Henry, L. (1962). C.R. Acad. Paris **255**, 76–78.

Castaing, R. and Henry, L. (1964). J. Microscopie **3**, 133–152.

Chamberlain, O. (1960). Ann. Rev. Nucl. Sci. **10**, 161–192.

Chechulin, V.N. and Yavor, S.Ya. (1969). Zh. Tekh. Fiz. **39**, 1457–1462; Sov. Phys. Tech. Phys. **14**, 1093–1096.

Christofilos, N. (1950). Focusing system for ions and electrons, US Patent 2736799, reprinted in *The Development of High-Energy Accelerators* (Livingston, M.S., ed) pp. 270–280 (Dover, New York, 1966); re unpublished paper of 1950, see Courant *et al.* (1953).

Ćirić, D., Terzić, I. and Vukanić, J. (1976). J. Phys. E: Sci. Instrum. **9**, 844–846.

Cotte, M. (1938). Ann. Physique (Paris) **10**, 333–405.

Courant, E.D., Livingston, M.S. and Snyder, H.S. (1952). Phys. Rev. **88**, 1190–1196.

Courant, E.D., Livingston, M.S., Snyder, H.S. and Blewett, J.P. (1953). Phys. Rev. **91**, 202–203.

Czapski, S. and Eppenstein, O. (1924). *Grundzüge der Theorie der Optischen Instrumente nach Abbe*, 3rd ed. (Barth, Leipzig).

Daumenov, T.D., Sapargaliev, A.A. and Yakushev, E.M. (1978). Zh. Tekh. Fiz. **48**, 2447–2454; Sov. Phys. Tech. Phys. **23**, 1400–1404.

Davisson, C.J. and Calbick, C.J. (1931). Phys. Rev. **38**, 585; errata ibid. **42**, 580 (1932).

Dušek, H. (1959). Optik **16**, 419–445.

Dyachenko, V. (1935). Ukr. Fiz. Zap. **3**, 53–65.

Dyachenko, V. and Sakharov, I.I. (1935). Ukr. Fiz. Zap. **4**, 23–32.

Dyachenko, V. and Sakharov, I.I. (1937). Ukr. Fiz. Zap. **6**, 37–48 and 49–52.

Dyachenko, V.E. and Sakharov, I.I. (1938). Ukr. Fiz. Zap. **7**, 3–21 and 175–191.

Dymnikov, A.D. (1968). Zh. Tekh. Fiz. **38**, 1120–1125; Sov. Phys. Tech. Phys. **13**, 929–933.

Dymnikov, A.D. and Yavor, S.Ya. (1963). Zh. Tekh. Fiz. **33**, 851–858; Sov. Phys. Tech. Phys. **8**, 639–643.

Dymnikov, A.D., Fishkova, T.Ya. and Yavor, S.Ya. (1963a). Izv. Akad. Nauk SSSR (Ser. Fiz.) **27**, 1131–1134; Bull. Acad. Sci. USSR (Phys. Ser.) **27**, 1112–1115.

Dymnikov, A.D., Ovsyannikova, L.P. and Yavor, S.Ya. (1963b). Zh. Tekh. Fiz. **33**, 393–397; Sov. Phys. Tech. Phys. **8**, 293–296.

Fry, T.C. (1932). Am. Math. Monthly **39**, 199–209.

Funk, P. (1950). Acta Phys. Austriaca **4**, 304–308.

Gautier, P. and Latour, C. (1959). C.R. Acad. Sci. Paris **248**, 1637–1640.

Glaser, W. (1933a). Z. Physik **80**, 451–464.

Glaser, W. (1933b). Z. Physik **81**, 647–686.

Glaser, W. (1933c). Z. Physik **83**, 104–122.

Glaser, W. (1933d). Ann. Physik (Leipzig) **18**, 557–585.

Glaser, W. (1936). Z. Tech. Phys. **17**, 617–622.

Glaser, W. (1940). Z. Physik **117**, 285–315.

Glaser, W. (1950). Paris, pp. 158–164.

Glaser, W. (1952). *Grundlagen der Elektronenoptik* (Springer, Vienna).

Glaser, W. (1956). Elektronen- und Ionenoptik. *Handbuch der Physik* **33**, 123–395.

Glaser, W. and Bergmann, O. (1950). Z. Angew. Math. Phys. **1**, 363–379.

Glaser, W. and Bergmann, O. (1951). Z. Angew. Math. Phys. **2**, 159–188.

Glaser, A. and Henneberg, W. (1935). Z. Tech. Phys. **16**, 222–230.

Glaser, W. and Lammel, E. (1941). Ann. Physik (Leipzig) **40**, 367–384.

Glaser, W. and Lammel, E. (1943). Monatsh. Math. Phys. **50**, 289–297.

Glikman, L.G. and Yakushev, E.M. (1967). Zh. Tekh. Fiz. **37**, 2097–2099; Sov. Phys. Tech. Phys. **12**, 1544–1545.

Glikman, L.G., Kel'man, V.M. and Yakushev, E.M. (1967a). Zh. Tekh. Fiz. **37**, 13–16; Sov. Phys. Tech. Phys. **12**, 9–11.

Glikman, L.G., Kel'man, V.M. and Yakushev, E.M. (1967b). Zh. Tekh. Fiz. **37**, 1028–1034 and 1720–1725; Sov. Phys. Tech. Phys. **12**, 740–744 and 1261–1264.

Glikman, L.G., Karetskaya, S.P., Kel'man, V.M. and Yakushev, E.M. (1971). Zh. Tekh. Fiz. **41**, 330–335; Sov. Phys. Tech. Phys. **16**, 247–251.

Glushko, M.T. and Strashkevich, A.M. (1940). Zh. Tekh. Fiz. **10**, 1793–1799.

Glushko, M.T. and Strashkevich, A.M. (1941). Zh. Tekh. Fiz. **11**, 205–228.

Gratsiatos, J. (1940). Z. Physik **115**, 61–68.

Grivet, P. and Septier, A. (1960). Nucl. Instrum. Meth. **6**, 126–156 and 243–275.

Gullstrand, A. (1900). Nova Acta R. Soc. Sci. Upsal. **20**, 204 pp.

Gullstrand, A. (1906). K. Svensk. Vetenskapsakad. Hand. **41**, No. 3, 119 pp.

Gullstrand, A. (1908). K. Svensk. Vetenskapsakad. Hand. **43**, No. 2, 58 pp.

Gullstrand, A. (1915). K. Svensk. Vetenskapsakad. Hand. **55**, No. 1, 139 pp.

Gullstrand, A. (1924). K. Svensk. Vetenskapsakad. Hand. **63**, No. 13, 175 pp.

Hahn, E. (1965). Jenaer Jahrbuch, 107–145.

Hahn, E. (1971). Wiss. Z. Tech. Univ. Dresden **20**, 361–363.

Harting, E. and Read, F.H. (1976). *Electrostatic Lenses* (Elsevier, Amsterdam, Oxford and New York).

Hawkes, P.W. (1966). *Quadrupole Optics* (Springer, Berlin and New York).

Hawkes, P.W. (1967). Brit. J. Appl. Phys. **18**, 545–547.

Hawkes, P.W. (1970). *Quadrupoles in Electron Lens Design,* Adv. Electron. Electron Phys. Suppl. 7 (Academic Press, New York and London).

Hawkes, P.W., ed. (1982). *Magnetic Electron Lenses* (Springer, Berlin and New York).

Henneberg, W. (1935). Z. Physik **94**, 22–27.

Henneberg, W. and Recknagel, A. (1935). Z. Tech. Phys. **16**, 621–623.

Herzberger, M. (1931). *Strahlenoptik* (Springer, Berlin).

Herzberger, M. (1943). J. Opt. Soc. Am. **33**, 651–655.

Herzberger, M. (1958). *Modern Geometrical Optics* (Interscience, New York and London).

Hibi, T., Takahashi, S. and Takahashi, K. (1967). J. Electron Microsc. **16**, 225–231.

Hottenroth, G. (1936). Z. Physik **103**, 460–462.

Hottenroth, G. (1937). Ann. Physik (Leipzig) **30**, 689–712.

Hutter, R.G.E. (1945). J. Appl. Phys. **16**, 670–678.

Ivanov, R.D. and Abalmazova, M.G. (1966). Prib. Tekh. Eksp. No. 5, 192–194; Instrum. Exp. Tech. **9**, 1221–1222.

Kamke, E. (1977). *Differentialgleichungen, Lösungsmethoden und Lösungen*, vol. 1, 9th ed. (Teubner, Stuttgart).

Karetskaya, S.P., Kel'man, V.M. and Yakushev, E.M. (1970). Zh. Tekh. Fiz. **40**, 2563–2567; Sov. Phys. Tech. Phys. **15**, 2010–2013.

Karetskaya, S.P., Kel'man, V.M. and Yakushev, E.M. (1971a). Zh. Tekh. Fiz. **41**, 325–329; Sov. Phys. Tech. Phys. **16**, 244–246.

Karetskaya, S.P., Kel'man, V.M. and Yakushev, E.M. (1971b). Zh. Tekh. Fiz. **41**, 548–552; Sov. Phys. Tech. Phys. **16**, 425–428.

Kel'man, V.M. and Yavor, S.Ya. (1954). Zh. Tekh. Fiz. **24**, 1329–1332.

Kel'man, V.M. and Yavor, S.Ya. (1955). Zh. Tekh. Fiz. **25**, 1405–1411.

Kel'man, V.M. and Yavor, S.Ya. (1968). *Elektronnaya Optika*, 3rd ed. (Nauka, Leningrad).

Kel'man, V.M., Kaminskii, D.L. and Yavor, S.Ya. (1954). Zh. Tekh. Fiz. **24**, 1410–1427.

Kel'man, V.M., Fedulina, L.V. and Yakushev, E.M. (1971). Zh. Tekh. Fiz. **41**, 1489–1497, 1825–1831, 1832–1838 and 2016–2022; Sov. Phys. Tech. Phys. **16**, 1171–1176, 1444–1448, 1449–1454 and 1598–1602.

Kel'man, V.M., Fedulina, L.V. and Yakushev, E.M. (1972a). Zh. Tekh. Fiz. **42**, 297–302; Sov. Phys. Tech. Phys. **17**, 238–241.

Kel'man, V.M., Sekunova, L.M. and Yakushev, E.M. (1972b). Zh. Tekh. Fiz. **42**, 2279–2287; Sov. Phys. Tech. Phys. **17**, 1786–1791.

Kel'man, V.M., Sekunova, L.M. and Yakushev, E.M. (1973a). Zh. Tekh. Fiz. **43**, 1799–1806 and 1807–1817; Sov. Phys. Tech. Phys. **18**, 1142–1146 and 1147–1152.

Kel'man, V.M., Karetskaya, S.P., Manabaev, Kh.Kh., Fedulina, L.V. and Yakushev, E.M. (1973b). Zh. Tekh. Fiz. **43**, 2238–2245 and 2463–2471; Sov. Phys. Tech. Phys. **18**, 1418–1421 and 1552–1556.

Kel'man, V.M., Karetskaya, S.P., Fedulina, L.V. and Yakushev, E.M. (1979). *Elektronno-Opticheskie Elementy Prismennykh Spektrometrov Zaryazhennykh Chastits* (Nauka, Alma-Ata).

Kel'man, V.M., Karetskaya, S.P., Saichenko, N.Yu. and Fedulina, L.F. (1982). Zh. Tekh. Fiz. **52**, 2140–2145; Sov. Phys. Tech. Phys. **27**, 1318–1321.

King, N.M. (1964). Theoretical techniques of high-energy beam design. Prog. Nucl. Phys. **9**, 71–116.

Knoll, M. and Ruska, E. (1932a). Ann. Physik (Leipzig) **12**, 607–661.

Knoll, M. and Ruska, E. (1932b). Z. Physik **78**, 318–339.

Koch, W. (1967). Optik **25**, 523–534 and 535–550.

Kochanov, E.S. (1962). Zh. Tekh. Fiz. **32**, 294–302; Sov. Phys. Tech. Phys. **7**, 209–214.

Kochanov, E.S. (1963). Zh. Tekh. Fiz. **33**, 1301–1310; Sov. Phys. Tech. Phys. **8**, 969–975.

Kot, M.V. (1952). Uchen. Zap. Chernovitskii Gos. Univ. (Ser. Fiz.-Mat. Nauk) **4**, 73–94.

Laudet, M. (1953). Cah. Phys. No. 41, **7**, 73–80.

Laudet, M. (1955). J. Phys. Radium **16**, 118–124 and 908–916.

Laudet, M. (1956). Ann. Fac. Sci. Toulouse **20**, 111–230.

Leitner, M. (1942). *Die Abbildungsfehler der Elektrischen Zylinderlinse*. Draft Dissertation, Berlin.

Lichte, H. (1983). In *Proc. Int. Symp. Foundations of Quantum Mechanics in the Light of New Technology* (Kamefuchi, S., ed.) pp. 29–38 (Physical Society of Japan, Tokyo).

Luckey, D. (1961). Beam optics. In *Techniques of High Energy Physics* (Ritson, D.M., ed.) pp. 403–463 (Wiley-Interscience, New York and London).

Marton, L. (1946). Electron microscopy. Repts Prog. Phys. **10**, 204–252.

Melkich, A. (1947). *Ausgezeichnete astigmatische Systeme der Elektronenoptik*. Dissertation, Berlin 1944; Sitzungsber. Akad. Wiss. Wien, Abt. IIA, **155**, 393–438 and 440–471.

Metherell, A.J.F. (1971). Adv. Opt. Electron Microsc. **4**, 263–360.

Möllenstedt, G. and Düker, H. (1953). Naturwissenschaften **42**, 41.

Mulvey, T. (1984). In *Electron Optical Systems for Microscopy, Microanalysis and Microlithography* (Hren, J.J., Lenz, F.A., Munro, E. and Sewell, P.B., eds) pp. 15–27 (Scanning Electron Microscopy, AMF O'Hare).

Mulvey, T. and Wallington, M.J. (1973). Electron lenses. Repts Prog. Phys. **36**, 347–421.

Nicholl, F.H. (1938). Proc. Phys. Soc. (London) **50**, 888–898.

Ollendorff, F. (1955). *Elektronik des Einzelektrons* (Springer, Vienna).

Ollendorff, F. and Wendt, G. (1932). Z. Physik **76**, 655–659.

Picht, J. (1932). Ann. Physik (Leipzig) **15**, 926–964.

Picht, J. (1933a). Z. Tech. Phys. **14**, 239–241.

Picht, J. (1933b). Z. Instrumentenkde **53**, 274–282.

Picht, J. (1939a). Ann. Physik (Leipzig) **36**, 249–264.

Picht, J. (1939b). *Einführung in die Theorie der Elektronenoptik*, 1st ed. (Barth, Leipzig).

Picht, J. (1963). *Einführung in die Theorie der Elektronenoptik*, 3rd ed. (Barth, Leipzig).

Putz, F. (1951). *Über die Elektronenoptische Abbildung in Starken Elektrischen und Magnetischen Abbildungsfeldern*. Dissertation, Vienna.

Rabin, B.M. and Strashkevich, A.M. (1950). Zh. Tekh. Fiz. **20**, 1232–1240.

Rabin, B.M., Strashkevich, A.M. and Khin, L.S. (1951). Zh. Tekh. Fiz. **21**, 438–444.

Rayleigh, Lord [J.W. Strutt] (1886). Phil. Mag. **21**, 466–476.

Recknagel, A. (1936). Z. Tech. Phys. **17**, 643–645.

Recknagel, A. (1937). Z. Physik **104**, 381–394.

Regenstreif, E. (1947). Ann. Radioél. **2**, 348–358.

Regenstreif, E. (1951). Ann. Radioél. **6**, 51–83 and 114–154.

Regenstreif, E. (1966). C.R. Acad. Sci. Paris **B263**, 1205–1207 and 1297–1299.

Regenstreif, E. (1967). C.R. Acad. Sci. Paris **B264**, 596–598.

Riecke, W.D. (1982). Practical lens design. In *Magnetic Electron Lenses* (Hawkes, P.W., ed.) pp. 164–357 (Springer, Berlin and New York).

Riecke, W.D. (1984). In *Electron Optical Systems for Microscopy, Microanalysis and Microlithography* (Hren, J.J., Lenz, F.A., Munro, E. and Sewell, P.B., eds) pp. 1–14 (Scanning Electron Microscopy, AMF O'Hare).

Rose, H. (1966/67). Optik **24**, 36–59 and 108–121.

Rose, H. (1972). Optik **36**, 19–36.

Ruska, E. (1934a). Z. Physik **87**, 580–602.

Ruska, E. (1934b). Z. Physik **89**, 90–128.

Rusterholz, A.A. (1950). *Elektronenoptik* (Birkhäuser, Basel).

Schiske, P. (1957). Optik **14**, 34–45.

Schriber, S.O. and Taylor, L.S., eds (1987). *Proc. 2nd Int. Conf. Charged Particle Optics*, Nucl. Instrum. Meth. Phys. Res. **A258**, 289–598 (North-Holland, Amsterdam).

Septier, A. (1953). Ann. Radioél. **9**, 374–410.

Septier, A. (1954). C.R. Acad. Sci. Paris **239**, 402–404.

Septier, A. (1961). Adv. Electron. Electron Phys. **14**, 85–205.

Shpak, E.V. and Yavor, S.Ya. (1964). Zh. Tekh. Fiz. **34**, 1037–1039; Sov. Phys. Tech. Phys. **9**, 803–805.

Steffen, K.G. (1965). *High Energy Beam Optics* (Wiley-Interscience, New York and London).

Steffen, K. (1985). Basic course on accelerator optics. In *CERN Accelerator School, General Accelerator Physics* (Bryant, P. and Turner, S., eds) vol. 1, pp. 25–63 (CERN, Geneva, Report 85-19).

Strashkevich, A.M. (1940a). J. Physics (USSR) **3**, 507–523.

Strashkevich, A.M. (1940b). Kiev. Tech. Inst. Kozhevenno-obuvnoi Prom. Sb. Nauch.-Issled. Rabot. **3**, 249–260.

Strashkevich, A.M. (1940c). Zh. Tekh. Fiz. **10**, 91–111.

Strashkevich, A.M. (1952a). Zh. Tekh. Fiz. **22**, 487–497 and 1848–1856.

Strashkevich, A.M. (1952b). Uch. Zap. Chernovits. Gos. Univ., Ser. Fiz.-Mat. Nauk **4**, 53–58.

Strashkevich, A.M. (1954). Zh. Tekh. Fiz. **24**, 274–286.

Strashkevich, A.M. (1955). Nauk. Zap. Chernivet'sk Derzhavn. Univ. (Ser. Fiz.-Mat.) **12**, 189–197.

Strashkevich, A.M. (1959). *Elektronnaya Optika Elektrostaticheskikh Polei ne Obladayushchikh Osevoi Simmetriei* (Fizmatgiz, Moscow).

Strashkevich, A.M. (1962). Zh. Tekh. Fiz. **32**, 1142–1152; Sov. Phys. Tech. Phys. **7**, 841–847.

Strashkevich, A.M. (1966). *Elektronnaya Optika Elektrostaticheskikh Sistem* (Energiya, Moscow and Leningrad).

Strashkevich, A.M. and Glushko, M.T. (1940). Kiev. Tech. Inst. Kozhevenno-obuvnoi Prom. Sb. Nauch.-Issled. Rabot. **3**, 239–248.

Strashkevich, A.M. and Yurchenko, N.P. (1952). Uch. Zap. Chernovits. Gos. Univ., Ser. Fiz.-Mat. Nauk **4**, 99–112.

Sturrock, P.A. (1951). C.R. Acad. Sci. Paris **233**, 401–403.

Sturrock, P.A. (1955). *Static and Dynamic Electron Optics* (University Press, Cambridge).

Tanaka, K. (1981). Optik **58**, 351–358 and **60**, 73–80.

Tanaka, K. (1982). Optik **62**, 211–214.

Tanaka, K. (1983). Optik **64**, 13–24 and 89–111.

Tanaka, K. (1986). Paraxial theory in optical design in terms of Gaussian brackets. Prog. Opt. **23**, 63–111.

Thomas, L.H. (1938). Phys. Rev. **54**, 580–588.

Vukanić, J., Terzić, I., Aničin, B. and Ćirić, D. (1976). J. Phys. E: Sci. Instrum. **9**, 842–843.

Wallington, M.J. (1970). J. Phys. E: Sci. Instrum. **3**, 599–604.

Wallington, M.J. (1971). J. Phys. E: Sci. Instrum. **4**, 1–8.

Whittaker, E.T. and Watson, G.N. (1927). *A Course of Modern Analysis*, 4th ed. (University Press, Cambridge).

Wollnik, H. (1987). *Optics of Charged Particles* (Academic Press, Orlando and London).

Ximen, J.-y., Zhou, L-.w. and Ai, K.-c. (1983). Optik **66**, 19–34.

Yakushev, E.M. and Sekunova, L.M. (1986). Adv. Electron. Electron Phys. **68**, 337–416.

Yavor, S.Ya. (1955). Zh. Tekh. Fiz. **25**, 779–790.

Yavor, S.Ya. (1962). In *Proc. Symp. Electron Vacuum Phys., Budapest* pp. 125–137.

Yavor, S.Ya. (1968). *Fokusirovka Zaryazhennykh Chastits Kvadrupol'nymi Linzami* (Atomizdat, Moscow).

Yavor, S.Ya. and Szilagyi, M. (1960). Zh. Tekh. Fiz. **30**, 927–932; Sov. Phys. Tech. Phys. **5**, 872–876.

Yavor, S.Ya., Shpak, E.V. and Minima, R.M. (1960). Zh. Tekh. Fiz. **30**, 395–404; Sov. Phys. Tech. Phys. **5**, 369–377.

Zworykin, V.K., Morton, G.A., Ramberg, E.G., Hillier, J. and Vance, A.W. (1945). *Electron Optics and the Electron Microscope* (Wiley, New York and Chapman and Hall, London).

Part IV, Chapters 21–31

Further discussion or alternative presentations of the basic ideas of this chapter are to be found in Carathéodory (1937), Sturrock (1952) and Kas'yankov (1958a) and aberrations in general are surveyed in Hawkes (1967a) and Lenz (1982a). The historical article by Kanaya (1985) also takes a broad view.

It has not been possible to find space for an account of the interesting new approach to aberrations in the language of Lie algebra, explained in detail by Dragt and Forest (1986). One of the attractions claimed for this approach is the ease with which inter-relations between aberration coefficients are recovered; we therefore remind the reader that this has long been known to be an advantage of the eikonal method and that various alternative ways of establishing such relations have been explored by Meads (1963) and Sivkov (1971). The 'symplecticity', which is a central feature of the Lie method, is powerfully exploited by Wollnik and Berz (1985) and examined critically by Rose (1987).

Many papers have been devoted to each of the aberrations analysed in Chapters 24 and 25. On spherical aberration, see Rebsch and Schneider (1937), Becker and Wallraff (1938), Rebsch (1938), Plass (1942), Sugiura and Suzuki (1943), Vlasov (1944), Sorokin and Timofeev (1948), Liebmann (1951), Kanaya (1951a, 1952a), Seman (1953a,b), Kas'yankov (1955), Grümm (1956), Archard (1958), Petrie (1962), Der-Shvarts and Makarova (1966, 1967), Septier (1966), Barnes and Openshaw (1968), Brookes *et al.* (1968), Der-Shvarts (1971), Hanszen *et al.* (1972a,b), Lyubchik and Mokhnatkin (1972), Ishikawa (1978), Suzuki and Ishikawa (1978) and van Gorkum and Spanjer (1986). Meyer (1956) considers the fifth-order spherical aberration present when the third-order aberration has been corrected. Measurements and calculations for lens models are to be found in Part VII, Chapters 35 and 36.

Coma is further considered by Kanaya (1951d), Lenz (1954), Seman (1959c), Rose (1971a,b), Moses (1972, 1973), Rose and Moses (1973) and Lenz et al. (1982).

For astigmatism and field curvature, see Becker and Wallraff (1939, 1940), Kas'yankov (1950, 1952, 1967), Kanaya (1951c), Seman (1959c), Dutova and Kas'yankov (1963), Taganov and Kas'yankov (1964, 1965, 1967), Taganov (1966) and Gurbanov and Kas'yankov (1966).

The Petzval curvature is discussed by Goddard (1946), Chiang (1956), Seman (1968), Kas'yankov et al. (1970a,b) and Lenz (1986).

Distortion is the subject of papers by Hillier (1946b), Rang (1948), Mulvey and Jacob (1949), Kanaya (1951b, 1952b), Kanaya and Kato (1951), Liebmann (1952a), Wegmann (1953, 1954), De and Saha (1954), Seman (1959a), Kynaston and Mulvey (1962, 1963), Dutova and Kas'yankov (1963), Reisner (1970), Marai and Mulvey (1975, 1977), Lambrakis et al. (1977), Alshwaikh and Mulvey (1977), Elkamali and Mulvey (1977, 1979, 1980), Tsuno and Harada (1981a,b) and Tsuno et al. (1980b).

A complete list of the fifth-order aberration coefficients of round lenses is to be found in Hawkes (1965a) and later in Li and Ni (1988) and in Ai and Szilagyi (1988). As well as the paper by Meyer already cited, see also Archard (1960) and U (1957).

For further theoretical work on chromatic aberration (Chapter 26), see Glaser (1940b), Scherzer (1941), Kanaya (1951a, 1952a), Liebmann (1952b), Katagiri (1953), Morito (1954, 1957), Vandakurov (1955a), Watanabe and Morito (1955), Schiske (1956), Seman (1959b) and Brookes et al. (1968) and as usual, the references for Chapters 35 and 36.

To Chapter 27, we may add Hawkes (1983b, 1984b), Hanszen et al. (1972), Brouwer and Walther (1967) and the use by Heritage (1973) and Lewis et al. (1986) of aberration polynomials.

The papers by Ximen (1957) and Ximen et al. (1983) are relevant to Chapter 28.

The general texts by Steffen (1965), Banford (1966), Carey (1987) and Wollnik (1987) are all useful in connection with Chapter 28 as are the conference proceedings edited by Wollnik (1981) and Schriber and Taylor (1987). For very full bibliographies, see Hawkes (1966, 1970a).

A few extra details on cylindrical lens theory (Chapter 30) are to be found in Bertein (1950, 1951), Laudet (1953) and Vandakurov (1955b).

Parasitic aberrations (Chapter 31) have a voluminous literature, in which few authors pay much attention to the work of their predecessors. The following list contains analyses of lens imperfections of various sorts and descriptions of many kinds of stigmator: Scherzer (1946), Hillier (1946a), Bertein (1947a–c, 1948b, 1949), Bertein and Regenstreif (1947, 1949), Bruck (1947), Bruck and Grivet (1947, 1950), Glaser (1948), Cotte (1949, 1950),

Grivet et al. (1949), Inoue (1950), Rabin et al. (1951), Regenstreif (1951a,b), Recknagel and Haufe (1952/53), Leisegang (1953, 1954), Hahn (1954, 1959, 1966), Lenz and Hahn (1953), Kanaya (1953, 1955, 1958, 1962), Sakaki and Maruse (1954), Morito (1955), Stoyanov (1958a,b), Riecke (1958, 1964, 1966/67, 1972, 1976, 1982), Kanaya and Ishikawa (1958, 1959), Kas'yankov (1959), Katagiri (1960a,b), Meyer (1961), Vlasov and Shakhmatova (1962), Watanabe and Someya (1963), Ximen and Chen (1964), Ximen and Xi (1964), Tadano et al. (1966), Reisner and Schuler (1967), Amboss and Jennings (1970), Yanaka and Shirota (1970) and Monastyrskii and Kolesnikov (1983).

Adams, A. and Read, F.H. (1972a). J. Phys. E: Sci. Instrum. **5**, 150–155.

Adams, A. and Read, F.H. (1972b). J. Phys. E: Sci. Instrum. **5**, 156–160.

Ade, G. (1973). *Der Einfluss der Bildfehler dritter Ordnung auf die elektronenmikroskopische Abbildung und die Korrektur dieser Fehler durch holographische Rekonstruktion.* PTB-Ber. APh-3, 114 pp.

Ade, G. (1982). Optik **63**, 43–54.

Ai, K.-c. and Szilagyi, M. (1988). Optik **79**, 33–40.

Alshwaikh, A. and Mulvey T. (1977). EMAG, pp. 25–28.

Amboss, K. (1959). *Electron Optics: aberrations of air-cored magnetic lenses.* Thesis, London.

Amboss, K. and Jennings, J.C.E. (1970). J. Appl. Phys. **41**, 1608–1616.

Archard, G.D. (1953). J. Sci. Instrum. **30**, 352–358.

Archard, G.D. (1958). Rev. Sci. Instrum. **29**, 1049–1050.

Archard, G.D. (1960). Brit. J. Appl. Phys **11**, 521–522.

Banford, A.P. (1966). *The Transport of Charged Particle Beams* (Spon, London).

Baranova, L.A. and Yavor, S.Ya (1986). *Elektrostaticheskie Elektronnye Linzy* (Nauka, Moscow).

Barnes, R.L. and Openshaw, I.K. (1968). J. Phys. E: Sci. Instrum. **1**, 628–630.

Barton, D. and Fitch, J.P. (1972). Repts Prog. Phys. **35**, 235–314.

Becker, H. and Wallraff, A. (1938). Arch. Elektrotech. **32**, 664–675.

Becker, H. and Wallraff, A. (1939). Arch. Elektrotech. **33**, 491–505.

Becker, H. and Wallraff, A. (1940). Arch. Elektrotech. **34**, 43–48.

Bertein, F. (1947a). C.R. Acad. Sci. Paris **224**, 106–107 and 560–562.

Bertein, F. (1947b). C.R. Acad. Sci. Paris **225**, 801–803.

Bertein, F. (1947c). C.R. Acad. Sci. **225**, 863–865.

Bertein, F. (1947d). Ann. Radioél. **2**, 249–252.

Bertein, F. (1947e). Ann. Radioél. **2**, 379–408.

Bertein, F. (1948a). Ann. Radioél. **3**, 49–62.

Bertein, F. (1948b). J. Phys. Radium **9**, 104–112.

Bertein, F. (1949). C.R. Acad. Sci. Paris **229**, 291–293.

Bertein, F. (1950). C.R. Acad. Sci. Paris **231**, 766–767, 1134–1136 and 1448–1449.

Bertein, F. (1951). J. Phys. Radium **12**, 595–601 and 25A–31A.

Bertein, F. and Regenstreif, E. (1947). C.R. Acad. Sci. Paris **224**, 737–739.

Bertein, F. and Regenstreif, E. (1949). C.R. Acad. Sci. Paris **228**, 1854–1856.

Bertein, F., Bruck, H. and Grivet, P. (1947). Ann. Radioél. **2**, 249–252.

Bonshtedt, B.E. (1964). Radiotekh. Elektron. **9**, 844–850; Radio Eng. Electron. **9**, 686–692.

Born, M. and Wolf, E. (1959). *Principles of Optics* (Pergamon, Oxford and New York); 7th ed., 1988.

Broglie, L. de (1950). *Optique Electronique et Corpusculaire* (Hermann, Paris).

Brookes, K.A., Mulvey, T. and Wallington, M.J. (1968). Rome, vol. 1, pp. 165–166.

Brouwer, W. (1957). *The Use of Matrix Algebra in Geometrical Optics*. Dissertation, Delft.

Brouwer, W. (1964). *Matrix Methods in Optical Instrument Design* (Benjamin, New York and Amsterdam).

Brouwer, W. and Walther, A. (1967). Design of optical instruments. In *Advanced Optical Techniques* (van Heel, A.C.S., ed.) pp. 571–631 (North Holland, Amsterdam).

Bruck, H. (1947). C.R. Acad. Sci. Paris **224**, 1628–1629 and 1818–1820.

Bruck, H. and Grivet, P. (1947). C.R. Acad. Sci. Paris **224**, 1768–1769.

Bruck, H. and Grivet P. (1950). Rev. Opt. **29**, 164–170.

Bruck, H., Remillon, R. and Romani, L. (1948). C.R. Acad. Sci. Paris **226**, 650–652.

Bruns, H. (1895). Abh. K. Sächs. Ges. Wiss., Math.-Phys. Kl. **21**, 321–436 ; also published as *Das Eikonal* (Hirzel, Leipzig).

Burfoot, J. (1953). Proc. Phys. Soc. (London) **B66**, 775–792.

Burfoot, J. (1954a). Proc. Phys. Soc. (London) **B67**, 523–528.

Burfoot, J. (1954b). London, pp. 105–109.

Busch H. and Brüche, E., eds (1937). *Beiträge zur Elektronenoptik* (Barth, Leipzig).

Carathéodory, C. (1937). *Geometrische Optik*. Ergebnisse der Mathematik und ihrer Grenzgebiete, vol. 5 (Springer, Berlin).

Carey, D.C. (1987). *The Optics of Charged Particle Beams* (Harwood, Chur, London and New York).

Chako, N. (1957). Trans. Chalmers Univ. Technol. Gothenburg, Nr. 191, 50 pp.

Chiang, M.-y. (1956). Acta Phys. Sin. **12**, 439–446.

Cotte, M. (1949). C.R. Acad. Sci. Paris **228**, 377–378.

Cotte, M. (1950). Paris, pp. 155–157.

Czapski, S. and Eppenstein, O. (1924). *Grundzüge der Theorie der Optischen Instrumente nach Abbe*, 3rd ed. (Barth, Leipzig).

Daumenov, T.D., Sapargaliev, A.A. and Yakushev, E.M. (1978). Zh. Tekh. Fiz. **48**, 2447–2454; Sov. Phys. Tech. Phys. **23**, 1400–1404.

De, M.L. and Saha, D.K. (1954). Ind. J. Phys. **28**, 263–268.

Der-Shvarts, G.V. (1954). Zh. Tekh. Fiz. **24**, 859–870.

Der-Shvarts, G.V. (1970). Elektrofiz. Elektrokhim. Met. Obrabotki, vyp. 4, pp. 7–10.

Der-Shvarts, G.V. (1971), Radiotekh. Elektron. **16**, 1305–1306; Radio Eng. Electron. Phys. **16**, 1240–1241.

Der-Shvarts, G.V. and Makarova I.S. (1966). Radiotekh. Elektron. **11**, 89–93; Radio Eng. Electron. Phys. **11**, 72–75.

Der-Shvarts, G.V. and Makarova I.S. (1967). Radiotekh. Elektron. **12**, 168–171; Radio Eng. Electron. Phys. **12**, 161–163.

Der-Shvarts, G.V. and Makarova I.S. (1972). Izv. Akad. Nauk SSSR (Ser. Fiz.) **36**, 1304–1311; Bull. Acad. Sci. USSR (Phys. Ser.) **36**, 1164–1170.

Der-Shvarts, G.V. and Makarova I.S. (1973). Radiotekh. Elektron. **18**, 2374–2378; Radio Eng. Electron. Phys. **18**, 1722–1725.

Dodin, A.L. and Nesvizhskii, M.B. (1981). Zh. Tekh. Fiz. **51**, 897–901; Sov. Phys. Tech. Phys. **26**, 539–541.

Douglas, D.R. and Dragt, A.J. (1983). IEEE Trans. **NS-30**, 2442–2444.

Dragt, A.J. (1979). IEEE Trans. **NS-26**, 3601–3603.

Dragt, A.J. (1982). J. Opt. Soc. Amer. **72**, 372–379.

Dragt, A.J. and Forest, E. (1983). J. Math. Phys. **24**, 2734–2744.

Dragt, A.J. and Forest, E. (1986). Lie algebraic theory of charged-particle optics and electron microscopes. Adv. Electron. Electron Phys. **67**, 65–120.

Dragt, A.J., Forest, E. and Wolf, K.B. (1986). Foundations of a Lie algebraic theory of geometrical optics. In *Lie Methods in Optics* (Mondragón, J.S. and Wolf, K.B., eds) pp. 105–157 (Springer, Berlin).

Dušek, H. (1959). Optik **16**, 419–445.

Dutova, K.P. and Kas'yankov, P.P. (1963). Izv. Akad. Nauk. SSSR (Ser. Fiz.) **27**, 1127–1130; Bull. Acad. Sci. USSR (Phys. Ser.) **27**, 1108–1111.

Elkamali, H.H. and Mulvey, T. (1977). EMAG, pp. 33–34.

Elkamali, H.H. and Mulvey, T. (1979). EMAG, pp. 63–64.

Elkamali, H.H. and Mulvey, T. (1980). The Hague, vol. 1, 74–75.

Finsterwalder, S. (1982). Abh. Math.-Phys. Kl. K. Bayr. Akad. Wiss. **17**, 517–587.

Fitch, J.P. (1979). In *Symbolic and Algebraic Computation* (Ng, E.W., ed.) pp. 30–41. Vol. 72 of Lecture Notes in Computer Science (Springer, Berlin and New York).

Fitch, J.P. (1985). J. Symbol. Comput. **1**, 211–227.

Focke, J. (1965). Prog. Opt. **4**, 1–36.

Funk, P. (1936). Monatsh. Math. Phys. **43**, 305–316.

Funk, P. (1937). Monatsh. Math. Phys. **45**, 314–319.

Glaser, W. (1933a). Z. Physik **81**, 647–686.

Glaser, W. (1933b). Z. Physik **83**, 104–122.

Glaser, W. (1933c). Ann. Physik (Leipzig) **18**, 557–585.

Glaser, W. (1935). Z. Physik **97**, 177–201.

Glaser, W. (1936a). Z. Tech. Phys. **17**, 617–622.

Glaser, W. (1936b). Z. Physik **104**, 157–160.

Glaser, W. (1937). In *Beiträge zur Elektronenoptik* (Busch, H. and Brüche, E., eds) pp. 24–33 (Barth, Leipzig).

Glaser, W. (1938). Z. Physik **109**, 700–721.

Glaser, W. (1940a). Z. Physik **116**, 19–33 and 734–735.

Glaser, W. (1940b). Z. Physik **116**, 56–67.

Glaser, W. (1942/43). Z. Physik **120**, 1–15.

Glaser, W. (1948). Öst. Ing.-Arch. **3**, 39–46.

Glaser, W. (1949). Ann. Physik (Leipzig) **4**, 389–408.

Glaser, W. (1952). *Grundlagen der Elektronenoptik* (Springer, Vienna).

Glaser, W. and Schiske, P. (1953). Z. Angew. Phys. **5**, 329–339.

Goddard, L.S. (1946). Proc. Camb. Philos. Soc. **42**, 127–131.

Gorkum, A.A. van and Spanjer, T.G. (1986). Optik **72**, 134–136.

Goto, E. and Soma, T. (1977). Optik **48**, 255–270.

Gratsiatos, J. (1936). Z. Physik **102**, 641–651.

Gratsiatos, J. (1937). Z. Physik **107**, 382–386.

Grinberg, G.A. (1942). Dokl. Akad. Nauk SSR **37**, 172–178 and 261–268.

Grinberg, G.A. (1943a). Dokl. Akad. Nauk. SSR **38**, 78–81.

Grinberg, G.A. (1943b). Zh. Tekh. Fiz **13**, 361–388 (Russian version of Grinberg 1942, 1943a).

Grinberg, G.A. (1948). *Izbrannye Voprosy Mathematicheskoi Teorii Elektricheskikh i Magnitnykh Yavlenii* (Academy of Sciences Press, Moscow).

Grinberg, G.A. (1957a). Opt. Spektrosk. **3**, 673.

Grinberg, G.A. (1957b). Zh. Tekh. Fiz. **27**, 2425–2431; Sov. Phys. Tech. Phys. **2**, 2259–2265.

Grivet, P., Bertein, F. and Regenstreif, E. (1949). Delft, pp. 86–88.

Grümm, H. (1956). Optik **13**, 92–93.

Gullstrand, A. (1915). Kungl. Svensk. Vetenskapsakad. Handl. **63** (suppl.), No. 13, 175 pp.

Gurbanov, G.G. and Kas'yankov, P.P. (1966). Izv. Akad. Nauk SSSR (Ser. Fiz.) **30**, 735–738; Bull. Acad. Sci. USSR (Phys. Ser.) **30**, 762–765.

Hahn, E. (1954). Jenaer Jahrb. Pt I, 63–75.

Hahn, E. (1959). Jenaer Jahrb. Pt I, 86–114.

Hahn, E. (1966). Jenaer Jahrb. 145–172.

Hamilton, W.R. (1931). *The Mathematical Papers of Sir William Rowan Hamilton* (Conway, A.W. and Synge, J.L., eds, University Press, Cambridge).

Hanszen, K.-J., Ade, G. and Lauer, R. (1972a). Optik **35**, 567–590.

Hanszen, K.-J., Lauer, R. and Ade, G. (1972b). Optik **36**, 156–159.

Hardy, D.F. (1967). *Combined Magnetic and Electrostatic Quadrupole Electron Lenses.* Dissertation, Cambridge.

Harrington, R.F. (1968). *Field Computation by Moment Methods* (Macmillan, New York and Collier-Macmillan, London).

Hawkes, P.W. (1964). Prague, vol. 1, pp. 5–6.

Hawkes, P.W. (1965a). Phil. Trans. Roy. Soc. (London) **A257**, 479–522.

Hawkes, P.W. (1965b). Optik **22**, 340–368.

Hawkes, P.W. (1965c). Optik **22**, 543–551.

Hawkes, P.W. (1965d). Optik **23**, 244–250.

Hawkes, P.W. (1966). *Quadrupole Optics* (Springer, Berlin and New York).

Hawkes, P.W. (1966/67a). Optik **24**, 60–78 and 95–107.

Hawkes, P.W. (1966/67b). Optik **24**, 252–262 and 275–282.

Hawkes, P.W. (1967a). Lens aberrations. In *Focusing of Charged Particles* (Septier, A., ed) vol. 1, pp. 411–468 (Academic Press, New York and London).

Hawkes, P.W. (1967b). J. Microscopie **6**, 917–932.

Hawkes, P.W. (1968). Optik **27**, 287–304.

Hawkes, P.W. (1970a). *Quadrupoles in Electron Lens Design* (Academic Press, New York and London), Supplement 7 to Adv. Electron. Electron Phys.

Hawkes, P.W. (1970b). Optik **31**, 213–219.

Hawkes, P.W. (1970c). Optik **31**, 302–314.

Hawkes, P.W. (1970d). Optik **31**, 592–599.

Hawkes, P.W. (1970e). Optik **32**, 50–60.

Hawkes, P.W. (1970f). J. Microscopie **9**, 435–454.

Hawkes, P.W. (1977a). Optik **48**, 29–51.

Hawkes, P.W. (1977b). HVEM Kyoto, pp. 57–60

Hawkes, P.W. (1980a). Methods of computing optical properties and combating aberrations for low-intensity beams. In *Applied Charged Particle Optics* (Septier, A., ed.) vol. A, pp. 45–157; Adv. Electron. Electron Phys. Suppl. 13A (Academic Press, New York and London).

Hawkes, P.W. (1980b). Optik **56**, 293–320.

Hawkes, P.W. (1983a). Optik **63**, 129–156 and **65**, 227–251.

Hawkes, P.W. (1983b). EMAG, pp. 471–474.

Hawkes, P.W. (1984a). Optik **66**, 379–380.

Hawkes, P.W. (1984b). Budapest, vol. 1, pp. 23–24.

Hawkes, P.W. (1985a). Optik **70**, 115–123.

Hawkes, P.W. (1985b). Optik **70**, 140–142.

Hawkes, P.W. (1986). Ultramicroscopy **20**, 189–194.

Hawkes, P.W. (1987). J. Phys. E: Sci. Instrum. **20**, 234–235.

Hawkes, P.W. and Cosslett, V.E. (1962). Brit. J. Appl. Phys. **13**, 272–279.

Hearn, A.C. (1985). *REDUCE User's Manual* (Rand Publication CP78, Rand Corp., Santa Monica, CA 90406).

Heel, A.C.S. van (1949). In *La Théorie des Images Optiques* (Fleury, P., ed.) pp. 32–67 (Editions Revue d'Optique, Paris).

Heel, A.C.S. van (1964). *Inleiding in de Optica*, 5th ed. (Nijhoff, The Hague).

Heritage, M.B. (1973). In *Image Processing and Computer-aided Design in Electron Optics* (Hawkes, P.W., ed.) pp. 324–338 (Academic Press, London and New York).

Herzberger, M. (1931). *Strahlenoptik* (Springer, Berlin).

Herzberger, M. (1958). *Modern Geometrical Optics* (Interscience, New York and London).

Hillier, J. (1946a). J. Appl. Phys. **17**, 307–309.

Hillier, J. (1946b). J. Appl. Phys. **17**, 411–419.

Hillier, J. and Ramberg, E.G. (1947). J. Appl. Phys. **18**, 48–71.

Inoue, Y. (1950). Paris, pp. 199–200.

Jandeleit, O. and Lenz, F. (1959). Optik **16**, 87–107.

Janse, J. (1971). Optik **33**, 270–281.

Kanaya, K. (1951a). Bull. Electrotech. Lab. **15**, 86–91.

Kanaya, K. (1951b). Bull. Electrotech. Lab. **15**, 91–94.

Kanaya, K. (1951c). Bull. Electrotech. Lab. **15**, 193–198.

Kanaya, K. (1951d). Bull. Electrotech. Lab. **15**, 199–202.

Kanaya, K. (1952a). Bull. Electrotech. Lab. **16**, 25–30.

Kanaya, K. (1952b). Bull. Electrotech. Lab. **16**, 135–142.

Kanaya, K. (1952c). Bull. Electrotech. Lab. **16**, 184–192.

Kanaya, K. (1953). J. Electronmicrosc. **1**, 7–12.

Kanaya, K. (1955). Res. Electrotech. Lab. No. 548, 70 pp.

Kanaya, K. (1958). Bull. Electrotech. Lab. **22**, 615–622.

Kanaya, K. (1962). Bull. Electrotech. Lab. **26**, 161–172.

Kanaya, K. (1985). In *The Beginnings of Electron Microscopy* (Hawkes, P.W., ed.) pp. 317–386 (Academic Press, Orlando and London).

Kanaya, K. and Ishikawa, A. (1958). Bull. Electrotech. Lab. **22**, 641–646.

Kanaya, K. and Ishikawa, A. (1959). J. Electronmicrosc. **7**, 13–15.

Kanaya, K. and Kato, A. (1951). Bull. Electrotech. Lab. **15**, 827–833.

Kasper, E. (1982). Magnetic field calculation and the determination of electron trajectories. In *Magnetic Electron Lenses* (Hawkes, P.W., ed.) pp. 57–118 (Springer, Berlin and New York).

Kas'yankov, P.P. (1950). Zh. Tekh. Fiz. **20**, 1426–1434.

Kas'yankov, P.P. (1952). Zh. Tekh. Fiz. **22**, 80–83.

Kas'yankov, P.P. (1955). Zh. Tekh. Fiz. **25**, 1639–1648.

Kas'yankov, P.P. (1956a). Dokl. Akad. Nauk. SSSR **108**, 813–816; Sov. Phys. Dokl. **1**, 367–371.

Kas'yankov, P.P. (1956b). *Teoriya Elektromagnitnykh Sistem s Krivolineinoi Os'yu* (University Press, Leningrad).

Kas'yankov, P.P. (1957). Opt. Spektrosk. **3**, 169–179.

Kas'yankov, P.P. (1958a). Dokl. Akad. Nauk SSSR **120**, 497–500; Sov. Phys. Dokl. **3**, 573–576.

Kas'yankov, P.P. (1958b). Zh. Tekh. Fiz. **28**, 915–918; Sov. Phys. Tech. Phys. **3**, 854–857.

Kas'yankov, P.P. (1959). Izv. Akad. Nauk SSSR (Ser. Fiz.) **23**, 711–715; Bull. Acad. Sci. USSR (Phys. Ser.) **23**, 706–710.

Kas'yankov, P.P. (1967). In *Chislennye Metody Rascheta Elektronno-Opticheskykh Sistem* (Marchuk, G.I., ed.) pp. 62–67 (Nauka, Novosibirsk).

Kas'yankov, P.P. and Dutova, K.P. (1961). Izv. Akad. Nauk SSSR (Ser. Fiz.) **25**, 665–667; Bull. Acad. Sci. USSR (Phys. Ser.) **25**, 680–682.

Kas'yankov, P.P., Cheremisina, N.S. and Rynkevich, N.P. (1970b). Opt.-Mekh. Prom. **37**, No. 11, 67; Sov. J. Opt. Technol. **37**, 757–758.

Kas'yankov, P.P., Cheremisina, N.S. and Rynkevich, N.P. (1970b). In *Metody Rascheta*

Elektronno-Opticheskykh Sistem (Marchuk, G.I., ed.) pp. 62–67 (Nauka, Novosibirsk).

Katagiri, S. (1953). J. Electronmicrosc. **1**, 13–18.

Katagiri, S. (1960). J. Electronmicrosc. **8**, 13–16.

Katagiri, S. (1960). J. Electronmicrosc. **9**, 119.

Kel'man, V.M. and Yavor, S.Ya (1961). Zh. Tekh. Fiz. **31**, 1439–1442; Sov. Phys. Tech. Phys. **6**, 1052–1054.

Kel'man, V.M., Fedulina, L.V. and Yakushev, E.M. (1971). Zh. Tekh. Fiz. **41**, 1489–1497, 1832–1838 and 2016–2022; Sov. Phys. Tech. Phys. **16**, 1171–1176, 1449–1454 and 1598–1602.

Kel'man, V.M., Sapargaliev, A.A. and Yakushev, E.M. (1972). Zh. Tekh. Fiz. **42**, 2001–2009; Sov. Phys. Tech. Phys. **17**, 1607–1611.

Kel'man, V.M., Sapargaliev, A.A. and Yakushev, E.M. (1973). Zh. Tekh. Phys. **43**, 52–60; Sov. Phys. Tech. Phys. **18**, 33–37.

Korringa, J. (1942). *Onderzoekingen op het Gebied der Algebraïsche Optiek*. Dissertation, Delft.

Kulikov, Yu.V. (1971). Radiotekh. Elektron. **16**, 654–655; Radio Eng. Electron. Phys. **16**, 715–717.

Kulikov, Yu.V. (1972). Radiotekh. Elektron. **17**, 373–375; Radio Eng. Electron. Phys. **17**, 286–288.

Kulikov, Yu.V. (1973). Radiotekh. Elektron. **18**, 2379–2383; Radio Eng. Electron. Phys. **18**, No. 11, 286–288.

Kulikov, Yu.V. (1975). Radiotekh. Elektron. **20**, 1249–1254; Radio Eng. Electron. Phys. **20**, No. 6, 93.

Kulikov, Yu.V., Monastyrskii, M.A. and Feygin, Kh.I. (1978). Radiotekh. Elektron. **23**, 167–175; Radio Eng. Electron. Phys. **23**, No. 1, 120.

Kuyatt, C.E. (1978). J. Vac. Sci. Technol. **15**, 861–864.

Kuyatt, C.E., DiChio, D. and Natali, S.V. (1974). Rev. Sci. Instrum. **45**, 1275–1280.

Kynaston, D. and Mulvey, T. (1962). Philadelphia, vol. 1, D-2.

Kynaston, D. and Mulvey, T. (1963). Brit. J. Appl. Phys. **14**, 199–206.

Lambrakis, E., Marai, F.Z. and Mulvey, T. (1977). EMAG, pp. 35–38.

Laudet, M. (1953). Cah. Phys. **41**, 73–80.

Laudet, M. (1955a). Ann. Fac. Sci. Toulouse (4) **20**, 111–230.

Laudet, M. (1955b). J. Phys. Radium **16**, 118–124 and 908–916.

Lauer, R. (1982). Characteristics of triode electron guns. Adv. Opt. Electron Microsc. **8**, 137–206.

Leisegang, S. (1953). Optik **10**, 5–14.

Leisegang, S. (1954). Optik **11**, 49–60.

Leitner, H. (1942). *Die Abbildungsfehler der Elektrischen Zylinderlinse.* Dissertation (draft), Berlin.

Lenz, F. (1954). London, pp. 86–88.

Lenz, F. (1956). Stockholm, pp. 48–51.

Lenz, F. (1957). Optik **14**, 74–82.

Lenz, F. (1982a). Properties of electron lenses. In *Magnetic Electron Lenses* (Hawkes, P.W., ed.) pp. 119–161 (Springer, Berlin and New York).

Lenz, F. (1982b). Mikroskopie **39**, 359–373.

Lenz, F. (1986). Kyoto, vol. 1, pp. 287–288.

Lenz, F. and Hahn, M. (1953). Optik **10**, 15–27.

Lenz, F., Speidel, R. and Kuhm, J. (1982). Optik **62**, 401–411.

Lewis, G.N., Paik, H., Mioduszewski, J. and Siegel, B.M. (1986). J. Vac. Sci. Technol. **B4**, 116–119.

Li, Y. and Ni, W.-x. (1988). Optik **78**, 45–47

Liebmann, G. (1951). Proc. Phys. Soc. (London) **B64**, 972–977.

Liebmann, G. (1952a). Proc. Phys. Soc. (London) **B65**, 94–108.

Liebmann, G. (1952b). Proc. Phys. Soc. (London) **B65**, 188–192.

Lyubchik, Ya.G. and Mokhnatkin, A.V. (1972). Radiotekh. Elektron. **17**, 2234–2237; Radio Eng. Electron. Phys. **17**, 1795–1798.

Marai, F.Z. and Mulvey, T. (1975). EMAG, pp. 43–44.

Marai, F.Z. and Mulvey, T. (1977). Ultramicroscopy **2**, 187–192.

Marschall, H.J. (1939). Telefunken-Röhre **16**, 190–197.

Meads, P. (1963). *The Theory of Aberrations of Quadrupole Focusing Arrays.* Thesis, University of California; UCRL-10807.

Melkich, A. (1947). Sitzungsber. Akad. Wiss. Wien, Math-Nat. Kl., Abt. IIa, **155**, 393–438 and 439–471.

Meyer, W.E. (1956). Optik **13**, 86–91.

Meyer, W.E. (1961). Optik **18**, 69–91 and 101–114.

Monastyrskii, M.A. (1978). Zh. Tekh. Fiz. **48**, 1117–1122 and 2228–2234; Sov. Phys. Tech. Phys. **23**, 624–627 and 1275–1278.

Monastyrskii, M.A. (1980). Zh. Tekh. Fiz. **50**, 1939–1947; Sov. Phys. Tech. Phys. **25**, 1129–1133.

Monastyrskii, M.A. and Kolesnikov, S.V. (1983). Zh. Tekh. Fiz. **53**, 1668–1677; Sov. Phys. Tech. Phys. **28**, 1029–1034.

Monastyrskii, M.A. and Kulikov, Yu.V. (1976). Radiotekh. Elektron. **21**, 2251–2254; Radio Eng. Electron. Phys. **21**, No. 10, 148–150.

Monastyrskii, M.A. and Kulikov, Yu.V. (1978). Radiotekh. Elektron. **23**, 644–648; Radio Eng. Electron. Phys. **23**, No. 3, 137–140.

Mondragón, J.S. and Wolf, K.B., eds (1986). *Lie Methods in Optics* (Springer, Berlin).

Morito, N. (1954). J. Appl. Phys. **25**, 986–993.

Morito, N. (1955). Hitachi Hyoron **37**, 817–822.

Morito, N. (1957). J. Electronmicrosc. **5**, 1–2.

Moses, R.W. (1966). Rev. Sci. Instrum. **37**, 1370–1372.

Moses, R.W. (1970). Rev. Sci. Instrum. **41**, 729–740.

Moses, R.W. (1971a). Rev. Sci. Instrum. **42**, 828–831.

Moses, R.W. (1971b). Rev. Sci. Instrum. **42**, 832–839.

Moses, R.W. (1971c). EMAG, pp. 88–89.

Moses, R.W. (1972). Manchester, pp. 86–87.

Moses, R.W. (1973). In *Image Processing and Computer-aided Design in Electron Optics* (Hawkes, P.W., ed.) pp. 250–272 (Academic Press, London and New York).

Moses, R.W. (1974). Proc. Roy. Soc. (London) **A339**, 483–512.

Mulvey, T. and Jacob, L. (1949). Nature **163**, 525–526.

Munro, E. (1970). Grenoble, vol. 2, pp. 55–56.

Munro, E. (1971). EMAG, pp. 84–87.

Munro, E. (1972). Manchester, pp. 22–23.

Munro, E. (1973). In *Image Processing and Computer-aided Design in Electron Optics* (Hawkes, P.W., ed.) pp. 284–323 (Academic Press, London and New York).

Munro, E. (1988). J. Vac. Sci. Technol. **B6**, 941–948.

Nesvizhskii, M.B. (1986). Radiotekh. Elektron. **31**, 162–168; Sov. J. Commun. Technol. Electron. **31**, No. 5, 153–160.

Ng, E.W., ed. (1979). *Symbolic and Algebraic Computing*. Lecture Notes in Computer Science, vol. 72 (Springer, Berlin and New York).

Orloff, J. (1983a). Optik **63**, 369–372.

Orloff, J. (1983b). Optik **65**, 369–371.

Ovsyannikova, L.P. and Yavor, S.Ya (1965). Zh. Tekh. Fiz. **35**, 940–946; Sov. Phys. Tech. Phys. **10**, 723–726.

Petrie, D.P.R. (1962). Philadelphia, vol. 1, KK-2.

Plass, G.N. (1942). J. Appl. Phys. **13**, 49–55 and 524.

Rabin, B.M., Strashkevich, A.M. and Khin, L.S. (1951). Zh. Tekh. Fiz. **21**, 438–444.

Ramberg, E.G. (1939). J. Opt. Soc. Am. **29**, 79–83.

Rang, O. (1948). Optik **4**, 251–257.

Rang, O. (1949a). Optik **5**, 518–530.

Rang, O. (1949b). Phys. Bl. **5**, 78–80.

Rauh, H. (1971). Z. Naturforsch. **26a**, 1667–1675.

Read, F.H., Adams, A. and Soto-Montiel, J.R. (1971). J. Phys. E: Sci. Instrum. **4**, 625–632.

Rebsch, R. (1938). Ann. Physik (Leipzig) **31**, 551–560.

Rebsch, R. (1940). Z. Physik **116**, 729–733.

Rebsch, R. and Schneider, W. (1937). Z. Physik **107**, 138–143.

Recknagel, A. (1941). Z. Physik **117**, 67–73.

Recknagel, A. and Haufe, G. (1952/3). Wiss. Z. Techn. Hochschule Dresden **2**, 1–10.

Regenstreif, E. (1951a). Ann. Radioél. **6**, 244–267 and 299–317.

Regenstreif, E. (1951b). C.R. Acad. Sci. Paris **232**, 1918–1920 and **233**, 854–856.

Reisner, J.H. (1970). EMSA **28**, pp. 350–351.

Reisner, J.H. and Schuler, J.J. (1967). EMSA, pp. 226–227.

Renau, A. and Heddle, D.W.O. (1986). J. Phys. E: Sci. Instrum. **19**, 284–288 and 288–295.

Renau, A. and Heddle, D.W.O. (1987). J. Phys. E: Sci. Instrum. **20**, 235–236.

Rheinfurth, M. (1955). Optik **12**, 411–416.

Riecke, W.D. (1958). Berlin, vol. 1, pp. 189–194.

Riecke, W.D. (1964). Prague, vol. A, pp. 7–8.

Riecke, W.D. (1966/7). Optik **24**, 397–426.

Riecke, W.D. (1972). Optik **36**, 66–84, 288–308 and 375–398.

Riecke, W.D. (1976). Instrument operation for microscopy and microdiffraction. In *Electron Microscopy in Materials Science* (Valdrè, U. and Ruedl, E., eds) pp. 19–111 (CEE, Luxembourg).

Riecke, W.D. (1982). In *Magnetic Electron Lenses* (Hawkes, P.W., ed.) pp. 163–357 (Springer, Berlin and New York).

Riedl, H. (1937). Z. Physik **107**, 210–216.

Rogowski, W. (1937). Arch. Elektrotech. **31**, 555–593.

Rose, H. (1967). Optik **25**, 587–597.

Rose, H. (1967/8). Optik **26**, 289–298.

Rose, H. (1968). Optik **27**, 466–474 and 497–514.

Rose, H. (1968/69). Optik **28**, 462–474.

Rose, H. (1971a). Optik **33**, 1–24.

Rose, H. (1971b). Optik **34**, 285–311.

Rose, H. (1987). Nucl. Instrum. Meth. **A258**, 374–401.

Rose, H. and Moses, R.W. (1973). Optik **37**, 316–336.

Rose, H. and Petri, U. (1971). Optik **33**, 151–165.

Sakaki, Y. and Maruse, S. (1954). J. Electronmicrosc. **2**, 8–9.

Scherzer, O. (1933). Z. Physik **80**, 193–202.

Scherzer, O. (1936a). Z. Physik **101**, 23–26.

Scherzer, O. (1936b). Z. Physik **101**, 593–603.

Scherzer, O. (1937). Berechnung der Bildfehler dritter Ordnung nach der Bahnmethode. In *Beiträge zur Elektronenoptik* (Busch, H. and Brüche, E., eds) pp. 33–41 (Barth, Leipzig).

Scherzer, O. (1941). Z. Physik **118**, 461–466.

Scherzer, O. (1946). Phys. Bl. **2**, 110.

Scherzer, O. (1947). Optik **2**, 114–132.

Schiske, P. (1956). Optik **13**, 502–505.

Schriber, S.O. and Taylor, L.S., eds (1987). *Proc. 2nd Int. Conf. Charged Particle Optics*, Nucl. Instrum. Meth. Phys. Res. **A258**, 289–598 (North-Holland, Amsterdam).

Seman, O.I. (1951). Dokl. Akad. Nauk SSSR **81**, 775–778.

Seman, O.I. (1953a). Zh. Eksp. Tekh. Fiz. **24**, 581–588.

Seman, O.I. (1953b). Dokl. Akad. Nauk SSSR **93**, 443–445.

Seman, O.I. (1954). Dokl. Akad. Nauk SSSR **96**, 1151–1154.

Seman, O.I. (1955a). Trudy Inst. Fiz. Astron. Akad. Nauk Eston. SSR, No. 2, 3–29.

Seman, O.I. (1955b). Trudy Inst. Fiz. Astron. Acad. Nauk Eston. SSR, No. 2, 30–49.

Seman, O.I. (1958a). Uch. Zap. Rostov. Gos. Univ. **68**, No. 8, 63–75.

Seman, O.I. (1958b). Uch. Zap. Rostov. Gos. Univ. **68**, No. 8, 77–90.

Seman, O.I. (1958c). Radiotekh. Elektron. **3**, 283–287; Radio Eng. Electron. **3**, 402–409.

Seman, O.I. (1959a). Radiotekh. Elektron. **4**, 1213–1214; Radio Eng. Electron. **4**, No. 7, 227–230.

Seman, O.I. (1959b). Radiotekh. Elektron. **4**, 1702–1707; Radio Eng. Electron. **4**, No. 10, 235–244.

Seman, O.I. (1959c). Opt. Spektrosk. **7**, 113–115; Opt. Spectrosc. **7**, 68–69.

Seman, O.I. (1968). Radiotekh. Elektron. **13**, 907–912; Radio Eng. Electron. Phys. **13**, 784–788.

Septier, A. (1963). C.R. Acad. Sci. Paris **256**, 2325–2328.

Septier, A. (1966). The struggle to overcome spherical aberration in the electron microscope. Adv. Opt. Electron Microsc. **1**, 204–274.

Shao, Z. (1987a). Optik **75**, 152–157.

Shao, Z. (1987b). EMSA **45**, pp. 134–135.

Shao, Z. and Crewe, A.V. (1987). J. Appl. Phys. **62**, 1149–1153.

Shi, C.-h. (1956). Acta Sci. Nat. Univ. Pekin **2**, 457–465.

Shimoyama, H. (1982). J. Electron Microsc. **31**, 8–17.

Shpak, E.V. and Yavor, S.Ya. (1964). Zh. Tekh. Fiz. **34**, 2003–2007; Sov. Phys. Tech. Phys. **9**, 1540–1543.

Shpak, E.V. and Yavor, S.Ya. (1965). Zh. Tekh. Fiz. **35**, 947–950; Sov. Phys. Tech. Phys. **10**, 727–729.

Shtepa, N.I. (1952). Zh. Tekh. Fiz. **22**, 216–226.

Siegbahn, K. (1946). Phil. Mag. **37**, 162–184.

Sivkov, Yu.P. (1971). Particle Acc. **2**, 243–249.

Smirnov, N.A., Monastyrskii, M.A. and Kulikov, Yu.V. (1979). Zh. Tekh. Fiz. **49**, 2590–2595; Sov. Phys. Tech. Phys. **24**, 1462–1465.

Smith, T. (1921/22). Trans. Opt. Soc. (London) **23**, 311–322.

Soma, T. (1977). Optik **49**, 255–262.

Sorokina, V.V. and Timofeev, P.V. (1948). Zh. Tekh. Fiz. **18**, 509–516.

Steffen, K.G. (1965). *High Energy Beam Optics* (Wiley-Interscience, New York and London).

Stephan, W.G. (1947). *Practische Toepassingen op het Gebied der Algebraïsche Optica*. Proefschrift, Delft.

Stoyanov, P.A. (1955a). Zh. Tekh. Fiz. **25**, 625–635.

Stoyanov, P.A. (1955b). Zh. Tekh. Fiz. **25**, 2537–2541.

Stoyanov, P.A. (1958a). Berlin, vol. 1, pp. 61–66.

Strashkevich, A.M. (1959). *Elektronnaya Optika Elektrostaticheskikh Polei ne obladayushchikh Osevoi Simmetriei* (Gos. Izd. Fiz-Mat. Lit., Moscow).

Strashkevich, A.M. (1965). Zh. Tekh. Fiz. **35**, 177–183; Sov. Phys. Tech. Phys. **10**, 147–151.

Strashkevich, A.M. (1966). *Elektronnaya Optika Elektrostaticheskikh Sistem* (Energiya, Moscow).

Strashkevich, A.M. and Gluzman, N.G. (1954). Zh. Tekh. Fiz. **24**, 2271–2284.

Strashkevich, A.M. and Pilat, I.M. (1951). Izv. Akad. Nauk SSSR (Ser. Fiz.) **15**, 448–466.

Strashkevich, A.M. and Pilat, I.M. (1952). Uchen. Zap. Chernovitskii Gos. Univ. (Ser. Fiz-Mat. Nauk) **4**, 113–122.

Sturrock, P.A. (1949). Delft, 89–93.

Sturrock, P.A. (1951a). Proc. Roy. Soc. (London) **A210**, 269–289.

Sturrock, P.A. (1951b). Phil. Trans. Roy. Soc. (London) **A243**, 387–429.

Sturrock, P.A. (1951c). C.R. Acad. Sci. Paris **233**, 146–147 and 243–245.

Sturrock, P.A. (1952). Phil. Trans. Roy. Soc. (London) A245, 155–187.

Sturrock, P.A. (1955). *Static and Dynamic Electron Optics* (University Press, Cambridge).

Sugiura, Y. and Suzuki, S. (1943). Proc. Imp. Acad. Japan 19, 293–302.

Suzuki, S. and Ishikawa, A. (1978). Toronto, vol. 1, pp. 24–25.

Tadano, B., Kimura, H. and Onuma, Y. (1966). Hitachi Rev. 15, 340–344.

Taganov, I.N. (1966). Radiotekh. Elektron 11, 1329–1330; Radio Eng. Electron Phys. 11, 1160–1162.

Taganov, I.N. and Kas'yankov, P.P. (1964). Opt.-Mekh. Prom. 31, No. 11, 14–16.

Taganov, I.N. and Kas'yankov, P.P. (1965). Opt.-Mekh. Prom. 32, No. 12, 21–23.

Taganov, I.N. and Kas'yankov, P.P. (1967). In *Chislennye Metody Rascheta Elektronno-Opticheskikh Sistem* (Marchuk, G.I., ed.) pp. 11–22 (Nauka, Novosibirsk).

Tretner, W. (1950). Optik 7, 242.

Tretner, W. (1954). Optik 11, 312–326; errata ibid. 12, 293–294 (1955).

Tretner, W. (1956). Optik 13, 516–519.

Tretner, W. (1959). Optik 16, 155–184.

Tsuno, K. and Harada, Y. (1981a). J. Phys. E: Sci. Instrum. 14, 313–319.

Tsuno, K. and Harada, Y. (1981b). J. Phys. E: Sci. Instrum. 14, 955–960.

Tsuno, K., Ishida, Y. and Harada, Y. (1980a). EMSA, pp. 280–281.

Tsuno, K., Arai, Y. and Harada, Y. (1980b). The Hague, vol. 1, pp. 76–77.

U, M.-zh. [Wu, M.-z.] (1957). Sci. Sin. 6, 833–846; Acta Phys. Sin. 13, 181–206.

Vandakurov, Yu.V. (1955a). Zh. Tekh. Fiz. 25, 1412–1425.

Vandakurov, Yu.V. (1955b). Zh. Tekh. Fiz. 25, 2545–2555.

Vandakurov, Yu.V. (1956a). Zh. Tekh. Fiz. 26, 1599–1610; Sov. Phys. Tech. Phys. 1, 1558–1569.

Vandakurov, Yu.V. (1956b). Zh. Tekh. Fiz. 26, 2578–2594; Sov. Phys. Tech. Phys. 1, 2491–2507.

Vandakurov, Yu.V. (1957). Zh. Tekh. Fiz. 27, 1850–1862; Sov. Phys. Tech. Phys. 2, 1719–1733.

Velzel, C.H.F. (1987). J. Opt. Soc. Am. A4, 1342–1348.

Velzel, C.H.F. and Meijere, J.L.F. de (1988). J. Opt. Soc. Am. A5, 246–250 and 251–256.

Verster, J.L. (1963). Philips Res. Repts 18, 465–605.

Vlasov, A.G. (1944). Izv. Akad. Nauk SSSR (Ser. Fiz.) 8, 235–239.

Vlasov, A.G. and Shakhmatova, I.P. (1962). Zh. Tekh. Fiz. 32, 695–705; Sov. Phys. Tech. Phys. 7, 507–514.

Voit, H. (1939). Z. Instrumentenkde **59**, 71–82.

Watanabe, A. and Morito, N. (1955). Optik **12**, 166–172 and 564.

Watanabe, M. and Someya, T. (1963). Optik **20**, 99–108.

Wegmann, L. (1953). Helv. Phys. Acta **26**, 448–449.

Wegmann, L. (1954). Optik **11**, 153–170.

Wollnik, H., ed. (1981). *Proc. 1st Conf. Charged Particle Optics.* Nucl. Instrum. Meth. Phys. Res. **187**, 314pp.

Wollnik, H. (1987). *Optics of Charged Particles* (Academic Press, Orlando and London).

Wollnik, H. and Berz, M. (1985). Nucl. Instrum. Meth. Phys. Res. **A238**, 127–140.

Ximen, J.-y. (1957). Acta Phys. Sin. **13**, 339–356.

Ximen, J.-y. and Chen Q.-s. (1964). Acta Electron. Sin. No. 4, 72–84.

Ximen, J.-y. and Xi Z.-h. (1964). Acta Electron. Sin. No. 3, 24–35.

Ximen, J.-y., Zhou, L.-w. and Ai, K.-c. (1983). Optik **66**, 19–34.

Yanaka, T. and Shirota, K. (1970). Grenoble, vol. 2, pp. 59–60.

Yavor, S.Ya. (1968). *Fokusirovka Zaryazhennykh Chastits Kvadrupol'nymi Linzami* (Atomizdat, Moscow).

Yavor, S.Ya., Dymnikov, A.D. and Ovsyannikova, L.P. (1964). Zh. Tekh. Fiz. **24**, 99–104; Sov. Phys. Tech. Phys. **9**, 76–80.

Zhou, L.-w., Ai, K.-c. and Pan, S.-c. (1983). Acta Phys. Sin. **32**, 376–392.

Zworykin, V.K., Morton, G.A., Ramberg, E.G., Hillier, J. and Vance, A.W. (1945). *Electron Optics and the Electron Microscope* (Wiley, New York and Chapman and Hall, London).

Part V, Chapter 32

We draw attention to the survey by Ritz (1979), to the earlier papers on deflection aberrations by Grümm and Spurny (1956), Werner (1963), Hutter (1967) and Schürmann and Haussmann (1967) and to other work by Munro (1975, 1980), Soma (1979) and Tsunagari *et al.* (1986), who consider parasitic aberrations. Papers on this subject frequently appear in the proceedings of the Electron, Ion and Laser Beam Technology conferences, published in *J. Vac.Sci.Technol.* and in those of the Microcircuit Engineering meetings now published in *Microelectronic Engineering*.

Amboss, K. and Wolf, E.D. (1971). In *Rec. 11th Symp. Electron Ion Laser Beam Technol.* (Thornley, R.F.M., ed.) pp. 195–204 (San Francisco Press, San Francisco).

Chu, H.C. and Munro, E. (1981). J. Vac. Sci. Technol. **19**, 1054–1057.

Chu, H.C. and Munro, E. (1982a). Optik **61**, 121–145.

Chu, H.C. and Munro, E. (1982b). Optik **61**, 213–236.

Crewe, A.V. and Parker, N.W. (1976). Optik **46**, 183–194.

Ding, S.Q. (1982). Hamburg, vol. 1, pp. 315–316.

Dodin, A.L. (1983). Radiotekh. Elektron. **28**, 357–361; Radio Eng. Electron. Phys. **28**, No. 2, 114–118.

Glaser, W. (1938). Z. Physik **111**, 357–372.

Glaser, W. (1941). Z. Physik **117**, 412.

Glaser, W. (1949). Ann. Physik (Leipzig) **4**, 389–408.

Glaser, W. (1952). *Grundlagen der Elektronenoptik* (Springer, Vienna).

Glaser, W. (1956). Elektronen- und Ionenoptik. In *Handbuch der Physik* **33**, 123–395.

Goto, E. and Soma, T. (1977). Optik **48**, 255–270.

Grümm, H. and Spurny, H. (1956). Öst. Ing.-Arch. **10**, 104–106.

Haantjes, J. and Lubben, G.J. (1957). Philips Tech. Repts **12**, 46–68.

Haantjes, J. and Lubben, G.J. (1959). Philips Res. Repts **14**, 65–97.

Hawkes, P.W. (1988). Optik, to be published.

Hosokawa, T. (1980). Optik **56**, 21–30.

Hutter, R.G.E. (1947). J. Appl. Phys. **18**, 740–758.

Hutter, R.G.E. (1948). Adv. Electron. **1**, 167–218.

Hutter, R.G.E. (1967). IEEE. Trans. **ED-14**, 694–699.

Kaashoek, J. (1968). Philips Res. Repts Suppl. No. 11, 114 pp.

Kanaya, K. and Kawakatsu, H. (1961a). Bull. Electrotech. Lab. **25**, 241–252.

Kanaya, K. and Kawakatsu, H. (1961b). J. Electronmicrosc. **10**, 218–221.

Kanaya, K. and Kawakatsu, H. (1962). Bull. Electrotech. Lab. **26**, 241–250.

Kanaya, K., Kawakatsu, H. and Tanaka, K. (1961). Bull. Electrotech. Lab. **25**, 481–494.

Kanaya, K., Kawakatsu, H., Okazaki, I. and Takizawa, T. (1963). Bull. Electrotech. Lab. **27**, 401–417.

Kanaya, K., Kawakatsu, H., Okazaki, I. and Takizawa, T. (1964). J. Electronmicrosc. **13**, 80–86.

Kawakatsu, H. and Kanaya, K. (1961). J. Electronmicrosc. **10**, 119–124.

Kern, D.P. (1979). J. Vac. Sci. Technol. **16**, 1686–1691.

Knauer, W. (1981). J. Vac. Sci. Technol. **19**, 1042–1047.

Koops, H. (1972). Optik **36**, 93–110.

Koops, H. (1973). J. Vac. Sci. Technol. **10**, 909–912.

Koops, H. and Bernhard, W. (1975). J. Vac. Sci. Technol. **12**, 1141–1145.

Kuroda, K. (1980). Optik **57**, 251–258.

Lenc, M. and Lencová, B. (1988). Optik **78**, 127–131.

Lencová, B. (1981). Optik **58**, 25–35.

Lencová, B. (1988). Optik **79**, 1–12.

Li, Y. (1981). Acta Phys. Sin. **30**, 1155–1164.

Li, Y. (1983). Optik **63**, 213–216.

Li, Y. and Ximen, J.-y. (1982a). Optik **61**, 315–332.

Li, Y. and Ximen, J.-y. (1982b). Acta Phys. Sin. **31**, 604–614.

Munro, E. (1974). Optik **39**, 450–466.

Munro, E. (1975). J. Vac. Sci. Technol. **12**, 1146–1150.

Munro, E. (1980). In *Microcircuit Engineering* (Ahmed, H. and Nixon, W.C., eds) pp. 513–534 (University Press, Cambridge).

Munro, E. and Chu, H.C. (1982a). Optik **60**, 371–390.

Munro, E. and Chu, H.C. (1982b). Optik **61**, 1–16.

Ohiwa, H. (1970). *Elimination of Third Order Aberrations in Electron Beam Scanning Systems*. Thesis, Tokyo.

Ohiwa, H. (1978). J. Vac. Sci. Technol. **15**, 849–852.

Ohiwa, H. (1979). Optik **53**, 63–68.

Ohiwa, H., Goto, E. and Ono, A. (1971). Trans. Inst. Electron. Commun. Eng. Japan **54–B**, 730–737; Electron. Commun. Japan **54–B**, No. 12, 44–51.

Owen, G. (1981). J. Vac. Sci. Technol. **19**, 1064–1068.

Owen, G. and Nixon, W.C. (1973). J. Vac. Sci. Technol. **10**, 983–986.

Picht, J. (1943). Ann. Physik (Leipzig) **43**, 53–72.

Picht, J. and Himpan (1941). Ann. Physik (Leipzig) **39**, 409–435, 436–477 and 478–501.

Plies, E. (1982). Siemens Forsch. Entwickl. Ber. **11**, 38–45 and 83–90.

Rao, V.R.M. and Nixon, W.C. (1981). J. Vac. Sci. Technol. **19**, 1037–1041.

Ritz, E.F. (1979). Adv. Electron. Electron Phys. **59**, 299–357.

Schürmann, J. and Haussmann, G. (1967). Z. Angew. Phys. **22**, 235–239.

Smith, M.R. and Munro, E. (1986). Optik **74**, 7–16.

Smith, M.R. and Munro, E. (1987). J. Vac. Sci. Technol. **B5**, 161–164.

Soma, T. (1977). Optik **49**, 255–262.

Soma, T. (1979). Optik **53**, 281–284.

Tang, T.T. (1986). Optik **74**, 43–47.

Wang, C.C.T. (1966). J. Appl. Phys. **37**, 5007–5008.

Wang, C.C.T. (1967a). J. Appl. Phys. **38**, 3991–3994.

Wang, C.C.T. (1967b). J. Appl. Phys. **38**, 4938–4944.

Wang, C.C.T. (1967c). IEEE Trans. **ED-14**, 357–365.

Wang, C.C.T. (1971). IEEE Trans. **ED-18**, 258–274.

Wendt, G. (1939). Telefunkenröhre **15**, 100–136.

Wendt, G. (1942a). Z. Physik **118**, 593–517.

Wendt, G. (1942b). Z. Physik **119**, 423–462.

Wendt, G. (1947). Ann. Physik (Leipzig) **1**, 83–94.

Wendt, G. (1953). Onde Elec. **33**, 93–106.

Werner, H. (1963). Exp. Tech. Phys. **11**, 32–50.

Ximen, J.-y. (1977). Acta Phys. Sin. **26**, 34–53.

Ximen, J.-y. (1978). Acta Phys. Sin. **27**, 247–259.

Ximen, J.-y. (1981). Optik **59**, 237–249; Chinese J. Sci. Instrum. **2**, 1–11.

Ximen, J.-y. (1986). *Aberration Theory in Electron and Ion Optics.* Adv. Electron. Electron Phys. Suppl. 17.

Ximen, J.-y. and Li, Y. (1982). Optik **62**, 287–297.

Part VI, Chapters 33 and 34

For some further development in numerical methods, see Munro (1987) and Ströer (1987). An early attempt to assess the combined effect of aberrations of different kinds is to found in Harte (1973). The use of a Russian computer algebra language (ANALITIK) for deriving aberration coefficients is described by Narylkov and Lyubchik (1982).

Barton, D. and Fitch, J.P. (1972). Repts Prog. Phys. **35**, 235–314.

Berz, M. and Wollnik, H. (1987). Nucl. Instrum. Meth. Phys. Res. **A258**, 364–373.

Buchberger, B., Collins, G.E., Loos, R. and Albrecht, R., eds (1982). *Computer Algebra, Symbolic and Algebraic Computation.* Computing, Suppl. 4 (Springer, Vienna and New York).

Burfoot, J. (1952). Brit. J. Appl. Phys. **3**, 22–24.

Busch, H. (1927). Arch. Elektrotech. **18**, 583–594.

Calmet, J. (1982). *Computer Algebra.* Lecture Notes in Computer Science **144** (Springer, Berlin and New York).

Chu, H.C. and Munro, E. (1982). Optik **61**, 121–145 and 213–236.

Crewe, A.V., Eggenberger, D.N., Wall, J. and Welter, L.M. (1968). Rev. Sci. Instrum. **39**, 576–583.

Davenport, J.H., Siret, Y. and Tournier, E. (1988). *Computer Algebra* (Academic Press, London and San Diego).

Dodin, A.L. and Nesvizhskii, M.B. (1981). Radiotekh. Elektron. **26**, 2651–2657; Radio Eng. Electron Phys. **26**, No. 12, 151–157.

Fitch, J.P. (1985). J. Symbol. Comput. **1**, 211–227.

Fox, L. and Goodwin, E.T. (1949). Proc. Cambridge Philos. Soc. **45**, 373–388.

Gill, S. (1951). Proc. Cambridge Philos. Soc. **47**, 96–108.

Glaser, W. (1933). Ann. Physik (Leipzig) **18**, 557–585.

Glaser, W. (1935). Z. Physik **97**, 177–201.

Glaser, W. (1952). *Grundlagen der Elektronenoptik* (Springer, Vienna).

Harte, K.J. (1973). J. Vac. Sci. Technol. **10**, 1098–1101.

Hauke, R. (1977). *Theoretische Untersuchungen rotationssymmetrischer Elektronenstrahlerzeugungssysteme unter Berücksichtigung von Raumladung.* Dissertation, Tübingen.

Hawkes, P.W. (1977). Optik **48**, 29–51.

Hawkes, P.W. (1983). Optik **63**, 129–156 and **65**, 227–251.

Hearn, A.C. (1985) *REDUCE User's Manual* (Rand, Santa Monica).

Hoch, H., Kasper, E. and Kern, D. (1976). Optik **46**, 463–473.

Householder, A.S. (1964). *The Theory of Matrices in Numerical Analysis* (Blaisdell, New York).

Hulzen, J.A. van and Calmet, J. (1982). Computing, Suppl. 4, 221–243.

Jacobs, D., ed. (1977). *The State of the Art in Numerical Analysis* (Academic Press, London and New York).

Jennings, J.C.E. and Pratt, R.G. (1955). Proc. Phys. Soc. (London) **B68**, 526–536.

Kasper, E. (1982). In *Magnetic Electron Lenses* (Hawkes, P.W., ed.) pp. 57–118 (Springer, Berlin and New York).

Kasper, E. (1984). In *Electron Optical Systems for Microscopy, Microanalysis and Microlithography* (Hren, J.J., Lenz, F.A., Munro, E. and Sewell, P.B., eds) pp. 63–73 (Scanning Electron Microscopy, AMF O'Hare).

Kasper, E. (1985). Optik **69**, 117–125.

Kasper, E. (1987a). Nucl. Instrum. Meth. Phys. Res. **A258**, 466–479.

Kasper, E. (1987b). Optik **77**, 3–12.

Kern, D. (1978). *Theoretische Untersuchungen an rotationssymmetrischen Strahlerzeugungssystemen mit Feldemissionsquelle.* Dissertation, Tübingen.

Lawson, C.L. and Hanson, R.J. (1974). *Solving Least-Squares Problems* (Prentice-Hall, Englewood Cliffs, NJ).

Levenberg, K. (1944). Quart. Appl. Math. **2**, 164–168.

Manning, W.F. and Millman, J. (1938). Phys. Rev. **53**, 673.

MIT (1983). *The MACSYMA Reference Manual* (Massachusetts Institute of Technology, Cambridge).

Moses, R.W. (1973). In *Image Processing and Computer-aided Design in Electron Optics* (Hawkes, P.W., ed.) pp. 250–272 (Academic Press, London and New York).

Munro, E. (1973). In *Image Processing and Computer-aided Design in Electron Optics* (Hawkes, P.W., ed.) pp. 284–323 (Academic Press, London and New York).

Munro, E. (1987). Nucl. Instrum. Meth. Phys. Res. **A258**, 443–461.

Munro, E. and Chu, H.C. (1982). Optik **60**, 371–390 and **61**, 1–16.

Narylkov, S.G. and Lyubchik, Ya.G. (1982). Radiotekh. Elektron. **27**, 1602–1605 (not translated).

Nesvizhskii, M.B. (1986). Radiotekh. Elektron. **31**, 162–168; Radio Eng. Electron. Phys. **31**, No. 1, 153–160.

Ng, E.W., ed. (1979). *Symbolic and Algebraic Computation*. Lecture Notes in Computer Science **72** (Springer, Berlin and New York).

Niemitz, H.P. (1980). *Theoretische Untersuchung von elektrostatischen Drei-Elektroden-Filterlinsen*. Dissertation, Tübingen.

Norman, A.C. (1982). In *Computer Algebra* (Calmet, J., ed.) pp. 237–248 (Springer, Berlin and New York).

Norman, A.C. and Davenport, J.H. (1979). In *Symbolic and Algebraic Computation* (Ng, E.W., ed.) pp. 398–407 (Springer, Berlin and New York).

Numerov, B. (1923). Publ. Obs. Astrophys. Central Russie **2**, 188–288.

Pavelle, R., ed. (1985). *Applications of Computer Algebra* (Kluwer, Dordrecht).

Pavelle, R., Rothstein, M. and Fitch, J. (1981). Sci. Am. **245**, No. 6, 136–152.

Ralston, A. (1960). In *Mathematical Methods for Digital Computers* (Ralston, A. and Wilf, H.S., eds) vol. 1, pp. 95–109 (Wiley, New York).

Rand, R.H. (1984). *Computer Algebra in Applied Mathematics: an Introduction to MACSYMA* (Pitman, Boston and London).

Rand, R.H. and Armbruster, D. (1987). *Perturbation Methods, Bifurcation Theory and Computer Algebra* (Springer, New York and Berlin).

Rayna, G. (1987). *REDUCE* (Springer, New York and Berlin).

Romanelli, M.J. (1960). In *Mathematical Methods for Digital Computers* (Ralston, A. and Wilf, H.S., eds) vol. 1, pp. 110–120 (Wiley, New York).

Scherle, W. (1983). Optik **63**, 217–226.

Scherle, W. (1984). Optik **67**, 307–314.

Scherzer, O. (1936). In *Beiträge zur Elektronenoptik* (Busch, H. and Brüche, E., eds) pp. 33–41 (Barth, Leipzig).

Soma, T. (1977). Optik **49**, 255–262.

Stoer, J. (1979). *Einführung in die Numerische Mathematik,* vol. I (Heidelberger Taschenbücher **105**, Springer, Berlin and New York).

Störmer, C. (1907). Arch. Math. Naturvid. **28**, No. 2.

Ströer, M. (1987). Optik **77**, 15–25.

Szilagyi, M. (1977). Optik **48**, 215–224.

Szilagyi, M. (1984). In *Electron Optical Systems for Microscopy, Microanalysis and Microlithography* (Hren, J.J., Lenz, F.A., Munro, E. and Sewell, P.B., eds) pp. 75–84 (Scanning Electron Microscopy, AMF O'Hare).

Whittaker, E.T. and Watson, G.N. (1950). *A Course of Modern Analysis,* 4th ed. (University Press, Cambridge).

Index

INDEX

Abbe sine condition 219
aberration correction 11, 365, 857–878
aberration disc
 – integral properties 555
 – numerical determination 547–558
 – prisms 1077
aberration integrals 534
 – for deflection systems 511
aberration matrices 301, 418–424
aberration minimization 301, 877
 – and deflection 508
aberration polynomials 12, 300, 396
aberration studies
 – historical development of 8, 11, 299
aberration theory
 – for curvilinear systems 1053
aberrations and system symmetry
 – exceptions 322
 – for $N = 2$ 325
 – for $N = 3$ 329
 – for $N = 4$ 330
 – for $N = 5$ and 6 333
 – general case 315
 – rotational symmetry 334
 – with a symmetry plane 324
aberrations
 – differential equations for 537–542
 – Fourier analysis of 550
aberrations of cylindrical lenses 466–469
aberrations of deflection systems 497–516

aberrations of foil lenses 680
aberrations of gradient 419, 442
aberrations of microwave cavity lenses 876
aberrations of mirrors and cathode lenses 425–433
 – cartesian form 432
 – parametric form 425
 – structure 430
aberrations of quadrupoles and octopoles 434–465
 – quadruplets 461
aberrations of round lenses 339–424
 – and Seman's technique 356, 386
 – in reduced coordinates 384
aberrations, parasitic 470–479
 – numerical determination 475
 – isoplanatic approximation 477
acceleration potential 19, 21
accelerators, optics of 647
acceptance 994
achromatic quadrupoles 461, 862
addition rules for aberrations
 – round lenses 420
 – quadrupoles 464
alpha filter 1095, 1099
alternating direction implicit methods 173
alternating gradient focusing 1062
analytic continuation 79
analytic function 70
analytical electron microscopy (AEM) 10
anamorphotic systems 802
angle characteristic 237

angular emission distribution 924
angular magnification 217
anisotropic aberrations
— astigmatism 370
— chromatic distortion 414
— coma 366
— distortion 379
anisotropic magnetic circuit 181
annular systems 894, 895, 897
anode 908
antisymmetric multiplets 286, 805
aperture aberrations 337, 350
— of quadrupoles and octopoles 439, 440, 453
— sign of 458
aperture field 143, 193
aperture
— of quadrupoles and octopoles 440
aperture-lenses 631
apertures, influence on brightness 977
Archard–Deltrap corrector 861
arc lenses 810
astigmatic difference for quadrupoles 281
astigmatic objects and images 280
astigmatic tube lenses 813
astigmatism 337, 520
— axial 470, 474
— of quadrupoles and octopoles 439, 441
astigmatism of round lenses, asymptotic 397
— and pupil magnification 407

— addition rules for 420
astigmatism of round lenses, real 349, 369–378
— aberration figure 369
— and aperture position 383
— circle of least confusion 374
— formulae for 376, 385
— thin-lens formula 378
asymptotes 226, 239
asymptotic aberration coefficients
— of magnetic lenses 705
— of quadrupoles and octopoles 434, 458
— of round lenses 393–408
— integrals 399
asymptotic aberrations
— dependence on object and aperture position 406, 424
— integrals for the bell-shaped model 701
asymptotic cardinal elements 226, 233
— focal length 228
— image focus 228, 244
— object focus 228
— of magnetic lenses 698
— of mirrors 270
— principal plane 228
asymptotic image formation 225
asymptotic object 225
asymptotic properties of thermionic guns 957
average brightness 971
axial angular momentum 24, 30, 38, 44, 210
axial astigmatism 470, 474

INDEX

axial conductors 870
axial fields or potentials, optimization 561
axial focusing (prisms) 1068
axial harmonic 79, 80, 82, 836
azimuthal Fourier series expansion 73
Barber's rule 1065
barrel distortion 379
basis-curve 881, 887
beam choppers 876
beam entropy 1011
beam matching 994
beam temperature 1007
　– longitudinal 1008
　– transverse 1007
beam transport systems 917
beam-shaping techniques 848, 850
bell-shaped models 253, 706, 801
beta-ray spectrometers 365, 685, 897
binormal 1041
Biot–Savart law 98, 829
biplanar lenses 811
blade focus 895
Boersch effect 917, 1004–1016, 1036
　– analytical calculations 1013
　– dimensional analysis 1016
　– energetic Boersch effect 1004
　– spatial Boersch effect 1004, 1012
　– thermodynamic limit 1009
Bohm–Aharonov effect 56
boundary conditions for magnetostatic fields 96

boundary-element method 125–158, 944
boundary-value problem 61, 94–106
　– and finite-difference method 170
　– electrostatic fields 94
　– for paraxial trajectories 535
　– general theory 94
　– magnetic fields 101
BR-product 28
brightness 917, 971–988
　– and emittance 996
　– influence of apertures 977
　– Lenz's theory 980
　– maximum brightness 975
　– measurement 985
　– of field emission guns 1021
　– of thermionic guns 1030
brightness function 971
　– generalized 972
brightness measurement
　– lens method 986
　– two-aperture method 986
Busch
　– and lens action 6
　– focal length formula 259, 472, 521
calculation of aberration coefficients
　– round lenses, trajectory method 341
　– round lenses, eikonal method 344
Caledonian quadruplet 878
CAMAL 301, 568
canonical equation 42, 973, 990
canonical formalism 41

INDEX

canonical momentum 37, 44, 1053
carbon emitters 913, 1036
cardinal elements
  – of quadrupole systems 281
  – of rotationally symmetric systems 225–260
cardinal elements of round lenses
  – approximate formulae 257
  – asymptotic 226
  – real 242
cartesian theory of mirror optics 271, 432
Castaing–Henry analyser 262, 797, 1058, 1095, 1099
cathode 908, 912
cathode lens 262, 425, 630, 799
cathode-ray tubes 823
catoptric systems 261
Cauchy–Riemann equations 70
caustic 57, 879–904
  – concept 879
  – higher-order focusing 893
  – intensity distribution 891
  – of a round lens 882
  – of astigmatic lenses 886
characteristic function 11, 49, 52, 298
  – first-order perturbation 307
  – general 46
chromatic aberration 298
  – differential equations for 541
  – numerical calculation 555
  – of deflection systems 507
  – of quadrupoles 460
  – of round lenses 409–417

chromatic aberration coefficients, see also: axial chromatic aberration 413
chromatic aberration coefficients 409–417
  – asymptotic coefficients of round lenses 415
  – of dual-channel sytems 516
  – real coefficients of round lenses 412
chromatic aberration
  – of magnification 414
  – of rotation 414
circular optic axis 1059
closed current loops 828
collineation 235
coma 337
coma of quadrupoles and octopoles 441
coma of round lenses, asymptotic 397
  – addition rules for 421
  – and pupil magnification 407
coma of round lenses, real 349, 365–369
  – aberration figure 365
  – and aperture position 383
  – formulae 367, 385
  – thin-lens formula 368
coma-free condition 508
coma-free lens 773
coma-free point 383, 701
computer algebra 301, 565–571
computer programs for ion optics 1098
condenser lens 207

INDEX

condenser-objective 691, 751
cone-edge focus 895
cone focus 895
conformal mapping 70, 685
congruences 56
conservation
 – of energy 18, 30
 – of momentum 24
contracurrent systems 261
control electrode 908
convergence 419
 – of lenses 216
convolutional models
 – for magnetic lenses 711
coordinate rotation 209
coordinate systems 22
correction of astigmatisms 517
crossed-field device 1097
crossed lenses 810
cross focus 895
crossover 983
cryolenses 688, 783
cubic spline 189
current loops 828
 – with ferrite shield 831
curvature vector 1042
curved cathodes 429
curved optic axis 1037
curvilinear coordinate systems 1040
curvilinear systems 1039–1099
cyclic reduction methods 173
cylindrical lens 290–293, 466–469, 685
cylindrical mirror analyser 901
Darmstadt project 862

defect function 559
deflection 483
deflection aberration (parasitic) 475
deflection coil fields
 – axial harmonics 836
 – saddle coils 834
 – toroidal coils 834
deflection modes 850
deflection systems 481, 823
 – aberrations of 497–516, 545
 – addition of aberrations 508, 545
 – alternative concepts 844
 – electric system 494
 – electrostatic system 484, 491
 – field models for magnetic deflection 828
 – hybrid system 495
 – in electron microscopes 823
 – large-angle deflection 504
 – magnetic system 484, 494, 828
 – parasitic aberrations 509
 – pure deflection, aberrations 498
 – rotation invariant 495
 – sensitivity 494
 – with magnetic lenses, aberrations 504, 570
differential equations for the aberrations 537–542
differentiation and interpolation
 – one-dimensional case 188
 – two-dimensional case 194
dilatation factor 20
diode 914
diode approximation 936
diode-field model 945

dipole field 88
Dirichlet problem
 – general case 109
 – planar case 117
 – two-dimensional case 113
 – three-dimensional form 149
disc of least confusion
 – for astigmatism 374
 – for spherical aberration 354
distortion 337
 – and triple-polepiece lenses 782
 – of deflection systems 504
 – of quadrupoles and octopoles 440, 442
distortion-free orthogonal systems 802
distortion of round lenses, asymptotic 397
 – addition rules 420
 – and pupil magnification 407
 – elimination 779, 782
distortion of round lenses, real 349, 378–382
 – aberration figures 379
 – and aperture position 384
 – formulae 380, 385
 – thin-lens formulae 382
divergence factor 954, 959, 961
divergent lens 216, 230
double focusing 1058
doublet
 – cardinal elements of 234
drift space 283
dual-channel deflection systems 848
dual-channel system 511, 515

Dušek matrix 258, 283
dynamic focusing and correction 841
dynamic programming 677, 773
EBL, see electron lithography systems
EFF, see fringing field effects 1072
effective length
 – of a quadrupole 803
effective potential 40, 45
eigenvalue problems
 – for paraxial trajectories 535
eikonal function 11, 50
eikonal method 298, 511
einzel lens 202, 630, 655–679
 – figure of merit 675
 – hyperbolic lenses 674
 – Kanaya–Baba model 666
 – measurements and exact calculations 671
 – multielectrode lenses 673
 – optimum design 675
 – Regenstreif's model 656
 – Schiske's model 659
 – Shimoyama's analysis 671
 – Wendt's theory 669
EL3 851
electrical input signals 493
electromagnetic potentials 63
electron bombardment ion sources 685
electron emission 907
 – field electron emission 927
 – theory 918
 – thermionic emission 921, 1034
electron guns 905–1036

- design 1017
- field emission guns 914, 1020, 1035
- general features 907
- hybrid emission guns 1022
- LaB_6 guns 913, 1034
- Lauer's model 948
- Pierce guns 1030
- point source model 1017
- thermionic guns 907, 948, 1028, 1034

electron lithography systems 483, 495, 509, 825, 844, 1004

electron mirror microscope 796

electron mirrors 261, 425, 796–798
- and aberration correction 870
- cartesian representation 271, 432
- in energy analysis 797, 1058
- parametric representation 264, 425
- quadratic transformation 274

electrostatic lenses 202, 629–686
- astigmatism and field curvature 376
- coma 367
- distortion 381
- quantitative investigations 671
- spherical aberration 359

electrostatic potential 19
- general definition 63
- series expansion 82

electrostatic principle 33

electrostatic prisms 1080, 1097
- aberrations 1056

elliptic differential equation
- general form 159

- self-adjoint form 162, 179

embedding material 9

emission microscope 262, 907

emittance 917
- and brightness 996
- general concepts 989–1003
- two-dimensional 991

emittance diagram 998
- and aberrations 1000

emittance ellipses 992
- acceptance ellipse 995
- drift 993
- focusing 993
- transfer 993

emittance matching 995

energy analysis 797
- literature 1098

energy filters 630

energy flux densities
- in thermionic guns 959

entropy, and Boersch effect 1011

equations of motion 267

equidensities 892

equivalent solenoid model 712

Euler–Lagrange equations 1047, 1053
- of fields 65
- of trajectories 36, 39, 44

expanding spherical-mesh grid 168

exponential model
- for magnetic lenses 709

extended paraxial domain 61, 73

FDM (see finite-difference method) 159

FEM (see finite-element method) 175
Fermat's principle 50
ferrite shields 831
field calculation 59–198
– electron sources with pointed cathodes 941
– hybrid methods 969
field curvature of round lenses, asymptotic 397
– addition rules 420
– and pupil magnification 407
field curvature of round lenses, real 337, 349, 369–378
– aberration figure 369
– and aperture position 384
– circle of least confusion 374
– formulae 375, 385
– thin-lens formula 378
field equations 62
field emission gun 10, 185, 914, 1019, 1020, 1035
– diode 939
– field model 945
field-emission microscopy 8, 799, 900
field emission, theory of 927
field-interpolation techniques 188–198
fifth-order aberrations
– of deflection systems 503
– of round lenses 334, 1066
filter lens 546
finite-difference method 159–174, 301, 944, 1026
finite-element method 175–187, 301

– infinite elements 184
Finsterwalder condition 384
five-electrode lens 655
five-point formulae 160
focal length, see real, asymptotic focal length
focus, see also real, asymptotic focus
– blade focus 895
– cone focus 895
– cone-edge focus 895
– cross focus 895
– line focus 895
focusing in first order 1068
foil lenses 679, 864, 865
form factor of magnetic lenses 689
fourfold errors 501
fourfold symmetry in deflection systems 500
Fourier analysis of the aberrations 550
Fourier integral kernels 125
Fourier series expansion 505
Fourier–Bessel series expansions 91
Fowler–Nordheim equation 928
Fox–Goodwin–Numerov method 526, 534
Fredholm equation 110
fringing field effects 1068, 1069, 1070
– extended fringing fields 1072
gas discharge electron sources 907
gauge transformation 53
Gaussian approximation
– cylindrical lenses 290–293
– deflection systems 487
– mirrors 261–275
– quadrupoles 276–289

– round lenses 225–260
Gaussian brackets 285
gauze lenses 300, 679
generalized brightness function 972
generalized model
  – for magnetic lenses 713
geometric aberration 298
  – differential equations for 537
  – of deflection systems 483, 505, 506
  – of quadrupoles and octopoles 434
  – of round lenses 339
  – of systems with a curved optic axis 1053
Glaser's bell-shaped model
  – for round lenses 253, 567, 661, 695
  – for quadrupoles 801, 804
Glaser–Robl model 638
g-ray (definition) 215, 217
G- and $\overline{G}$-rays (definition) 226
Green's theorem 107
grid electrode 908
grid, for finite differences 159
grid lenses 679
Grigson coils 498
Grinberg's theory 11
Grinberg's model 639
Grivet–Lenz model 640, 707
Gullstrand's classification 323
Hahn's model 723
Hamiltonian 41, 42, 48, 919, 990
Hamilton's central equation 47
Hamilton–Jacobi equation 48

Hamilton–Jacobi theory 46
Hermite interpolation
  – one-dimensional case 189
  – two-dimensional case 194
Herschel's condition 219
high-frequency lenses 872
high-voltage microscope 9
  – pulsed-beam proposal 876
higher-order focusing 893
history of electron optics 6–12
hollow beam 963, 1000
holography 11, 877
Householder transformation 544
h-ray (definition) 214, 217
Hutter's model 638
hybrid emission gun 916
hybrid field calculation methods 170, 184
hyperbolic lens 674
hyperemittance 991, 1011
Ichinokawa analyser 1095, 1099
ideal deflection 490
image converter 263
immersion lens 202, 630, 631–655
  – single aperture 631
  – three or more electrodes 647–655
  – two-electrode lens 635–647
immersion objective 263
improper integral
  – evaluation 141
initial value problem
  – ordinary differential equations 526
  – paraxial trajectories 533

integral equation 107–124
 – and interpolation 136
 – for magnetic deflection coils 119
 – for multipoles 121
 – for parasitic aberrations 122
 – for rotationally symmetric potentials 117, 118
 – for scalar potentials 107
 – for unconventional lenses 119
 – general theory 107
 – numerical solution 131
intensity considerations for caustics 891
interface conditions 97
 – general case 111
 – two-dimensional case 116
interferometer 262
intermediate lens 207
inverse problem for magnetic lenses 737
inversion of the principal planes 253
ion microprobes 801, 806
ion optics, unified theory 1080–1099
 – Nakabushi's studies 1085
isophotes 893
isoplanatic approximation 477, 886
isotropic aberration terms in deflection systems 501
iterative solution techniques (in the FDM) 170
Judd's rule 1065
Kanaya–Baba model
 – for bipotential lenses 640
 – for einzel lenses 666
Kasper's diode model 943
Kerst–Serber equations 1062

kinematic function 21
kinetic energy 19
kinetic momentum 17, 20, 36, 1046, 1052
kinetic potential 36
Lagrange bracket 57, 309
Lagrange density 65
Lagrange differential invariant (see also Lagrange bracket) 220
Lagrange equations 36
Lagrange formalism 35
Lagrange–Helmholtz relation 1048
Lagrangian 35, 989, 1046
Lambert's law 924
laminated lenses 10, 181, 779
lancet cathode 912
Langmuir–Child law 962, 1031
lanthanum hexaboride cathode 913, 1034
Laplace's equation 63, 1043
Larmor frequency 26, 210, 472
lateral aberration 500, 501, 1056
lateral magnification 939
Lauer's triode model 948
least-squares-fit method in electron optics 543, 551
Legendre transform 40, 45, 990
lens design (see also electrostatic, magnetic lenses)
 – for analytical instruments 693
 – for three-dimensional reconstruction 694
lens-mirror transition 630
Lenz's brightness theory 980, 996
Lie algebra and aberrations 301, 1065

Liebmann's curves 733
Liebmann's method (see finite difference method) 159
line focus 895
Liouville's theorem 972, 981, 989
load characteristic 286, 809
longitudinal
 – magnification 218
 – temperature 1008
Lorentz equation 17
Lorentz force 17
Luneberg lens 679
MACSYMA 565
magnetic deflection systems
 – aberrations 497
 – field calculation 119
 – field models 828
 – saddle coils 484
 – sensitivity 494
 – toroidal coils 105, 484
magnetic flux function 24, 67
magnetic flux potential 101, 104, 181
magnetic lenses 202, 687–795
 – astigmatism and field curvature 377
 – coma 368
 – convolutional models 711
 – distortion 381
 – exponential model 709
 – field calculation 102, 155, 175
 – field models 695
 – form factor 689
 – generalized model 713
 – Glaser's bell-shaped model 695
 – Grivet–Lenz model 707
 – Hahn's mapping 723
 – laminated lenses 779
 – lens with low specimen field 790
 – measurements and universal curves 726
   – unsaturated lenses 733
   – saturated lenses 737
 – minilenses 776
 – modes of operation 691
 – notation 694
 – optimization 772
 – other bell-shaped models 706
 – pancake lenses 776
 – permanent-magnet lenses 787
 – power-law model 710
 – practical design 691
 – rectangular model 718
 – scaling rules 689
 – single-polepiece lenses 776
 – spherical aberration coefficient 363
 – superconducting lenses 687
 – supertwin lens 792
 – symmetric 687, 695
 – Tretner's analysis 755
 – triple-polepiece lens 782
 – twin lens 792
 – unconventional 104, 776
 – unsymmetric 687, 718
magnetic prisms 1058–1079, 1096
 – aberrations 1054, 1073
 – axial focusing 1068
 – dispersion 1067
 – focusing in first order 1063

- fringing field effects 1070
- optimization 1076
- radial focusing 1063
- with circular axis 1059

magnetic sector fields: see magnetic prisms 1058–1079

magnetostatics, boundary conditions 96

magnification
- angular 217
- complex 216
- longitudinal 218
- transverse 216

mass spectrometer 71, 1079, 1099
material coefficients 98
material equations 63
maximum brightness 974
Maxwell's equations 62
Maxwell's fish-eye 679
mean energy 923, 1005
mechanical aberration, see parasitic aberration
meridional rays 211
merit function 559
microwave cavity lenses 876
minilenses 776
mirror electron microscope 262
mirrors 261, 425, 796
Möllenstedt analyser 1095, 1099
monochromators 630
Monte Carlo method 965, 1004, 1005
moving objective lens (MOL) 508, 839
multielectrode lenses 647
multipole correctors 859

multipole fields 1043
- series expansions 88

multipole system
- field calculation 121
- in ion optics 1098

Munro–Chu formulae for deflection aberrations 509

Murphy–Good–Young equation 928
Neill parabola 884
Neumann problems
- general case 111

Newton's lens equation 231, 236, 246

Newtonian field 242, 252, 698
nine-point formulae 165
nodal point 232, 1022
normalized brightness 976, 1011
Nottingham effect 930
numerical differentiation 190
Numerov method (see Fox–Goodwin–Numerov method) 526, 534
objective lens 207, 226, 242
- spherical aberration 354

omega filters 1079, 1095, 1099
optimization procedures 1021
- damped least-squares method 562
- for axial distributions 561, 771
- for magnetic prisms 1076
- general considerations 558

ordinary differential equations, numerical solution 526–533
orthogonal trajectories 49, 52
orthogonality condition
- for quadrupole lenses 279
orthomorphic systems 802, 810

oscilloscope 483, 484, 491, 498
osculating cardinal elements 246
pancake-lenses 776
parasitic aberration 298, 470, 517
    – classification 319, 472
    – numerical determination 475
    – of deflection systems 509
paraxial deflection 496, 520
paraxial distortion 474, 496
paraxial ray equation
    – derivation 207, 211
    – for momentum 223
    – of combined systems 509
    – of deflection systems 487
    – of mirrors, cartesian form 271
    – of quadrupoles 277
    – of round lenses 207, 213
    – Picht's transformation 214
    – transformations of 222
paraxial trajectory equations 1047
partial-integration rule 291
permanent-magnet lenses 202, 787
perturbation
    – characteristic function 307
    – eikonal 499, 1055
    – first-order 307
    – general formalism 297, 303
    – operator 305
    – second-order 310
perturbations of the rotational symmetry 122
perveance 984
Petzval coefficient
    – asymptotic 401
    – real 376

phase condition, for HF-lenses 872
phase shifts 338
phase space 989, 919, 1011
Picht's transformation 214, 472
Pierce systems 962, 1030
pincushion distortion 379
pinhole lens 718
pivot-point 491
planar fields 69, 90, 117, 290
pocket-handkerchief distortion 379
Poincaré's integral invariant 54, 57
point characteristic function 49, 305
point source: see virtual source 1017
pointed filament 912, 934
Poisson equation
    – scalar form 73, 170, 953
    – vector form 63, 64
polepiece design 691
    – material 687, 753
polynomial forms of asymptotic aberration coefficients 12, 396, 416, 439
potential barrier 920
potential energy 19
potential minimum 955, 961
power law model for magnetic lenses 710
predictor-corrector method 529
principal normal 1041
principal planes, see real, asymptotic principal planes
principal ray 491
probe-forming lens 207
    – spherical aberration 354
projective transformation 235
projector lens 207

pseudo-stigmatic systems 802
pupils 349
quadrupole coefficients 489
quadrupole field 89
    – bell-shaped model 801, 804
    – rectangular model 801, 803
quadrupole lens 276–289, 801–822
quadrupole multiplets 284, 461
    – anamorphotic 802
    – antisymmetric 286, 436
    – distortion-free 802
    – orthomorphic 802
    – pseudo-stigmatic 802
    – regular 802
    – symmetric 436
quadrupole-octopole correctors 859
quadrupoles
    – antisymmetric multiplets 286, 805
    – arc lenses 810
    – astigmatic tube lenses 813
    – biplanar lenses 811
    – crossed lenses 810
    – doublets 805
    – ion microanalysers 806
    – isolated 805
    – load characteristic 809
    – paraxial equations 277
    – quadruplets 809
    – radial lenses 819
    – Russian quadruplet 810
    – special geometries 810
    – transaxial lenses 819
    – triplets 806
quasistatic approximation 61

radial focusing (prisms) 1063
radial lenses 819
radial series expansion
    – evaluation 192
    – for integral kernels 126
    – for fields 78
rare-earth polepieces 753
raster pattern 483
real aberration coefficients
    – of round lenses 339
    – of quadrupoles 454
real and asymptotic aberrations 11
real cardinal elements 242, 697
    – focal length 244
    – image focal plane 243
    – image focus 243
    – principal planes 244
        – inversion of 253
    – object focus 242
Rebsch–Schneider measure of lens performance 769
reciprocal magnification 395
rectangular model
    – for magnetic lenses 717
recurrence relation
    – for integral kernels 127
REDUCE 301, 565, 568
reduced brightness (see normalized brightness) 976, 981
reduced coordinates 231, 339, 384–386
reduced magnetic scalar potential 100
reference sphere 350
    – and astigmatism 375
    – and coma 366

- and distortion 376
- and field curvature 375
- and spherical aberration 356
refractive index 51
Regenstreif's model 656
regular systems 802
relative refractive index 419
relativistic kinematics 17
relativistic mass 17
relativistic proper-time element 29
reluctance 63
resolving power 1068
Richardson–Dushman equation 922
Richtstrahlwert 971
rotating coordinate system: see rotating frame
rotating frame 208, 267
rotation-invariant deflection 495
rotationally symmetric fields
   - general relations 67
   - series expansions 85
rotationally symmetric system 202
   - general systems 38
   - geometrical aberrations 339–392
   - paraxial properties 202
   - static systems 44
Runge–Kutta method 527
Ruska–Riecke lens 691
Russian quadruplet 286, 810
saddle coil systems 484, 834
saturated lenses 737
saturation 66, 689
scalar magnetic potential 78
   - general definition 64

- series expansion 83
scalar potential
   - series expansion 73, 78
scaling rules 33
scanning electron microscope (SEM) 10, 483, 498, 823
scanning transmission electron microscope (STEM) 10, 821
scatter-plots 802
Scherzer's theorem 11, 363, 857
Schiske's model 659
Schottky emission 931, 975
Schwarz's alternating method 185
SCOFF, see fringing field effects 1070
second-zone objective 691, 751
sector fields, magnetic 1058–1079
self-adjoint elliptic equations, FEM solution 179
Seman's technique 11, 356, 387
sensitivity of deflectors
   - complex 497
   - real 494
series expansions 1049
   - for potentials and fields 73–93
   - with curved optic axis 1042
seven-electrode lens 655
sextupole correctors 863
shaped-beam systems 514
shaped-beam technique 483, 851
sharp cutoff fringing field 1070
shielding cryolens 688
shift matrices 422
Shimoyama's model 671
shooting-method 536
sign convention 230

single aperture
- field of 143, 631
- optics of 632
single octopoles as corrector 862
single-polepiece lenses 776
skew optic axis 1039
skew rays 211
SLOR (see successive line overrelaxation method) 173
Smith–Helmholtz formula 218
SOC, see sphere-on-orthogonal cone model 941
solid angle potential 99
Sommerfeld–Bethe model 918
SOR (see successive overrelaxation method) 171
space charge 911, 953, 963, 975, 982
- and aberration correction 864
- counting method 965, 969
- semianalytic method 965
spectral radiance 923
sphere-on-orthogonal-cone model 941
spherical aberration coefficient
- finite magnification 737
- the bell-shaped model 701
- the rectangular model 718
spherical aberration of round lenses, asymptotic 396
- addition rules 421
- and pupil magnification 407
spherical aberration of round lenses, real 349, 350–365
- and aperture position 383
- disc of least confusion 354
- formulae 356, 385

- forward and backward 356
- sign 352, 363
- thin-lens formula 365
spherical cathode 934
spherical diode 953
spherical-mesh grid 168
spiral distortion 379
spline 136
spot diagram 526, 802
step-width adjustment 531
stigmatic image 214
stigmator 471, 479, 516–521, 899
Störmer equation 45
strong focusing 276, 1062
Stufenfeld 723
Sturrock's units 52, 384–386
successive line overrelaxation method 173
successive overrelaxation method 171
superconducting lenses 9, 202, 687, 753, 783
- yokes 101
- zone plate 865
superposition, of aperture fields 146, 193
supertwin lens 792
surface charge density 96
surface current density 97
surface normal 94
synklysmotron 875
tangential vector 1041
telescopic-condenser mode 691
television tube 484, 823, 498
thermal-field (TF) emission 927

thermionic electron guns 908, 974, 1034
- conventional forms 908, 1028
- diodes 939
- field model 945
- short focus 1029
- telefocus 1029
- triode guns 908, 948
thermionic emission 921
thermodynamics, and Boersch effect 1006
thin-lens approximation 257, 365
threefold astigmatism 475
toroidal coil systems 484, 834
torsion 1042
total energy distribution 923, 929
trace space 989, 998
trajectory equations 27, 1052
- arc length representation 27
- cartesian representation 30
- in curvilinear systems 1045, 1052
- parametric representation 27
- relativistic proper-time representation 29
trajectory method 11, 297, 341
transaxial lens 286, 819
transfer matrices 226, 231, 292, 1065, 1069
- of a drift space 283
- of quadrupole systems 280
transmission factor 919, 921
transverse axial field strength 489
transverse chromatic error 507
transverse magnification 216

transverse momentum distribution 924, 931
transverse temperature 1007
Tretner and ultimate lens performance 365, 755
- electrostatic lenses 763
- magnetic lenses 766
triode gun 908, 948
triple-polepiece projector lenses 782
truncated power series algebra 1095
tubular lens 901
tunnel effect 914, 925
twelve-pole stigmator 520
twin lenses 792
two different symmetries in deflection systems 498
two-electrode lenses 635
ultimate lens performance 755
- early studies 769
- Glaser's aberration-free lens 770
- Tretner's analysis 365, 755
ultramicrotome 9
unconventional lenses 10, 119, 687, 776
unipotential lens 202, 630, 655
unsaturated lenses, properties 733
unsymmetric magnetic lenses, properties 718
variable-axis immersion lens (VAIL) 843
variable shaped-beam technique 851
variable-axis lens (VAL) 825, 839
variational principle
- for fields 65, 175
- for paraxial equations 212

– for trajectories 35
– for trajectories with curved optic axis 1045
– time-independent form 43
vector potential 51
    – general definition 63
    – series expansions 75, 80, 83
vector scan technique 850
vertical landing 839
virtual cathode 961
virtual source 914, 937, 940, 1021
wave aberration 473, 518
weak-lens approximation 257, 771
wehnelt electrode 908, 914
Wendt's theory 669
Wien filter 494, 1097
work function 921, 1024
working distance 483
Wronskian 217, 392
Yada's combined lens 771
yoke design 691
zoom lenses 647

AUG 1 7 1989